Lecture Notes in Computer Science 17089

The series Lecture Notes in Computer Science (LNCS), including its subseries Lecture Notes in Artificial Intelligence (LNAI) and Lecture Notes in Bioinformatics (LNBI), has established itself as a medium for the publication of new developments in computer science and information technology research, teaching, and education.

LNCS enjoys close cooperation with the computer science R & D community, the series counts many renowned academics among its volume editors and paper authors, and collaborates with prestigious societies. Its mission is to serve this international community by providing an invaluable service, mainly focused on the publication of conference and workshop proceedings and postproceedings. LNCS commenced publication in 1973.

Luca Pasa · Alessandra Lintas · Igor V. Tetko ·
Alessio Micheli · Nicolò Navarin ·
Alessandro E. P. Villa · Domenico Tortorella ·
Mirko Polato
Editors

Artificial Neural Networks and Machine Learning – ICANN 2026

35th International Conference on Artificial Neural Networks
Padua, Italy, September 14–17, 2026
Proceedings, Part I

Springer

Editors
Luca Pasa
University of Padova
Padua, Italy

Igor V. Tetko
Helmholtz Munich
Munich, Germany

Nicolò Navarin
University of Padova
Padua, Italy

Domenico Tortorella
University of Pisa
Pisa, Italy

Alessandra Lintas
University of Lausanne
Lausanne, Switzerland

Alessio Micheli
University of Pisa
Pisa, Italy

Alessandro E. P. Villa
University of Lausanne
Lausanne, Switzerland

Mirko Polato
Università degli studi di Torino
Turin, Italy

ISSN 0302-9743 ISSN 1611-3349 (electronic)
Lecture Notes in Computer Science
ISBN 978-3-032-38397-6 ISBN 978-3-032-38398-3 (eBook)
https://doi.org/10.1007/978-3-032-38398-3

This Springer imprint is published by the registered company Springer Nature Switzerland AG
The registered company address is: Gewerbestrasse 11, 6330 Cham, Switzerland

Preface

The 35th International Conference on Artificial Neural Networks (ICANN 2026) was held in Padua, Italy on September 14–17, 2026. It serves as a global platform for presenting the latest breakthroughs in artificial intelligence, machine learning, neural networks, deep learning, and brain-inspired computing. Organized annually by the European Neural Network Society (ENNS), ICANN has, since its inception in 1992, evolved into one of Europe's most prestigious conferences at the intersection of Artificial Intelligence and Neuroscience. It offers researchers, practitioners, and industry professionals a unique opportunity to share innovations, discuss theoretical advances, and explore practical applications of neural computation. The conference strongly values the synergy between theoretical progress and impactful real-world applications, and actively encourages contributions that demonstrate how artificial neural networks are being used to address pressing societal and technological challenges.

We received 814 submissions for the main research track and the eight special sessions, of which 295 papers were accepted (36%). In addition, there were 89 papers submitted to workshops, which were mainly invited contributions (of which 57 were accepted). Each paper was assigned at least three reviewers; an average of 4 reviews per paper was obtained, totaling 2728 reviews. The review process followed a double-blind peer-review model, and area chairs oversaw quality and fairness. The papers submitted to the main research track and the special sessions were presented over three days. Additionally, one day was devoted to the workshops.

The conference featured six keynote talks by distinguished invited speakers spanning a wide spectrum of topics, reflecting both emerging scientific challenges and the growing convergence of theoretical insights, real-world applications, and ethical considerations in AI and neuroscience. These talks addressed the frontiers of artificial intelligence, computational neuroscience, complex networks, and dynamical systems, showcasing the diverse landscape ICANN 2026 aimed to represent: Pierre Baldi, UC Irvine, "The Theory of Synaptic Neural Balance"; Ginestra Bianconi, Queen Mary University of London, "Topology Shapes Dynamics of Higher-Order Networks"; José María Delgado García, Seville, "Functional States Corresponding to Different Types of Associative and Social Learning Tasks"; Massimiliano Pontil, Italian Institute of Technology, Genova, "Learning Linear Evolution Operators: Theory and Applications"; Ausra Saudargiene, Lithuanian University of Health Sciences, "Toward Digital Brain Twins of Neurodegenerative Disorders"; and Marco Gori, University of Siena; "Generative System Dynamics in Recurrent Neural Networks".

Both ENNS and the ICANN 2026 Organizing Committee are committed to supporting participation by students and early-career researchers, also by providing travel grant awards. By enabling broader participation, ICANN 2026 continued its efforts towards fostering a more inclusive and representative research environment, encouraging the exchange of ideas across cultures, disciplines, and career stages.

We extend our sincere thanks to all the authors, workshop and tutorial organizers, and participants whose scientific contributions and engagement made ICANN 2026 such a dynamic and inspiring event. Crafting a rich and interdisciplinary program would not have been possible without the dedication and significant commitment of the Area Chairs, the Program Committee, and all the members of the Organizing Committee. We are also deeply grateful to the many volunteers and session chairs, whose behind-the-scenes work ensured the smooth running of the conference, as well as to all reviewers. We thank Springer for their continued collaboration in publishing the conference proceedings, and Microsoft for providing access to the CMT conference management system and offering technical assistance throughout the organization process. We deeply appreciate the ongoing support and guidance of the European Neural Network Society (ENNS), whose leadership and vision continue to shape the ICANN community.

July 2026

Luca Pasa
Alessandra Lintas
Igor V. Tetko
Alessio Micheli
Nicolò Navarin
Alessandro E. P. Villa
Domenico Tortorella
Mirko Polato

Organization

General Chairs

Alessio Micheli	University of Pisa, Italy
Nicolò Navarin	University of Padua, Italy
Alessandro E.P. Villa	University of Lausanne, Switzerland

Program Chairs

Luca Pasa	University of Padua, Italy
Alessandra Lintas	University of Lausanne, Switzerland
Igor V. Tetko	Helmholtz Munich, Germany

Submissions Chair

Linas Petkevičius	Vilnius University, Lithuania

Workshops and Special Sessions Chairs

Luca Oneto	University of Genoa, Italy
Valerie Vaquet	Bielefeld University, Germany

Social Media Chairs

Giovanna Maria Dimitri	University of Milan, Italy
Alessandro Dipalma	University of Pisa, Italy

Diversity and Inclusion Chair

Roseli S. Wedemann	Universidade do Estado do Rio de Janeiro, Brazil

Communication Chairs

Kristína Malinovská	Comenius University, Slovakia
Davide Rigoni	University of Padua, Italy
Matteo Zavatteri	Fondazione Bruno Kessler, Italy

Publication Chairs

Domenico Tortorella	University of Pisa, Italy
Mirko Polato	University of Turin, Italy

COI Chair

Daniele Zambon	Università della Svizzera italiana, IDSIA, Switzerland

Volunteers Chair

Giovanni Donghi	University of Padua, Italy

Local Arrangements Chairs

Merylin Monaro	University of Padua, Italy
Andrea Zangrossi	University of Padua, Italy

Steering Committee

Angelo Cangelosi	University of Manchester, UK
Igor Farkaš	Comenius University, Slovakia
Matthias Kerzel	University of Hamburg, Germany
Sebastian Otte	University of Lübeck, Germany
Michael Wand	Dalle Molle Institute for Artificial Intelligence, Switzerland
Erkki Oja	Aalto University, Finland
Włodzisław Duch	Nicolaus Copernicus University, Poland
Věra Kůrková	Czech Academy of Sciences, Czechia
Stefan Wermter	University of Hamburg, Germany

Area Chairs

Adrian Bors	University of York, UK
Alberto Cabri	University of Milan, Italy
Alberto Testolin	University of Padova, Italy
Alessandra Bertoldo	University of Padova, Italy
Alessandro Betti	IMT Alti Studi Lucca, Italy
Alessandro D'Ausilio	University of Ferrara and Italian Institute of Technology, Italy
André Artelt	Bielefeld University, Germany
Andrea Cossu	University of Pisa, Italy
Antonio Longa	University of Trento, Italy
Ausra Saudargiene	Lithuanian University of Health Sciences, Lithuania
Barbara Hammer	Bielefeld University, Germany
Barbara Toniella Corradini	Università degli Studi di Siena, Italy
Benjamin Paaßen	Bielefeld University, Germany
Claudio Gallicchio	University of Pisa, Italy
Daniele Castellana	Università di Firenze, Italy
Danilo Croce	University of Rome, Tor Vergata, Italy
Fabio Aiolli	University of Padova, Italy
Fabrizia Devito	Università degli Studi di Bari Aldo Moro, Italy
Francesc Serratosa	Universitat Rovira i Virgili, Spain
Franco Scarselli	University of Siena, Italy
Giuseppe Marra	KU Leuven, Belgium
Gustav Šír	Czech Technical University, Czechia
Hui Liu	University of Bremen, Germany
Igor Farkaš	Comenius University, Slovakia
Johannes Brinkrolf	Bielefeld University, Germany
Juan-Pablo Ortega	Nanyang Technological University, Singapore
Kohulan Rajan	Friedrich Schiller University Jena, Germany
Kristína Malinovská	Comenius University, Slovakia
Luca Bergamin	University of Padua, Italy
Luca Calatroni	Università degli Studi di Genova, Italy
Luca Pedrelli	Scuola Superiore Sant'Anna, Italy
Marc Bianciotto	Sanofi, France
Marcello Sanguineti	University of Genova, Italy
Marco Cardia	University of Pisa, Italy
Marco Fiorucci	University of Padova, Italy
Marco Lippi	University of Florence, Italy
Marco Podda	University of Pisa, Italy
Marika Kaden	UAS Mittweida, Germany

Matthias Kerzel	Universität Hamburg, Germany
Michael Wand	IDSIA USI-SUPSI, Switzerland
Michele Rossi	University of Padova, Italy
Mihai Andries	IMT Atlantique, France
Monica Bianchini	University of Siena, Italy
Ola Engkvist	AstraZeneca, Sweden
Pablo Barros	Sony, Belgium
Paolo Frazzetto	University of Valencia, Spain
Petra Bevandic	University of Zagreb, Croatia
Pietro Novelli	Italian Institute of Technology, Genova, Italy
Povilas Daniušis	Vytautas Magnus University, Lithuania
Qilong Gu	Microsoft, USA
Roseli Suzi Wedemann	Universidade do Estado do Rio de Janeiro, Brazil
Sandro Ridella	University of Genoa, Italy
Shirui Pan	Griffith University, Australia
Shuo Wang	University of Birmingham, UK
Simone Scardapane	Sapienza University, Italy
Vincenzo Lomonaco	LUISS Guido Carli, Italy
Vittorio Cuculo	Università degli Studi di Modena e Reggio Emilia, Italy
Yibo Yang	KAUST, South Korea
Zhiqi Shao	University of Sydney, Australia

Abstracts of Invited Talks

The Theory of Synaptic Neural Balance

Pierre Baldi

UC Irvine, USA

Abstract. We develop a general theory of synaptic neural balance and how it can emerge or be enforced in neural networks. For a given regularizer, a neuron is said to be in balance if the total cost of its input weights is equal to the total cost of its output weights. The basic example is provided by feedforward networks of ReLU units trained with L2 regularizers, which exhibit balance after proper training. The theory explains this phenomenon and extends it in several directions. The first direction is the extension to bilinear and other activation functions. The second direction is the extension to more general regularizers, including all Lp regularizers. The third direction is the extension to non-layered architectures, recurrent architectures, convolutional architectures, as well as architectures with mixed activation functions. Gradient descent on the error function alone does not converge in general to a balanced state, where every neuron is in balance, even when starting from a balanced state. However, gradient descent on the regularized error function ought to converge to a balanced state, and thus network balance can be used to assess learning progress. The theory is based on two local neuronal operations: scaling which is commutative, and balancing which is not commutative. Given any initial set of weights, when local balancing operations are applied to each neuron in a stochastic manner, global order always emerges through the convergence of the stochastic balancing algorithm to the same unique set of balanced weights. The reason for this is the existence of an underlying strictly convex optimization problem where the relevant variables are constrained to a linear, only architecture-dependent, manifold. Simulations show that balancing neurons prior to learning, or during learning in alternation with gradient descent steps, can improve learning speed and final performance. Finally, we show how synaptic balancing is intimately connected to questions of neural overparameterization and algebraic geometry, and discuss the relevance of synaptic neural balance for neurobiological or neuromorphic neural networks.

Topology Shapes Dynamics of Higher-Order Networks

Ginestra Bianconi

Queen Mary University of London, UK

Abstract. Complex systems like the brain, climate, and next-generation artificial intelligence rely on interactions that extend beyond simple pairwise relationships. This talk reveals the critical role of higher-order networks, structures that capture multi-body interactions, in shaping the dynamics of such systems.

By integrating algebraic topology with non-linear dynamics, the research highlights how topological signals, dynamical variables defined on nodes, edges, triangles, and other higher-order structures, drive phenomena such as topological synchronization, pattern formation, and triadic percolation. The surprising result that emerges from this research is that topological operators including the Topological Dirac operator, offer a common language for treating complexity, AI algorithms, and quantum physics.

These findings not only advance the understanding of the underlying mechanisms in neuroscience and climate science but also pave the way for innovative machine learning algorithms inspired by theoretical physics.

Functional States Corresponding to Different Types of Associative and Social Learning Tasks

José María Delgado García

Siviglia, Spain

Abstract. The most important thing that I have learned across my scientific life is that the complexity of brain functions can only be approached by multidisciplinary, comparative, and live approaches. The availability of genetically manipulated mammals (mice and rats) and of sophisticated electrophysiological and pharmacological techniques, which can be applied in behaving animals during the acquisition of new motor and cognitive abilities, have largely facilitated my scientific approach in the past 20 years. In this regard, our group has studied the contribution of cortical, subcortical and cerebellar circuits to associative (Pavlovian, instrumental), social, brain–machine, and decision-making learning paradigms. For this, we have recorded unitary firing rates, local field potentials, and activity dependent changes in synaptic strength in cortical and subcortical neurons during the respective acquisition processes. The main output of our studies is that learning is the result of the activity of wide cortical and subcortical circuits activating particular functional properties of involved synaptic nodes and that brain functions during learning processes have to be studied at the very moment of the acquisition process.

Learning Linear Evolution Operators: Theory and Applications

Massimiliano Pontil

Italian Institute of Technology, Genova, Italy

Abstract. Dynamical systems are central to science and engineering, with applications in climate modeling, molecular dynamics, robotics, neuroscience, finance, and beyond. Remarkably, despite their diversity, many such systems can be understood within the unifying framework of linear evolution operators. The key idea is to study how functions of the state evolve over time, instead of tracking the full state. This transforms a nonlinear problem into a linear one and enables spectral analysis to uncover global system dynamics. This perspective has a long history, rooted in foundational work by Markov, Koopman, and von Neumann. However, while conceptually powerful, linear evolution operators are often computationally intractable in high-dimensional settings. Consequently, over the past two decades, significant effort has focused on data-driven methods, yet their theoretical guarantees remain poorly understood. This talk presents a framework for placing data-driven approaches to dynamical systems on a firm statistical foundation. By formulating the problem of learning linear evolution operators from a statistical perspective, we develop novel learning algorithms backed by finite-sample learning guarantees, which show promising results in challenging real-world applications in molecular dynamics, climate modeling, and robotics. The algorithms provably learn evolution operators and their spectra, and leverage modern deep learning architectures to learn representations of dynamics efficiently and reliably. Finally, if time permits, I will discuss how linear operators play a central role in reinforcement learning and statistical inference, including uncertainty quantification and causality.

Toward Digital Brain Twins of Neurodegenerative Disorders

Ausra Saudargiene

Lithuanian University of Health Sciences, Lithuania

Abstract. EBRAINS is a European digital brain research infrastructure for integrating multimodal neuroscience data, computational models, simulation tools, and AI-assisted clinical research workflows within a shared ecosystem. By enabling interoperable data analysis pipelines and biologically informed large-scale simulations, EBRAINS supports a transition from descriptive neuroscience toward mechanistic and predictive models of brain disorders. Within the EBRAINS ecosystem, The Virtual Brain (TVB) provides a framework for constructing individualized whole-brain models by integrating patient-specific anatomical connectivity, empirical neurophysiological recordings, and biophysically interpretable neural mass models. This capacity to link empirical data with mechanistic simulations is particularly relevant for neurodegenerative disorders, which are characterized by progressive neuronal dysfunction and loss, often accompanied by widespread alterations in brain network dynamics. Parkinson's disease, one of the most common neurodegenerative disorders, is marked by dopaminergic neuronal loss and heterogeneous motor and non-motor symptoms. In Parkinson's disease, TVB-based modeling is especially promising as both dopaminergic medication and deep brain stimulation modulate large-scale neural dynamics, yet the mechanisms underlying individual therapeutic responses remain incompletely understood. Integrating EEG-derived biomarkers with mechanistic whole-brain modeling may therefore support the development of patient-specific digital brain models and personalized neuromodulation strategies.

We combine EEG-derived biomarkers, including spectral connectivity, synchronization measures, burst dynamics, and neuronal avalanche statistics, with dopamine-sensitive neural mass modeling to characterize patient-specific brain dynamics across medication and deep-brain stimulation states. The results suggest that neuromodulation shifts individuals through different dynamical regimes, including pathological synchronization, asynchronous activity, bistability, and near-critical states. By linking AI-assisted feature extraction and model inversion with mechanistic whole-brain simulations, EBRAINS The Virtual Brain Twin can help infer latent physiological parameters such as excitability, synaptic gain, dopaminergic modulation, and network coupling. This interaction between AI, empirical biomarkers, and digital twin modeling offers a pathway toward predictive, patient-specific models of neurodegenerative disorders and more personalized neuromodulation strategies.

Generative System Dynamics in Recurrent Neural Networks

Marco Gori

University of Siena, Italy

Abstract. In this talk, I discuss an alternative view of learning from interaction, one that casts time in the leading role. I then show that this alternative path is a candidate to replace transformer-based architectures in carrying out generative processes. The emphasis is on mechanisms that favor generative processes arising simply from the initial conditions of the neural network. I prove that linear units enable universal generation capabilities, which can be extended to piecewise-linear units. Finally, I present evidence of the underlying symbolic structure that emerges from the proposed generation process.

Contents

Neuroscience: Cognitive Neuroscience, Brain-Inspired Computing, Digital Twins and Virtual Brain Models

Training Dataset Factors Determining Brain-Like Representations in the AlexNet Model for Object Classification

Nobuhiko Wagatsuma[1(✉)], Naoyuki Ohta[1], Akinori Hidaka[2], and Hiroshi Tamura[3,4]

[1] Toho University, 2–2-1 Miyama, Funabashi, Chiba 274-8510, Japan
nwagatsuma@is.sci.toho-u.ac.jp

[2] Tokyo Denki University, Ishizaka, Hatoyama-machi, Hiki-gun, Saitama 350–0394, Japan

[3] The University of Osaka, 1–4 Yamadaoka, Suita, Osaka 565–0871, Japan

[4] Center for Information and Neural Networks, 1–4 Yamadaoka, Suita, Osaka 565–0871, Japan

Abstract. Deep convolutional neural networks (DCNNs) trained for object classification often develop hierarchical representations that partially resemble the primate ventral visual stream. However, which properties of the training datasets are critical for the emergence of such brain-like representations, particularly in higher network layers, remains unclear. In this study, we systematically examined the effects of object class number and image resolution on the representational properties of AlexNet models trained using a (i) low-resolution ImageNet dataset with 1,000 object classes or (ii) high-resolution dataset with 10 object classes. Representational dissimilarity matrices were computed from each network layer and quantitatively compared with recorded neural representations from macaque visual areas V1 and V4, and the inferotemporal cortex (IT). Our results indicate that early convolutional layers show robust correspondence with V1, irrespective of the training dataset. In contrast, the fully connected layers trained on datasets with a limited number of object classes exhibited a pronounced reduction in representational similarity to V4 and IT, even when image resolution was preserved. Notably, reduction in image resolution alone did not substantially impair the correspondence between higher network layers and higher visual cortical areas. These findings indicate that the number of object classes in the training dataset is a key determinant of brain-like representations in the higher layers of DCNNs, exerting a stronger influence than image resolution. Moreover, these results demonstrate that task structure and dataset diversity critically shape the organization of internal representations in artificial vision models, with important implications for computational neuroscience and biologically inspired deep learning.

Keywords: AlexNet model · deep convolutional neural network · training dataset factors · visual cortex · object recognition

L. Pasa et al. (Eds.): ICANN 2026, LNCS 17089, pp. 3–14, 2027.
https://doi.org/10.1007/978-3-032-38398-3_1

1 Introduction

Recent models based on deep convolutional neural networks (DCNNs) form a powerful class of tools for addressing specific challenges such as object classification [1,2] through training on large-scale datasets such as ImageNet [3]. Trained DCNN models, exemplified by the AlexNet model [4,5], have led to substantial advances in object classification performance in computer vision (Fig. 1). The overall architecture of DCNNs is often regarded as reflecting the hierarchical organization of the primate visual system used for object perception [6–8]. Notably, previous studies have reported that an AlexNet model trained on the ImageNet dataset may acquire mechanisms for object classification that are, at least in part, analogous to those of the primate ventral visual stream involved in object recognition [9,10].

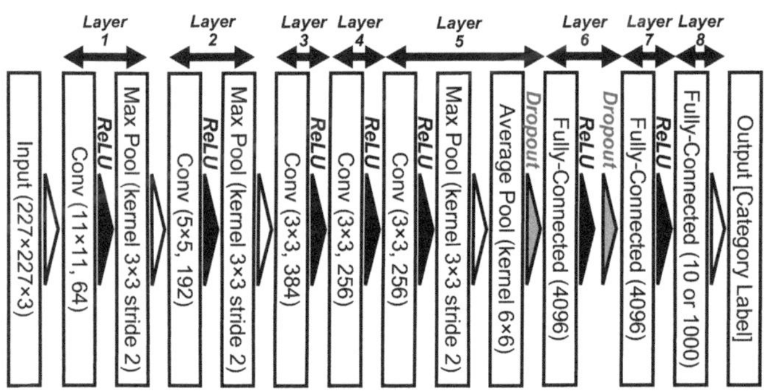

Fig. 1. Architecture of the AlexNet model [5] used in this study. We used various resized datasets (RGB color images, 227 × 227 pixels) as input for the network model.

In an ImageNet-trained AlexNet model, neurons in the lower layers tend to exhibit preferences for orientation and spatial frequency, mirroring key response properties of neurons in the primary visual cortex (V1) [6] and resembling Gabor filter–like characteristics [4]. Previous studies have further demonstrated that, in an ImageNet-trained AlexNet model, representations in the intermediate and higher layers show strong correlations with neural responses in the intermediate visual area V4 and inferotemporal cortex (IT) [9,10]. In contrast, a previous study reported that the artificial representations in the fully connected (FC) layers of AlexNet trained on the CIFAR-10 dataset [11], which contains fewer object categories and lower image resolution than ImageNet, differ substantially from the neural response properties observed in macaque V1, V4, and IT [12]. These observations suggest that the resolution of the training data, the number of object labels, or a combination of both may play a critical role in determining model neuron responses in the FC layers of the AlexNet model. However, it remains unclear which specific dataset properties are responsible for shaping these representational characteristics.

In this study, we examined how the resolution of the training dataset and number of object classes influence the representational properties underlying object recognition in the AlexNet model. We achieved this through quantitative comparisons with neuronal response properties in the primate visual cortex. Specifically, AlexNet models were trained on either (i) a low-resolution dataset obtained by downsampling ImageNet while preserving all 1,000 object classes (the low-resolution 1,000-class dataset) or (ii) a high-resolution dataset constructed by selecting 10 object classes from ImageNet (the high-resolution 10-class dataset). We quantitatively compared the representational properties of model neurons across the layers of AlexNet models trained on each dataset with the neuronal response properties of macaque V1, V4, and IT [9,13]. Our analyses revealed that the FC layers trained on the high-resolution 10-class dataset exhibited a pronounced reduction in representational similarity to V4 and IT, whereas the model trained on the low-resolution 1,000-class dataset substantially preserved the correspondence between higher network layers and higher visual cortical areas. These findings suggest that the number of object labels in the training dataset may contribute to the emergence of representational properties in the FC layers of AlexNet that more closely resemble those observed in higher visual cortical areas.

2 Materials and Methods

2.1 Training Datasets

In this study, we aimed to characterize the artificial representations of model neuron populations across layers of AlexNet models trained on either a low-resolution 1,000-class dataset or a high-resolution 10-class dataset. To this end, we quantitatively compared the artificial representations learned by these AlexNet models with the neural response properties of primate V1, V4, and IT [14].

To examine how the number of object classes and image resolution in a training dataset affect the representational properties of a trained DCNN, we trained AlexNet using either the low-resolution 1,000-class dataset (Fig. 2(a)) or the high-resolution 10-class dataset (Fig. 2(b)). Both datasets were constructed from subsets of the ImageNet dataset [3].

The low-resolution 1,000-class dataset was generated by substantially reducing the image resolution of the ImageNet dataset to degrade the texture and material information in the images (Fig. 2(a)). Specifically, the images comprising the ImageNet dataset were first downsampled by a factor of one-tenth in both the horizontal and vertical dimensions, yielding low-resolution images. This procedure substantially degraded the original image quality, particularly attenuating fine visual details such as textures and patterns dominated by high-frequency components. These low-resolution images were then upsampled to 227×227 pixels, the input resolution of the AlexNet model, resulting in a 1,000-class image dataset with an effective image quality that is similar to that of low-resolution

Fig. 2. Examples of the (a) low-resolution 1,000-class dataset (RGB images, 22 × 22 pixels) and (b) high-resolution 10-class dataset (RGB images, 227 × 227 pixels) used to train the AlexNet models in this study. All images were resized to 227 × 227 pixels prior to input into the AlexNet models (Fig. 1).

datasets such as CIFAR-10 [11]. Image downsampling and upsampling were performed using the OpenCV resize function in Python using bilinear interpolation.

The high-resolution 10-class dataset comprised approximately 12,000 training images, with about 1,200 images per object class, along with an additional 500 images reserved for validation (Fig. 2(b)). The dataset included 10 object classes: airplane, automobile, bird, cat, dog, frog, truck, monkey, fish, and motorcycle. Among these, seven categories (airplane, automobile, bird, cat, dog, frog, and truck) directly corresponded to classes in the CIFAR-10 dataset [11]. In contrast, the remaining categories (monkey, fish, and motorcycle) do not have direct counterparts in CIFAR-10 and were therefore treated as non-overlapping classes. Importantly, image resolution was not manipulated in the high-resolution 10-class dataset.

2.2 AlexNet Object Classification Model

The structure of the AlexNet model used in this study is shown in Fig. 1. The model follows the architecture proposed by Krizhevsky [5], which differs slightly from the original AlexNet model [4] employed in the 2012 ImageNet Large Scale Visual Recognition Challenge [3]. The network architecture consists of five convolutional (Conv) layers, three max-pooling layers, one average-pooling layer, and three FC layers. The model was adapted for either the low-resolution 1,000-class dataset or the high-resolution 10-class dataset by configuring the output layer to include 1,000 or 10 output model neurons, respectively, corresponding to the number of classification classes. A rectified linear unit (ReLU) [15] was used as the nonlinear activation function in all Conv layers. In layers 1, 2, and 5, the activated feature maps were subjected to 3 × 3 max pooling with a stride of two pixels. The final output layer, which follows the third FC layer, computes the probability that a given input image belongs to each object class. To reduce overfitting, two dropout layers were incorporated into the network. Model training and all simulations were performed using the PyTorch framework [16] on a ZOTAC GeForce RTX 3070 Ti GPU.

We trained the AlexNet model using either the low-resolution 1,000-class dataset or the high-resolution 10-class dataset, as defined in the previous section, to investigate the mechanisms underlying object classification. All training and validation images were resized to 227 × 227 pixels using the OpenCV resize function with bilinear interpolation. Model training was performed using stochastic gradient descent with a cross-entropy loss function. The initial learning rate was set to 0.01 and reduced three times over the course of training. The network was trained for 90 epochs with a batch size of 256 images. All training parameters followed those used in the original AlexNet implementation [4].

2.3 Physiological Experiments for Recording the Neural Responses of V1, V4, and IT to Natural Object Surface Images

In this study, we analyzed the neuronal responses from three visual cortical regions (V1, V4, and IT) of *Macaca* fuscata to natural object surface images [14] (Fig. 3). These neural responses were used as quantitative metrics to assess their correspondence with the responses of the model neurons in AlexNet models trained on either the low-resolution 1,000-class dataset or the high-resolution 10-class dataset. The details of the surgery, neural recording, and experimental procedures are provided in [14].

In this analysis, we used the firing rates of single unit spiking activity recorded from these visual cortices as references to examine the artificial representations across layers of the AlexNet models trained on the two datasets (Fig. 1). The mean firing rates were standardized according to a Gaussian distribution with a mean of zero and variance of one for each V1, V4, and IT neuron. The stimulus set of natural object surface images includes eight types of natural objects, comprising 64 images in total (Fig. 3): stones (#1–8), tree bark (#9–16), leaves (#17–24), flowers (#25–32), fruits and vegetables (#33–40), butterfly wings (#41–48), feathers (#49–56), and skins and furs (#57–64). Neuronal responses to the 64 images were recorded from V1 (691 neurons from two animals), V4 (494 neurons from two animals), and IT (294 neurons from three animals).

2.4 Data Analyses Based on Representational Dissimilarity Matrices (RDMs)

Representational dissimilarity matrices (RDMs) have been used to investigate the relationship between the artificial representations in the DCNN model and neural representations in V1, V4, and IT [14]. To obtain the RDMs for neural representations, we computed the representational dissimilarity of the neuronal responses in V1, V4, and IT for all pairs of natural object surface images as follows:

$$RD_v(i,j) = 1 - R_v(i,j) \tag{1}$$

where R_v is the Pearson's correlation coefficient, using the standardized firing rates of neuron populations in a visual cortical area v, for two input images i

Fig. 3. Natural object surfaces used for recording neuronal responses in M. fuscata V1, V4, and IT taken from [14]. A stimulus set in that physiological study consisted of 64 images of eight types of natural objects: stones (St, #1–8), tree barks (Ba, #9–16), leaves (Le, #17–24), flowers (Fl, #25–32), fruits and vegetables (FV, #33– 40), butterfly wings (BW, #41–48), feathers (Fe, #49–56), and skins and furs (SF, #57–64).

and j. The index RD is 0 if the response patterns of the neuron populations are identical for two input images. In contrast, RD increases with increasing representational dissimilarity in the response patterns of a neuron population for two stimuli. The RD_v for all 2016 pairs of natural object surface images are then used as the individual elements of the RDMs.

In this study, we obtained RDMs with 64 × 64 elements, where each element of an RDM quantifies the representational dissimilarity of a neuron population for two natural object surface images. These RDMs are symmetrical, featuring zeros along the main diagonal. We also computed the RDM_l for the model neuron activities in layer l of the AlexNet models trained on either the low-resolution 1,000-class dataset or the high-resolution 10-class dataset. To quantify the similarity between neural and artificial representations underlying object classification, we used Pearson's correlation coefficient (r) to compare the RDMs derived from macaque V1, V4, and IT with those obtained from each layer of the corresponding models.

3 Results

3.1 Accuracy of the AlexNet Models Trained on the Low-Resolution 1,000-Class Dataset or High-Resolution 10-Class Dataset

In this study, we aimed to elucidate how dataset characteristics shape the artificial representations of model neuron populations across different layers of a DCNN trained for object classification. To this end, we quantitatively compared the artificial representations of the AlexNet models trained on either the low-resolution 1,000-class dataset or the high-resolution 10-class dataset with neural representations observed in *Macaca* V1, V4, and IT [14]. First, to assess whether the AlexNet model trained on the low-resolution 1,000-class dataset was capable of robust object discrimination, we evaluated the model using 50,000 natural images from the corresponding validation set. This model achieved a top-5 accuracy of 63.3% on these images. In addition, the model trained on the high-resolution 10-class dataset achieved a top-1 accuracy of 75.4% on its validation image set. Together, these results indicate that both AlexNet models were adequately trained for classifying natural images into distinct object classes.

3.2 Relationship Between Neural Representations in Visual Cortical Areas and Artificial Representations Learned from Low-Resolution 1,000-Class and High-Resolution 10-Class Datasets

The RDMs computed for the artificial representations of the layers in an AlexNet model trained either on the low-resolution 1,000-class dataset or the high-resolution 10-class dataset are shown in Figs. 4(a) and (b), respectively. In the RDMs, elements with smaller values, shown as darker colors, indicate greater similarity between the response properties of pairs of object surface images i and j. The FC layers of the model trained on the low-resolution 1,000-class dataset (Fig. 4(a)) exhibited an overall tendency toward dissimilarity values that are lower than those of the model trained on the high-resolution 10-class dataset (Fig. 4(b)). The RDMs computed from the neural representations in macaque visual cortical areas reported in prior studies [9,12,13] were used to assess the similarity between neural and artificial representational structures.

3.3 Dataset-Dependent Correspondence Between Neural Representations in Visual Cortex and Artificial Representations

To examine how the characteristics of the training dataset modulate the artificial representations acquired by the AlexNet model, we computed the correlations between RDMs derived from the neural responses in macaque V1, V4, and IT and RDMs based on AlexNet model neuron responses. Figure 5(a) shows the correlation (r_{V1}) between the RDM of V1 and the RDMs of AlexNet model trained on either the low-resolution 1,000-class dataset (solid black line) or the high-resolution 10-class dataset (black dashed line), plotted as a function of model layer. For comparison, the gray solid line indicates the correlation between the RDM derived from V1 neural responses and the RDM computed from the AlexNet model trained on the standard ImageNet dataset, as reported in [9]. Across all models, the artificial representations in the early Conv layers were highly correlated with the neural representations in V1. The magnitude of r_{V1} exhibited by each model tended to decrease with increases in the model layers. Notably, the AlexNet model trained on the high-resolution 10-class dataset exhibited a more rapid qualitative decline in correlation strength compared with those of the other models.

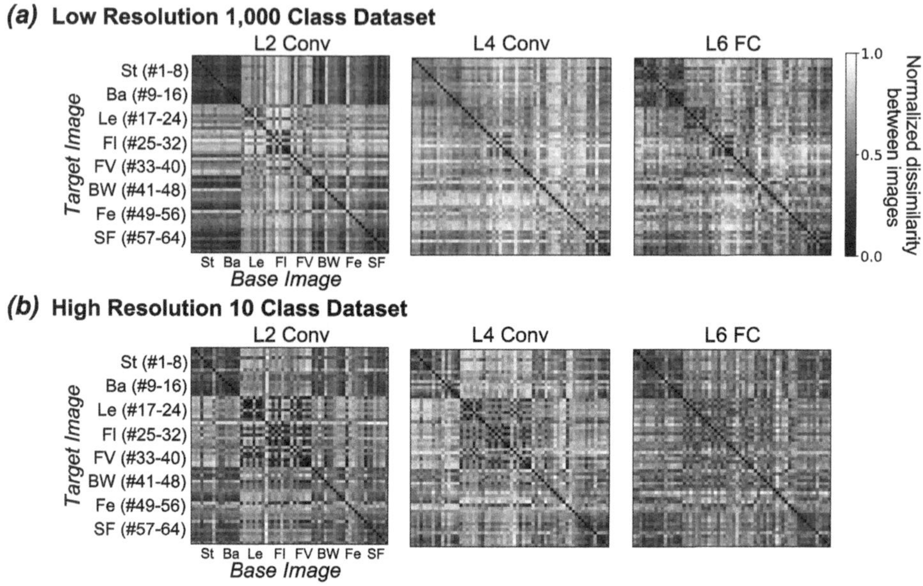

Fig. 4. RDMs derived from model neuron responses to natural object surface images in the Conv and FC layers of an AlexNet model trained on the (a) low-resolution 1,000-class dataset and (b) high-resolution 10-class dataset. Each element quantifies the dissimilarity between the model neuron activity patterns evoked by a pair of natural object surface images.

Figure 5(b) presents the correlation between the RDM derived from the neural representations in V4 and the RDMs computed from three AlexNet models trained on different datasets (r_{V4}). For the AlexNet model trained on the low-resolution 1,000-class dataset, r_{V4} (solid black line) increased from Conv layers 1 to 3 and continued to rise with increasing layer levels. High r_{V4} values were also preserved in the FC layers. This pattern was similar to that observed in the AlexNet model trained on ImageNet (gray line). In contrast, within the FC layers, the AlexNet model trained on the high-resolution 10-class dataset exhibited markedly lower r_{V4} values (black dashed line) than the other two models.

Figure 5(c) shows the correlation between the IT cortex responses and each layer of the three models (r_{IT}). For models trained on the standard ImageNet and low-resolution 1,000-class datasets (gray and solid black lines, respectively), strong correlations were maintained across the FC layers, reaching levels similar to the r_{V4} values shown in Fig. 5(b). In contrast, the model trained on the high-resolution 10-class dataset (black dashed line in Fig. 5(c)) exhibited substantially lower correlations in the FC layers relative to the other two models. These findings indicate that, in shaping the representational characteristics acquired in the FC layers of AlexNet, the number of object classes may play a more critical role than image resolution. In contrast, relative to the standard ImageNet-trained AlexNet model, the model trained on the high-resolution 10-class dataset showed

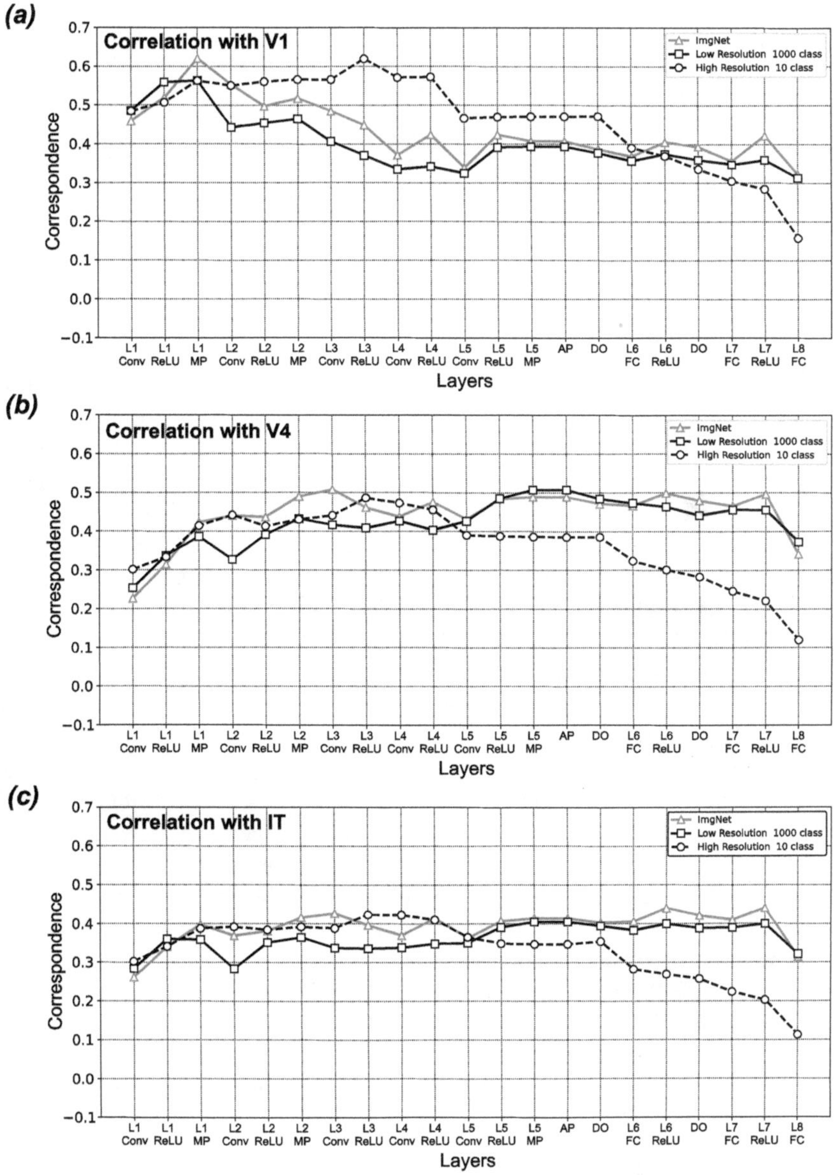

Fig. 5. Correlation between the artificial representations in AlexNet and neural representations in V1, V4, and IT. Correlation coefficients (r_v) for AlexNet models trained on the low-resolution 1,000-class dataset, the high-resolution 10-class dataset, and standard ImageNet are shown as solid black, dashed black, and gray lines, respectively. The x-axis shows layer l of the AlexNet model (Fig. 1). (a) Correlation with V1 (r_{V1}) (b) Correlation with V4 (r_{V4}). (c) Correlation with IT (r_{IT}).

a pronounced reduction in the correlation with neuronal responses in V4 and IT at the FC layers.

4 Discussion

In this study, we investigated how key characteristics of the training dataset such as the number of object classes and image resolution shape the representational characteristics that emerge across the layers of the AlexNet model. These effects were quantitatively evaluated through direct comparisons with the neuronal response properties of the macaque visual cortex [14]. In the FC layers, the neuronal responses of the model trained on the low-resolution 1,000-class dataset exhibited correlation with those of macaque V1, V4, and IT that is similar in magnitude to that observed for the standard ImageNet-trained AlexNet model with 1,000 object classes. In contrast, the correlation of the neuronal responses of the FC layers of the model trained on the high-resolution 10-class dataset with those of V4 and IT was markedly lower than that of the standard ImageNet-trained AlexNet. Taken together, these results suggest that, for the FC layers of the AlexNet model to acquire artificial representations analogous to those observed in higher-order visual cortical areas of macaques, the number of object classes in the training dataset may play a more important role than image spatial resolution. In addition, the activities of neurons in V4 and IT may contribute to the neural representations of the textures and materials of natural objects [17,18]. The pronounced correspondence between neural representations in V4 and IT and the representations in the FC layers of AlexNet trained on the low-resolution 1,000-class dataset suggests that greater diversity of object categories in the training set may play an important role in shaping higher-level visual representations, including their selectivity for texture and material properties [19].

In contrast to the FC layers, the early layers of all models examined in this study exhibited strong correlations with V1. This observation suggests that model neurons exhibiting preferences analogous to those of V1 neurons can emerge in the early layers of AlexNet largely independently of image resolution or the number of classes in the training dataset. In primates, neurons in V1 are well known to respond selectively to the orientation of visual stimuli presented within their receptive fields [6]. Consistent with this property, model neurons in the early layers of AlexNet models trained on the ImageNet, low-resolution 1,000-class, and high-resolution 10-class datasets may acquire orientation-selective response properties to input images, resembling those of orientation-selective neurons in V1.

Our findings suggest that the number of object classes in the training dataset may be an important factor in the emergence of brain-like representations in the higher layers of DCNNs, potentially exerting a stronger influence than image resolution. However, the total number of training images differs substantially between the low-resolution 1,000-class dataset and the high-resolution 10-class dataset, and this difference may also contribute to the observed representational characteristics in the AlexNet model. Further investigations systematically controlling for the total number of training images will be necessary to more precisely clarify the mechanisms underlying the acquisition of representational properties that resemble those observed in visual cortical areas.

Acknowledgments. This work was partly supported by KAKENHI grants JP23H03697, JP25K15850 and JP25K15199 and a grant from a Toho University Grant for Research Initiative Program (TUGRIP).

References

1. Simonyan, K., Zisserman, A.: Very deep convolutional networks for large-scale image recognition. arXiv 1409, pp. 1556 (2014)
2. He, K., Zhang, X., Ren, S., Sun, J.: Deep residual learning for image recognition. In: Proceedings of the 2016 IEEE Conference on Computer Vision and Pattern Recognition (CVPR), vol. 1, pp. 770–778 (2016)
3. Russakovsky, O., et al.: ImageNet large scale visual recognition challenge. Int. J. Comput. Vis. **115**, 211–252 (2015)
4. Krizhevsky, A., Sutskever, I., Hinton, G.E.: ImageNet classification with deep convolutional neural networks. In: Proceedings of the 25th International Conference on Neural Information Processing Systems, vol. 1, pp. 1097–1105 (2012)
5. Krizhevsky, A.: One weird trick for parallelizing convolutional neural networks. arXiv 1404, pp. 5997 (2014)
6. Hubel, D.H., Wiesel, T.N.: Receptive fields and functional architecture of monkey striate cortex. J. Physiol. **195**, 215–243 (1968)
7. Felleman, D.J., Van Essen, D.C.: Distributed hierarchical processing in the primate cerebral cortex. Cereb. Cortex **1**, 1–47 (1991)
8. LeCun, Y., Bengio, Y., Hinton, G.: Deep learning. Nature **521**, 436–444 (2015)
9. Wagatsuma, N., Hidaka, A., Tamura, H.: Analysis based on neural representation of natural object surfaces to elucidate the mechanisms of a trained AlexNet model. Front. Comput. Neurosci. **16**, 979258 (2022)
10. Khaligh-Razavi, S.-M., Kriegeskorte, N.: Deep supervised, but not unsupervised models may explain IT cortical representation. PLoS Comput. Biol. **10**, e1003915 (2014)
11. Krizhevsky, A.: Learning multiple layers of features from tiny images (2009). https://www.cs.toronto.edu/~kriz/learning-features-2009-TR.pdf
12. Wagatsuma, N., Yamaguchi, M., Hidaka, A., Tamura, H.: Analysis on artificial representations of a trained alexNet model using the CIFAR-10 dataset. Commun. Comput. Inf. Sci. **2296**, 90–101 (2024)
13. Wagatsuma, N., Hidaka, A., Tamura, H.: Correspondence between monkey visual cortices and layers of a saliency map model based on a deep convolutional neural network for representations of natural images. eNeuro **8** (2020). ENEURO.0200-20.2020
14. Tamura, H., Otsuka, H., Yamane, Y.: Neurons in the inferior temporal cortex of macaque monkeys are sensitive to multiple surface features from natural objects. bioRxiv 10.1101/086157 (2016)
15. Nair, V., Hinton, G.E.: Rectified linear units improve restricted boltzmann machines. In: Proceedings of the 27th International Conference on Machine Learning, vol. 27, pp. 807–814 (2010)

16. Paszke, A., Gross, S., Massa, F., Lerer, A., Bradbury, J., Chanan, G.: PyTorch: an imperative style, high-performance deep learning library. In: Proceedings of the 33th International Conference on Neural Information Processing Systems, vol. 721, pp. 8024–8035 (2019)
17. Goda, N., Tachibana, A., Okazawa, G., Komatsu, H.: Representation of the material properties of objects in the visual cortex of nonhuman primates. J. Neurosci. **37**, 2660–2673 (2014)
18. Komatsu, H., Goda, N.: Neural mechanisms of material perception: quest on Shitsukan. Neuroscience **392**, 329–347 (2018)
19. Wagatsuma, N., Ito, K., Hidaka, A.: Layer-specific feature preference in a trained alexNet model revealed by stylized image analysis. Sci. Rep. (2026). https://doi.org/10.1038/s41598-026-53152-9

A Predictive Coding Model for Memory Replay in the Neocortex

Lucie Fontaine[1,2,3(✉)] and Frédéric Alexandre[1,2,3]

[1] Centre Inria de l'université de Bordeaux, Talence, France
frederic.alexandre@inria.fr
[2] Laboratoire Bordelais de Recherche en Informatique (LaBRI), Talence, France
[3] Institut des Maladies Neurodégénératives (IMN), Bordeaux, France
lucie.fontaine@inria.fr

Abstract. The neocortex and the hippocampus are two complementary learning systems which interact during memory construction and consolidation. The hippocampus stores episodic memories coming from the neocortex, and later replays them back to the neocortex to transform them into semantic memory during memory consolidation. It was found that the hippocampus is also involved in imagining novel, never experienced before episodes, suggesting the idea that hippocampal replay is a generative process. We propose a predictive coding model of the neocortex and model its interactions with the hippocampus during memory replay. Our work suggests that the quality of replay depends on the separability of the representations, which can be controlled by the width and depth of the network.

Keywords: Predictive coding · Memory · Representation learning

1 Introduction

According to McClelland and McNaughton [13], the hippocampus and the neocortex are complementary learning systems (CLS). The necortex is responsible for semantic memory, that is the general knowledge that we have about the world, whereas hippocampus stores episodic memories, which correspond to an individual's emotional and sensory experiences, [11]. After storage in the hippocampus, an episodic memory can be recalled from a corrupted version of it. Furthermore, episodic memories are replayed during rest or sleep for memory consolidation. Memory replay corresponds to the spontaneous reactivation of the activity associated to an episodic memory in the hippocampus and its subsequent reinstantiation in the neocortex.

According to the CLS theory, the reinstatement of new memories interleaves them with existing memories to integrate them into the semantic memory of the neocortex. This principle has inspired successful applications in reinforcement learning and supervised continual learning tasks, where the learning of

L. Pasa et al. (Eds.): ICANN 2026, LNCS 17089, pp. 15–26, 2027.
https://doi.org/10.1007/978-3-032-38398-3_2

new tasks interferes with the knowledge of previously learned tasks [5,14]. Subsequently, experimental evidence suggested that the hippocampus is engaged in mental simulation and imagination, beyond episodic memory [20,25]. This has inspired researchers from NeuroAI to implement hippocampal replay as a generative process, in the form of "generative replay", which exhibits some advantages over experience replay [22,24].

Despite the importance of memory replay for continual learning, its implementation in the neocortex is not well understood. Because of the similarity of architecture in all its areas, researchers hypothesized that a common algorithm underlies computations in the neocortex [8,9]. Principles of organization in the neocortex have emerged from empirical studies in the visual cortex, which have shown that this region is arranged hierarchically, with forward connections from lower to higher areas, and backward connections from higher to lower areas [8]. Predictive coding networks (PCN), originally proposed by Rao and Ballard [18], were later shown by Friston [8] to align closely with the neocortex's architecture and information processing dynamics, establishing predictive coding as a compelling framework for understanding cortical computation. Since the foundational work of Friston [8], numerous computational models have demonstrated that PCN can effectively account for a wide range of functional properties observed along the visual cortical hierarchy [3,6,12,19].

Located at the apex of the neocortical hierarchy [2], the hippocampus receives input from the entorhinal cortex, which combines representations from different high-level neocortical areas of different sensory modalities, through forward connections (Fig. 1a). Thus, Bogacz and colleagues [19,23] modelled the hippocampus as the top level of a predictive coding network. However, by modelling the hippocampus and the neocortex as auto-associative memory systems within the same predictive coding framework, they overlooked the complementary roles these structures play in learning. In a previous study [7], we re-evaluated the role of the neocortex in auto-associative memory. We reproduced the result of Salvatori et al. [19] that PCNs can store training images as memories (Fig. 1c), but showed that this is done by overfitting an overparameterized network to a few training images. This work showed that while the neocortex might not be able to store patterns like the hippocampus, it integrates episodic memory in a semantic form during memory replay.

Memory replay can be decomposed into two phases, which we call hippocampal replay and cortical replay respectively (Fig. 1b). Memory replay is initiated in the hippocampus by hippocampal replay, which generates an episode in the entorhinal cortex. Then, cortical replay transfers the episode from the entorhinal cortex to the different neocortical areas through backward connections. In this paper, we propose to model the two phases of memory replay using a 2-layer PCN, focusing on the visual pathway. Experience and generative replay are two modes of hippocampal replay. Experience replay corresponds to the reactivation of a stored pattern of activations in the hippocampus, whereas generative replay activates a novel pattern of activations in the hippocampus. In both cases, the pattern of activations in the hippocampus is transfered to the neocortex during the subsequent cortical replay phase. Whereas hippocampal replay is modelled at a coarse, computational level, we model cortical replay at a more fine grained, algorithmic level. Our experiments study the effect of the depth and width of the PCN on the representations and on replay. Results are presented for the MNIST dataset, and discussed in light of additional experiments using Fashion-MNIST and CIFAR-10.

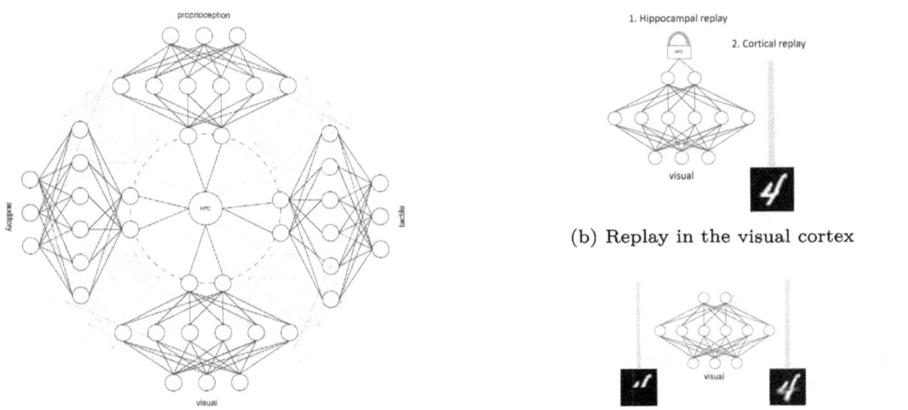

(a) Neocortical hierachy

(b) Replay in the visual cortex

(c) Pattern completion in the visual cortex

Fig. 1. (a) The high-level representations in neocortical areas corresponding to different sensory modalities are combined in the entorhinal cortex (represented as the dotted circle) which is the input of the hippocampus (HPC), located at the apex of the neocortical hierarchy. (b) After storage of an episodic memory in the hippocampus, it can be replayed in the hippocampus (hippocampal replay) before being reinstantiated in the neocortex, including the visual cortex (cortical replay). (c) The visual cortex can complete an image based on semantic memory [7].

2 Related Works

Models of Memory Replay. Replay in the neocortex was modelled computationnally to study memory consolidation, using artificial neural network, such as Boltzmann machines [10] or Variational Autoencoders [21], which are not faithful to the architecture and information processing of the neocortex.

Image Generation in PCN. Predictive coding networks (PCN) were used for the generation of inputs, but without specifying how generative processing in the neocortex can coexist with the presence of sensory inputs, or its relation to the hippocampus. On the one hand, Monte Carlo predictive coding [15] was proposed for learning probability distributions of sensory inputs, arguing that classical PCN demonstrated limited performance in generative tasks. On the other hand, Ororbia and Kifer [17] generated images from classical PCNs, but only a few samples were shown, leaving aside a large part of the image space which is covered by the generative model.

3 Methods

Building upon our previous study [7], we propose a minimal predictive coding model of the visual hierarchy to account for memory replay in the visual cortex.

3.1 Model

Our model is a PCN based on a hierarchical generative model [8] with $L = 2$ layers

$$\begin{aligned} p_{\boldsymbol{\theta}_0}(\boldsymbol{h_0} \mid \boldsymbol{h_1}) &= \mathcal{N}(\boldsymbol{h_0}; \boldsymbol{W_0}\boldsymbol{h_1}, \boldsymbol{\Sigma_0}) \\ p_{\boldsymbol{\theta}_1}(\boldsymbol{h_1} \mid \boldsymbol{h_2}) &= \mathcal{N}(\boldsymbol{h_1}; f(\boldsymbol{W_1}\boldsymbol{h_2} + \boldsymbol{b_1}), \boldsymbol{\Sigma_1}) \\ p(\boldsymbol{h_2}) &= \mathcal{N}(\boldsymbol{h_2}; \boldsymbol{0}, \boldsymbol{\Sigma_2}) \end{aligned}$$

where $\boldsymbol{h_0}, \boldsymbol{h_1}, \boldsymbol{h_2}$ are the states, $\boldsymbol{\theta}_0 = \boldsymbol{W_0}$ and $\boldsymbol{\theta}_1 = (\boldsymbol{W_1}, \boldsymbol{b_1})$ are the parameters and f is a nonlinearity. The bias and nonlinearity are not required in level 0, which is the input level, because an image can be represented as a linear combination of basis functions [16]. The covariances $\boldsymbol{\Sigma}_i$ of the likelihoods at each level i are set to $\boldsymbol{I}$, except for the input level 0 during replay, as explained in Sect. 3.3. Therefore, the prior on the latent state of level 2 is a centered isotropic multivariate Gaussian with no parameter. Recognition is assumed to be deterministic, such that

$$q_{\boldsymbol{\phi}}(\boldsymbol{h_1}, \boldsymbol{h_2} \mid \boldsymbol{x}) = \delta(\boldsymbol{h_1} - \boldsymbol{\phi_1})\delta(\boldsymbol{h_2} - \boldsymbol{\phi_2})$$

where $\boldsymbol{\phi} = (\boldsymbol{\phi_1}, \boldsymbol{\phi_2})$ is an estimate of the causes of the input $\boldsymbol{x}$.

The variational free energy for an input $\boldsymbol{x}$ is

$$\mathcal{L}(\boldsymbol{\theta}, \boldsymbol{\phi}; \boldsymbol{x}) = \frac{1}{2}\xi_0^\top \xi_0 + \frac{1}{2}\xi_1^\top \xi_1 + \frac{1}{2}\xi_2^\top \xi_2 + \text{constant}$$

where we have defined

$$\xi_0 = \boldsymbol{x} - \boldsymbol{W_0}\boldsymbol{\phi}_1 \tag{1}$$

$$\xi_1 = \boldsymbol{\phi}_1 - f(\boldsymbol{W_1}\boldsymbol{\phi}_2 + \boldsymbol{b_1}) \tag{2}$$

$$\xi_2 = \boldsymbol{\phi}_2. \tag{3}$$

3.2 Training Algorithm

Let us consider a dataset $\boldsymbol{X} = \{\boldsymbol{x}^{(i)}\}_{i=1}^N$ of N i.i.d. samples of a continous variable $\boldsymbol{x}$. The log likelihood can be written $\log p_{\boldsymbol{\theta}}(\boldsymbol{x}^{(1)}, ..., \boldsymbol{x}^{(N)}) = \sum_{i=1}^N \log p_{\boldsymbol{\theta}}(\boldsymbol{x}^{(i)})$. Therefore, the variational free energy of the dataset $\boldsymbol{X}$ is

$$\mathcal{L}(\boldsymbol{\theta}; \boldsymbol{X}) = \sum_{i=1}^N \mathcal{L}(\boldsymbol{\theta}, \boldsymbol{\phi}; \boldsymbol{x}^{(i)}),$$

which can be estimated based on minibatches

$$\mathcal{L}(\boldsymbol{\theta}; \boldsymbol{X}) \approx \mathcal{L}^M(\boldsymbol{\theta}; \boldsymbol{X}^M) = \frac{N}{M} \sum_{i=1}^{M} \mathcal{L}(\boldsymbol{\theta}, \boldsymbol{\phi}^{(i)}; \boldsymbol{x}^{(i)}) \tag{4}$$

where the minibatch $\boldsymbol{X}^M = \{\boldsymbol{x}^{(i)}\}_{i=1}^M$ of size M is randomly drawn from the dataset $\boldsymbol{X}$.

Given a minibatch $\boldsymbol{X}^M$, the latent states $\boldsymbol{\phi}_0^{(i)}$ and $\boldsymbol{\phi}_1^{(i)}$ are updated during T iterations for each datapoint $\boldsymbol{x}^{(i)}$ to minimize the variational free energy $\mathcal{L}(\boldsymbol{\theta}, \boldsymbol{\phi}^{(i)}; \boldsymbol{x}^{(i)})$. For a vector $\mathbf{v}$, we note diag[$\mathbf{v}$] the diagonal matrix with $\mathbf{v}$ in its diagonal. The update rules for the two layers in our model are

$$\Delta\boldsymbol{\phi}_1 = \alpha(\boldsymbol{W}_0^\top \boldsymbol{\xi}_0 - \boldsymbol{\xi}_1) \tag{5}$$

$$\Delta\boldsymbol{\phi}_2 = \alpha(\boldsymbol{W}_1^\top \mathrm{diag}\left[f'(\boldsymbol{W}_1\boldsymbol{\phi}_2 + \boldsymbol{b}_1)\right] \boldsymbol{\xi}_1 - \boldsymbol{\xi}_2). \tag{6}$$

where α is the inference rate.

Then, the parameters $\boldsymbol{\theta}_0 = \boldsymbol{W}_0$ and $\boldsymbol{\theta}_1 = (\boldsymbol{W}_1, \boldsymbol{b}_1)$ are updated once to minimize the estimate $\mathcal{L}^M(\boldsymbol{\theta}; \boldsymbol{X}^M)$ given in Eq. 4. Thus, the learning rules can be calculated from the sum of gradients $\sum_{i=1}^M \nabla_{\boldsymbol{\theta}_l} \mathcal{L}(\boldsymbol{\theta}; \boldsymbol{x}^{(i)})$

$$\Delta\boldsymbol{W}_0 = \beta \sum_{i=1}^{M} \boldsymbol{\xi}_0^{(i)} \boldsymbol{\phi}_1^{(i)\top}$$

$$\Delta\boldsymbol{W}_1 = \beta \sum_{i=1}^{M} \left[\boldsymbol{\xi}_1^{(i)} \odot f'(\boldsymbol{W}_1 \boldsymbol{\phi}_2^{(i)} + \boldsymbol{b}_1)\right] \boldsymbol{\phi}_2^{(i)\top}$$

$$\Delta\boldsymbol{b}_1 = \beta \sum_{i=1}^{M} \boldsymbol{\xi}_1^{(i)} \odot f'(\boldsymbol{W}_1 \boldsymbol{\phi}_2^{(i)} + \boldsymbol{b}_1).$$

where $\odot$ is the element-wise product and β is the learning rate.

We train the model on a dataset $\boldsymbol{X}$ for multiple epochs, and reduce the learning rate by a multiplicative factor $\gamma < 1$ at each epoch of training to prevent instability issues.

3.3 Mapping to the Brain

Figure 2 shows how this algorithm can be implemented in a neural network mapped onto the visual pathway of the brain, with only local computations. The three levels in the network correspond, from bottom to top, to the lateral geniculate nucleus (LGN) in thalamus (and not the retina because it doesn't receive feedback connections from LGN), the visual cortex (VC) and the entorhinal cortex (EC). This mapping allows us to study memory replay after training the model.

At inference time, the latent states are updated until convergence or until a maximum number of iterations $T_{\max}$ is reached in order to study the converged representations. Convergence occurs when the relative change in the norm of the latent state ϕ_l is smaller than a threshold ϵ, i.e.

$$\frac{\|\nabla_{\phi_l}\mathcal{L}(\boldsymbol{\theta}, \boldsymbol{\phi}; \boldsymbol{x})\|}{\|\phi_l\|} < \epsilon.$$

During perception, the network is driven by the input image in LGN. While the LGN is set to the image, the VC and EC converge to hierarchical predictive representations of the image following the inference rules 5 and 6 respectively. The representation in VC is predictive of the image, as the prediction $\boldsymbol{W_0\phi_1}$ is a reconstruction of the image. The representations in EC can be stored in a memory buffer representing the hippocampus and later replayed during experience replay.

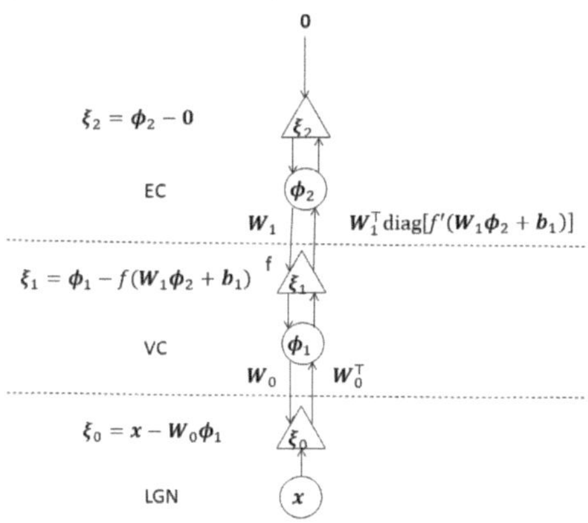

Fig. 2. PCN with 3 levels mapped to lateral geniculate nucleus (LGN), visual cortex (VC) and entorhinal cortex (EC). Each node corresponds to multiple neurons. Circle nodes correspond to units ϕ_l and triangle nodes correspond to error units $\boldsymbol{\xi}_l$. Each connection between nodes corresponds to a fully connected network, with excitatory connections in the upward direction and inhibitory connections in the downward direction. We consider the upmost prediction to be 0, to have a standard normal distribution as prior.

Hippocampal replay is modelled as the retrieval of a pattern from the hippocampal memory buffer in the case of experience replay and as a sampling process in the case of generative replay. Thus, during memory replay, hippocampal replay first outputs the EC representation of an image retrieved from the memory buffer in the case of experience replay or a sample generated in the latent space of EC in the case of generative replay (as described in the next paragraph). Then, during cortical replay, the network is driven by the representation in EC obtained from hippocampal replay. While setting the EC level to the corresponding representation, the VC converges to the replayed representation following the inference rule 5. As the network can also be presented with an input in LGN during memory replay, the representation in VC is protected from ascending input in LGN by setting the inverse covariance, also called precision, $\boldsymbol{\Sigma}_{\mathbf{0}}^{-1}$ in the LGN to 0, preventing the prediction errors in LGN to influence the activity in VC. In this way, attention is focused on the representation in EC, and not on the current input. Then, the prediction $\boldsymbol{W}_{\mathbf{0}}\boldsymbol{\phi}_{\mathbf{1}}$ based on the replayed representation $\boldsymbol{\phi}_{\mathbf{1}}$ in VC corresponds to the replayed image.

During generative replay, sampling of the latent space of EC is conditioned on the class. Indeed, we fit a multivariate Gaussian distribution to each class in the latent space of EC, by estimating the mean and covariance of training samples belonging to each class. Then, samples from a given class can be generated by sampling the corresponding Gaussian distribution.

During perception, the network can also perform pattern completion when presented with a corrupted (noisy or incomplete) input. Following Salvatori et al. [19], while the LGN level is initialized to the corrupted input, the VC and EC converge following the same dynamics as in regular perception (Eqs. 5 and 6), whereas the corrupted part of LGN converges following the inference rule

$$\Delta\boldsymbol{\phi}_0 = -\alpha\boldsymbol{\phi}_0.$$

obtained by calculating the gradient of the variational free energy of the input with respect to the corrupted part of LGN. Then, the prediction $\boldsymbol{W}_{\mathbf{0}}\boldsymbol{\phi}_{\mathbf{1}}$ based on the recalled representation $\boldsymbol{\phi}_{\mathbf{1}}$ in VC corresponds to the recalled image.

3.4 Experiments

The model is trained with 3 random seeds on the MNIST dataset, which contains images of handwritten digits from 0 to 9. We normalize the images before training. The original training set of 60,000 images is split into a training set of size 50,000 and a validation set of size 10,000. The validation set is used to evaluate the model during training and hyperparameter tuning. After training, we evaluate the model on the original test set of 10,000 images.

Hyperparameter values are chosen based on empirical trials informed by the predictive coding literature. In addition, the number of hidden units in level 1 is obtained by minimizing the variational free energy on the validation set for a PCN with $L = 1$ using grid search.

The choice of the number of hidden units in level 2 is more complex and results from a trade-off between different metrics. In our simulations, we study the influence of the number of hidden units in level 2 on the learned representations and on memory replay, both quantitatively and qualitatively. Quantitatively, we evaluate the predictive performance of the model, the quality of the experience replays and the pattern completion performance using the reconstruction, replay and completion errors, based on the mean squared error (MSE). The MSE between two flattened images $\boldsymbol{x}$ and $\hat{\boldsymbol{x}}$ is

$$\mathrm{MSE}(\boldsymbol{x}, \hat{\boldsymbol{x}}) = \frac{1}{N_{\text{pixels}}} \sum_{i=1}^{N_{\text{pixels}}} (x_i - \hat{x}_i)^2.$$

The reconstruction and completion errors are computed between the original and reconstructed images and between the original and recalled images respectively, averaged over the validation set, whereas the replay error is computed between the original and replayed images, averaged over the training set. In addition, we evaluate the linear separability of the latent manifolds corresponding to the different classes in the two hidden levels of the model using the classification accuracy of a simple multinomial logistic regression. For comparison, these measures are also computed for a one-layer PCN with the same width in level 1.

On the qualitative side, we examine examples of reconstructions of images from the validation set, as well as examples of experience replay corresponding to images from the training set. Additionnally, we show examples of images generated by generative replay and visualize their hierarchical representations. Whereas the quantitative results are shown for the 3 random seeds, the qualitative results are shown for only one of the seeds.

4 Results

The reconstruction error, completion error, replay error and classification accuracy are plotted against the number of hidden units in level 2 for 3 random seeds in Fig. 3. These measures are evaluated after training the models until convergence of all layers. We also include as a baseline the average values of the same metrics for a one-layer PCN with the same number of hidden units in level 1, except for the replay error because our implementation of replay requires at least two layers (see Sect. 3.3).

On the one hand, the second hidden layer in our model does not improve the predictive power of its representations (Fig. 3, left). Indeed, the same model with the top layer removed has a lower reconstruction error than any of the models with two hidden layers. The reconstruction error on the validation set increases before decreasing as the width of level 2 is increased. In addition, the completion error increases with the number of units in the second hidden level, but it is always lower than for a model with only one hidden layer, suggesting that adding a second layer is still beneficial to the completion performance, contrarily to the reconstruction performance.

On the other hand, the quality of replay and the linear separability of the classes in the second hidden level increase with the number of units in that level, even though the linear separability remains below that of a one-layer model (Fig. 3, right). In addition, we observe that if the top level is sufficiently large, the linear separability of the classes in this level surpasses that of the level below.

In sum, the width of the top level improves the quality of replay and the linear separability of the classes in that level, but degrades the completion performance on the validation set. Therefore, choosing the number of units in level 2 based on these metrics is not straightforward, and we will turn to the qualitative evaluation.

In Fig. 4a, we visualize examples of experience replay obtained with models of different widths. As shown previously [7], setting the width of the top

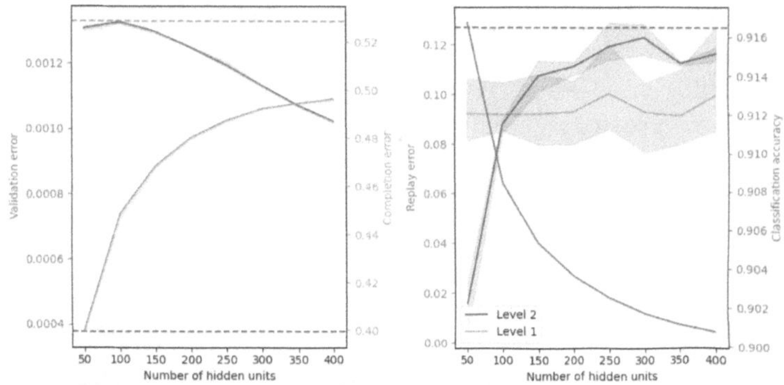

Fig. 3. Quantitative evaluation of our model according to different metrics, depending on the number of hidden units in level 2. The lines represent the averages over 3 random seeds, while the shaded areas represent the confidence interval. Dotted lines represent the average values for the same model, but with level 2 removed.

level to the number of classes (i.e. $n_2 = 10$) results in replayed images that are blurry and that do not retain the details of the original images. This issue is solved by increasing the width of the top level to 30. Increasing it further to 100 improves the sharpness of the replayed images, but at the cost of the completion performance. Indeed, while a width of 30 enables the network to semantically complete the bottom half of images taken from the validation set that were masked, a higher width of 100 makes the completion uninformative of the classes of the masked digits (Fig. 4b). Thus, the trade-off between the replay fidelity and completion performance could lead us to choose a width of 30.

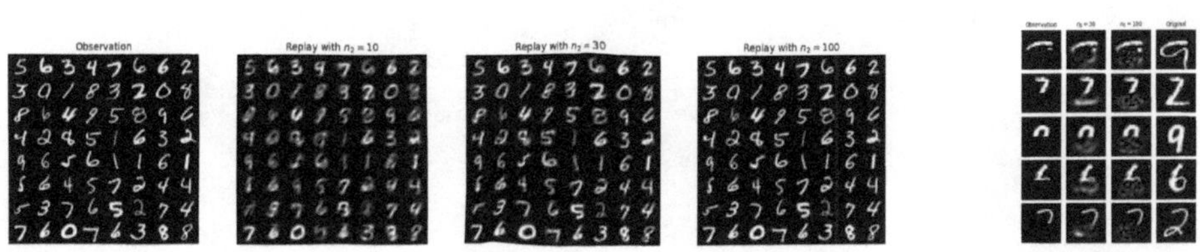

(a) Replay

(b) Pattern completion

Fig. 4. Replay and pattern completion by a PCN with different top level widths n_2. (a) A minibatch of images from the training set is replayed. (b) Five images from the validation set are masked and completed by the models.

Examples of reconstructions of images from the validation set and replayed images obtained by generative replay are shown for a model with $n_2 = 30$ units in level 2 (Fig. 5). Figure 5a shows that the representations learned by the model are perfectly predictive of images it has never seen during training. In Fig. 5c, we plotted hierarchical representations obtained with generative replay, over hierarchical representations inferred from images of the validation set. The representations of the different classes are well separated in all three levels, including the top level which we previously found to have overlapping clusters due to an insufficient width [7], and the representations generated by generative replay mostly fall within the right clusters in all levels. Consequently, most of the images generated for each class are realistic examples of their classes, even though some of them are blurry (Fig. 5b).

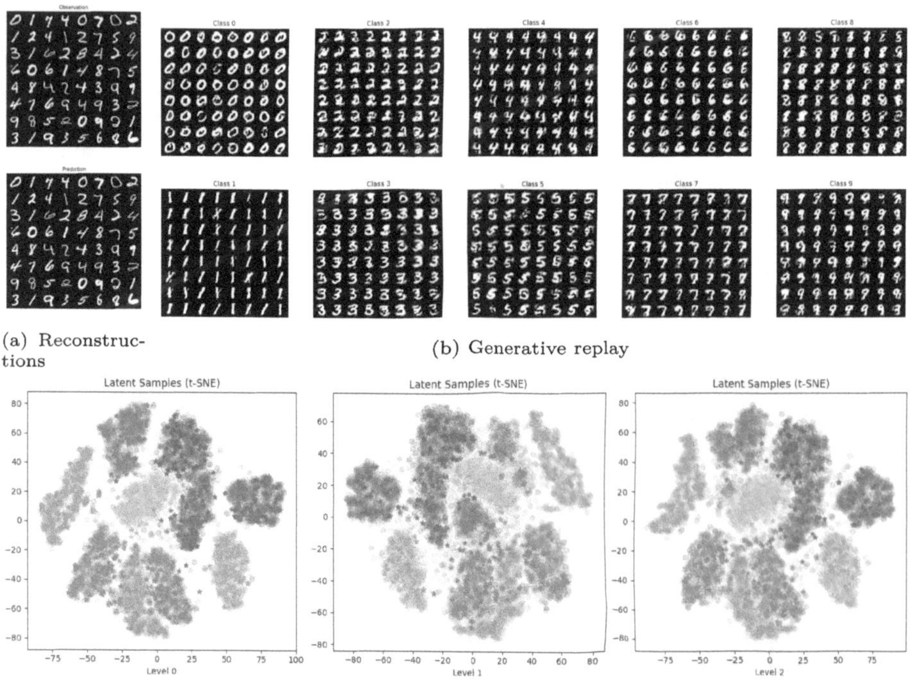

(a) Reconstructions

(b) Generative replay

(c) Latent and input space

Fig. 5. Visualizations for a model with $n_2 = 30$ units in the top level. (a) Reconstructions for a random mini-batch from the validation set. (b) Images generated by replay for each class. (b) Hierarchical representations of the images generated by generative replay (star-shaped markers) and of the images of the validation set (transparent circle markers), visualized in 2D using t-SNE. Each image is represented as one data point in each of the three subplots, colored according to the class.

5 Discussion

We proposed a model of hippocampo-neocortical interactions involved in memory replay using a PCN. Our work shows that predictive coding accounts for cortical replay, i.e. the transfer of an episodic memory from the hippocampus to the neocortex. To this purpose, we modelled the two modes of hippocampal replay, experience and generative replay, in a minimal way, by mapping the top level of the PCN onto the output of the hippocampus, the entorhinal cortex.

However, some of the images generated by our model were found to be out-of-distribution, probably because the simple Gaussian distribution we used does not capture the complex, non-linear geometry of the latent manifolds. This issue could be solved using other latent density models or using a Riemannian metric [1]. However, additional experiments with Fashion-MNIST and CIFAR-10 show that the complexity of the latent manifolds is increased with higher data complexity, making image generation more challenging. Indeed, class separability decreases across all layers when the dataset becomes more complex. However, for a given dataset, this separability tends to increase as we move up the network's hierarchy and can surpass that of a one-layer PCN, especially for more complex datasets such as CIFAR-10. Therefore, our model could be adapted to more complex datasets by increasing the network depth. Nonetheless, in order to provide a more complete account of memory replay, future work should aim at modelling the hippocampal formation with its different components to understand hippocampal replay at a more detailed level.

This work also contributes to understanding the effect of depth and width on the representations learned by PCNs. Regarding the width, we showed that a larger level-2 width improves the replay quality and the linear separability of the classes. Indeed, expanding the dimensionality of the representations is known to increase their separability [4], which could cause the replay of memorized patterns to improve due to the decreased interference between them. Regarding the depth, we observed that adding a second layer does not help in making the representations of the MNIST classes more linearly separable. This could be due to the simplicity of the dataset, which is already separable at the input level (Fig. 5c). Indeed, additional experiments show that the second layer improves the linear separability of the classes only for the most complex dataset, CIFAR-10.

References

1. Arvanitidis, G., Hansen, L.K., Hauberg, S.: Latent space oddity: on the curvature of deep generative models. arXiv preprint arXiv:1710.11379 (2017)
2. Barron, H.C., Auksztulewicz, R., Friston, K.: Prediction and memory: a predictive coding account. Prog. Neurobiol. **192**, 101821 (2020)
3. Brucklacher, M., Bohté, S.M., Mejias, J.F., Pennartz, C.M.: Local minimization of prediction errors drives learning of invariant object representations in a generative network model of visual perception. Front. Comput. Neurosci. **17**, 1207361 (2023)
4. Cayco-Gajic, N.A., Silver, R.A.: Re-evaluating circuit mechanisms underlying pattern separation. Neuron **101**(4), 584–602 (2019)
5. Chaudhry, A., et al.: On tiny episodic memories in continual learning. arXiv preprint arXiv:1902.10486 (2019)
6. Dora, S., Bohte, S.M., Pennartz, C.M.: Deep gated hebbian predictive coding accounts for emergence of complex neural response properties along the visual cortical hierarchy. Front. Comput. Neurosci. **15**, 666131 (2021)

7. Fontaine, L., Alexandre, F.: Semantic and episodic memories in a predictive coding model of the neocortex. In: 2025 International Joint Conference on Neural Networks (IJCNN), pp. 1–9. IEEE (2025)
8. Friston, K.: Learning and inference in the brain. Neural Netw. **16**(9), 1325–1352 (2003)
9. Hawkins, J., Lewis, M., Klukas, M., Purdy, S., Ahmad, S.: A framework for intelligence and cortical function based on grid cells in the neocortex. Front. Neural Circuits **12**, 121 (2019)
10. Káli, S., Dayan, P.: Off-line replay maintains declarative memories in a model of hippocampal-neocortical interactions. Nat. Neurosci. **7**(3), 286–294 (2004)
11. Kumaran, D., Hassabis, D., McClelland, J.L.: What learning systems do intelligent agents need? complementary learning systems theory updated. Trends Cogn. Sci. **20**(7), 512–534 (2016)
12. Li, T.E., Tang, M., Bogacz, R.: Predictive coding model detects novelty on different levels of representation hierarchy. Neural Comput. **37**(8), 1373–1408 (2025)
13. McClelland, J.L., McNaughton, B.L., O'Reilly, R.C.: Why there are complementary learning systems in the hippocampus and neocortex: insights from the successes and failures of connectionist models of learning and memory. Psychol. Rev. **102**(3), 419 (1995)
14. Mnih, V., et al.: Human-level control through deep reinforcement learning. Nature **518**(7540), 529–533 (2015)
15. Oliviers, G., Bogacz, R., Meulemans, A.: Learning probability distributions of sensory inputs with monte carlo predictive coding. PLoS Comput. Biol. **20**(10), e1012532 (2024)
16. Olshausen, B.A., Field, D.J.: Emergence of simple-cell receptive field properties by learning a sparse code for natural images. Nature **381**(6583), 607–609 (1996)
17. Ororbia, A., Kifer, D.: The neural coding framework for learning generative models. Nat. Commun. **13**(1), 2064 (2022)
18. Rao, R.P., Ballard, D.H.: Predictive coding in the visual cortex: a functional interpretation of some extra-classical receptive-field effects. Nat. Neurosci. **2**(1), 79–87 (1999)
19. Salvatori, T., et al.: Associative memories via predictive coding. Adv. Neural. Inf. Process. Syst. **34**, 3874–3886 (2021)
20. Schacter, D.L., Addis, D.R., Hassabis, D., Martin, V.C., Spreng, R.N., Szpunar, K.K.: The future of memory: remembering, imagining, and the brain. Neuron **76**(4), 677–694 (2012)
21. Spens, E., Burgess, N.: A generative model of memory construction and consolidation. Nat. Hum. Behav., 1–18 (2024)
22. Stoianov, I., Maisto, D., Pezzulo, G.: The hippocampal formation as a hierarchical generative model supporting generative replay and continual learning. Prog. Neurobiol. **217**, 102329 (2022)
23. Tang, M., Salvatori, T., Millidge, B., Song, Y., Lukasiewicz, T., Bogacz, R.: Recurrent predictive coding models for associative memory employing covariance learning. PLoS Comput. Biol. **19**(4), e1010719 (2023)
24. van de Ven, G.M., Siegelmann, H.T., Tolias, A.S.: Brain-inspired replay for continual learning with artificial neural networks. Nature Commun. **11** (2020)
25. Wikenheiser, A.M., Redish, A.D.: Decoding the cognitive map: ensemble hippocampal sequences and decision making. Curr. Opin. Neurobiol. **32**, 8–15 (2015)

White-Matter-Microstructure-Informed Whole-Brain Models Reveal Localized Excitation-Inhibition Imbalance in Schizophrenia

Kailin Zhu[1(✉)], Georg Reich[2], Xiaojun Zhou[1], and Trang-Anh E. Nghiem[1]

[1] Hertie Institute for AI in Brain Health, Faculty of Medicine, University of Tübingen, Tübingen, Germany
zklin20@gmail.com , trang-anh-estelle.nghiem@uni-tuebingen.de
[2] Neural Information Processing Group, Technical University of Berlin, Berlin, Germany
gr@ni.tu-berlin.de

Abstract. Providing early diagnosis and personalized treatment for psychiatric disorders like schizophrenia remains challenging, due to important interpersonal differences and still elusive neuronal mechanisms. Whole-brain network models show promising results with clinical relevance for individualized treatment recommendations in neurological disorders. However, their applicability to psychiatry is still limited as models fail to account for inter-individual differences in the correlation structure of brain dynamics. What physiological mechanisms should models incorporate to better account for individual profiles of brain dynamics in schizophrenia patients and healthy controls? Our study compares various metrics of white matter structure and microstructure to inform connection weights between regions. To do so, we inferred regional parameters of whole-brain mean-field models with The Virtual Brain simulator to account for empirical functional connectivity from resting-state functional magnetic resonance imaging of schizophrenia patients and healthy controls. We found that using global fractional anisotropy or apparent diffusion coefficient of white matter fibers to inform the weights in neural mass models can drastically improve model performance. The data-model correlations of simulated and empirical data were significantly improved (from 0.2 to 0.7) over using the number or density of fibers as in many state-of-the-art methods. This approach allows us to uncover personalized maps of excitation-inhibition imbalance, hypothesized to underlie symptoms in schizophrenia. These maps prove meaningful in that they can predict diagnosis better than model-independent neuroimaging benchmarks. Our findings highlight the importance of white matter microstructure in whole-brain modeling. The novel white-matter-informed models reveal mechanisms that can cause altered brain dynamics in schizophrenia and could inform treatment in personalized psychiatry. Our code is available at https://github.com/Comp-Neuropsy-Nghiem-Lab/TVB-schz.git.

L. Pasa et al. (Eds.): ICANN 2026, LNCS 17089, pp. 27–39, 2027.
https://doi.org/10.1007/978-3-032-38398-3_3

Keywords: The Virtual Brain · apparent diffusion coefficient · fractional anisotropy · personalized psychiatry

1 Introduction

Schizophrenia affects more than 23 million people worldwide [16]. Symptoms are wide-ranging across domains including hallucinations, delusions, cognitive control, and social interaction. Important inter-subject variability in symptoms and biomarkers has crucially hampered our ability to detect biomarkers of schizophrenia for early diagnosis as well as to develop widely effective treatments. Moreover, specifically targeting treatments to limit secondary effects is still challenging due to the fundamental gap in our knowledge of the cellular mechanisms causing altered brain function behind symptoms. What cellular mechanisms, in which brain regions, underlie altered brain dynamics in each schizophrenia patient? At the neuronal level, imbalance between excitation (E) and inhibition (I) due to loss of inhibitory conductance has been hypothesized to underlie symptoms [1]. As an alternative hypothesis, dysfunction of the dopamine system has been proposed - though recent evidence suggests a causal association between E/I imbalance and dopaminergic dysregulation in schizophrenia [6]. White matter structure and microstructure, as measured by diffusion-weighted magnetic resonance imaging (dwMRI), has also been observed to be altered schizophrenia [11]. However, it remains unclear whether cellular-level E/I imbalance or circuit-level white-matter could explain brain dynamical differences in schizophrenia. Further, it is still unclear which brain regions may be affected by these mechanistic alterations. The multi-domain nature of schizophrenia symptoms suggests that multiple brain regions may be affected, with distinct altered areas associated with different symptoms across patients. For instance, hallucinations may be caused by imbalance in sensory areas, with the most common hallucinations, auditory verbal i.e. hearing voices, possibly associated with E/I imbalance among language processing areas in the temporal lobe [8] and the salience network [10]. The default-mode network, associated with self-referential and social cognition, has also been found to be affected in schizophrenia among other disorders, including in its interactions with the salience network [9]. However, it remains difficult to map E/I balance at the individual level *in vivo*, as modern methods in magnetic resonance spectroscopy to probe E and I neurotransmitter concentrations need to be targeted to a single hypothesized region at a time.

To address this crucial challenge, personalized whole-brain models have emerged as a powerful paradigm to probe brain mechanisms behind brain dynamics at the individual level [4,7,15,20]. Briefly, whole-brain models simulate brain activity based on differential equations for each region, with connection weights and delays between regions informed by individualized white-matter data. Whole-brain models have proven clinically relevant for E/I imbalance disorders like epilepsy, by helping localize the epileptic focus in each patient to guide surgery [7]. However, inferring parameters like inhibitory conductance, controlling E/I imbalance, at the regional level has remained computationally challenging, until new developments introduced E-to-E and E-to-I inter-regional coupling inference [15]. Yet, these novel tools remain to be tested in their ability to account for individual brain dynamics in patient populations. Moreover, decades of literature on whole-brain modeling have never systematically investigated the effects of white-matter-microstructure in shaping functional connectivity (FC). In most models, when specified, the number or density of fibers is used to inform connection weights between regions. However, microstructural white-matter properties like generalized fractional anisotropy (gFA) and apparent diffusivity coefficient (ADC) may also modulate inter-regional interactions, which whole-brain modeling has never accounted for.

In this manuscript, for the first time, we systematically compare metrics of white matter structure and microstructure to inform personalized simulations of brain activity in schizophrenia and controls. To do so, we infer regional parameters of whole-brain mean-field models with The Virtual Brain (TVB) to account for individual FC from resting-state functional magnetic resonance imaging (fMRI) data. First, we compare models informed with different white-matter metrics in their fit to empirical FC. Then, we uncover the relevant aspects of white-matter data to capture FC, by shuffling white-matter matrices across regions and participants and testing whether data-model correlations worsen. Next, we reveal which regions are affected by E/I imbalance in schizophrenia by investigating inferred E/I parameter maps. Finally, we demonstrate that inferred E/I maps are meaningful by leveraging them for diagnostic classification and comparing accuracy with metrics independent of our model. Our approach provides a white-matter-microstructure-informed platform to model brain activity at the individual level in health and pathology, allowing us to introduce and validate tools to map E/I balance and derive mechanistic inductive biases supporting machine-learning-based diagnostics.

2 Materials and Methods

Data. The dataset [19] consists of structural and functional connectomes from 27 (age 41 $\pm$ 9.6) schizophrenic patients and 27 matched healthy controls (age 35 $\pm$ 6.8 years). All patients met the DSM-IV criteria for a diagnosis of schizophrenic or schizoaffective disorders and were medicated. All data is from eyes-open resting-state fMRI. A Desikan-Killiany parcellation into 83 and 129 areas was applied to the data. Structural connectivity (SC) is quantified by white-matter metrics including the number of fibers i.e. streamlines between regions, the density of streamlines normalized by their length and regional surface area, as well as the gFA and average ADC defined by the anisotropy of inter-regional streamline directions and water diffusivity respectively. Functional connectivity was computed as pairwise Pearson correlation between the fMRI signal of regions, after detrending, smoothing, removal of artifacts, and exclusion of the first four time points (Fig. 1.

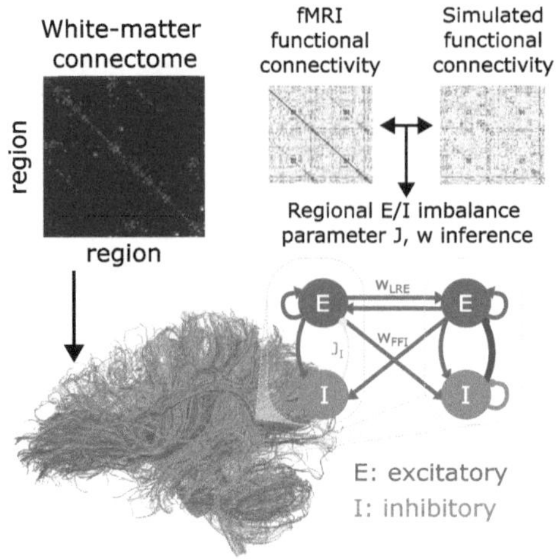

Fig. 1. Diagram of the model and inference procedure. After obtaining the individual connectome from white-matter data, we construct a personalized whole-brain dynamical model using reduced Wong–Wang excitatory–inhibitory equations for each brain region. Then by fitting simulated FC to empirical FC, we infer regional and inter-regional E/I parameters for every individual.

TVB Simulations. We used The Virtual Brain simulator to construct personalized whole-brain network models and simulate regional synaptic activity and FC [4]. Each brain region i is modeled as an E/I population governed by a reduced Wong–Wang neural mass model [2,15], capturing slow N-methyl-D-aspartate (NMDA)-mediated synaptic dynamics relevant for resting-state fMRI and widely used in whole-brain FC modeling. The E and I firing rates $H_i^{(E)}$, $H_i^{(I)}$ and corresponding synaptic gating variables $S_i^{(E)}$ and $S_i^{(I)}$ are given by

$$\begin{aligned}
\dot{S}_i^{(E)} &= -\frac{S_i^{(E)}}{\tau_E} + (1 - S_i^{(E)})\gamma_E H_i^{(E)} + \sigma\eta_i, \\
\dot{S}_i^{(I)} &= -\frac{S_i^{(I)}}{\tau_I} + \gamma_I H_i^{(I)} + \sigma\eta_i, \\
H_i^{(E)} &= \frac{a_E I_i^{(E)} - b_E}{1 - \exp\Big(-d_E(a_E I_i^{(E)} - b_E)\Big)}, \\
H_i^{(I)} &= \frac{a_I I_i^{(I)} - b_I}{1 - \exp\Big(-d_I(a_I I_i^{(I)} - b_I)\Big)}, \\
I_i^{(E)} &= w_p J_{\text{NMDA}} S_i^{(E)} - J_i S_i^{(I)} + W_E I_0 + J_{\text{NMDA}} c_i^{\text{LRE}} + I_{\text{ext}}, \\
I_i^{(I)} &= J_{\text{NMDA}} S_i^{(E)} - S_i^{(I)} + W_I I_0 + \lambda J_{\text{NMDA}} c_i^{\text{FFI}}.
\end{aligned} \tag{1}$$

Inter-regional couplings are defined by

$$\begin{aligned} c_i^{\mathrm{LRE}} &= \sum_{j=1}^{N} w_{ij}^{\mathrm{LRE}} C_{ij} S_j^{(E)}(t-\tau_{ij}), \\ c_i^{\mathrm{FFI}} &= \sum_{j=1}^{N} w_{ij}^{\mathrm{FFI}} C_{ij} S_j^{(E)}(t-\tau_{ij}), \end{aligned} \tag{2}$$

The parameters $\tau_{(E,I)}$ and $\gamma_{(E,I)}$ control synaptic time scales and saturation, while J_{NMDA} and w_p determine excitatory coupling and local recurrence, and J_i denotes local inhibitory feedback. The inputs $I_i^{(E)}$ and $I_i^{(I)}$ combine local excitation–inhibition, background drive $W_{(E,I)} I_0$, external input I_{ext}, and delayed inter-regional coupling. Between regions, coupling strength and delays are given by the normalized empirical structural connectivity is C_{ij} and delays τ_{ij} derived assuming a constant conduction speed of 3.0 m/s, while w_{ij}^{LRE} and w_{ij}^{FFI} denote long-range excitation and feedforward inhibition weights from region j to i. The term $\sigma\eta_i$ represents additive Gaussian noise.

TVB Inference. To maximize the Pearson correlation between simulated and empirical FC, we infer regional and inter-regional parameters J_i, w^{LRE}, and w^{FFI} at the individual level. Regional parameters are adjusted using feedback inhibition control[2]. At each optimization step, regional I feedback parameters J_i are updated according to $\Delta J_i = \alpha_{\mathrm{FIC}} \cdot (\langle S_i^{(I)}\rangle \langle S_i^{(E)}\rangle - r_{\mathrm{target}} \langle S_i^{(I)}\rangle)$, with $\langle S^{(E)}\rangle$ the synaptic gating activity variable for the excitatory population. This ensures that the average firing rate over time does not diverge or saturate in each region. Simultaneously, the inter-region coupling weights to E and I populations, w^{LRE} and w^{FFI}, are adjusted by update rules based on the difference between simulated and empirical FC following $\Delta w_{ij}^{\mathrm{LRE}} = \alpha_{\mathrm{EIB}} \cdot (\mathrm{FC}_{ij}^{\mathrm{target}} - \mathrm{FC}_{ij}^{\mathrm{pred}}) \cdot \mathrm{RMSE}_i$ and $\Delta w_{ij}^{\mathrm{FFI}} = -\alpha_{\mathrm{EIB}} \cdot (\mathrm{FC}_{ij}^{\mathrm{target}} - \mathrm{FC}_{ij}^{\mathrm{pred}}) \cdot \mathrm{RMSE}_i$ [12]. Simulated FC was computed by hemodynamic response function convolution [12].

White-Matter Matrix Shuffling. To interpret our findings, we use a permutation approach with two analyses. First, we randomly permute participant's labels such that each person's FC is simulated using another person's SC matrix. Second, we shuffle across the regions in each individual to test whether regional specificity has influence on model fit of empiral FC. We randomly shuffle region labels and rearrange the SC matrix for each participant according to the shuffled region labels, these are used to fit their FC matrix. This ensures that overall graph properties are preserved but the regional specificity was destroyed. For both shuffling analyses, models are inferred again with these different SC matrices.

Diagnostic Classification. To study whether inferred model parameters were meaningful for diagnosis, we use stochastic gradient descent (SGD) classifiers from the python *scikit-learn* library. As inputs, we provide vectors of model-inferred J_i, w_{ij}^{LRE}, and w_{ij}^{FFI} parameters for each participant. Using these, the classifiers are trained to discriminate between schizophrenia and control diagnostic labels. The dataset is randomly split into training and test sets of equal size. Performance was then evaluated using test accuracy across $n = 100$ random splits. For comparison, model-independent inputs consisting of concatenated FC and SC matrices are given to the classifiers. We use the hinge, log loss, modified Huber, and perceptron loss functions. The maximum iteration number is set to 5000 and the stopping criterion is given by $tol = 5 * 10^{-3}$. To prevent overfitting, L2 regularization is applied.

Statistical Testing. All p-values are obtained from non-parametric two-sided Wilcoxon signed-rank tests. We use the Holm-Sidak correction or the Nadeau-Bengio correction for multiple comparisons over metrics, pair of metrics, or loss functions wherever applicable.

3 Results

3.1 Informing Whole-Brain Models with White Matter ADC and gFA Drastically Improves Data-Model Fit

Table 1. E/I tuning model Pearson FC correlation (mean ± std)

	number	density	gFA	ADC
83 regions (ctrl)	0.16±0.05	0.28±0.06	**0.70±0.06**	**0.71±0.06**
83 regions (schz)	0.16±0.05	0.26±0.04	**0.71±0.05**	**0.69±0.05**
129 regions (ctrl)	0.14±0.04	0.23±0.05	**0.65±0.07**	**0.68±0.06**
129 regions (schz)	0.13±0.03	0.20±0.04	**0.64±0.06**	**0.67±0.07**

To systematically test to what extent integrating white-matter microstructure into whole-brain modeling helps account for the data, we simulate brain activity using, as inter-regional weight matrices C_{ij}, successively only the number of fibers, only the density of fibers, only the ADC, and only the gFA. All other parameters are either kept identical across simulations or inferred at the individual level for direct comparison purposes. This procedure is applied to each participant individually: simulations are run successively with each of their white-matter metrics as the SC weight matrix, and model parameters are inferred to match their personal FC matrix. We find that data-model correlations are considerably larger ($r \approx 0.7$) for models informed by ADC and gFA compared to the state-of-the-art models informed by the number and density of fibers ($r \approx 0.2$) (Table 1). Using a Wilcoxon rank-sum test, we found the difference is statistically significant ($^{***}p < 10^{-8}$ for all metric pairs except between ADC and gFA). No significant differences are found between data-model correlations in patients vs controls ($p > 0.05$). Results are replicated across both 83-region and 129-region brain parcellations. The results emphasize the role of white-matter microstructure in sculpting resting-state FC and the importance of informing whole-brain models with ADC or gFA metrics to capture brain activity in both health and schizophrenia.

3.2 The Regional Specificity, but Individual Specificity of ADC and gFA Graphs, are Relevant for Data-Model Fit

To uncover the aspects of ADC and gFA supporting model fit, we perform a permutation analysis across all participants as well as brain regions.

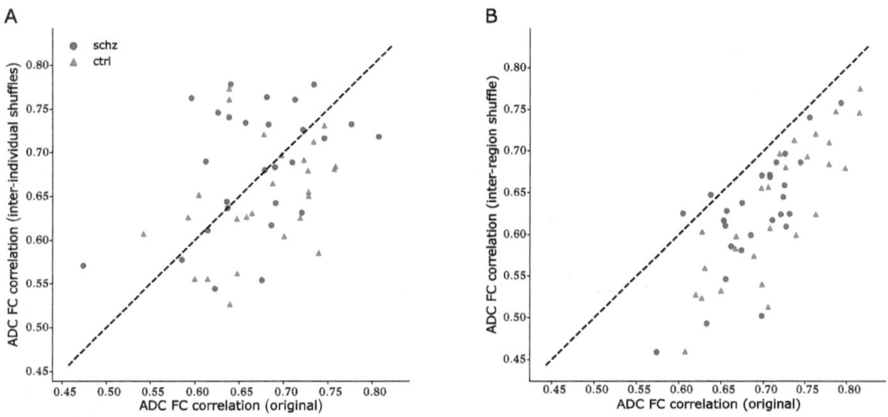

Fig. 2. Correlations between empirical and simulated FC, informed by ADC, for original vs **(A)** inter-individually and **(B)** inter-regionally shuffled weights. Each point represents one subject. The dashed diagonal line is the identity. Points below the diagonal indicate worsened model performance after shuffling.

Surprisingly, for ADC-informed models, shuffling the white-matter matrices across the 54 participants yields no significant difference for any metric in model performance (Fig. 2, $p = 0.7$). The same non-significant effects are observed upon shuffling gFA, number, and density matrices ($p > 0.05$ for all metrics). In contrast, shuffling across brain regions in each individual and thus, disrupting regional specificity in structural connectivity matrices, we observe a significant reduction in model performance. This is the most prominent in ADC, number of fibers, and density ($^{***}p < 0.0001$). However, for the models informed by gFA, inter-regional shuffling worsens fit significantly for schizophrenia patients but not controls ($^{*}p = 0.02$ and $p = 0.2$ respectively). Corrected Wilcoxon-test of the differences between shuffled and non-shuffled showed no significant differences in any metric between the two groups. The results show that regional specificity of ADC and gFA matrices matters but not individual specificity.

3.3 Inferred E/I Parameter Maps Reveal Putative Imbalance Among Posterior Cingulate and Paracentral Regions

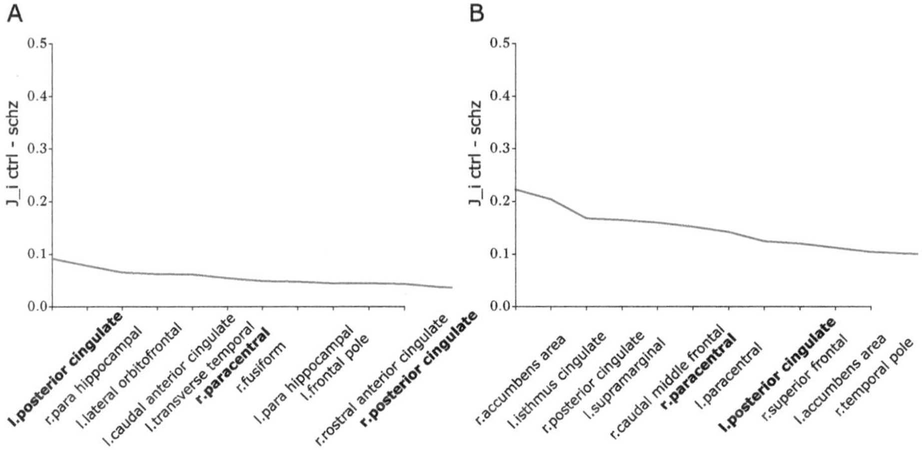

Fig. 3. Regional J_i differences between control and schizophrenia groups. The top ten regions are ranked in decreasing order of J_i differences for (**A**) ADC - and (**B**) gFA - informed models. Regions appearing in both are highlighted in bold.

To localize regions of E/I imbalance, we further investigate the inferred J_i parameters and their difference between controls and schizophrenia, averaged across participants in each group. Regions that appear among the ten with the largest difference for both ADC and gFA included the left and right posterior cingulate cortex (PCC), an area of the default-mode network, and paracentral region, a sensorimotor integration area overlapping with the language processing system, which seems robust across all metrics except the number of fibers for the posterior cingulate cortex (Fig. 3). Notably, the largest difference observed for gFA was in the accumbens, followed by the isthmus cingulate, which are not found for other metrics. For ADC, the most prominent difference is observed in the PCC. The results show that the PCC and paracentral regions are the most consistently affected by E/I imbalance in schizophrenia according to our model.

3.4 Using Inferred E/I Maps Tends to Improve Diagnostic Classification over Using Model-Independent Metrics

To investigate whether E/I maps can underpin symptoms and test their utility for diagnosis, we compare diagnostic classification from model-inferred E/I balance parameters vs empirical FC and SC matrices. Using the w^{LRE} to inform the classifier yields a mean accuracy of 0.73 for ADC across random data splits. This is a significant improvement over the test scores obtained using combined empirical FC and SC data which accounts for a test score of 0.69 ($p<0.05$). However, significance did not survive Nadeau–Bengio correction (paired t-tests, all $p>0.05$) (Fig. 4). The results show that inferred E/I maps can capture clinically relevant features of SC and FC, and hence could support diagnostics.

4 Discussion

In this work, we compared whole-brain models informed by various white-matter metrics in their ability to account for resting-state fMRI in schizophrenia and controls. We found that models informed by gFA and ADC microstructural metrics consistently outperformed models informed by state-of-the-art number and density of fibers at capturing empirical FC. The regional specificity, but not the individual variability of white-matter matrices was crucial to reproduce subject FC in simulations, which suggests that in our models, E/I imbalance accounts for most of inter-subject and schizophrenia-control differences. E/I imbalance maps revealed reduced inhibition in the PCC and paracentral areas of the default-mode and sensorimotor networks in schizophrenia. Maps were also meaningful in that they were relevant for diagnostic classification. The findings support that whole-brain models informed with white-matter microstructure can capture the structure of fMRI correlations and meaningfully map E/I imbalance underlying brain dynamical differences and potentially symptoms in schizophrenia.

First, we find that informing models with gFA or ADC yields significantly improved data-model correlations ($r \approx 0.7$) compared to using the number or density of fibers as commonly used in the literature ($r \approx 0.2$). Here, our findings contribute a systematic comparison of these metrics to the literature, when different past works may have used different white-matter metrics to fit various datasets. A correlation between empirical and simulated FC of $r \approx 0.7$ improves on even recent works [13]. In particular, fitting regional and inter-regional inhibition parameters supports model performance, consistent with published work that found up to $r \approx 0.7$ by inferring regional inhibition [20], including on the same data or with the same pipeline [3, 12]. Together with our results, this supports the importance of regional E/I balance in shaping FC. To our knowledge, our research is the first to apply this approach to patient data, allowing us to observe that the methods are also relevant to mapping pathological E/I imbalance towards uncovering root causes of symptoms and guiding treatment.

Second, we reveal that shuffling white-matter matrices across participants did not significantly worsen model performance. Even shuffling between participants with schizophrenia and controls did not worsen fit to data, surprisingly as white-matter microstructure was shown to be affected in schizophrenia [11]. Our results suggest, by contrast, that resting-state dynamical alterations in schizophrenia are not explained by underlying white-matter microstructure. However, shuffling between regions gave rise to small, although significant, changes in data-model fit. This is due to inter-regional E/I balance parameters playing a comparatively crucial part in shaping FC, with the potential confound of overfitting due to heavy parametrization of inter-regional E-to-E and E-to-I weights. Taken together, these results open doors to using white-matter microstructure from one dataset to fit FC from other data where white matter data are unavailable, hence increasing the applicability of personalized whole-brain modeling beyond the state of the art where individual diffusion data is needed.

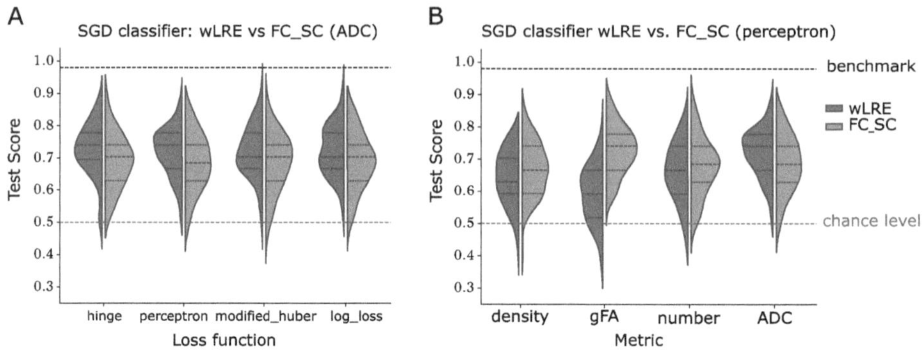

Fig. 4. Diagnostic classification accuracy using E/I balance model parameters. **(A)** Accuracy across different loss functions, comparing SGD classifier informed by simulated w_{LRE} and empirical FC and SC. The models were informed by ADC. The red line indicates chance level and the black line the benchmark performance in the field [14]. Each plot shows the distribution of test scores for 100 splits. **(B)** Accuracy across different metrics for perceptron loss.

Our models are heavily parametrized in that inferred regional and inter-regional maps importantly contribute to model-data fit over white-matter specifics. Hence, it is crucial to ensure that inferred E/I imbalance parameter maps are meaningful. A first way to tackle this validation question is to investigate regions uncovered by the model as affected by E/I imbalance in schizophrenia. This approach revealed that the PCC and paracentral areas showed on average reduced inhibition in schizophrenia. Dynamical differences consistent with E/I imbalance were also observed in the PCC in autism spectrum disorder, which shares social interaction symptoms with schizophrenia [17]. The PCC is also associated with autobiographical memory, potentially related to delusion symptoms domains. The paracentral area is a sensorimotor region among areas involved in language and salience processing, relevant to auditory verbal hallucinations and found to be affected in schizophrenia [18]. Thus, the regions found by our model are in line with the literature, which supports the validity of our method. Our model produces a putative explanation for dynamical differences in these regions, by suggesting they could be due to reduced inhibition compared to excitation.

Another way to validate our approach is to use model-derived E/I balance parameters to predict diagnosis, as the model was not informed with diagnostic data. We find that the inferred regional and inter-regional parameters controlling E/I balance can equal or outperform model-independent SC and FC data for diagnostic classification. Here, the ADC appears from our analyses as a white-matter metric able to both capture resting-state FC and diagnostic-relevant information. State-of-the-art diagnostic prediction methods perform similarly or better than E/I balance maps [5,14,18], which could be because they take into account dynamical aspects of fMRI signals beyond average FC over entire time series. Thus, future work inferring E/I balance maps as dynamical, time-varying parameters from patient data may yield diagnostic improvements. Overall, our method may be understood as a way to analyze data, combining information from SC, FC, and E/I balance biophysics to generate inductive biases for clinical machine learning tasks like diagnostics. By amplifying data directions relevant that are diagnostically relevant and associated with hypothesized causal mechanisms, our white-matter-microstructure informed modeling may be useful to replace or add to existing inductive biases for clinical machine learning tasks.

Importantly, our work has several limitations. First, we provide a proof of principle on a small dataset of 27 schizophrenia patients and 27 control participants [19]. It remains to be seen if results generalize to larger and more heterogeneous cohorts such as including early psychosis. Investigation in larger cohorts would also enable the clustering of patients into disorder subtypes potentially corresponding to different E/I imbalance maps. Moreover, the effects we identify are observed in resting state and do not necessarily generalize to brain function in tasks tied to symptoms. While we chose to focus on resting-state data as it is currently the most readily available in public datasets from schizophrenia patients, extensions of this approach to tasks tied to symptoms may provide more insights into the causal role of E/I balance in altered brain function.

In conclusion, we developed personalized and white-matter-microstructure informed models of brain dynamics in health and schizophrenia. Our results provide the first systematic investigation of the role of white matter in whole-brain modeling, emphasizing the importance of ADC and gFA metrics. This approach reveals, for the first time, maps of E/I imbalance in a patient population, with reduced inhibition in distinct brain regions across functional networks. Inferred imbalance maps are meaningful in that they can support clinical machine learning tasks like diagnostic classification. The findings create a fundamentally novel bridge between cellular-scale E/I imbalance mechanisms hypothesized in schizophrenia and brain-wide dynamics that can be recorded *in vivo* in patients. Personalized white-matter microstructure informed models could therefore be relevant to simulate disorder progression for early diagnosis and to optimize intervention protocols toward individualized treatment recommendations.

Disclosure of Interests. The authors have no competing interests.

References

1. Anticevic, A., Lisman, J.: How can global alteration of excitation/inhibition balance lead to the local dysfunctions that underlie schizophrenia? Biol. Psychiat. **81**(10), 818–820 (2017)
2. Deco, G., et al.: Identification of optimal structural connectivity using functional connectivity and neural modeling. J. Neurosci. **34**(23), 7910–7916 (2014). https://doi.org/10.1523/JNEUROSCI.4423-13.2014. https://www.jneurosci.org/lookup/doi/10.1523/JNEUROSCI.4423-13.2014
3. Deng, N.: Large-Scale Brain Simulation to Characterize Neural Circuits of Schizophrenia. McGill University, Canada (2024)
4. Goldman, J.S., et al.: A comprehensive neural simulation of slow-wave sleep and highly responsive wakefulness dynamics. Front. Comput. Neurosci. **16**, 1058957 (2023)
5. Gutiérrez-Gómez, L., et al.: Stable biomarker identification for predicting schizophrenia in the human connectome. NeuroImage: Clin. **27**, 102316 (2020). https://doi.org/10.1016/j.nicl.2020.102316. https://pmc.ncbi.nlm.nih.gov/articles/PMC7334612/
6. Howes, O.D., Shatalina, E.: Integrating the neurodevelopmental and dopamine hypotheses of schizophrenia and the role of cortical excitation-inhibition balance. Biol. Psychiat. **92**(6), 501–513 (2022)
7. Jirsa, V.: Personalised virtual brain models in epilepsy. The Lancet Neurology **22**(5), 443–454 (2023)
8. Kaur, A., et al.: Structural and functional alterations of the temporal lobe in schizophrenia: a literature review. Cureus **12**(10) (2020)
9. Menon, V.: Large-scale brain networks and psychopathology: a unifying triple network model. Trends Cogn. Sci. **15**(10), 483–506 (2011)
10. Palaniyappan, L., Liddle, P.F.: Does the salience network play a cardinal role in psychosis? an emerging hypothesis of insular dysfunction. J. Psychiatry Neurosci. **37**(1), 17–27 (2012)
11. Pavan, T., et al.: White matter microstructure alterations in early psychosis and schizophrenia. Transl. Psychiatry **15**(1), 179 (2025). https://doi.org/10.1038/s41398-025-03397-1. https://www.nature.com/articles/s41398-025-03397-1
12. Pille, M., Martin, L., Richter, E., Perdikis, D., Schirner, M., Ritter, P.: Fast and Easy Whole-Brain Network Model Parameter Estimation with Automatic Differentiation (2025). https://doi.org/10.1101/2025.11.18.689003. https://www.biorxiv.org/content/10.1101/2025.11.18.689003v1, iSSN: 2692-8205 Pages: 2025.11.18.689003 Section: New Results
13. Plüss, R., Villota, H., Orio, P.: Hemispheric-Specific Coupling Improves Modeling of Functional Connectivity Using Wilson–Cowan Dynamics. In: Talevi, A., Rosa Cota, V. (eds.) Computational Neuroscience. LNCS, vol. 2734, pp. 109–121. Springer Nature Switzerland, Cham (2026). https://doi.org/10.1007/978-3-032-14664-9_10
14. Raeisi, Z., Mehrnia, M., Ahmadi Lashaki, R., Abedi Lomer, F.: Enhancing schizophrenia diagnosis through deep learning: a resting-state FMRI approach. Neural Comput. Appl. **37**(20), 15277–15309 (2025)
15. Schirner, M., Deco, G., Ritter, P.: Learning how network structure shapes decision-making for bio-inspired computing. Nature Commun. **14**(1), 2963 (2023). https://doi.org/10.1038/s41467-023-38626-y. https://www.nature.com/articles/s41467-023-38626-y

16. Solmi, M., et al.: Incidence, prevalence, and global burden of schizophrenia-data, with critical appraisal, from the global burden of disease (GBD) 2019. Mol. Psychiatry **28**(12), 5319–5327 (2023)
17. Strock, A., Nghiem, T.A.E., Lecea, A.D., Hassan, A., Ryali, S., Menon, V.: Explainable AI techniques for dynamic functional brain imaging: validation and analysis of e/i imbalance in autism. bioRxiv, pp. 2025–02 (2025)
18. Supekar, K., et al.: Robust and replicable functional brain signatures of 22q11. 2 deletion syndrome and associated psychosis: a deep neural network-based multi-cohort study. Mol. Psychiatry **29**(10), 2951–2966 (2024)
19. Vohryzek, J., et al.: Structural and functional connectomes from 27 schizophrenic patients and 27 matched healthy adults (2020). https://doi.org/10.5281/zenodo.3758534. https://zenodo.org/records/3758534
20. Zhang, S., et al.: In vivo whole-cortex marker of excitation-inhibition ratio indexes cortical maturation and cognitive ability in youth. Proc. Natl. Acad. Sci. **121**(23), e2318641121 (2024)

Dopamine-Receptor Dynamics in Basal Ganglia Reinforcement Learning: A Mechanistic Model of Probabilistic Reversal Learning

Giulia Piermaria[1(✉)], Silvana Pelle[1], Miriam Schirru[1,2], Fahima Nekka[2,3], and Mauro Ursino[1]

[1] Department of Electrical, Electronic and Information Engineering Guglielmo Marconi, University of Bologna, Cesena, Italy
giulia.piermaria3@unibo.it

[2] Faculté de Pharmacie, Université de Montréal, Montreal, Canada

[3] Centre de recherches mathématiques, Université de Montréal; Centre for Applied Mathematics in Bioscience and Medicine (CAMBAM), McGill University, Montreal, Canada

Abstract. Biologically inspired reinforcement learning (RL) models of the basal ganglia (BG) often treat dopamine (DA) as an ad hoc modulatory signal, without explicitly accounting for DA kinetics and D1/D2 receptor dynamics. Here we present an extended mechanistic BG model that integrates DA release, reuptake, and degradation with striatal D1 and D2 receptor occupancy, and couples receptor binding/unbinding to pathway-specific modulation of Go/NoGo activity and Hebbian synaptic plasticity. First, single-trial simulations under naïve conditions reproduce plausible temporal patterns: reward-evoked DA peaks increase D1/D2 occupancy, enhance Go activity, and potentiate corticostriatal synapses for the selected action, whereas punishment-evoked DA dips reduce occupancy, suppress Go activity, and favor potentiation of NoGo synapses. Second, we evaluate the model on a two-choice probabilistic reversal learning task (80/20 contingencies; 40 acquisition + 40 reversal trials; 50 simulated subjects). The model captures acquisition and post-switch adaptation without relying on engineered RL update rules, as learning emerges from the interaction of DA-dependent pathway modulation and thresholded Hebbian plasticity. Finally, sensitivity analyses show that excessive exploratory noise markedly impairs performance, reduced D1 sensitivity preferentially disrupts reversal, and reduced D2 sensitivity in NoGo strongly compromises both acquisition and reversal, highlighting mechanistic constraints on flexibility in non-stationary environments. In perspective, the model can be valuable for testing alternative hypotheses about receptor dynamics and exploring therapeutic strategies for targeting D1 and D2 receptor occupancy.

Keywords: basal ganglia · dopamine kinetics · D1/D2 receptor occupancy · Hebbian learning · probabilistic reversal learning · exploration–exploitation

L. Pasa et al. (Eds.): ICANN 2026, LNCS 17089, pp. 40–52, 2027.
https://doi.org/10.1007/978-3-032-38398-3_4

1 Introduction

Reinforcement learning (RL) is a fundamental aspect of cognition, enabling training from the environment without a teacher, based on a history of rewards and punishments. In a non-stationary environment, where the probabilities of rewards and punishments can abruptly change, an essential requisite of RL is not only the learning capacity (often summarized with the term "exploitation"), but also the ability to modify previous choices (often named "exploration"). Such an exploration/exploitation trade-off is an essential aspect of biological adaptivity to external contingencies. Several algorithms for RL have been proposed in recent years, but are often scarcely inspired by the biology field. Nevertheless, modeling biological RL can be highly valuable, enriching our knowledge of brain function, providing a better understanding of several neurological deficits, and inspiring innovative artificial intelligence algorithms.

In our brain, one of the most important structures implicated in RL is the basal ganglia (BG), a cluster of subcortical nuclei involved in action selection. Learning in the BG occurs for each selected action by altering the balance between two distinct, competitive branches [1,2]: the Go pathway, which favors action selection, and the NoGo pathway, which brakes action selection. Unexpected rewards following a given action reinforce its Go path and weaken the NoGo, thus favoring a future selection of the same action. Conversely, unexpected punishments reinforce the corresponding NoGo path and weaken the Go path, thereby jeopardizing future selection. Mechanistically, these rewards/punishments are realized through phasic changes (transient peaks or dips, respectively) of the neurotransmitter dopamine (DA), while the tonic DA level sets the working point of the overall structure [3]. Actually, DA alterations are implicated in several neurological disorders, such as Parkinson's, attention-deficit hyperactivity disorder (ADHD), Tourette syndrome, and depression [4,5].

The effect of DA on RL in the BG depends on the presence of two DA receptor classes: D1-and D2, which differentially modulate the Go and NoGo pathways. Projection neurons in the Go pathway express D1 dopamine receptors (D1Rs) and are excited by dopamine peaks. Conversely, projection neurons in the NoGo pathway mainly express D2 dopamine receptors (D2Rs), which exert an opposite effect, exciting neurons during dopamine dips. Another important distinction is that D1R and D2R exhibit different affinity (higher for D2R and lower for D1R) and possibly different dynamics.

Although a few biologically inspired neurocomputational models of the BG have been proposed in recent years [6–9], they generally do not explicitly incorporate the dynamics of D1R and D2R, nor do they discuss their roles in RL. Rather, in these models, DA is an external input that modifies some network properties. A recent model by Romero Pinto and Uchida [10] includes D1R and D2R within an RL algorithm, but their learning equations do not explicitly reproduce the physiological structures of the BG nuclei or the neurons' activity.

In recent years, a neurocomputational model that incorporates the different nuclei in the BG and their connections has been developed [11,12]. The model was able to simulate several RL tasks in both healthy subjects and Parkinsonian patients, including probabilistic and deterministic learning under direct and reversal conditions, and to simulate bradykinesia [13] and aberrant learning in Parkinsonism [14]. This model, however, did not include an explicit description of DA dynamics, D1R and D2R occupancy, and their role in BG during learning.

In this work, we present an extended version of the model that incorporates dopamine, D1R, and D2R dynamics and their effect on learning. We show that the new model can produce reliable temporal patterns in the main quantities in response to rewards and punishments, in accordance with Hebbian learning, and correctly simulate a two-choice probabilistic reversal learning task. In addition, a sensitivity analysis is performed on parameters related to DA dynamics and on the sensitivity of D1R and D2R to illustrate how variations in these parameters affect subject performance. In perspective, this model can be useful for assessing pathological conditions characterized by alterations in DA signaling and for testing therapeutic interventions targeting D1R and D2R.

2 Method

2.1 Qualitative Model Description

The model represents human BG based on previous works [12]. Figure 1 shows a schematic representation of the model.

The network presents several units, each simulating a group of neurons with similar functions. Each unit's dynamics can be described by a first-order low-pass filter that mimics the integrative properties of the synapse and membrane, and by a sigmoidal relationship for the output activity, indicating upper and lower saturation values.

The model includes both a sensory representation, which codes the input stimulus, and a motor representation in the cortex, which encodes possible actions. In addition, the model includes the striatum, further subdivided into Go and NoGo pathways, the Globus Pallidus pars externa (Gpe), the Globus Pallidus pars interna (Gpi), and the Thalamus (T). Every unit in these structures has a 1-to-1 relationship with neurons in the motor cortex, representing the possible actions. Conversely, the cholinergic area (Chl, see also Fig. 2) and the subthalamic nucleus (STN) are modeled with a single Unit, which affects the overall structure. In the current model version, only two actions are included (as shown in Fig. 1), corresponding to the two possible choices in our task.

To simulate exploration, Gaussian noise with a zero mean value and a specified standard deviation (SD), denoted as parameter n, has been added to the two neurons in the motor cortex, thereby influencing the winner-takes-all dynamics.

The neural units are connected to realize the three main pathways responsible for BG functioning: direct (via Go neurons), indirect (via NoGo neurons), and hyperdirect (via the STN). The balance between these pathways modulates Gpi inhibition of the thalamus, either permitting or blocking a coded response by motor neurons. At steady state, the Cortex, Striatum, and Thalamus are inhibited, whereas the Gpi and Gpe exhibit basal activity, consistent with physiological data [15]. Following sufficient stimulation, action selection emerges through

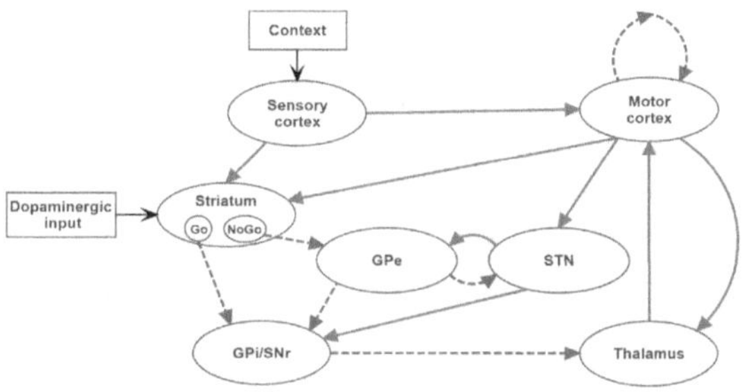

Fig. 1. Block diagram of the different parts included in the model. Green continuous lines represent excitatory synapses; red dashed lines are inhibitory synapses. Only the synapses from the sensory and motor cortices to the striatum are trained with Hebbian mechanisms. (Color figure online)

a Winner-Takes-All mechanism in the Motor Cortex, with the direct pathway facilitating the selected response via Gpi inhibition and the indirect pathway suppressing competing responses via Gpe inhibition. The hyperdirect pathway helps resolve intense conflict among cortical units by blocking responses and providing additional time for the cortex to select a winning neuron. In detail, the STN receives a conflict level as input, and it sends excitation to the Gpi, providing additional thalamic inhibition.

The Hebbian rule is used to train the network in response to rewards and/or punishments, modifying synaptic weights that enter the Striatum based on the activity of the Go and NoGo neurons. Let y_i^A and y_j^B denote the activity of neurons i and j in layers A and B, respectively:

$$\Delta w_{ij}^{AB} = \gamma(y_j^B - \theta^{PRE})(y_i^A - \theta^{POST}) \quad (1)$$

where Δw_{ij}^{AB} is the variation of the synapse between the pre-synaptic neuron j in layer B ($B = S$ or C) and the post-synaptic neuron i, in layer A ($A = G$ or N), y represents neuronal activity (normalized between 0 and 1), θ^{PRE} and θ^{POST} are the thresholds for the pre-synaptic and post-synaptic activity, and γ is the learning factor. Moreover, we assume that synapse values cannot exceed an upper saturation value or fall below a lower saturation.

Compared with previous models [12], in which tonic and phasic dopamine were treated as external inputs, here we introduced an explicit description of DA concentration production, striatal D1 and D2 receptor occupancy, and the dependence of Go and NoGo activities on receptor occupancy. These new equations are reported below.

2.2 Dopamine Dynamics and D1R, D2R Occupancy

Dopamine Dynamics - Following Dreyer et al. [16], in our model DA concentration is described as a function of release, reuptake, and degradation (Fig. 2).

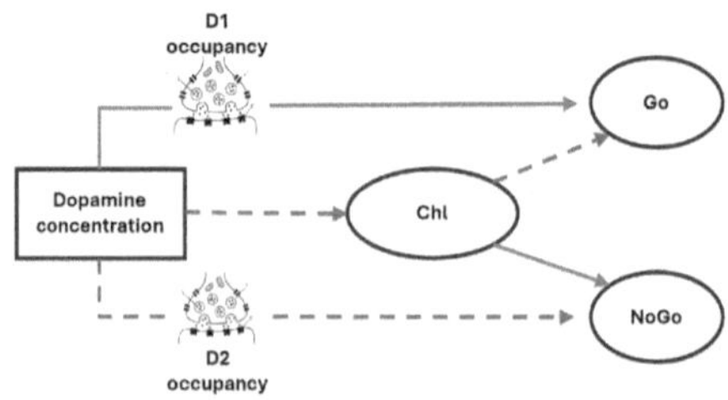

Fig. 2. Block diagram focusing on DA circuit. Arrows indicate the excitatory and inhibitory synapses (Fig. 1 for details).

It is defined as:

$$\frac{dC_{DA}(t)}{dt} = I_{DA}^{tonic} + I_{DA}^{phasic}(t) - \frac{V_{max}C_{DA}(t)}{k_m + C_{DA}(t)} - k_{rem}C_{DA}(t) \tag{2}$$

where reuptake is described with a Michaelis-Menten term (V_{max} is the maximal reuptake velocity and k_m the concentration at half the maximum) and k_{rem} is a removal constant. I_{DA}^{tonic} is the **tonic dopamine** production rate:

$$I_{DA}^{tonic} = \eta \cdot v_{tonic} \tag{3}$$

where v_{tonic} is the average tonic firing rate of dopaminergic neurons, which is an input for the model. The term η represents the amount of DA release per action potential, accounting for the density of dopamine terminals ρ, volume fraction α_{vf}, Avogadro's number N_A, vesicle release probability P_r, and the number of DA molecules released per vesicle η_0:

$$\eta = \rho \frac{Pr\ \eta_0}{\alpha_{vf} N_A} \tag{4}$$

As for the phasic dopamine release, $I_{DA}^{phasic}(t) = 0$ when there is no response. When there is a reward or a punishment in a specific window of time, we have:

$$I_{DA}^{phasic}(t) = \eta \cdot v_{phasic} \tag{5}$$

where $v_{phasic} > 0$ during a reward and $v_{phasic} < 0$ during a punishment.

The expressions for v_{phasic} depend on how much a reward (or punishment) is expected, with greater changes in case of unexpected rewards (or unexpected punishments, respectively). We followed expressions similar to those used in another work to produce phasic dopamine changes [17] adapted to the new equations. However, since, of course, the average firing rate cannot decrease below zero during a punishment, we set $v_{phasic} \geq -v_{tonic}$.

Receptor Occupancy - The following equation describes how dopamine receptor occupancy (for D1R or D2R) changes over time as a function of DA concentration. The two processes described are the binding of dopamine to the receptor and the unbinding or dissociation of dopamine from the receptor:

$$\frac{dO_j(t)}{dt} = (1 - O_j(t)) \cdot k_{on,j}[C_{DA}(t)] - k_{off,j}\, O_j(t), \quad j = 1, 2 \tag{6}$$

where:

- $O_j(t)$ is the fraction of receptors occupied by DA at time t ($j = 1$ or 2 as to D1 or D2 receptor type).
- $k_{on,j}$ is the rate of DA bound to the receptor.
- $k_{off,j}$ is the rate of DA unbound from the receptor.

Dependence of the Go and NoGo Pathway from D1 and D2 Receptor Occupancy. As a last step, we need to specify how the activity of projection neurons in the Go and NoGo pathways is affected by receptor occupancy. To this end, we modified the equations in the previous model, which specify the neuron inputs. In the following, as in Eq. 1, the synapses w are identified by two superscripts, specifying the target and the donor layers (S, C, Go, NoGo, Gpe, Gpi), and two subscripts, distinguishing post and pre synaptic neurons (i, j with i, j = 1, 2, ..., N). y is neuron activity, with N being the number of action channels (N is set to 2 in this work since we simulate a two-choice task), and S is a Nx1 vector containing the input stimulus.

The inputs to projection neurons in the Go pathway and their output are described by the following differential equation:

$$\tau \frac{du_i^G}{dt} = -u_i^G + \sum_{j=1}^{N} w_{ij}^{GS} s_j + w_{ii}^{GC} y_i^C + \alpha \cdot O_1 \cdot (y_i^G - \theta^G) + w^{GH} y^H \tag{7}$$

$$y_i^G = \zeta(u_i^G) \tag{8}$$

According to Eq. 7, inputs to the Go come from the whole stimulus S and the corresponding neuron in the motor cortex, C, both excitatory. Furthermore, the activity of each neuron in the Go is modulated by D1 receptor occupancy, O_1, and by the cholinergic interneuron activity (y^H, see also Fig. 2).

$\zeta(u_i^G)$ in Eq. 8 represents a sigmoidal relationship, providing neuron output.

Dopamine is excitatory ($\alpha > 0$) if the Go activity is above threshold (θ^G), inhibitory otherwise, thus realizing the contrast enhancement effect [18]. The cholinergic interneurons are always inhibitory ($w^{GH} < 0$) to the Go.

The inputs to projection neurons in the NoGo part of the striatum are, for i=1,..., N:

$$\tau \frac{du_i^N}{dt} = -u_i^N + \sum_{j=1}^{N} w_{ij}^{NS} s_j + w_{ii}^{NC} y_i^C + \beta \cdot O_2 + w^{NH} y^H \tag{9}$$

$$y_i^N = \zeta(u_i^N) \tag{10}$$

Every neuron of the NoGo receives excitatory input from the whole stimulus S and excitatory projection from the corresponding neuron in the cortex C. Furthermore, the activity of each neuron in the NoGo is modulated by the D2 receptor occupancy, O_2, and by the cholinergic interneuron activity (y^H). Dopamine provides inhibition (β<0) to all the NoGo neurons, while the cholinergic interneuron provides excitation (w^{NH}>0). Parameters in Equations (2–5) have been taken from [16], while parameters in Equation (6) have been taken from [10,16]. Parameters in Eqs. 7 and 9 have been assigned to obtain a response of Go and NoGo neurons comparable to another work [12]. All other equations in the model are the same as in a recent paper [12].

3 Results

3.1 Single Simulations in Naïve conditions

The first simulations have been performed starting from the network under naïve conditions, i.e., with the two actions equally probable. This is obtained using identical synaptic weights for the two action pathways.

Figure 3 illustrates the model response to a reward. The neuron in the motor cortex associated with action 1 wins the competition, and this choice is rewarded, leading to a DA peak and increased D1R and D2R occupancy. As a consequence, the winner Go neuron increases its activity, resulting in Hebbian potentiation, whereas the loser Go neuron and both NoGo neurons are inhibited, resulting in Hebbian depotentiation. The two vertical dashed black lines in both Figs. 3 and 4 represent the period when the Hebb rule is applied to change synapses in the Go and NoGo pathways. Figure 4 illustrates the dynamics following a punishment, assuming an initially naïve network configuration. In this case as well, Action 1 wins, but is causally punished. This results in a dopamine dip and consequent reduction in receptor occupancy. Consequently, Go neurons are inhibited (Hebbian depotentiation), and the winner NoGo neuron increases its activity, resulting in Hebbian potentiation.

3.2 Simulation of a Two-Choice Probabilistic Reversal Learning Task

Finally, to test the model, we simulated a two-choice probabilistic task with reversal, as described in [19]. The subject must choose between two possible actions. The first leads to an 80% probability of rewards, while the second is associated with a 20% probability. In this work, the acquisition and the reversal stages are kept distinct. In particular, after completion of the acquisition stage, which consists of 40 epochs, the reversal phase follows, in which probabilities are oppositely associated. For each test, we simulated 50 different subjects, which differ only in the noise realization superimposed on the motor cortex neurons. As in [19], we considered that a subject passes the task (independently in the direct acquisition and in the reversal phase) if the participant meets the criterion of responding correctly, for at least eight consecutive trials (the correct response

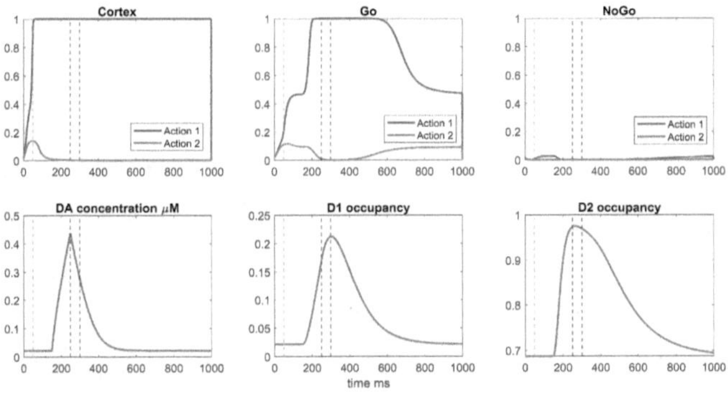

Fig. 3. Temporal pattern of normalized neuron activity in the Cortex and in the Go and NoGo pathways (upper panels), DA concentration, and D1R/D2R occupancy (bottom panels), during rewarded simulation, starting from a naïve network. The Cortex, Go, and NoGo pathways are represented by two neurons, encoding one of the two possible actions (blue line: Action 1; red line: Action 2). Action 1 wins the competition (at the instant signaled by the vertical green line), and the network is rewarded. Phasic DA release occurs 100 ms after the choice. The two vertical dashed black lines represent the period when the Hebb rule is applied to change synapses in the Go and NoGo pathways. The threshold for potentiation/depotentiation in the Hebb rule is 0.5 (Eq. 1). The winning Go neuron shows strong activation and potentiation, whereas the losing Go neuron and both NoGo neurons are inhibited and undergo depotentiation.

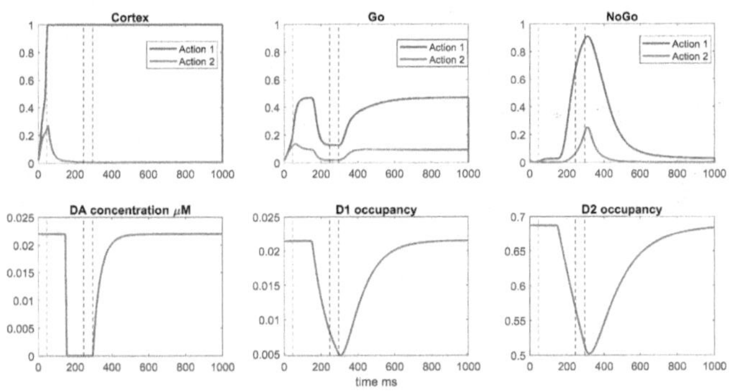

Fig. 4. Same format and conventions as Fig. 3, except that Action 1 is now punished, and the DA concentration falls. The winner, Go neuron, is inhibited by the dopamine fall, and both Go neurons exhibit depotentiation. The winner NoGo is active and exhibits potentiation, whereas the loser NoGo neuron remains poorly active and exhibits modest depotentiation.

is the one rewarded in 80% of cases during that phase). The upper left panel in Fig. 5 compares model performance obtained with nominal parameter values to the behavioral performance reported by Cools et al. [19], showing that control subjects can pass both direct and reversal phases in the majority of cases. The upper right panel shows the results of a noise SD sensitivity analysis. Results show that a high noise level (signifying much exploration by the subject) has a strong negative impact on the subject's performance.

Finally, the two bottom panels report a sensitivity analysis on parameters α and β in Equations (7) and (9), which represent the strength of dopaminergic modulation of Go and NoGo to D1 and D2 receptor occupancy, respectively. As expected, performance decreases when this modulation is reduced; however the effect is much greater for the NoGo pathway (parameter β).

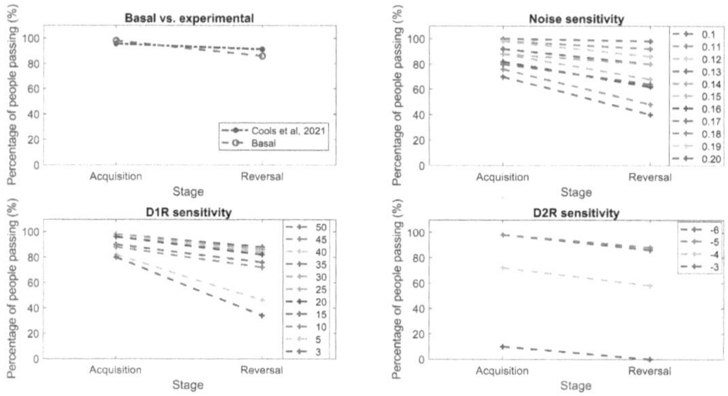

Fig. 5. Percentage of subjects passing a two-choice probabilistic task during the acquisition phase (40 trials) and the reversal phase (40 trials) according to the criterion defined in Cools et al. [19]. Upper left panel: comparison between model simulation with basal parameter values and the results reported in [19]. The other three panels show the results of a sensitivity analysis obtained by varying the noise n (upper right), decreasing the Go pathway sensitivity to D1R occupancy (left bottom), and reducing the NoGo pathway sensitivity to D2R occupancy (right bottom). The basal value of noise SD (n) was 0.12, and α and β basal values were 40 and 6 respectively.

4 Discussion

This study extends biologically grounded models of the BG by including an explicit description of DA kinetics – release, reuptake, and degradation - together with D1/D2 receptor occupancy dynamics, within a mechanistic neural architecture for action selection and RL. Unlike many neurocomputational BG models, in which DA is represented as an external modulatory input, our approach links DA phasic/tonic variations to receptor binding/unbinding and, in turn, to pathway-specific modulation of Go/NoGo activity. This work connects a model of BG dynamics with pharmacodynamics description of DA and receptor kinetics, which are often considered separately [6,16]. DA has been firmly established as a key regulator of learning and motivation: changes in its dynamics are recognized as important factors in the etiology of several neurological pathologies involving dysfunction in action selection, reduced motivation, and mood. Hence, developing a model that integrates neural and DA into a single theoretical framework may have serious implications, not only for deepening our understanding of brain function but also for developing new tools for clinical support and inspiring new ideas for treatment strategies.

Functional Implications – The present preliminary simulations yield some interesting results. First, we have shown that the neural BG model, integrated with DA dynamics, can simulate a two-choice probabilistic task quite well, both in the direct and reversal phases, ascribing learning to a simple Hebbian mechanism involving Go and NoGo projection neurons in the striatum. By explicitly modeling receptor occupancy, the present framework provides a principled account of how transient DA fluctuations can reshape striatal plasticity by differentially engaging pathways. In our simulations, reward-related DA peaks increase D1/D2 occupancy, promoting Go activation and potentiation of the selected action. In contrast, punishment-related DA dips reduce occupancy, favoring NoGo-mediated suppression. This mechanism offers an interpretable mapping between DA transients, receptor state, pathway excitability, and Hebbian weight changes, yielding temporally plausible response patterns and coherent learning dynamics.

Biological RL Requirements – A key requirement for biological RL is adaptation in non-stationary contexts, such as reversal learning. The model successfully reproduces performance in a two-choice probabilistic reversal task, capturing learning both during acquisition and after contingency switching. Importantly, learning emerges naturally from the interaction between DA-dependent pathway modulation and a simple Hebbian rule with pre-/post-synaptic thresholds. This allows the network to learn from experience without implementing an explicit engineered RL update rule.

Sensitivity Analysis – Sensitivity analyses show that excessive exploratory noise impairs both acquisition and reversal learning, suggesting a computational interpretation of "over-exploration" as a direct cause of impaired performance in probabilistic settings. Second, reducing sensitivity to D1 receptor occupancy selectively affects reversal more strongly, consistent with the idea that flexible switching may rely on efficient Go reconfiguration. Third, reducing sensitivity to D2 receptor occupancy in NoGo produces a severe impairment across both phases, indicating that appropriate NoGo gating may be a dominant requirement not only for suppressing maladaptive actions after reversal, but also for stabilizing learning during acquisition.

Limitations and Open Questions – A primary limitation concerns the uncertainty in kinetic parameters governing receptor binding/unbinding (e.g., k_{on}/k_{off}). We adopted fast dynamics (i.e., we used upper-bound estimates reported in the literature), i.e., we simulated transient times within one second in both the on and off phases. Still, values in the literature span more than an order of magnitude, and data by some authors – particularly for D2 – suggest dissociation rates compatible with occupancy transients lasting seconds to minutes.

Actually, the main benefit of the present model, compared with previous versions that lack an explicit receptor description, lies in the ability to investigate the relationship between receptor kinetics and BG problem-solving capacity. This has important theoretical consequences: slow occupancy dynamics would imply that striatal DA receptors are unable to track rapid DA changes, so the BG act as temporal integrators of reinforcement, potentially smoothing rapid phasic signals into longer-lasting eligibility-like traces. Conversely, faster dynamics, as in our approach, imply the capacity to respond quickly to immediate environmental changes. Testing these hypotheses is a natural next step and one that our framework is well-suited to address, because it allows systematic manipulation of receptor kinetics while keeping the neural circuit unchanged.

A second open question concerns the different roles of Go and NoGo (and, correspondingly, D1 vs D2 signaling) across different task regimes – e.g., rapid reversal, gradual shifts, or volatility-driven exploration - where differences in these pathways may lead to distinct pieces of behavior. Such distinctions are particularly relevant for interpreting pharmacological manipulations and for designing interventions that preferentially target one receptor class.

Clinical and Translational Perspectives - DA dysfunction is implicated in multiple disorders characterized by altered action selection, motivation, and learning. By providing a unified model in which DA kinetics, receptor occupancy, circuit dynamics, and synaptic plasticity interact, this framework can support in silico experimentation on pathological conditions and treatment hypotheses [20]. Promising immediate directions include modeling Parkinson's disease (altered tonic/phasic DA and reuptake), and ADHD-related alterations in dopaminergic signaling.

Another important related subject of study is the effect of receptor-specific agonist/antagonist manipulations on differential effects on acquisition versus reversal and on the exploration/exploitation balance. Results of these manipulations can help the design of innovative treatment procedures.

References

1. Kumar, K.S., Maithreye, R., Gupte, N., Chakravarthy, V.S.: Expanding the Go/NoGo depiction of the action of basal ganglia pathways. In: The 2011 International Joint Conference on Neural Networks (IJCNN 2011), pp. 3207–3212. San Jose, CA, USA (2011). https://doi.org/10.1109/IJCNN.2011.6033646
2. Berthet, P., Hellgren-Kotaleski, J., Lansner, A.: Action selection performance of a reconfigurable basal ganglia inspired model with Hebbian-Bayesian Go-NoGo connectivity. Front. Behav. Neurosci. **6**, 65 (2012). https://doi.org/10.3389/fnbeh.2012.00065

3. Gilbertson, T., Steele, D., et al.: Tonic dopamine, uncertainty and basal ganglia action selection. Neuroscience **466**, 109–124 (2021). https://doi.org/10.1016/j.neuroscience.2021.08.022
4. Meder, D., Herz, D.M., Rowe, J.B., Lehéricy, S., Siebner, H.R.: The role of dopamine in the brain – lessons learned from Parkinson's disease. Neuroimage **190**, 79–93 (2019). https://doi.org/10.1016/j.neuroimage.2018.11.021
5. Ijomone, O.K., Oria, R.S., Ijomone, O.M., Aschner, M., Bornhorst, J.: Dopaminergic perturbation in the aetiology of neurodevelopmental disorders. Mol. Neurobiol. **62**(2), 2420–2434 (2025). https://doi.org/10.1007/s12035-024-04418-8
6. Frank, M.J.: Dynamic dopamine modulation in the basal ganglia: a neurocomputational account of cognitive deficits in medicated and nonmedicated Parkinsonism. J. Cogn. Neurosci. **17**(1), 51–72 (2005). https://doi.org/10.1162/0898929052880093
7. Mikhael, J.G., Bogacz, R.: Learning reward uncertainty in the basal ganglia. PLoS Comput. Biol. **12**(9), e1005062 (2016). https://doi.org/10.1371/journal.pcbi.1005062
8. Stewart, T.C., Bekolay, T., Eliasmith, C.: Learning to select actions with spiking neurons in the basal ganglia. Front. Neurosci. **6**, 2 (2012). https://doi.org/10.3389/fnins.2012.00002
9. Blackwell, K.T., Doya, K.: Enhancing reinforcement learning models by including direct and indirect pathways improves performance on striatal dependent tasks. PLoS Comput. Biol. **19**(8) (2023). https://doi.org/10.1371/journal.pcbi.1011385
10. Romero Pinto, S., Uchida, N.: Tonic dopamine and biases in value learning linked through a biologically inspired reinforcement learning model. Nat. Commun. **16**, 7529 (2025). https://doi.org/10.1038/s41467-025-62280-1
11. Baston, C., Ursino, M.: A biologically inspired computational model of basal ganglia in action selection. Comput. Intell. Neurosci. **2015**, 187417 (2015). https://doi.org/10.1155/2015/187417
12. Ursino, M., Pelle, S., et al.: Valence-dependent dopaminergic modulation during reversal learning in Parkinson's disease: a neurocomputational approach. Neurobiol. Learn. Mem. **215**, 107985 (2024). https://doi.org/10.1016/j.nlm.2024.107985
13. Baston, C., Contin, M., et al.: A mathematical model of levodopa medication effect on basal ganglia in Parkinson's disease: an application to the alternate finger tapping task. Front. Hum. Neurosci. **10**, 280 (2016). https://doi.org/10.3389/fnhum.2016.00280
14. Ursino, M., Baston, C.: Aberrant learning in Parkinson's disease: a neurocomputational study on bradykinesia. Eur. J. Neurosci. **47**(12), 1563–1582 (2018). https://doi.org/10.1111/ejn.13960
15. van Albada, S.J., Robinson, P.A.: Mean-field modeling of the basal ganglia-thalamocortical system. I. Firing rates in healthy and Parkinsonian states. J. Theor. Biol. **257**(4), 642–663 (2009). https://doi.org/10.1016/j.jtbi.2008.12.018
16. Dreyer, J.K., Herrik, K.F., Berg, R.W., Hounsgaard, J.D.: Influence of phasic and tonic dopamine release on receptor activation. J. Neurosci. **30**(42), 14273–14283 (2010). https://doi.org/10.1523/JNEUROSCI.1894-10.2010
17. Schirru, M., Véronneau-Veilleux, F., Nekka, F., Ursino, M.: Phasic dopamine changes and Hebbian mechanisms during probabilistic reversal learning in striatal circuits: a computational study. Int. J. Mol. Sci. **23**(7), 3452 (2022). https://doi.org/10.3390/ijms23073452
18. Nicola, S.M., Surmeier, J., Malenka, R.C.: Dopaminergic modulation of neuronal excitability in the striatum and nucleus accumbens. Annu. Rev. Neurosci. **23**, 185–215 (2000). https://doi.org/10.1146/annurev.neuro.23.1.185

19. Cools, R., Barker, R.A., Sahakian, B.J., Robbins, T.W.: Enhanced or impaired cognitive function in Parkinson's disease as a function of dopaminergic medication and task demands. Cereb. Cortex **11**(12), 1136–1143 (2001). https://doi.org/10.1093/cercor/11.12.1136
20. Da Cunha, C., Gomez-A, A., Blaha, C.D.: The role of the basal ganglia in motivated behavior. Rev. Neurosci. **23**(5–6), 747–767 (2012). https://doi.org/10.1515/revneuro-2012-0063

A Dynamic Neural Field Model of Chronotopic Map Formation

Paulo Barbosa[1(✉)], Flora Ferreira[2,3], Estela Bicho[4], and Wolfram Erlhagen[1]

[1] Centre of Mathematics, University of Minho, Guimarães, Portugal
pjbarbosa98@gmail.com, wolfram.erlhagen@math.uminho.pt
[2] Centre of Mathematics, University of Porto, Porto, Portugal
flora.ferreira@fep.up.pt
[3] School of Economics and Management, University of Porto, Porto, Portugal
[4] ALGORITMI Research Centre, University of Minho, Guimarães, Portugal
estela.bicho@dei.uminho.pt

Abstract. Quantifying temporal information is essential for interacting with dynamic environments, yet the neural mechanisms underlying time representation remain elusive. Neurophysiological and behavioral studies suggest two principal coding strategies: monotonic ramping activity, in which firing rates increase continuously to encode elapsed time (rate code), and duration-tuned responses (place code), where neurons respond selectively to preferred durations and form topographically organized maps in higher cortical regions. Here, we present a Dynamic Neural Field model that transforms rate-coded into place-coded temporal representations. The model combines a neural integrator with self-organized learning of feed-forward connections to a neural field, where the position of localized activity patterns encodes event duration. In line with experimental findings, the resulting chronotopic maps represent both absolute and relative time and exhibit strong context dependence. The maps adapt to different ranges of experienced durations, consistent with the notion of a scalable population code for time. The neural tuning dynamically reorganizes under nonstationary input statistics while maintaining stable representations. Moreover, the learned temporal representations may generalize across non-temporal input features (e.g., spatial location), providing a neural basis for categorical temporal decisions. We discuss these results with respect to our goal of endowing autonomous systems with human-like temporal cognition.

Keywords: Dynamic Neural Fields · Time Perception · Chronotopic Maps

1 Introduction

In a dynamic world, our ability to sense the passage of time is essential. It helps us identify the duration and temporal relationship between sensory events and

L. Pasa et al. (Eds.): ICANN 2026, LNCS 17089, pp. 53–64, 2027.
https://doi.org/10.1007/978-3-032-38398-3_5

anticipate actions. However, how the brain measures and represents temporal information remains unclear, as there are neither dedicated sensory organs nor well-defined neural pathways for time processing.

One possible neural mechanism to robustly encode temporal information is ramping activity defined by a consistently increasing firing rate of a neural population that peaks at the end of a timed interval [23,33]. Particularly compelling evidence comes from an experiment in which monkeys were trained to first measure different sample intervals defined by visual cues and immediately afterward reproduce them with the onset of a saccadic eye movement to a predefined target [17]. The recorded neural activity increased with the duration of the sample interval, and the peak value at the end of the measurement period predicted the slope of the ramping activity toward a fixed decision threshold during the production period. The idea that ramping activity reflects a temporal integration process has been exploited in several neurocomputational models of timed behavior [11,27,31]. Importantly, the processing of time by a neural population may go hand in hand with the representation of other environmental information, like, for instance, the location and color of an object or stimulus frequency [4,13]. This integrated view might explain why ramping activity in timing tasks has been described in a wide variety of brain areas [29].

A conceptually distinct coding strategy for temporal information is represented by neural populations tuned to different time intervals. Converging evidence from neurophysiological and behavioral experiments supports the notion that neurons in different brain areas respond selectively to a narrow range of stimulus durations with a peak at the preferred duration [16,22]. The response of an array of neurons with distinct, though overlapping, tuning curves would represent the most likely perceived duration. This type of code is commonly referred to as a place code. Consistent with this view, neuroimaging studies report the existence of timing-tuned neural responses that are topographically mapped such that duration preferences change gradually across the cortical surface, forming so-called chronotopic maps [14,24]. Interestingly, recent studies analyzing timing selective neural response along the visual pathway report the coexistence of monotonic rate and place coding schemes. The observation that duration-tuned responses gradually become dominant across successive stages of visual processing has been interpreted as evidence for temporal representations that become independent of sensory event attributes [6,15]. However, the question of how duration-tuned responses can be derived from monotonic early visual responses is still an unsolved problem which, to the best of our knowledge, has not yet been investigated computationally.

In this paper, we present a model based on the theoretical framework of Dynamic Neural Fields (DNFs) that addresses the problem of converting neural population activity from rate code to place code. It is based on our previous work on precisely-timed behavior, where we introduced a neural integrator model representing elapsed time in the amplitude of self-stabilized neural activation patterns or bumps [30]. The neural integrator has been applied to explain experimental data and to endow an autonomous robot with the capacity of adaptive motor timing for fluent human-robot interactions [31,32]. Here, we extend this work by adapting an existing DNF model of self-organized map formation in the somatosensory cortex [10] to the temporal domain. The unsupervised learning process establishes inter-field connections that map bumps of different amplitudes onto specific positions of a DNF in a spatially ordered manner. We show in model simulations that the developing topographic duration map dynamically adapts to different ranges of time intervals while abstracting the temporal representations from spatial properties of the input like, for instance, location. The promising first results are discussed in the context of our ongoing work. The focus is on exploiting the assumed complementary role of rate and space coding of event duration to qualitatively explain experimental data and to endow autonomous robots with a human-like time perception capacity.

2 Dynamic Neural Fields

DNFs provide a coarse-grained description of the mean firing-rate activity across a population of interconnected neurons, typically defined over continuous feature dimensions. Their recurrent interactions give rise to a wide range of activity patterns, such as stationary bumps and traveling waves [7], which have been used to model a range of neural phenomena [3,12,18]. We consider two distinct DNF models, described below.

2.1 The Amari Neural Field Model

First, we consider the classical neural field formulation introduced by Amari [2]. Let $\Omega \subset \mathbb{R}$ denote a compact, connected spatial domain. The mean population activity $u(x,t)$ at position $x \in \Omega$ and time $t \geq 0$ is governed by the following dynamics:

$$\frac{\partial u(x,t)}{\partial t} = -u(x,t) + \int_{\Omega} w(|x-y|)\mathcal{H}(u(y,t))\,dy + I_{\text{EXT}}(x,t), \tag{1}$$

where $\mathcal{H}$ denotes the Heaviside step function with threshold 0, $I_{\text{EXT}}(x,t)$ is the localized external input to the network, and w is the lateral synaptic coupling kernel. We assume an isotropic connectivity profile:

$$w(|x-y|) = w_e(|x-y|) - w_i(|x-y|) \tag{2}$$

where the excitatory and inhibitory components are given by Gaussian functions:

$$w_e(x) = A_{\text{exc}}\, e^{-x^2/(2\sigma_{\text{exc}}^2)}, \tag{3a}$$

$$w_i(x) = A_{\text{inh}}\, e^{-x^2/(2\sigma_{\text{inh}}^2)} + g_{\text{inh}}, \tag{3b}$$

with $A_{exc} > A_{inh} > 0$, $\sigma_{inh} > \sigma_{exc} > 0$, and $g_{inh} > 0$.

The model support the existence of self-stabilized bumps which are triggered by transient inputs. However, the stereotyped bump shape does not reflect input duration.

2.2 The Neural Integrator Model

The neural integrator model is a two-population extension of the Amari model [30]. It supports the existence of bumps with a continuum of amplitudes representing elapsed time. The dynamics of the coupled u- and v-populations are governed by:

$$\frac{\partial u(x,t)}{\partial t} = -u(x,t) + v(x,t) + \int_{\Omega} w(|x-y|)\mathcal{H}(u(y,t))\,dy + I_{\text{EXT}}(x,t), \quad (4a)$$

$$\frac{\partial v(x,t)}{\partial t} = -v(x,t) + u(x,t) - \int_{\Omega} w(|x-y|)\mathcal{H}(u(y,t))\,dy. \quad (4b)$$

Behavioral studies suggest that variations in stimulus intensity may influence perceived duration [20]. Since the bump amplitude of the integrator model is also modulated by stimulus strength, we normalize the input to unambiguously represent the integration of time. As illustrated in Fig. 1, the model exhibits a strong linear relationship between duration and amplitude (Pearson $r = 0.999$).

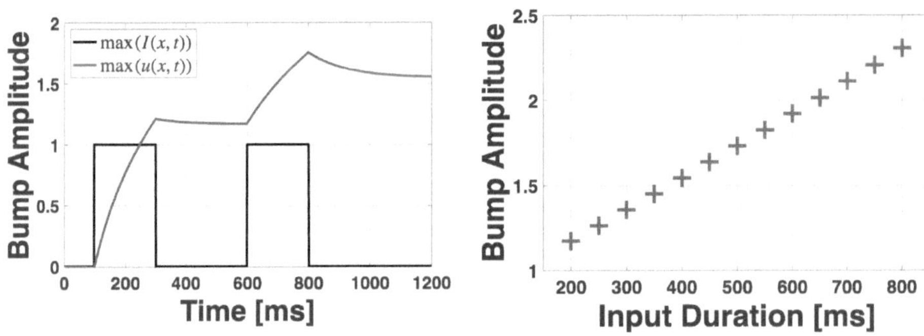

Fig. 1. (Left) Temporal evolution of the bump amplitude in response to two sequentially presented inputs of equal duration: the first applied from 100 to 300 milliseconds and the second from 600 to 800 milliseconds. (Right) Bump amplitude as a function of sample interval length, revealing an approximately linear relationship.

3 Learning a Topographic Event Duration Map

The model architecture sketched in Fig. 2 consists of two modules: a temporal integration module and a topographic map formation module. The former integrates stimuli of varying duration and produces a normalized scalar signal. The latter takes this signal as input and generates a localized activity pattern representing a place code of the specific duration.

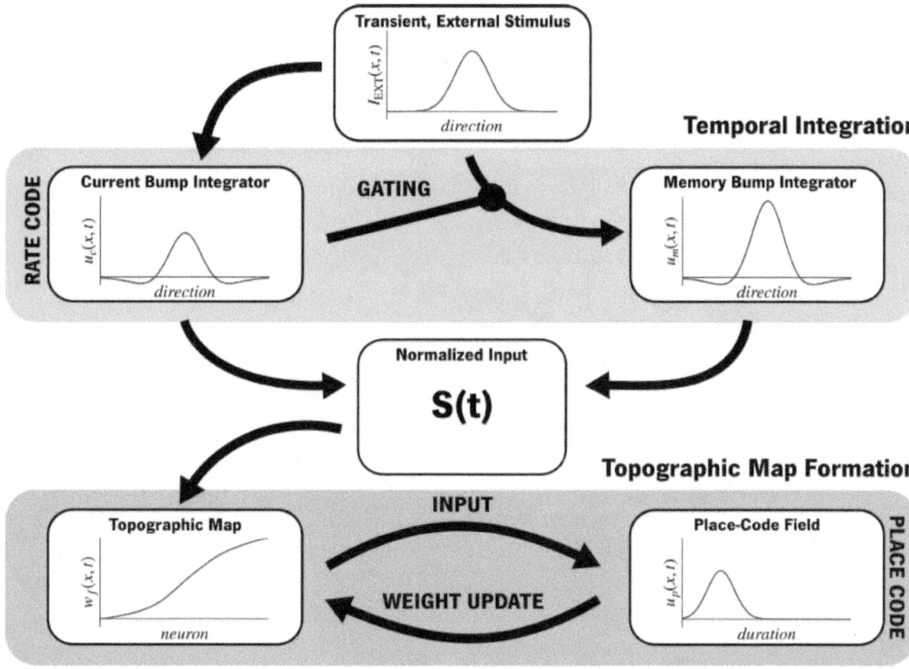

Fig. 2. Schematic view of the proposed architecture for learning topographic event duration maps. The temporal integration module converts transient stimuli into a normalized duration signal, which is then passed to the topographic map formation module to generate a spatially organized, place-coded representation of event duration.

3.1 Temporal Integration

The module consists of two integrator-type DNFs. The current activity bump, u_c, represents the temporal integration of the normalized input in each trial. It is reset to zero at the onset of each new trial. Second, an associated memory bump, u_m, tracks and memorizes in its amplitude the longest duration of all inputs presented at the specific position. This is achieved for simplicity with a hand-coded gating signal which opens the input integration whenever the amplitude of the current bump exceeds the amplitude of the memory bump:

$$\phi(x,t) = \mathcal{H}\big(\max_{x \in \Omega} (u_c(x,t)) - \max_{x \in \Omega} (u_m(x,t)) \big). \tag{5}$$

The input to the u- Eq. (4a) of the memory bump integrator is thus given by $\phi(x,t) \cdot I_{\mathrm{EXT}}(x,t)$. The implementation of the comparison using properties of the neural ramping dynamics as shown in monkey experiments [17] is left for future studies.

The supra-threshold activity of the memory bump is used for a divisive input normalization [5]. At the end of each trial, the input signal to the topographic map formation module is defined as the ratio of the total supra-threshold activity of the current integrator to that of the memory integrator

$$S(t) = \frac{\int_{\Omega} [u_c(x,t)]_+ \, dx}{\varepsilon + \int_{\Omega} [u_m(x,t)]_+ \, dx}, \tag{6}$$

where $0 < \varepsilon \ll 1$ is a small constant introduced to prevent division by zero and $[\cdot]_+$ denotes rectification. The ratio is bounded between 0 and 1, since the integrated supra-threshold activity of the memory bump is as least as large as that of the current bump.

3.2 Topographic Map Formation

The formation of chronotopic maps is implemented using the principles of the DNF Self-Organizing Map (DNF-SOM), originally proposed to explain the development of ordered maps in the somatosensory cortex with input from the spatially distributed mechanoreceptor of the hand [10].

At the end of each trial, the input to a neuron at position $x \in \Omega$ of the place-code field u_p is given by the function:

$$I(x,t) = 1 - \|w_f(x,t) - S(t)\|_1, \tag{7}$$

where $w_f(x,t) : \Omega \times \mathbb{R}_0^+ \to [0,1]$ represents the feed-forward weights at time t and $\|\cdot\|_1$ denotes the $\mathcal{L}_1$ norm. The neural activity of u_p then evolves according to the Amari dynamics (Eq. 1). The location of the resulting bump pattern encodes event duration. Once the field activity reaches a stationary state, the feed-forward weights are updated through a variant of Oja's Hebbian learning rule:

$$\frac{\partial w_f(x,t)}{\partial t} = \gamma\big(S(t) - w_f(x,t)\big) \int_{\Omega} w_e(|x-y|)\mathcal{H}(u_p(y,t))\, dy, \tag{8}$$

where γ is the constant learning rate governing the adaptation speed of the feed-forward weights. Starting with small random values of a uniform distribution, we assume an adiabatic regime in which weight updates evolve much more slowly than neural activity, i.e., $\gamma \ll 1$. In this regime, small values of γ ensure stable but slow convergence, whereas excessively large values may induce instability or oscillatory behavior, preventing the formation of a stable topographic map. Similarly to Kohonen's SOM algorithm, the feed-forward weights are attracted toward the current input [19]. However, instead of a hand-coded neighborhood function which is assumed to decrease over time with a certain rate, the neural activity pattern in the post-synaptic field guides the Hebbian learning process. Moreover, it is well known that strong recurrent interactions in neural populations suppress noisy and weakly correlated inputs which contributes to the development of a smooth and continuous topographic map. Under mild constraints imposed on the lateral connectivity kernel, the resulting DNF-SOM has been shown to be convergent and stable [9].

4 Experimental Results

This section illustrates how the proposed architecture transforms a rate code into a place code, corresponding to absolute and relational representations of perceived time [1], and evaluates its ability to adapt to changing input distributions, exhibit temporal rescaling, and abstract temporal information from non-temporal features such as spatial location.

In the experiments, we assume normalized Gaussian inputs of varying durations centered at a certain location of the integrator fields, representing a specific value of continuous dimension direction.

The place-code field has a finite length, discretized into N neurons. This means that the outermost neurons do not have neighbors on one side with which they can interact like in real cortical networks. This will cause distortions of the topographic mapping at the shortest and longest durations. To minimize this boundary effect, we have extended in the model simulations the range of sample durations beyond the limits that the developing chronotopic map is supposed to represent (e.g., 500 - 1000 ms).

4.1 Temporal Rescaling and Map Reorganization

In the first numerical experiment, we trained the model using a single cue located at direction $\theta = -45°$ over 3000 iterations. During the first 1500 iterations, stimulus durations were sampled uniformly from 500 – 1000 ms. During the remaining iterations, the range was expanded to 500 – 2000 ms.

Figure 3 (A-C) compares the rate- and place coded responses at iterations 1500 and 3000 for a 900 ms sample. The rate code represents, in both cases, the 900 ms interval but the place-code representation differs significantly. At iteration 1500, the population response is localized close to the right border of the field whereas at iteration 3000 it appears shifted to the left. Panel D compares the mapping of the different ranges of sample durations on the normalized relative field position, showing a strongly linear relationship (Pearson r = 0.997 and 0.996 at iterations 1500 and 3000, respectively). It is important to stress however that for a given stationary temporal context (e.g., 500 - 1000 ms), the read-out of the bump position provides absolute duration information as encoded by the bump amplitude of the neural integrator.

Moreover, the chronotopic map successfully reorganizes in response to changes in the input distribution. To quantify this reorganization capacity, we computed the network's distortion over the training trials (Panel E), defined as the quantization error of the self-organizing map (for details, see [8]). It took about 500 iterations to reach the same low distortion level as before the expansion of the duration range.

4.2 Chronotopic Maps from Different Input Positions

In the second experiment, we considered two visual cues presented at distinct spatial field positions, namely $\theta_1 = -45°$ and $\theta_2 = +45°$. The corresponding stimulus-duration ranges were 500 - 1000 ms and 1000 - 2000 ms, respectively.

After learning, in separate blocks, the feed-forward connections from each input position to the tuning field, we tested the network with a fixed 500 ms interval for the visual cue at θ_1 and subsequently with a fixed 1000 ms interval for the visual cue at θ_2. These intervals correspond to the minimum durations of the respective ranges being trained.

The different rate amplitudes faithfully reflect the difference in input duration (Fig. 4 - A, B). By contrast, the strongly overlapping place-codes reveal that the same neural population encodes different intervals depending on the context, since both tested intervals represent the minimum within the respective duration ranges. This finding is consistent with the idea that place codes constitute relational representations of elapsed time.

Compared to the first experiment, the temporal tuning is more abstract since it generalizes across stimulus locations [15]. Such abstract chronotopic maps may provide a neural substrate for learning to quantify intervals as short, intermediate or long depending on the relative bump position in the field. It is important to notice that behavioral experiments strongly suggest that absolute and relative temporal representations coexist in the brain to guide performance [1].

5 Discussion

Neural tuning and topographic organization of input features and quantities are common throughout sensory and cognitive processing in the brain [28]. While pathways with approximately topological connections for the transmission of sensory information from retinal, auditory, or somatosensory receptors to specific brain areas have been described [25], the emergence of topographic maps for temporal features, such as the duration of a sensory event, remains an open question.

We have presented a DNF model that implements neuro-plausible processing mechanisms supporting the emergence of duration-tuned neurons. Two main assumptions are guiding our work. First, the dynamics of neural responses to input features, such as, for instance, position or color, may simultaneously represent the passage of time. The assumption of a decentralized neural process for interval timing is supported by the observation of ramping activity in many brain regions involved in timing tasks [29]. In the model, the summed bump amplitude of the neural integrator reflects event duration. Secondly, we assume that normalized bumps with different amplitudes representing a whole range of durations are orderly mapped in contiguous portions of a neural field. The self-organizing inter-field connections are established online with a Hebbian-type learning rule [9]. The computational process shares with Kohonen's SOM [19] that the feed-forward weights are shifted towards the input. However, the field implementation adds biological realism since the evolving activation pattern in the post-synaptic field avoids 1) the *"argmax"* function that scans all neurons to find the perfect mathematical winner, and 2) the hand-coded neighborhood function of the Kohonen algorithm. The stability conditions of the DNF-SOM are known [9] whereas the convergence of SOM depends heavily on the dynamic coupling of the chosen neighborhood function and decay rate.

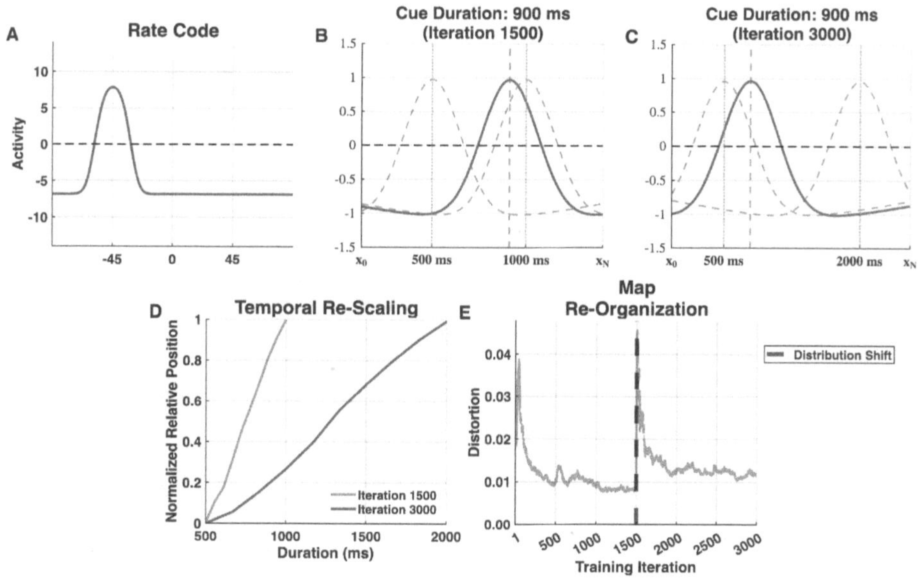

Fig. 3. (A–C) Bumps (solid line) representing the rate-coded (A) and the place-coded responses to a fixed 900 ms stimulus at iterations 1500 (B) and 3000 (C) when at iteration 1500 the temporal context changed. Rate codes remain invariant, whereas place codes shift with the experienced duration range. The dashed lines indicate the bump positions representing the minimum and the maximum duration in each temporal context. (D) Normalized positions of place-coded responses across cue durations at iterations 1500 and 3000, illustrating temporal rescaling. (E) Map distortion during training. After the change in input distribution, distortion briefly increases and then returns to a low steady state, indicating successful reorganization.

The model simulations show that a neural summation code for duration can be transformed into a place code of neurons with different preferred durations. Why is the nervous system using both coding schemes, given that bump amplitude and bump position may convey the same information? One hypothesis is a difference in the level of abstraction [15]. Consistent with experimental evidence, the model predicts that the duration-tuning may become independent of input features like position (Fig. 2) or the specific sensory modality, for instance, representing the visual and/or auditory duration of an event. Moreover, the tuning in chronotopic maps is known to be highly context dependent. The neural representation of the same physical duration presented in the context of different, but overlapping, duration distributions appears to be spatially shifted according to the relative position within each context (Fig. 4) [21,24]. In the model, this relative scaling is achieved due to the input normalization by the accumulated evidence of the neural integrator dynamically tracking the longest tested duration.

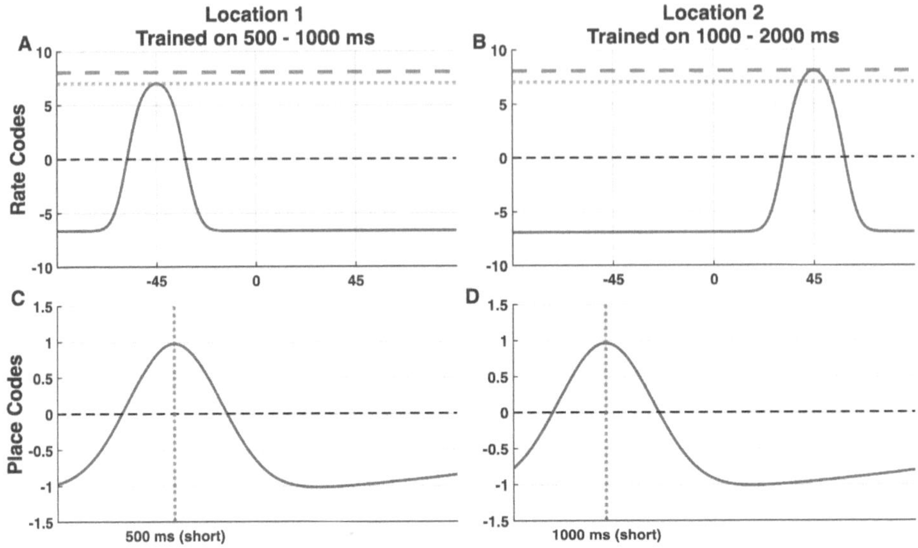

Fig. 4. Context-dependent rate and place coding for the shortest trained durations. (A, B) Rate-coded responses to the minimum duration in each trained range: 500 ms at Location 1 (500 - 1000 ms) and 1000 ms at Location 2 (1000 - 2000 ms). The amplitudes differ because the absolute durations differ. (C, D) Corresponding place-coded responses occupy the same field position in both cases, indicating that the place code preserves the same relational classification.

A robust finding of contextual effects on time perception, which we are currently investigating, is the central-tendency bias [26]. Estimates of time are systematically pulled toward the mean of the duration range. This can be explained with the space coding of the DNF model by assuming that the chronotopic map appears to be pre-activated to a subthreshold level by the distribution of time intervals with a peak at the mean duration. Due to the recurrent interactions in the network, the position of the evolving bump of the current input appears to be attracted to some extent towards the peak position of the pre-activation.

Following our ultimate goal to endow autonomous robots with a human-like time perception capacity, we plan to test both coding schemes for time perception in human-robot interactions. In previous work, we have used ramping activity in an object hand-over task requiring an active monitoring of the passage of time by the robot for precise and adaptive motor timing that best serves the human user [32]. In many daily collaborative tasks, there will be sub-tasks that do not require any direct interaction. The temporal decision of an assistive robot on what to do next should thus take into account the predicted duration of an ongoing action of the user. Duration-tuned neurons may provide an abstract neural substrate for a memory store of (sub-) task durations, which can be used as a basis for learning to associate durations classified as "short" or "long" (depending on their relative position in the field) with predictive contextual cues and specific actions.

Acknowledgments. This study was supported by Fundação para a Ciência e a Tecnologia (FCT) through a doctoral scholarship with reference **UI/BD/153737/2022** (https://doi.org/10.54499/UI/BD/153737/2022) and the project **UID/00013/2025** (https://doi.org/10.54499/UID/00013/2025).

Disclosure of Interests. The authors have no competing interests to declare that are relevant to the content of this article.

Code Availability. The developed source code and model parameters used for the numerical simulations and figures can be found at https://github.com/PJBarbosa98/Chronotopic-Maps/.

References

1. Akdoğan, B., Wanar, A., Gersten, B.K., Gallistel, C.R., Balsam, P.D.: Temporal encoding: relative and absolute representations of time guide behavior. J. Exp. Psychol. Anim. Learn. Cogn. **49**(1), 46 (2023)
2. Amari, S.I.: Dynamics of pattern formation in lateral-inhibition type neural fields. Biol. Cybern. **27**(2), 77–87 (1977)
3. Bressloff, P.C., Cowan, J.D., Golubitsky, M., Thomas, P.J., Wiener, M.C.: Geometric visual hallucinations, euclidean symmetry and the functional architecture of striate cortex. Philos. Trans. R. Soc. Lond. B Biol. Sci. **356**(1407), 299–330 (2001)
4. Brody, C.D., Hernández, A., Zainos, A., Romo, R.: Timing and neural encoding of somatosensory parametric working memory in macaque prefrontal cortex. Cereb. Cortex **13**(11), 1196–1207 (2003)
5. Carandini, M., Heeger, D.J.: Normalization as a canonical neural computation. Nat. Rev. Neurosci. **13**(1), 51–62 (2012)
6. Centanino, V., Fortunato, G., Bueti, D.: The neural link between stimulus duration and spatial location in the human visual hierarchy. Nat. Commun. **15**(1), 10720 (2024)
7. Coombes, S.: Waves, bumps, and patterns in neural field theories. Biol. Cybern. **93**(2), 91–108 (2005)
8. Cover, T.M.: Elements of Information Theory. John Wiley & Sons (1999)
9. Detorakis, G., Chaillet, A., Rougier, N.P.: Stability analysis of a neural field self-organizing map. J. Math. Neurosci. **10**(1), 1–20 (2020). https://doi.org/10.1186/s13408-020-00097-6
10. Detorakis, G.I., Rougier, N.P.: A neural field model of the somatosensory cortex: formation, maintenance and reorganization of ordered topographic maps. PLoS ONE **7**(7), e40257 (2012)
11. Durstewitz, D.: Neural representation of interval time. NeuroReport **15**(5), 745–749 (2004)
12. Ermentrout, G.B., Cowan, J.D.: A mathematical theory of visual hallucination patterns. Biol. Cybern. **34**(3), 137–150 (1979)
13. Ferreira, F., Wojtak, W., Sousa, E., Louro, L., Bicho, E., Erlhagen, W.: Rapid learning of complex sequences with time constraints: a dynamic neural field model. IEEE Trans. Cogn. Dev. Syst. **13**(4), 853–864 (2020)
14. Harvey, B.M., Dumoulin, S.O., Fracasso, A., Paul, J.M.: A network of topographic maps in human association cortex hierarchically transforms visual timing-selective responses. Curr. Biol. **30**(8), 1424–1434 (2020)

15. Hendrikx, E., Paul, J.M., van Ackooij, M., van der Stoep, N., Harvey, B.M.: Visual timing-tuned responses in human association cortices and response dynamics in early visual cortex. Nat. Commun. **13**(1), 3952 (2022)
16. Heron, J., Aaen-Stockdale, C., Hotchkiss, J., Roach, N.W., McGraw, P.V., Whitaker, D.: Duration channels mediate human time perception. Proc. R. Soc. B Biol. Sci. **279**(1729), 690–698 (2012)
17. Jazayeri, M., Shadlen, M.N.: A neural mechanism for sensing and reproducing a time interval. Curr. Biol. **25**(20), 2599–2609 (2015)
18. Jirsa, V.K., Jantzen, K.J., Fuchs, A., Kelso, J.A.S.: Neural field dynamics on the folded three-dimensional cortical sheet and its forward EEG and MEG. In: Insana, M.F., Leahy, R.M. (eds.) IPMI 2001. LNCS, vol. 2082, pp. 286–299. Springer, Heidelberg (2001). https://doi.org/10.1007/3-540-45729-1_30
19. Kohonen, T.: Self-organizing Maps, vol. 30. Springer Science & Business Media (2012)
20. Matthews, W.J., Stewart, N., Wearden, J.H.: Stimulus intensity and the perception of duration. J. Exp. Psychol. Hum. Percept. Perform. **37**(1), 303 (2011)
21. Mello, G.B., Soares, S., Paton, J.J.: A scalable population code for time in the striatum. Curr. Biol. **25**(9), 1113–1122 (2015)
22. Merchant, H., Pérez, O., Zarco, W., Gámez, J.: Interval tuning in the primate medial premotor cortex as a general timing mechanism. J. Neurosci. **33**(21), 9082–9096 (2013)
23. Merchant, H., Zarco, W., Pérez, O., Prado, L., Bartolo, R.: Measuring time with different neural chronometers during a synchronization-continuation task. Proc. Natl. Acad. Sci. **108**(49), 19784–19789 (2011)
24. Protopapa, F., et al.: Chronotopic maps in human supplementary motor area. PLoS Biol. **17**(3), e3000026 (2019)
25. Sereno, M.I., Sood, M.R., Huang, R.S.: Topological maps and brain computations from low to high. Front. Syst. Neurosci. **16**, 787737 (2022)
26. Shi, Z., Church, R.M., Meck, W.H.: Bayesian optimization of time perception. Trends Cogn. Sci. **17**(11), 556–564 (2013)
27. Simen, P., Balci, F., deSouza, L., Cohen, J.D., Holmes, P.: A model of interval timing by neural integration. J. Neurosci. **31**(25), 9238–9253 (2011)
28. Tsouli, A., et al.: The role of neural tuning in quantity perception. Trends Cogn. Sci. **26**(1), 11–24 (2022)
29. Wittmann, M.: The inner sense of time: how the brain creates a representation of duration. Nat. Rev. Neurosci. **14**(3), 217–223 (2013)
30. Wojtak, W., Coombes, S., Avitabile, D., Bicho, E., Erlhagen, W.: A dynamic neural field model of continuous input integration. Biol. Cybern. **115**(5), 451–471 (2021). https://doi.org/10.1007/s00422-021-00893-7
31. Wojtak, W., Ferreira, F., Bicho, E., Erlhagen, W.: Neural field model for measuring and reproducing time intervals. In: Tetko, I.V., Kurková, V., Karpov, P., Theis, F. (eds.) ICANN 2019. LNCS, vol. 11727, pp. 327–338. Springer, Cham (2019). https://doi.org/10.1007/978-3-030-30487-4_26
32. Wojtak, W., Ferreira, F., Louro, L., Bicho, E., Erlhagen, W.: Adaptive timing in a dynamic field architecture for natural human-robot interactions. Cogn. Syst. Res. **82**, 101148 (2023)
33. Zhou, J., Benson, N.C., Kay, K.N., Winawer, J.: Compressive temporal summation in human visual cortex. J. Neurosci. **38**(3), 691–709 (2018)

From NEST Benchmark to FPGA Silicon: Quantifying the Cost of STDP and Demonstrating Efficient Kernel Offloading

Maxime Carriere(✉)

Brain Language Laboratory, Freie Universität Berlin, Berlin, Germany
carriere.maxim@gmail.com

Abstract. Spiking Neural Networks trained with Spike-Timing Dependent Plasticity (STDP) are expensive to simulate on conventional CPUs at scale, but the precise conditions under which that cost becomes critical have not been systematically quantified. This paper makes two contributions. First, a benchmark of NEST 3.6 across a wide parameter space shows when STDP synapses become expensive relative to static synapses: overhead is negligible for small networks but grows to a 3.75× slowdown at $N = 8{,}000$ ($p = 0.1$), and per-synapse cost exhibits a pattern consistent with memory-hierarchy effects above $N \approx 800$. Multi-core scaling leaves the STDP overhead ratio largely unchanged. Second, a pipelined HLS co-processor on the Xilinx Zynq XC7Z020 shows that the core STDP weight-update computation can be offloaded to low-resource FPGA hardware: 100 M syn/s at 500 ns per spike and 1.73 W board power. Hardware correctness is validated against a Python reference implementing the same event-driven rule (MAE = 1.91×10^{-3}, dominated by quantisation); the larger residual against NEST (MAE = 1.10×10^{-2}) reflects a difference in update-timing semantics, not a hardware error. The main integration bottleneck is identified and quantified, and a batched-transfer scheme is outlined as a plausible path toward closing it.

Keywords: STDP · Spiking Neural Networks · FPGA · NEST Simulator · Neuromorphic Computing · HLS · Hardware Acceleration

1 Introduction

Spiking Neural Networks (SNNs) are a biologically motivated computing paradigm in which neurons communicate through discrete electrical pulses and learning rules operate on spike timing rather than continuous activations [14]. They are of growing interest for neuromorphic embedded systems, closed-loop brain-machine interfaces, and fundamental neuroscience modelling. A central challenge is efficient synaptic plasticity: the mechanism by which connection strengths are updated during learning or simulation.

Spike-Timing Dependent Plasticity (STDP) [1,15,22] is the dominant biologically observed learning rule. For fixed connection probability and firing rate,

L. Pasa et al. (Eds.): ICANN 2026, LNCS 17089, pp. 65–77, 2027.
https://doi.org/10.1007/978-3-032-38398-3_6

the total synaptic update rate scales quadratically with network size: each spike triggers a weight update at every outgoing synapse, so larger or denser networks rapidly exhaust the memory bandwidth and arithmetic capacity of general-purpose processors. Software simulators such as NEST [9] serialise these updates on CPUs, which is adequate for small networks but increasingly costly at scale, because each spike can trigger weight updates across an entire row or column of the synapse table. NEST is also used in our broader modelling context for large-scale, biologically grounded spiking network simulations [2].

Despite this known bottleneck, a quantitative characterisation of *when* STDP overhead becomes critical—across network sizes, densities, firing rates, and core counts—has not been published. Such a baseline is useful for hardware–software co-design: it identifies the regime where a software-only approach is insufficient and bounds the expected benefit of dedicated hardware. This matters most for applications that need on-line plasticity under real-time or power constraints—closed-loop brain–machine interfaces [5,16], adaptive neuromorphic edge devices [10], and large-scale plastic simulations in computational neuroscience [2,13]. In these settings the per-spike cost of STDP, not inference, sets the achievable scale.

On the hardware side, prior FPGA STDP kernels rely on custom RTL targeting high-end devices [20,23]; HLS-based deployment on a mid-range Zynq SoC with validation against a reference simulator has not been demonstrated. The Zynq SoC platform is particularly relevant here: an embedded ARM processor communicates spike events to programmable logic over AXI-Lite, a setup that is flexible and reproducible but also makes host–accelerator transfer overhead explicit and measurable.

Contributions. Two things are shown separately:

1. A controlled benchmark of NEST 3.6 STDP synapses versus static synapses across 1,950 parameter configurations, showing the software-side scaling bottleneck and a per-synapse cost pattern consistent with memory-hierarchy effects above $N \approx 800$ neurons.
2. A pipelined HLS co-processor on the Xilinx Zynq XC7Z020 demonstrating that the core STDP update computation can be offloaded to FPGA hardware, validated against a Python reference implementing the same event-driven formulation. Comparison with NEST highlights a difference in update-timing semantics. End-to-end integration with NEST is an open engineering challenge, discussed explicitly.

Section 2 reviews STDP and related work. Section 3 presents the NEST benchmark. Section 4 describes the FPGA design. Section 5 reports validation and performance. Section 6 discusses limitations and future work. Section 7 concludes.

2 Background and Related Work

2.1 The STDP Learning Rule

STDP modifies synaptic weights W_{ij} based on the relative timing of pre- and post-synaptic spikes [1,18,22]. Each synapse maintains a pre-synaptic trace x_{pre}

decaying with time constant τ_+ and a post-synaptic trace x_{post} decaying with τ_-. Weights are updated at every spike:

$$\text{post-synaptic spike at } j: \quad \Delta W_{ij} = +A_+ \, x_{\text{pre},i}(t) \quad \text{(LTP)} \tag{1}$$

$$\text{pre-synaptic spike at } i: \quad \Delta W_{ij} = -A_- \, x_{\text{post},j}(t) \quad \text{(LTD)} \tag{2}$$

In the additive variant of the rule ($\mu_\pm = 0$), the learning amplitudes are $A_+ = \lambda$ and $A_- = \lambda\alpha$, where λ is the learning rate and $\alpha > 1$ gives asymmetric LTD/LTP magnitudes. Weights are bounded: $W_{ij} \in [0, W_{\max}]$. Both traces share the same time constant $\tau_\pm = 20\,\text{ms}$ in the standard parameterisation. The total update rate is

$$\text{Updates/s} = \nu \cdot p \cdot N^2 \tag{3}$$

where ν is the mean firing rate and p the connection probability. For fixed ν and p, this yields quadratic growth with network size, which drives the bottleneck analysed in Sect. 3.

The rule used here is the widely adopted pair-based form; higher-order variants such as triplet STDP [21] and more recent kernel-driven supervised formulations [4] improve biological fidelity at additional cost, but share the same per-synapse, event-driven update structure analysed in this paper.

2.2 Related Work

FPGA-based implementations of SNNs have been studied extensively across neuron models, learning rules, and system architectures, as reviewed recently by Karamimanesh et al. [12]. However, much of this literature targets inference, fixed-weight simulation, or custom RTL implementations rather than on-line plasticity on modest FPGA platforms. For example, Pani et al. [20] present a custom RTL platform for real-time SNN simulation on high-end Xilinx devices; Cheung et al. [6] target large-scale FPGA acceleration without on-chip plasticity; and Mitchell et al. [17] demonstrate a small, low-cost event-driven FPGA platform for SNNs but do not address on-line STDP offloading. At the system level, Kauth et al. [13] and Trensch and Morrison [23] focus on larger simulation architectures and system-level constraints—communication, reproducibility, and multi-node execution—rather than a standalone on-line STDP kernel. On digital many-core neuromorphic hardware, Jin et al. [11] implement STDP on SpiNNaker using a deferred, pre-synaptic-sensitive scheme; notably, they identify a large fixed per-row processing cost that favours long synaptic rows—a memory-access effect consistent with the per-synapse cost pattern we observe on CPU (Sect. 3). Custom neuromorphic ASICs have demonstrated on-chip STDP at far greater scale and energy efficiency: Chen et al. [3] integrate 4,096 neurons and 1M synapses with on-chip STDP in 10-nm FinFET CMOS at 3.8 pJ/SOP, and Intel's Loihi [8] supports on-chip learning across 128 neuromorphic cores (≈130,000 neurons). These dedicated implementations define the efficiency frontier for STDP hardware. The approach proposed here targets a complementary regime: a reconfigurable FPGA platform consuming below 2 W, synthesised directly from C++ via HLS, and validated against a standard neuroscience reference simulator (NEST), prioritising reproducibility, accessibility, and rapid prototyping over peak energy efficiency or transistor-count scale. To our knowledge, no prior work combines all three of: (i) a systematic CPU-side STDP cost benchmark, (ii) an HLS-based STDP kernel on a mid-range SoC, and (iii) validation against recorded NEST spike trains.

Table 1. Benchmark parameter space (1,950 configurations, 3 trials each).

Parameter	Values
Network size N	10, 50, 100, 200, 500, 1 000, 2 000, 4 000, 8 000
Connection probability p	0.05, 0.10, 0.20
Poisson input rate ν	10, 20, 40 Hz
CPU cores	1, 2, 4, 8
Simulation duration	20 s biological time
Synapse delay	0.1 ms; STDP: $\tau_{\pm} = 20$ ms, $\alpha = 1.2$

3 Quantifying STDP Overhead in NEST

3.1 Experimental Setup

NEST is a widely used event-driven reference simulator for large-scale spiking networks [9,19]. The benchmark compares NEST 3.6 STDP synapses against static synapses across a full factorial sweep. Table 1 lists all parameters. Neurons follow leaky integrate-and-fire dynamics with alpha-function synaptic currents and a 400 pA bias current.

A key methodological choice is setting the learning rate to $\lambda = 0$: trace computation and weight-update code still execute at every spike, but $\Delta W = 0$ by construction. This prevents STDP from altering network activity (pilot runs with $\lambda = 0.01$ produced 11.6 % fewer spikes on average), ensuring that any timing difference reflects purely the cost of running the STDP machinery. Identical random seeds are used for each paired static/STDP trial; firing rates were verified to be identical across all 1,950 configurations (maximum deviation < 0.01 %).

All simulations ran on a desktop workstation running Arch Linux (x86-64, kernel 6.9.8), equipped with an Intel Core i9-10900X (10 cores, 20 threads; 3.70 GHz base, 4.70 GHz boost); each configuration used 1, 2, 4, or 8 cores of that machine. Wall-clock time is measured with a high-resolution system timer around each simulation call. NEST's internal simulation timer agreed to within 0.04 % across all configurations.

3.2 Overhead Grows Super-Linearly with Network Size

Figure 1A shows STDP synapse overhead relative to static synapses as a function of N. Below $N \approx 200$ the overhead is negligible ($< 10\,\%$). A sharp transition occurs around $N \sim 1{,}000$, consistent with the per-synapse cost change discussed in Sect. 3.3: for $p = 0.1$ the overhead reaches 275 % at $N = 8{,}000$ ($3.75\times$); for $p = 0.2$ it reaches 369 %; the global maximum is 475 % ($N = 8{,}000$, $p = 0.2$, 10 Hz, 2 cores). In absolute terms, STDP at $N = 8{,}000$ adds over 15 min of extra runtime per 20-second simulation.

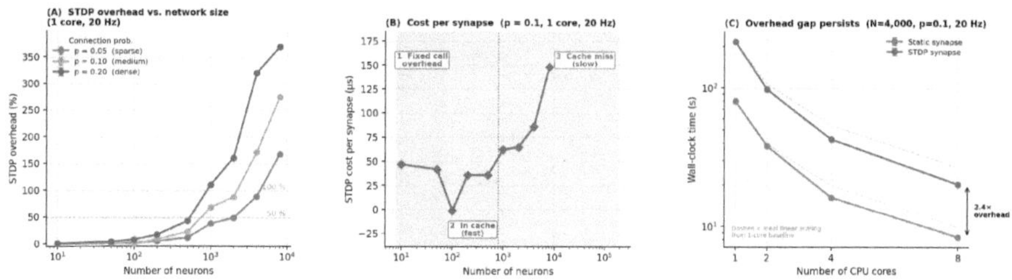

Fig. 1. STDP computational cost in NEST 3.6 (1 core, 20 Hz, $p = 0.1$ unless noted). (A) STDP synapse overhead (%) vs. N for three connection probabilities; dashed lines mark 50 % and 100 % thresholds. (B) Per-synapse overhead vs. N; the non-flat curve is consistent with memory-hierarchy effects: a fast zone where the synapse table fits in cache ($N = 100$–800, green) and a slower zone above $N \approx 800$ (red) where cost approximately doubles. (C) Wall-clock time vs. CPU-core count at $N = 4{,}000$; both models scale similarly (dashed = ideal linear scaling), leaving the $\sim 2.4\times$ overhead ratio unchanged at all core counts. (Color figure online)

3.3 Per-synapse Cost: Consistent with Memory-Hierarchy Effects

If per-synapse update cost were constant, the overhead-per-synapse curve would be flat. Instead, Fig. 1B shows three regimes consistent with memory-hierarchy effects:

For small networks ($N < 100$), cost is $\approx 60\,\mu s$ per synapse, likely dominated by fixed function-call overhead not amortised over a large synapse table. For medium-sized networks ($100 \leq N < 800$), cost falls to $\approx 36\,\mu s$ per synapse—the lowest observed value, consistent with the ~ 1 MB synapse table fitting in cache. For large networks ($N \geq 800$), cost rises to $\approx 63\,\mu s$ per synapse, approximately double the medium regime; the table grows to ~ 3 MB, consistent with increased cache pressure and more frequent main-memory accesses.

The inflection near $N \sim 800$ is consistent with a memory-hierarchy transition. Direct confirmation via cache-miss profiling is left as future work (Sect. 6). An FPGA holding synaptic state in BRAM offers predictable access latency for networks that fit within the available on-chip memory, making it a natural platform for this regime.

3.4 Multi-core Parallelism Does Not Close the Gap

Figure 1C shows wall-clock time for static and STDP synapses at $N = 4{,}000$ across core counts. Both models scale similarly: going from 1 to 8 cores reduces static synapse time from 80 s to 8.4 s (9.5×) and STDP time from 217 s to 20.3 s (10.7×). Because both models scale by approximately the same factor, the overhead ratio remains 2.41–2.71× at every core count, motivating an architectural approach that reduces the per-synapse update cost itself.

Table 2. FPGA resource utilisation after implementation (Xilinx XC7Z020, 100 MHz).

Resource	Used	Available	Utilisation
LUTs	5,995	53,200	11.3 %
FFs	8,225	106,400	7.7 %
BRAMs (36k)	27	140	19.3 %
DSP48s	7	220	3.2 %
Clock	100 MHz	—	WNS +0.495 ns
Power	1.73 W	—	—

4 FPGA Co-processor Design

4.1 Architecture Overview

The co-processor implements the pair-based additive STDP rule (Eqs. 1–2) in event-driven form using Vitis HLS 2023.1 targeting the Xilinx Zynq XC7Z020. The host ARM processor communicates spike events and neuron identifiers over AXI-Lite, a lightweight memory-mapped interface to the programmable logic, which executes all weight arithmetic. This setup is convenient for prototyping but makes the host–accelerator interface overhead explicit in the end-to-end measurements reported later (Sect. 5).

For a pre-synaptic spike from neuron i: the kernel reads all post-synaptic traces $x_{\mathrm{post},j}$, applies LTD to each weight W_{ij}, and updates $x_{\mathrm{pre},i}$. For a post-synaptic spike from j: LTP is applied across all $x_{\mathrm{pre},i}$, and $x_{\mathrm{post},j}$ is updated. The inner loop is pipelined to process one synapse per clock cycle, limited only by BRAM read/write bandwidth.

4.2 Fixed-Point Arithmetic and Exponential LUT

Weights use Q1.15 (16-bit, 1 integer and 15 fractional bits, resolution 3.05×10^{-5} over $[0, 2)$); traces use Q2.14 (16-bit, range $[0, 4)$). A 16-bit hardware multiplier handles the multiply-accumulate natively.

Exponential trace decay is applied as $x \leftarrow x\, e^{-\Delta t/\tau}$, where Δt is the elapsed time since the previous spike on that neuron. The factor $e^{-\Delta t/\tau}$ is evaluated via a 2,048-entry look-up table stored in dual-port BRAM, pre-computed at synthesis time, avoiding a runtime exponential in hardware. In the validated 50×50 configuration, the weight matrix is also stored in dual-port BRAM, allowing simultaneous read/write access within the one-synapse-per-cycle pipeline.

4.3 Resource Utilisation

Table 2 summarises post-implementation resources and timing at 100 MHz.

Resource use is low across all categories, leaving substantial resource headroom on the XC7Z020. The remaining BRAM budget would support a 256×256 synapse matrix (65,536 synapses) without architectural change.

Table 3. Pairwise accuracy: mean absolute error (MAE) across 2,500 synapses after a 35,309-spike replay.

Comparison	Source of difference	MAE
FPGA vs. Python online reference	Q1.15 quantisation	1.91×10^{-3}
FPGA vs. NEST	Update-timing semantics + quantisation	1.10×10^{-2}

5 Validation and Performance

5.1 Experimental Setup

The co-processor is validated using a 10-second spike train recorded from a NEST 3.6 simulation of a 50×50 all-to-all network (2,500 synapses, 35,309 spikes). Both the 50 pre-synaptic and 50 post-synaptic neurons fired at a mean rate of approximately 35 Hz (17,710 pre-synaptic and 17,599 post-synaptic spikes over 10 s). With 2,500 synapses, each pre-synaptic spike updates a full row of 50 synapses, giving on the order of 10^5 weight updates per second. The same spike stream is replayed on (i) the FPGA, (ii) a Python online reference implementing the identical event-driven formulation, and (iii) NEST. Final weight matrices are compared pairwise.

5.2 Accuracy

Table 3 and Fig. 2 report pairwise accuracy after the full replay.

Figure 2A shows that FPGA and NEST final weights cluster near the identity line, with residual scatter arising from the difference in update semantics. By contrast, the much smaller MAE of 1.91×10^{-3} against the Python reference is consistent with Q1.15 rounding (3.05×10^{-5} per step, accumulated over ~35k spikes), confirming hardware correctness. The larger residual of 1.10×10^{-2} against NEST reflects a difference in update semantics rather than a hardware fault. The FPGA implements a fully on-line event-driven rule, with LTP and LTD applied immediately at spike time, whereas in NEST LTD is applied at pre-spike delivery using the post-synaptic trace at delivery time. The error distributions in Fig. 2B illustrate this distinction: FPGA–Python errors fall almost entirely below the quantisation step; the broader FPGA–NEST distribution reflects the semantics gap.

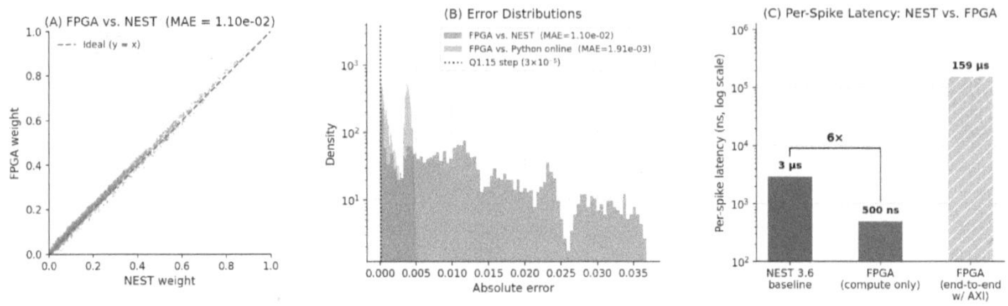

Fig. 2. FPGA validation and performance. (A) FPGA vs. NEST final weights for all 2,500 synapses; points cluster near the identity line (MAE $= 1.10 \times 10^{-2}$), with scatter from the update-timing semantics difference. (B) Absolute error distributions: FPGA–Python errors (red) are concentrated below the Q1.15 step (3×10^{-5}, dotted line), confirming hardware correctness; the broader FPGA–NEST distribution (blue) reflects the semantics gap. (C) Per-spike latency: the compute kernel achieves 500 ns vs. 3 μs for NEST (6×); the hatched bar shows end-to-end latency including AXI-Lite transfers (159 μs), with DMA batching as the identified path to closing this gap.

5.3 Throughput and Latency

Figure 2C summarises per-spike latency. The FPGA kernel runs at 500 ns per spike (50 cycles at 100 MHz), corresponding to 100 M syn/s for a 2,500-synapse network and a 6× reduction in compute latency relative to NEST (3 μs per spike).

End-to-end latency including AXI-Lite host transfers is currently 159 μs per spike, showing that the present integrated system is dominated by host–accelerator interface overhead rather than FPGA kernel compute. The current integrated system therefore does not yet achieve end-to-end acceleration relative to NEST; the gap and a plausible solution path are discussed in Sect. 6.

The firing rate also determines the sustained transfer load: with N neurons firing at mean rate ν, the host issues νN spike transfers per second. At 159 μs per un-batched AXI-Lite transaction the link saturates at approximately 6,300 transfers/s. The validation workload ($\nu \approx 35$ Hz, 100 neurons, $\approx 3{,}500$ transfers/s) sits below this ceiling, but higher firing rates or larger networks would approach it—at which point the bus, not the compute kernel, becomes the binding constraint. DMA batching (Sect. 6) addresses this directly: packing B spikes per transaction raises the sustainable transfer rate by a factor of B, moving the saturation point well beyond the regime considered here.

5.4 Scalability

BRAM utilisation for a 50×50 network (19.3 %) projects approximately linearly to ∼74 % for 256×256 synapses (65,536 connections), within the 140 available BRAM tiles. The current architecture is expected to preserve one-synapse-per-cycle throughput for larger networks that still fit within the BRAM budget.

The benchmark in Sect. 3 identifies $N \approx 1{,}000$–$8{,}000$ as the regime where STDP overhead becomes severe on CPU (2.4–3.75× slowdown). A single XC7Z020 holds a 256×256 dense network in BRAM; the more relevant target is biologically realistic sparse connectivity ($p \approx 0.1$), at which a sparse-storage adaptation of the weight matrix would allow the same BRAM budget to accommodate substantially larger neuron counts. The present 50×50 dense prototype is therefore a compute-kernel feasibility study rather than a deployment-scale system; the architectural path to the benchmark's high-overhead regime is multi-BRAM partitioning and, at larger scales, multi-FPGA designs (Sect. 6).

6 Limitations and Future Work

Scope. This paper separately examines two things: (i) the software-side scaling cost of STDP in NEST, and (ii) the compute efficiency of a dedicated STDP kernel on FPGA hardware. End-to-end acceleration of a running NEST simulation is not claimed; the host–accelerator interface overhead (159 μs per spike) means the current integrated system is slower than NEST. Isolating compute from integration is a useful first step toward a full co-processor design.

Host-Transfer Bottleneck and DMA Batching. A plausible next step is DMA burst transfer, replacing per-spike AXI-Lite transactions with bursts that carry many spike events in a single memory transaction. The current 159 μs/spike is dominated by per-transaction handshake and setup overhead on the single-beat AXI-Lite link; grouping B spikes per transaction amortises this fixed cost across B events. Treating the transfer cost as a fixed per-transaction term amortised over B spikes plus the 500 ns compute kernel, $B = 256$ yields an estimated end-to-end cost on the order of 1–1.5 μs/spike, compared with NEST's 3 μs/spike—a projected ∼2× improvement. This is an optimistic estimate: it treats the 159 μs as essentially fixed overhead, whereas in practice each batched transaction still moves more payload (spike identifiers and timing data) across the ARM–PL boundary, so a residual per-spike transfer cost remains that batching cannot eliminate. The achievable speedup therefore depends on the ratio of fixed handshake cost to per-spike payload cost, and on the spike-arrival statistics that determine realistic batch sizes; the estimate above should be read as an upper bound rather than a guaranteed figure. Host–accelerator communication has similarly been identified as a primary system-level constraint in FPGA-based neuroscience simulation frameworks [13].

Memory Access as the Central Constraint. More broadly, STDP performance is constrained less by arithmetic than by the location and movement of synaptic state. On the CPU side, the per-synapse cost increase beyond $N \approx 800$ reflects a transition from cache-resident to memory-limited operation; on the FPGA side, the kernel is efficient because weights and traces reside in on-chip BRAM, with the bottleneck shifting to host–accelerator communication. Both point to the same conclusion: future acceleration depends primarily on improving memory locality and data movement, not only on reducing compute latency. This mirrors recent work on the NEST core itself, where cache-efficient spike delivery yields substantial speedups at scale [24].

Memory-Access Profiling. The per-synapse cost pattern is consistent with memory-hierarchy effects above $N \approx 800$, but other microarchitectural effects (e.g. TLB pressure) could contribute. Direct measurement of cache-miss rates via hardware performance counters would confirm or refine this.

Scale and Topology. Hardware validation used a 50×50 all-to-all network; larger scales and biologically realistic sparse connectivity remain untested. The trade-off is between predictable on-chip access and limited on-chip capacity: the XC7Z020 accommodates moderate networks within BRAM, while larger models require more aggressive partitioning, external-memory-aware designs, or multi-FPGA architectures—a constraint of on-chip capacity, not a fundamental limitation of the FPGA approach. A natural scaling path is therefore modular: the present single-device kernel becomes one compute node, and a 50×50 (or, with the available BRAM headroom, 256×256) network maps to one cluster representing an abstraction of a cortical column or area, with multiple FPGAs interconnected to form the full model. This mirrors established multi-FPGA SNN platforms—the cluster-of-columns abstraction of Wang et al. [25], the Zynq-based multi-node compute node of Trensch and Morrison [23], and large-scale FPGA simulators such as NeuroFlow [6]—in which inter-node communication and memory-access patterns, rather than per-node compute, are the dominant system constraints [13]. Such scaling is feasible but shifts the design focus to these system-level constraints rather than per-node implementation details. Our contribution at this level is not the cluster architecture itself but a compact, NEST-validated STDP plasticity kernel suitable as the per-node learning engine.

Beyond Additive STDP. Extensions to triplet or reward-modulated rules would require additional trace variables but are compatible with the HLS pipeline approach used here. Voltage-based rules [7] (in NEST, `clopath_synapse`) remain spike-triggered but make each update depend on the postsynaptic membrane potential, so the accelerator would need a continuous neuronal variable in addition to spike events—amplifying the host-transfer bottleneck of Sect. 5 unless neuron state is co-located on-chip. Should such a rule be adopted, quantifying this trade-off would be a natural next step.

Practical Implications. Beyond the specific measurements, the benchmark functions as a decision tool for practitioners building STDP-based SNN simulations: the regime boundaries identified here—negligible overhead below $N \approx 200$, memory-hierarchy transition near $N \approx 800$, and a 2.4–3.75× slowdown by $N = 8{,}000$—indicate when a software-only NEST workflow is adequate and when the engineering investment of a hardware co-processor begins to pay off. More broadly, the finding that STDP performance is bounded by memory locality and data movement rather than arithmetic—consistent across both the CPU benchmark and the FPGA results—suggests that acceleration effort is best directed at the memory system and host–accelerator data path, generalising beyond the specific rule and platform studied here.

7 Conclusion

This paper has examined the computational cost of STDP in SNNs from two angles. A benchmark of NEST 3.6 across 1,950 configurations shows that STDP synapse overhead grows from negligible to 3.75× at $N = 8{,}000$ (and up to 5.75× in the full sweep), that the per-synapse cost pattern is consistent with memory-hierarchy effects above $N \approx 800$, and that multi-core parallelism does not reduce the overhead ratio. A pipelined HLS co-processor on the Xilinx XC7Z020 shows that the core STDP update computation can be offloaded to FPGA hardware: 100 M syn/s at 500 ns per spike and 1.73 W, with hardware correctness confirmed against a matched Python reference. The remaining bottleneck—AXI-Lite host-transfer overhead—is quantified, with DMA batching proposed as the next integration step. Together, the benchmark and the hardware design provide a quantitative foundation for further work towards FPGA co-processing of STDP in neuromorphic simulation pipelines. All simulation scripts, benchmark data, HLS sources, and figure-generation code are publicly available at https://github.com/MaximeCarriere/fpga_stdp.

Acknowledgments. Research funding was provided by the European Research Council, Advanced Grant "Material constraints enabling human cognition (MatCo)" (ERC-2019-ADG 883811).

Disclosure of Interests. The authors has no competing interests to declare that are relevant to the content of this article.

References

1. Bi, G.Q., Poo, M.M.: Synaptic modifications in cultured hippocampal neurons: dependence on spike timing, synaptic strength, and postsynaptic cell type. J. Neurosci. **18**(24), 10464–10472 (1998)
2. Carriere, M., Dobler, F., Plesser, H. E., Feledyn, A., Tomasello, R., Wennekers, T., Pulvermüller, F.: A brain-constrained neural model of cognition and language with NEST: transitioning from the Felix framework. Cogn. Neurodyn. **20**(1) (2026). https://doi.org/10.1007/s11571-026-10415-5

3. Chen, G.K., Kumar, R., Sumbul, H.E., Knag, P.C., Krishnamurthy, R.K.: A 4096-neuron 1M-synapse 3.8-pJ/SOP spiking neural network with on-chip STDP learning and sparse weights in 10-nm FinFET CMOS. IEEE J. Solid-State Circ. **54**(4), 992–1002 (2019). https://doi.org/10.1109/JSSC.2018.2884336
4. Chen, G., Lin, X., Wang, G.: TSTKD: triple-spike train kernel-driven supervised learning algorithm. Pattern Recogn. **164**, 111608 (2025)
5. Cheslet, J., Bernert, M., Beaubois, R., Yvert, B., Lévi, T.: Real-time spike sorting using an optimized STDP spiking neural network on FPGA. In: 2024 International Joint Conference on Neural Networks (IJCNN), pp. 1–7. IEEE (2024)
6. Cheung, K., Schultz, S.R., Luk, W.: A large-scale spiking neural network accelerator for FPGA systems. In: Villa, A.E.P., Duch, W., Érdi, P., Masulli, F., Palm, G. (eds.) ICANN 2012. LNCS, vol. 7552, pp. 113–120. Springer, Heidelberg (2012). https://doi.org/10.1007/978-3-642-33269-2_15
7. Clopath, C., Büsing, L., Vasilaki, E., Gerstner, W.: Connectivity reflects coding: a model of voltage-based STDP with homeostasis. Nat. Neurosci. **13**(3), 344–352 (2010). https://doi.org/10.1038/nn.2479
8. Davies, M., et al.: Loihi: a neuromorphic manycore processor with on-chip learning. IEEE Micro **38**(1), 82–99 (2018). https://doi.org/10.1109/MM.2018.112130359
9. Gewaltig, M.O., Diesmann, M.: NEST (NEural Simulation Tool). Scholarpedia **2**(4), 1430 (2007)
10. He, J., et al.: A visual cortex-inspired edge neuromorphic hardware architecture with on-chip multi-layer STDP learning. Comput. Electr. Eng. **120**, 109806 (2024)
11. Jin, X., Rast, A., Galluppi, F., Davies, S., Furber, S.: Implementing spike-timing-dependent plasticity on SpiNNaker neuromorphic hardware. In: 2010 International Joint Conference on Neural Networks (IJCNN), pp. 1–8. IEEE (2010). https://doi.org/10.1109/IJCNN.2010.5596372
12. Karamimanesh, M., Abiri, E., Shahsavari, M., Hassanli, K., van Schaik, A., Eshraghian, J.K.: Spiking neural networks on FPGA: a survey of methodologies and recent advancements. Neural Netw. **186**, 107256 (2025). https://doi.org/10.1016/j.neunet.2025.107256
13. Kauth, K., Stadtmann, T., Sobhani, V., Gemmeke, T.: neuroAIx-framework: design of future neuroscience simulation systems exhibiting execution of the cortical microcircuit model 20× faster than biological real-time. Front. Comput. Neurosci. **17**, 1144143 (2023). https://doi.org/10.3389/fncom.2023.1144143
14. Maass, W.: Networks of spiking neurons: the third generation of neural network models. Neural Netw. **10**(9), 1659–1671 (1997)
15. Markram, H., Lübke, J., Frotscher, M., Sakmann, B.: Regulation of synaptic efficacy by coincidence of postsynaptic APs and EPSPs. Science **275**, 213–215 (1997)
16. Martis, L., Leone, G., Raffo, L., Meloni, P.: Low-power FPGA-based spiking neural networks for real-time decoding of intracortical neural activity. IEEE Sens. J. **24**(24), 42448–42459 (2024)
17. Mitchell, J.P., Schuman, C.D., Potok, T.E.: A small, low cost event-driven architecture for spiking neural networks on FPGAs. In: Proceedings of the International Conference on Neuromorphic Systems (ICONS). ACM (2020)
18. Morrison, A., Aertsen, A., Diesmann, M.: Spike-timing-dependent plasticity in balanced random networks. Neural Comput. **19**(6), 1437–1467 (2007). https://doi.org/10.1162/neco.2007.19.6.1437
19. Nordlie, E., Gewaltig, M.O., Plesser, H.E.: Towards reproducible descriptions of neuronal network models. PLoS Comput. Biol. **5**(8), e1000456 (2009)

20. Pani, D., Meloni, P., Tuveri, G., Palumbo, F., Massobrio, P., Raffo, L.: An FPGA platform for real-time simulation of spiking neuronal networks. Front. Neurosci. **11**, 90 (2017)
21. Pfister, J.P., Gerstner, W.: Triplets of spikes in a model of spike timing-dependent plasticity. J. Neurosci. **26**(38), 9673–9682 (2006)
22. Song, S., Miller, K.D., Abbott, L.F.: Competitive Hebbian learning through spike-timing-dependent synaptic plasticity. Nat. Neurosci. **3**, 919–926 (2000)
23. Trensch, G., Morrison, A.: A system-on-chip based hybrid neuromorphic compute node architecture for reproducible hyper-real-time simulations of spiking neural networks. Front. Neuroinform. **16**, 884033 (2022). https://doi.org/10.3389/fninf.2022.884033
24. Vogelsang, J., Kunkel, S.: Scalability via cache-efficiency on manycore exascale systems. In: Presented at the NEST Conference 2026, 16–17 June 2026, Virtual (2026). https://events.hifis.net/event/3408/. Accessed 19 June 2026
25. Wang, R.M., Thakur, C.S., van Schaik, A.: An FPGA-based massively parallel neuromorphic cortex simulator. Front. Neurosci. **12**, 213 (2018). https://doi.org/10.3389/fnins.2018.00213

Synaptic Plasticity Shapes Complex Intermittency in an Izhikevich Spiking Neural Network

Marco Cafiso[1,2](✉) and Paolo Paradisi[2,3]

[1] Department of Physics 'E. Fermi', University of Pisa, Largo Bruno Pontecorvo 3, 56127 Pisa, Italy
marco.cafiso@isti.cnr.it

[2] Institute of Information Science and Technologies 'A. Faedo', ISTI-CNR, Via G. Moruzzi 1, 56124 Pisa, Italy

[3] BCAM-Basque Center for Applied Mathematics, Alameda de Mazarredo 14, 48009 Bilbao, BASQUE COUNTRY, Spain

Abstract. Neural avalanches are a hallmark of cortical criticality, a dynamical regime thought to optimize information processing, dynamic range, and computational flexibility. Despite growing evidence linking synaptic plasticity to critical avalanche dynamics, the mechanisms by which spike-timing-dependent plasticity (STDP) modulates the temporal complexity of collective neural events, and how this modulation depends on the statistical structure of sensory inputs, remain poorly understood. Here, we investigate this problem in a biologically plausible spiking network model in which a scale-free Izhikevich cortical network is reciprocally coupled to a sensory layer driven by different structured signals across multiple frequencies and sensory population sizes. Avalanche temporal complexity is quantified within the Event-Driven Diffusion Scaling (EDDiS) framework. We found that STDP systematically reduces long-time Hurst exponents of avalanche birth–death dynamics, while scaling indices remain robust across input type and frequency, revealing a form of universality in plasticity-driven temporal reorganization. Avalanches always show two scaling regimes: short-time superdiffusion shifting to long-time subdiffusion, driven by neuronal refractoriness and adaptation. Larger sensory populations boost STDP efficacy and sharpen complexity signatures, revealing finite-size effects in critical scaling.

Keywords: Synaptic Plasticity · Brain-Inspired Computing · Neural Avalanches · Temporal Complexity · Diffusion Scaling

1 Introduction

The emergence of complex intermittent temporal dynamics in cortical networks has attracted considerable interest in computational and systems neuroscience. A central concept in this context is that of "scale-free" neural avalanches, cascades of collective neural activity whose size and duration distributions follow

L. Pasa et al. (Eds.): ICANN 2026, LNCS 17089, pp. 78–90, 2027.
https://doi.org/10.1007/978-3-032-38398-3_7

power laws, first reported experimentally by Beggs and Plenz [4]. This has motivated the hypothesis that the cortex operates near a critical point, a regime that is thought to maximize information transmission, dynamic range, and computational capacity [6,25]. Understanding how this critical organization is established and maintained in the presence of learning and plasticity, therefore, constitutes a fundamental open question in theoretical neuroscience. Spike-timing-dependent plasticity (STDP) [16] is a primary biological mechanism through which neural circuits self-organize in response to experience. STDP is a local, unsupervised learning rule that reshapes network connectivity based on the relative timing of pre- and post-synaptic activity [17]. Theoretical studies have demonstrated that STDP can drive networks toward critical avalanche dynamics [14,24,31], suggesting a deep connection between learning, network structure, and the emergence of scale-free avalanches. Nevertheless, the precise mechanisms by which STDP modulates the temporal complexity of avalanches, and how this modulation interacts with structured sensory input, remain poorly understood. To fill this gap, we investigate here how structured sensory inputs drive synaptic plasticity and shape the emergence of temporal self-organization and complexity in a cortical circuit. In particular, we employ a network architecture that consists of a module composed of Izhikevich neurons, designed to replicate the dynamical properties of a cortical area, reciprocally coupled to a smaller sensory module that receives externally applied input currents. The Izhikevich neuron model was selected as the fundamental building block of this architecture based on three complementary strengths: computational efficiency, biological plausibility, and parametric flexibility [12,13].

The synaptic plasticity mechanism implemented is a soft-bound variant of the STDP rule [26], in which the magnitude of potentiation and depression is scaled by the proximity of the synaptic weight to its upper and lower bounds, respectively, naturally preventing unbounded weight growth. Plasticity was restricted to excitatory synapses, while inhibitory synaptic strengths were held fixed throughout all simulations. To study the effects of synaptic plasticity on temporal self-organization we exploit tools coming from the Temporal Complexity (TC) framework, also known as Intermittency-Driven Complexity (IDC) [19]. The TC framework is applied to those systems which display complex intermittent behaviour, i.e., systems that display self-organized metastable states, whose transitions are marked by crucial events. A key quantity in TC/IDC analysis is the probability density function (PDF) of inter-event times (IETs), which are the time duration between two consecutive crucial events. However, direct estimation of the IET-PDF is often unreliable due to contamination by noise-related secondary events [3,20]. A more robust characterization is instead obtained through scaling analyses of diffusion processes constructed from the event sequence [1], as systematically implemented in the Event-Driven Diffusion Scaling (EDDiS) algorithm [19]. This framework has been successfully applied to brain data in both wake and sleep conditions [2], as well as in spiking neural models [5].

2 Methods

2.1 Neural Network Model

This work employs a two-layer spiking neural network, in which each neuron is modelled as an Izhikevich unit [13]. The first layer, the Sensory Network (SN), serves as the input stage of the architecture and consists of N_s fully connected excitatory neurons. The second layer, the Cortical Network (CN), models the dynamics of a cortical neural ensemble. This layer is composed of $N_C = N_{exc} + N_{inh} = 500$ neurons arranged in a scale-free (SF) topology [5,9], including both excitatory and inhibitory cells, with an inhibitory-to-excitatory ratio of 1:4, consistent with neurophysiological evidence from studies of the cerebral cortex [7]. The choice of an SF topology is motivated by the well-established observation that neuronal connectivity in the human brain departs significantly from random organization; rather, it is characterized by highly connected hub nodes, a structural feature consistently documented across human connectome studies [10]. To develop the SF topology, we follow the algorithm presented in this study [5] with a power-law exponent of $\alpha = 2.5$. The SN is built by selecting the N_s highest out-degree excitatory nodes from an SF network of $N_{exc} + N_s$ neurons, where N_s is a free parameter. These sensory neurons are then rewired into an all-to-all subnetwork and connected via full feedforward projections onto the CN. The following set of equations governs the i-th neuron:

$$\frac{dV_i}{dt} = 0.04V_i^2 + 5V_i + 140 - u_i + I_i^{DC} + I_i^{syn}, \tag{1}$$

$$\frac{du_i}{dt} = a(bV_i - u_i),$$

where V_i is the membrane potential of the i-th neuron, and u_i is its membrane recovery variable. When a spike occurs, the after-spike reset rule is applied as follows: if $V_i \geq 30$ mV then $V_i \leftarrow c$ and $u_i \leftarrow u_i + d$. To model a cortical network, neuronal parameters were set to produce regular spiking in excitatory neurons and fast spiking in inhibitory neurons (see Table 1 for the full parameter values list). I_i^{DC} denotes the external input current received by the i-th sensory neuron; this term is $I_i^{DC} = 0$ for neurons in the CN. The term I_i^{syn}, on the other hand, describes the chemical synaptic current that arrive to each post-synaptic neuron i, and it is modelled as a sum of two exponentials [22]:

$$I_i^{syn} = \frac{E_0 - V_i}{D_i} \sum_j g_{ij} \frac{e^{-\frac{t-t_j}{\tau_{decay}}} - e^{-\frac{t-t_j}{\tau_{rise}}}}{\tau_{decay} - \tau_{rise}}. \tag{2}$$

Here, E_0 denotes the synaptic reversal potential, which takes different values for excitatory and inhibitory synapses (see Table 1), D_i is the in-degree value of node i, and t_j refers to the time of the most recent spike emitted by the j-th pre-synaptic neuron. The constants τ_{rise} and τ_{decay} represent the synaptic time constants governing the rise and decay phases of the synaptic response, respectively, while g_{ij} denotes the synaptic conductance strength of the connection from neuron j to neuron i. Finally, a soft-bound STDP rule [26] drives plasticity exclusively at excitatory synapses, while inhibitory weights remain fixed throughout all simulations. When a post-synaptic neuron j fires at time t_{post} and its pre-synaptic counterpart i fires at time t_{pre}, the synaptic conductance is updated according to $g_{ij} \to g_{ij} + \Delta g_{ij}$, where:

$$\Delta g_{ij} = \begin{cases} A_+(g_{max} - g_{ij})e^{-\frac{t_{post}-t_{pre}}{\tau_+}} & \Delta t > 0 \\ -A_-(g_{ij} - g_{min})e^{\frac{t_{post}-t_{pre}}{\tau_-}} & \Delta t \leq 0 \end{cases} \tag{3}$$

Here, A_+ and A_- are the amplitudes of synaptic potentiation and depression, respectively. τ_+ and τ_- set the width of the STDP learning window for each process, and g_{min} and g_{max} define the lower and upper bounds on the synaptic conductance. The full set of STDP parameter values is provided in Table 1. All the dynamical equations were integrated using a fourth-order Runge-Kutta scheme with a fixed time step of $h = 0.01$ ms.

2.2 Input Currents

Three input signal current types are considered: (i) square wave signal, (ii) Poisson and (iii) power-law telegraph signals. This selection allows a direct comparison between a strictly periodic signal (square wave) and stochastic signals with either simple (Poisson) or complex (power-law) IET statistics. Each signal alternates between $I_{low} = 0$ and $I_{high} = 8$ at each event occurrence.

The Poisson and power-law telegraph signals (TS) were constructed by sampling IETs from their respective distributions [1,8,19]:

$$\psi_P(\tau) = r_p\, e^{-r_p \tau} \; ; \quad \psi_{PL}(\tau) = \frac{\mu - 1}{T} \frac{1}{(T+\tau)^{\mu}} \; ; \tag{4}$$

with event rate r_p and mean IET $\langle \tau_P \rangle = 1/r_p$ for the Poisson events, and μ and T the power-law exponent and the timescale of complex events, respectively. For $\mu > 2$, the mean complex IET is finite: $\langle \tau_{PL} \rangle = T/(\mu - 2)$. In this work, we set $\mu = 2.5$.

Three input frequencies were tested relative to the characteristic timescale of the excitatory neuron recovery variable ($a_{exc} = 0.02$ ms^{-1}, giving $\tau_{a_{exc}} = 1/a_{exc} = 50$ ms, or equivalently 20 Hz): a sub-resonant frequency (10 Hz), the resonant frequency (20 Hz), and a supra-resonant frequency (50 Hz). For the stochastic signals, parameters were matched to the square wave frequency f. Specifically, since $r_p = 2f$ and $T = \langle \tau_P \rangle \cdot (\mu - 2)$, a square wave at $f = 10$ Hz gives $r_p = 20$ Hz and $T = 0.025$ s; at $f = 20$ Hz, $r_p = 40$ Hz and $T = 0.0125$ s; and at $f = 50$ Hz, $r_p = 100$ Hz and $T = 0.005$ s. For our simulations, input signals of duration $t_{\text{signal}} = 10$ s were generated.

Table 1. Values of constant parameters used in the neural network model.

Parameters	Symbol	Value
Neuron Parameters		
Time-Scale Recovery Variable Excitatory	a_{exc}	0.02 ms^{-1}
Sensitivity Recovery Variable Excitatory	b_{exc}	0.2 $(\text{ms} \cdot \text{mV})^{-1}$
After-Spike Reset Membrane Potential Excitatory	c_{exc}	−65 mV
After-Spike Reset Recovery Variable Excitatory	d_{exc}	8
Time-Scale Recovery Variable Inhibitory	a_{inh}	0.1 ms^{-1}
Sensitivity Recovery Variable Inhibitory	b_{inh}	0.2 $(\text{ms} \cdot \text{mV})^{-1}$
After-Spike Reset Membrane Potential Inhibitory	c_{inh}	−65 mV
After-Spike Reset Recovery Variable Inhibitory	d_{inh}	2
Synaptic Parameters		
Synaptic Decay Phase Time Constant	τ_{decay}	1.7 ms
Synaptic Rise Phase Time Constant	τ_{rise}	0.2 ms
Excitatory Reversal Potential	$E_{0,exc}$	0 mV
Inhibitory Reversal Potential	$E_{0,inh}$	−75 mV
STDP Parameters		
Maximum Synaptic Potentiation	A_+	0.05
Maximum Synaptic Depression	A_-	0.05
Temporal STDP Window Potentiation	τ_+	20 ms
Temporal STDP Window Depression	τ_-	20 ms
Maximum Synaptic Conductance	g_{max}	6 nS
Minimum Synaptic Conductance	g_{min}	0 nS

2.3 Temporal Complexity Analyses

Temporal Complexity (TC) analysis characterizes the scaling behaviour of intermittent transition events in a time series, identifying whether their statistics follow power-law distributions indicative of complex, self-organizing dynamics [19]. Specifically, the method adopted in this study is called Event-Driven Diffusion Scaling analysis (EDDiS) [19]. The EDDiS method extracts two scaling indices, the second moment scaling and the probability density function (PDF) self-similarity index, from diffusion processes built from the event sequence. The diffusion processes were computed using a Continuous Time Random Walk (CTRW) model [18,29], where motion is triggered only by the occurrence of events, leaving the particle at rest between consecutive events. In the present study, we decided to use a single diffusion process generated using the Asymmetric Jump (AJ) walking rule, motivated by its intrinsic robustness in evaluating scaling behaviours [8]. This rule consists of a unit forward jump at each event occurrence, and therefore corresponds to the counting process generated by the event sequence: $X(t) = \#\{n : t_n < t\}^1$. The methods used to extract and analyze the crucial events are reported below.

Neural Avalanches. Neural avalanches are a hallmark of intermittent, large-scale coordinated activity across brain regions [4,27], and stand as one of the most studied manifestations of complex, self-organized dynamics in the brain. In neural networks, avalanches are identified through a firing counting procedure applied across the entire network using non-overlapping time windows of duration Δt_{av}. A birth time is defined as the onset of the first time window in which the number of firing events surpasses a prescribed threshold N_{av}, immediately following a window in which the count fell below it. Conversely, a death time marks the end of the last time window exceeding N_{av} before the activity drops back below threshold in the subsequent window. The avalanche duration τ_{pn} is then given by the difference between each death time and its corresponding birth time. Here, $\Delta t_{av} = 1$ time-step (0.01 ms), and N_{av} is set to the 35th percentile of the neural coincidence size distribution. A coincidence is defined as any instant in which at least $N_c = 2$ neurons fire simultaneously, with coincidence size referring to the number of neurons involved. This narrow time window is chosen to capture the true onset of coincident activity, and it is particularly appropriate in our setting because the sensory network delivers the same input to all neurons, leading to a high firing rate.

Detrended Fluctuation Analysis. The Detrended Fluctuation Analysis (DFA) is a method used to evaluate the long-range temporal correlations of a signal (see, e.g., [21]), by estimating the second-moment scaling exponent H defined as:

$$F^2(\Delta t) = \langle (\Delta X(t, \Delta t) - \Delta X_{trend}(t, \Delta t))^2 \rangle_t \sim \Delta t^{2H}. \tag{5}$$

Here $\langle \cdot \rangle_t$ denotes a temporal average over all time positions t for a given lag Δt, $\Delta X(t, \Delta t) = X(t + \Delta t) - X(t)$, and $\Delta X_{trend}(t, \Delta t)$ a proper local trend of the time series. The exponent H essentially coincides with the classical Hurst exponent [11] and characterizes long-range correlations. Specifically, $H < 0.5$ (subdiffusion) indicates anti-correlated (anti-persistent) dynamics, $H = 0.5$ is the hallmark of uncorrelated (white-noise) fluctuations, and $H > 0.5$ (superdiffusion) indicates positively correlated (persistent) dynamics. All DFA computations were carried out using the `MFDFA` function from the Python package `MFDFA` [23].

Diffusion Entropy. The Diffusion Entropy (DE) analysis quantifies the self-similarity exponent δ of a diffusion process via the Shannon entropy of its probability density function (PDF) [1,8,19]. The PDF self-similarity index δ is defined as:

$$p(x, \Delta t)\, dx = \frac{1}{\Delta t^{\delta}} F\left(\frac{x}{\Delta t^{\delta}}\right) dx\ . \tag{6}$$

---
[1] Here $\#A$ denotes the cardinality of the set A, which is also denoted as $card(A)$ or $|A|$.

Here $p(x, \Delta t)$ is the PDF of the diffusion process $\Delta X(t, \Delta t)$, which is computed as an average over t. Analogously to the DFA exponent H, the index δ characterizes the diffusive behaviour of the process, but without requiring detrending. In particular, $\delta = 0.5$ corresponds to normal diffusion (no correlations), whereas $\delta < 0.5$ and $\delta > 0.5$ indicate anomalous subdiffusion and superdiffusion, associated with anti-persistent and persistent correlations, respectively.

3 Results

3.1 Simulation Results

The primary objective of this work is to examine how STDP shapes the temporal self-organization of cortical networks under different input regimes. Each simulation followed a two-phase protocol: (i) the *Baseline phase* (without STDP), a $t_{\mathrm{sim},1} = 60$ s simulation with frozen synaptic weights; and (ii) the *Learning phase* (with STDP), a $t_{\mathrm{sim},2} = 120$ s simulation with active synaptic plasticity. Input signals were delivered continuously and repeatedly across both phases. The extended duration of the learning phase was chosen to ensure that synaptic weight dynamics reached a stable steady state. All simulations were carried out using the *Brian2* neural simulator [28]. Figure 1 depicts the time course of the mean excitatory synaptic weight $\langle g_{exc} \rangle$ over the full simulation period $t_{sim} = 180$ s. Across all input conditions, the overall magnitude of weight potentiation is systematically greater for $N_s = 15$ than for $N_s = 10$. In the $N_s = 15$ case, the strongest potentiation is driven by the 50 Hz square wave signal, reaching a steady-state value of $\langle g_{exc} \rangle \simeq 4.6$ nS, followed by the 20 Hz square wave ($\langle g_{exc} \rangle \simeq 4.1$ nS); all remaining input signals converge to approximately $\langle g_{exc} \rangle \simeq 4$ nS. In the $N_s = 10$ regime, the 50 Hz square wave again elicits the largest weight increase ($\langle g_{exc} \rangle \simeq 4.1$ nS), while the remaining signals settle near $\langle g_{exc} \rangle \simeq 3.8$ nS. Notably, the 20 Hz square wave yields the weakest mean potentiation for $N_s = 10$ ($\langle g_{exc} \rangle \simeq 3.7$ nS), in stark contrast to its relatively strong effect observed at $N_s = 15$.

3.2 Temporal Complexity Results

Figure 2 displays the DFA and DE analyses of avalanche birth–death events for $N_s = 10$. Avalanche statistics were computed over the entire simulation time ($t_{sim,1} = 60$ s) in the baseline phase, whereas in the learning phase, they were estimated only over the last 60 s, i.e. after the average synaptic weight evolution had reached a stationary regime. Panel a) presents the DFA and DE curves averaged by signal type (across frequencies). These curves primarily distinguish the baseline (no STDP) from the learning (with STDP) phase, with negligible differences across signal classes. Consistently, panel b) reveals a superdiffusive short-time regime ($H > 0.5$) with median $H \approx 1.25$ across all signal types and phases, while at long times the learning phase yields systematically lower H values than baseline, with only marginal signal-dependent variation. Specifically, for square wave signals, H drops from $H \approx 0.25$ (baseline) to $H \approx 0.1$ (learning), whereas for Poisson and power-law telegraph signals, it decreases from $H \approx 0.3$

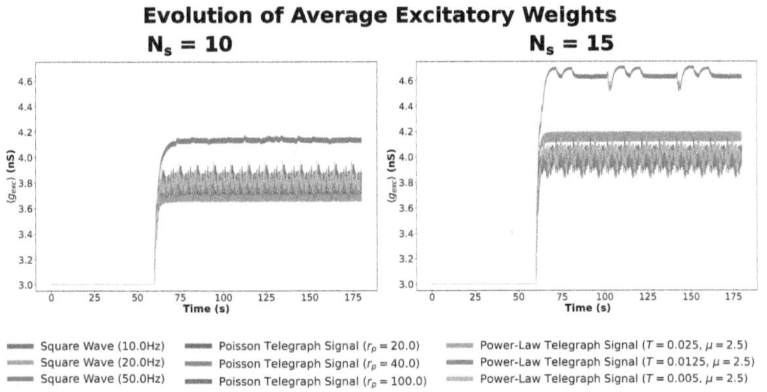

Fig. 1. Evolution of the average excitatory synaptic weight over the full simulation period for all input conditions. Results are shown separately for sensory network sizes $N_s = 10$ (left) and $N_s = 15$ (right).

(baseline) to $H \approx 0.2$ (learning). The DE results confirm the same two-regime scaling behaviour, with median $\delta \approx 0.85$ (superdiffusive, $\delta > 0.5$) at short times and $\delta \approx 0.45$ (subdiffusive, $\delta < 0.5$) at long times Panels c) and d) present the same quantities grouped by frequency (averaged over signal types), reinforcing the conclusion that avalanche birth–death dynamics are largely insensitive to input frequency. Panel d) again exhibits a robust short-time regime with $H \approx 1.25$ across all frequencies and phases, followed by a long-time regime in which STDP shifts H from ≈ 0.3 (baseline) to ≈ 0.2 (learning). The DE analysis remains consistent with this picture, yielding $\delta \approx 0.85$ at short times and $\delta \approx 0.45$ at long times, with no discernible modulation by frequency or phase. Figure 3 demonstrates that the same qualitative picture holds for $N_s = 15$, with STDP-related differences in DFA becoming more pronounced. Specifically, panel b) reports short-time superdiffusion with median $H \approx 1.2$ across all signal types and phases, while long-time H is again reduced in the learning phase relative to baseline, with minor signal-dependent variation (for square waves, from $H \approx 0.3$ to $H \approx 0.1$; for Poisson and power-law telegraph signals, from $H \approx 0.4$ to $H \approx 0.3$). Panel d) reveals the same phase-dependent shift across frequencies (short-time $H \approx 1.2$ in all cases; long-time H declining from ≈ 0.4 in baseline to ≈ 0.3 in learning). For $N_s = 15$, modest phase-dependent differences also emerge in the long-time DE exponent: the median δ for square wave signals decreases from $\delta \approx 0.5$ to $\delta \approx 0.4$, while telegraph signals exhibit no marked variation. Across frequencies, δ undergoes a transition at $f = 10$ Hz and $f = 50$ Hz from $\delta \approx 0.5$ to $\delta \approx 0.4$, whereas at $f = 20$ Hz it remains near $\delta \approx 0.5$ in both phases.

4 Discussion

The present work investigates TC in a biologically plausible spiking neural network comprising an Izhikevich cortical network (CN) coupled to a sensory network (SN), subjected to a soft-bound STDP rule. To further enhance biological

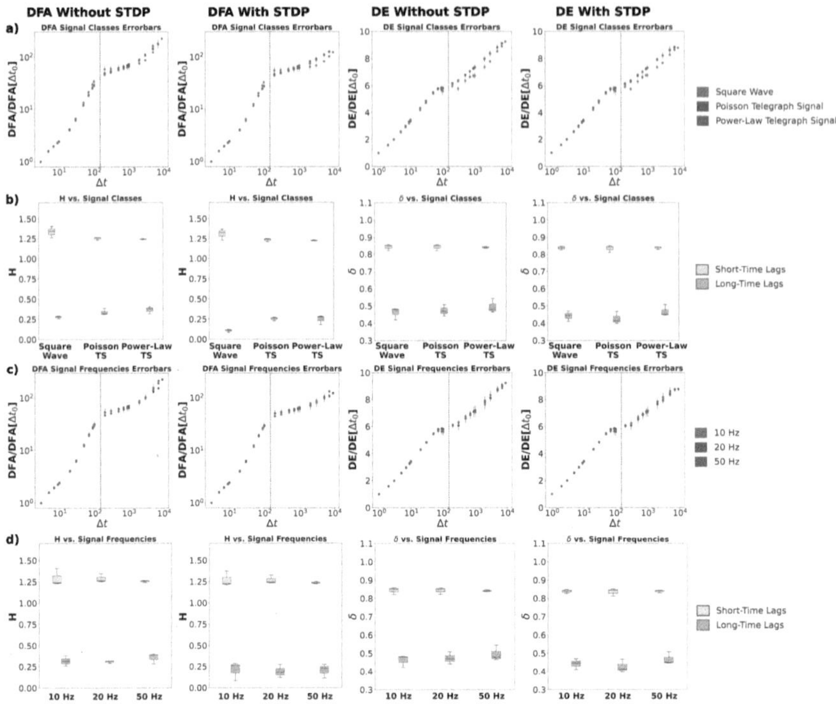

Fig. 2. EDDiS analysis of avalanches birth-death events for $N_s = 10$. (a) DFA (left two panels) and DE (right panels) curves stratified by input signal type (square wave: red; Poisson telegraph: blue; power-law telegraph: green); the vertical dashed line marks the boundary between short- and long-time lag regimes. (b) Boxplots of the short- and long-time scaling exponents H (left two panels) and δ (right panels) across the three signal classes. (c) DFA (left two panels) and DE (right panels) curves stratified by input frequency (10 Hz: red; 20 Hz: blue; 50 Hz: green); the vertical dashed line again marks the short-/long-time lag boundary. (d) Boxplots of H (left two panels) and δ (right panels) across the three frequency conditions, for both short- and long-time lag regimes. All DFA and DE curves are normalized by their respective values at the smallest evaluated time lag, Δt_0. (Color figure online)

plausibility, the CN connectivity follows an SF topology. A central finding is the strong sensitivity of avalanche dynamics to STDP-induced synaptic modifications. Avalanche birth–death events showed pronounced STDP-driven changes in TC indices: long-time Hurst exponents systematically decreased from $H \approx 0.25$–0.4 at baseline to $H \approx 0.1$–0.3 during learning. This reflects the tight coupling between avalanche dynamics and network connectivity, which STDP continuously reshapes by modulating the effective coupling strength, thereby altering the propagation and termination timescales of collective bursts. This interpretation is consistent with theoretical work linking synaptic plasticity to critical avalanche dynamics [14, 24, 31]. The frequency invariance of temporal complexity indices for avalanche events, with H and δ remaining stable across 10, 20, and

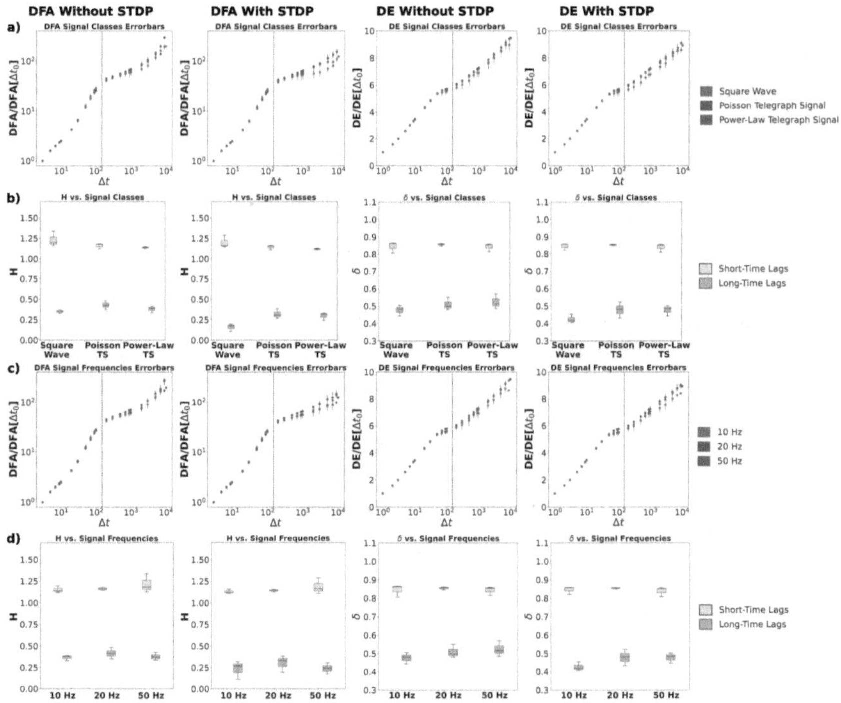

Fig. 3. Same as Fig. 2, but for $N_s = 15$.

50 Hz inputs, suggests that the network's complexity signature is determined by event statistics and dynamical phases rather than event rate. Yet synaptic weight evolution revealed clear frequency-dependent STDP efficacy. The 50 Hz square wave produced the largest mean weight increases ($\langle g_{exc} \rangle \approx 4.6$ nS for $N_s = 15$, Fig. 1), while the 20 Hz input, whose inter-stimulus interval (50 ms) matches the adaptation recovery timescale $\tau_{a_{exc}} = 50$ ms, yielded a more complex modification profile strongly sensitive to sensory population size. Notably, small sensory populations ($N_s = 10$) drove only modest synaptic modifications, whereas larger ones ($N_s = 15$) led to more pronounced weight updates, suggesting that the resonance between input periodicity and intrinsic neuronal timescales can only be fully expressed when the collective sensory drive exceeds a critical threshold. The enhanced STDP efficacy and more structured avalanche TC features observed at $N_s = 15$ relative to $N_s = 10$ indicate that sensory population size modulates learning-induced self-organization. Larger sensory populations provide stronger, more consistent drive to the cortical network, boosting pre- and post-synaptic correlations and amplifying STDP-induced weight changes. The sharper phase-dependent differences in avalanche dynamics at $N_s = 15$ further suggest that sufficient input strength is required to stabilize emerging scaling laws, consistent with theoretical predictions that self-organized criticality is subject to finite-size effects, with critical exponents converging only asymptotically with system

size [15,30]. The increasing structure of complexity metrics with N_s suggests the $N_C = 500$ cortical network, while capturing the qualitative phenomenology, may still operate in a finite-size regime where fluctuations remain non-negligible.

5 Concluding Remarks

The central result is that avalanche birth–death events exhibit pronounced STDP-dependent modulation, with Hurst exponents systematically decreasing during learning, reflecting strong coupling to the network's evolving connectivity. Across all input conditions, neural avalanches consistently display a double-scaling regime: short-time superdiffusion transitioning to long-time subdiffusion, the latter likely rooted in the refractory and adaptation dynamics of the Izhikevich neuron model. Simulations at larger network sizes would clarify whether this STDP-modulated long-time scaling reflects a genuine thermodynamic-limit behaviour or a finite-size transient.

Acknowledgments. This work was supported by the Next-Generation-EU programme under the funding schemes PNRR-PE-AI scheme (M4C2, investment 1.3, line on AI) FAIR "Future Artificial Intelligence Research", grant id PE00000013, Spoke-8: Pervasive AI.

References

1. Akin, O., Paradisi, P., Grigolini, P.: Perturbation-induced emergence of Poisson-like behavior in non-Poisson systems. J. Stat. Mech.: Theory Exp, P01013 (2009). https://doi.org/10.1088/1742-5468/2009/01/P01013
2. Allegrini, P., et al.: Sleep unconsciousness and breakdown of serial critical intermittency: new vistas on the global workspace. Chaos Solitons Fract. **55**, 32–43 (2013). https://doi.org/10.1016/j.chaos.2013.05.019
3. Allegrini, P., Menicucci, D., Bedini, R., Gemignani, A., Paradisi, P.: Complex intermittency blurred by noise: theory and application to neural dynamics. Phys. Rev. E **82**, 015103 (2010). https://doi.org/10.1103/PhysRevE.82.015103
4. Beggs, J.M., Plenz, D.: Neuronal avalanches in neocortical circuits. J. Neurosci. **23**(35), 11167–11177 (2003). https://doi.org/10.1523/JNEUROSCI.23-35-11167.2003
5. Cafiso, M., Paradisi, P.: Temporal complexity of a Hopfield-type neural model in random and scale-free graphs. In: Proceedings of the 16th International Joint Conference on Computational Intelligence - NCTA, pp. 438–448. INSTICC, SciTePress (2024). https://doi.org/10.5220/0013007600003837
6. Chialvo, D.R.: Emergent complex neural dynamics. Nat. Phys. **6**, 744–750 (2010). https://doi.org/10.1038/nphys1803
7. DeFelipe, J.: Neocortical neuronal diversity: chemical heterogeneity revealed by colocalization studies of classic neurotransmitters, neuropeptides, calcium-binding proteins, and cell surface molecules. Cerebral Cortex **3**(4), 273–289 (1993). https://doi.org/10.1093/cercor/3.4.273
8. Grigolini, P., Palatella, L., Raffaelli, G.: Asymmetric anomalous diffusion: an efficient way to detect memory in time series. Fractals **9**(04), 439–449 (2001). https://doi.org/10.1142/S0218348X01000865

9. Grinstein, G., Linsker, R.: Synchronous neural activity in scale-free network models versus random network models. **102**, 9948–9953 (2005). https://doi.org/10.1073/pnas.0504127102
10. van den Heuvel, M.P., Sporns, O.: Network hubs in the human brain. Trends Cogn. Sci. **17**(12), 683–696 (2013). https://doi.org/10.1016/j.tics.2013.09.012
11. Hurst, H.: Long-term storage capacity of reservoirs. Trans. Am. Soc. Civil Eng. **116**(1), 770–799 (1951). https://doi.org/10.1061/TACEAT.0006518
12. Izhikevich, E.: Which model to use for cortical spiking neurons? IEEE Trans. Neural Netw. **15**(5), 1063–1070 (2004). https://doi.org/10.1109/TNN.2004.832719
13. Izhikevich, E.M.: Simple model of spiking neurons. IEEE Trans. Neural Netw. **14**(6), 1569–1572 (2003). https://doi.org/10.1109/TNN.2003.820440
14. Khoshkhou, M., Montakhab, A.: Spike-timing-dependent plasticity with axonal delay tunes networks of Izhikevich neurons to the edge of synchronization transition with scale-free avalanches. Front. Syst. Neurosci. **13** (2019). https://doi.org/10.3389/fnsys.2019.00073
15. Klaus, A., Yu, S., Plenz, D.: Statistical analyses support power law distributions found in neuronal avalanches. PLOS ONE **6**(5), 1–12 (2011). https://doi.org/10.1371/journal.pone.0019779
16. Markram, H., Gerstner, W., Sjöström, P.J.: Spike-timing-dependent plasticity: a comprehensive overview. Front. Synaptic Neurosci. **4** (2012). https://doi.org/10.3389/fnsyn.2012.00002
17. Markram, H., Lübke, J., Frotscher, M., Sakmann, B.: Regulation of synaptic efficacy by coincidence of postsynaptic APs and EPSPs. Science **275**(5297), 213–215 (1997). https://doi.org/10.1126/science.275.5297.213
18. Montroll, E.: Random walks on lattices. Proc. Symp. Appl. Math. **16**, 193–220 (1964). https://doi.org/10.1090/psapm/016
19. Paradisi, P., Allegrini, P.: Intermittency-driven complexity in signal processing. In: Barbieri, R., Scilingo, E.P., Valenza, G. (eds.) Complexity and Nonlinearity in Cardiovascular Signals, pp. 161–195. Springer, Cham (2017). https://doi.org/10.1007/978-3-319-58709-7_6
20. Paradisi, P., Cesari, R., Donateo, A., Contini, D., Allegrini, P.: Scaling laws of diffusion and time intermittency generated by coherent structures in atmospheric turbulence. Nonlin. Processes Geophys. **19**(1), 113–126 (2012). https://doi.org/10.5194/npg-19-113-2012
21. Peng, C.K., Buldyrev, S.V., Havlin, S., Simons, M., Stanley, H.E., Goldberger, A.L.: Mosaic organization of DNA nucleotides. Phys. Rev. E **49**, 1685–1689 (1994). https://doi.org/10.1103/PhysRevE.49.1685
22. Roth, A., van Rossum, M.C.W.: Modeling synapses. In: De Schutter, E. (ed.) Computational Modeling Methods for Neuroscientists, pp. 139–160. MIT Press, Cambridge, MA (2009). https://doi.org/10.7551/mitpress/9780262013277.003.0007
23. Rydin Gorjão, L., Hassan, G., Kurths, J., Witthaut, D.: MFDFA: efficient multifractal detrended fluctuation analysis in Python. Comput. Phys. Commun. **273**, 108254 (2022). https://doi.org/10.1016/j.cpc.2021.108254
24. Scarpetta, S., de Candia, A.: Neural avalanches at the critical point between replay and non-replay of spatiotemporal patterns. PLOS ONE **8**(6), 1–9 (2013). https://doi.org/10.1371/journal.pone.0064162
25. Shew, W.L., Yang, H., Yu, S., Roy, R., Plenz, D.: Information capacity and transmission are maximized in balanced cortical networks with neuronal avalanches. J. Neurosci. **31**(1), 55–63 (2011). https://doi.org/10.1523/JNEUROSCI.4637-10.2011

26. Song, S., Miller, K.D., Abbott, L.F.: Competitive Hebbian learning through spike-timing-dependent synaptic plasticity. Nat. Neurosci. **3**, 919–926 (2000). https://doi.org/10.1038/78829
27. Sorrentino, P., et al.: Brain fingerprint is based on the aperiodic, scale-free, neuronal activity. Neuroimage **277**, 120260 (2023). https://doi.org/10.1016/j.neuroimage.2023.120260
28. Stimberg, M., Brette, R., Goodman, D.F.: Brian 2, an intuitive and efficient neural simulator. Elife **8**, e47314 (2019). https://doi.org/10.7554/eLife.47314
29. Weiss, G.H., Rubin, R.J.: Random walks: theory and selected applications, pp. 363–505 (1982). https://doi.org/10.1002/9780470142769.ch5
30. Yadav, A.C., Quadir, A., Jafri, H.H.: Finite-size scaling of critical avalanches. Phys. Rev. E **106**, 014148 (2022). https://doi.org/10.1103/PhysRevE.106.014148
31. Yang, J., Feng, P., Wu, Y.: Neuronal avalanche dynamics regulated by spike-timing-dependent plasticity under different topologies and heterogeneities. Cogn. Neurodyn. **18**(6), 1307–1321 (2024). https://doi.org/10.1007/s11571-023-09966-8

Input Manifold Structure Drives Reservoir Partitioning: Insights into Task-Related Brain Activation and the Connectivity Hypothesis

Peter Ford Dominey[1(✉)] and Paul Triebkorn[2]

[1] Université Bourgogne Europe, Inserm, CAPS UMR 1093, 21000 Dijon, France
peter-ford.dominey@u-bourgogne.fr
[2] Aix Marseille Univ, INSERM, INS, Inst Neurosci Syst, Marseille, France

Abstract. A major open question in neuroscience concerns how brain regions acquire their functional specificity. Is it due to intrinsic properties of particular brain regions, or to some more global cause? The connectivity hypothesis proposes that such regional brain functional specificity is based on extrinsic connectivity of the region, rather than on its internal properties. While neuroscience research can contribute to this argument, it is technically difficult to directly address this question in the human brain. Artificial neural networks allow us to more directly test this hypothesis and demonstrate that functional partitioning in a recurrent neural network is an emergent consequence of structured input manifolds interacting with network connectivity, forcing the reservoir into discrete, non-overlapping state attractors. This is related to recent studies that characterize how the functional computing dynamics of reservoirs can be systematically changed in response to control inputs that modify the reservoir dynamics. The first contribution of the current research is to investigate how structure that is directly embedded in the input can serve to reconfigure the network. This is of interest in the context of reservoir computing and the use of control inputs to increase reservoir expressivity. The second contribution is to demonstrate how these observations in reservoir computing provide a testable model for the connectivity hypothesis. This allows us to extend the connectivity hypothesis and make specific predictions for neuroscience related to specific patterns of functional connectivity and dynamics. This contributes to the growing research on input control of reservoir function, and importantly opens the door to a novel way of conceptualizing task-related brain dynamics, as a consequence of input-driven task structure reconfiguring the computational organization of the brain.

Keywords: connectivity · connectome · reservoir computing · input manifold

1 Introduction

A hallmark in human cognitive function is the ability to perform a variety of distinct cognitive tasks. This poses the question of how neural networks can spontaneously adapt and switch between different tasks, and how brains display task-related regional

L. Pasa et al. (Eds.): ICANN 2026, LNCS 17089, pp. 91–102, 2027.
https://doi.org/10.1007/978-3-032-38398-3_8

specificity. Recent research in recurrent neural networks has demonstrated how recurrent networks can be reconfigured into diverse dynamics regimes for performing distinct cognitive functions by input biases [1]. These studies indicate that neural computation involves a rich interaction between structure in the input, and structure in the network. They show that one method that allows the same network to reconfigure itself for different tasks is to use structure in the different context coding inputs, in addition to task inputs, to drive the network into different task-specific functional regimes [1]. These observations in theoretical neuroscience are pertinent in understanding a major open question in neuroscience.

Indeed, understanding how different brain networks are activated for different cognitive functions is one of the major open problems in neuroscience today [2]. The visual word form area (VWFA) is an interesting example. This area is active in particular for the processing of written words [3]. Written words have arrived fairly recently in the time-scale of the evolution of the modern human brain. Thus it is supposed that the VWFA emerges not because of some intrinsic sensitivity to words, but rather because of its existing connectivity, at the junction between detailed visual processing and language areas [2, 4]. More generally, this leads to the connectivity hypothesis, that brain connectivity provides a framework whereby specific structure in the input will be routed to, and processed by specific subnetworks, thus giving rise to specialized areas such as the VWFA [2, 4].

In the current research, inspired by theoretical work that shows how reservoir dynamics can be shaped by input structure [1], we examine whether input structure can produce region specificity as in the VWFA. Previous research has shown that distinct input structures will generate distinct and stable patterns of functional connectivity [5]. Here we extend this work by testing the hypothesis that structured input manifolds will generate distinct and stable patterns of network activation. The first contribution of the current approach is to demonstrate how stable biases can be embedded directly in the input signal, rather than as a separate bias input [1]. The input thus at the same time includes the data for the task, and a configuring bias that drives the network into a particular state that is appropriate for the computations required by the task. This is important because it defines a new computational paradigm for exploiting multitasking and expressivity in reservoir computing [1].

The second contribution is to demonstrate how this approach provides a new method to understand how specific regional neural activity in the human brain, such as the VWFA, can arise during input driven cognitive processing. We predict that this interaction between structure in the input and network connectivity will be manifested in three distinct and complimentary manners that are used in human neuroscience: (A) mean input-driven neural activation [6, 7], (B) input-driven functional connectivity [8, 9], and (C) input-driven phase-space dynamics [10, 11].

The current research thus sets out to test the hypothesis that these three key aspects of network activation that can be observed in human neuroscience can be explained, at least in part, in terms of the interaction between coherent structure that is present in the input, and the anatomical structure of the network. In order to do this, we use biologically constrained reservoir networks with structured inputs to produce distinct task related activations, functional connectivity and dynamics. Recent research has demonstrated that different input manifolds can generate different patterns of functional connectivity, using reservoirs that followed a local connectivity rule [5]. Here we extend this research, using more structured connectivity, based on the human connectome, that has been demonstrated to simulate human brain activity during narrative processing [12].

2 Methods

2.1 Connectome Reservoir

The reservoir architecture is illustrated in Fig. 1. The reservoir model is based on a classic echo state network with leaky integrator tanh units. A set of recurrently connected nodes – the reservoir – is stimulated by inputs. This produces a dynamic reverberation of activation throughout the reservoir as information propagates through the recurrent connections.

The basic discrete-time, tanh-unit echo state network with N reservoir units and K inputs is characterized by the state update equation:

$$\mathbf{x}(t+1) = (1-\alpha)\mathbf{x}(t) + \alpha \cdot f(\mathbf{W}x(t) + \mathbf{W}_{in}u(t)) \tag{1}$$

where $\mathbf{x}(n)$ is the N-dimensional reservoir state, f is the tanh function, $\mathbf{W}$ is the $N \times N$ reservoir weight matrix, $\mathbf{W}in$ is the $N \times K$ input weight matrix, $\mathbf{u}(n)$ is the K dimensional input signal, α is the leaking rate. The matrix elements of W and W_{in} are drawn from a random distribution.

The reservoir was instantiated using easyesn, a python library for recurrent neural networks using echo state networks (https://pypi.org/project/easyesn/) [13]. We used a reservoir of 1000 neurons, with input and output dimensions of 300. The W and W_in matrices are initialized with uniform distribution of values from −0.5 to 0.5. The leak rate was 0.2, as in a previous related study [14]. The reservoir is relatively robust to changes in these values, as long as the reservoir dynamics are neither diverging nor collapsing.

Human connectome matrices were constructed using data from the Human Connectome Project (HCP) [15] to create reservoir connectivity matrices that are defined by human brain connectivity. We chose to use these connectome based connectivity matrices for the reservoir because they have been demonstrated to produce human-like input-driven dynamics [12]. It is possible that our results will extend to other network topologies, but this remains a question to be investigated. We used minimally preprocessed MRI data of 20 subjects to construct connectivity matrices [16]. Tractography, based on diffusion MRI, was performed using the MRtrix3 software package [17] to obtain the connection strength between brain regions. The resulting 15 million streamlines were then used to generate the 1000 x 1000 connectivity matrices according to the Schaefer 1000 area parcellation [18]. The connectome reservoir was then implemented by multiplying the W matrix by the connectome matrix, as schematized in Fig. 1. Inputs are provided to a limited number (150 × 2) units in visual and auditory sensory areas of the left and right hemisphere, as illustrated in Fig. 1. Thus, 300 neurons are directly input driven, and 700 are only driven indirectly, and referred to as associative neurons. Connectome reservoirs constructed in this manner have been demonstrated to display anatomically distributed temporal processing hierarchies [12] as observed in the human brain [19], and so are suitably structured for our analysis of the interaction between input structure and networks structure.

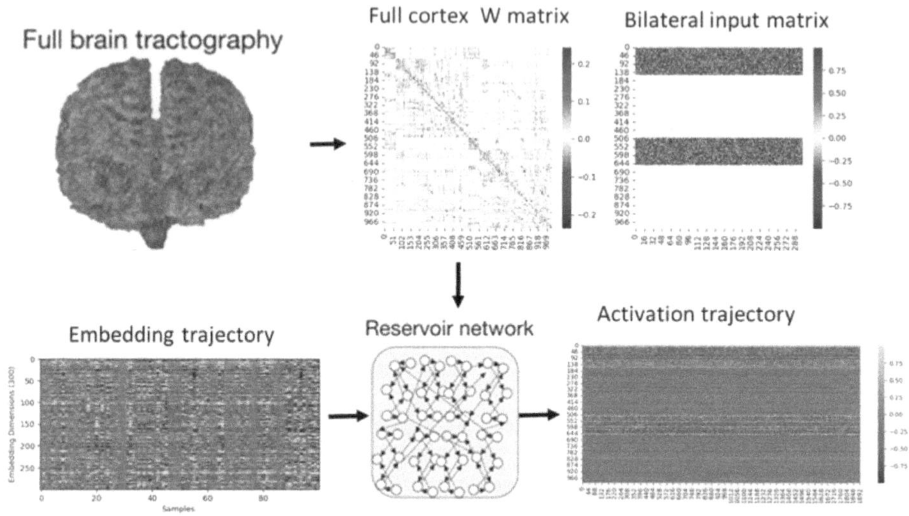

Fig. 1. Connectome Reservoir. Full brain connectome is used to generate a 1000×1000 connectivity matrix corresponding to the full brain Schaefer parcellation, used as the connectivity matrix for a reservoir. W_matrix – note the squares along the diagonal that correspond to left and right hemisphere connectivity. N-word narrative text from human fMRI experiments is used to generate word embeddings from a language model that are input to the reservoir. Bilateral input is limited to a set of 150 sensory neurons in each of the left and right hemispheres. This generates an Nx1000 trajectory of reservoir neuron activations, which can be analyzed as outlined in Fig. 1.

We do note the following limitations: the connectome data does not specify the directionality of connections, nor their excitatory or inhibitory nature. We use a combination of excitatory and inhibitory connections, and the issue of directionality is addressed by multiplying the symmetric connectome matrix with native reservoir W matrix. Additionally, neurons are tanh with uniform parameters across regions. This level of granularity is sufficient to allow simulation of regional specificity in human neural temporal processing hierarchies as a function of network connectivity [12, 20]. In any case, the important thing here is that we have a defined connectivity structure against which we can test the impact of input manifold structure.

2.2 Input Manifold Geometry

To investigate how input geometry interacts with network topology, we generate two distinct high-dimensional input structures, hereafter referred to as Input Manifold 1 and Input Manifold 2 from two embedding models, en_core_web_lg and fr_core_news_lg from https://spacy.io/models. Each of the two embedding models was used to generate a sequence of embedding vectors from a narrative text. The text was from the Sherlock television series that has been transcribed and is available in the Narratives Dataset, which includes fMRI from subjects listening to different narratives, along with ancillary data including text transcripts [21]. These two embedding sequences are our two input manifolds,

It is important to note that we did not choose to use such embedding sequences as a proxy for linguistic input. Rather, we chose them because of their inherent properties as embeddings from distinct embedding spaces, with sufficient statistical differences to drive the network to distinct state spaces. We thus treat these two embedding sequences as abstract geometric entities rather than linguistic signals, so that we can investigate the reservoir's capacity for functional specialization driven solely by the spatiotemporal properties of the input. It is also important to note that we specifically do not want to use transformer models [22] to generate the embeddings, because we want to look at integration properties of the reservoir with well characterized embeddings that do not change with context.

We performed PCA on these these two manifolds in order to characterize the structural differences between them. Figure 2 demonstrates that the two input manifolds provide systematic and distinct input bias through three related geometric mechanisms. First, the Manifold Separation shown in the PCA projection (left) reveals that the two embedding models occupy distinct regions of the vector space, ensuring that the reservoir will be driven by statistically distinct input regimes. Second, the mean embeddings for the two manifolds are uncorrelated, and third, the mean variability in the two embedding spaces are also uncorrelated. Together, we can be confident that these features should be sufficient to provide the baseline control signal required to shift the RNN's operating point into separate dynamical states, if this is possible.

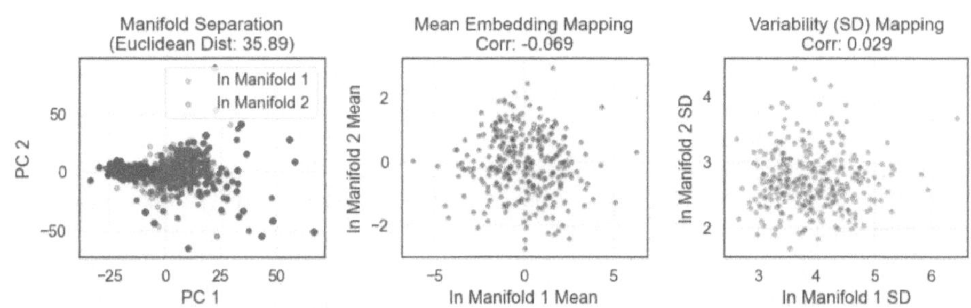

Fig. 2. Left. The two input manifolds, consisting of embedding trajectories, were analysed by PCA. The two patterns display clear differences in their distance along PC1, and their dispersion along both PCs. Middle, mean activations across the 300D embeddings for the two manifolds are correlated. Right, mean variability (standard deviations) across the 300D embeddings for the two manifolds are uncorrelated

2.3 Approach

Given this infrastructure of structured reservoir connectivity, and structured input manifolds we can now perform experiments to test the predictions from our three-point hypothesis: distinct input manifolds will produce (A) mean activation patterns that reliably differ, (B) distinct patterns of functional connectivity, and (C) distinct spatiotemporal dynamics. If so, this contributes to explaining how input structure influences task related brain activation in human fMRI, helping to demonstrate how the connectivity hypothesis can explain regional specificity as observed in the VWFA. In addition it makes a link between these neuroscience questions, and fundamental work in input driven manipulation of reservoir computation fundamentals [1].

3 Results

3.1 Input Manifold Drives Network Activation Geometry

In order to determine if driving the reservoir with distinct input manifolds would produce different spatial patterns of mean activation, we ran the population of reservoirs on the two input manifolds of 1800 time-steps for each, and then compute the mean activation for each of the neurons. In order to eliminate false responses related directly to the input drive, we restricted the analysis to the 700/1000 neurons that do not receive direct input.

not receive direct input.

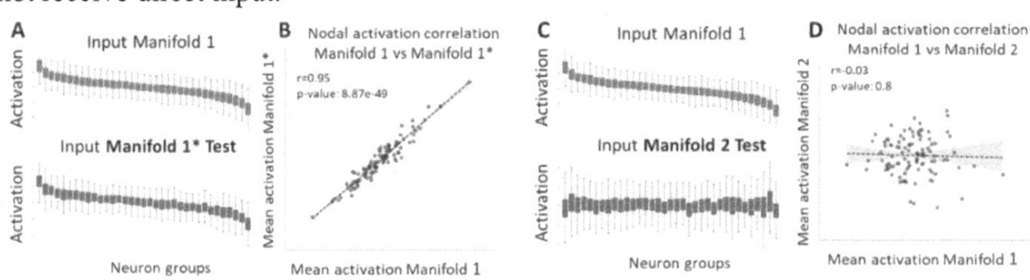

Fig. 3. Input-driven neural hierarchy. A. Input Manifold 1 produces a specific pattern of activations, here sorted from highest to lowest. Input Manifold 1* - A second input generated from manifold 1 drives the reservoirs to the same activation pattern. B. Highly significant correlation between activation patterns for input manifold 1 and 1*. C. Input Manifold 2 produces an activation pattern that is unrelated to that from Input Manifold 1. D. Neural activity from the two input manifilds are uncorrelated.

The mean activations of the neurons driven by input manifold 1 were ranked from strongest to weakest. The neurons are grouped into bins of 10 for better visualization. This activation profile is illustrated in Fig. 3A. The first observation is that this ranking is coherent across the population of reservoirs. The error bars show the standard deviation across the 20 connectome instances. To confirm the consistency of the reservoir's internal representations across independent runs, we calculated the Intraclass Correlation Coefficient (ICC) for the 20 instances of the sorted activation profiles. We observed excellent reliability for the mean representation (ICC(3,k) = 0.974, 95% CI [0.97, 0.98], $p < 0.001$). These results demonstrate that while individual runs contain stochastic noise, the underlying manifold structure is highly conserved across all instances. This is a significant result, as it argues for the connectivity hypothesis: input structure interacts with connectivity to produce specific regional activation patterns, just what we see here. We then ran the same reservoirs on a new input sequence from manifold 1 (indicated with *). As seen in Fig. 3A and B, there we get a highly correlated activation pattern, again in support of the connectivity hypothesis.

In order to determine how this spatial activation pattern depends on the specific input manifold structure, we then plotted the activation of the same neuron groups in the same ordering, with their activation driven by input manifold 2. There we observe that the activation pattern is unrelated to that for input manifold 1, Fig. 3B. This dissociation between activation patterns for the two input manifolds is further revealed in the scatter-plot and correlation where each point represents the neural activation for the two input manifolds in the 2d space, Fig. 3C. The two activation profiles are uncorrelated. This analysis allows us to conclude that the interaction between a structured input manifold and the connectivity structure produces distinct patterns of mean activation across 20 human connectome reservoirs. From a biological perspective, this is coherent with the hypothesis that task related activation patterns seen in human fMRI can be explained by these same kinds of interactions between input structure and brain architecture.

Previous research has shown that such results are highly dependent on the connectivity structure [5, 12, 20, 23, 24]. To evaluate the effects of connectivity structure we tested the ability of the networks to predict the input manifold based on network activation with intact and shuffled connectomes. To exhaustively evaluate this measure, 20 reservoir instances were each tested with 20 separate human connectome connectivity matrices. In each of these 400 runs, the network was exposed to alternating inputs from the two manifolds. A decoder (Logistic Regression with L2 regularization) was trained on data from 15 subjects (300 runs) and tested on 5 completely novel subjects (100 runs). The intact networks achieved perfect decoding accuracy, corresponding to the clear separation see in Fig. 3. In contrast, the shuffled connectome networks were at 0.55 (chance) accuracy. This argues that structure in both the input manifold and the network connectivity are required to produce a coherent input driven activation geometry.

3.2 Input Manifold Shapes Functional Connectivity

Recent neuroscience research has shown that functional connectivity is modulated by task-related input [8, 9]. Here we can analyze how functional connectivity is stable for a given input manifold, and dissociated for different input manifolds. We characterize functional connectivity as the pairwise correlations between all pairs of neurons over a given time period. This generates an NxN matrix where N is the number of neurons in the network, in our case the 700 neurons that are not directly driven by the input. Each cell represents the correlation between the two neurons, i.e. their functional connectivity. We can then collapse the NxN matrix into an N dimensional vector that is the sum of the functional connections. If the input structure has a reliable impact on functional connectivity, then we would predict that this representation of the functional connectivity will be correlated when the reservoir is driven by inputs from the same input manifold, and that they will be significantly less correlated when the reservoir receives non-similar inputs from two different input manifolds.

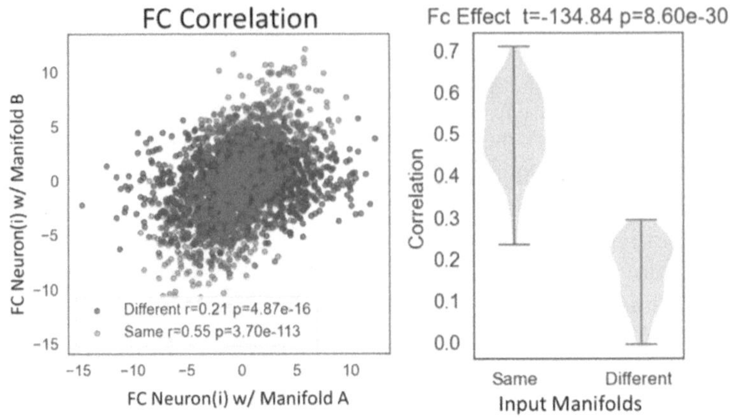

Fig. 4. Correlations between functional connectivity (FC) for networks exposed to same or different input manifolds. Each point is the FC for neuron i in 1–700 for the two input manifolds. Same (Red) Here the two FC matrices were generated from activation of the networks with inputs from the same manifold (A and B are the same manifolds). The correlation between the two FC networks is r = 0.55. Different (Blue) Here the two FC networks were generated from network activations from the same networks exposed to inputs from the two manifolds (A and B are different manifolds). In this case the correlation between the two FC matrices is much lower, r = 0.21. Right panel. Displays the functional connectivity correlations for the 20 networks when exposed to inputs from same vs. different manifolds. The correlation between the FC matrices when reservoirs are exposed to different input manifolds is significantly lower than for the FC matrices when the reservoirs are exposed to inputs from the same manifold.

The first observation is that when inputs from a given manifold are presented to the reservoirs, there is a reliable and consistent resulting functional connectivity across the 20 reservoirs. This is revealed by the high correlation between the FC vectors for two instances of the same input manifold input to the reservoirs. The difference between functional connectivity correlation for same vs different inputs to the network can be considered a metric for an input driven functional connectivity effect. Figure 4 displays this effect. There we see that indeed, network functional connectivity depends on structure in the input. Functional connectivity for networks driven by the same type of input structures displays significantly greater correlation that for networks driven by different types of input structure. The panel on the right show the difference between the correlation scores (Pearson-r statistic correlation values). We observe that there is a highly significant increase in the correlations for functional connectivity driven by inputs from the same input manifold.

This provides support for the connectivity hypothesis, and also extends it from regional specificity to functional connectivity. It provides an ANN model to provide an explanation for human research that indicates that different input driven cognitive tasks result in different functional connectivity networks [8, 9].

3.3 Network Dynamics

A final prediction from the connectivity hypothesis is that the interaction of a structured input manifold with network structure should drive the network into a particular dynamic state. In order to test this, we stimulated the reservoir with input that alternated between the two input manifolds from Sect. 3.1. PCA revealed two highly separable clusters illustrated as the blue stars for manifold 1 inputs, and red stars for manifold 2 in Fig. 5. We then stimulated a reservoir with a different subject connectome and projected the resulting activation trajectory into the original PCA space, with the same color coding as filled circles. The PCA issued from one subject's connectome reliably separated the novel data from a reservoir with a different connectome. Again, to exhaustively evaluate this effect, 20 reservoir instances were each tested with 20 separate human connectome connectivity matrices with intact and shuffled weight matrices. The intact networks showed a reliable separation that was significantly greater than for the shuffled connection matrices.

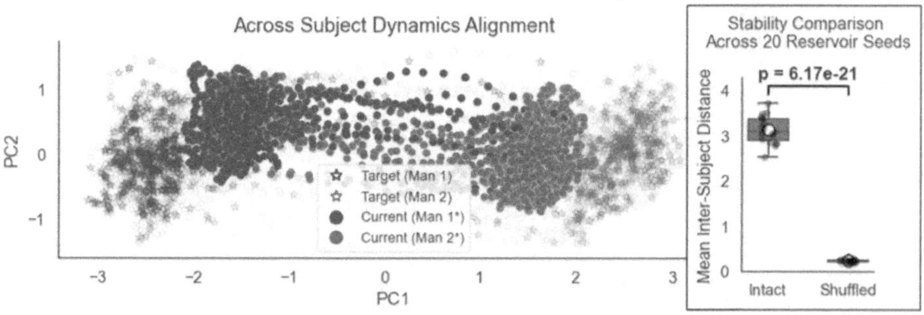

Fig. 5. Left. PCA applied to reservoir dynamics produced by an alternating manifold 1 – manifold 2 input. Blue and red stars illustrate input manifold labels. Blue and red circles display how these same principle components reliably separated network dynamics for reservoir activation from a network with a different subject connectome. Right. Separation distances for 20 reservoir instances each using the 20 connectome matrices with intact vs. shuffled matrices.

These results extend the connectivity hypothesis into the domain of network dynamics: that functionally specific network dynamics develop their characteristic form because of pre-existing connectivity, and importantly, because of the interaction with structure in the input.

4 Interpretation and Discussion

Here we set out to characterize how regular structure encoded in the input can interact with anatomical structure to produce characteristic mean activations, functional connectivity and network dynamics that are specific to the input patterns. Driving the network with distinct input structures can shift the dynamics in computationally pertinent manner [1]. This supports the hypothesis that multiple aspects of network activation that can be observed in human neuroscience can be explained, at least in part, in terms of the interaction between coherent structure that is present in the input, and the anatomical structure of the network.

4.1 Neuroscience Interpretations and Predictions

This is coherent with the Connectivity Hypothesis in human neuroscience which states that "the specialization of a given brain region is largely shaped by how it connects and communicates with the rest of the brain" [4]. It extends the Connectivity Hypothesis such that it is not the connectivity alone, but its interaction with structure that is encoded in the input, that drives the specialization of a given brain region. Here we demonstrate how driving reservoirs with different structured input manifolds produces reliably different specialization of the network. We do not claim to demonstrate emergence of the VWFA. Rather, we provide evidence about how task-related regional specialization can emerge from the interaction of input and network structures.

Importantly, we also extend the connectivity hypothesis to make predictions that can be tested in neuroscience. The first prediction is that during task specific processing that produces regional specific activation, the interaction of input and connectivity will also be revealed by specific patterns of functional connectivity. Recent neuroscience research is consisten with this prediction, and has demonstrated that functional connectivity is modified in a systematic way based on the engaged cognitive task [8, 9]. Here we demonstrated that structure in the input can drive biologically constrained reservoirs into different regimes of functional connectivity, as observed in human neuroscience. As illustrated in Sect. 3.2 and Fig. 4.

Finally, we extend the connectivity hypothesis to dynamics. The PCA analysis if network activation driven by different input manifolds illustrates how the dynamics of the network closely follow the input structure, and predict that similar observations should be made neuroscience. Interestingly, there is evidence in favor of this prediction, from observations of human task-related brain dynamics [10, 11].

Regarding the connectivity hypothesis there is an important caveat to make clear. Saygin and colleagues note that connectivity precedes function and allows brain regions to develop specificity, but show that in early development the specificity may not yet be present [2]. Because we do not model developmental changes, our modeling directly displays the specificity, and how development can be addressed in more detail remains as a topic for future research.

4.2 Conclusions

We believe that this research is pertinent because it creates a link between state of the art research in network neuroscience [25, 26], and computational biology, and neuroscience via reservoir computing. Recent research has shown that providing homogenous control inputs to reservoirs can be used to shift the systems into a small finite set of dynamic regimes, each of which has well defined computational properties [1]. This opens the possibility for a more general setup, with heterogeneous inputs that could provide an even richer dynamic computational repertoire. At the same time, recent research has characterized how characteristics of structure in the input can shape responses in cortex [27, 28]. The key point is that structure in the input can have an important impact on network dynamics. In parallel, research in human neuroscience has proposed the connectivity hypothesis, stressing that specificity of regions in a network/brain are due to their connectivity.

In this context, the current research creates a hopefully productive synthesis, exploring how it is an interaction between network connectivity and input structure that produces regional specificity. We thus provide a computational model for exploring the connectivity hypothesis. At the same time, we propose that this provides an infrastructure to extend the network flexibility produced with baseline control.

Acknowledgments. Reservoir simulations were performed using the easyesn toolkit https://github.com/kalekiu/easyesn. Narrative transcripts [29] were obtained from the Narratives dataset https://fcon_1000.projects.nitrc.org/indi/retro/Narratives.html [21]. Data were provided [in part] by the Human Connectome Project, WU-Minn Consortium (Principal Investigators: David Van Essen and Kamil Ugurbil; 1U54MH091657) funded by the 16 NIH Institutes and Centers that support the NIH Blueprint for Neuroscience Research; and by the McDonnell Center for Systems Neuroscience at Washington University.

Disclosure of Interests. The authors have no competing interests to declare that are relevant to the content of this article.

References

1. Ogawa, S., Fumarola, F., Mazzucato, L.: Multitasking via baseline control in recurrent neural networks. Proc. Natl. Acad. Sci. **120**(33), e2304394120 (2023)
2. Saygin, Z.M., et al.: Connectivity precedes function in the development of the visual word form area. Nat. Neurosci. **19**(9), 1250–1255 (2016)
3. Dehaene, S., Cohen, L.: The unique role of the visual word form area in reading. Trends Cogn. Sci. **15**(6), 254–262 (2011)
4. Li, J., et al.: Innate connectivity patterns drive the development of the visual word form area. Sci. Rep. **10**(1), 18039 (2020)
5. Dominey, P.F.: Effects of input structure and topology on input-driven functional connectivity stability. In: Wand, M., Malinovská, K., Schmidhuber, J., Tetko, I.V. (eds.) International Conference on Artificial Neural Networks. Springer (2024). https://doi.org/10.1007/978-3-031-72359-9_7
6. Price, C.J.: A review and synthesis of the first 20 years of PET and fMRI studies of heard speech, spoken language and reading. Neuroimage **62**(2), 816–847 (2012)
7. Amalric, M., Dehaene, S.: A distinct cortical network for mathematical knowledge in the human brain. Neuroimage **189**, 19–31 (2019)
8. Cole, M.W., et al.: The functional relevance of task-state functional connectivity. J. Neurosci. **41**(12), 2684–2702 (2021)
9. Elton, A., Gao, W.: Task-related modulation of functional connectivity variability and its behavioral correlations. Hum. Brain Mapp. **36**(8), 3260–3272 (2015)
10. Meer, J.N.v.d., et al.: Movie viewing elicits rich and reliable brain state dynamics. Nat. Commun. **11**(1), 5004 (2020)
11. Venkatesh, M., Jaja, J., Pessoa, L.: Brain dynamics and temporal trajectories during task and naturalistic processing. Neuroimage **186**, 410–423 (2019)

12. Triebkorn, P., Jirsa, V., Dominey, P.: Modeling the impact of white matter connectivity on brain network processing time scales – a computational connectomic study. Commun. Biol. (2025)
13. Thiede, L.A., Zimmermann, R.S.: EasyESN: a library for recurrent neural networks using echo state networks (2017). https://github.com/kalekiu/easyesn
14. Dominey, P.F.: Narrative event segmentation in the cortical reservoir. PLoS Comput. Biol. **17**(10), e1008993 (2021). https://doi.org/10.1371/journal.pcbi.1008993.PMID:34618804; PMCID:PMC8525778
15. Van Essen, D.C., et al.: The WU-Minn human connectome project: an overview. Neuroimage **80**, 62–79 (2013)
16. Glasser, M.F., et al.: The minimal preprocessing pipelines for the human connectome project. Neuroimage **80**, 105–124 (2013)
17. Tournier, J.-D., et al.: MRtrix3: a fast, flexible and open software framework for medical image processing and visualisation. Neuroimage **202**, 116137 (2019)
18. Schaefer, A., et al.: Local-global parcellation of the human cerebral cortex from intrinsic functional connectivity MRI. Cereb. Cortex **28**(9), 3095–3114 (2018)
19. Chien, H.-Y.S., Honey, C.J.: Constructing and forgetting temporal context in the human cerebral cortex. Neuron **106**(4), 675–686. e11 (2020)
20. Chang, C.H., et al.: Emergence of a temporal processing gradient from naturalistic inputs and network connectivity. Proc. Natl. Acad. Sci. **122**(28), e2420105122 (2025)
21. Nastase, S.A., et al.: The "Narratives" fMRI dataset for evaluating models of naturalistic language comprehension. Sci. Data **8**(1), 250 (2021)
22. Devlin, J., et al.: BERT: pre-training of deep bidirectional transformers for language understanding. In: Proceedings of the 2019 Conference of the North American Chapter of the Association for Computational Linguistics: Human Language Technologies, Volume 1 (Long and Short Papers) (2019)
23. Chaudhuri, R., Bernacchia, A., Wang, X.-J.: A diversity of localized timescales in network activity. elife. **3**, e01239 (2014)
24. Dominey, P.F., Ellmore, T.M., Ventre-Dominey, J.: Effects of connectivity on narrative temporal processing in structured reservoir computing. In: 2022 International Joint Conference on Neural Networks (IJCNN), p. 1–8. IEEE (2022)
25. Kawai, Y., Park, J., Asada, M.: A small-world topology enhances the echo state property and signal propagation in reservoir computing. Neural Netw. **112**, 15–23 (2019)
26. Suárez, L.E., et al.: Learning function from structure in neuromorphic networks. Nat. Mach. Intell. **3**(9), 771–786 (2021)
27. Tang, D., et al.: Stimulus type shapes the topology of cellular functional networks in mouse visual cortex. Nat. Commun. **15**(1), 5753 (2024)
28. Bányai, M., et al.: Stimulus complexity shapes response correlations in primary visual cortex. Proc. Natl. Acad. Sci. **116**(7), 2723–2732 (2019)
29. Baldassano, C., et al., Discovering event structure in continuous narrative perception and memory. Neuron **95**(3), 709–721. e5 (2017)

Statistical Dynamics of Continuous Memory Neural Networks: the Effects of Gradient and Curl Forces

Roseli S. Wedemann[1(✉)], Angel R. Plastino[2], and Frank Quing Poi Chen[1]

[1] Instituto de Matemática e Estatística, Universidade do Estado do Rio de Janeiro, Rua São Francisco Xavier 524, 20550-900 Rio de Janeiro, RJ, Brazil
roseli@ime.uerj.br

[2] CeBio y Departamento de Ciencias Básicas, Universidad Nacional del Noroeste de la Provincia de Buenos Aires (UNNOBA), CONICET, Roque Saenz Peña 456, B6000 Junin, Buenos Aires, Argentina
arplastino@unnoba.edu.ar

Abstract. Associative memory neural networks (NNs) are essential ingredients for developing computational models for mental phenomena, including neurosis, delusions, creativity, and the interplay between unconsciousness and consciousness, and also for solving diverse problems in AI. Previous work suggests that power-law, q-MaxEnt probability densities, that optimize S_q entropies associated with nonextensive statistical mechanics, may emerge from the dynamics of artificial, associative memory NNs. These densities have also been detected in experimental observations of brain dynamics. Here, we propose and analyze a Fokker-Planck (FP) representation of the dynamics of continuous NNs, in the presence of both gradient and curl force contributions to the NNs' phase-space flow. Gradient forces are typical of NNs with symmetric synaptic connections, while curl forces can describe asymmetric connections. We investigate power-law, nonlinear FP descriptions of the time evolution of NNs that lead to q-MaxEnt stationary solutions, and analyze features of the related flows of energy, information and entropy. This work extends and generalizes previous studies, in which we considered curl forces within a particular form for the drift field (which we here generalize) or conducted an analysis of the energy, entropy, and information flows in nonlinear FP descriptions of NN models without curl forces.

Keywords: Associative memory · Dynamics of continuous neural networks · Curl forces · Generalized statistical mechanics · S_q entropies

1 Introduction

Memory functioning is an essential element in the description of brain and mental phenomena, as well as in the treatment of many tasks in artificial intelligence

L. Pasa et al. (Eds.): ICANN 2026, LNCS 17089, pp. 103–114, 2027.
https://doi.org/10.1007/978-3-032-38398-3_9

(AI). Both pathological and normal mental processes studied by psychology, psychoanalysis, psychiatry and neuroscience commonly describe phenomena such as neurosis, delusions, disorganized thought, psychotic thinking, schizophrenia, creativity and the interaction between consciousness and unconsciousness, based on aspects involving broader, distant, looser or narrower associations and unfocusing of attention [1,2,8,17,18,22,26]. In AI, there has been recent understanding relating Hopfield-like networks, such as restricted Boltzmann machines (BMs), to deep learning architectures such as transformers [9,10,16]. Associative memory, artificial neural network (NN) models, like the Hopfield model [6] have thus played a key role in the theoretical study of many of these mental phenomena and artificial devices [2,9,10,16,17,22,26].

There has been intense recent research dedicated to the study of a generalization of the Boltzmann-Gibbs (BG) thermostatistics referred to as nonextensive statistical mechanics, or S_q-thermostatistics, based on the S_q, power-law entropic functionals proposed by Tsallis [19]. It has been applied to the study of many types of complex systems in many fields, including neuroscience and AI [1,11,12,17,19,22–24]. For instance, in theoretical studies of mental processes presented in [17,22,26,28], memory access has been simulated with discrete versions of Hopfield-like models, using the Generalized Simulated Annealing (GSA) algorithm, based on the the S_q entropic form [19,20], which recovers the standard BM when $q = 1$. When implementing these algorithms, the network reaches minimum energy states that can be shallower or deeper depending on how the annealing schedule is tuned. The process of reaching shallower attractor states was interpreted in [17,22] as being able to make more loose or creative associations and the authors found that the GSA memory NN was significantly more creative than the BM. When GSA is applied to discrete-state NNs, it notably provides a dynamical mechanism that generates maximum S_q entropy (q-MaxEnt) distributions [17,22,26], such as the distributions of avalanche sizes that occur during memory retrieval [17]. These distributions have q-exponential forms and are consistent with power-law distributions observed in results from numerical simulations of theoretical models [13,22,26], as well as in experiments in neuroscience (see [1] and [17] and references therein), and they are central to the S_q-thermostatistical formalism. These observations provide a strong motivation for investigating the application of the S_q thermostatistical framework to the study of the dynamics of NN models.

The S_q-thermostatistical, theoretical framework has also been successfully applied to the analysis of the dynamics of continuous NN models. This is particularly relevant, as biological neurons in animals (including humans) are typically described by continuous state variables [3,6,8]. Dynamical models of NNs with continuous variables are represented by sets of coupled, ordinary differential equations and, in the presence of noise, can be treated by the Fokker-Planck formalism. Such a treatment of the dynamics of continuous NNs has revealed interesting properties of these systems, such as the limit cycles that occur in the presence of asymmetric synapses [11,12,23–25]. Treatment of asymmetric connections in NNs is also relevant as they occur in biological and artificial NNs.

In this contribution, we explore a general theoretical framework for the nonlinear Fokker-Planck representation of the dynamics of continuous memory models, such as the celebrated Cohen-Grossberg models [3]. We consider models that include both gradient and curl forces, and exhibit a more general form than the ones we studied in previous works. By "forces", we mean the right-hand sides of the coupled differential equations of motion governing the model (see Eq. (2)). Gradient forces arise from a potential function, and occur in memory NNs with symmetrical synaptic connections, whereas curl forces do not derive from a potential and provide a formalism for describing NNs with asymmetric connections [12]. We obtain the form for the nonlinear Fokker-Planck equation (NLFPE) appropriate for these models, and determine a set of relations between the curl forces and the potential function that guarantee the existence of stationary solutions of the q-MaxEnt form. We derive the flows of energy, information and entropy, which differ from those observed in systems where these forces are absent [29]. We derive an H theorem for the NNs' NLFPE, in terms of the behavior of a free-energy-like quantity that characterizes the irreversible nature of the time-dependent solutions. NLFPEs, such as the one investigated here, are specially valuable as they provide a tenable dynamical mechanism that can generate the q-MaxEnt distributions observed in physical and artificial complex systems, particularly in biology and in brain networks, that may exhibit spatial disorder and long-range interactions [19]. Our approach generalizes, within the framework of the S_q formalism and the corresponding NLFPE, work recently presented in [4,7,30], which is based on the standard, linear Fokker-Planck treatment. In particular, we extend previous works, in which we considered curl forces within more restricted forms of the NNs' phase-space flow than the one addressed here [11,12,23], or in which the energy, entropy, and information flows were analyzed in NLFPEs describing NN models without curl forces [29].

2 S_q-Entropy Based Statistical Mechanics

The S_q-based, statistical mechanical, theoretical formalism is being used currently to study a vast range of complex systems in many different fields and has attracted increasingly intense attention in recent years [19]. The S_q power-law, nonadditive entropy [19] is given by

$$S_q[\mathcal{P}] = (q-1)^{-1} \int \left[1 - \left(\frac{\mathcal{P}(\boldsymbol{u})}{\mathcal{P}_c}\right)^{q-1}\right] \mathcal{P}(\boldsymbol{u})\, d\boldsymbol{u}\,, \tag{1}$$

where $q \in \mathbb{R}$ is a dimensionless parameter, and $\mathcal{P}_c$ is a constant with the same dimension as the normalized probability density $\mathcal{P}(\boldsymbol{u})$. This density is defined on an N-dimensional configuration space, whose points are represented by the vectors $\boldsymbol{u} \in \mathbb{R}^N$ and where $d\boldsymbol{u}$ is the volume element. The entropy S_q coincides with the BG one, $S_{BG} = -\int \mathcal{P} \ln(\mathcal{P}/\mathcal{P}_c) d\boldsymbol{u}$, and $\mathcal{P}$ is expressed by an exponential function when $q \to 1$. Essential to the S_q-statistical theory is the q-exponential function, defined as $\exp_q(x) = [1 + (1-q)x]_+^{\frac{1}{(1-q)}}$, where "+" indicates that the function is equal to zero when the quantity inside the brackets is negative. Importantly, $\exp_q(x)$ describes the shape of the densities that are derived from the constrained optimization of S_q, which are called q-MaxEnt densities [15,19].

The S_q-statistics, and the associated q-MaxEnt distributions, have been remarkably useful for the theoretical interpretation of data observed in experimental studies and in numerical results obtained with computer simulations of various types of complex systems, in a vast variety of subject areas [17,19,22,26]. Systems that seem to be adequate candidates for a description based on S_q-thermostatistics, have features characterized by classes that include [19]: (1) systems with long-range interactions that exhibit quasi-equilibrium (or meta-equilibrium); (2) systems located at the edge of chaos; and (3) systems whose behavior allows a description based on Fokker-Planck equations (FPEs), with power-law diffusion terms [5,14,19]. The study of the theoretical relationships between nonlinear Fokker-Planck dynamics and S_q-statistical mechanics has attracted considerable research efforts in recent years [5,14], as they are central to some of the most studied, and better understood applications of the S_q-based statistical theory (see [19,25] and references therein).

3 Continuous NN Models for Associative Memory

Let us consider the deterministic dynamics of a continuous memory NN that can be expressed by a set of coupled, ordinary differential equations of motion as

$$du_i/dt = \mathcal{F}_i(u_1, \ldots, u_N), \quad i = 1, \ldots, N\,, \tag{2}$$

where u_i characterizes the state of the ith neuron in a network with N neurons. The u_i and functions $\mathcal{F}_i(\boldsymbol{u})$ are then components of the N-dimensional vectors $\boldsymbol{u} \in \mathbb{R}^N$ and $\boldsymbol{\mathcal{F}}(\boldsymbol{u}) \in \mathbb{R}^N$. Vector $\boldsymbol{u}$ represents the global state of the NN, and $\boldsymbol{\mathcal{F}}(\boldsymbol{u})$ describes the phase-space flux field according to which the global state evolves. The equations of motion can thus be cast as $d\boldsymbol{u}/dt = \boldsymbol{\mathcal{F}}$, and we refer to the quantities $\mathcal{F}_i$ as "forces". Important models for continuous memory NNs obey equations of motion in which the $\mathcal{F}_i$ can be derived from a potential energy function $\Omega(u_1, \ldots, u_N)$. The deterministic dynamics associated with Ω is such that the state of the network always moves "downhill" in the landscape determined by Ω, i.e., $\boldsymbol{\mathcal{F}}$ has a positive projection in the direction of $-\boldsymbol{\nabla}\Omega$, where $\boldsymbol{\nabla} = \left(\frac{\partial}{\partial u_1}, \ldots, \frac{\partial}{\partial u_N}\right)$. Note that $\boldsymbol{\mathcal{F}}$ and $-\boldsymbol{\nabla}\Omega$ do not have to be parallel. In fact, a more general form for the network dynamics associated with a potential function is given by

$$\frac{d\boldsymbol{u}}{dt} = \boldsymbol{\mathcal{F}} = -\boldsymbol{\mathcal{A}}\boldsymbol{\nabla}\Omega, \tag{3}$$

where $\boldsymbol{\mathcal{A}}(\boldsymbol{u})$ is, for all $\boldsymbol{u}$, an invertible, positive definite matrix. An important illustration of the general network dynamical system (3) is given by the family of models for continuous memory NNs introduced by Cohen and Grossberg [3,27], which are governed by the dynamical equations

$$\frac{du_i}{dt} = a_i(u_i)\left[b_i(u_i) - \sum_{j=1}^{N} c_{ij} d_j(u_j)\right], \quad i = 1, \ldots, N, \tag{4}$$

where the c_{ij}'s are constant parameters representing synaptic weights, and the $a_i(u_i)$'s, $b_i(u_i)$'s, and $d_i(u_i)$'s are functions of u_i that comply with $a_i(u_i)d_i'(u_i) \geq 0$. Each particular choice for the $a_i(u_i)$'s, $b_i(u_i)$'s, and $d_i(u_i)$'s characterizes a particular model within the Cohen-Grossberg family. If the synaptic weights are symmetric, $c_{ij} = c_{ji}$, the model admits the *potential function* (also known as an energy function), given by [3]

$$\Omega = -\sum_{i=1}^{N} \int_0^{u_i} b_i(u_i) d_i'(u_i) du_i + \frac{1}{2} \sum_{j,k=1}^{N} c_{jk} d_j(u_j) d_k(u_k). \tag{5}$$

The dynamical Eq. (4) when expressed in terms of the gradient of Ω, can be cast in the form (3), where the elements $\mathcal{A}_{ij}$ of matrix $\boldsymbol{\mathcal{A}}(\boldsymbol{u})$ are given by

$$\mathcal{A}_{ij} = \delta_{ij} \frac{a_i(u_i)}{d_i'(u_i)}, \tag{6}$$

where δ_{ij} stands for Kronecker's delta. For particular choices of $a_i(u_i)$, $b_i(u_i)$, and $d_i(u_i)$, the equations of motion of the Cohen-Grossberg model (4) reduce to those of the well-known continuous, Hopfield NN model.

In general, for a potential function to exist, the synapses connecting neurons have to be symmetric. In contrast, asymmetric connections are associated with equations of motion in which the forces have components that do not arise from a potential function [11, 12, 23]. We refer to these components as "curl forces", in analogy with the situation in Newtonian dynamics, in which a three-dimensional force field that cannot be derived from a potential function has a non-vanishing curl, or rotor. To develop a general formalism for continuous NN dynamics that allows the analysis of effects due to possible asymmetries in synapses, one has to consider a more general form for the forces determining the network's dynamics,

$$\boldsymbol{\mathcal{F}} = \boldsymbol{\mathcal{B}} - \boldsymbol{\mathcal{A}} \nabla \Omega, \tag{7}$$

that includes the curl force field $\boldsymbol{\mathcal{B}}(\boldsymbol{u})$, not derivable from a potential function. The detailed connection between the curl field and the asymmetries in the neurons' interactions can be complex, and must be analyzed in a case-by-case basis. In general, if the force on a neuron involves a summation of components due to each of the other neurons (as occurs in the Cohen-Grossberg model) and the dynamics arises from a potential, then the potential will involve a double summation like the one appearing in (5). This kind of summation can always be written in terms of a symmetric matrix $\{c_{ij}\}$. Therefore, a system with asymmetric interactions cannot be derived from a potential and will necessarily involve forces that do not arise from a potential [12]. Some illustrative examples can be found in [11,12]. Curl forces representing asymmetric interactions have also been considered in [21], although through a formulation different from ours and in the context of linear FPEs. We shall here consider a statistical ensemble of systems governed by a dynamics driven by (7). The time-dependent, phase-space density describing the ensemble is governed by an appropriate FPE. We shall consider the general framework of power-law NLFPEs that admit power-law-like, q-exponential stationary densities. Our first task is to determine a set of relations linking the curl field $\boldsymbol{\mathcal{B}}(\boldsymbol{u})$ and the potential function Ω, that guarantee the existence of this type of stationary solutions.

4 Power-Law Nonlinear Fokker-Planck Equations for Continuous Memory NNs with Curl Forces

To briefly review the main concepts behind the Fokker-Planck treatment, it is convenient to consider first the linear case. Then we proceed to discuss the more general, nonlinear one, which is the main focus of the present contribution. An individual realization of the dynamical memory NN evolves according to equations of the form (2) (an important example of which are Eq. (4)). When studying complex dynamical systems with high dimensionality, instead of following the behavior of a single instance of the system, it is more feasible and illuminating to study the behavior of a statistical ensemble of identical copies of the system (see, for instance [4,7,12,30]). The statistical ensemble is described by a time-dependent probability density in phase space $\mathcal{P}(u_1, \cdots, u_N, t)$ governed by the Liouville partial differential equation $(\partial\mathcal{P}/\partial t) + \boldsymbol{\nabla} \cdot (\mathcal{P}\boldsymbol{\mathcal{F}}) = 0$. The effects of noise, which are often relevant in biological NNs, can be accounted for by incorporating a diffusion (Laplacian) term to the evolution equation. One then obtains the linear FPE, $\frac{\partial\mathcal{P}}{\partial t} - \mathcal{D}\nabla^2\mathcal{P} + \boldsymbol{\nabla} \cdot (\mathcal{P}\boldsymbol{\mathcal{F}}) = 0$, where $\mathcal{D}$ is a diffusion constant. In the FPE, the term that involves $\boldsymbol{\mathcal{F}}$ is referred to as the *drift* term and we then refer to $\boldsymbol{\mathcal{F}}$ as the the phase-space flux, or the *drift* field. When linear FPEs have stationary solutions, these generally have an exponential shape and are the well-known Boltzmann distributions, $\mathcal{P}_{BG}(\boldsymbol{u}) = (1/\mathcal{Q}) \exp\left[-\frac{1}{\mathcal{D}}\Omega(\boldsymbol{u})\right]$, where $\mathcal{Q} = \int \exp\left[-\Omega(\boldsymbol{u})/\mathcal{D}\right] d\boldsymbol{u}$ is a normalization constant called the partition function. It is normally assumed that the potential $\Omega(\boldsymbol{u})$ leads to a convergent $\mathcal{Q}$. The probability density $\mathcal{P}_{BG}$ optimizes the entropy S_{BG} under the constraints of normalization and the mean value of the potential, $\langle\Omega\rangle = \int \Omega(\boldsymbol{u})\,\mathcal{P}(\boldsymbol{u})\,d\boldsymbol{u}$.

In this work, we explore features of a more general type of FPE, the power-law nonlinear ones, NLFPEs. These equations have been used recently for the study of diverse aspects of the dynamics of complex systems. When NLFPEs admit stationary solutions, these exhibit a q-exponential, power-law form. These q-exponential distributions optimize an S_q entropy under the constraints of normalization and the mean value of $\Omega(\boldsymbol{u})$. In the context of biological NNs, the nonlinear Fokker-Planck dynamics provides a possible mechanism leading to the diverse power laws observed both in experimental studies and in numerical simulations. In previous studies, the nonlinear Fokker-Planck dynamics of continuous memory NNs was explored mainly for networks whose phase-flow is derived from a potential. As already explained, in important models for continuous memory NNs with symmetric synapses, the field $\boldsymbol{\mathcal{F}}$ is related to Ω through Eq. (3). In this work, we consider more general scenarios where the drift field incorporates curl force terms, that do not arise from a potential. Some scenarios of this kind were considered in [11,12,23,25] with particular cases that, in our present formulation, correspond to $\boldsymbol{\mathcal{A}}$ constant and proportional to the identity matrix. Of special relevance for our purposes is that the curl (non-gradient) flows considered in [11,12,23,25] can be used to describe the effects of asymmetric interactions. Here, we explore the effects resulting from the incorporation of curl forces, described by the more general field given by (7), with a not necessarily diagonal $\boldsymbol{\mathcal{A}}$ whose elements may depend on the u_i. As already mentioned, we want to determine a set of connections between Ω and the curl forces $\boldsymbol{\mathcal{B}}(\boldsymbol{u})$ for which the NLFPE describing the network admits q-exponential, stationary solutions. We start by considering NLFPEs of the form

$$\frac{\partial \mathcal{P}}{\partial t} = \mathcal{D}\left(\frac{2-q}{1-q}\right)\boldsymbol{\nabla}\cdot\left\{\boldsymbol{\mathcal{A}}\mathcal{P}\boldsymbol{\nabla}\left[\left(\frac{\mathcal{P}}{\mathcal{P}_0}\right)^{1-q}\right]\right\} - \boldsymbol{\nabla}\cdot\left[\mathcal{P}\boldsymbol{\mathcal{F}}\right], \tag{8}$$

with $(2-q)\mathcal{D} > 0$. Evolution equations of the form (8) are known to be the appropriate power-law NLFPEs for continuous networks that obey a dynamics derivable from a potential, in which the forces are those given by Eq. (3) [12,27, 29]. The field $\boldsymbol{\mathcal{F}}(\boldsymbol{u})$, which describes the deterministic global network's dynamics, corresponds to the drift forces in the Fokker-Planck context. It may seem that Eq. (8) has a singularity for $q = 1$, but it can be verified that this is not the case, because the factor $(1-q)^{-1}$ cancels out when the gradient operator acts on $\mathcal{P}^{1-q}$. The FPE (8) is a continuity equation that can be cast as

$$\frac{\partial \mathcal{P}}{\partial t} + \boldsymbol{\nabla}\cdot\boldsymbol{\mathcal{J}} = 0, \tag{9}$$

where the effective probability density current $\boldsymbol{\mathcal{J}}$ is

$$\boldsymbol{\mathcal{J}} = \mathcal{P}\left\{\boldsymbol{\mathcal{F}} - \mathcal{D}\left(\frac{2-q}{1-q}\right)\boldsymbol{\mathcal{A}}\boldsymbol{\nabla}\left[\left(\frac{\mathcal{P}}{\mathcal{P}_0}\right)^{1-q}\right]\right\}. \tag{10}$$

It can be verified that, if the forces $\boldsymbol{\mathcal{F}}(\boldsymbol{u})$ arise from an Ω, according to (3), the FPE (8) admits q-exponential, stationary solutions of the form

$$\mathcal{P}_{st}(\boldsymbol{u}) = C\left[1-(1-q)\beta\Omega(\boldsymbol{u})\right]_+^{1/(1-q)}, \tag{11}$$

where the parameters C and β are related through

$$(2-q)\mathcal{D}\beta C^{(1-q)} = 1. \tag{12}$$

As an example of the above Fokker-Planck treatment, we can mention the FPE associated with the Cohen-Grossberg models. The power-law, evolution NLFPE specifically associated with the Cohen-Grossberg NN models is [27]

$$\frac{\partial \mathcal{P}}{\partial t} = D\left(\frac{q-2}{q-1}\right)\sum_i \frac{\partial}{\partial u_i}\left\{\alpha_i(u_i)\,\mathcal{P}(\boldsymbol{u},t)\,\frac{\partial}{\partial u_i}\left[\left(\frac{\mathcal{P}}{\mathcal{P}_c}\right)^{1-q}\right]\right\}$$

$$+ \sum_i \frac{\partial}{\partial u_i} \left[\alpha_i(u_i) \, \mathcal{P}(\boldsymbol{u}, t) \, \frac{\partial \Omega}{\partial u_i} \right] . \tag{13}$$

The above equation has precisely the form (8), with the drift field $\boldsymbol{\mathcal{F}}$ appearing in (3), the potential Ω equal to (5), and the matrix elements of $\boldsymbol{\mathcal{A}}$ given by (6).

We now assume that the FPE (8) admits a normalizable, stationary solution $\mathcal{P}_{st}(\boldsymbol{u})$ of the form given by equations (11)-(12), but with the more general drift field (7), that includes the curl force component $\boldsymbol{\mathcal{B}}(\boldsymbol{u})$. We shall show that this assumption leads to a structural relationship linking the matrix $\boldsymbol{\mathcal{A}}(\boldsymbol{u})$ with the drift field $\boldsymbol{\mathcal{F}}(\boldsymbol{u})$, which, in turn, implies some connections between $\boldsymbol{\mathcal{A}}(\boldsymbol{u})$, $\boldsymbol{\mathcal{B}}(\boldsymbol{u})$, and Ω. When the q-exponential, stationary solution exists, we have

$$\begin{aligned} &\boldsymbol{\nabla} \cdot \left\{ (2-q) \, \boldsymbol{\mathcal{A}} \mathcal{P}_{st} \left[\mathcal{D} C^{(1-q)} \beta \, (\boldsymbol{\nabla} \Omega) \, + \, (2-q)^{-1} \, \boldsymbol{\mathcal{A}}^{-1} \boldsymbol{\mathcal{F}} \right] \right\} = \\ &\boldsymbol{\nabla} \cdot \left\{ \boldsymbol{\mathcal{A}} \mathcal{P}_{st} \left[(\boldsymbol{\nabla} \Omega) \, + \, \boldsymbol{\mathcal{A}}^{-1} \boldsymbol{\mathcal{F}} \right] \right\} = 0 \, , \end{aligned} \tag{14}$$

where we have used (11)-(12) in (8), with $\partial \mathcal{P} / \partial t = 0$. We shall use the last equality in (14) to establish a relation connecting the matrix $\boldsymbol{\mathcal{A}}$ with the curl field $\boldsymbol{\mathcal{B}}$ and Ω. Substituting $\boldsymbol{\mathcal{F}}(\boldsymbol{u})$ in the second line of (14) by the expression in the right-hand side of (7), one gets $\boldsymbol{\nabla} \cdot (\mathcal{P}_{st} \, \boldsymbol{\mathcal{B}}) = 0$, leading to the fundamental relations linking $\boldsymbol{\mathcal{A}}$ and $\boldsymbol{\mathcal{B}}$ with Ω. It can be shown, after some calculations, that the condition $\boldsymbol{\nabla} \cdot (\mathcal{P}_{st} \, \boldsymbol{\mathcal{B}}) = 0$ is satisfied, if the field $\boldsymbol{\mathcal{B}}$ and Ω satisfy

$$\boldsymbol{\nabla} \cdot \boldsymbol{\mathcal{B}} = 0; \qquad \boldsymbol{\mathcal{B}} \cdot (\boldsymbol{\nabla} \Omega) = 0 \, . \tag{15}$$

Equations (15) summarize a set of conditions relating the drift field and the potential, which guarantee that the NLFPE for the network, including curl forces, admits stationary solutions of the q-exponential, power-law-like form. It is remarkable that relations (15) are q-invariant: they hold for any q-value, in spite of the fact that the nonlinear structure of the NLFPEs depends drastically on q. According to the relations (15), the field $\boldsymbol{\mathcal{B}}$ is divergenceless and orthogonal to the gradient of the potential. That is, the lines of force of $\boldsymbol{\mathcal{B}}$ are lines of constant Ω. Our present considerations prove that conditions (15) are valid in scenarios more general than those previously analyzed in [11,12,23,25] for particular systems where $\boldsymbol{\mathcal{A}}$ is proportional to the identity matrix.

5 Energy and Information Flow: Effects of the Gradient and the Curl Forces

We now investigate, for the FPE (8), the time evolution of the average potential energy $\langle \Omega \rangle$, and of an entropic-information measure $S_{\tilde{q}}$, where, based on previous experience with the Fokker-Planck treatment of network dynamics, we adopt $\tilde{q} = 2 - q$, which leads to the relevant (entropic) information measure. We consider a NN with both gradient and curl forces. We are interested in both the global changes of $\langle \Omega \rangle$ and $S_{\tilde{q}}$, as well as their local changes, and the concomitant density flows and creation rates. The continuity equation (9) leads to

$$\frac{d \langle \Omega \rangle}{dt} = - \int \Omega \Big(\boldsymbol{\nabla} \cdot \boldsymbol{\mathcal{J}} \Big) d\boldsymbol{u} \quad \text{and} \, ,$$

$$\frac{dS_{\tilde{q}}}{dt} = \frac{1}{q-1} \int \left[1 + (q-2) \left(\frac{\mathcal{P}}{\mathcal{P}_c} \right)^{1-q} \right] \left(\nabla \cdot \mathcal{J} \right) du. \tag{16}$$

Defining the quantities

$$\omega = \Omega \mathcal{P} \, , \tag{17}$$

$$\mathcal{J}_\omega = \Omega \mathcal{J} \, , \tag{18}$$

$$R_\omega = (\nabla \Omega) \cdot (\mathcal{J} - \mathcal{P}\mathcal{B}) \, , \tag{19}$$

$$\sigma = \frac{\mathcal{P}}{1-q} \left[1 - \left(\frac{\mathcal{P}}{\mathcal{P}_c} \right)^{1-q} \right] , \tag{20}$$

$$\mathcal{J}_\sigma = \frac{1}{q-1} \left[1 + (q-2) \left(\frac{\mathcal{P}}{\mathcal{P}_c} \right)^{1-q} \right] \mathcal{J} \, , \tag{21}$$

$$R_\sigma = \left\{ \nabla \left[\frac{q-2}{1-q} \left(\frac{\mathcal{P}}{\mathcal{P}_c} \right)^{1-q} \right] \right\} \cdot (\mathcal{J} - \mathcal{P}\mathcal{B}) \, , \tag{22}$$

and, by recourse to (9), one can derive the evolution equations

$$\frac{\partial \omega}{\partial t} + \nabla \cdot \mathcal{J}_\omega = R_\omega ; \qquad \frac{\partial \sigma}{\partial t} + \nabla \cdot \mathcal{J}_\sigma = R_\sigma . \tag{23}$$

These are continuity equations with source terms. The quantities ω, $\mathcal{J}_\omega$, and R_ω can be regarded as the phase-space effective density, density current, and local "production" rate of energy, respectively. On the other hand, σ, $\mathcal{J}_\sigma$, and R_σ can be interpreted as the phase-space effective density, density current, and local "production" rate of entropy. If the entropy $S_{\tilde{q}}$ is interpreted as an information measure, the current $\mathcal{J}_\sigma$ describes the flow of information between different regions of phase space. If the entropy creation rate R_σ vanishes everywhere, the total amount of information, as measured by $S_{\tilde{q}}$, remains constant. The creation or destruction of information is determined by the production rate R_σ, which describes local sources or sinks of information in different regions of phase-space. A remarkable fact that transpires from equations (18)-(19) and (21)-(22) is that, while the curl forces $\mathcal{B}$ have an effect on the density currents of energy and information, those forces do not affect the corresponding production rates. Indeed, in (18) and (21) $\mathcal{J}$ depends on $\mathcal{B}$ (see (7) and (10)), and this dependence on $\mathcal{B}$ is canceled out in the right-hand sides of (19) and (22).

Equations (9) and (23) determine the co-evolution of the phase-space densities of probability ($\mathcal{P}$), energy (ω), and entropy-information (σ). The concepts behind equations (23) are inspired on, and constitute a generalization of, a recent work by Fagerholm et al., who explored the co-evolution of the probability and information densities in a system described by a linear FPE, with constant diffusion and uniform constant drift, in [4]. Here we applied the notions advanced in [4] to more general scenarios, corresponding to NLFPEs associated with models for continuous memory NNs, such as the Cohen-Grossberg ones. In our treatment we considered the effects of both gradient and curl forces. We shall now use our analysis of information flow to derive an H-theorem for the NLFPE (8). Let us consider the free-energy-like quantity $\mathcal{H} = \langle \Omega \rangle - \mathcal{D} S_{\tilde{q}}$. According to equations (23), the local rate of change of $\mathcal{H}$ is given by the quantity

$$R_\omega - \mathcal{D} R_\sigma = -(\boldsymbol{\mathcal{J}} - \mathcal{P}\boldsymbol{\mathcal{B}})^\top \boldsymbol{\mathcal{A}} (\boldsymbol{\mathcal{J}} - \mathcal{P}\boldsymbol{\mathcal{B}}) \leq 0, \qquad (24)$$

where $\top$ denotes the transpose. The above inequality holds because the matrix $\boldsymbol{\mathcal{A}}$ is positive definite. Equality in (24) holds for the stationary density. Integrating the quantity $R_\omega - \mathcal{D} R_\sigma$ over all phase-space, and taking into consideration the inequality in (24), one obtains the H-theorem $d\mathcal{H}/dt = \int (R_\omega - \mathcal{D} R_\sigma) d\boldsymbol{u} \leq 0$, where equality holds for the stationary density (11). This implies that the stationary solution (11), which is a q-exponential, is the density that minimizes $\mathcal{H}$ under the normalization constraint. In other words, the probability density that optimizes entropy $S_{\tilde{q}}$, under the constraints imposed by normalization and the mean potential energy $\Omega(\boldsymbol{u})$, is the stationary one.

6 Conclusions

We have explored a power-law, nonlinear Fokker-Planck description of continuous models for memory NNs, such as the Cohen-Grossberg ones. We considered NN models whose phase-space flows have gradient and curl contributions (here referred to as "forces") and are of a more general form than the ones we studied in previous works. Gradient forces, arising from a potential, are typical of NNs with symmetric synaptic connections, while curl forces may describe the presence of asymmetric connections. We obtained relevant forms of the NLFPEs describing the models, that admit power-like, q-MaxEnt stationary solutions. We proved that this occurs when the potential and the curl forces satisfy an appropriate set of relations. We analyzed, emphasizing the role of the curl forces, the flows of energy and information corresponding to time-dependent solutions of the above mentioned NLFPEs, and obtained an H-theorem. The present developments contribute to elucidate possible dynamical mechanisms whereby power-law densities optimizing the S_q entropy may emerge. These probability densities are often observed in the study of natural and artificial complex systems. They have been observed in the context of neuroscience, in connection with both numerical simulations of NNs, and experimental studies of brain activity. Our present developments provide new theoretical tools that may be useful when applying the S_q-based thermostatistical formalism to neuroscience [1] and AI [9,10,16].

Acknowledgments. We acknowledge financial support from the Brazilian National Research Council (CNPq), the Brazilian agency that funds graduate studies (CAPES) and from the Rio de Janeiro State Research Foundation (FAPERJ).

References

1. Abramov, D.M.: Complexity and psychopathology: from mechanistic science to a physical understanding of the mental conditions based on chaos and complexity. Front. Psychiatry **17**, 1764422 (2026). https://doi.org/10.3389/fpsyt.2026.1764422

2. de Carvalho, L.A.V., Mendes, D.Q., Wedemann, R.S.: Creativity and delusions: the dopaminergic modulation of cortical maps. In: Sloot, P.M.A., Abramson, D., Bogdanov, A.V., Dongarra, J.J., Zomaya, A.Y., Gorbachev, Y.E. (eds.) ICCS 2003. LNCS, vol. 2657, pp. 511–520. Springer, Berlin, Heidelberg (2003). https://doi.org/10.1007/3-540-44860-8_53
3. Cohen, M.A., Grossberg, S.: Absolute stability of global pattern formation and parallel memory storage by competitive neural networks. IEEE Trans. Syst. Man Cybern. **13**, 815–826 (1983). https://doi.org/10.1109/TSMC.1983.6313075
4. Fagerholm, E.D., Scott, G., Leech, R., Turkheimer, F.E., Friston, K.J., Brázdil, M.: Mapping conservative Fokker-Planck entropy in neural systems. J. Phys. D Appl. Phys. **58**(14), 145401 (2025). https://doi.org/10.1088/1361-6463/adb318
5. Frank, T.D.: Nonlinear Fokker-Planck Equations: Fundamentals and Applications. Springer, Berlin, Heidelberg (2005). https://doi.org/10.1007/b137680
6. Hopfield, J.J.: Neurons with graded responses have collective computational properties like those of two-state neurons. Proc. Natl. Acad. Sci. **81**, 3088–3092 (1984). https://doi.org/10.1073/pnas.81.10.3088
7. Jirsa, V., Sheheitli, H.: Entropy, free energy, symmetry and dynamics in the brain. J. Phys. Complexity **3**(1), 015007 (2022). https://doi.org/10.1088/2632-072X/ac4bec
8. Kandel, E.: Psychiatry, Psychoanalysis, and the New Biology of Mind. American Psychiatric Association Publishing, Inc., Washington D.C. (2005). https://doi.org/10.1176/appi.books.9781615377527
9. Krotov, D.: A new frontier for Hopfield networks. Nat. Rev. Phys. **5**(7), 366–367 (2023). https://doi.org/10.1038/s42254-023-00595-y
10. Krotov, D., Hopfield, J.J.: Large associative memory problem in neurobiology and machine learning. In: International Conference on Learning Representations (ICLR 2021) (2021). https://openreview.net/forum?id=X4y_10OX-hX
11. de Luca, V.T.F., Wedemann, R.S., Plastino, A.R.: Neuronal asymmetries and Fokker-Planck dynamics. In: Kůrková, V., Manolopoulos, Y., Hammer, B., Iliadis, L., Maglogiannis, I. (eds.) ICANN 2018. LNCS, vol. 11141, pp. 703–713. Springer, Cham, Switzerland (2018). https://doi.org/10.1007/978-3-030-01424-7_69
12. de Luca, V.T.F., Wedemann, R.S., Plastino, A.R.: The nonlinear Fokker-Planck equation with nongradient drift forces and an anisotropic potential. Chaos **35**(9), 093115 (2025). https://doi.org/10.1063/5.0280921
13. Papa, A.R.R., da Silva, L.: Earthquakes in the brain. Theory Biosci. **116**, 321–327 (1997)
14. Plastino, A.R., Plastino, A.: Non-extensive statistical mechanics and generalized Fokker-Planck equation. Phys. A **222**(1), 347–354 (1995). https://doi.org/10.1016/0378-4371(95)00211-1
15. Plastino, A.R., Tsallis, C., Wedemann, R.S., Haubold, H.J.: Entropy optimization, generalized logarithms, and duality relations. Entropy **24**(12), 1723.1–13 (2022)
16. Ramsauer, H., et al.: Hopfield networks is all you need. In: International Conference on Learning Representations (ICLR 2021) (2021). https://openreview.net/forum?id=tL89RnzIiCd
17. Siddiqui, M., Wedemann, R.S., Jensen, H.J.: Avalanches and generalized memory associativity in a network model for conscious and unconscious mental functioning. Phys. A **490**, 127–138 (2018). https://doi.org/10.1016/j.physa.2017.08.011
18. Taylor, J.G.: A neural model of the loss of self in schizophrenia. Schizophrenia Bull. **37**(6), 1229–1247 (2011). https://doi.org/10.1093/schbul/sbq033

19. Tsallis, C.: Introduction to Nonextensive Statistical Mechanics, Approaching a Complex World. Springer, Cham, Switzerland, 2nd. edn. (2023). https://doi.org/10.1007/978-3-030-79569-6
20. Tsallis, C., Stariolo, D.A.: Generalized simulated annealing. Phys. A **233**, 395–406 (1996). https://doi.org/10.1016/S0378-4371(96)00271-3
21. Wang, J., Xu, L., Wang, E.: Potential landscape and flux framework of nonequilibrium networks: robustness, dissipation, and coherence of biochemical oscillations. Proc. Natl. Acad. Sci. **105**(34), 12271–12276 (2008). https://doi.org/10.1073/pnas.0800579105
22. Wedemann, R.S., Donangelo, R., de Carvalho, L.A.V.: Generalized memory associativity in a network model for the neuroses. Chaos **19**(1), 015116–(1–11) (2009). https://doi.org/10.1063/1.3099608
23. Wedemann, R.S., Plastino, A.R.: Asymmetries in synaptic connections and the nonlinear Fokker-Planck formalism. In: Villa, A.E.P., Masulli, P., Pons Rivero, A.J. (eds.) ICANN 2016. LNCS, vol. 9886, pp. 19–27. Springer, Cham, Switzerland (2016). https://doi.org/10.1007/978-3-319-44778-0_3
24. Wedemann, R.S., Plastino, A.R.: q-Maximum entropy distributions and memory neural networks. In: Lintas, A., Rovetta, S., Verschure, P.F.M.J., Villa, A.E.P. (eds.) ICANN 2017. LNCS, vol. 10613, pp. 300–308. Springer, Cham, Switzerland (2017). https://doi.org/10.1007/978-3-319-68600-4_35
25. Wedemann, R.S., Plastino, A.R., Tsallis, C.: Curl forces and the nonlinear Fokker-Planck equation. Phys. Rev. E **94**(6), 062105 (2016). https://doi.org/10.1103/PhysRevE.94.062105
26. Wedemann, R.S., de Carvalho, L.A.V., Donangelo, R.: Access to symbolization and associativity mechanisms in a model of conscious and unconscious processes. In: Samsonovich, A.V., Jóhannsdóttir, K.R. (eds.) Biologically Inspired Cognitive Architectures 2011, Frontiers in Artificial Intelligence and Applications, vol. 233, pp. 444–449. IOS Press, Amsterdam, Netherlands (2011). https://doi.org/10.3233/978-1-60750-959-2-444
27. Wedemann, R.S., Plastino, A.R.: Nonlinear Fokker-Planck approach to the cohen-grossberg model. In: Lintas, A., Enrico, P., Pan, X., Wang, R., Villa, A. (eds.) Advances in Cognitive Neurodynamics (VII), pp. 61–72. Springer, Singapore (2021). https://doi.org/10.1007/978-981-16-0317-4_7
28. Wedemann, R.S., Plastino, A.R.: Associative memory models for mental processes: connections with q-statistical mechanics. Brainiacs J. Brain Imaging Comput. Sci. **6**(1), Edoc X41D74D73 (2025). https://doi.org/10.48085/X41D74D73
29. Wedemann, R.S., Plastino, A.R.: A Fokker-Planck perspective on the flow of information in continuous memory neural networks. In: Senn, W., et al. (eds.) ICANN 2025. LNCS, vol. 16068, pp. 658–669. Springer, Cham, Switzerland (2026). https://doi.org/10.1007/978-3-032-04558-4_53
30. Yan, H., Zhao, L., Hu, L., Wang, X., Wang, E., Wang, J.: Nonequilibrium landscape theory of neural networks. Proc. Natl. Acad. Sci. **110**(45), E4185–E4194 (2013). https://doi.org/10.1073/pnas.1310692110

Group Resonance Network: Learnable Prototypes and Multi-subject Resonance for EEG Emotion Recognition

Renwei Meng(✉)

Anhui University, Hefei, China
R32314095@stu.ahu.edu.cn

Abstract. Electroencephalography (EEG)-based emotion recognition remains challenging in cross-subject settings due to severe inter-subject variability. Existing methods mainly learn subject-invariant features, but often under-exploit stimulus-locked group regularities shared across subjects. To address this issue, we propose the *Group Resonance Network (GRN)*, which integrates individual EEG dynamics with offline group resonance modeling. GRN contains three components: an individual encoder for band-wise EEG features, a set of learnable group prototypes for prototype-induced resonance, and a multi-subject resonance branch that encodes PLV/coherence-based synchrony with a small reference set. A resonance-aware fusion module combines individual and group-level representations for final classification. Experiments on SEED and DEAP under both subject-dependent and leave-one-subject-out protocols show that GRN consistently outperforms competitive baselines, while additional analyses confirm the effects of prototype learning, PLV/coherence resonance, and leakage-safe reference construction.

Keywords: EEG · Emotion Recognition · Cross-Subject Generalization · Prototype Learning · Neural Synchrony

1 Introduction

Electroencephalogram (EEG)-based emotion recognition is a key technique for affective brain-computer interfaces and human-machine interaction, and has been widely evaluated on public benchmarks such as DEAP, SEED, DREAMER, and AMIGOS [14,15,20,33]. Although deep models, including CNNs, temporal-spatial networks, and graph-based methods, have achieved strong performance, they remain vulnerable to severe inter-subject variability in cross-subject settings [6,17,29].

Existing cross-subject methods mainly address this issue through transfer learning, domain adaptation, and subject-invariant representation learning [2,5,10,18,25,26,34,35]. However, most of them treat each subject as an isolated domain and under-exploit the *shared, stimulus-locked* neural patterns that may

L. Pasa et al. (Eds.): ICANN 2026, LNCS 17089, pp. 115–126, 2027.
https://doi.org/10.1007/978-3-032-38398-3_10

emerge when different subjects experience the same affective stimulus. Neuroscience studies on inter-subject correlation and hyperscanning suggest that brain responses can exhibit measurable synchrony across individuals, often quantified by phase locking value (PLV) and coherence [1,3,8,11,12,16,21–24]. These findings indicate that group-level reference structure may provide useful cues for robust affect decoding.

To this end, we propose the *Group Resonance Network (GRN)* for EEG emotion recognition. GRN models *offline* group resonance by computing PLV/coherence-based synchrony between a target subject and a small reference set, while simultaneously learning a set of *learnable prototypes* that summarize group-consistent affective structure [4,28,31]. Our contributions are threefold: (i) we introduce stimulus-locked multi-subject resonance tensors for EEG emotion recognition; (ii) we combine learnable prototypes with resonance-aware fusion for group-level affect modeling; and (iii) we provide extensive analyses on component effects, reference alignment, statistical significance, and computational cost.

2 Related Work

2.1 EEG Emotion Recognition and Cross-Subject Learning

EEG emotion recognition has been widely studied on public benchmarks such as DEAP, SEED, DREAMER, and AMIGOS [14,15,20,33]. Representative models include compact CNNs, temporal-spatial CNNs, graph neural networks, and recent Transformer-based variants [6,17,19,27,29,30]. To address cross-subject variability, prior work mainly adopts transfer learning, domain adaptation, adversarial alignment, or contrastive subject-invariant learning [2,5,10,13,18,25,26,34,35]. These methods reduce domain discrepancy, but rarely exploit group-level stimulus-locked structure explicitly.

2.2 Inter-subject Synchrony and Prototype Learning

Common stimuli can induce synchronized neural responses across subjects, as reflected by inter-subject correlation and related shared-response analyses [7,8,12,22,23]. Hyperscanning research further measures inter-brain coupling using PLV and coherence [1,9,11,21], grounded in electrophysiological connectivity analysis [3,16,24]. Separately, prototype-based and metric learning methods represent latent structure with representative vectors [4,28,32]; in EEG, prototype learning has also been linked to domain adaptation on SPD representations [31]. GRN connects these directions by integrating learnable prototypes with synchrony-derived offline resonance.

3 Methodology

3.1 Overview and Definition of Resonance

Given an EEG segment $X \in \mathbb{R}^{C\times T}$ and label y, GRN predicts the emotional state by combining an individual embedding, a prototype-induced group embedding, and a synchrony-based multi-subject embedding. In this paper, *resonance* denotes stimulus-locked cross-subject alignment: two EEG responses are considered resonant if their phase or spectral coupling is high after being aligned by the onset and duration of the same affective stimulus. Thus, resonance is not a causal interaction as in online hyperscanning; it is an offline similarity/synchrony measure computed under shared stimulus timing.

Figure 1 summarizes the architecture. The method contains four stages: individual encoding, learnable group prototype attention, multi-subject resonance tensor construction, and resonance-aware fusion.

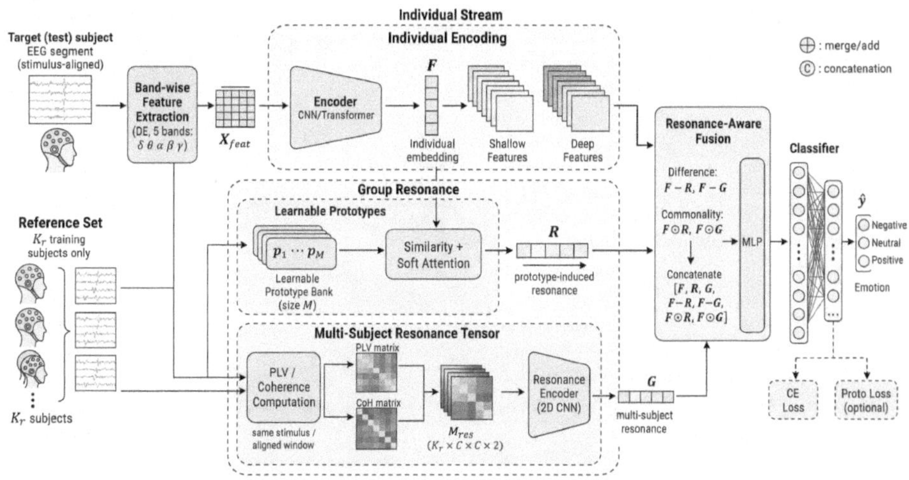

Fig. 1. Overall GRN pipeline. The model integrates an individual representation F, a prototype-induced group representation R, and a multi-subject resonance representation G for emotion classification.

3.2 Individual Encoding

We first extract band-wise features over five frequency bands $(\delta, \theta, \alpha, \beta, \gamma)$, denoted as X_{feat}. An encoder Enc(·) maps the feature tensor to a compact individual embedding:

$$F = \text{Enc}(X_{feat}) \in \mathbb{R}^d. \tag{1}$$

The encoder is implemented as a lightweight convolutional-temporal backbone, but the proposed resonance modules are compatible with other EEG encoders such as graph networks or CNN-Transformer hybrids.

3.3 Learnable Group Prototypes

Instead of selecting a fixed prototype subject, GRN maintains M trainable group prototypes:

$$\mathcal{P} = \{p_m\}_{m=1}^{M}, \quad p_m \in \mathbb{R}^d. \tag{2}$$

For each sample, cosine similarity is computed between F and each prototype:

$$s_m = \frac{F^\top p_m}{\|F\|_2 \|p_m\|_2}. \tag{3}$$

The attention weight over prototypes is obtained by temperature-scaled softmax:

$$\alpha_m = \frac{\exp(s_m/\tau)}{\sum_{j=1}^{M} \exp(s_j/\tau)}. \tag{4}$$

The prototype-induced resonance embedding is then

$$R = \sum_{m=1}^{M} \alpha_m p_m. \tag{5}$$

This design turns prototypes into trainable group anchors, allowing the model to learn latent affective patterns that are shared across subjects rather than depending on a single hand-selected reference.

3.4 Multi-subject Resonance Tensor

For each training fold, a reference set $\mathcal{S} = \{X^{(k)}\}_{k=1}^{K_r}$ is sampled strictly from training subjects. In LOSO, the held-out test subject is never used as a reference. Moreover, resonance is computed only between samples aligned to the same stimulus timeline.

For a target sample and reference $X^{(k)}$, PLV captures phase synchrony:

$$\mathrm{PLV}_{ij}^{(k)} = \left| \frac{1}{N} \sum_{n=1}^{N} e^{\mathrm{i}(\phi_i(n) - \phi_j^{(k)}(n))} \right|, \tag{6}$$

where $\phi_i(n)$ and $\phi_j^{(k)}(n)$ are analytic phases estimated by the Hilbert transform. Coherence measures frequency-domain coupling:

$$\mathrm{CoH}_{ij}^{(k)}(f) = \frac{|P_{ij}^{(k)}(f)|^2}{P_{ii}(f) P_{jj}^{(k)}(f)}, \tag{7}$$

where $P_{ij}^{(k)}(f)$ is the cross-spectral density and $P_{ii}(f)$, $P_{jj}^{(k)}(f)$ are power spectral densities. PLV and band-averaged CoH are stacked into

$$\mathbf{M} \in \mathbb{R}^{K_r \times C \times C \times 2}. \tag{8}$$

A shallow resonance encoder compresses $\mathbf{M}$ into

$$G = \mathrm{ResEnc}(\mathbf{M}) \in \mathbb{R}^d. \tag{9}$$

Fig. 2 illustrates the construction process.

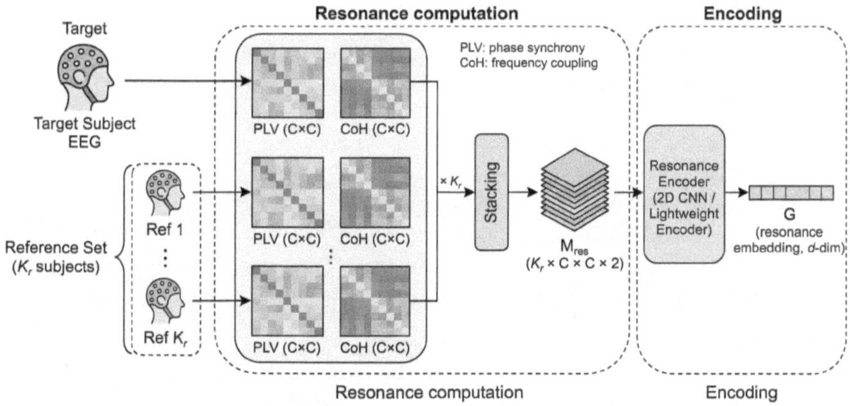

Fig. 2. Multi-subject resonance tensor construction. For each training-only reference subject, PLV and CoH matrices are computed under stimulus alignment and stacked into **M**.

3.5 Resonance-Aware Fusion and Optimization

GRN fuses F, R, and G by explicitly modeling both difference and commonality:

$$D_R = F - R, \quad D_G = F - G, \quad C_R = F \odot R, \quad C_G = F \odot G. \tag{10}$$

The final representation is

$$H = \mathrm{MLP}([F \| R \| G \| D_R \| D_G \| C_R \| C_G]), \tag{11}$$

followed by a softmax classifier. The training loss is

$$\mathcal{L} = \mathcal{L}_{cls} + \lambda \mathcal{L}_{proto}, \quad \mathcal{L}_{proto} = \sum_{m=1}^{M} \alpha_m \|F - p_m\|_2^2. \tag{12}$$

The prototype term encourages embeddings to remain close to the prototypes receiving high attention, stabilizing the learned group anchors. The additional resonance computation scales as $O(K_r C^2 N)$ per sample before encoding; since K_r is small and fixed, the overhead is bounded and can be precomputed for offline benchmarks.

4 Experiments

4.1 Datasets, Protocols, and Implementation

We evaluate GRN on SEED and DEAP. SEED is a three-class benchmark with Negative, Neutral, and Positive labels; DEAP is evaluated as binary Valence and Arousal classification by thresholding ratings at 5.0. We report subject-dependent (SD) results and subject-independent (SI) results using Leave-One-Subject-Out (LOSO). For every LOSO fold, references are sampled only from the training subjects. Default hyperparameters are $d = 256$, $M = 8$, $K_r = 3$, batch size 64, Adam optimizer with learning rate 10^{-4} and weight decay 10^{-4}, and early stopping with patience 10.

4.2 Main Results

Tables 1 and 2 report the main accuracy results. GRN achieves the best performance on both datasets without changing the standard evaluation protocols.

Table 1. Accuracy (%) on SEED under SD and SI protocols.

Method	SD	SI (LOSO)
DGCNN	90.51 ± 3.46	79.45 ± 10.11
ST-DADGAT	94.87 ± 2.51	83.59 ± 6.82
FCAnet	96.93 ± 2.28	84.16 ± 5.93
LATN	–	86.15 ± 6.74
DVIE-Net	97.01 ± 1.93	86.34 ± 4.18
GRN	$\mathbf{97.42 \pm 1.61}$	$\mathbf{87.90 \pm 3.85}$

Table 2. Accuracy (%) on DEAP under SI protocol.

Method	Valence	Arousal
DGCNN	83.27 ± 4.30	82.36 ± 4.30
ST-DADGAT	86.61 ± 3.10	85.12 ± 5.80
FCAnet	88.92 ± 3.90	86.41 ± 3.40
DVIE-Net	89.27 ± 3.28	88.51 ± 3.76
GRN	$\mathbf{90.35 \pm 2.95}$	$\mathbf{89.40 \pm 3.20}$

Near-peer methods such as LATN and DVIE-Net also reduce inter-subject variability through transfer or prototype-style alignment, which explains their competitive performance. GRN differs by combining learned group anchors with explicit PLV/CoH synchrony, leading to consistent but moderate gains. A paired t-test across LOSO folds shows that GRN improves over DVIE-Net on SEED SI ($p = 0.018$) and over the strongest DEAP baseline on Valence and Arousal ($p < 0.05$). Table 3 further reports macro-F1 and AUC.

4.3 Ablation and Component Analysis

Table 4 evaluates each component on SEED SI. Both group prototypes and multi-subject resonance improve the individual encoder, and removing the difference/commonality terms degrades performance. The PLV-only and CoH-only variants show that the two synchrony measures are complementary.

Table 3. Additional SI metrics for GRN. AUC is macro one-vs-rest on SEED and binary on DEAP.

Task	Acc. (%)	Macro-F1 (%)	AUC
SEED	87.90	87.56	0.941
DEAP-Valence	90.35	90.12	0.922
DEAP-Arousal	89.40	89.08	0.911

Table 4. Ablation on SEED SI (LOSO).

Variant	SI Acc. (%)
Individual only (remove R, G)	84.60
Capacity-matched individual encoder	85.02
+ Learnable prototypes only (add R, no $\mathbf{M}$)	86.40
+ Multi-subject resonance only (add G, no prototypes)	86.85
Full GRN w/o difference/commonality fusion	86.92
Full GRN w/o prototype regularizer	87.35
Full GRN with PLV only	87.22
Full GRN with CoH only	86.98
Full GRN	**87.90**

Table 5 confirms that the component trend also holds on DEAP. Prototype attention analysis shows that learned prototypes form class-biased group anchors: the dominant prototype assignment purity reaches 0.81 on SEED, while the mean attention entropy decreases from 1.72 to 1.18 during training, indicating increasingly selective prototype usage.

Table 5. Ablation on DEAP SI.

Variant	Valence	Arousal
Individual only	87.84	86.71
+ Learnable prototypes only	89.23	88.19
+ Multi-subject resonance only	89.41	88.55
Full GRN	**90.35**	**89.40**

4.4 Stimulus Alignment, Sensitivity, and Cost

Since SEED and DEAP assign labels through stimuli, the resonance branch must be interpreted as a stimulus-locked feature rather than a general-purpose hyperscanning signal. Table 6 verifies this constraint: when references are stimulus-misaligned, the synchrony signal collapses toward the individual-only baseline.

This supports that GRN exploits temporal synchrony rather than leakage from held-out subjects.

Table 6. Reference alignment controls on SEED SI.

Reference condition	SI Acc. (%)
Aligned training references	**87.90**
Stimulus-misaligned training references	84.83
Random class-agnostic references	84.65
Label-shuffled reference pairing	84.41
Individual only	84.60

Table 7 shows that performance is stable across K_r and M. The small variation suggests that GRN does not rely on a narrow hyperparameter choice; the best setting is used for all main experiments. Table 8 reports model size and runtime and does not provide the hyperparameter setting.

Table 7. Sensitivity on SEED SI (LOSO).

K_r	1	3	5
Acc. (%)	87.10	**87.90**	87.85

M	4	8	12
Acc. (%)	87.45	**87.90**	87.88

Table 8. Model size and runtime on SEED SI. Runtime is measured per batch.

Model	Params (M)	Runtime (ms)
Individual encoder	1.42	4.1
Capacity-matched encoder	1.86	4.7
GRN	**1.89**	**5.6**

4.5 Visualization Analysis

To further inspect class-wise behavior, Fig. 3 reports the confusion matrices on SEED. GRN maintains balanced recognition under SI and becomes strongly diagonal under SD. The training curves in Fig. 4 show stable convergence without severe overfitting.

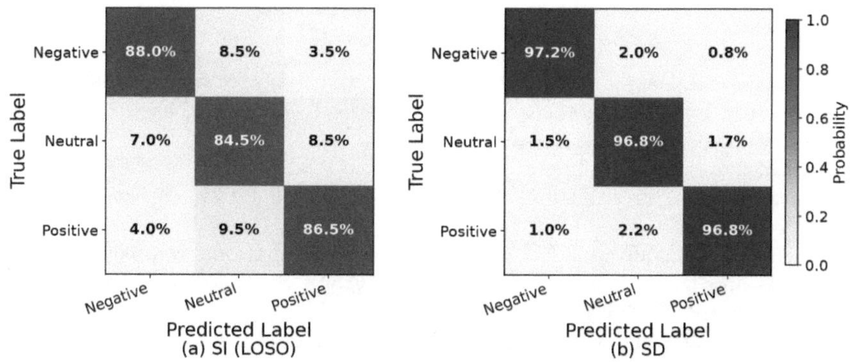

Fig. 3. Confusion matrices on SEED: (a) SI (LOSO) and (b) SD.

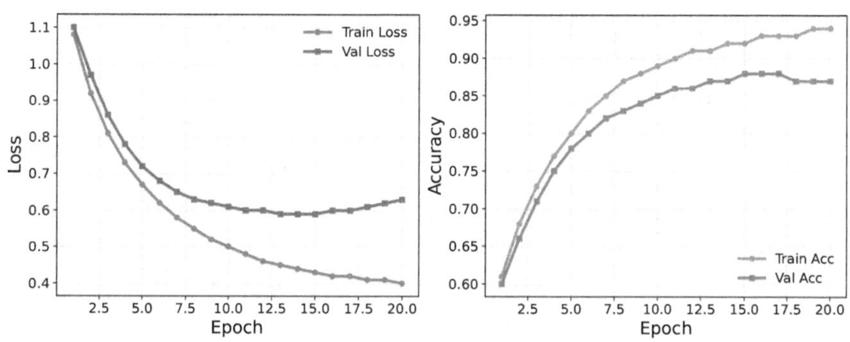

Fig. 4. Training/validation curves on SEED SI: loss and accuracy vs. epoch.

5 Conclusion

We presented GRN, a group-aware framework for EEG emotion recognition. GRN combines individual EEG encoding, learnable group prototypes, and PLV/CoH-based multi-subject resonance tensors under leakage-safe stimulus alignment. The resonance-aware fusion module jointly models subject-specific differences and group-level commonality. Experiments on SEED and DEAP show consistent improvements over competitive baselines, while additional analyses verify the contributions of prototype learning, PLV/CoH synchrony, fusion interactions, and reference alignment. The method is most suitable for offline affective benchmarks with shared stimulus timelines; extending it to finer-grained affect labels and weaker stimulus-label coupling is an important direction for future work.

References

1. Babiloni, F., Astolfi, L.: Social neuroscience and hyperscanning techniques: past, present and future. Neurosci. Biobehav. Rev. **44**, 76–93 (2014). https://doi.org/10.1016/j.neubiorev.2012.07.006
2. Bao, G., et al.: Two-level domain adaptation neural network for EEG-based emotion recognition. Front. Hum. Neurosci. **14**, 605246 (2021). https://doi.org/10.3389/fnhum.2020.605246
3. Bastos, A.M., Schoffelen, J.M.: A tutorial review of functional connectivity analysis methods and their interpretational pitfalls. Front. Syst. Neurosci. **9**, 175 (2016). https://doi.org/10.3389/fnsys.2015.00175
4. Caron, M., Misra, I., Mairal, J., Goyal, P., Bojanowski, P., Joulin, A.: Unsupervised learning of visual features by contrasting cluster assignments. In: Advances in Neural Information Processing Systems (2020)
5. Chai, X., Wang, Q., Zhao, Y., Liu, X., Bai, O., Li, Y.: Unsupervised domain adaptation techniques based on auto-encoder for non-stationary EEG-based emotion recognition. Comput. Biol. Med. **79**, 205–214 (2016). https://doi.org/10.1016/j.compbiomed.2016.10.019
6. Ding, Y., Robinson, N., Zhang, S., Zeng, Q., Guan, C.: TSception: capturing temporal dynamics and spatial asymmetry from EEG for emotion recognition. IEEE Trans. Affect. Comput. (2023). https://doi.org/10.1109/TAFFC.2022.3169001
7. Dmochowski, J.P., Bezdek, M.A., Abelson, B.P., Johnson, J.S., Schumacher, E.H., Parra, L.C.: Audience preferences are predicted by temporal reliability of neural processing. Nat. Commun. **5**, 4567 (2014). https://doi.org/10.1038/ncomms5567
8. Dmochowski, J.P., Sajda, P., Dias, J., Parra, L.C.: Correlated components of ongoing EEG point to emotionally laden attention. Front. Hum. Neurosci. **6**, 112 (2012). https://doi.org/10.3389/fnhum.2012.00112
9. Dumas, G., Nadel, J., Soussignan, R., Martinerie, J., Garnero, L.: Inter-brain synchronization during social interaction. PLoS ONE **5**(8), e12166 (2010). https://doi.org/10.1371/journal.pone.0012166
10. Ganin, Y., et al.: Domain-adversarial training of neural networks. J. Mach. Learn. Res. **17**, 1–35 (2016)
11. Hakim, U., et al.: Quantification of inter-brain coupling: a review of current methods used in haemodynamic and electrophysiological hyperscanning studies. Neuroimage **280**, 120354 (2023). https://doi.org/10.1016/j.neuroimage.2023.120354
12. Hasson, U., Nir, Y., Levy, I., Fuhrmann, G., Malach, R.: Intersubject synchronization of cortical activity during natural vision. Science **303**(5664), 1634–1640 (2004). https://doi.org/10.1126/science.1089506
13. He, H., Wu, D.: Transfer learning for brain-computer interfaces: a Euclidean space data alignment approach. IEEE Trans. Biomed. Eng. **67**(2), 399–410 (2020). https://doi.org/10.1109/TBME.2019.2913914
14. Katsigiannis, S., Ramzan, N.: DREAMER: a database for emotion recognition through EEG and ECG signals from wireless low-cost off-the-shelf devices. IEEE J. Biomed. Health Inform. **22**(1), 98–107 (2018). https://doi.org/10.1109/JBHI.2017.2688239
15. Koelstra, S., et al.: DEAP: a database for emotion analysis using physiological signals. IEEE Trans. Affect. Comput. **3**(1), 18–31 (2012). https://doi.org/10.1109/T-AFFC.2011.15
16. Lachaux, J.P., Rodriguez, E., Martinerie, J., Varela, F.J.: Measuring phase synchrony in brain signals. Hum. Brain Mapp. **8**(4), 194–208 (1999). https://doi.org/10.1002/(SICI)1097-0193(1999)8:4<194::AID-HBM4>3.0.CO;2-C

17. Lawhern, V.J., Solon, A.J., Waytowich, N.R., Gordon, S.M., Hung, C.P., Lance, B.J.: EEGNet: a compact convolutional neural network for EEG-based brain-computer interfaces. J. Neural Eng. **15**(5), 056013 (2018). https://doi.org/10.1088/1741-2552/aace8c
18. Li, Y., Zheng, W., Zong, Y., Cui, Z., Zhang, T., Zhou, X.: A bi-hemisphere domain adversarial neural network model for EEG emotion recognition. IEEE Trans. Affect. Comput. **12**(2), 494–504 (2021). https://doi.org/10.1109/TAFFC.2018.2885474
19. Liu, R., et al.: ERTNet: an interpretable transformer-based framework for EEG emotion recognition. Front. Neurosci. **18**, 1320645 (2024). https://doi.org/10.3389/fnins.2024.1320645
20. Miranda-Correa, J.A., Abadi, M.K., Sebe, N., Patras, I.: AMIGOS: a dataset for affect, personality and mood research on individuals and groups. IEEE Trans. Affect. Comput. **12**(2), 479–493 (2021). https://doi.org/10.1109/TAFFC.2018.2884461
21. Montague, P.R., et al.: HyperScanning: simultaneous fMRI during linked social interactions. Neuroimage **16**(4), 1159–1164 (2002). https://doi.org/10.1006/nimg.2002.1150
22. Nastase, S.A., Gazzola, V., Hasson, U., Keysers, C.: Measuring shared responses across subjects using intersubject correlation. Soci. Cognit. Affect. Neurosci. **14**(6), 667–685 (2019). https://doi.org/10.1093/scan/nsz037
23. Nummenmaa, L., Glerean, E., Viinikainen, M., Jaaskelainen, I.P., Hari, R., Sams, M.: Emotions promote social interaction by synchronizing brain activity across individuals. Proc. Natl. Acad. Sci. **109**(24), 9599–9604 (2012). https://doi.org/10.1073/pnas.1206095109
24. Nunez, P.L., et al.: EEG coherency. i: statistics, reference electrode, volume conduction, laplacians, cortical imaging, and interpretation at multiple scales. Electroencephalogr. Clin. Neurophysiol. **103**(5), 499–515 (1997). https://doi.org/10.1016/S0013-4694(97)00066-7
25. Pan, S.J., Tsang, I.W., Kwok, J.T., Yang, Q.: Domain adaptation via transfer component analysis. IEEE Trans. Neural Netw. **22**(2), 199–210 (2011). https://doi.org/10.1109/TNN.2010.2091281
26. Shen, X., Liu, X., Hu, X., Zhang, D., Song, S.: Contrastive learning of subject-invariant EEG representations for cross-subject emotion recognition. IEEE Trans. Affect. Comput. **14**(3), 2496–2511 (2023). https://doi.org/10.1109/TAFFC.2022.3164516
27. Si, X., et al.: Temporal aware mixed attention-based convolution and transformer network for cross-subject EEG emotion recognition. Comput. Biol. Med. **181**, 108973 (2024). https://doi.org/10.1016/j.compbiomed.2024.108973
28. Snell, J., Swersky, K., Zemel, R.S.: Prototypical networks for few-shot learning. In: Advances in Neural Information Processing Systems (2017)
29. Song, T., Zheng, W., Song, P., Cui, Z.: EEG emotion recognition using dynamical graph convolutional neural networks. IEEE Trans. Affect. Comput. **11**(3), 532–541 (2020). https://doi.org/10.1109/TAFFC.2018.2817622
30. Song, Y., Zheng, Q., Liu, B., Gao, X.: EEG conformer: convolutional transformer for EEG decoding and visualization. IEEE Trans. Neural Syst. Rehabil. Eng. **31**, 710–719 (2023). https://doi.org/10.1109/TNSRE.2022.3230250
31. Wang, Y., Qiu, S., Ma, X., He, H.: A prototype-based SPD matrix network for domain adaptation EEG emotion recognition. Pattern Recogn. **110**, 107626 (2021). https://doi.org/10.1016/j.patcog.2020.107626

32. Weinberger, K.Q., Saul, L.K.: Distance metric learning for large margin nearest neighbor classification. J. Mach. Learn. Res. **10**, 207–244 (2009)
33. Zheng, W.L., Lu, B.L.: Investigating critical frequency bands and channels for EEG-based emotion recognition with deep neural networks. IEEE Trans. Auton. Ment. Dev. **7**(3), 162–175 (2015). https://doi.org/10.1109/TAMD.2015.2431497
34. Zheng, W.L., Lu, B.L.: Personalizing EEG-based affective models with transfer learning. In: Proceedings of the International Joint Conference on Artificial Intelligence, pp. 2732–2738 (2016)
35. Zheng, W.L., Zhang, Y.Q., Zhu, J.Y., Lu, B.L.: Transfer components between subjects for EEG-based emotion recognition. In: International Conference on Affective Computing and Intelligent Interaction, pp. 917–922 (2015). https://doi.org/10.1109/ACII.2015.7344684

OAG-CAE: A Supervised Autoencoder Framework for Brain-Age Prediction from Whole-Brain Functional Connectivity

Yikang Cao[1,2], Marco Castellaro[2], Marco Zorzi[1,3], and Alessandra Bertoldo[1,2](✉)

[1] Padova Neuroscience Center, University of Padova, Padova, Italy
marco.zorzi@unipd.it
[2] Department of Information Engineering, University of Padova, Padova, Italy
alessandra.bertoldo@unipd.it
[3] Department of General Psychology, University of Padova, Padova, Italy

Abstract. Functional brain aging is characterized by large-scale network reorganization, yet predicting brain age from whole-brain functional connectivity remains challenging under the high-dimensional, limited-sample conditions typical of neuroimaging studies. To address this limitation, a supervised deep-learning framework for brain-age prediction was developed based on an Orthogonal-Auxiliary-Guided Convolutional Autoencoder (OAG-CAE). Resting-state fMRI data from 599 adults were used to construct whole-brain functional connectivity matrices spanning cortical, subcortical, and cerebellar regions. OAG-CAE was designed to encourage the separation of age-relevant and age-irrelevant latent features through orthogonality regularization and auxiliary supervision. The age-related latent representation was then used by a multi-scale convolutional regressor for brain-age estimation. The model achieved stable performance under five-fold cross-validation (MAE $= 6.110 \pm 0.539$ years, Pearson's $r = 0.850$, $R^2 = 0.723 \pm 0.048$), outperforming conventional regression and machine-learning baselines. Ablation analyses further showed that convolutional encoding, auxiliary supervision, and the final convolution-based prediction module each contributed to performance gains. Together, these results suggest that OAG-CAE provides a feasible and methodologically grounded framework for modeling age-related patterns in whole-brain functional connectivity and may offer a useful basis for neuroimaging-based brain-age prediction.

Keywords: Brain-age prediction · Functional connectivity · Autoencoder · Deep learning · Healthy aging

1 Introduction

Normal aging is associated with progressive changes in brain function, cognitive decline, and increased vulnerability to neurological disorders [10,11]. Contemporary theories such as the Scaffolding Theory of Aging and Cognition (STAC)

L. Pasa et al. (Eds.): ICANN 2026, LNCS 17089, pp. 127–136, 2027.
https://doi.org/10.1007/978-3-032-38398-3_11

and its revised version (STAC-r) suggest that these changes reflect large-scale functional reorganization rather than isolated local decline [21,23]. Consistent with this view, resting-state fMRI studies have reported reduced within-network coherence, particularly in the default mode network, together with increased cross-network coupling involving somatosensory and subcortical systems [8]. These patterns have often been interpreted as reflecting compensatory or adaptive network reorganization during aging [24]. However, most previous work has focused primarily on cortical systems, leaving the contributions of subcortical and cerebellar circuits comparatively underexplored despite their important roles in cognition, attention, and motor control [22,25].

Brain-age prediction provides a useful framework for characterizing such aging-related brain changes from neuroimaging data. In particular, deep learning models applied to resting-state functional connectivity (FC) can capture nonlinear and distributed aging signatures that may be difficult to detect using conventional methods [1]. Their ability to model multiscale relationships in high-dimensional connectivity data makes them especially attractive for studying distributed functional reorganization across cortical, subcortical, and cerebellar systems. At the same time, prediction from fMRI remains challenging because FC features are high-dimensional relative to typical sample sizes [13]. Classical dimensionality-reduction approaches such as principal component analysis (PCA) are computationally efficient and have been widely used in brain-age studies [18,27], but their linear nature limits their ability to represent the nonlinear and hierarchical organization of functional brain aging.

Autoencoder-based methods offer a more flexible alternative by learning compact latent representations directly from the data [19,29]. Rather than relying on predefined linear bases, they can capture coordinated network-level patterns and reduce noise in the input space, making them well-suited for modeling functional brain aging as a distributed process. Recent work has shown that autoencoders and related architectures can recover latent features that covary systematically with age and improve brain-age prediction performance [19,20]. While FC provides a denoised and meaningful representation of interactions between brain regions, latent representations learned by conventional autoencoders may still entangle age-related and age-irrelevant sources of variance without explicit guidance [19,29].

To address these limitations, a supervised brain-age prediction framework was proposed based on an Orthogonal-Auxiliary-Guided Convolutional Autoencoder (OAG-CAE). The model is designed to learn age-relevant latent representations from whole-brain FC while reducing potential overlap with irrelevant variance. A nonlinear convolution-based regressor is then used to map these latent features to brain-age estimates. This study evaluates the framework using whole-brain resting-state FC and shows that OAG-CAE achieves stable performance compared with selected conventional baselines, while benefiting from auxiliary supervision and latent-space regularization.

2 Methods

2.1 Dataset and Participants

Resting-state fMRI data from 599 participants from the Human Connectome Project Aging (HCP-A) cohort who passed preprocessing and quality-control screening were analyzed. Participants ranged in age from 36 to 95 years and included 262 males and 337 females. The data were obtained through the Aging, Brain, Cognition, and Connectome (AABC)/HCP-A data framework under the applicable data use terms [2,9]. Imaging data were first preprocessed using fMRIPrep [6] and then further processed with XCP-D, a post-processing pipeline for functional MRI data. Whole-brain functional connectivity was computed by combining 246 cortical and subcortical regions from the Brainnetome atlas with 32 cerebellar regions from the Spatially Unbiased Infratentorial Template (SUIT) atlas, yielding 278 regions in total [5,7]. Regional mean BOLD time series were extracted from all regions, and FC was computed as the Pearson correlation between each pair of regional time series, resulting in a 278-by-278 FC matrix for each participant. Correlation coefficients were Fisher z-transformed before model training. To avoid data leakage, normalization parameters were estimated on the training set only and then applied to the validation and test sets.

2.2 OAG-CAE for Brain-Age Prediction

To address the high-dimensional, low-sample-size challenge of neuroimaging data [26], this study proposes a supervised autoencoder framework termed the Orthogonal-Auxiliary-Guided Convolutional Autoencoder (OAG-CAE). As shown in Fig. 1, OAG-CAE adopts a convolutional autoencoder backbone [17] to learn compact latent representations from the input functional connectivity (FC) matrix. Rather than treating the latent space as a single undifferentiated representation, the model was designed to encourage a structured separation between age-related and age-irrelevant variability. This was achieved through two complementary mechanisms: auxiliary age supervision, which encouraged the latent representation to encode age-informative structure, and orthogonality regularization, which discouraged overlap between the age-related and age-irrelevant latent components [12]. Specifically, orthogonality regularization was implemented as a sample-wise cosine orthogonality loss. For each mini-batch, the two latent components were first centered by subtracting their batch-wise means and then L2-normalized along the feature dimension. The loss was computed as:

$$\mathcal{L}_{orth} = \frac{1}{B} \sum_{i=1}^{B} \left| \cos(\tilde{z}_i^{age}, \tilde{z}_i^{noise}) \right|, \tag{1}$$

where B denotes the mini-batch size, and $\tilde{z}_i^{age}$ and $\tilde{z}_i^{noise}$ denote the centered age-related and age-irrelevant latent vectors for sample i, respectively. This loss penalizes both positive and negative alignment between the two latent components and encourages them to encode less redundant information.

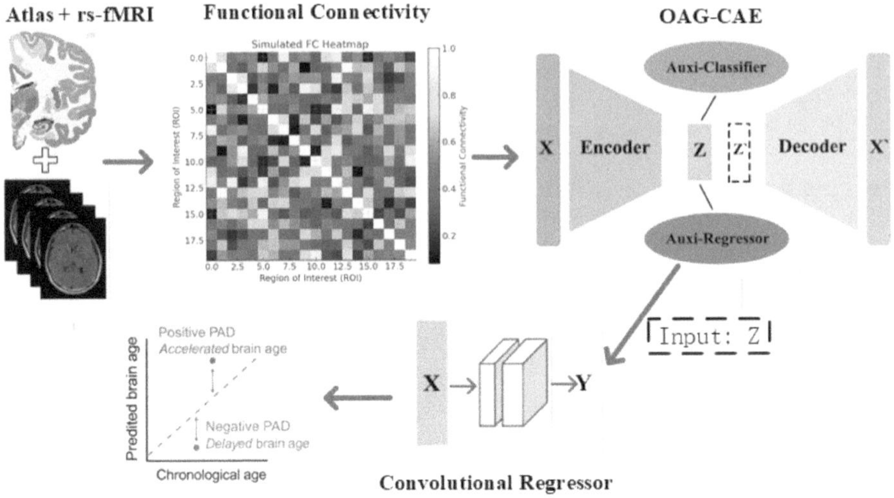

Fig. 1. Schematic illustration of the proposed OAG-CAE framework for brain-age prediction. The model encodes the input FC matrix into two latent components: an age-related representation (Z) and an age-irrelevant representation (Z'). Orthogonality regularization is used to discourage overlap between the two components, and Z is used by the convolutional regressor to predict brain age.

To further stabilize age-related feature learning, the age-related latent component is jointly optimized with complementary auxiliary regression and classification objectives. These auxiliary heads served as training-time supervision signals and were not used for final inference; the reported brain-age predictions were generated by the convolutional regressor from the age-related latent component. This multi-task supervision provides additional constraints on latent-space organization and encourages the model to retain information that is both predictive of chronological age and robust across subjects. Compared with conventional autoencoders, which may mix age-related variation with irrelevant sources of variance, this design provides additional supervision and regularization to guide representation learning toward age-relevant structure.

The age-related latent representation is then fed into a nonlinear regression network for brain-age prediction. The regressor consists of convolutional blocks operating along the latent dimension together with a lightweight feature-wise attention mechanism, enabling adaptive reweighting of informative latent features. A trainable bias-shift term and post-hoc linear calibration are further applied to reduce systematic prediction bias. Importantly, the use of a nonlinear regressor does not contradict the constrained latent-space design. In the present framework, auxiliary supervision and orthogonality regularization encourage age-relevant information to be organized in the latent space, whereas the nonlinear read-out allows residual structure in the latent-age relationship to be captured. This is desirable given the heterogeneous and potentially nonlinear trajectories of brain aging.

Model performance was evaluated using five-fold cross-validation stratified by age and gender. In each outer fold, the data were split into training, validation, and test subsets, with the validation set nested within the training data for model selection. FC was first computed as a 278-by-278 matrix for each participant. For convolutional models, the FC matrix was used in matrix form with the diagonal and redundant triangular elements masked. Both the age-related and age-irrelevant latent components were 32-dimensional. The OAG-CAE module was optimized using AdamW, a decoupled weight-decay variant of Adam [14], with a learning rate of 0.0003 and a weight decay of 0.002. Full-batch training was used within each fold. Training was performed for up to 2000 epochs, with validation MAE evaluated every 20 epochs and early stopping applied after 10 validation checks without improvement. The representation-learning loss combined reconstruction, orthogonality, auxiliary regression, and auxiliary classification terms, with weights of 0.1, 0.4, 0.3, and 0.7, respectively. The additional nonlinear prediction model was then trained on the age-related latent component using MAE loss. MAE was also used as the primary evaluation metric. All experiments were implemented in PyTorch, and the same five-fold split was used across all models. The implementation details and model code are available at: https://github.com/YKCIAO/OAG-CAE.

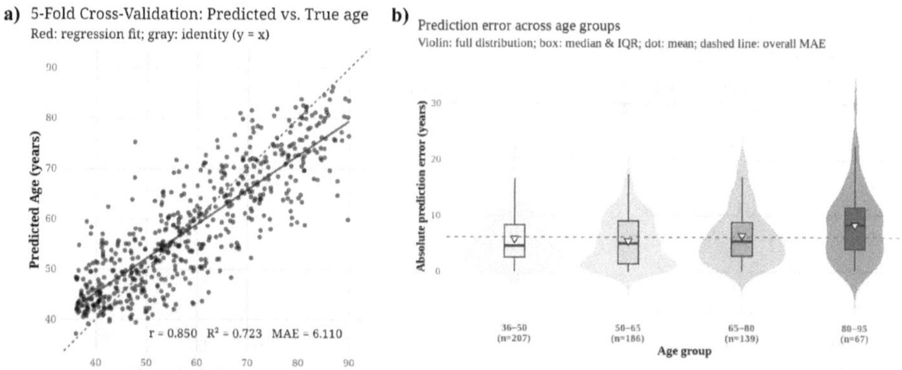

Fig. 2. (a) Scatter plot of brain-age prediction versus chronological age. The red line shows the regression fit, and the gray dashed line shows the identity line (y = x). (b) Absolute prediction error across four age groups, shown as violin and boxplots. White dots indicate mean error, and the dashed line marks the overall MAE. (Color figure online)

Table 1. Brain-age prediction performance of baseline models and the proposed OAG-CAE. "Not stable" indicates unstable convergence across folds under the same training protocol.

Model	MAE (Mean ± SD)	R^2 (Mean ± SD)
Proposed Model		
OAG-CAE	6.110 ± 0.539	0.723 ± 0.048
Traditional ML Baselines		
Linear Regression	6.942 ± 0.450	0.678 ± 0.042
Ridge Regression	6.842 ± 0.450	0.678 ± 0.042
Lasso Regression	6.858 ± 0.503	0.660 ± 0.036
SVR	6.741 ± 0.452	0.677 ± 0.040
XGBoost	8.661 ± 0.361	0.478 ± 0.047
Deep Learning Baselines		
Convolutional Neural Network	not stable	–

3 Results

The proposed OAG-CAE achieved stable brain-age prediction performance under the five-fold cross-validation setting. As shown in Fig. 2a, predicted brain age was strongly correlated with chronological age (Pearson's r = 0.850, p < 0.001), with a mean absolute error of 6.110 ± 0.539 years across the five cross-validation folds. Pearson's r was computed directly between predicted and chronological age, whereas R^2 was used as the coefficient-of-determination metric in the model-comparison tables. The fitted regression line was close to the identity line, indicating limited systematic bias. Figure 2b further shows that absolute prediction errors varied across age groups, with relatively higher variability in the 50–65 and 80–95 groups.

This study next compared OAG-CAE with several baseline models commonly used in brain-age prediction. For a fair comparison, all baseline models were trained on the same whole-brain FC information and evaluated using the same five-fold data splits as OAG-CAE. Vector-based models used the upper-triangular FC features, whereas convolutional models used the matrix-form FC representation. As shown in Table 1, OAG-CAE achieved the best overall performance (MAE = 6.110 ± 0.539 years, R^2 = 0.723 ± 0.048), outperforming the selected conventional regression and machine-learning baselines. Among the latter, Linear Regression, Ridge Regression, Lasso Regression, and Support Vector Regression (SVR) showed weaker performance, whereas XGBoost performed worst.

An ablation study further confirmed the contribution of each model component (Table 2). The basic AE with linear regression performed worst, while replacing it with a CAE substantially improved performance, indicating the benefit of convolutional representation learning for FC data. Auxiliary supervision yielded additional gains, with regression-based supervision contributing more strongly than classification alone, and the combination of both producing further improvement. The full OAG-CAE, which integrates convolutional encoding, auxiliary supervision, and convolution-based regression, achieved the best results overall. Together, these findings show that each component contributes to predictive performance, while the complete architecture provided the best performance among the tested variants.

Table 2. Ablation results for model variants on the brain-age prediction task.

Model	Aux. cls.	Aux. reg.	Predictor	Five-fold MAE	Five-fold R^2
AE	No	No	Linear	8.352 ± 0.759	0.508 ± 0.073
CAE	No	No	Linear	7.431 ± 0.404	0.600 ± 0.052
CAE	Yes	No	Linear	7.108 ± 0.554	0.628 ± 0.056
CAE	No	Yes	Linear	6.615 ± 0.508	0.686 ± 0.045
CAE	Yes	Yes	Linear	6.468 ± 0.580	0.694 ± 0.035
CAE	Yes	Yes	SVR	6.741 ± 0.391	0.677 ± 0.035
CAE	Yes	Yes	Convolutional	6.110 ± 0.539	0.723 ± 0.048

4 Discussion

The results of the proposed OAG-CAE support the use of supervised nonlinear representation learning for modeling distributed aging-related patterns in high-dimensional functional connectivity data [30]. In particular, the combination of convolutional feature encoding and auxiliary supervision appears to improve the extraction of age-relevant latent representations, while also stabilizing performance across folds, consistent with previous work showing that autoencoder-based regression frameworks can yield age-informative latent representations for aging prediction [19,28,29]. The ablation results further indicate that each component of the framework contributes meaningfully to prediction accuracy, highlighting the value of additional supervision and regularization in brain-age modeling.

The broad age range of the HCP-A cohort (36–95 years) introduces substantial inter-individual variability, making accurate prediction particularly challenging. The scatter plot suggests a residual age-related prediction bias, with younger participants tending to be overestimated and older participants underestimated. This common brain-age bias likely reflects regression-to-the-mean effects under broad age-range training, even when bias-reduction strategies are applied [3,16]. Thus, individual brain-age gaps should be interpreted cautiously, especially at the extremes of the age distribution. In the present healthy cohort, a negative brain-age gap in older adults should not be interpreted directly as evidence of better brain health; rather, it reflects a lower predicted age relative to chronological age within the model's learned normative reference pattern. Clinical interpretation would require external validation, bias correction, and association with independent cognitive or clinical outcomes.

Although multimodal approaches may achieve higher predictive accuracy [15], the present FC-only design was adopted to focus on functional rather than structural aging-related variation, allowing the analysis to target age-sensitive patterns in large-scale functional connectivity [1]. Latent-space compression may also help reduce overfitting while preserving informative functional variation [4,18,20].

Several limitations should be noted. First, the current evaluation relied on cross-validation only, and external validation is needed to assess out-of-sample generalizability. Second, although OAG-CAE outperformed the selected conventional baselines, further comparisons with recent deep-learning and autoencoder-based architectures are required. The unstable CNN baseline likely reflects the difficulty of training an unconstrained CNN on high-dimensional FC inputs under the present sample-size conditions, rather than a general limitation of CNN-based approaches. Third, although orthogonality regularization and auxiliary supervision encouraged a more structured latent space, future work should assess latent-space organization and identify the functional connections most relevant to brain-age prediction.

5 Conclusion

This study presents OAG-CAE, a supervised autoencoder framework for brain-age prediction from whole-brain functional connectivity. Rather than relying on unconstrained latent representations, the proposed approach combines orthogonality regularization, auxiliary supervision, and nonlinear prediction to encourage age-relevant latent-space organization under high-dimensional and limited-sample conditions. In this sense, the main contribution of the framework lies not only in predictive performance, but also in providing a more structured strategy for learning brain-age representations from functional connectivity data.

The empirical results support the methodological value of this design: convolutional encoding improves representation learning for FC data, auxiliary objectives further support age-related latent structure, and the full model achieved the best performance among the tested variants. More broadly, OAG-CAE suggests a feasible route for combining supervised prediction with constraints that promote more structured latent representations in neuroimaging-based aging research. This framework may therefore serve as a useful basis for future work on brain-age modeling in both healthy and pathological populations.

References

1. Azzam, M., et al.: A review of artificial intelligence-based brain age estimation and its applications for related diseases. Brief. Funct. Genom. **24**, elae042 (2024). https://doi.org/10.1093/bfgp/elae042
2. Bookheimer, S.Y., et al.: The lifespan human connectome project in aging: an overview. Neuroimage **185**, 335–348 (2019). https://doi.org/10.1016/j.neuroimage.2018.10.009
3. Borkar, K., Chaturvedi, A., Vinod, P.K., Bapi, R.S.: Ayu-Characterization of healthy aging from neuroimaging data with deep learning and rsfMRI. Front. Comput. Neurosci. **16** (2022). https://doi.org/10.3389/fncom.2022.940922
4. Calesella, F., Testolin, A., De Filippo De Grazia, M., Zorzi, M.: A comparison of feature extraction methods for prediction of neuropsychological scores from functional connectivity data of stroke patients. Brain Info. **8**(1), 8 (2021). https://doi.org/10.1186/s40708-021-00129-1
5. Diedrichsen, J.: A spatially unbiased atlas template of the human cerebellum. Neuroimage **33**(1), 127–138 (2006). https://doi.org/10.1016/j.neuroimage.2006.05.056
6. Esteban, O., et al.: fMRIPrep: a robust preprocessing pipeline for functional MRI. Nat. Methods **16**(1), 111–116 (2019). https://doi.org/10.1038/s41592-018-0235-4
7. Fan, L., et al.: The human Brainnetome Atlas: a new brain atlas based on connectional architecture. Cereb. Cortex **26**(8), 3508–3526 (2016). https://doi.org/10.1093/cercor/bhw157
8. Goelman, G., Dan, R., Bezdicek, O., Jech, R., Ekstein, D.: Directed functional connectivity of the default-mode-network of young and older healthy subjects. Sci. Rep. **14**(1), 4304 (2024). https://doi.org/10.1038/s41598-024-54802-6
9. Harms, M.P., et al.: Extending the human connectome project across ages: imaging protocols for the lifespan development and aging projects. Neuroimage **183**, 972–984 (2018). https://doi.org/10.1016/j.neuroimage.2018.09.060
10. Hedden, T., Gabrieli, J.D.E.: Insights into the ageing mind: a view from cognitive neuroscience. Nat. Rev. Neurosci. **5**(2), 87–96 (2004). https://doi.org/10.1038/nrn1323
11. Hou, Y., et al.: Ageing as a risk factor for neurodegenerative disease. Nat. Rev. Neurol. **15**(10), 565–581 (2019). https://doi.org/10.1038/s41582-019-0244-7
12. Hu, D., et al.: Disentangled-multimodal adversarial autoencoder: application to infant age prediction with incomplete multimodal neuroimages. IEEE Trans. Med. Imag. **39**(12), 4137–4149 (2020). https://doi.org/10.1109/TMI.2020.3013825
13. Liu, S., Xu, L., Xu, G., Wang, Y., Zhang, G., He, L.: A systematic review: brain age gap as a promising early diagnostic biomarker for Alzheimer's disease. J. Neurol. Sci. **475**, 123563 (2025). https://doi.org/10.1016/j.jns.2025.123563
14. Loshchilov, I., Hutter, F.: Decoupled weight decay regularization. In: International Conference on Learning Representations (2017). https://api.semanticscholar.org/CorpusID:53592270
15. Luo, N., Shi, W., Yang, Z., Song, M., Jiang, T.: Multimodal fusion of brain imaging data: methods and applications. Mach. Intell. Res. **21**(1), 136–152 (2024). https://doi.org/10.1007/s11633-023-1442-8
16. Luo, Y., Chen, W., Qiu, J., Jia, T.: Accelerated functional brain aging in major depressive disorder: evidence from a large scale fMRI analysis of Chinese participants. Transl. Psych. **12**(1), 397 (2022). https://doi.org/10.1038/s41398-022-02162-y

17. Masci, J., Meier, U., Cireşan, D., Schmidhuber, J.: Stacked convolutional auto-encoders for hierarchical feature extraction. In: Honkela, T., Duch, W., Girolami, M., Kaski, S. (eds.) Artificial Neural Networks and Machine Learning – ICANN 2011, pp. 52–59. Springer, Berlin Heidelberg, Berlin, Heidelberg (2011). https://doi.org/10.1007/978-3-642-21735-7_7
18. Monti, R.P., et al.: Interpretable brain age prediction using linear latent variable models of functional connectivity. PLoS ONE **15**(6), 1–25 (2020). https://doi.org/10.1371/journal.pone.0232296
19. Mouches, P., Wilms, M., Rajashekar, D., Langner, S., Forkert, N.: Unifying brain age prediction and age-conditioned template generation with a deterministic autoencoder. In: Heinrich, M., Dou, Q., de Bruijne, M., Lellmann, J., Schläfer, A., Ernst, F. (eds.) Proceedings of the Fourth Conference on Medical Imaging with Deep Learning. Proceedings of Machine Learning Research, vol. 143, pp. 497–506. PMLR (2021). https://proceedings.mlr.press/v143/mouches21a.html
20. Ouyang, J., Zhao, Q., Adeli, E., Zaharchuk, G., Pohl, K.M.: Self-supervised learning of neighborhood embedding for longitudinal MRI. Med. Image Anal. **82**, 102571 (2022). https://doi.org/10.1016/j.media.2022.102571
21. Park, D.C., Reuter-Lorenz, P.: The adaptive brain: aging and neurocognitive scaffolding. Annual Rev. Psychol. **60**, 173–196 (2009). https://doi.org/10.1146/annurev.psych.59.103006.093656
22. Patil, A.U., Madathil, D., Huang, C.M.: Healthy aging alters the functional connectivity of creative cognition in the default mode network and cerebellar network. Front. Aging Neurosci. **13**, 607988 (2021). https://doi.org/10.3389/fnagi.2021.607988
23. Reuter-Lorenz, P.A., Park, D.C.: How does it STAC Up? Revisiting the scaffolding theory of aging and cognition. Neuropsychol. Rev. **24**(3), 355–370 (2014). https://doi.org/10.1007/s11065-014-9270-9
24. Sala-Llonch, R., Bartrés-Faz, D., Junqué, C.: Reorganization of brain networks in aging: a review of functional connectivity studies. Front. Psychol. **6** (2015). https://doi.org/10.3389/fpsyg.2015.00663
25. Stoodley, C.J., Schmahmann, J.D.: Functional topography in the human Cerebellum: a meta-analysis of neuroimaging studies. Neuroimage **44**(2), 489–501 (2009). https://doi.org/10.1016/j.neuroimage.2008.08.039
26. Tang, Y., Chen, D., Li, X.: Dimensionality reduction methods for brain imaging data analysis. ACM Comput. Surv. **54**(4) (2021). https://doi.org/10.1145/3448302
27. Vanni, F., Cinetto, S., De Filippo De Grazia, M., Zorzi, M., Filippini, N.: Obesity accelerates brain ageing: a multimodal imaging study. Brain Commun. **7**(6), fcaf389 (2025). https://doi.org/10.1093/braincomms/fcaf389
28. Wilms, M., et al.: Invertible modeling of bidirectional relationships in neuroimaging with normalizing flows: application to brain aging. IEEE Trans. Med. Imag. **41**(9), 2331–2347 (2022). https://doi.org/10.1109/TMI.2022.3161947
29. Zhao, Q., Adeli, E., Honnorat, N., Leng, T., Pohl, K.M.: Variational AutoEncoder for Regression: application to brain aging analysis. In: Shen, D., et al. (eds.) Medical Image Computing and Computer Assisted Intervention – MICCAI 2019, pp. 823–831. Springer International Publishing, Cham (2019). https://doi.org/10.1007/978-3-030-32245-8_91
30. Zhou, T., Thung, K.H., Liu, M., Shi, F., Zhang, C., Shen, D.: Multi-modal neuroimaging data fusion via latent space learning for Alzheimer's disease diagnosis. In: Rekik, I., Unal, G., Adeli, E., Park, S.H. (eds.) Pred. Intell. Med., pp. 76–84. Springer International Publishing, Cham (2018). https://doi.org/10.1007/978-3-030-00320-3_10

Representational Drift in Memory Under Incremental Learning with Local Synaptic Mechanisms

Jayani Hewavitharana(✉), Pedro Henrique da Costa Avelar, Peter Giese, and Kathleen Steinhofel

King's College London, London, UK
{jayani.hewavitharana,pedro_henrique.da_costa_avelar,karl.giese,
kathleen.steinhofel}@kcl.ac.uk

Abstract. Memory representations in the brain are formed incrementally and are therefore susceptible to representational drift when learning is disrupted by input interference. In this work, we use a brain-inspired neural network with local synaptic learning rules to investigate how two synaptic mechanisms, long-term potentiation (LTP) and multi-innervated spine formation (MIS), shape memory representations under incremental learning with controlled interference. Through analyses of representational dispersion, pre-activation dynamics, and output sequence structure, we show that representational trajectories are greatly influenced by the characteristics of synaptic mechanisms underlying memory encoding. LTP supports stable, recoverable encoding through selective reinforcement and normalization, while MIS drives gradual drift through continuous weight redistribution. These findings provide useful insights into conditions that control variability in memory engrams.

Keywords: Memory representations · Representational drift · Incremental learning · Computational neuroscience · Cognitive models

1 Introduction

Memory formation in the brain is a sequential process. As sensory experiences emerge, the brain integrates each new event into an existing representational framework through synaptic plasticity, the ability of synaptic connections to strengthen or weaken in response to neural activity [3,12]. At the cellular level, memories are stored as groups of neurons known as memory engrams that are selectively activated during learning [11,18,22]. Only a small fraction of neurons participate in any single engram, making the representations sparse and allowing a large number of distinct memories [19].

Engrams formed close together in time, or in response to similar inputs tend to share overlapping neural populations, reflecting the influence of both temporal proximity and input similarity [11]. This raises the question of how memory

L. Pasa et al. (Eds.): ICANN 2026, LNCS 17089, pp. 137–148, 2027.
https://doi.org/10.1007/978-3-032-38398-3_12

engrams evolve when the input stream is not uniform. In realistic learning conditions, a sequence of related experiences can be interrupted by out-of-distribution stimuli arising from distractions, context switches, or incidental events. Such interference introduces competing synaptic demands that may drive representational drift, a gradual shift in the neural population codes underlying stored memories [5]. While memory interference has been studied in terms of representational overlap [4], the specific contribution of input interference to engram drift and the role of the underlying synaptic mechanism remains poorly understood.

Artificial neural networks (ANNs) face closely related challenges as brain-inspired models that learn incrementally as online, batch, or continual learning. Non-stationary inputs can cause previously learnt representations to change leading to representational drift and catastrophic forgetting [6,17,20]. Understanding how synaptic mechanisms determine the variability of memory engrams under interference offers biological insights for this challenge as ANNs move toward sparser, more plasticity-based architectures for efficiency [1,7,10].

In this work, we use a simple ANN with local learning rules to model the evolution of memory representations during incremental learning over a sequence of related inputs with controlled interference. We study two biologically motivated synaptic mechanisms, long-term potentiation (LTP) [14], which is prominent in younger brains, and multi-innervated spine formation (MIS) [2], a compensatory mechanism for reduced plasticity in ageing brains, and examine how each produce different representational dynamics. Through analyses of engram similarity, pre-activation density, and output sequence structure, we show that representational trajectory largely depends on the consistency of synaptic connections.

2 Methods

Our study uses two independent components for input and memory encoding. The input encoding stage transforms raw inputs into sparse neural activation patterns. The memory encoding stage forms the primary component of this study, providing the representational space to analyse how memory traces evolve under incremental learning and varied interference conditions. In this section, we describe the structure and function of the two components[1]

2.1 Input Encoding

We use the MNIST handwritten digit dataset [13] as the source of raw input, representing simple visual stimuli for downstream memory formation. MNIST provides limited variability within classes and reasonable separation across classes, making it well-suited for studying memory representations under incremental learning and their response to class-inconsistent interference in the input stream.

To convert image data into neural activations, we employ an autoencoder with a standard encoder-decoder architecture and a k-softmax operator [9,15,23]

[1] Implementation - https://github.com/JayaniH/brain-inspired-learning.

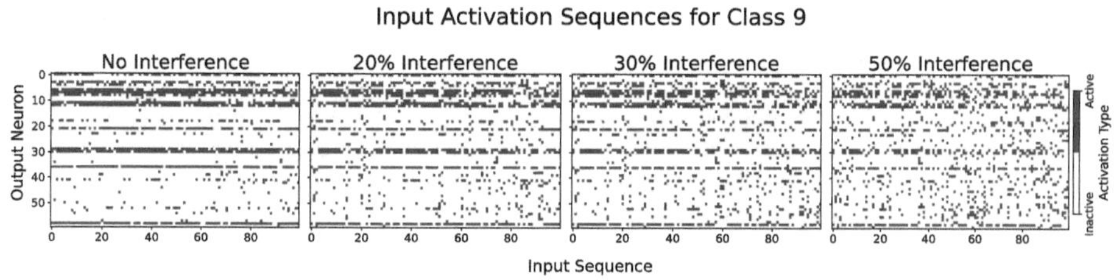

Fig. 1. The sequence of input activations for different interference levels.

applied at the bottleneck layer. We extract binary latent codes ($\boldsymbol{z} \in \{0,1\}^d$) retaining only the top-k activations after the k-softmax layer in the bottleneck, yielding sparse input representations with k active nodes ($\sum_i \boldsymbol{z}_i = k$). We generate 60-dimensional latent vectors to balance representational capacity with a tractable input space. We set $k = 12$, corresponding to an activation rate of 20%, consistent with observed sparse coding levels in the brain [19] [16].

We remove duplicates and select a subset of $n = 100$ latent vectors per digit class to construct an internally consistent input set, prioritising high within-class similarity. For each class, we use a prototype vector representing the k most frequently active bit positions and retain the top-n overlapping candidate vectors. This produces a set of representations with high within-class consistency while preserving distinctions between classes. Therefore, any changes observed in the output representations can be attributed to explicit interference rather than variability in the inputs themselves, providing a stable baseline for isolating representational drift. The use of small subsets enables granular observation of learning dynamics at early stages, avoiding generalisation.

We consider the ten class-specific subsets as separate sets of input representations and conduct independent learning simulations, allowing us to assess the consistency and robustness of memory encoding dynamics across varied inputs. To establish a baseline, incremental learning is carried out on each subset without cross-class interference. To investigate the effect of external disturbances, we introduce controlled interference by replacing a proportion of each subset with latent codes sampled randomly from the remaining classes. We examine interference levels of 20%, 30%, and 50% replacement, corresponding to progressively noisier learning contexts, as illustrated in Fig. 1.

2.2 Memory Encoding

We model memory encoding using an artificial neural network (ANN) consisting of a binary input layer ($\boldsymbol{z} \in \{0,1\}^d$) fully connected to an output layer of equal dimensionality ($\boldsymbol{h} \in \mathbb{R}^d$) representing the connectivity between hippocampal regions [8]. The symmetric structure provides a direct mapping between inputs and memory representations, and both layers contain $d = 60$ units, consistent with the dimensionality of the latent codes produced by input encoding.

The synaptic weight matrix ($W \in \mathbb{R}^{d \times d}$) captures the connection strengths between input and output neurons. Weights are positively initialised from a normal distribution and rescaled so that the total outgoing weight of each input unit sums to one, providing an unbiased starting condition. The output signals $\boldsymbol{h}$ are computed as the weighted sum of the input activations (Eq. (1)).

$$\boldsymbol{h} = \boldsymbol{z} \cdot W \tag{1}$$

As the inputs are sparse binary vectors, only weights originating from active input units ($z_i = 1$) contribute to the output signal and to subsequent weight updates, as formalised in Eq. (2). This reduces the fully connected network to a functionally sparse one during learning.

$$w_{ij} = \begin{cases} w_{ij} & \text{if } z_i = 1 \\ 0 & \text{if } z_i = 0 \end{cases} \tag{2}$$

To represent memory engrams as discrete activation patterns, the output h is thresholded to produce a binary vector $\boldsymbol{y} \in \{0, 1\}^d$, as defined in Eq. (3).

$$y_j = \begin{cases} 1 & \text{if } h_j \geq T \\ 0 & \text{if } h_j < T \end{cases} \tag{3}$$

The threshold T is set to the lower bound that ensures that all output units remain inactive across each input vector in the complete dataset given the initial weights. This guarantees that any output activity reflects learning effects rather than the initial weight configuration.

2.3 Incremental Learning

We simulate incremental learning by presenting inputs $\{\boldsymbol{z}^a\}_{a=1}^n$ to the model sequentially and updating the weights continuously. Each input produces an output representation $\boldsymbol{y}$, and the accumulation of weight changes introduces dependencies between successive outputs, such that the representations formed for later inputs are shaped by those formed earlier. In simulations that incorporate interference, inputs are presented in the same sequential order as the baseline condition, although some are replaced by inputs drawn from other classes.

We employ brain-inspired local learning in which weight updates are restricted to the synapses of a single output neuron per iteration. For a given input, output neurons are iteratively selected via fitness-proportionate sampling weighted by signal magnitude h_j, reflecting the biological principle that strongly potentiated neurons compete more successfully for engram participation [11]. The selected neuron's synapses are then updated according to the local learning rules defined later in the section. This selection-and-update cycle continues until exactly k output neurons exceed the threshold T to form a complete engram representation. This procedure is repeated for each input in the sequence using the updated weights.

As learning progresses, some output neurons accumulate sufficient synaptic weights to exceed T before any weight update occurs for a given input. This pre-activation effectively reduces the learning effort required per input, since fewer than k neurons need to be activated through weight updates. When the number of pre-activated neurons exceeds k for a given input, a k-winner-take-all mechanism is applied, retaining only the k strongest activations and bypassing the iterative update process entirely.

We increment T by δ after each input to control convergence and produce interpretable learning dynamics. It gradually raises the requirement for engram participation, preventing output neurons from becoming saturated early in the learning sequence. We experimentally selected $\delta = 0.001$ for minimal increment. Smaller values drive rapid output saturation without allowing the majority of the inputs to contribute to the memory trajectory, while larger values slow down the learning processes without visible impact on the output representations.

We independently implement two synaptic mechanisms, LTP and MIS, for local learning as defined below, each representing a distinct mode of synaptic modification observed in the brain.

Long-Term Potentiation (LTP). Under the LTP rule, the active weights w_j (Eq. (2)) associated with the selected output unit h_j are scaled by a potentiation factor $\alpha > 1$, as defined in Eq. (4). We set $\alpha = 1.2$ to produce moderate potentiation in the lower range of biological observations [21].

$$w_j \leftarrow \alpha \, w_j \tag{4}$$

To prevent unbounded weight growth due to repeated amplification, we renormalise the outgoing weights from each input neuron after each input so that their sum equals one. This preserves the relative changes induced by learning while preventing runaway excitation and serves as a functional reflection of homeostatic mechanisms such as synaptic scaling.

Multi-innervated Spine Generation (MIS). Our MIS rule abstracts the redistribution of synaptic weight rather than explicit amplification. A fitness-proportionately selected weight w_{ij} of h_j is updated by transferring the weight from a neighbouring synapse w_{il} originating at the same input neuron z_i as formalised in Eq. (5). The donor synapse is chosen from the immediate neighbours $(\{j-1, j+1\})$ with equal probability.

$$\begin{aligned} w_{ij} &\leftarrow w_{ij} + w_{il} \\ w_{il} &\leftarrow 0 \end{aligned} \tag{5}$$

As MIS transfers weight, the total outgoing weight per input neuron is automatically conserved. However, over successive learning steps, connectivity progressively concentrates onto fewer synapses, making the network incrementally sparser and reducing to each input neuron having a single connection of unit weight.

3 Results

3.1 Representational Consistency and Drift

We analyse the dispersion of the generated representations across all classes to characterise engram consistency and the drift induced by interference. Figure 2 shows the distribution of engrams projected into two dimensions, where similarity is measured by the number of overlapping active neurons between representations.

In the absence of interference, both learning rules produce visible engram clusters, indicating a coherent within-class representation structure throughout incremental learning. This is expected when learning a consistent input stream, as there is no mechanism to disrupt the developing representational structure. However, MIS produces less self-contained clusters, suggesting a degree of representational variability through weight redistribution even under noiseless conditions. This points to a structural property in MIS as synaptic weight concentrates onto fewer connections over time, memory encoding becomes increasingly sensitive to small variations in the active input units, making drift more likely.

As interference increases, within-class coherence progressively degrades for both mechanisms, reflecting the noise introduced by out-of-distribution inputs. This degradation is particularly more pronounced under MIS, which exhibits greater sensitivity to interference-induced drift. The weight redistribution process appears to amplify the effect of cross-class inputs by redirecting connectivity in ways that are inconsistent with the within-class representational structure. In contrast, LTP benefits from renormalisation after each input, which partially absorbs the effect of interfering inputs and preserves greater cluster coherence at higher interference levels.

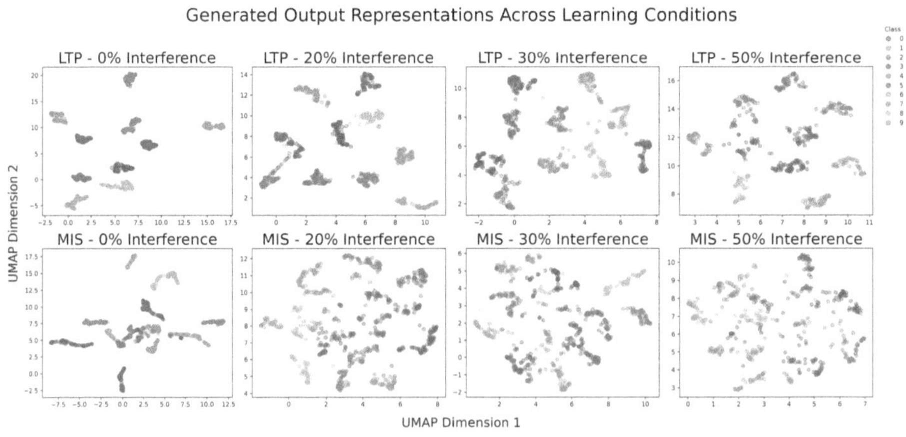

Fig. 2. Two-dimensional projection of generated memory representations for both learning mechanisms under varying interference levels. Interference conditions show only outputs corresponding to within-class inputs.

3.2 Output Neuron Pre-activation

Output pre-activation reflects the cumulative influence of prior learning on current encoding. Larger pre-activation counts indicate that the existing network weight configuration supports the incoming representation. However, pre-activation carries competing implications as it may either stabilise representations by anchoring new engrams to frequently established ones or immediately preceding ones that are inconsistent with the current input.

To quantify pre-activation behaviour, we measure three metrics for each learning simulation: mean pre-activation count μ, their mean increment between consecutive outputs μ_Δ where positive increments indicate propagation of knowledge along the learning trajectory, and fluctuation magnitude σ_Δ capturing the stability of knowledge propagation. These are defined respectively in Eq. (6), Eq. (7) and Eq. (8) where $n = 100$ is the number of inputs and c_a denotes the pre-activation count for the a-th output. Table 1 reports the mean-aggregated results across all ten digit classes for each interference level for fixed initial weights.

$$\mu = \frac{1}{n}\sum_{a=1}^{n} c_a \tag{6}$$

$$\mu_\Delta = \frac{1}{n-1}\sum_{a=2}^{n}(c_a - c_{a-1}) \tag{7}$$

$$\sigma_\Delta = \sqrt{\frac{1}{n-2}\sum_{a=2}^{n}\left[(c_a - c_{a-1}) - \mu_\Delta\right]^2} \tag{8}$$

Table 1. Mean pre-activation count, mean increment, and fluctuation magnitude across interference levels, aggregated over all digit classes.

Learning Mechanism	Interference Level (%)	Mean Pre-activation	Mean Increment	Fluctuation Magnitude
LTP	0	6.43 ± 0.46	0.075 ± 0.017	2.76 ± 0.45
	20	4.93 ± 0.46	0.075 ± 0.021	4.09 ± 0.35
	30	4.11 ± 0.35	0.067 ± 0.014	4.25 ± 0.20
	50	2.97 ± 0.31	0.016 ± 0.019	4.07 ± 0.50
MIS	0	10.90 ± 0.21	0.121 ± 0.011	2.30 ± 0.25
	20	9.96 ± 0.16	0.119 ± 0.013	3.16 ± 0.22
	30	9.69 ± 0.24	0.107 ± 0.011	3.05 ± 0.25
	50	9.25 ± 0.15	0.094 ± 0.019	3.02 ± 0.20

MIS consistently produces more pre-activations than LTP across all interference levels suggesting that the progressive sparsification under MIS causes the network to generalise strongly from immediately prior inputs. As interference increases, mean pre-activation declines for both mechanisms, indicating that noisy input streams disrupt the output structure that drives consistent pre-activation. This decline is considerably steeper for LTP, suggesting that it is more sensitive to input perturbation and actively adapts its weight configuration in response to changing inputs, supporting greater representational flexibility.

The mean increment μ_Δ decreases with higher interference for both mechanisms, reflecting a slower propagation of knowledge over the learning sequence. MIS maintains a higher positive trend than LTP at all levels, consistent with its greater overall pre-activation density. Fluctuation magnitude σ_Δ increases under moderate interference before stabilising at higher levels, suggesting that the learning trajectory becomes more irregular with initial interference but settles as pre-activations reach a lower, more consistent range.

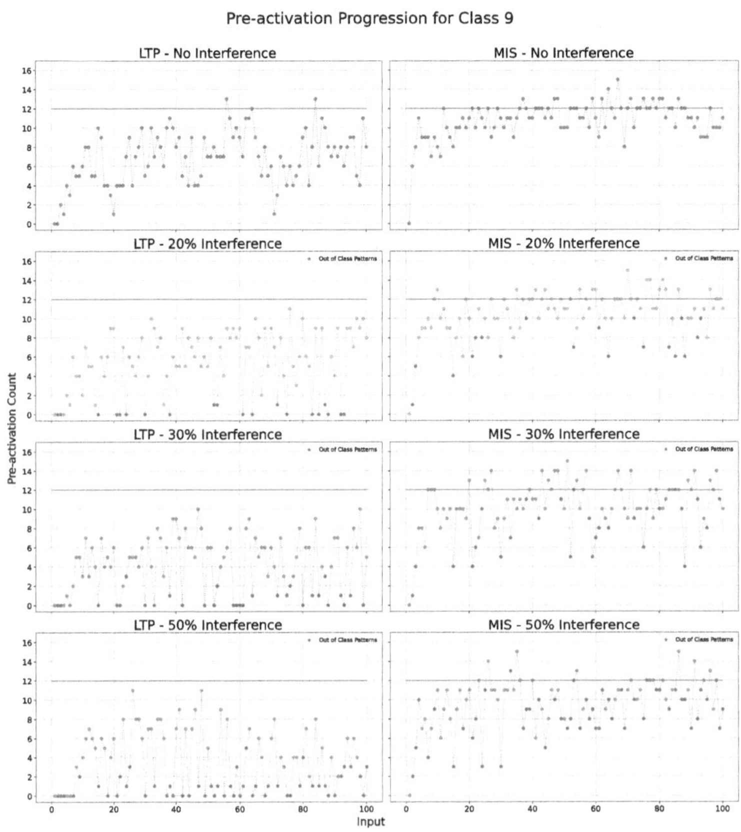

Fig. 3. Trajectory of pre-activated output neuron count throughout incremental learning for digit class 9. Red points indicate out-of-class interference inputs and the red horizontal line marks the threshold count required to form a memory engram. (Color figure online)

The trends are consistent for all classes and Fig. 3 shows the trajectories for digit class 9 as an example. Under LTP, out-of-class inputs cause sharp drops in pre-activation followed by recovery, producing high fluctuation which indicates active reorganisation of the network in response to interference. Under MIS, the same perturbations produce smaller changes, reflecting a more rigid representational structure that is less responsive to input variation. This limits the network's ability to distinguish within-class from out-of-class inputs which is a key factor in the greater drift observed under MIS.

The co-efficient of variation of the pre-activation means and fluctuations across five weight initialisations is $< 5\%$ for all conditions, while the increments show a variation $< 10\%$ for all conditions except LTP with the largest interference level where the variation is 24.5%.

3.3 Output Pattern Sequence

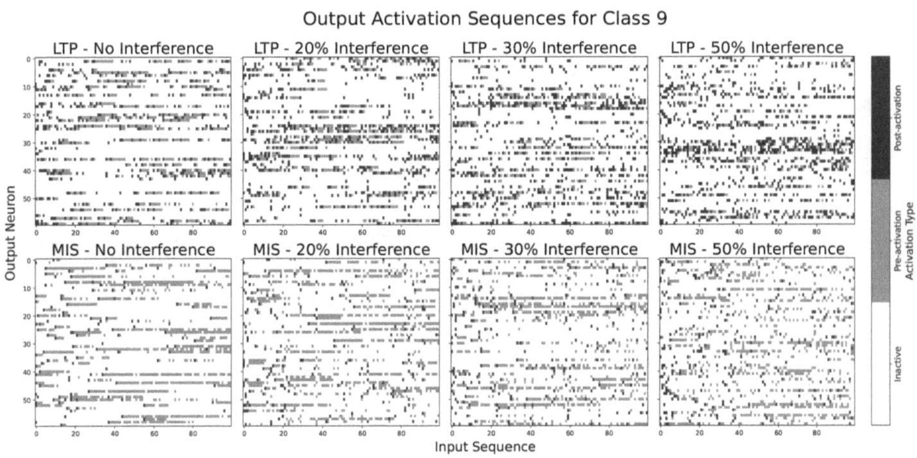

Fig. 4. Sequences of output representations under varying interference conditions, displayed as heatmaps. Yellow indicates pre-activated neurons and blue indicates neurons activated specifically by the current input. (Color figure online)

Finally, we examine how the sequence of output engrams evolves during incremental learning and how it is affected by interference. We observe similar behaviour for all classes and Fig. 4 shows class 9 as an example of the output sequences for both learning mechanisms.

Under MIS, pre-activations dominate across all interference conditions, consistent with the high pre-activation density reported in the previous section, whereas with LTP, pre-activations are comparatively limited. The consistency of pre-activated neurons across the learning trajectory is reflected in the mean pairwise overlap ratio, which declines relative to baseline by 16.5%, 19.2%, and

32.5% for MIS, and by 1.5%, 12.6%, and 33.3% for LTP, for the three interference levels, respectively. Although LTP maintains noticeably higher consistency at low to moderate interference, the two mechanisms converge toward comparable levels of disruption at the highest interference level. This pattern is consistent with the redistributive nature of MIS, where increasing sparsity leaves fewer synapses available to absorb further redistribution, gradually reducing adaptability. Under LTP, the smaller set of pre-activated neurons appears to anchor a more stable core of connections, though this advantage diminishes as interference increases.

These observations indicate that representational stability under incremental learning is not directly dependent on the quantity of pre-activations, but rather on whether they remain anchored to a consistent set of synaptic connections. LTP supports this anchoring through controlled potentiation and renormalisation, while MIS hinders it through continual weight redistribution that becomes increasingly inflexible as the network sparsity grows.

4 Discussion

This study examined how memory representations evolve under incremental learning in the presence of input interference, using a brain-inspired neural network with local synaptic learning rules. LTP converges on a stable set of repeatedly reinforced connections that persist across inputs and absorb moderate interference without significant reorganisation. MIS, by contrast, continuously redistributes weight between synapses, preventing the consolidation of a stable representational structure. From a neuroscientific perspective, this provides insights to synaptic conditions that support stable engram formation.

The progressive degradation of representational consistency with increasing interference demonstrates that even moderate proportions of out-of-distribution inputs are sufficient to vary incrementally formed engrams. Importantly, interference actively reshapes the synaptic trajectory in a way that accumulates over the input sequence. LTP exhibits a responsive and recoverable dynamic, where the extent of drift is limited, and the network partially recovers on subsequent within-class inputs. MIS lacks an equivalent stabilising effect, accumulating the effects of each interfering input into a progressively changing connectivity structure from which recovery becomes limited. This suggests that stable biological memory uses not only plasticity mechanisms that strengthen relevant connections but also complementary homeostatic processes that bound the extent of synaptic change.

The conditions studied here resemble those encountered in online and continual learning, where ANNs must acquire new representations without interfering with previously learnt ones. The LTP results suggest that local potentiation combined with a normalisation constraint can support representational stability under moderate interference. In contrast, the MIS results highlight a risk in sparsification-based learning rules. While sparse connectivity is associated with increased representational capacity, progressive weight redistribution can make the network strained as it approaches a maximally sparse configuration.

Although these findings offer meaningful insights into the dynamics of engram stability and drift, they are derived from an intentionally simple model that only uses excitatory neurons to prioritise interpretability over scale. While this design enables analysis of the mechanisms under study, it leaves open the question of how these dynamics generalise to larger networks, higher-dimensional inputs, and more complex interference patterns. Future work could extend this framework to deeper architectures or recurrent networks, where temporal dependencies in the input stream play a more prominent role in shaping representations. Further, studying LTP and MIS in combination, or alongside inhibitory plasticity (LTP), may reveal interactions that moderate the drift observed here in isolation. Finally, incorporating a retrieval component would allow drift to be analysed in terms of its functional consequences for memory recall, connecting the representational findings more directly to behaviour and to the broader goal of developing biologically grounded artificial learning systems.

References

1. Atashgahi, Z., Pieterse, J., Liu, S., Mocanu, D.C., Veldhuis, R., Pechenizkiy, M.: A brain-inspired algorithm for training highly sparse neural networks. Mach. Learn. **111**(12), 4411–4452 (2022). https://doi.org/10.1007/s10994-022-06266-w
2. Aziz, W., et al.: Multi-input synapses, but not LTP-strengthened synapses, correlate with hippocampal memory storage in aged mice. Curr. Biol. **29**(21), 3600-3610.e4 (2019). https://doi.org/10.1016/j.cub.2019.08.064
3. Bliss, T.V.P., Lømo, T.: Long-lasting potentiation of synaptic transmission in the dentate area of the anaesthetized rabbit following stimulation of the perforant path. J. Physiol. **232**(2), 331–356 (1973). https://doi.org/10.1113/jphysiol.1973.sp010273
4. Chanales, A.J.H., Dudukovic, N.M., Richter, F.R., Kuhl, B.A.: Interference between overlapping memories is predicted by neural states during learning. Nat. Commun. **10**(1), 5363 (2019). https://doi.org/10.1038/s41467-019-13377-x
5. Driscoll, L.N., Pettit, N.L., Minderer, M., Chettih, S.N., Harvey, C.D.: Dynamic reorganization of neuronal activity patterns in parietal cortex. Cell **170**(5), 986-999.e16 (2017). https://doi.org/10.1016/j.cell.2017.07.021
6. Gomez-Villa, A., Goswami, D., Wang, K., Bagdanov, A.D., Twardowski, B., van de Weijer, J.: Exemplar-free continual representation learning via learnable drift compensation. In: Leonardis, A., Ricci, E., Roth, S., Russakovsky, O., Sattler, T., Varol, G. (eds.) Computer Vision – ECCV 2024, pp. 473–490. Springer Nature Switzerland, Cham (2025). https://doi.org/10.1007/978-3-031-72667-5_27
7. Hassabis, D., Kumaran, D., Summerfield, C., Botvinick, M.: Neuroscience-inspired artificial intelligence. Neuron **95**(2), 245–258 (2017). https://doi.org/10.1016/j.neuron.2017.06.011
8. Hewavitharana, J., Anand, A., Giese, P., Moretti Ierardi, C., Steinhofel, K.: Brain inspired learning for neural networks. In: Iliadis, L., Maglogiannis, I., Kyriacou, E., Jayne, C. (eds.) Engineering Applications of Neural Networks, pp. 59–72. Springer Nature Switzerland, Cham (2025). https://doi.org/10.1007/978-3-031-96196-0_5
9. Hinton, G.E., Salakhutdinov, R.R.: Reducing the dimensionality of data with neural networks. Science **313**(5786), 504–507 (2006). https://doi.org/10.1126/science.1127647

10. Jiao, L., et al.: Brain-inspired learning, perception, and cognition: a comprehensive review. IEEE Trans. Neural Netw. Learn. Syst. **36**(4), 5921–5941 (2025). https://doi.org/10.1109/TNNLS.2024.3401711
11. Josselyn, S.A., Tonegawa, S.: Memory engrams: recalling the past and imagining the future. Science **367**(6473), eaaw4325 (2020). https://doi.org/10.1126/science.aaw4325
12. Kitamura, T., et al.: Engrams and circuits crucial for systems consolidation of a memory. Science **356**(6333), 73–78 (2017). https://doi.org/10.1126/science.aam6808
13. Lecun, Y., Bottou, L., Bengio, Y., Haffner, P.: Gradient-based learning applied to document recognition. Proc. IEEE **86**(11), 2278–2324 (1998). https://doi.org/10.1109/5.726791
14. LYNCH, M.A.: Long-term potentiation and memory. Physiol. Rev. **84**(1), 87–136 (2004). https://doi.org/10.1152/physrev.00014.2003
15. Masci, J., Meier, U., Cireşan, D., Schmidhuber, J.: Stacked convolutional auto-encoders for hierarchical feature extraction. In: Honkela, T., Duch, W., Girolami, M., Kaski, S. (eds.) Artificial Neural Networks and Machine Learning – ICANN 2011, pp. 52–59. Springer, Berlin, Heidelberg (2011). https://doi.org/10.1007/978-3-642-21735-7_7
16. Mayford, M.: The search for a hippocampal engram. Philos. Trans. Royal Soc. B: Biolog. Sci. **369**(1633), 20130161 (2014). https://doi.org/10.1098/rstb.2013.0161
17. McCloskey, M., Cohen, N.J.: Catastrophic interference in connectionist networks: the sequential learning problem. In: Psychology of Learning and Motivation, vol. 24, pp. 109–165. Elsevier (1989). https://doi.org/10.1016/S0079-7421(08)60536-8
18. Miry, O., Li, J., Chen, L.: The quest for the hippocampal memory engram: from theories to experimental evidence. Front. Behav. Neurosci.**14** (2021). https://doi.org/10.3389/fnbeh.2020.632019
19. Rao-Ruiz, P., Yu, J., Kushner, S.A., Josselyn, S.A.: Neuronal competition: microcircuit mechanisms define the sparsity of the engram. Curr. Opin. Neurobiol. **54**, 163–170 (2019). https://doi.org/10.1016/j.conb.2018.10.013
20. Rymarczyk, D., Van De Weijer, J., Zieliński, B., Twardowski, B.: ICICLE: interpretable class incremental continual learning. In: 2023 IEEE/CVF International Conference on Computer Vision (ICCV), pp. 1887–1898. IEEE, Paris, France (2023). https://doi.org/10.1109/ICCV51070.2023.00181
21. Schulz, P.E.: Long-term potentiation involves increases in the probability of neurotransmitter release. Proc. Natl. Acad. Sci. U.S.A. **94**(11), 5888–5893 (1997). https://doi.org/10.1073/pnas.94.11.5888, https://pmc.ncbi.nlm.nih.gov/articles/PMC20876/
22. Tonegawa, S., Liu, X., Ramirez, S., Redondo, R.: Memory engram cells have come of age. Neuron **87**(5), 918–931 (2015). https://doi.org/10.1016/j.neuron.2015.08.002
23. Xie, S.M., Ermon, S.: Reparameterizable subset sampling via continuous relaxations. In: Kraus, S. (ed.) Proceedings of the Twenty-Eighth International Joint Conference on Artificial Intelligence, IJCAI 2019, Macao, China, August 10-16, 2019, pp. 3919–3925. ijcai.org (2019). https://doi.org/10.24963/IJCAI.2019/544

A Tool for Formal Verification of Probabilistic Spiking Neural Networks via Weight-Discretized Quotient Abstractions

Nikan Zandian Jazi(✉), Elisabetta De Maria, and Christopher Leturc

Université Côte d'Azur, CNRS, I3S, Nice, France
{zandian,edemaria,leturc}@i3s.unice.fr

Abstract. Spiking Neural Networks (SNNs) model biological neural dynamics more faithfully than classical artificial networks, but their stochastic, event-driven computation demands probabilistic models for which deterministic abstractions are inadequate. Formal verification via probabilistic model checking then faces a fundamental barrier: the *state space explosion problem*, where the transition system grows exponentially with the number of neurons. General-purpose quotient model abstractions can in principle mitigate this growth by partitioning membrane potentials into equivalence classes, but a naïve application to SNNs discards synaptic weight information, limiting the properties that can be verified. This paper introduces a *weight-discretized quotient model abstraction* that maps continuous synaptic weights to a compact integer range while preserving the relative contribution of each synapse, and presents CogSpike, a unified workbench that integrates probabilistic SNN design, simulation, and PRISM-based formal verification within a single tool chain. The discretization is accompanied by formal correctness guarantees: a Threshold Preservation theorem ensures that firing configurations with sufficient margin are preserved, and an Asymptotic Silence theorem guarantees that spurious firing cannot persist. A case study on contralateral inhibition demonstrates the full design–simulate–verify workflow and confirms that the discretized model preserves Winner-Takes-All dynamics with a 9.3× state space reduction.

Keywords: Spiking Neural Networks · Formal Verification · Model Checking · PRISM · Quotient Abstraction · Weight Discretization

1 Introduction

Spiking Neural Networks (SNNs), the third generation of artificial neural networks [14], are modelled as directed graphs whose nodes represent neurons and whose edges represent synaptic connections that can be either *excitatory* (positive weight) or *inhibitory* (negative weight). Unlike rate-coded deep networks, SNNs communicate via discrete, asynchronous spikes whose precise timing encodes information alongside aggregate firing rates, making them a natural

L. Pasa et al. (Eds.): ICANN 2026, LNCS 17089, pp. 149–159, 2027.
https://doi.org/10.1007/978-3-032-38398-3_13

candidate for studying how real neural circuits process and transmit information. Among SNN neuron models, the Leaky Integrate-and-Fire (LIF) formulation provides analytical tractability, while biophysically detailed models such as Hodgkin–Huxley [10] and computationally efficient alternatives such as Izhikevich neurons [11] reproduce a wider repertoire of biological firing patterns at higher computational cost.

Understanding how the brain computes requires studying the *temporal dynamics* of neural circuits—how spikes propagate, interact, and give rise to emergent behaviours in small but functionally relevant topologies such as chains, convergent motifs, and recurrent loops. Modelling these dynamics demands a delicate balance: biological neurons are inherently stochastic—ion-channel noise, unreliable vesicle release, and variable axonal delays inject randomness at every stage [10,15], so probabilistic models are necessary.

One promising avenue encodes the network as a Discrete-Time Markov Chain (DTMC): probabilistic model checking can then determine whether the DTMC satisfies the behavioural properties expected by neuroscientists, such as tonic spiking under sustained input or silence without stimulation [5]. However, this approach faces a fundamental barrier: the *state space explosion problem*, where the DTMC state space grows exponentially with network size, rendering verification intractable beyond a handful of neurons.

This paper addresses that challenge. We propose CogSpike, a unified tool for probabilistic spiking neural networks that integrates three tightly coupled capabilities within a single framework: (i) *simulation* of probabilistic LIF-based SNN dynamics, (ii) *formal modelling* of the same networks as DTMCs for the PRISM model checker [13] (a tool that computes the probability with which a temporal-logic property holds over a DTMC, up to floating-point precision), and (iii) *automated model checking* of behavioural properties expressed in Probabilistic Computation Tree Logic (PCTL) [8]. The underlying neuron model employs a weight-discretized quotient abstraction that overcomes the limitations of naïve quotient models [1], which discard synaptic weight information when partitioning states into equivalence classes.

Concretely, the contributions are threefold: (1) a *weight discretization scheme* that maps continuous synaptic weights to a finite discrete range while preserving threshold feasibility and relative synaptic contributions; (2) a *Threshold Preservation* theorem ensuring that firing patterns with sufficient margin are preserved by the discretization, and an *Asymptotic Silence* theorem guaranteeing that the discretized model cannot sustain spurious firing; and (3) **CogSpike**[1], a unified workbench integrating SNN design, simulation, and formal verification, whose code generator produces a PRISM representation isomorphic to the simulation engine, enabling automated formal modelling and model checking. The complete formal proofs, additional derivations, and an empirical scaling study across seven canonical topologies are provided in the extended version of this paper [18]. The remainder surveys related work (Sect. 2) and background (Sect. 3), presents the

[1] All code and experiments are available at https://github.com/QuietRocket/CogSpike.

weight-discretized abstraction (Sect. 4) and the CogSpike workbench (Sect. 5), and reports a contralateral-inhibition case study (Sect. 6).

2 Related Work

Biological neurons exhibit significant trial-to-trial variability even under identical stimulation [6]. Classical approaches capture this stochasticity through three mechanisms: *escape noise*, which introduces a probabilistic firing threshold; *diffusive noise*, modelling stochastic spike arrivals via synaptic bombardment; and *slow noise*, which adds fluctuations to neuronal parameters [6]. These formulations underpin large-scale analyses of noisy integrate-and-fire networks [3], but they operate in continuous state spaces and do not yield discrete-state models amenable to exhaustive formal verification. Among existing simulators, Brian 2 [16] supports stochastic firing thresholds via escape noise (SDE integration), while Nengo [2], NEST [7], and BindsNET [9] are limited to noise injection at the input level. Crucially, none of these platforms support probabilistic model checking.

Our model takes a different route: rather than adding continuous noise to a differential equation, we discretize the membrane potential into threshold levels and assign each level an explicit firing probability, yielding a finite-state probabilistic model that maps directly onto a DTMC. This enables exhaustive formal verification via model checking, a capability that is fundamentally unavailable with continuous-noise formulations.

Concerning the use of formal verification for Spiking Neural Networks, De Maria et al. [5] pioneered the modelling of SNNs as timed automata, formalizing Leaky Integrate-and-Fire (LIF) neurons with parameter learning. Their work established key biological properties, e.g., tonic spiking, integrator behaviour, and excitability, as formal verification targets, and introduced the Advice Back-Propagation (ABP) algorithm for supervised parameter inference driven by model-checking counter-examples rather than continuous gradients. A formal approach to model and verify *neuronal archetypes*, which are primitive micro-circuits such as contralateral inhibition or parallel composition, was introduced in [4], allowing macroscopic network properties to be composed from formally verified building blocks. More recently, Yao et al. [17] introduced the Refractory-evolve Probabilistic LI&F (RP-LI&F) neuron model, unifying discrete-time refractory dynamics with probabilistic spike generation. Their contract-based verification approach translates SNN topologies into Discrete-Time Markov Chains (DTMCs) and specifies behavioural properties using Probabilistic Computation Tree Logic (PCTL), enabling rigorous assume/guarantee contracts.

These approaches establish the feasibility of formal SNN verification but share the *state space explosion problem*: the state space grows exponentially with network size. General-purpose quotient abstractions [1] can mitigate this in principle, but a naïve application to SNNs discards synaptic weight information;

moreover, no existing tool unifies probabilistic SNN simulation and formal verification. The present work addresses both through a weight-discretized quotient abstraction integrated into a single workbench.

3 Preliminaries

In this section, we introduce important background on the LIF model, probabilistic model checking, and quotient model abstraction.

3.1 Spiking Neural Network Model

An SNN is modelled as a directed graph $G = (V, E)$, where directed edges represent unidirectional synaptic connections and $V = V_{\text{in}} \cup V_{\text{proc}} \cup V_{\text{out}}$ partitions into input, processing, and output neurons, and $E \subseteq V \times V$ represents directed synaptic connections with integer weights $w_e \in [-100, 100]$.

Each neuron n follows Leaky Integrate-and-Fire (LIF) dynamics [5,10]. Let $r \in [0, 1]$ be the leak factor and T be the firing threshold. At each discrete time step t, the membrane potential p integrates incoming weighted spikes and decays toward rest:

$$p_n(t+1) = \max\left(0,\ r \cdot p_n(t) + \sum_{i \in \text{In}(n)} w_{i,n} \cdot y_i(t)\right) \tag{1}$$

where $y_i(t) \in \{0, 1\}$ is the spike event of presynaptic neuron i at time t. When $p_n(t+1) \geq T$, neuron n emits a spike ($y_n(t+1) = 1$) and resets to zero.

Firing is *probabilistic*: the potential maps to a discrete threshold level $L \in \{0, \ldots, k-1\}$ (k configurable, 1–10), each level yielding a firing probability. Optionally, neurons implement a three-state refractory machine—Normal, Absolute (ARP), and Relative (RRP)—with firing probability during RRP scaled by α.

3.2 Model Checking and Temporal Logics

Definition 1. *A* Discrete-Time Markov Chain *(DTMC) is a tuple* $\mathcal{D} = (S, s_0, \mathbf{P})$ *where* S *is a finite set of states,* $s_0 \in S$ *is the initial state, and* $\mathbf{P} : S \times S \to [0, 1]$ *is the transition probability matrix satisfying* $\sum_{s' \in S} \mathbf{P}(s, s') = 1$ *for all* $s \in S$ *[1].*

Behavioural properties are expressed in *Computation Tree Logic* (CTL) [1], built from the path operators $\mathbf{X}$ (*next*), $\mathbf{U}$ (*until*), $\mathbf{F}$ (*finally*), and $\mathbf{G}$ (*globally*). *Probabilistic CTL* (PCTL) [8] replaces CTL's path quantifiers with a probabilistic operator $P_{\bowtie p}[\psi]$, asserting that the probability of satisfying ψ meets the bound $\bowtie p$, $\bowtie \in \{>, \geq, =, <, \leq\}$. For instance, $P_{\geq 1}[\mathbf{F}(y_n = 1)]$ asserts that neuron n fires with probability one and $P_{\geq 1}[\mathbf{G}(y_n = 0)]$ permanent silence.

The role of a *probabilistic model checker* is to compute, given a DTMC $\mathcal{D}$ and a PCTL property ϕ, the probability with which ϕ is satisfied from the initial state, up to floating-point precision. In the context of SNN verification, this serves two purposes: (i) *validating model correctness*, i.e., confirming that the formal DTMC encoding faithfully reproduces expected neural behaviours (e.g., tonic spiking under sustained input, silence without input); and (ii) *studying the temporal dynamics* of small but functionally relevant neural configurations, such as quantifying spike propagation probabilities across chains or characterizing inhibitory gating in convergent motifs. PRISM [13] is a probabilistic model checker supporting DTMCs. Models are specified as parallel compositions of *modules* with local integer variables and guarded probabilistic transitions, synchronizing via shared labels. The global state space is the Cartesian product of all module state spaces. In our encoding every neuron module synchronises on a single global step ("tick") action: on each step a module reads its presynaptic neighbours' current spike outputs and updates simultaneously, so a spike crosses one synapse per step. PRISM offers an *explicit* engine (enumerates reachable states) and a *symbolic BDD* engine (uses Binary Decision Diagrams via the CUDD library), whose efficiency depends on variable ordering.

3.3 Quotient Model Abstraction

Quotient model abstraction [1,12] reduces the DTMC state space by partitioning *probabilistically bisimilar* states—those yielding identical firing probabilities and, under every input, transitioning to equivalent successor classes. This reduces the per-neuron state space from $|P_{\max} - P_{\min} + 1|$ values to $k+1$ threshold classes. However, a naïve application to SNNs treats all synapses uniformly, collapsing weight information: it cannot distinguish strong from weak excitatory or inhibitory contributions. The weight discretization scheme in Sect. 4 addresses this limitation.

4 Weight-Discretized Quotient Abstraction

The quotient model of Sect. 3.3 abstracts membrane potentials into equivalence classes but treats all synapses uniformly. This section introduces a *weight discretization scheme* that resolves this limitation while preserving the relative contribution of each synapse.

4.1 Weight Discretization Function

Definition 2. *Given a weight range* $[-w_{\max}, w_{\max}]$ *(typically* $w_{\max} = 100$*) and a discretization parameter* $W \in \mathbb{N}^+$*, the* weight discretization function $\delta_W : \mathbb{R} \to \mathbb{Z}$ *is*

$$\delta_W(w) = \text{round}(w \cdot W/w_{\max}) = \lfloor w \cdot W/w_{\max} \rceil ,$$

mapping original weights to the discrete range $[-W, W] \subset \mathbb{Z}$.

The function preserves relative weight magnitudes: for $W = 3$, strong excitatory ($w = 100$) maps to $\delta_3(100) = 3$, medium ($w = 67$) to $\delta_3(67) = 2$, weak ($w = 33$) to $\delta_3(33) = 1$, and inhibitory ($w = -50$) to $\delta_3(-50) = -2$. Weights are stored as static integer constants in the generated PRISM model, contributing no additional state variables.

The *weighted contribution* for neuron n with discretized incoming weights $\{w_1^d, \ldots, w_m^d\}$ replaces the binary class evolution of the original quotient model:

$$C_n = \sum_{i=1}^{m} w_i^d \cdot y_i \tag{2}$$

where $y_i \in \{0, 1\}$ is the spike output of presynaptic neuron i and m is the fan-in.

4.2 Threshold Calibration

To ensure that the discretized neuron preserves the same firing difficulty as the original, the threshold must be recalibrated to the discrete weight domain.

Definition 3. *The* discretized threshold *for a neuron with original threshold T is*

$$T_d = \lceil T \cdot W/w_{\max} \rceil .$$

The use of the ceiling function $\lceil \cdot \rceil$ (rather than rounding) ensures $T_d \geq T \cdot W/w_{\max}$, so the discretized neuron is *at least as hard* to fire as the original. This conservative calibration underlies the Asymptotic Silence guarantee (Theorem 2).

4.3 Multiplicative Leak

In the discretized model, the membrane potential decays via the same multiplicative leak as the original LIF dynamics (Eq. 1): the discretized update rule is

$$p'_n = \lfloor r \cdot p_n \rfloor + C_n \tag{3}$$

where C_n is the weighted contribution from Eq. 2 and $r \in [0, 1]$ is the leak factor. Since $r < 1$, the floor operation guarantees that $\lfloor r \cdot p_n \rfloor < p_n$ for all $p_n > 0$, ensuring strict decay in the absence of input. This preserves strict isomorphism between the simulation engine and the PRISM model.

4.4 Key Properties of the Discretization

This subsection presents the two main formal guarantees.

The following theorem ensures that discretization does not suppress any firing configuration that was possible in the original model.

Theorem 1 (Threshold Preservation). *Let $\mathcal{N}$ be a neuron with weights $\{w_1, \ldots, w_m\}$ and threshold T. If $\mathcal{N}$ can fire with margin (i.e., $\exists \mathbf{y} \in \{0,1\}^m$ such that $\sum_{i=1}^{m} w_i \cdot y_i \geq T + w_{\max}(m/2+1)/W$), then the discretized neuron $\mathcal{N}'$ with weights $\{\delta_W(w_1), \ldots, \delta_W(w_m)\}$ and threshold T_d can also fire.*

The proof is constructive and yields a practical rule for *choosing* W: a single-step firing with margin $\gamma = \sum_i w_i y_i - T$ is preserved by any $W \geq w_{\max}(m/2 + 1)/\gamma$, where m is the fan-in. Weights within $w_{\max}\, m/(2W)$ of threshold may flip—the gray zone that W trades against the state-space reduction.

Conversely, the next theorem provides a safety guarantee: the discretization cannot sustain spurious firing.

Theorem 2 (Asymptotic Silence). *Let $\mathcal{N}'$ be a discretized neuron with potential $p_t < T_d$ and leak factor $r < 1$. If the input is zero for all $t' \geq t$ (i.e., $C_n = 0$ henceforth), then $\mathcal{N}'$ will never fire.*

The guarantee is *asymptotic* rather than single-step: δ_W rounds each weight by up to 1/2, so for fan-in $m \geq 2$ the discretized contribution can exceed its scaled original by up to $m/2$, and an isolated subthreshold step is not guaranteed silent. Once the supra-threshold drive ceases, however, the strict decay $\lfloor r \cdot p \rfloor < p$ ($r < 1$) sends the potential monotonically to the absorbing value $0 < T_d$, so the neuron stays silent for the entire remaining trajectory; inhibitory input only hastens this.

4.5 Biological Property Preservation

The discretized model preserves the core LIF properties formalized by De Maria et al. [5]:

- **Tonic spiking.** Under constant input C_{in}, the neuron has non-zero firing probability iff the net gain per step overcomes the multiplicative decay, i.e., $C_{\text{in}} > T_d \cdot (1 - r)$.
- **Integrator.** The probability of immediate firing on simultaneous inputs reaches 1.0 iff $\sum \delta_W(w_i) \geq T_d$.
- **Excitability.** The expected inter-spike interval decreases monotonically as input strength increases, since stronger input yields higher net accumulation per step.

5 The CogSpike Workbench

Building on the abstraction of Sect. 4, CogSpike turns the design–simulate–verify loop into a single tool. Whereas the simulators surveyed in Sect. 2 provide stochastic dynamics but no path to formal verification, CogSpike unifies *probabilistic* SNN *design*, *simulation*, and *verification* in a single desktop workbench. The tool is implemented in Rust with an immediate-mode GUI (egui), and its core design principle is strict isomorphism: the PRISM code generator produces a DTMC representation that is isomorphic to the simulation engine, that is, both share the same mathematical model, namely the LIF dynamics of Eq. 1, the three-state refractory machine, and the probabilistic firing logic, so that verification results faithfully analyse simulation behaviour.

The workbench provides:

1. A **visual graph editor** for constructing SNN topologies as directed graphs with configurable synaptic weights, input spike generators (periodic, Poisson, burst, custom), and multi-generator combination modes (OR, AND, XOR).
2. An **isomorphic simulation engine** that executes the LIF dynamics with configurable model complexity: three presets—Deterministic (1 threshold level, no refractory), Fast (4 levels, no refractory), and Full (10 levels, ARP/RRP enabled)—allow trading biological fidelity for computational tractability. Results are visualized via raster plots, membrane potential traces, and aggregate firing statistics.
3. **Automated PRISM code generation** that translates the SNN graph into a DTMC model. The generator supports both *precise* and *weight-discretized quotient* abstraction modes (Sect. 4), and synthesizes PCTL properties for reachability, safety, and liveness verification. Per-neuron potential bounds are computed from *fan-in analysis*—bounding each neuron's reachable potential from the signs and magnitudes of its incoming weights—to minimize the state space.
4. A **verification bridge** that invokes PRISM as a background process with configurable engines (explicit, sparse, MTBDD) and solver options, parsing results inline alongside the simulation output.

6 Case Study: Contralateral Inhibition

To demonstrate the complete design–simulate–verify workflow, we apply CogSpike to a *contralateral inhibition* network, a neuronal archetype in which competing neurons mutually suppress each other until a single winner emerges [4]. Figure 1 shows the 9-neuron topology constructed in the CogSpike workbench: 3 input neurons (S1–S3) providing constant excitation ($w = +100$), 3 competing processing neurons (N1–N3) connected by mutual inhibitory synapses, and 3 output neurons (O1–O3). The inhibitory weights are *asymmetric*: N1 delivers $w = -100$ to N2 and N3 while receiving only $w = -70$ in return, predetermining N1 as the winner. The model uses $k = 4$ threshold levels, threshold $T = 80$, and leak factor $r = 0.5$.

6.1 State Space Reduction

To quantify the impact of weight discretization on this topology, we compare the reachable state space of the *precise* PRISM model against the *weight-discretized* model ($W = 6$), both using the non-refractory configuration ($k = 4$ threshold levels, no ARP/RRP).

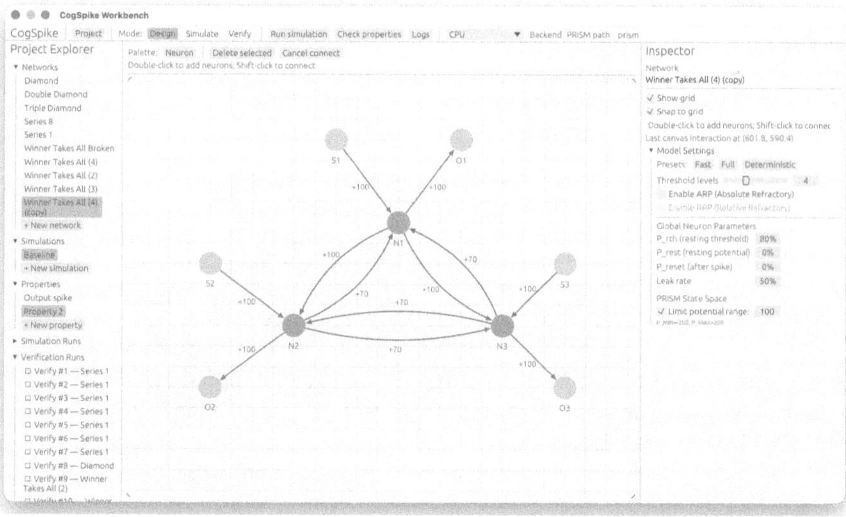

Fig. 1. The CogSpike workbench showing the contralateral inhibition topology: input neurons (orange, S1–S3), competing processing neurons (blue, N1–N3) with mutual inhibitory connections (red edges), and output neurons (purple, O1–O3). (Color figure online)

The precise model yields 3,603 reachable states and 9,917 transitions, while the $W = 6$ discretized model reduces this to just 387 states and 1,000 transitions, that is, a **9.3**× *state* reduction and a **9.9**× *transition* reduction. This compression is especially notable given the recurrent inhibitory connections of the network. The saving is not specific to this topology: across seven canonical feedforward motifs the per-neuron reduction compounds at roughly 17× per neuron (for $W = 3$), growing exponentially with network size. The full topology-by-topology measurements appear in the extended version [18].

6.2 Formal Verification of Winner-Takes-All Dynamics

Table 1 reports the PCTL verification results on both models. Three properties characterize classical Winner-Takes-All behaviour: the predetermined winner (N1) must fire infinitely often, and both losers (N2, N3) must eventually become permanently silent.

Both models verify all three WTA properties with identical probabilities: the asymmetric weight advantage makes N1 the persistent winner, while both losers are driven to permanent silence. The discretized model preserves these properties exactly with only 387 states—a 9.3× reduction from 3,603—demonstrating substantial compression without loss of verification fidelity.

Table 1. PCTL verification results on both the precise model (3,603 states, 9,917 transitions) and the discretized $W = 6$ model (387 states, 1,000 transitions).

PCTL Property	Precise	Disc. W=6	Interpretation
$P_{=?}[\mathbf{GF}(y_{\mathrm{N1}} = 1)]$	1.0	1.0	Winner fires infinitely often
$P_{=?}[\mathbf{FG}(y_{\mathrm{N2}} = 0)]$	1.0	1.0	Loser N2 eventually silent
$P_{=?}[\mathbf{FG}(y_{\mathrm{N3}} = 0)]$	1.0	1.0	Loser N3 eventually silent

7 Conclusion

Formal verification of spiking neural networks must balance faithfulness to the SNN model against the combinatorial explosion of exhaustive state-space exploration. We presented a weight-discretized quotient abstraction that maps continuous synaptic weights to a compact integer range while preserving threshold feasibility and preventing spurious sustained firing, maintaining the core LIF properties of tonic spiking, integrator behaviour, and excitability. We also introduced CogSpike, a unified workbench that integrates probabilistic SNN design, simulation, and PRISM-based formal verification. A case study on contralateral inhibition demonstrated the full design–simulate–verify workflow, formally proving Winner-Takes-All dynamics in a 9-neuron network. The $W = 6$ discretized model preserves all verified properties with identical probabilities while reducing the state space from 3,603 to 387 states—a 9.3× reduction.

Future work will pursue automated W selection from fan-in analysis and a PCTL-guided parameter-learning facility, in which property violations trigger targeted corrective updates until the key specifications are satisfied with probability close to 1.

Disclosure of Interests. The authors have no competing interests to declare that are relevant to the content of this article.

References

1. Baier, C., Katoen, J.P.: Principles of Model Checking, MIT Press (2008)
2. Bekolay, T., et al.: Nengo: a Python tool for building large-scale functional brain models. Front. Neuroinform. **7**, 48 (2014)
3. Brunel, N., Hakim, V.: Fast global oscillations in networks of integrate-and-fire neurons with low firing rates. Neural Comput. **11**(7), 1621–1671 (1999)
4. De Maria, E., et al.: On the use of formal methods to model and verify neuronal archetypes. Front. Comp. Sci. **16**(3), 163404 (2022). https://doi.org/10.1007/S11704-020-0029-6
5. De Maria, E., Di Giusto, C., Laversa, L.: Spiking neural networks modelled as timed automata: with parameter learning. Nat. Comput. **19**(1), 135–155 (2020). https://doi.org/10.1007/S11047-019-09727-9
6. Gerstner, W., Kistler, W.M.: Spiking Neuron Models: Single Neurons, Populations. Cambridge University Press, Plasticity (2002)

7. Gewaltig, M.O., Diesmann, M.: NEST (NEural simulation tool). Scholarpedia **2**(4), 1430 (2007)
8. Hansson, H., Jonsson, B.: A logic for reasoning about time and reliability. Formal Aspects Comput. **6**(5), 512–535 (1994)
9. Hazan, H., et al.: BindsNET: a machine learning-oriented spiking neural networks library in Python. Front. Neuroinform. **12**, 89 (2018). https://doi.org/10.3389/fninf.2018.00089
10. Hodgkin, A.L., Huxley, A.F.: A quantitative description of membrane current and its application to conduction and excitation in nerve. J. Physiol. **117**(4), 500–544 (1952)
11. Izhikevich, E.M.: Simple model of spiking neurons. IEEE Trans. Neural Netw. **14**(6), 1569–1572 (2003)
12. Katoen, J., Kemna, T., Zapreev, I.S., Jansen, D.N.: Bisimulation minimisation mostly speeds up probabilistic model checking. In: Tools and Algorithms for the Construction and Analysis of Systems, 13th International Conference, TACAS 2007, pp. 87–101 (2007). https://doi.org/10.1007/978-3-540-71209-1_9, https://doi.org/10.1007/978-3-540-71209-1_9
13. Kwiatkowska, M.Z., Norman, G., Parker, D.: PRISM 4.0: verification of probabilistic real-time systems. In: Computer Aided Verification - 23rd International Conference, CAV 2011, Snowbird, UT, USA, July 14-20, 2011. Proceedings, pp. 585–591 (2011). https://doi.org/10.1007/978-3-642-22110-1_47
14. Maass, W.: Networks of spiking neurons: the third generation of neural network models. Neural Netw. **10**(9), 1659–1671 (1997)
15. Nguyen, D.A., Tran, X.T., Iacopi, F.: A review of algorithms and hardware implementations for spiking neural networks. J. Low Power Electron. Appl. **11**(2), 23 (2021)
16. Stimberg, M., Brette, R., Goodman, D.F.M.: Brian 2, an intuitive and efficient neural simulator. eLife **8**, e47314 (2019)
17. Yao, Z., De Maria, E., de Simone, R.: Probabilistic spiking neural networks: formal verification and simulation. In: Formenti, E., Manzoni, L. (eds.) Unconventional Computation and Natural Computation - 22nd International Conference, UCNC 2025, Nice, France, September 1-5, 2025, Proceedings, pp. 100–114. Lecture Notes in Computer Science, Springer (2025). https://doi.org/10.1007/978-3-032-15641-9_8
18. Zandian Jazi, N., De Maria, E., Leturc, C.: A formal tool for verification of probabilistic spiking neural networks based on quotient abstractions (extended version). arXiv:2606.20674 [cs.NE] (2026). https://arxiv.org/abs/2606.20674, extended version with full proofs and a topology-dependent scaling analysis, last accessed 2026-06-23

Analytical Approach to a Large-Scale Cognitive Multi-area Language Model

Agata Feledyn[1,2](✉), Tilo Schwalger[2,3], and Thomas Wennekers[4]

[1] Freie Universität Berlin, Berlin, Germany
[2] Bernstein Center for Computational Neuroscience Berlin, 10115 Berlin, Germany
agata.feledyn@fu-berlin.de, schwalger@math.tu-berlin.de
[3] Institute of Mathematics, Technical University Berlin, 10623 Berlin, Germany
[4] School of Engineering, Computing and Mathematics, University of Plymouth, Plymouth, UK
twennekers@plymouth.ac.uk

Abstract. Understanding how semantic representations emerge from large recurrent cortical networks demands models that are both biologically grounded and analytically tractable. MatCo is a 12-area neural field model of language-related cortex widely used in cognitive studies, but its high-dimensional parameter space and substantial simulation cost hinder systematic exploration. We map the model to a set of Wilson–Cowan equations and derive closed-form conditions for saddle-node and Hopf bifurcations primarily in terms of the excitatory and inhibitory loop gains. This leads to an intuitive bifurcation diagram and allows to identify parameter regimes that support stationary input-driven, oscillatory, and bistable attractor dynamics without exhaustive simulation. Applying the analysis to a single model area, we show variability in local connectivity leads to heterogeneous firing rates within a single assembly. Drawing on homeostatic regulation in the brain and to enable tractable analysis, we normalise inputs to each column, yielding homogeneous assembly dynamics and aligning simulations with theoretical predictions. This reveals regimes from transient to sustained reverberant activity, and shows that robust inter-areal propagation requires operating near the saddle-node boundary, where recurrent amplification recruits associative areas across modalities. The framework provides a tool for parameter selection and dynamical interpretation of MatCo, and offers a transferable approach for related large-scale models.

Keywords: Language model · Associative memory · Neural dynamics · Mathematical analysis

1 Introduction

Large-scale cortical models are powerful tools for investigating how cognitive functions such as language emerge from distributed neural activity across interconnected brain areas. The MatCo model is a 12-area network that includes core

L. Pasa et al. (Eds.): ICANN 2026, LNCS 17089, pp. 160–171, 2027.
https://doi.org/10.1007/978-3-032-38398-3_14

language-related regions of the human neocortex to address precisely this question. It has been used in studies addressing symbol grounding, word and concept representation, language-attention interactions, and visual cortex engagement during speech in blind subjects [1–3]. MatCo is primarily a cognitive model that incorporates substantial physiological and anatomical detail, linking cognitive and neural mechanisms in the spirit of Hebb's cell assembly theory [4–6]. In this regard, it shares the broader aim of large-scale models such as Lansner's approach [7] or SPAUN [8]. Recent modelling efforts that pursue hyper-realistic artificial cortices currently address largely complementary questions from the neurophysiological and high-performance computing end [9,10]. Both traditions approach Hebb's neuropsychological programme from different directions.

Large-scale cortical models are difficult to control. Their extensive parameter spaces give rise to a wide range of possible spatio-temporal dynamics, which makes the interpretation of simulated activity challenging. Identifying regimes relevant to a specific research question often relies on prior experience, heuristic reasoning, or trial and error. This problem is further exacerbated by high computing resource requirements, which limit parameter searches and exploration of the dynamic repertoire of the model. Therefore, in the present work, we use analytical methods from dynamical systems and bifurcation theory to guide parameter choices in the MatCo model and gain insight into its dynamics. The analysis exploits the recurrent excitatory-inhibitory loop structure of local circuits within the model, which gives rise to distinct fixed points and oscillatory regimes depending on inputs and parameters. The mathematical framework for coupled excitatory-inhibitory populations was first developed by Wilson and Cowan [11] and has since been widely applied [12]. We extend this framework to the MatCo model, where activation patterns representing words and their meanings are stored associatively across recurrent and inter-areal connections, providing a tractable analytical entry point into how semantic memory is encoded in the brain.

2 Methods

2.1 The MatCo Model

The MatCo model comprises twelve neural fields representing cortical areas involved in language processing (Fig. 1A–B) and organised into four functional systems: visual, auditory, articulatory, and hand-motor – each with hierarchically arranged primary, secondary, and associative areas connected according to neuroanatomical data [13]. Each area is a 25×25 grid of excitatory populations (or units) with cyclic boundaries. Units receive lateral input from an 11×11 intra-areal neighbourhood and, for connected areas, from a corresponding inter-areal neighbourhood (Fig. 1C), together defining the dendritic receptive field (RF) of the unit. Each area also contains a layer of 25×25 local inhibitory neuronal populations, in which each inhibitory unit receives input from a 5×5 excitatory neighbourhood and projects back to a single excitatory population. Each excitatory-inhibitory pair in the model represents a population of neurons

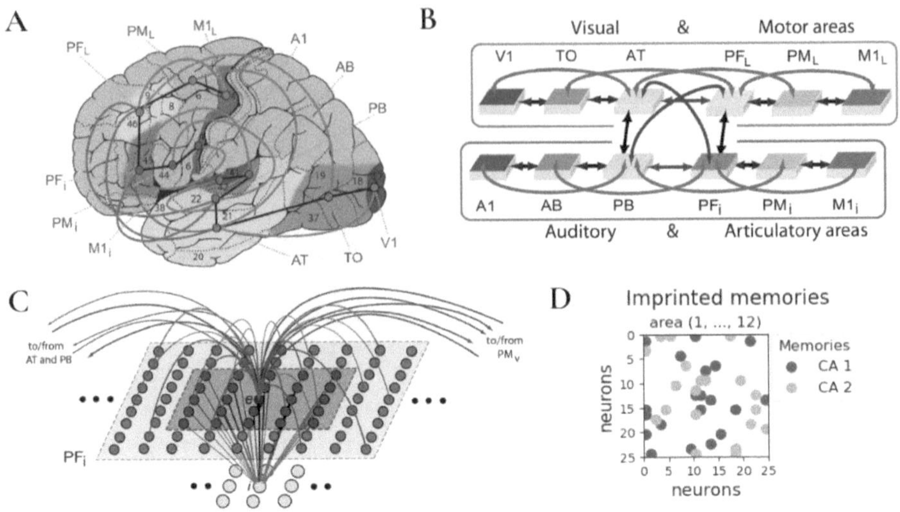

Fig. 1. Model architecture. (**A**) Cortical areas involved in language processing forming the anatomical basis of the model. (**B**) Twelve areas organised into four functional systems; arrows indicate neuroanatomically constrained between-area connectivity. (**C**) Connectivity of a single area: excitatory populations (red) receive topographic input within and between areas; local inhibitory populations (grey) receive input from a 5×5 excitatory neighbourhood and project to a single excitatory population. (**D**) Two imprinted cell assemblies (CA 1, blue; CA 2, orange), each of 20 units per area.(Color figure online)

spanning approximately 250–500 μm, corresponding to a single cortical column. In addition, each area includes a single slow global inhibitory unit that sums total excitatory activity within the area and inhibits all excitatory units proportionally. External input is delivered to one of the primary areas V1, A1, $\mathrm{M1_i}$, or $\mathrm{M1_L}$. Synaptic weights are set by one-shot rather than online Hebbian learning [1,13], giving precise control over connectivities and direct comparability with theoretical predictions. Prior to a simulation, P cell assemblies CA_μ ($\mu = 1, \ldots, P$) are imprinted, each comprising the same 20 random units per area for simplicity with excitatory synaptic weights set as defined later (Fig. 1D).

The membrane potential, x_i^a, of an excitatory unit i in area a evolves as:

$$\tau_E \frac{dx_i^a}{dt} = -\, x_i^a + J^{\mathrm{rec}} \sum_{j=1}^{N_E} w_{ij}^a\, f^E\left(x_j^a\right) + J^{\mathrm{ffb}} \sum_{\substack{b=1\\ b\neq a}}^{A} \sum_{j=1}^{N_E} w_{ij}^{ab}\, f^E\left(x_j^b\right) \\ - J^{\mathrm{linh}} \sum_{j=1}^{N_I} w_{ij}^{\mathrm{EI}}\, f^I\left(y_j^a\right) - J^{\mathrm{ginh}}\, \varphi_S^a + I_i^a + \sigma \xi_i^a \tag{1}$$

where $N_A = N_I = 25 \times 25$ are the numbers of excitatory and inhibitory populations. I_i^a is external input. ξ_i^a are i.i.d. uniform noise drawn from an interval with mean zero and variance one. σ scales the impact of the noise. τ_E is the excitatory membrane time constant and J^{rec}, J^{ffb}, J^{linh} and J^{ginh} are scaling factors for excitatory within-area, between-area, local inhibitory, and global inhibitory inputs. w_{ij}^{a} are recurrent excitatory weights within area a, w_{ij}^{ab} are excitatory synaptic weights from unit j in area b to unit i in area a (zero for non-connected pairs), and w_{ij}^{EI} are weights from inhibitory unit j to excitatory unit i in area a.

The activity of inhibitory unit i in area a evolves as:

$$\tau_{lI}\frac{dy_i^a}{dt} = -y_i^a + \sum_{j=1}^{N_E} w_{ij}^{\text{IE}}\, f^E(x_j^a) \tag{2}$$

where y_i^a is the membrane potential of inhibitory population i in area a, τ_I is the inhibitory time constant, and w_{ij}^{IE} are synaptic weights from excitatory population j to inhibitory population i within area a.

The slow global inhibition φ_S^a of area a evolves as:

$$\tau_S\frac{d\varphi_S^a}{dt} = -\varphi_S^a + \sum_{i=1}^{N_E} f^E(x_i^a) \tag{3}$$

where τ_S sets the time-scale. φ_S^a sums the total excitatory activity in area a.

The excitatory activation function f^E is piecewise linear with a lower threshold at 0, slope of 1 and saturation at 1. The inhibitory activation function f^I is piecewise linear with a lower threshold at 0, slope of 1, and no upper saturation. Euler integration is used for the simulations ($\Delta t = 0.25$ ms).

Excitatory connectivities, w_{ij}^{ab}, are defined as follows: We consider P memory items labelled $\mu = 1, \ldots, P$. Units in the network can be selective to one of the items, or none. The set of neural units selective for item μ form a global cell assembly called CA_μ. If excitatory unit i of area a and excitatory unit j of area b both belong to the same cell assembly and areas a and b are connected, we set $w_{ij}^{ab} = \bar{w}^{EE} > 0$; otherwise, $w_{ij}^{ab} = 0$. This includes the within-area case $a = b$.

The connectivity w_{ij}^{IE} between excitatory units j and inhibitory units i is defined under two schemes explored in this paper. In the neighbourhood scheme, $w_{ij}^{IE} = \bar{w}^{IE} \cdot g(d_{ij})$ if excitatory unit j falls within the 5×5 neighbourhood centred on inhibitory unit i, and zero otherwise. Here $g(d_{ij})$ is a Gaussian function of the distance d_{ij} between units i and j (defined similarly to [1]). In the one-to-one scheme, $w_{ij}^{IE} = \bar{w}^{IE}$ if $i = j$ and zero otherwise. In both schemes, each inhibitory population projects back to its local single excitatory population, and the inhibitory-to-excitatory weights are set to a fixed value $\bar{w}^{EI} > 0$.

Parameters held constant throughout the study are: $\tau_E = 2.5$ ms, $\tau_{lI} = 5.0$ ms, $\tau_S = 12$ ms, $P = 2$, $\bar{w}^{EE} = \bar{w}^{IE} = \bar{w}^{EI} = 1.0$. The noise strength is set to $\sigma = .02$ in Sect. 3.2, and 0 in Sects. 3.3 – 3.4 for a cleaner analysis.

2.2 Wilson-Cowan Dynamics

The Wilson-Cowan equations describe the mean-field dynamics of coupled excitatory and inhibitory neural populations [11]. We adopt an ODE formulation adjusted to the specifics of MatCo model:

$$\tau_E \dot{x} = -x + J_{EE} f^E(x) - J_{EI} f^I(y) + I \tag{4}$$

$$\tau_I \dot{y} = -y + J_{IE} f^E(x) \ . \tag{5}$$

Here, x and y are the membrane potentials of excitatory and inhibitory populations, I is external input, and J_{EE}, J_{EI}, J_{IE} are coupling strengths. These equations will be used to approximate the local dynamics of an excitatory-inhibitory pair within the MatCo model (see Sect. 3.2) yielding fixed point and stability diagrams against which the full model can be tested. The slow inhibition could be included in (4) and (5) but this is not required for the rest of the paper.

3 Results

3.1 Fixed Point and Stability Analysis

We perform a linear stability analysis of fixed points, x^*, y^*, of (4) and (5) by requiring $\dot{x} = \dot{y} = 0$. This straightforwardly yields $y^* = J_{IE} f^E(x^*)$ and

$$x^* - I = (J_{EE} - J_{EI} J_{IE}) f^E(x^*) =: C f^E(x^*) \ . \tag{6}$$

As a function of x^*, the left-hand side (LHS) of (6) is a line shifting rightwards for increasing I, and the right-hand side (RHS) is f^E scaled by $C := J_{EE} - J_{EI} J_{IE}$ (Fig. 2A). If inhibition dominates, i.e. C is negative or small $C < 1$, only one fixed point exists, which is suppressed ($C < 0$) or amplified ($0 < C < 1$). Above a critical value of C, three fixed points co-exist – a zero-activity background state, an unstable intermediate, and an upper state with $f^E(x^*) = 1$ – giving rise to saddle-node bifurcations and hysteresis (Fig. 2B).

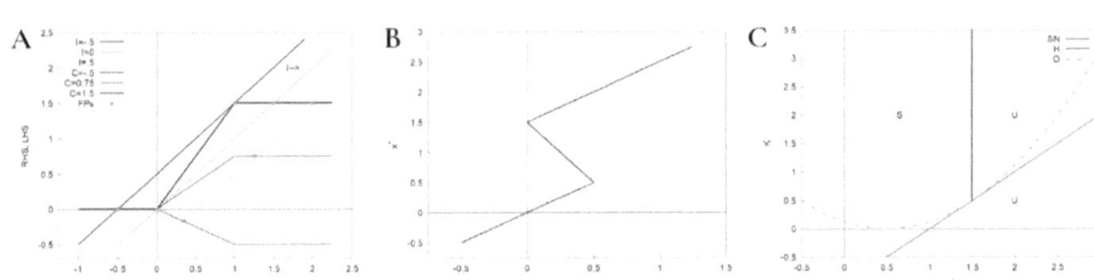

Fig. 2. Wilson-Cowan model analysis for a piecewise linear rate function. (A) Graphical representation of the fixed point equation (6) for three values of $C = J - K$. **(B)** Fixed point diagram with hysteresis ($C = 1.5$). **(C)** Stability diagram (s:stable, u:unstable).

To analyse stability, we consider perturbations $(\Delta x, \Delta y) \sim e^{\lambda t}$ with $\lambda = \alpha + i\omega$ around a fixed point. $\alpha < 0$ implies exponential relaxation to the fixed point (stable), whereas $\alpha > 0$ implies divergence (unstable). For $\omega > 0$, trajectories oscillate about the fixed point, with amplitude decaying or growing depending on α. The case $\alpha = 0$ is critical, marking the transition from stability to instability.

Inserting the Ansatz into (4)–(5) and linearising $f^E(x^* + \Delta x) \approx f^E(x^*) + f'^E(x^*)\Delta x$ yields a system of linear equations with non-trivial solutions only if the characteristic polynomial $D(\lambda)$ of the linearised equations vanishes:

$$D(\lambda) = \tau_E \tau_I \lambda^2 + (\tau_E + (1 - J)\tau_I)\lambda + (1 - J + K) \ . \tag{7}$$

Here, $J := J_{EE}\beta$ and $K := J_{EI}J_{IE}\beta$ are loop gains with $\beta := f'^E(x^*)$ characterising the impact of a perturbation on the membrane potential via the excitatory and inhibitory loops. For our rate function, $f'^E = 1$ on the rising branch and zero elsewhere, so on the whole rising branch $J = J_{EE}$ and $K = J_{EI}J_{IE}$.

Setting $\alpha = 0$ in $D(\lambda) = 0$ yields two bifurcation lines in J-K space, depending on whether ω is also zero or not (shown in Fig. 2C):

$$SN: \quad D(\lambda = 0) = 0 \longrightarrow K = J - 1 \tag{8}$$

$$H: \quad D(i\omega = 0) = 0 \longrightarrow J = 1 + \frac{\tau_E}{\tau_I} \quad \text{and} \quad \omega^2 = \frac{1 - J + K}{\tau_E \tau_I} \tag{9}$$

The first condition indicates a saddle-node bifurcation (SN), the second a degenerate Hopf bifurcation (H), where a family of marginally stable local limit cycles exist at the bifurcation point. Oscillation amplitudes increase in the unstable regime until bounded by the upper or lower threshold in f^E. This is not a local phenomenon and, therefore, not covered by a linear stability analysis. Values below the SN-line are unstable; activation grows until it reaches the upper fixed point. Above the SN-line and close to the H-line, solutions are oscillatory with frequencies given by ω in (9): ω is zero where the SN- and H-lines meet and increases with K along the H-line. For arbitrary J and K the solutions $\lambda = \alpha + i\omega$ of the quadratic polynomial (7) determine the decay or growth rate of perturbations, as well as frequencies if $\omega \neq 0$. This indicates that to the left of the H-line and above the SN-line solutions are stable and relax to the only fixed point in that region. Near either bifurcation line $\alpha \to 0$, so that relaxation times $1/\alpha$ can greatly exceed τ_E and τ_I. Finally, considering $D(\alpha \neq 0, i\omega = 0) = 0$ we can determine that along the O-line in Fig. 2C oscillation frequencies become zero. This line, however, does not indicate a bifurcation because $\alpha \neq 0$.

3.2 Mapping the MatCo Model to the Wilson-Cowan Dynamics

Parameters such as inputs, time constants, thresholds, and rate functions correspond directly between the MatCo and Wilson-Cowan models. Regarding coupling terms, we first consider a single area and set $J^{ffb} = 0$ in (Eq. 1). We focus on the activity of an assembly in a single area. The total recurrent excitatory synaptic efficacy received by a unit i can then be identified with the Wilson-Cowan coupling parameter as $J^i = J^{rec} \sum_j w_{ij}$, where the sum runs over assembly units j within a recurrent RF of the postsynaptic cell. This weight is specific to unit i as it depends on how many assembly units fall within its RF. An example distribution for CA 1 is shown in Fig. 3C.

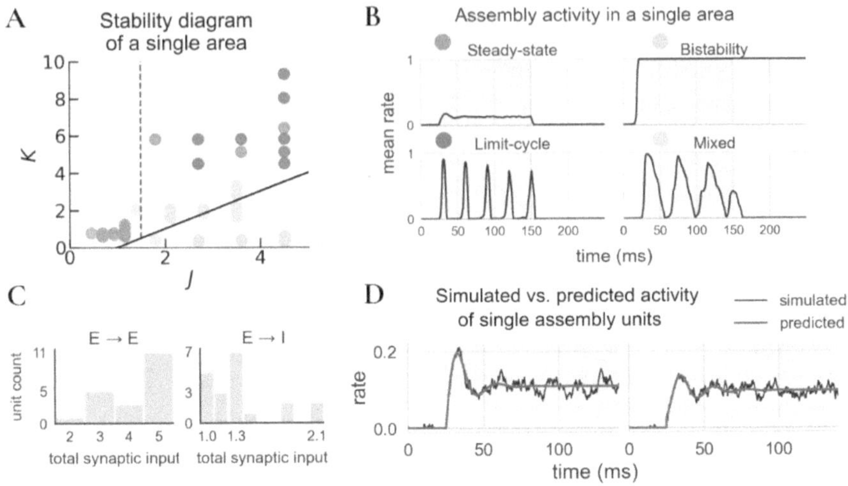

Fig. 3. A cell assembly in a single area has a broad distribution of effective excitatory and inhibitory connectivities, resulting in heterogeneous dynamics. (**A**) Stability diagram for CA 1 in area A1. Each dot denotes one cortical column, positioned by its effective connection strengths J^i and K^i. Colours represent the same CA population with differently scaled J and K. The dashed and solid lines indicate Hopf and saddle node bifurcations, respectively. (**B**) Mean firing rates for the parameter settings shown in panel A during transient stimulation (grey shading). (**C**) Distributions of total excitatory input to E-units and I-units in CA 1, showing substantial variability in effective coupling strength. (**D**) Simulated vs. theory-predicted activity for single selected units of the CA corresponding to the blue case in panel A).(Color figure online)

The inhibitory term is mapped analogously. For each excitatory population i, we define the inhibitory loop gain as $J_{IE}^{(i)} J_{EI}^{(i)}$. Here, $J_{IE}^{(i)} = J^{linh} \sum_j w_{ij}^{IE}$ denotes the total excitatory weights from assembly units onto the corresponding inhibitory population j. The sum of excitatory weights is taken over the RF of the inhibitory population. $J_{EI}^{(i)}$ denote the feedback from that inhibitory population j to the excitatory population i. As Fig. 3 shows a single assembly does not generally occupy one well-defined dynamical regime (pink circles), although this can be reached using suitable coupling strengths (blue, red, and grey circles).

3.3 Homogeneous Single-Area Dynamics

The broad spread of effective couplings in Fig. 3 arises from unequal afferent input into assembly units and may capture genuine cortical heterogeneity, although dynamic and homeostatic mechanisms can still constrain assemblies toward similar functional regimes [14]. We technically enforce homogeneity by normalising incoming weights per unit, dividing each afferent weight by the total incoming synaptic weight. Normalisation collapses the assembly columns to a single operating point while preserving their connectivity (Fig. 4A).

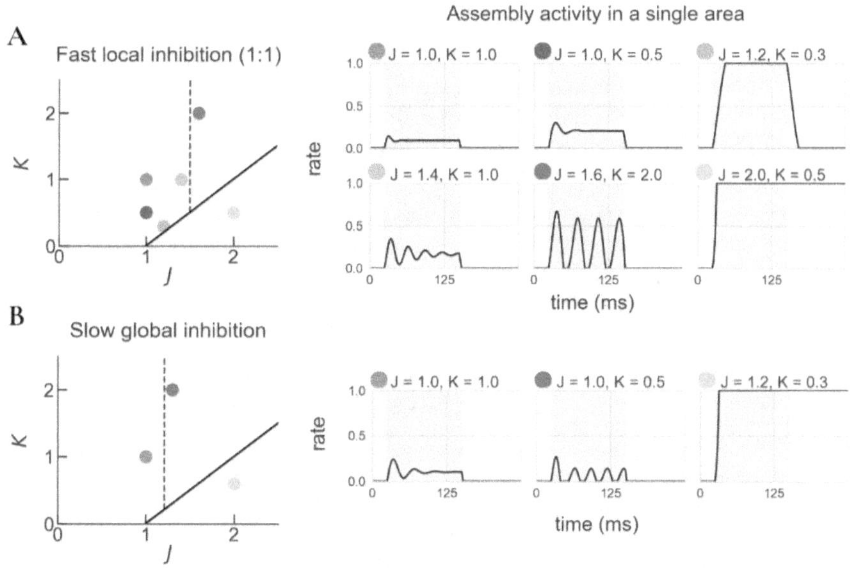

Fig. 4. Single area dynamics of a homogenous local assembly. (**A**) Fast local inhibition with one-to-one E-I-E coupling. Left: stability diagram for CA 1, colours indicate different J and K parameters for the assembly. Right: overlaid rate traces of all CA 1 units for representative parameter settings with transient stimulation (grey shading). (**B**) As in the panel A, but for slow global inhibition.(Color figure online)

Inhibitory synapses could be normalised in the same way, given that inhibitory circuits constrain excitatory firing via homeostatic and inhibitory plasticity [15]. However, inhibitory receptive fields in the original MatCo model are unrealistically wide, spanning several columns, while inhibitory projections are local, within a single column [16] or neighbouring columns [17].

We therefore impose one-to-one E–I–E loops where each excitatory population activates a dedicated inhibitory population and vice-versa. We compare this to slow global inhibition, where a single inhibitory unit receives excitation from all excitatory populations in an area and feeds back proportional inhibition, approximating an area-wide modulatory signal.

Under a transient stimulus to a single cell assembly, both inhibitory architectures admit a Wilson–Cowan description. Because all excitatory populations fire uniformly, they experience effectively identical inhibitory drive through local E–I loops or global feedback. The resulting single-area dynamics can therefore be predicted using a 2-population Wilson–Cowan framework, with inhibition scaled by J^{linh} for the local and J^{ginh} in the global case, and with the two architectures differing in inhibitory timescale (fast τ_{lI} vs. slow τ_S).

The resulting dynamics are shown in Fig. 4 (noise-free). After normalisation, all columns within the stimulated assembly behave identically, so the traces are single-population responses rather than averages. Each parameter setting corresponds to a dynamical regime. In the stable regime, activity is input-driven and decays after stimulus offset; closer to the saddle-node bifurcation, it amplifies and slows due to reverberation. Near the H-line on the stable side, responses show damped oscillations; beyond it, sustained oscillations emerge. Crossing the SN-line yields a persistent upper fixed point, so activity remains elevated after stimulus removal, and the system becomes bistable. The same regimes arise for global inhibition (Fig. 4B), but slower inhibition shifts the H-line to lower J, permitting oscillations where fast inhibition would stabilise fixed points.

3.4 Multi-area Dynamics and Cortical Propagation

We consider the full 12-area network to examine how local regimes shape inter-areal activity. Excitatory input is normalised separately for recurrent and inter-areal projections, preserving MatCo anatomy while equalising total input across areas, placing all areas in a common stability framework (Fig. 5A).

We use a configuration with 80% intra- and 20% inter-areal excitation, consistent with empirical estimates of local versus long-range synaptic input [18]. To test propagation across regimes, we sample trajectories perpendicular and parallel to the SN line that cross the Hopf boundary (Fig. 5A). Transient stimulation of a cell assembly in A1 mimics auditory input; propagation to V1 reflects cross-modal semantic retrieval. We measure propagation efficiency by taking the peak CA rate in V1 (Fig. 5B).

In the stable regime, propagation from A1 to V1 is strongest near the saddle-node bifurcation, it increases with proximity to the SN line and higher loop gain J (Fig. 5C,D). Crossing the Hopf boundary enhances propagation, especially near the SN line (Fig. 5D). Stable responses relax to a fixed point, whereas oscillatory regimes sustain limit cycles that amplify signals, yielding stronger, more persistent V1 activation. Near the SN boundary, reduced damping allows inputs to accumulate across cycles; farther away, stronger damping limits this gain. Thus, propagation is boosted by both critical amplification near SN and oscillatory regeneration beyond Hopf, with strongest coupling when both act.

4 Discussion

This paper presents an analytical treatment of a system-scale multi-area neural network. The MatCo model has until now been explored exclusively through simulations, which require expertise, computing resources, and tedious parameter searches. We map it to an adjusted set of Wilson–Cowan equations yielding particularly simple fixed point and stability criteria. This can substantially reduce the effort required to navigate parameter spaces and, in many cases, replace the need for simulations.

Several technical constraints of the present analysis can be relaxed without altering results qualitatively, so other large-scale models with recurrent excitatory-inhibitory loops structure may benefit from similar treatment. Local inhibitory-to-inhibitory couplings and both inhibitory architectures (fast-local

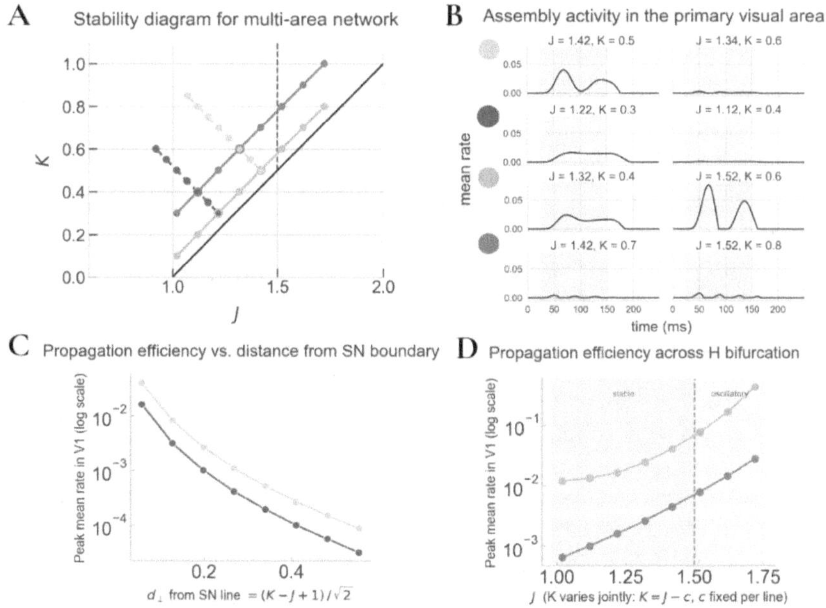

Fig. 5. Multi-area propagation depends on proximity to a local bifurcation. (**A**) Stability diagram of the multi-area model (intra-/inter-areal excitatory drive fixed at a 4:1 ratio, inputs normalised per assembly unit). Coloured dots mark four parameter trajectories in the (J, K) plane, used in B–D: two (pink) run perpendicular to the SN line at varying distance from it; two (green) run parallel to the SN line, at fixed distance, while crossing the H boundary (dashed). (**B**) Mean simulated rate of excitatory population in V1 following transient stimulation of A1 (grey shading). Each trace represent different example of J and K parameters shown in A. (**C**) Propagation efficiency (peak mean V1 rate) as a function of distance from the SN line, for the two pink trajectories in A. (**D**) Propagation efficiency as a function of J along the two green trajectories in A, crossing from the stable (blue) to oscillatory (orange) regime at the H boundary.(Color figure online)

and slow-global) can be incorporated simultaneously; the analysis is more involved but remains exactly solvable, with identical fixed point diagrams and minor changes to the stability diagrams. The analytic simplicity of the diagrams follows from the choice of piecewise linear rate functions. Replacing f^E with a smooth sigmoid leaves the fixed point and stability diagrams (Fig. 2A–C) qualitatively similar, but loop gains J and K then depend on the slope f'^E at the fixed point, making damping constants α and frequencies ω fixed point dependent. Replacing f^I with sigmoid changes the fixed point diagram (Fig. 2A,B) substantially, admitting additional fixed points and bifurcations (see e.g. [19]). A threshold-linear f^I is nonetheless physiologically reasonable, as inhibitory population rates tend to grow proportionally with excitatory drive and saturate less sharply.

The heterogeneity of effective coupling strengths revealed without weight normalization (Sect. 3.2) is unlikely to be a pure model artefact: long-tailed, strongly inhomogeneous synaptic and effective connectivity has been reported in cortical circuits [20], and unequal recruitment of interacting columns is plausible under extended stimulation, for example in V1 or inferior temporal cortex during complex object viewing [21]. Yet such inhomogeneity must coexist with stereotyped stimulus-evoked responses within cortical columns and with coherent mesoscopic signals such as EEG and steady-state responses, which depend primarily on temporally aligned population activity, and can arise even with substantial heterogeneity in single-synapse strengths [22]. This suggests that biological circuits actively constrain effective coupling within assemblies, for instance via synaptic and intrinsic homeostatic plasticity, including synaptic scaling and related mechanisms that stabilize firing rates and excitation–inhibition balance [23]. We capture these constraints phenomenologically by normalizing input weights, leaving detailed biological mechanisms to future studies.

The present work naturally raises the question of a cortical "set point" under external input. We hypothesize that operating near both the H and SN bifurcation boundaries could confer dynamical flexibility, by supporting rich oscillatory dynamics [24] and prolonged integration windows [25], potentially favourable for the large-scale propagation and integration of linguistic information.

Acknowledgments. This study was funded by the European Research Council, European Union, Advanced Grant "Material Constraints Enabling Human Cognition" (ERC-2019-ADG 883811).

Disclosure of Interests. The authors have no competing interests to declare that are relevant to the content of this article.

References

1. Garagnani, M., Wennekers, T., Pulvermüller, F.: A neuroanatomically grounded Hebbian-learning model of attention-language interactions in the human brain. Eur. J. Neurosci. **27**(2), 492–513 (2008)
2. Tomasello, R., Garagnani, M., Wennekers, T., Pulvermüller, F.: Brain connections of words, perceptions and actions: a neurobiological model of spatio-temporal semantic activation in the human cortex. Neuropsychologia **98**, 111–129 (2017)
3. Tomasello, R., Wennekers, T., Garagnani, M., Pulvermüller, F.: Visual cortex recruitment during language processing in blind individuals is explained by Hebbian learning. Sci. Rep. **9**(1) (2019)
4. Hebb, D.O.: The Organization of Behavior: A Neuropsychological Theory. John Wiley & Sons, New York (1949)
5. Palm, G.: Neural Assemblies: An Alternative Approach to Artificial Intelligence, Studies of Brain Function, vol. 7. Springer-Verlag, Berlin, Heidelberg (1982)
6. Pulvermüller, F.: Words in the brain's language. Behav. Brain Sci. **22**(2), 253–336 (1999)
7. Lansner, A.: Associative memory models: from the cell-assembly theory to biophysically detailed cortex simulations. Trends Neurosci. **32**(3), 178–186 (2009)

8. Eliasmith, C., et al.: A large-scale model of the functioning brain. Science **338**(6111), 1202–1205 (2012)
9. Potjans, T.C., Diesmann, M.: The cell-type specific cortical microcircuit: relating structure and activity in a full-scale spiking network model. Cereb. Cortex **24**(3), 785–806 (2014)
10. Pronold, J., et al.: Multi-scale spiking network model of human cerebral cortex. Cereb. Cortex **34**(10) (2024)
11. Wilson, H.R., Cowan, J.D.: Excitatory and inhibitory interactions in localized populations of model neurons. Biophys. J. **12**(1), 1–24 (1972)
12. Destexhe, A., Sejnowski, T.J.: The Wilson-cowan model, 36 years later. Biol. Cybern. **101**(1), 1–2 (2009)
13. Tomasello, R., Garagnani, M., Wennekers, T., Pulvermüller, F.: A neurobiologically constrained cortex model of semantic grounding with spiking neurons and brain-like connectivity. Front. Comput. Neurosci. **12** (2018)
14. Gainey, M.A., Feldman, D.E.: Multiple shared mechanisms for homeostatic plasticity in rodent somatosensory and visual cortex. Philos. Trans. Royal Soc. B: Biolog. Sci. **372**(1715), 20160157 (2017)
15. Vogels, T.P., Sprekeler, H., Zenke, F., Clopath, C., Gerstner, W.: Inhibitory plasticity balances excitation and inhibition in sensory pathways and memory networks. Science **334**(6062), 1569–1573 (2011)
16. Levy, R.B., Reyes, A.D.: Spatial profile of excitatory and inhibitory synaptic connectivity in mouse primary auditory cortex. J. Neurosci. **32**(16), 5609–5619 (2012)
17. Matsumura, M., fen Chen, D., Sawaguchi, T., Kubota, K., Fetz, E.E.: Synaptic interactions between primate precentral cortex neurons revealed by spike-triggered averaging of intracellular membrane potentials<i>in vivo</i>. J. Neurosci. **16**(23), 7757–7767 (1996)
18. Markov, N.T., Ercsey-Ravasz, M., Essen, D.C.V., Knoblauch, K., Toroczkai, Z., Kennedy, H.: Cortical high-density counterstream architectures. Science **342**(6158) (2013)
19. Izhikevich, E.: Dynamical Systems in Neuroscience: The Geometry of Excitability and Bursting, vol. 25. MIT Press (01 2007)
20. Mizuseki, K., Buzsáki, G.: Preconfigured, skewed distribution of firing rates in the hippocampus and entorhinal cortex. Cell Rep. **4**(5), 1010–1021 (2013)
21. Hu, D., Sato, T., Rockland, K.S., Tanifuji, M., Tanigawa, H.: Relationship between functional structures and horizontal connections in macaque inferior temporal cortex. Sci. Rep. **15**(1) (2025)
22. Johnson, T.D., Gallagher, A.J., Coulson, S., Rangel, L.M.: Network resonance and the auditory steady state response. Sci. Rep. **14**(1) (2024)
23. Wen, W., Prada, A.M., Turrigiano, G.G.: Modular arrangement of synaptic and intrinsic homeostatic plasticity within visual cortical circuits. Proc. Natl. Acad. Sci. **122**(22) (2025)
24. Buzsáki, G., Wang, X.J.: Mechanisms of gamma oscillations. Annu. Rev. Neurosci. **35**, 203–225 (2012)
25. Meisel, C., Klaus, A., Kuehn, C., Plenz, D.: Critical slowing down governs the transition to neuron spiking. PLoS Comput. Biol. **11**(2), e1004097 (2015)

From Concepts to Relations: Generative Models as Relational Associative Memories

Giovanni Donghi[1,2](✉), Francesco Lollato[1], Vincenzo Spinello[1], Alessandro Sperduti[1,3], Cesare Alippi[2,4], and Nicolò Navarin[1]

[1] University of Padua, Via Trieste 63, 35121 Padova, Italy
[2] Università della Svizzera italiana, Via Buffi 13, 6900 Lugano, Switzerland
giovanni.donghi@phd.unipd.it
[3] University of Trento, Via Sommarive, 9, 38123 Povo, Italy
[4] Politecnico di Milano, Piazza Leonardo Da Vinci 32, 20133 Milano, Italy

Abstract. This work investigates whether modern generative models can function as relational associative memories: systems that retrieve conceptually related patterns by conditioning on both an input instance and a relation. While classical associative memory systems reconstruct stored patterns without generalization, generative models learn distributions over concepts and may therefore support relational transformations between them. We introduce a relational extension of MNIST, containing digit pairs (00–99) linked by arithmetic relations, and train generative models under relational queries. Experiments show that these models generate relation-consistent outputs from image inputs, with diffusion models capturing multi-valued relations. The models additionally exhibit generalization capability to unseen concept-relation pairs. This suggests their potential as associative memory systems capable of relational reasoning.

Keywords: Associative memories · Generative models · Relational reasoning

1 Introduction

Associative memory is a cognitive function that allows humans to retrieve information from incomplete stimuli and to use contextual cues to recall related items. Early artificial associative memories, most notably Hopfield Networks [9], captured aspects of this behavior via attractor dynamics, but they primarily reconstruct memorized instances without generalization. Later extensions, such as Dense Associative Memories [14] and modern Hopfield formulations [21], improved capacity but remained limited to an *auto-associative* function: reconstructing memorized instances from corrupt inputs. An evolution of this paradigm is *hetero-associative* memory, where the system retrieves an individual target pattern distinct from, but related to, the input clue. While extensions

L. Pasa et al. (Eds.): ICANN 2026, LNCS 17089, pp. 172–184, 2027.
https://doi.org/10.1007/978-3-032-38398-3_15

of previous architectures address this, [13,26], they assume a single relation type and a maximum number of neighbors, preventing them from encoding diverse relations and producing novel outputs.

We argue that to achieve a fully realized associative memory we must overcome these limitations, and memory systems must be relational and generative. This perspective aligns with the constructive view of biological memory, which posits that retrieval is not a playback of static inputs, but a generative process constrained by current cues [23]. Furthermore, the relational theory of hippocampal function suggests that flexible cognition relies on the ability to bind distinct elements through flexible relations rather than fixed associations [4].

Promisingly, recent work has reframed generative models, including Variational Autoencoders (VAEs) [12] and Denoising Diffusion Probabilistic Models (DDPMs) [6], as associative memory systems with genuine generalization capabilities [1,8,19]. These models learn distributions rather than individual patterns, enabling the generation of novel but conceptually similar outputs. However, these works focus primarily on unconditional or class-conditional recall and do not address the problem of learning relational structure between concepts.

We investigate whether these models can act as *relational-associative* memories: given an example of a concept and a relation, the task is to generate an example of the related concept. This extends ideas from conditional generative modeling [25], by conditioning not on labels but on relational transformations. To study this model in a controlled setting, we introduce a relational extension of MNIST, with both single-valued and multi-valued relations over digit pairs, and adapt VAE and DDPM architectures to accept relational queries. Our results show that both models can successfully retrieve conceptually appropriate outputs for single-valued relations, and that DDPM also handles multi-valued ones. Moreover, both models exhibit generalization to unseen concept–relation pairs, indicating their potential as associative memories endowed with relational reasoning capabilities.

2 Background and Related Works

Auto-Associative Memory. Classical associative memory models store patterns as stable fixed points of a dynamical system. Hopfield Networks [9] introduced the energy-based formulation in which memories correspond to attractors, enabling robust retrieval from corrupted inputs. Later extensions such as Dense Associative Memories [14] increased memory storage capacity by modifying the energy function, achieving super-linear or even exponential capacity [3]. Modern continuous-state formulations of Hopfield Networks [21] introduce an update rule that has been shown to be mathematically equivalent to the attention mechanism in transformers. This formulation enables their integration into deep learning architectures as differentiable memory modules to store and retrieve relevant information. These models, however, are primarily utilized for auto-associative tasks, such as retrieving memorized patterns or serving as internal memory layers for deep learning architectures, and lack the ability to generalize or generate novel content. Recursive architectures such as RAAM/LRAAM

[20,26] compress structured inputs into latent vectors but rely on a maximum pre-set number of related samples and do not support heterogeneous relations or generative retrieval.

Hetero-Associative Memory. A natural extension of auto-associative memory models was introduced with Bidirectional Associative Memories [13]. Such models consist of two interconnected layers and behave as a hetero-associative content addressable memory that, given an incomplete cue in one domain, iteratively retrieves the associated pattern in the other domain. Most recently, Entropic Hetero-Associative Memory (EHAM) [18] introduced a non-neural model that frames memory retrieval as a constructive rather than merely reproductive process. In this framework, hetero-associations are stored as discrete functions in a 4D memory register using Hebbian learning. This distributed representation leads to the missing cue problem: while an input cue can identify the relevant plane in the target domain, it lacks the specific information required to extract a unique object. To resolve this indeterminacy, EHAM employs costly iterative search-based inference strategies, while maintaining the limitation of handling a single relation type.

Generative Models as Associative Memories. Generative models provide a different perspective. VAEs [2,12] learn latent variables that capture the underlying structure of a dataset, while diffusion models [6,16,24] perform iterative denoising to model complex data distributions. Recent theoretical analyses show that diffusion models can be interpreted as associative memory systems, with denoising dynamics corresponding to retrieval in an energy landscape [1,8], and identify regime transitions from memorization to generalization [19]. These models store distributions rather than specific patterns, supporting abstraction and generalization properties unattained by classical associative memories. However, their current associative memory interpretations are still auto-associative, treating noise as corruption or masking, and do not consider variability across instances of the same concept, nor more importantly do they incorporate relational structure between concepts. Several works have also explored associative extensions of generative architectures. Associative VAEs [10] introduce structured latent spaces and cross-modality associations for the same instance, while conditional generative models [7,25] provide mechanisms for conditioning on auxiliary information, such as labels or continuous attributes. These components form the basis for conditioning generative models on relational information. To the best of our knowledge, no existing generative framework has been evaluated as *relational* associative memory.

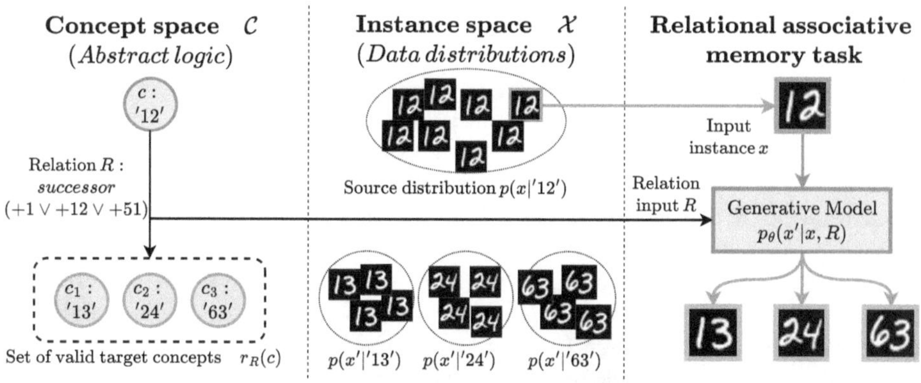

Fig. 1. Illustration of the relational associative memory task with an example from Relational MNIST-99 dataset. **Left:** concepts are associated to other concepts through relations, with possibly multiple associated concepts via the same relation. **Center:** each concept has an associated distribution in the space of instances. **Right:** for the relational associative memory task, a model is given an input instance drawn from the distribution of a source concept, a relation with an appropriate embedding, and is tasked to generate new instances according to the distributions of the related concept(s).

Limitations and Relation to the Present Work. Overall, classical architectures established the foundations of associative memory, but they remain constrained by their reconstructive nature: they excel at reconstructing a fixed patterns, but struggle to produce novel outputs, handle ambiguity or relations between concepts. This limitation is similarly present in hetero-associative frameworks like EHAM, where the system struggles to find a unique match among multiple valid associations. We propose that the solution lies in a paradigm shift from retrieval to generating from learned conditional distributions. Generative models can bypass the missing clue problem, replacing search with sampling conditioned on a relation index, additionally enabling the handling of multiple possible relations, each with multiple possible valid target concepts. Finally, a fully neural and generative relational-associative memory could also exhibit generalization capabilities beyond the instances and concepts observed during its training. The following section formalizes this task to rigorously test these capabilities.

3 Methods

We formulate relational associative memory as a conditional generation task. We consider examples drawn from different concepts. Given an input instance x representing a concept (i.e., a sample drawn from a probability distribution) and a relation R, the model must produce an output instance corresponding to the concept linked by R. Unlike standard conditional generative modeling [25], here the conditioning variable encodes a relational transformation.

The underlying assumption is that generative models, which learn structured latent spaces [6,12], may capture the relational structure of these transformations. Formally, let $\mathcal{X} \subseteq \mathbb{R}^n$ be a space of instances, $\mathcal{C}$ a finite set of concepts, and $\mathcal{R}$ a set of relations between concepts. Each concept $c \in \mathcal{C}$ is associated with a distribution $p(\cdot \mid c)$ over $\mathcal{X}$. A relation $R \subseteq \mathcal{C} \times \mathcal{C}$ is used to define $r_R(c) = \{c' \in \mathcal{C} \mid (c, c') \in R\}$. Given an input instance x drawn from $p(\cdot \mid c)$ and a relation R, the model is tasked to generate a target instance x' from a concept c' taken uniformly between the ones in $r_R(c)$, effectively approximating sampling from $p(x' \mid x, R) = \sum_{c \in \mathcal{C}} p(c \mid x) \frac{1}{|r_R(c)|} \sum_{c' \in r_R(c)} p(x' \mid c')$. In order to function as a relational associative memory, a model must therefore be able to simultaneously identify which concept an instance comes from, understand the relational structure that links concepts (with the potential challenge of multiple related concepts), and generate from the target distribution. Figure 1 illustrates this setting, with an example from the considered dataset.

3.1 Relational MNIST-99.

To evaluate relational reasoning over conceptually related images, we construct a dataset extending MNIST [15]: each integer in [0, 99] is represented by concatenating images of its digits. Each specific image represents an instance of the concept that is the number it represents (thus, we have 100 concepts). We then define six arithmetic relations $\{+1, -1, +12, -12, +51, -51\}$, mapping the concept of a number with the one resulting from the arithmetic operation if it lies in [0, 99], otherwise the relation is not present. In practice, for training and inference, a corresponding number image is used as target, and we exploit negative sampling using a black image to signal the absence of a relation, which we refer to as invalid target. The relation offsets were chosen to allow for diversity both in the target digits and the number of applicable relations for each digit pair. The final dataset contains 240k samples, uniformly covering all concept–relation pairs.

In addition to these single-valued relations, where the target concept is unique for a given query, we introduce also multi-valued ones: a *successor* relation grouping $\{+1, +12, +51\}$ and a *predecessor* relation grouping $\{-1, -12, -51\}$. These associate to a concept the set of all valid concepts (numbers) resulting from the respective single-valued arithmetic operations.

To further investigate the model's ability beyond memorization, we introduce a generalization setting over unseen concept-relation pairs, testing whether the model has learned the rule underlying each relation, rather than memorizing specific input–output mappings. While in the two experiments described above the model is exposed during training to instances of all possible concept-relation pairs, in this setting we partition the set of all such pairs into disjoint training, validation, and test subsets, containing respectively 70%, 15% and 15% of the pairs. As a result, the model is evaluated on concept–relation combinations that were not observed during training, while still involving concepts and relations seen individually. This generalization setting is defined only over the single-valued relations.

While a natural image dataset would offer greater visual complexity and variety, and more nuanced relations, they would introduce significant noise for the relational memory. We deliberately choose this controlled MNIST-based setting to decouple the difficulty of the visual perception and generative task by themselves from the challenge of relational reasoning. This ensures that any observed failures can be attributed directly to the relational associative mechanism, rather than to challenges in perception or generation. By fixing concepts as numbers and relations as unambiguous arithmetic operations, we can rigorously measure the model's ability to approximate $p(x' \mid x, R)$ without confounding factors.

3.2 Models

VAE Architecture. We extend the standard VAE [12] to accept a dual input: an image x and a relation index R. The relation is mapped to a learnable embedding vector and concatenated with the flattened image before being passed to the encoder. The encoder follows a standard design, consisting of fully connected layers with batch normalization, LeakyReLU activations, and dropout, producing the mean and log-variance of the latent distribution. The decoder mirrors this structure to reconstruct the output image from the latent representation. Crucially, the decoder receives no relational information, ensuring that the relational signal affects only the encoding stage.

DDPM Architecture. Our second architecture is a DDPM [6], adapted to operate as a relational associative memory. We use a standard U-Net [22] trained to predict the noise residual ϵ from the noisy target image x_t, conditioned on a source image x. The diffusion timestep t and the relation index R are embedded and summed into a global context vector, which is injected into each residual block.

4 Experimental Results

We present here empirical evidence of the ability of generative models to function as relational associative memories on the adapted Relational MNIST-99 dataset.

4.1 Training Details

For the VAE, we optimize the evidence lower bound (ELBO), consisting of binary cross-entropy reconstruction term and Kullback-Leibler divergence. We employ a β-VAE [5] train strategy: β is fixed at $1/128$ for the single-valued relations task, while it gradually increases from 0 to 1 for the multi-valued relations task. This choice is motivated by the need to give more weight to the KL term when multiple outputs are possible, encouraging diverse latent representations. The DDPM is trained to minimize the mean squared error between the predicted and true noise injected into the target image, with a linear noise schedule over 1000 steps. Both models are trained using Adam optimizer [11] with a batch size of 128 and early stopping based on validation loss.

4.2 Evaluation Protocol

We evaluate models according to three aspects. *Reconstruction quality*, measuring the fidelity of generated outputs, assessed both visually and quantitatively through four metrics: reconstruction loss as through pixel-wise MSE between generated and target images, Peak Signal-to-Noise Ratio (PSNR), Structural Similarity Index (SSIM), and Sharpness. *Correctness*, measuring the capacity of the model to generate images of the correct number given the query. For this, a MNIST-99 classifier was trained separately to automatically assess correctness and detect black-image outputs, achieving 99.3% test accuracy. We assess correctness separately on queries that have a *valid target* in 0–99, and on ones with *invalid target* (such as an image of 91 with relation +51) which were presented in training with a black output, and finally overall global accuracy. For multi-valued relations, we also consider *coverage* of all possible underlying single-valued operations, computing three metrics over 100 generations per query: junk rate (fraction of incorrect outputs), Wasserstein distance and Jensen-Shannon divergence (JSD) between the generated and true distributions. Both distances penalize not only the generation of incorrect classes, but also deviations of the generated distribution from the ground-truth distribution over the valid classes. To reduce noise in the generated distributions, we also apply a filtering strategy by retaining only the most frequently generated classes up to a cumulative mass q selected to minimize validation JSD.

4.3 Results and Discussion

Single-Valued Relations. Figure 2 illustrates some representative outputs for the single-valued relations. The VAE tends to generate smooth and often blurry samples, while the DDPM produces significantly sharper and more realistic images. This is reflected in the reconstruction quality metrics reported in Table 1: the VAE achieves better pixel-wise MSE and PSNR, as it tends to generate almost identical images for each target, while for these metrics the DDPM is penalized for producing a wider variety of images; on the other hand, the DDPM achieves higher SSIM and sharpness, indicating that its generated images are perceptually more realistic. Looking at the accuracy of generations in Table 1, we see that both models easily recognize invalid relations, while overall the VAE achieves higher accuracy. This could be due to the fact that the external classifier may favor the more prototypical digits generated by the VAE.

We emphasize that our aim is not to benchmark generative models against each other, but rather to demonstrate their use as associative memories. Both models are able to generate images of the correct target number given an input instance and relation, validating the central hypothesis of this work.

Multi-valued Relations. For the multi-valued relations task, Table 2 shows a clear gap between the two models. The VAE performs poorly: due to its tendency to reconstruct the mean of the target distribution, it fails to capture the diversity of multiple valid outputs, as seen in Fig. 3. In contrast, the stochastic nature

Table 1. Comparison of reconstruction metrics and accuracy between VAE and DDPM in the single-valued relations task.

Metric	VAE	DDPM
Reconstruction		
MSE ↓	**0.0541 ± 0.0001**	0.1066 ± 0.0001
PSNR (dB) ↑	**12.8060 ± 0.0092**	9.8821 ± 0.0042
SSIM ↑	0.4708 ± 0.0003	**0.6023 ± 0.0004**
Sharpness ↑	0.0712 ± 0.0001	**0.1321 ± 0.0001**
Accuracy (%) ↑		
Valid target	**95.71 ± 0.28**	87.09 ± 0.32
Invalid target	98.56 ± 0.38	**99.47 ± 0.14**
Global	**96.32 ± 0.21**	89.73 ± 0.26

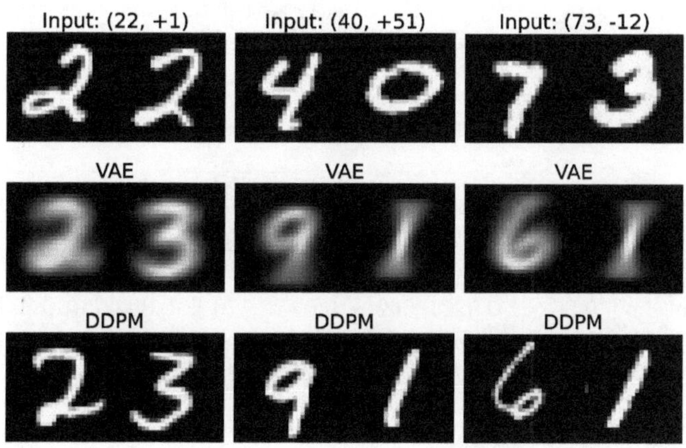

Fig. 2. Example outputs for single-valued relation queries. The first row shows the input, the second row the output generated by VAE, and the third the output generated by DDPM.

of the DDPM enables it to naturally handle multi-valued relations, generating a wider variety of valid outputs. The proposed filtering strategy improves all distribution-level metrics for both models, with a particularly strong effect for the DDPM.

Generalization on Single-Valued Relations. Finally, for the generalization task on unseen concept-relation pairs, quantitative results are reported in Table 3. The VAE outperforms the DDPM in terms of accuracy, especially in the valid targets subset, indicating that, when evaluated on unseen combinations, the VAE is able to apply the learned arithmetic rules, rather than relying on memorization of specific input–output mappings. Reconstruction quality met-

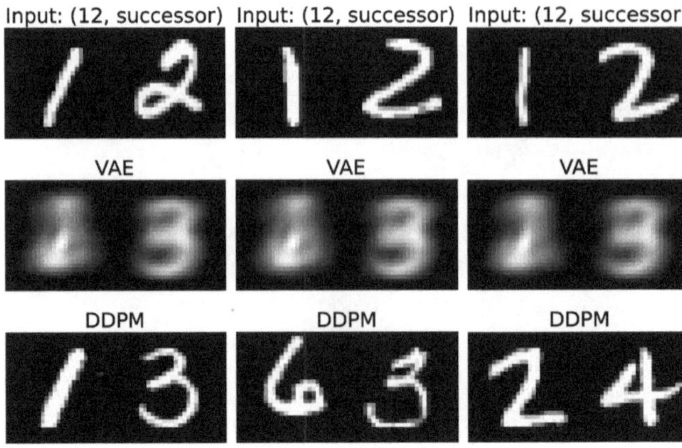

Fig. 3. Example outputs for multi-valued relation queries with the query (12, successor). The first row shows the input, the second row the output generated by VAE, and the third the output generated by DDPM.

Table 2. Average distribution-level metrics over all input pairs in the multi-valued relations task, comparing raw and filtered generations.

Metric	VAE		DDPM	
	Raw	Filtered	Raw	Filtered
Junk rate % ↓	54.18 ± 3.22	51.20 ± 3.74	14.15 ± 0.42	**0.04** ± **0.03**
Wasserstein ↓	0.661 ± 0.025	0.662 ± 0.025	0.170 ± 0.003	**0.086** ± **0.003**
JSD ↓	0.715 ± 0.020	0.707 ± 0.020	0.285 ± 0.004	**0.080** ± **0.003**

ric closely follow to those observed in the single-valued setting, and are thus omitted. Some generations are shown in Fig. 4.

To investigate the mechanism behind the VAE's superior generalization capacity, we analyzed its latent space: Fig. 5 shows a t-SNE visualization [17] of the latent representations obtained with the VAE's encoder for pairs (x, R) with resulting target number between 40 and 59 (to allow distinctions through colors). These embeddings appear to cluster according to the target number rather than input concept, indicating that the encoder successfully performs reasoning. In contrast to the DDPM, where the generation of the output is an iterative

Table 3. Comparison of accuracy between VAE and DDPM in the generalization task.

Accuracy (%) ↑	VAE	DDPM
Valid target	**83.55** ± **0.59**	67.45 ± 0.04
Invalid target	**96.76** ± **0.21**	93.40 ± 0.61
Global	**86.19** ± **0.47**	72.64 ± 0.14

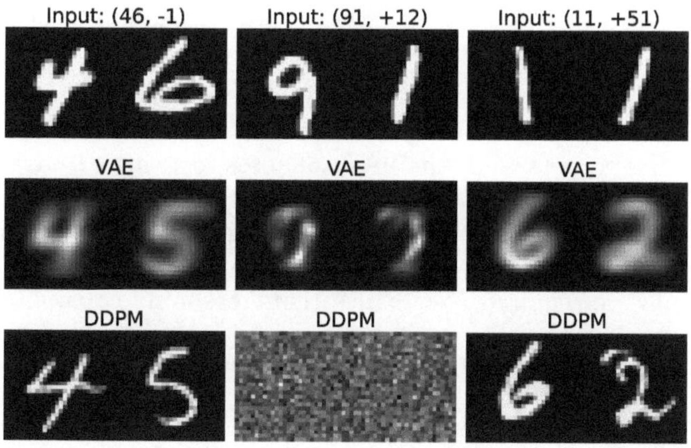

Fig. 4. Example outputs for single-valued relation queries in the generalization task. The first row shows the input, the second row the output generated by VAE, and the third the output generated by DDPM.

process distributed across time steps without a single static bottleneck representation, the VAE's latent space allows us to directly visualize this encoding or arithmetic rules. We argue that this ability to project diverse input-relation pairs onto a well-structured and consistent latent space, effectively solving the relation before decoding, is the key to the model's capacity to generalize rules to unseen

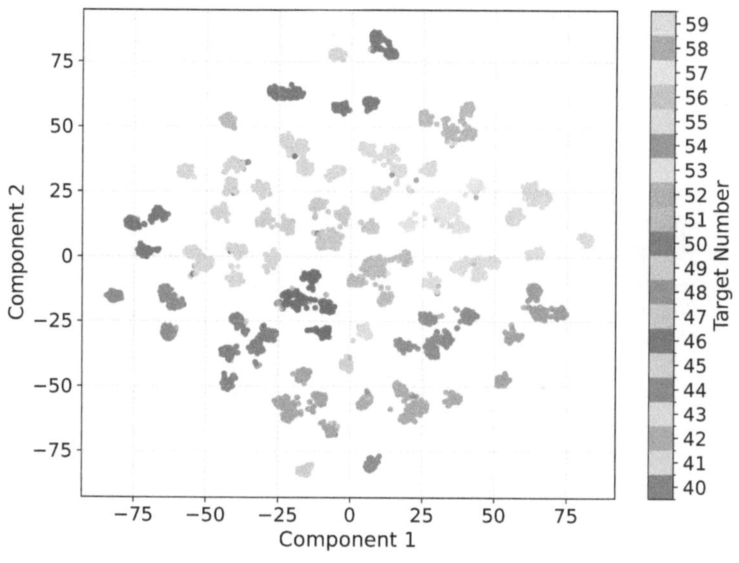

Fig. 5. t-SNE visualization of the VAE latent space for instance-relation pairs with target concepts 40–59.

combination, while also limiting its effectiveness in dealing with multi-valued relations.

In summary, our results show a trade-off between generation quality and rule abstraction. The DDPM proves superior for multi-valued relations, successfully generating diverse targets while the VAE collapses to a single blurry mean. However, this expressive power seems to lead the DDPM to have lower generalization capacity to unseen concept-relation pairs. This suggests that while diffusion models are more viable as relational associative memories for complex data, further mechanisms may be required to enhance their reasoning capabilities in unseen contexts.

5 Conclusions

This work examined whether modern generative models can operate as relational associative memories. Using a modified relational MNIST dataset, we showed that both VAEs and DDPMs can retrieve conceptually related outputs when conditioned on an input instance and an arithmetic operation. While both models successfully handle single-valued relations, their behavior diverges in more complex settings. In the multi-valued setting the VAE collapses towards mean reconstruction and fails to represent one-to-many mappings, while the DDPM is able to generate diverse and coherent output distributions. Conversely, while both model exhibit generalization capabilities over unseen concept-relation pairs, indicating that they also rely on the application of learned arithmetic rules, the VAE exhibits stronger generalization.

Despite the simple nature of the considered dataset, the results demonstrate that generative models could constitute a viable framework for relational associative memory, capable of integrating relational structure in their generative dynamics. Extending this framework to knowledge graphs and further exploring relational reasoning will be a future work.

Acknowledgments. This research was supported by: the European Union - NextGenerationEU as part of the Italian National Recovery and Resilience Plan (PNRR); the Italian Ministry of University and Research - NextGenerationEU, Mission 4 Component 2 Investment 1.3 - Call for tender No. 341 of March 15, 2022, project code PE0000013, Concession Decree No. 1555 of October 11, 2022, CUP C63C22000770006, project title "Future AI Research (FAIR) - Spoke 2 Integrative AI - Symbolic conditioning of Graph Generative Models (SymboliG)"; the Italian Ministry of University and Research under the PRIN program - Progetti di Rilevante Interesse Nazionale - PRIN 2022 (Secretary-General's Decree n. 1401 of 18/09/2024) - CUP C53C24000770006, project title "DEEP-GRAPH: Design and Theory of Deep Graph Learning".

Disclosure of Interests. The authors have no competing interests to declare that are relevant to the content of this article.

References

1. Ambrogioni, L.: In search of dispersed memories: generative diffusion models are associative memory networks. Entropy **26**(5) (2024)
2. Chen, X., et al.: Variational lossy autoencoder. In: International Conference on Learning Representations (2017)
3. Demircigil, M., Heusel, J., Löwe, M., Upgang, S., Vermet, F.: On a model of associative memory with huge storage capacity. J. Stat. Phys. **168**(2), 288–299 (2017)
4. Eichenbaum, H., Cohen, N.J.: From Conditioning to Conscious Recollection: Memory Systems of the Brain. Oxford University Press, New York (2001)
5. Higgins, I., et al.: beta-VAE: learning Basic Visual Concepts with a Constrained Variational Framework. In: International Conference on Learning Representations (2017)
6. Ho, J., Jain, A., Abbeel, P.: Denoising diffusion probabilistic models. In: Proceedings of the 34th International Conference on Neural Information Processing Systems. NIPS '20 (2020)
7. Ho, J., Salimans, T.: Classifier-Free Diffusion Guidance. arXiv:2207.12598 (2022)
8. Hoover, B., Strobelt, H., Krotov, D., Hoffman, J., Kira, Z., Chau, D.H.: Memory in plain sight: a survey of the uncanny resemblances between diffusion models and associative memories. In: Associative Memory & Hopfield Networks Workshop (NeurIPS 2023) (2023)
9. Hopfield, J.J.: Neural networks and physical systems with emergent collective computational abilities. Proc. Natl. Acad. Sci. **79**(8) (1982)
10. Jo, D.U., Lee, B., Choi, J., Yoo, H., Choi, J.Y.: Associative variational auto-encoder with distributed latent spaces and associators. In: Proceedings of the AAAI Conference on Artificial Intelligence, vol. 34, issue (07) (2020)
11. Kingma, D.P., Ba, J.: Adam: a method for stochastic optimization. In: International Conference on Learning Representations (ICLR) (2015)
12. Kingma, D.P., Welling, M.: Auto-encoding variational bayes. In: International Conference on Learning Representations (2014)
13. Kosko, B.: Bidirectional associative memories. IEEE Trans. Syst. Man Cybern. **18**(1), 49–60 (1988)
14. Krotov, D., Hopfield, J.J.: Dense associative memory for pattern recognition. In: Proceedings of the 30th International Conference on Neural Information Processing Systems. NIPS'16 (2016)
15. LeCun, Y., Bottou, L., Bengio, Y., Haffner, P.: Gradient-based learning applied to document recognition. Proc. IEEE **86**(11), 2278–2324 (1998)
16. Lim, S., Yoon, E., Byun, T., Kang, T., Kim, S., Lee, K., Choi, S.: Score-based generative modeling through stochastic evolution equations in hilbert spaces. In: Proceedings of the 37th International Conference on Neural Information Processing Systems. NIPS '23 (2023)
17. van der Maaten, L., Hinton, G.: Visualizing data using t-SNE. J. Mach. Learn. Res. **9**(86), 2579–2605 (2008)
18. Morales, R., Pineda, L.A.: The missing cue problem in hetero associative memory retrieval. Sci. Rep. **15**(1), 21850 (2025)
19. Pham, B., Raya, G., Negri, M., Zaki, M.J., Ambrogioni, L., Krotov, D.: Memorization to generalization: emergence of diffusion models from associative memory networks. In: New Frontiers in Associative Memories Workshop at ICLR (2025)
20. Pollack, J.B.: Recursive distributed representations. Artif. Intell. **46**(1) (1990)

21. Ramsauer, H., et al.: Hopfield networks is all you need. In: International Conference on Learning Representations (2020)
22. Ronneberger, O., Fischer, P., Brox, T.: U-Net: convolutional networks for biomedical image segmentation. In: Medical Image Computing and Computer-Assisted Intervention, MICCAI (2015)
23. Schacter, D.L., Addis, D.R.: The cognitive neuroscience of constructive memory: remembering the past and imagining the future. Philos. Trans. Royal Soc. B: Biol. Sci. **362**(1481), 773–786 (2007)
24. Sohl-Dickstein, J., Weiss, E., Maheswaranathan, N., Ganguli, S.: Deep unsupervised learning using nonequilibrium thermodynamics. In: ICML (2015)
25. Sohn, K., Yan, X., Lee, H.: Learning structured output representation using deep conditional generative models. In: Proceedings of the 29th International Conference on Neural Information Processing Systems - vol. 2. NIPS'15 (2015)
26. Sperduti, A.: Encoding labeled graphs by labeling RAAM. In: Proceedings of the 7th International Conference on Neural Information Processing Systems. NIPS'93 (1993)

Special Session: Spiking Neural Networks and Neuromorphic Computing

Equivalence of Approximation by Networks of Single and Multi-spike Neurons

Dominik Dold(✉) and Philipp Petersen

Faculty of Mathematics and Research Network Data Science @ Uni Vienna, University of Vienna, Kolingasse 14-16, 1090 Vienna, Austria
dominik.dold@univie.ac.at

Abstract. In a spiking neural network, is it enough for each neuron to spike at most once? In recent work, approximation bounds for spiking neural networks have been derived, quantifying how well they can fit target functions. However, these results are only valid for neurons that spike at most once, which is commonly thought to be a strong limitation. Here, we show that the opposite is true for a large class of spiking neuron models, including the commonly used leaky integrate-and-fire model with subtractive reset: for every approximation bound that is valid for a set of multi-spike neural networks, there is an equivalent set of single-spike neural networks with only linearly more neurons (in the maximum number of spikes) for which the bound holds. The same is true for the reverse direction too, showing that regarding their approximation capabilities in general machine learning tasks, single-spike and multi-spike neural networks are equivalent. Consequently, many approximation results in the literature for single-spike neural networks also hold for the multi-spike case.

Keywords: Spiking neural network · Spike encoding · Approximation bounds · Transference principle

1 Introduction

Is it sufficient for spiking neurons to spike at most once, or is there any benefit of spiking multiple times? How to best encode information in the temporal domain of spikes is still an open question. The possibilities range from encoding via exact spike times, such as in the latency of a neuron's spike response given a stimuli onset [8,27], to averaged quantities such as spike time rates [7]. Although using only single spikes might seem restrictive at first, it comes with several benefits. First, for exact spike times, analytical gradient-based learning rules are available for specific neuron models [2,9,13,14,17,24], as well as learning of synaptic delays [10] in addition to synaptic weights. Second, decoding outputs using lowest latency leads to a low time-to-solution, while the inherent sparsity of these spiking neural networks (SNNs) promises high energy efficiency [9,25].

L. Pasa et al. (Eds.): ICANN 2026, LNCS 17089, pp. 187–199, 2027.
https://doi.org/10.1007/978-3-032-38398-3_16

This makes single-spike SNNs quite attractive for edge applications, as recently proposed in the context of artificial intelligence (AI) onboard spacecraft [12,15].

For most practical purposes though, SNNs consisting of leaky integrate-and-fire (LIF) neurons (or variations thereof) with an unconstrained number of spikes per neuron are used nowadays. These SNNs are then commonly trained using the surrogate gradient method [5,19,21]. In fact, a common critique is that neurons which spike at most once restrict the expressivity of SNNs [5,26,28] and do not reflect biology well. This raises the question whether theoretical results derived for single-spike neurons apply to the more significant and relevant case of multi-spike neurons.

This work clarifies this open question, and we show a simple but extremely consequential result: both single-spike and multi-spike SNNs are equivalent when it comes to their approximation capabilities. Our result is inspired by recent work on the equivalence in approximation of fully-connected and convolutional neural networks (NNs) [22]. A formal statement of the result is given in Theorem 1, which is roughly summarised as follows:

1. All upper and lower bounds concerning approximation rates of multi-spike SNNs with $n \in \mathbb{N}$ neurons that spike at most $N_\mathrm{s} \in \mathbb{N}$ times translate to the same bounds concerning approximation rates of single-spike SNNs with at most $N_\mathrm{s} \cdot n$ neurons.
2. All upper and lower bounds concerning approximation rates of single-spike SNNs translate to the same bounds concerning approximation rates of multi-spike SNNs with at most $\alpha N_\mathrm{s} \cdot n$ neurons that spike at most N_s times. The prefactor satisfies $\alpha \leq \min\left(1, \frac{6}{\pi^2} + 1/\sqrt{N_\mathrm{s}}\right)$ for $N_\mathrm{s} \geq 1$ and $\alpha < \frac{6}{\pi^2} + 1/(2\sqrt{N_\mathrm{s}})$ for $N_\mathrm{s} \geq 8$.

Both the single-spike and multi-spike NNs have to be based on the same neuron model for this transference principle to hold, i.e., the single-spike neurons behave dynamically like multi-spike neurons that stop spiking after the first spike. The first part can be seen with the following direct argument: first, we replace each multi-spike neuron by N_s single-spike neurons, one per spike. Second, to ensure that the single-spike neurons spike at the correct time, their threshold is adjusted accordingly (Fig. 1A,B), and we are done. The central idea for showing the second statement is to combine the output of a small population of multi-spike neurons in such a way that all except one spike cancel out, allowing the population to mimic single-spike neurons (Fig. 1C,D). This result holds for a large family of spiking neuron models, including the widely used LIF neuron model with and without synaptic delays.

The result has far-reaching consequences, as every achievable approximation rate derived for single-spike SNNs automatically applies to multi-spike SNNs, and vice versa. Thus, our result directly strengthens recently reported approximation bounds for SNNs, such as covering-number based upper learning bounds derived for a simple spike-response model [20], and lower bounds of the approximation error for non-leaky integrate-and-fire neurons (nLIF) derived using a piecewise decomposition of SNNs [4]. Moreover, in [24,25], it was shown that for a simple spike-response model with linearly rising postsynaptic potential, an exact mapping between ReLU NNs and SNNs exists, i.e., where both NNs produce the exact same output for every input sample. Applying our result to this work, it follows that for each ReLU NN, there exists a multi-spike SNN (with the aforementioned neuron model) that satisfies the same approximation bounds. Generalisation error bounds for ReLU NNs are well known and can, for instance, be derived using covering numbers or Rademacher complexity, while interpolation error bounds can be derived via the number of affine pieces [23].

In the following, we first provide an overview of all assumptions for deriving Theorem 1. We then continue to stating our main theorem, proving it, and briefly discussing the main limitations of our result for completeness.

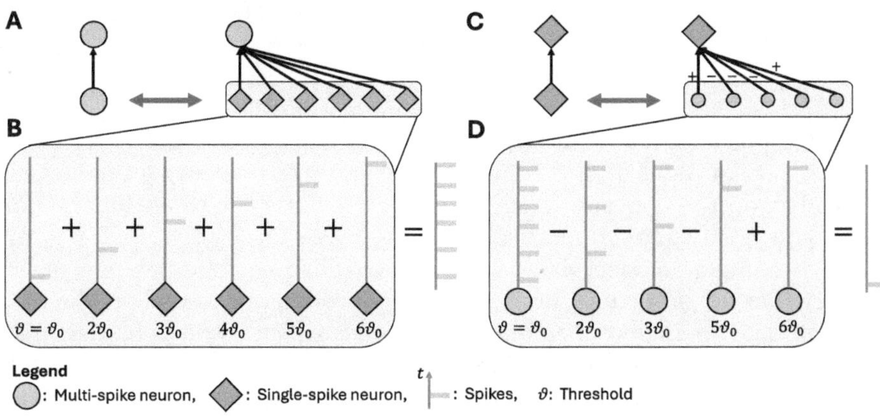

Fig. 1. Illustration of how to construct equivalent single-and multi-spike SNNs, shown here for multi-spike neurons that spike at most six times in a given finite time interval. **A.** A multi-spike neuron can be replaced by a population of single-spike neurons that produce an equivalent spike output. **B.** The kth single-spike neuron spikes at the kth spike time of the multi-spike neuron. This is achieved by setting the threshold of each neuron accordingly (alternatively, the reset potential could be adjusted). **C.** A single-spike neuron can also be replaced by a population of multi-spike neurons. **D.** Given any input, the first neuron produces a spike train. We add additional neurons to the population that gain identical input, but have increasing thresholds. For instance, the second neuron only spikes at every second spike time of the first neuron. By choosing synaptic weights accordingly, all but the first spike of the first neuron are cancelled out, producing an effective spike train with only one spike.

2 Background

2.1 Learning Task

First, we introduce the set of all spike trains consisting of at most $N \in \mathbb{N}$ spikes:

$$S_N := \bigcup_{k=1}^{N} \left\{ (t_1, \dots, t_k) \in [0,T]^k : t_1 \leq \dots \leq t_k \right\} \cup \{()\} \text{ with } T \in \mathbb{R} \cup \{\infty\}, \quad (1)$$

where () is the empty sequence representing no spiking. In particular, we define $S_\infty = \bigcup_{n\in\mathbb{N}} S_n$ as the set containing all spike trains of finite length. From this, we define the set of all functions $f : S_\infty^{d_\text{in}} \to S_{N_\text{s}}^{d_\text{out}}$ mapping a set of $d_\text{in} \in \mathbb{N}$ input spike trains to $d_\text{out} \in \mathbb{N}$ output spike trains with maximally $N_\text{s} \in \mathbb{N}$ spikes per neuron. To ensure that SNNs can learn these functions, we enforce that f is causal in time, i.e., $\min(f(X)) > \min(X)$ for all $X \in S_\infty^{d_\text{in}}$.

To cover most standard learning scenarios encountered in the SNN literature, we compose these spike-train functions with a spike-encoder and decoder. A spike-encoder is a function that translates the features of our data to spike trains, which then become the input to our SNN. Formally, we define encoders here as functions $E : \mathbb{R}^{d_\text{enc}} \to S_\infty^{d_\text{in}}$, where d_enc is the feature dimension of the data samples (e.g., the number of pixels of an image). This general formulation covers established encoders, such as rate-and latency encoding [5,9]. In case of rate-encoding, the input features are real, positive values representing spike rates, which are translated to spike trains that feature these rates. Similarly, for latency encoding, the inputs are the times-to-first-spike, which is translated into single-spike spike trains with this latency compared to a reference time.

Decoders do the opposite: take the output of the spike-based function and map it into a measurable space $\mathcal{O}$, $D : S_{N_\text{s}}^{d_\text{out}} \to \mathcal{O}$, e.g., $\mathcal{O} = \mathbb{R}^{d_\text{dec}}$ with $d_\text{dec} \in \mathbb{N}$. Common decoders either work with exact spike times (latency, relative spike times, exact spike time matching) or derived quantities such as the membrane potential (matching membrane trace, mean membrane potential, maximum membrane potential) or spike counts [5].

In summary, our results cover the case of general spike-based functions f extended using common encoders and decoders: $f \mapsto D \circ f \circ E$.

2.2 Spiking Neuron Models

This work applies at least to the family of spiking neuron models that satisfy the following three conditions:

1. The neuron is threshold-based, i.e., spikes are generated when the neuron's membrane potential u crosses a constant threshold value ϑ: $\{t : u(t) = \vartheta\}$.
2. After spiking, the neuron's membrane potential dynamics are unaffected apart from a constant offset $-\vartheta$.
3. Input currents are independent of a neuron's membrane potential, as for instance in case of current-based models (but not conductance-based ones when close to the reversal potentials).

There is a large set of models satisfying these conditions, including the widely used leaky integrate-and-fire (LIF) neuron model without refractory period, variations thereof, such as the nLIF model [9,17], and other models such as the simple spike response model [20,24]. For these models, the reset mechanisms that are commonly used also satisfy the second condition. For example, especially for machine learning applications, the following type of LIF model is often used [5,21]:

$$u(t + \mathrm{d}t) = -\beta u(t) + I(t) - s(t)\vartheta\,, \tag{2}$$

where $u(t)$ is the membrane potential of a single neuron at time t, $\beta > 0$ a constant, $I(t)$ is the time-dependent input to the neuron (e.g., currents caused by pre-synaptic spikes of other neurons), and $s(t) = \sum_i \delta(t - t_i)$ is a sum of Dirac delta functions representing the output spike times t_i of the neuron. Thus, whenever the neuron spikes, its immediate membrane potential is reduced by ϑ, as required by our second condition. Although not specifically included here in our definition, the presented results apply to spiking neuron models with synaptic delays.

Models that do not necessarily satisfy these conditions are those with refractory periods, adaptive thresholds [1], and spike generation without thresholds [6,11,18]. This does not necessarily mean that the presented theorem does not hold for these neuron models, but merely that the presented proof cannot be applied in these cases.

2.3 Network Architectures

The results in this work are stated for layered networks of single-and multi-spike neurons, allowing for connections between neurons of the same layer or neurons of different layers. We rule out autapses, i.e., recurrent connections of a neuron with itself. However, we further assume that network weights are bounded by B in absolute value, with $B > 0$. When talking about NNs in our main result, we always refer to this architecture.

2.4 Bounded Number of Spikes

Generally, we assume that spiking is constrained to a finite time interval T, leading to a maximum number N_s of times any multi-spike neuron can spike. This does match realistic neurons, which also can only fire a finite number of times in a given time interval. Moreover, for most spiking neuron models, with finite weights and non-zero membrane/synaptic time constants, the time-to-spike is finite, leading to an upper bound for the spike rate.

3 Results

Below, we state our main result: the equivalence of approximation of single-spike and multi-spike NNs (Theorem 1). It holds for any spiking neuron models with simple reset dynamics, various spike time decoders, and synaptic delays (see Sect. 2).

Theorem 1 (Transference principle). *Let $d_{in}, d_{out}, N_s \in \mathbb{N}$, $f : S_\infty^{d_{in}} \to S_{N_s}^{d_{out}}$ be time-causal, and $F = D \circ f$ with decoder D. Furthermore, set $\Omega \subseteq S_\infty^{d_{in}}$. Then, the following hold for each $p \in (0, \infty]$ and each $\epsilon \geq 0$:*

1. *If there is a multi-spike NN Ψ with $n \in \mathbb{N}$ neurons and L layers satisfying $\|F - D \circ \Psi\|_{L^p(\Omega)} \leq \epsilon$, then there is a decoder D' and a single-spike NN Φ with L layers and at most $N_s \cdot n$ neurons, such that $\|F - D' \circ \Phi\|_{L^p(\Omega)} \leq \epsilon$.*

2. *If there is a single-spike NN Φ with $n \in \mathbb{N}$ neurons and L layers satisfying $\|F - D \circ \Phi\|_{L^p(\Omega)} \leq \epsilon$, then there is a multi-spike NN Ψ with L layers and at most $\alpha N_s \cdot n$ neurons, $\alpha \leq min\left(1, \frac{6}{\pi^2} + 1/\sqrt{N_s}\right)$ for $N_s \geq 1$ and $\alpha < \frac{6}{\pi^2} + 1/(2 \cdot \sqrt{N_s})$ for $N_s \geq 8$, such that $\|F - D \circ \Psi\|_{L^p(\Omega)} \leq \epsilon$.*

In other words: given a multi-spike NN, we can always find an equivalent single-spike NN that performs similarly on a given learning task and vice versa – assuming that the dynamics of single and multi-spike neurons is the same (e.g., single-spike and multi-spike LIF with identical dynamics). This result is obtained without changing the neuron models or the underlying network architecture. Remarkably, for both directions, the number of neurons required to construct an equivalent NN increases, at most, linearly with the maximum number of spikes the multi-spike neurons can emit.

The following two extensions follow immediately from Theorem 1:

Corollary 1 (Lower bounds). *Theorem 1 extends also to lower bounds of approximation rates:*

1. *If for all multi-spike NNs Ψ with $n \in \mathbb{N}$ neurons and L layers it holds that $\|F - D \circ \Psi\|_{L^p(\Omega)} \geq \epsilon$, then for a decoder D' as given in Eq. (3) and all single-spike NNs Φ with L layers and at most $N_s \cdot n$ neurons, it holds that $\|F - D' \circ \Phi\|_{L^p(\Omega)} \geq \epsilon$.*
2. *If for all single-spike NNs Φ with $n \in \mathbb{N}$ neurons and L layers it holds that $\|F - D \circ \Phi\|_{L^p(\Omega)} \geq \epsilon$, then for all multi-spike NNs Ψ with L layers and at most $\alpha N_s \cdot n$ neurons, it holds that $\|F - D \circ \Psi\|_{L^p(\Omega)} \geq \epsilon$, with $\alpha \leq min\left(1, \frac{6}{\pi^2} + 1/\sqrt{N_s}\right)$ for $N_s \geq 1$ and $\alpha < \frac{6}{\pi^2} + 1/(2 \cdot \sqrt{N_s})$ for $N_s \geq 8$.*

Corollary 2 (Encoders). *Theorem 1 and Corollary 1 extend to common input spike-encoders as described in Sect. 2.*

Thus, through our result, generalization and approximation bounds obtained for single-spike neuron models can be immediately mapped to the multi-spike case, such as bounds obtained using covering numbers [20] and causal pieces [4]. As we phrased Theorem 1 for the rather general learning task of mapping from spike trains to spike trains, it applies both to static (e.g., image recognition) and temporal learning tasks (e.g., temporal sequence prediction) often studied with SNNs [3].

3.1 Proofs

To prove Theorem 1, we first show how to replace a single neuron in a network, from which the theorem follows immediately by replacing all neurons in a network accordingly.

Proof of the first part. The first part of Theorem 1 is straightforward. Assume a single multi-spike neuron with input weights $\boldsymbol{W}_{\text{in}} \in \mathbb{R}^{d_0}$, output weights $\boldsymbol{W}_{\text{out}} \in \mathbb{R}^{d_1}$, and input spike trains $s_{\text{in}} \in S_\infty^{d_0}$. Given the inputs, the neuron produces an output spike train $s_{\text{out}} \in S_{N_\text{s}}$ with at most N_s output spikes. We can replace this neuron by at most N_s single-spike neurons with the same membrane dynamics, with neuron k having threshold $k \cdot \vartheta$ for $k \in [1, N_\text{s}+1]$, and otherwise all neurons sharing the same input weights $\boldsymbol{W}_{\text{in}}$, output weights $\boldsymbol{W}_{\text{out}}$, neuron parameters (e.g., time constants), and inputs s_{in} as the multi-spike neuron. By *the same membrane dynamics*, we mean that the single-spike neurons behave identical to the multi-spike neurons, just that they stop spiking after the first spike.

Assume now that $s_{\text{out}} = (t_1, \dots, t_N)$ for $N \le N_\text{s}$. The first single-spike neuron has the lowest threshold, and hence will spike at time t_1 and remain silent afterwards. At t_1, the multi-spike neuron is reset and its membrane potential will rise by ϑ again until the second output spike time t_2. Thus, at time t_2, the second single-spike neuron's membrane potential reaches its threshold $2 \cdot \vartheta$ and thus spikes. Similarly, the third single-spike neuron spikes at t_3, the fourth at t_4, and so on. Single-spike neurons $k > N$ are silent, as the remaining input spikes that occurred after t_N do not drive the multi-and single-spike neurons across their thresholds.

Using this construction, any multi-spike neuron in Ψ can be replaced by N_s single-spike neurons, and the resulting NN Φ will produce the exact same output spike times as Ψ given any input spike trains. However, the output format is now different, since spike times are scattered across neurons. Assume that the $[(k-1) \cdot N_\text{s}, k \cdot N_\text{s}]$ single-spike neurons in the output layer of Φ represent the kth multi-spike neuron in the last layer of Ψ. Denote by $s_i \in S_1$ the output of the ith neuron in the output layer of Φ. By choosing the decoder $D' = D \circ A$ with

$$A(X)_k = s_{(k-1)N_\text{s}} \oplus \cdots \oplus s_{k \cdot N_\text{s}}\,, \tag{3}$$

for $X \in S_1^{d_{\text{out}} \cdot N_\text{s}}$, we then arrive at the second statement of Theorem 1. Here, we introduce a notation for appending sequences: $(t_0, \dots, t_K) \oplus (t_{K+1}, \dots, t_N) = (t_0, \dots, t_K, t_{K+1}, \dots, t_N)$, and $(t_0, \dots, t_K) \oplus () = (t_0, \dots, t_K)$.

In the above proof, we assumed no synaptic delays. Assume $\boldsymbol{\delta}_{\text{in}} \in \mathbb{R}_+^{d_0}$ are input delays, meaning that the spike times of the ith input spike train are all shifted by $\delta_{\text{in},i}$. Further assume that $\boldsymbol{\delta}_{\text{out}} \in \mathbb{R}_+^{d_1}$ are output delays, meaning that output spikes transmitted to the jth post-synaptic neuron are delayed by $\delta_{\text{out},j}$. As long as the single-spike neurons have the same input and output delays as the multi-spike neuron (similarly to how they share the same input and output weights), the above proof applies.

Proof of the Second Part. The second part of Theorem 1 can be derived using a similar idea. Instead of using multiple neurons to create a spike train, we use them to ensure that, in the relevant time period, this population of multi-spike neurons emits at most one effective spike, i.e., only one spike that affects its postsynaptic partners.

We take the same setup as before, but start from a single-spike neuron that produces outputs $s_{\text{out}} \in S_1$. We reproduce its input-output mapping using a population of at most N_s multi-spike neurons that have the same membrane dynamics as the single-spike neuron, as well as the same inputs, input weights, and input and output delays. In the following, without loss of generality and to simplify the notation, we drop the delays. Then, we set the threshold of the kth multi-spike neuron to $k \cdot \vartheta$, and its output weights to $\alpha_k \cdot \boldsymbol{W}_{\text{out}}$. All other neuron parameters are identical to the ones of the single-spike neuron.

Assume that given some input, the single-spike neuron spikes at time t_1. We choose as our first neuron in the multi-spike population one that spikes at times $s_1 = (t_1, \ldots, t_{N_\text{s}})$, where $t_2, \ldots, t_{N_\text{s}}$ are sorted and larger than t_1, but otherwise arbitrary. Then the second neuron in the multi-spike population, with threshold $2 \cdot \vartheta$, spikes only every second time $s_2 = \left(t_2, t_4, \ldots, t_{2 \cdot \lfloor N_\text{s}/2 \rfloor}\right)$, with $\lfloor \cdot \rfloor$ denoting rounding down. Similarly, the third neuron in the multi-spike population spikes only at times $s_3 = \left(t_3, t_6, \ldots, t_{3 \cdot \lfloor N_\text{s}/3 \rfloor}\right)$. Generally, since the first multi-spike neuron spikes N_s times in the time period $[0, T]$, the kth neuron spikes only $\lfloor N_\text{s}/k \rfloor$ times. To have the same effect on the postsynaptic neurons as the single-spike neuron, we thus need:

$$\sum_{i=1}^{N_\text{s}} \sum_{t' \in s_i} \alpha_i \boldsymbol{W}_{\text{out}} \delta(t - t') = \boldsymbol{W}_{\text{out}} \delta(t - t_1)\,, \tag{4}$$

which is equivalent to the condition $\boldsymbol{M\alpha} = \boldsymbol{e}_0$ with

$$M_{ij} = \begin{cases} 1 & \text{if } i \geq j \text{ and } (i+1)\%(j+1) = 0\,, \\ 0 & \text{otherwise}\,, \end{cases} \tag{5}$$

with % being the modulo operator, $\alpha = (\alpha_1, \ldots, \alpha_{N_\text{s}})$, and $\boldsymbol{e}_0 = (1, 0, \ldots, 0)$ the unit vector with all zeroes except the first element. For example, for $N_\text{s} = 6$, the linear equation is:

$$\begin{pmatrix} 1\,0\,0\,0\,0\,0 \\ 1\,1\,0\,0\,0\,0 \\ 1\,0\,1\,0\,0\,0 \\ 1\,1\,0\,1\,0\,0 \\ 1\,0\,0\,0\,1\,0 \\ 1\,1\,1\,0\,0\,1 \end{pmatrix} \begin{pmatrix} \alpha_1 \\ \alpha_2 \\ \alpha_3 \\ \alpha_4 \\ \alpha_5 \\ \alpha_6 \end{pmatrix} = \begin{pmatrix} 1 \\ 0 \\ 0 \\ 0 \\ 0 \\ 0 \end{pmatrix}. \tag{6}$$

In other words: we have to find appropriate weights α_i such that all spikes except the first one at time t_1 cancel each other. From this, it follows immediately that $\alpha_1 = 1$. Furthermore, since $\boldsymbol{M}$ is a lower-triangular matrix with non-zero diagonal elements, this linear equation always has a unique solution.

This way, we can replace every neuron in the single-spike NN Φ by at most N_s multi-spike neurons and appropriately chosen weights to get an equivalent multi-spike NN Ψ. For neurons in the output layer, we choose a decoder D that filters out the first spike of each neuron to conclude the proof. The solution for the weight factors is given by Lemma 1.

Lemma 1. *(Weight factors) Assume we want to replace a single-spike neuron with N_s multi-spike neurons. Let $n \in [1, N_s]$. Furthermore, let $n = \prod_{i=1}^{k} p_i^{m_i}$ be its prime factorization with $k \in \mathbb{N}$, $k > 0$, $m_i \in \mathbb{N}$, and $p_i \leq n$ being prime numbers. Choose $\alpha_1 = 1$. If for every $n > 1$, we choose the weight factors*

$$\alpha_n = \begin{cases} 0 & \text{if } \max\left(\{m_i : i \in [1,k]\}\right) > 1\,, \\ (-1)^k & \text{otherwise}\,, \end{cases} \tag{7}$$

then the vector $\boldsymbol{\alpha} = (\alpha_1, \ldots, \alpha_{N_s})$ solves $\boldsymbol{M\alpha} = \boldsymbol{e}_0$.

Lemma 1 is summarised as follows: a single-spike neuron can be replaced by $\eta(N_\mathrm{s})$ multi-spike neurons that spike at most N_s times in a given time window $[0, T]$, where $\eta(N_\mathrm{s})$ is the number of square-free numbers smaller or equal than N_s (i.e., we remove all neurons j with weight factor $\alpha_j = 0$). A number is square-free if in its prime factorization, all factors are unique, i.e., do not appear multiple times. $\eta(N_\mathrm{s})$ is upper bounded by $\eta(N_\mathrm{s}) < \zeta(2) \cdot N_\mathrm{s} + \sqrt{N_\mathrm{s}}$ for $N_\mathrm{s} \geq 1$ and $\eta(N_\mathrm{s}) < \zeta(2) \cdot N_\mathrm{s} + \frac{1}{2}\sqrt{N_\mathrm{s}}$ for $N_\mathrm{s} \geq 8$, with $\zeta(2) = 6/\pi^2$ [16]. In addition, we always have $\eta(N_\mathrm{s}) \leq N_\mathrm{s}$. Combined, this results in the estimate for the required number of neurons given in Theorem 1. Particularly, in the limit of large N_s, we have $\lim_{N_\mathrm{s} \to \infty} \eta(N_\mathrm{s})/N_\mathrm{s} < 6/\pi^2$.

Proof Idea. Before formalising the proof, we briefly illustrate the main idea. Assume we have N_s multi-spike neurons. The neuron with index 1 spikes exactly N_s times. The neuron with index 2 spikes at every second spike time of the neuron with index 1, the third at every third spike time of the neuron with index 1, and so on (see Fig. 2). We now ask: which neurons with index > 1 spike at the nth time? In the end, for these neurons we have to choose the outgoing weights in a way that the effect of all spikes cancels, meaning that this population of neurons produces *effectively* no spike at the nth time.

For example, assume $n = 30$. For a neuron to spike at the 30th spike time, its index has to divide n without remainder. In our example, this means that the neurons with index $\mathcal{D} = \{2, 3, 5, 6, 10, 15, 30\}$ spike at this time – all other neurons do **not** spike at the 30th time (Fig. 2). As we are looking for dividing factors, we can simplify by looking at the prime factorisation of $n = 30 = 1{\cdot}2{\cdot}3{\cdot}5$. $\mathcal{D}$ is then obtained by forming the products over all possible subsets, i.e., $\mathcal{D} = \{2, 3, 5, 2 \cdot 3, 2 \cdot 5, 3 \cdot 5, 2 \cdot 3 \cdot 5\}$. The assigned weights for a neuron with index m is given by $(-1)^{k_m}$, where k_m is the number of prime factors m composes to. Thus, we just have to count: we have $\binom{3}{1} = 3$ neurons whose indices are already a prime number (2,3,5), $\binom{3}{2} = 3$ neurons whose indices decompose into a product of two primes (6, 10, 15), and $\binom{3}{2} = 1$ neuron whose index decomposes into a product of three primes (30). Summing up these spikes with their corresponding weights, we get $\binom{3}{1}(-1)^1 + \binom{3}{2}(-1)^2 + \binom{3}{3}(-1)^3 = -3 + 3 - 1 = -1$. Adding the spike of the neuron with index 1 (and weight $\alpha_1 = 1$), all spikes cancel perfectly out $(-1 + 1 = 0)$.

In this example, $n = 30$ was a square-free number, meaning that no prime number appears multiple times in its factorisation. However, since all the neurons that were involved in cancelling the 30th spike time will also spike at all multiples of 30, they also cancel out at those times. For example, assume $n = 60 = 1{\cdot}2^2{\cdot}3{\cdot}5$. We already showed that the neurons with indices $\{2, 3, 5, 6, 10, 15, 30\}$ cancel out the 30th spike time of the neuron with index 1. Hence, they also cancel out the 60th spike time, meaning we do not need an additional neuron with index 60 at this time (i.e., we either set $\alpha_{60} = 0$, or we simply remove this neuron from the population altogether). This is true for all neurons whose index is not a square-free number. Hence, the number of neurons required to cancel out N_s spikes is given by the number of square-free numbers smaller or equal to N_s. In this work, we cite two standard bounds for this, with the second one being tighter but also only valid for larger values of N_s.

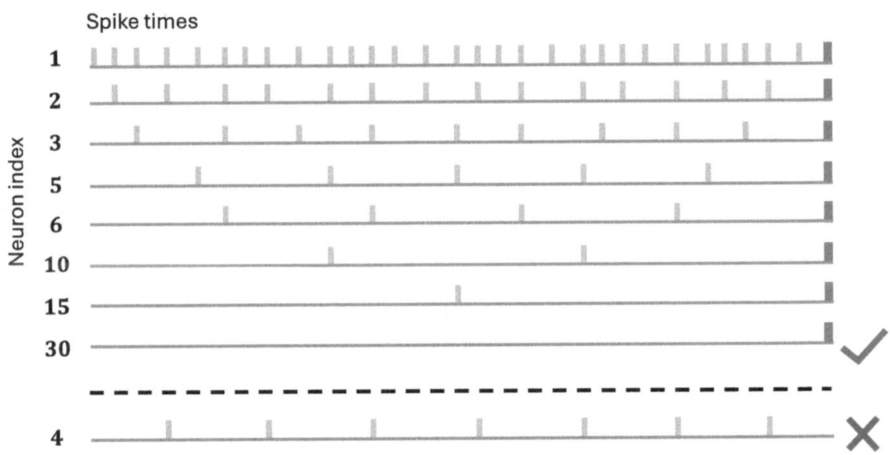

Fig. 2. Illustration of how to effectively cancel out spike times. We assume a population of neurons where the neuron with index i spikes at every i^{th} spike time of the first neuron with index 1. Given the nth spike time (here: $n = 30$), only those neurons whose index divides n without remainder also spike at this time. Thus, we only have to consider the outgoing weights of this subset of neurons for cancelling the nth spike.

Proof. Assume the nth spike time t_n, with $1 < n \leq N_\text{s}$. Let $n = \prod_{i=1}^{k} p_i^{m_i}$ be its prime factorization. First, we isolate all multiple factors by defining $\beta = \prod_{i=1}^{k} p_i^{m_i - 1}$, resulting in $n = \beta \cdot \prod_{i=1}^{k} p_i$. Note then that for any $C \subseteq \{1, \ldots, k\}$ (with $\bar{C}$ being its complement), we have $n = L_C \cdot \beta \prod_{i \in \bar{C}} p_i$, with $L_C = \prod_{i \in C} p_i$, which means that the neuron with index L_C, which emits a spike at every L_C^{th} spike time, also spikes at the nth spike time (n is a multiple of its period L_C). This is also true for neurons with indices whose prime factorization is not square-free, but according to Lemma 1, their weight factors are set to 0 and we can neglect them. As an example, assuming $n = 60$ again, one possible arrangement is $\beta = 2$ and $C = \{2, 3\}$, leading to $L_C = 2 \cdot 3 = 6$.

Thus, for the nth spike time, output spikes of all neurons with indices that decompose into products of subsets of the prime numbers $\{p_1, \ldots, p_k\}$ are summed up. There are, in total, $\sum_{i=1}^{k} \binom{k}{i}$ such combinations, and therefore $\sum_{i=1}^{k} \binom{k}{i}$ spikes that overlap. Following Lemma 1, the weight factor for each neuron's spike only depends on the number of primes its index factors in. Thus, the weighted sum of all neurons is given by $\sum_{i=1}^{k} \binom{k}{i}(-1)^i = \sum_{i=0}^{k} \binom{k}{i}(-1)^i 1^{k-i} - 1 = (1-1)^k - 1 = -1$, which perfectly cancels the spike of the first neuron with $\alpha_1 = 1$. Since this is valid for all values of n, the proof is complete.

Corollaries. Both Ψ and Φ produce the exact same spike times as output given the same spike trains as input, from which both corollaries follow immediately.

4 Limitations

We would like to stress that Theorem 1 is not universally valid for all spiking neuron models. Formally, we only prove it for models satisfying the requirements stated in Sect. 2 – although we do not rule out that the theorem holds true for other neuron models as well.

A major type of neurons not covered is those with a hard refractory period. In this case, proving the direction multi-spike → single-spike in Theorem 1 becomes very hard, if not impossible. For example, for the proof, we require that every single-spike neuron has, up to a constant offset, the same membrane trace as the multi-spike neuron until the threshold is crossed. However, during a hard refractory period, the multi-spike neuron's membrane potential is clamped to a constant value, while all single-spike neurons that haven't spiked yet continue to follow their inputs – thus diverging from the multi-spike neuron and potentially leading to differences in the spike times. The same applies for other effects that change a neuron's membrane dynamics after it spiked, such as adaptive thresholds.

Generally, Theorem 1 can still be shown to be valid as long as the effect induced by spiking on the membrane dynamics can be modelled via an ordinary self-connection, such as certain soft refractory period mechanisms.

Disclosure of Interests. The authors have no competing interests to declare that are relevant to the content of this article.

References

1. Brette, R., Gerstner, W.: Adaptive exponential integrate-and-fire model as an effective description of neuronal activity. J. Neurophysiol. **94**(5), 3637–3642 (2005)
2. Comsa, I.M., Potempa, K., Versari, L., Fischbacher, T., Gesmundo, A., Alakuijala, J.: Temporal coding in spiking neural networks with alpha synaptic function. In: ICASSP 2020-2020 IEEE International Conference on Acoustics, Speech and Signal Processing (ICASSP), pp. 8529–8533. IEEE (2020)

3. Cramer, B., Stradmann, Y., Schemmel, J., Zenke, F.: The Heidelberg spiking data sets for the systematic evaluation of spiking neural networks. IEEE Trans. Neural Netw. Learn. Syst. **33**(7), 2744–2757 (2020)
4. Dold, D., Petersen, P.C.: Causal pieces: analysing and improving spiking neural networks piece by piece. arXiv preprint arXiv:2504.14015 (2025)
5. Eshraghian, J.K., et al.: Training spiking neural networks using lessons from deep learning. Proc. IEEE **111**(9), 1016–1054 (2023)
6. FitzHugh, R.: Impulses and physiological states in theoretical models of nerve membrane. Biophys. J. **1**(6), 445–466 (1961)
7. Gerstner, W., Kistler, W.M., Naud, R., Paninski, L.: Neuronal Dynamics: From Single Neurons to Networks and Models of Cognition. Cambridge University Press, Cambridge (2014)
8. Gollisch, T., Meister, M.: Rapid neural coding in the retina with relative spike latencies. Science **319**(5866), 1108–1111 (2008)
9. Göltz, J., et al.: Fast and energy-efficient neuromorphic deep learning with first-spike times. Nat. Mach. Intell. **3**(9), 823–835 (2021)
10. Göltz, J., et al.: DelGrad: exact event-based gradients for training delays and weights on spiking neuromorphic hardware. Nat. Commun. **16**(1), 8245 (2025)
11. Hodgkin, A.L., Huxley, A.F.: A quantitative description of membrane current and its application to conduction and excitation in nerve. J. Physiol. **117**(4), 500 (1952)
12. Izzo, D., Hadjiivanov, A., Dold, D., Meoni, G., Blazquez, E.: Neuromorphic computing and sensing in space. In: Artificial Intelligence for Space: AI4SPACE, pp. 107–159. CRC Press (2023)
13. Kheradpisheh, S.R., Mirsadeghi, M., Masquelier, T.: BS4NN: binarized spiking neural networks with temporal coding and learning. Neural Process. Lett. **54**(2), 1255–1273 (2022)
14. Klos, C., Memmesheimer, R.M.: Smooth exact gradient descent learning in spiking neural networks. Phys. Rev. Lett. **134**(2), 027301 (2025)
15. Lunghi, P., Silvestrini, S., Dold, D., Meoni, G., Hadjiivanov, A., Izzo, D.: Energy efficiency analysis of spiking neural networks for space applications. Astrodynamics **9**(6), 909–932 (2025)
16. Moser, L., MacLeod, R.: The error term for the squarefree integers. Can. Math. Bull. **9**(3), 303–306 (1966)
17. Mostafa, H.: Supervised learning based on temporal coding in spiking neural networks. IEEE Trans. Neural Networks Learn. Syst. **29**(7), 3227–3235 (2017)
18. Nagumo, J., Arimoto, S., Yoshizawa, S.: An active pulse transmission line simulating nerve axon. Proc. IRE **50**(10), 2061–2070 (1962)
19. Neftci, E.O., Mostafa, H., Zenke, F.: Surrogate gradient learning in spiking neural networks: bringing the power of gradient-based optimization to spiking neural networks. IEEE Signal Process. Mag. **36**(6), 51–63 (2019)
20. Neuman, A.M., Dold, D., Petersen, P.C.: Stable learning using spiking neural networks equipped with affine encoders and decoders. J. Mach. Learn. Res. **26**(246), 1–49 (2025)
21. Pehle, C.G., Egholm Pedersen, J.: Norse-a deep learning library for spiking neural networks. Zenodo (2021)
22. Petersen, P., Voigtlaender, F.: Equivalence of approximation by convolutional neural networks and fully-connected networks. Proc. Am. Math. Soc. **148**(4), 1567–1581 (2020)
23. Petersen, P., Zech, J.: Mathematical theory of deep learning. arXiv preprint arXiv:2407.18384 (2024)

24. Stanojevic, A., Woźniak, S., Bellec, G., Cherubini, G., Pantazi, A., Gerstner, W.: An exact mapping from ReLU networks to spiking neural networks. Neural Netw. **168**, 74–88 (2023)
25. Stanojevic, A., Woźniak, S., Bellec, G., Cherubini, G., Pantazi, A., Gerstner, W.: High-performance deep spiking neural networks with 0.3 spikes per neuron. Nat. Commun. **15**(1), 6793 (2024)
26. Taylor, L., King, A., Harper, N.: Robust and accelerated single-spike spiking neural network training with applicability to challenging temporal tasks. arXiv preprint arXiv:2205.15286 (2022)
27. Thorpe, S., Fize, D., Marlot, C.: Speed of processing in the human visual system. Nature **381**(6582), 520–522 (1996)
28. Zenke, F.: Visualizing a joint future of neuroscience and neuromorphic engineering. Neuron **109**(4), 571–575 (2021)

Neuromuscular-Inspired Function Approximation for Neuromorphic Computing

Jordi Timmermans[1,2(✉)] and Lambert Schomaker[1,2]

[1] Bernoulli Institute for Mathematics, Computer Science and Artificial Intelligence,Department of Artificial Intelligence, University of Groningen, Groningen, The Netherlands

[2] Groningen Cognitive Systems and Materials Center (CogniGron), University of Groningen, Groningen, The Netherlands

j.j.m.a.timmermans@rug.nl

Abstract. Spiking neural network research remains focused on models that generate a spiking output. In neuromorphic and edge computing applications, a continuous output might be required. We propose extending a spiking neural network with a continuous, muscular-inspired output. More specifically, we filter each output spike by a muscle twitch-like signal, parametrized per output neuron. Similar to the antagonistic principle in the neuromuscular system, the model has outputs that generate positive signals, and outputs that generate negative signals. Rather than using the conventional backpropagation through time method to train the network, we use a more biologically plausible, and neuromorphically compatible training method. Using this model, which we call the twitch neural network (TwNN), and the biologically plausible learning rule, we show (practical) function fitting capabilities. We show that the model has basic function fitting capabilities by training it to estimate basic functions from static random inputs. Furthermore, we demonstrate that a single model is able to store different input/output time-function mappings. Namely, we show that the model can be trained to generate handwritten digit trajectories, conditioned on audio inputs. Even the basic version of the twitch neural network shows good function fitting performance, suggesting that architectural or training improvements can produce better performance, and accommodate a more diverse set of use cases.

Keywords: Spiking neural networks · Neuromuscular modelling · Function approximation · Analog control tasks

L. Pasa et al. (Eds.): ICANN 2026, LNCS 17089, pp. 200–211, 2027.
https://doi.org/10.1007/978-3-032-38398-3_17

1 Introduction

Spiking neural networks (SNNs) are mainly tested or benchmarked on discrete classification tasks, e.g., classifying MNIST [10,21] or generating a predetermined spike trains [19,25]. In classification tasks, the neural outputs are decoded through different decoding schemes. Alternatively, in spike-train fitting tasks, the network is trained to time spike events at exact points in time. Both the post-hoc output decoding in classification and the precise spike-timing requirement are artificial and lack biological plausibility. In neuromorphic computing applications where a continuous output signal is needed, different modeling approaches and evaluation methods would be needed. Continuous signal-generating spiking neural networks can be applied, for example, in robotic actuator control, prediction in time series and in synthetic data generation. Different studies have tried to approximate continuous functions using spiking neural networks. [16] utilized inter-spike-intervals (ISIs) to represent output values, requiring one output neuron per output value. [1] used first-order spikes as output stage. First-order filtered spikes however, cause discontinuities at time of spiking, and many spikes are required to fit the function properly. [15] utilized the summed membrane voltage of the output neurons to drive a robotic arm. The learning rule used here however is backpropagation through time, which is not practical for neuromorphic applications and does not scale well with larger timescales. To the best of our knowledge exact, continuous function approximation using a biologically plausible learning rule is not yet investigated. In previous studies [20,24] we showed that it is possible to utilize muscular-inspired outputs for control tasks. This only required the actuator to reach sufficient output work (W) or distance before a certain time t, while trajectory details were less important. In the current study we want to extend the task requirements by incorporating learning of timed spiking patterns. However, a precise timing of individual neuron spikes is less important than the overall output spiking pattern, where different neurons in an output pool can contribute to the required analog output pattern. As shown in [6] the timing of the whole pool (ensemble) is important, rather than that of each individual spike. Their solution however relied on knowledge of the dynamical system to analytically derive the neural network weights.

Under the assumption that spiking neural networks will be deployed efficiently and effectively on compatible hardware [8], the ability to energy-efficiently generate continuous output can unlock low-power solutions in different fields. For example: In forecasting, event-based inputs can be utilized to produce a prediction of the future. In neuromorphic robotics, the output of event-based sensors [11,23] can be used to generate desired smooth movement, avoiding abrupt jumps in control value.

Spiking neural networks can generate continuous outputs by filtering the output spikes. Filtering smears the spike out in time, generating a base function in time. Using a summation of base functions, complex functions can be generated. Similar to approximating a function with a Fourier series, the quality of fit will be determined by the number of different base function parametrizations in the ensemble. Different base functions can be utilized to approximate complex functions [7,9,14]. In the neuromuscular system, a motor neuron's action potential causes a muscle fiber to contract via a second-order overdamped impulse response, referred to as a *twitch*. The twitch serves as a biological base function for movement. In electronics this response is also cheaply realized by means of a double RC filter. By co-contracting, an ensemble of motor units [17] can generate a complex movement pattern. These patterns can vary in duration by difference in contraction time and by repeated contractions [2], and vary in force by means of motor unit recruitment [13].

Since filtering a spiking neural network's output spikes with a twitch base function produces only positive outputs, agonist and antagonist output populations are required to represent positive and negative signal components, analogous to antagonistic muscle pairs in biological motor control. Training such neural networks to produce desired output patterns in an energy-efficient, scalable manner requires suitable learning rules. Ideally, the learning rule is local by means of Hebbian plasticity [12], is limited in memory, is able to handle the temporal credit assignment problem, i.e., determining for each unit what the contribution was to the output, and does not lock layers in the backward pass such as in backpropagation.

In this work we:

i) Propose a twitch neural network (TwNN). We propose a trainable SNN whose output spikes are filtered by second-order overdamped impulse response filters, analogous to the neuromuscular system's twitches. The model proposes an approach to solve the problem of generating continuous outputs using an SNN [22], specifically in a neuromorphic computing context.

ii) Train the TwNN using a biologically-inspired, close to local learning rule. We use the Superspike method [25] combined with Direct Feedback Alignment (DFA) [18], avoiding the biologically implausible and computationally expensive backpropagation through time.

iii) Show the function estimation capabilities of the TwNN. We show viability of the TwNN by training it to approximate the sine, Gaussian and square pulse functions from a static random input. We subsequently show that it is possible to train the model to produce handwriting trajectories from speech input. The code for the model and implementations can be found on GitHub[1].

2 Methods

2.1 Model Structure

The spiking neural network models we use consist of fully connected layers of leaky integrate and fire neurons. The evolution of the synaptic input current is described in Eq. 1, and the membrane potential dynamic is described in Eq. 2. The weighted connection w_{ij} between the presynaptic neuron i and the postsynaptic neuron j weighs the incoming spikes S, and adds the weighted value to

[1] https://github.com/JordiTi/TwNN.

the synaptic current I. The current is added to the membrane potential V. Both the input current and membrane potential leak away, parametrized by τ_{syn} and τ_{mem}.

$$I_{i,t+1} = I_{i,t}e^{-dt/\tau_{syn}} + \sum_j w_{ij}S_{ij,t} \tag{1}$$

$$V_{i,t+1} = V_{i,t}e^{-dt/\tau_{mem}} + I_{i,t+1}(1 - e^{-dt/\tau_{mem}}) \tag{2}$$

When the membrane potential exceeds a predefined neural threshold Θ, the neuron emits a spike S and resets its membrane potential. The spiking dynamics are described in Eqs. 3 and 4.

$$S_{i,t} = H(V_{i,t} - V_{\Theta}) \tag{3}$$

$$V_{i,t+1} = V_{i,t}(1 - S_t) \tag{4}$$

Inspired by the neuromuscular system, the output neuron spikes are filtered with a second-order impulse response filter, yielding a *twitch*. These filters are parametrized by an amplitude A and two time constants τ_{rise}, τ_{fall}, which are randomly sampled from uniform distributions. The output twitch generation is described by Eqs. 5 and 6. $\mathcal{F}_2$ is the final twitch output signal per output neuron. The output of the model is the sum of all individual twitch outputs. An image of the output characteristics can be seen in Fig. 1.

$$\mathcal{F}_{1,t+1} = \mathcal{F}_{1,t}e^{-dt/\tau_{rise}} + (1 - e^{-dt/\tau_{rise}})S_tA \tag{5}$$

$$\mathcal{F}_{2,t+1} = \mathcal{F}_{2,t}e^{-dt/\tau_{fall}} + (1 - e^{-dt/\tau_{fall}})\mathcal{F}_{1,t+1} \tag{6}$$

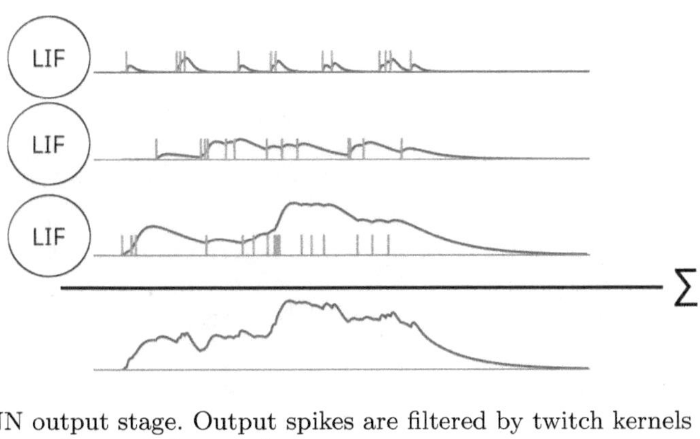

Fig. 1. TwNN output stage. Output spikes are filtered by twitch kernels and summed across agonist and antagonist populations.

2.2 Training

Superspike [25] is a three-factor learning rule with a local gradient. At every timestep t, presynaptic traces χ for each neuron are updated. Subsequently, the eligibility trace λ is calculated using Eq. 7, where α is a double exponential filter, and σ is the partial derivative of the negative side of the fast sigmoid.

$$\lambda_t = \alpha(\chi_{\mathbf{t}} \otimes \sigma'(V_t)) \tag{7}$$

The error vector E is calculated as in Eq. 8, where o_t denotes the model output.

$$E_t = o_t - \mathcal{F}_{i,t}/A_{max} \tag{8}$$

Credit assignment for the hidden layers is performed using DFA, i.e., the global error is fed back into the hidden layers via a random feedback matrix B that corrects the dimensionality difference, mapping the error to the number of neurons in the layer.

$$\delta = \mathbf{E} \cdot B \tag{9}$$

Finally the weights are updated using Eq. 10.

$$\Delta w_{hidden} = \lambda \odot \delta, \Delta w_{out} = \lambda \odot E \tag{10}$$

2.3 Basic Function Approximation

To show the viability of the concept of neuromuscular-inspired function fitting, we need to show that the TwNN model can learn to fit basic functions. We train TwNN to fit three different functions that together cover a broad range of approximation requirements: the Gaussian which is a universal probability density function approximator, the sine which is bipolar and used in Fourier approximations, and the square pulse which is discontinuous and has infinitely many spectral commponents. We will treat time in arbitrary time units.

A single Gaussian, a single oscillation of the sine function, and a single square pulse are generated on a time-axis of 500 arbitrary time units. The Gaussian is generated such that it is close to 0 at $t = 0$ and $t = 500$. The generated functions are centered on a time-axis with 1000 arbitrary time units, i.e., all functions started at $t = 250$, and ended at $t = 750$. To make the function approximation of similar difficulty without changing the model's hyperparameters, we normalized all functions such that the integral of their absolute value equaled 1.

We train a TwNN to fit the functions. The spiking neural network consists of two layers. The 50 input neurons generate a precomputed spiking pattern with a spiking probability of 10% per timestep per neuron. The 100 output neurons are LIF-neurons, characterized by parameters, chosen to be realistic for the problem at hand: time-constants $\tau_1 = 15$, $\tau_2 = 15$, and neuron thresholds $\Theta = 2$. All output filter parameters are uniquely drawn from a uniform distribution. The time constants are drawn from $T \sim U(20, 50)$. The sample space is picked such that the time constants were faster than the timescale of the functions, and superposition of outputs is required to fit the functions. The amplitudes are also uniquely drawn from a uniform distribution $M \sim U(\frac{A_{max}}{10}, A_{max})$, where A_{max} is the maximum amplitude. The weights are clipped such that $w \in [-1, 1]$, which avoids the need to use amplifiers in electronics, and promotes diversification in outputs.

All time-constants required for the Superspike learning-rule are set to $\tau = 15$. In the Superspike paper, for stability, Δw is added up to a running sum, and executed after a preset time, after which it is reset to 0. We do *not* implement the running sum. Instead, weight updates are performed every timestep. To get an impression of the hyperparameter landscape, we performed a hyperparameter sweep over the learning rate α, and the maximum amplitude A_{max}.

2.4 Approximation of Multiple Functions Triggered by Different Input Traces, by a Single Network

In the basic function approximation task, only a single input to output mapping is performed. To see whether the model and training method are capable of learning multiple mappings, we train the model to perform handwriting from speech. More specifically, we train the model to map 10 Mel-spectrograms of spoken digits, to their respective handwriting trajectories.

Ten audio files are extracted from the audio MNIST dataset [3], one for each digit, spoken by the same speaker (speaker 6) to prevent frequency bias. For each audio file, silence is trimmed, and all samples are truncated to the shortest sample's duration ($0.323s$). The audio files are converted to mel spectrograms with 8000 frequencies, logarithmically divided into 256 bins. The time-axis is divided into 970 chunks. The global maximum amplitude is chosen as reference amplitude. The sound level is divided by -100 to ensure all values lie between 0 and 1, which will be most practical when using it as an input to the TwNN.

The handwritten digits dataset is created from scratch. A handwriting trajectory can be split up into an x-component and y-component. Matching coordinates in time retrieves the original trajectory. A program is made that records the x and y values of drawings. With this program, we wrote each digit once. The x and y values of all digits were interpolated to equal length (800 timesteps), warping the time-axis so all digits share the same trajectory duration, which is not realistic but sufficient for demonstrating event-based input to continuous output mapping. After interpolating, there is still a discrepancy in timesteps between the audio spectrograms and the handwritten digit trajectories. The audio spectrograms have 970 timesteps, while the handwritten digit trajectories have 800. To center the trajectories, we appended 85 zeros to the front and back of the trajectories. An initial period where the audio provides input spikes, while the output requires no activity accommodated a warm-up period for the TwNN, i.e., a period where membrane potentials can stabilize, allowing neurons to be close to spiking when the writing trajectory commences. The audio spectrogram as well as the handwritten version of the digit 8 are depicted in Fig. 2.

To map audio spectrograms to handwritten digits, we expect that the TwNN requires more nonlinearity than in the base function estimation task. We therefore adopt a fully-connected three-layer feedforward TwNN with 256, 250 and

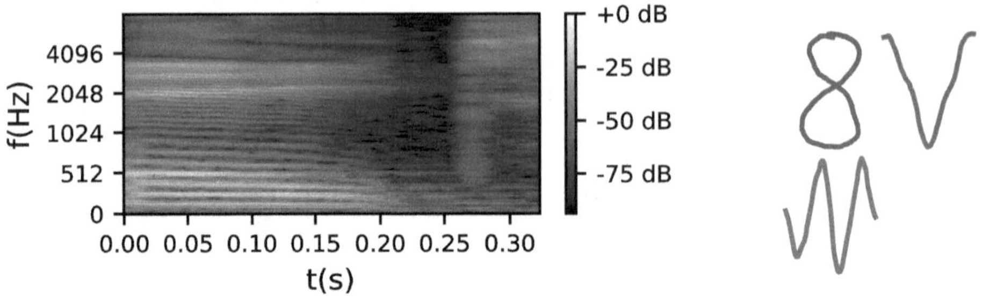

(a) (b)

Fig. 2. a) Spectrogram of the spoken digit "8". We can see an initial increase in amplitude between 0 and 0.2s representing "EY", we then see a period of silence between 0.2 and 0.25s, followed by another increase in amplitude, which represents the "T" sound. **b)** Handwriting of the digit "8". Top left: handwritten digit. Top right: y-trajectory as a function of time. Bottom left: x-trajectory as a function of time.

200 neurons in the respective input, hidden and output layer. All neuron and eligibility trace time-constants are set to 25, which is slower than the basic function approximation network, as the functions to fit are on a larger timescale (500 timesteps to 800 timesteps). Each input neuron is sensitive to one specific frequency bin. Every timestep, the sound level of the respective frequency bin is added to the non-leaky input neuron. When the neuron threshold of $\Theta = 25$ is reached, the neuron spikes and the membrane potential is reset. Each output twitch is parametrized by time constants τ_{rise}, τ_{fall}, and amplitude A. The time constants are sampled from a uniform distribution $T \sim U(20, 100)$. A is one of the hyper parameters that are varied, and is sampled from a different uniform distribution $M \sim U(\frac{A_{max}}{10}, A_{max})$. To see how many mappings of audio data to handwriting patterns we can save, we train 10 different models, incrementally adding one digit per model, i.e., one model trains on $\{0\}$, another trains on $\{0, 1\}$, and so on such that the last model trains on all 10 digits. To save computing resources, while being able to sufficiently assess performance of the model, only the x-trajectories are used in training and testing.

To evaluate model performance on the full handwriting task, the top-3 hyperparameter combinations from the x-trajectory task are used to train 5 model instances each. Each trajectory (x and y) is estimated by a separate model, so 2 trained TwNN's are required to produce one handwritten digit. Matching both estimations in time reconstructs the handwritten trajectory.

3 Results

3.1 Basic Function Approximation

A hyperparameter sweep is performed over the learning rate $\alpha \in \{0.001, 0.01, 0.1\}$, and the maximum amplitude $A_{max} \in \{0.01, 0.1\}$. For each parameter combination, 5 runs of 50,000 trials each are performed. The average errors per hyperparameter combination for each target function are plotted in Fig. 3. We can see that in general models with lower twitch amplitudes can learn to fit the functions better. The best combination is a high learning rate combined with outputs with low amplitudes. The mean model fits on the target functions are shown in Fig. 4. We can see that the model learned to estimate the sine and Gaussian well, while the model was not able to estimate the square pulse as well. Fitting straight horizontal lines and sharp drops or rises appears to be a problem for the model.

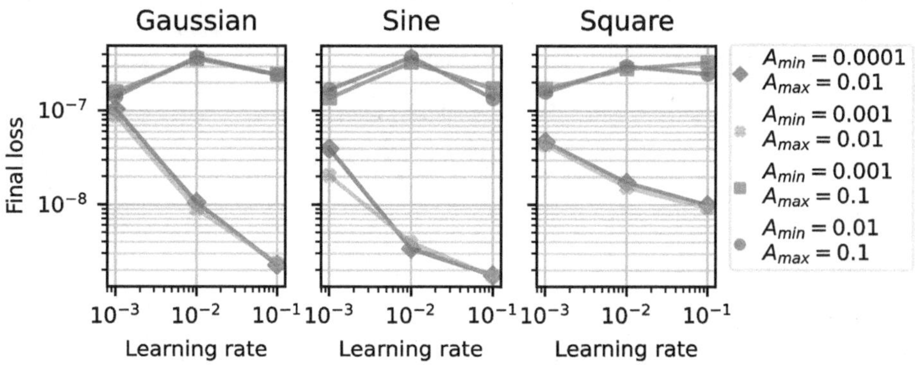

Fig. 3. Average losses for the final function estimations of the trained models. For each hyperparameter combinations, 5 models are trained. A_{min} and A_{max} are the minimum and maximum twitch amplitude respectively. Note that the vertical loss axis is in log scale.

3.2 Multiple Function Approximation

A hyperparameter sweep is performed over the learning rate of the hidden layer $\alpha_h \in \{0.001, 0.0001\}$, the learning rate of the output layer $\alpha_o \in \{0.001, 0.0001\}$, and the maximum amplitude $A_{max} \in \{1, 10, 20\}$. A single run of 50,000 trials each is performed per parameter combination. As an indication of the final model performance, we took the average over the last 1000 errors. The loss landscape can be seen in Fig. 5. We can see that models with some hyperparameter combinations perform well on one or two digits. The more digits are required to fit, the worse the performance gets. It does not seem like the model performances drop dramatically, indicating only a limited influence of catastrophic forgetting. Example fits of the 10-digit model are shown in Fig. 6.

4 Discussion

Research on continuous outputs for SNNs is limited. We developed a twitch neural network that allows generating continuous outputs. We trained the network in a biologically and neuromorphically plausible way.

The simple LIF neuron model can be extended with a variable threshold (ALIF) or time delay. When extended, the surrogate gradient of the Superspike rule is not viable anymore. Instead, (improvements of) the e-prop learning rule [4] can be used to compute surrogate gradients for the extended neuron functionality. Training such models encourages spike sparsity, and potentially improves performance in time-based tasks. On the other hand, they do have a higher computational cost and are more complex to implement in electronics.

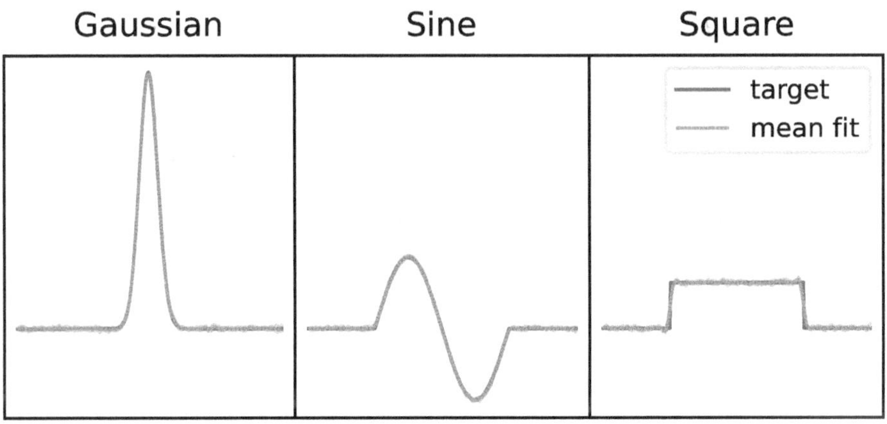

Fig. 4. Mean fit of 5 models trained to estimate basis functions from random static noise. The standard deviation is shown as a transparent band around the fit. The sine and Gaussian are fit best, while the square pulse is fit worst.

The use of a TwNN brings new challenges. We noticed learning instability in some parametrizations of the output. In SNNs with spiking outputs it is clear whether a neuron should spike or not, within a narrow time window. In TwNNs, one spike yielding a big-amplitude output twitch can cause a sudden long-lasting and large deviation from the target. This represents a big disturbance in the learning process. Using TwNNs mainly raises the questions of how to pick the right twitch parametrizations, and how to stably teach the network to generate desired functions. If we look at the biological neuromuscular system, there are two mechanisms that may give rise to stable learning in TwNNs, while also allow for learning a range of tasks. Firstly, the recruitment mechanism [13], where the neural system recruits muscle fibers depending on desired force, allows the system to tackle a variety of tasks. Secondly, in learning, infants usually learn via the proximodistal principle [5], i.e., first learning gross motor control, followed by fine motor control. A TwNN can be trained on the basis of such biology-inspired principle, i.e., first learning to use the large-twitch outputs, while correcting the remaining errors later in the learning process through learning small-twitch outputs for refinement. This may also naturally give rise to the recruitment mechanism. While both agonist and antagonist muscles co-contract to increase movement stability, they are also able to inhibit each other through inter-neurons. Inhibition saves energy and prevents muscles from tearing. In the TwNN, as we implemented it, there is no mutual inhibition. This allows the network to activate unnecessary twitches. Making use of inhibitory connections can potentially increase energy-efficiency.

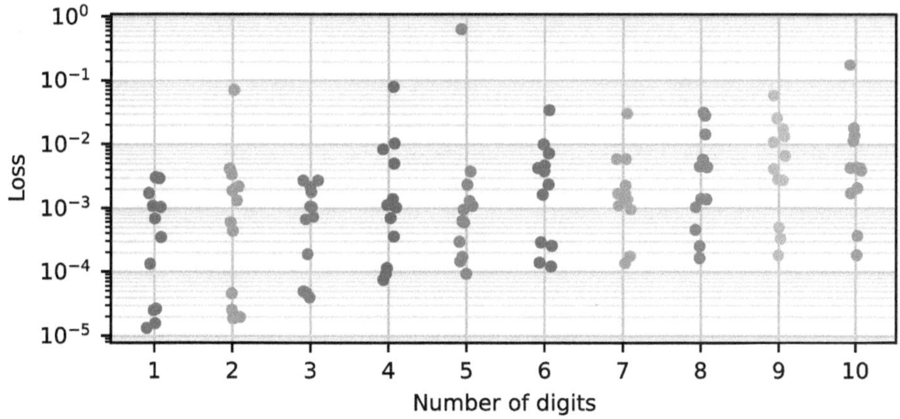

Fig. 5. Models with different hyperparameter combinations are trained to write a digit from an audio signal. Each model hyper parameter combination is used to fit a predetermined number of digits, increasing from only 0, until 0 to 9.

In this work, the task is to map an audio spectrogram to a constant output function. To perform a simple real-time speach-to-handwriting task, two challenges need to be addressed. Firstly the model needs to learn when to start and stop writing, often referred to as "PEN UP" and "PEN DOWN" states. Secondly the model needs to retain history of speech to distinguish between words with similar sounds (e.g., "seven" and "eleven").

5 Conclusion

We proposed a twitch neural network, to approximate various continuous functions from a spiking input. The TwNN is trained using a biologically plausible learning rule that allows local learning, and prevents layer-locking during training. We first showed that it is possible to approximate the Gaussian, sine and square pulse functions, of which the Gaussian and sine are known for their function approximation capabilities and the square contains infinitely many spectral components. We found that the Gaussian and sine functions were approximated accurately. The horizontal lines of the square function however, created a bottleneck for a good fit. Subsequently, we showed that the TwNN is also capable of saving multiple patterns. We showed this by training the network to map 10 Mel spectrograms of spoken digits, to handwriting trajectories of the corresponding

Fig. 6. Fits of 5 TwNN models, trained using the best hyperparameters from the exploratory search. The models are trained to map spoken digits to handwriting. The target is plotted, together with the mean output of the 5 trained models. Each individual model output is also plotted in transparent curves. Similar to the fitting attempts on the square pulse, we see that fitting straight lines is challenging.

spoken digits. The TwNN shows basic function fitting capabilities. Its performance and/or energy-efficiency can be improved by more biologically-inspired learning, a different network architecture and/or more complex neuron models.

Acknowledgments. This project was funded by the CogniGron Center, University of Groningen. We thank the Center for Information Technology of the University of Groningen for their support and for providing access to the Hábrók high performance computing cluster

Disclosure of Interests. The authors have no competing interests to declare that are relevant to the content of this article.

References

1. Abbott, L.F., DePasquale, B., Memmesheimer, R.M.: Building functional networks of spiking model neurons. Nat. Neurosci. **19**(3), 350–355 (2016)
2. Adrian, E.D., Bronk, D.W.: The discharge of impulses in motor nerve fibres: part ii. the frequency of discharge in reflex and voluntary contractions. J. Physiol. **67**(2), i3 (1929)
3. Becker, S., Vielhaben, J., Ackermann, M., Müller, K.R., Lapuschkin, S., Samek, W.: Audiomnist: Exploring explainable artificial intelligence for audio analysis on a simple benchmark. J. Franklin Inst. (2023). https://doi.org/10.1016/j.jfranklin.2023.11.038
4. Bellec, G., et al.: A solution to the learning dilemma for recurrent networks of spiking neurons. Nat. Commun. **11**(1), 3625 (2020)
5. Berthier, N.E., Clifton, R.K., McCall, D.D., Robin, D.J.: Proximodistal structure of early reaching in human infants. Exp. Brain Res. **127**(3), 259–269 (1999)
6. Boerlin, M., Machens, C.K., Denève, S.: Predictive coding of dynamical variables in balanced spiking networks. PLoS Comput. Biol. **9**(11), e1003258 (2013)

7. Buhmann, M.D.: Radial basis functions. Acta Numer **9**, 1–38 (2000)
8. Davies, M., et al.: Loihi: a neuromorphic manycore processor with on-chip learning. IEEE Micro **38**(1), 82–99 (2018)
9. DeVore, R.A.: Nonlinear approximation. Acta Numer **7**, 51–150 (1998)
10. Diehl, P.U., Cook, M.: Unsupervised learning of digit recognition using spike-timing-dependent plasticity. Front. Comput. Neurosci. **9**, 149773 (2015)
11. Gallego, G., et al.: Event-based vision: a survey. IEEE Trans. Pattern Anal. Mach. Intell. **44**(1), 154–180 (2020)
12. Hebb, D.O.: The Organization of Behavior: A Neuropsychological Theory. Psychology press (2005)
13. Henneman, E.: Relation between size of neurons and their susceptibility to discharge. Science **126**(3287), 1345–1347 (1957)
14. Hornik, K.: Approximation capabilities of multilayer feedforward networks. Neural Netw. **4**(2), 251–257 (1991)
15. Huebotter, J., Lanillos, P., van Gerven, M., Thill, S.: Spiking neural networks for continuous control via end-to-end model-based learning. Neuromorphic Comput. Eng. **6**(2), 024004 (2026)
16. Iannella, N., Back, A.D.: A spiking neural network architecture for nonlinear function approximation. Neural Netw. **14**(6–7), 933–939 (2001)
17. Liddell, E.G.T., Sherrington, C.S.: Reflexes in response to stretch (myotatic reflexes). In: Proceedings of the Royal Society of London. Series B, Containing Papers of a Biological Character, vol. 96, issue 675, pp. 212–242 (1924)
18. Nøkland, A.: Direct feedback alignment provides learning in deep neural networks. Adv. Neural. Inf. Process. Syst. **29** (2016)
19. Ponulak, F., Kasiński, A.: Supervised learning in spiking neural networks with resume: sequence learning, classification, and spike shifting. Neural Comput. **22**(2), 467–510 (2010)
20. Schomaker, L., Timmermans, J., Banerjee, T.: Non-linear adaptive control inspired by neuromuscular systems. Bioinspiration Biomimetics **18**(4), 046015 (2023)
21. Stanojevic, A., Woźniak, S., Bellec, G., Cherubini, G., Pantazi, A., Gerstner, W.: High-performance deep spiking neural networks with 0.3 spikes per neuron. Nat. Commun. **15**(1), 6793 (2024)
22. Sun, T., Bohté, S.: Uncertainty-aware spiking neural networks for regression. In: 2025 International Conference on Neuromorphic Systems (ICONS), pp. 195–201. IEEE (2025)
23. Tayarani-Najaran, M.H., Schmuker, M.: Event-based sensing and signal processing in the visual, auditory, and olfactory domain: a review. Front. Neural Circ. **15**, 610446 (2021)
24. Timmermans, J., Schomaker, L.: Trainable movement control using spikes and muscle-twitch dynamics. Front. Neurorobotics **20** (2026). https://doi.org/10.3389/fnbot.2026.1761767
25. Zenke, F., Ganguli, S.: SuperSpike: supervised learning in multilayer spiking neural networks. Neural Comput. **30**(6), 1514–1541 (2018)

Introducing Metrics for Tuning and Evaluating Decision Making Processes Based on Spiking Neural Networks

Paul Quentin Schauer[1], Dominik Krenzer[1], Ferney Beltran-Velandia[1,2](✉), and Martin Bogdan[1,2]

[1] Leipzig University, Augustusplatz 10, Leipzig, Germany
[2] Center for Scalable Data Analytics and Artificial Intelligence ScaDS.AI, Dresden/Leipzig, Humboldtstrasse 25, Leipzig, Germany
beltran@informatik.uni-leipzig.de

Abstract. Perceptual decision-making tasks often rely on models that capture competitive neural dynamics. A set of quantitative metrics is introduced for evaluating decision-making dynamics in spiking neural network implementations of the Wang model. These new metrics, such as stability and clarity, capture complementary aspects of evidence accumulation and competitive interactions between excitatory populations. The metrics are applied in a systematic grid search to identify parameter regimes that optimize decision performance. The results demonstrate that the metrics reveal characteristic boundaries in parameter space and enable robust comparison across model configurations, supporting tuning and interpretation of neural decision-making models.

Keywords: Spiking Neural Networks · Perceptual Decision Making · Hyperparameter Optimization · Wang Model

1 Introduction

Decisions accompany us every day, no matter where we are. Do you walk to work or take the train? Do you wear a T-shirt or a sweater? Such decisions are based on experience and information, some of which we gather from our surroundings. In order to better understand such decisions, numerous experiments have already been conducted in the field of perception-based decision-making (PDM) [4]. A classic paradigm for the experimental investigation of PDMs is the random dot motion (RDM) task [6]. In this experiment, rhesus monkeys observe a field of moving dots, some of which move in a common direction while the rest move randomly within the field. The monkeys then indicate their decision regarding the perceived direction of movement by making a saccadic eye movement toward one of two target dots. It was shown that certain neurons in the lateral intraparietal cortex (LIP) of the monkeys exhibit a ramp-shaped activity during decision-making processes. This is interpreted as evidence accumulation, which

L. Pasa et al. (Eds.): ICANN 2026, LNCS 17089, pp. 212–223, 2027.
https://doi.org/10.1007/978-3-032-38398-3_18

represents a central mechanism in many decision-making models [8–10]. In order to further examine the observations from these experiments, attempts have been made to simulate PDM behavior using spiking neural networks (SNNs). A well-known model for simulating PDM is the Wang model [10]. The network (Fig. 1) consists of two excitatory populations $E1$ and $E2$, each representing a decision alternative. These two populations compete via a common inhibitory population. Within the excitatory populations, there are strong recurrent connections that stabilize their own activity. A decision bias is realized via the two different stimuli on the two populations $E1$ and $E2$. In addition, each population is subject to background noise. The shared connection with the inhibitory population means that increased activity in one group suppresses the activity of the other group. In this way, the network implements a winner-takes-all principle. However, the emergence and stability of this behavior depends on the choice of model parameters. While tuning parameters is already challenging in SNNs due to their nonlinear dynamics and high-dimensional parameter spaces [2,7], this challenge is further amplified in attractor-based perceptual decision-making models such as the Wang model [10], where multiple interacting dynamical properties jointly shape the network's behavior, making systematic evaluation difficult. To address this, a set of intuitive metrics is introduced that enable systematic evaluation and comparison of the Wang model while providing a quantitative description of its decision dynamics, including decision reliability, strength of competition, inhibitory suppression, and temporal stability. Moreover, it is demonstrated how these metrics can be used to optimize model parameters through grid search.

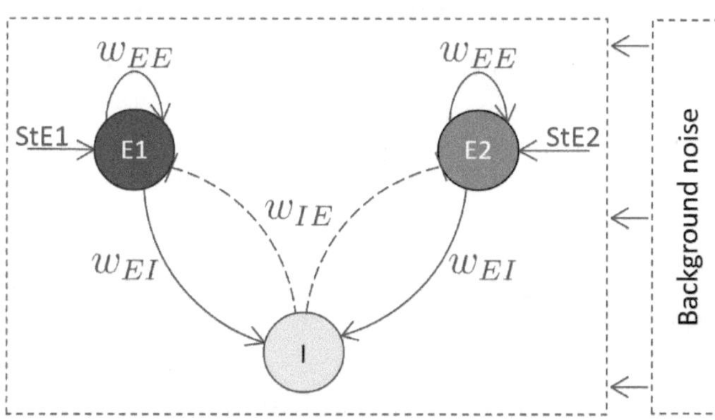

Fig. 1. Variation of the Wang model [10]. Two excitatory populations ($E1$, $E2$), having recurrent connections (w_{EE}), compete via a common inhibitory population (I) under background noise. $E1$ and $E2$ receive distinct stimuli ($StE1$, $StE2$). The inhibitory population receives input from both excitatory population (w_{EI}) and provides feedback inhibition (w_{IE}).

2 Methodology

This section introduces novel metrics for decision-making processes in models with two excitatory, mutually competing populations [10]. It is also demonstrated how these metrics can be applied to optimize the model's hyperparameters.

2.1 Metrics Definition

The proposed metrics are based on the population activity of the two excitatory populations $E1$ and $E2$. Both of these populations receive external stimulus inputs (Fig. 2) as well as stochastic background activity. The stimulus input is modeled as a stochastic input process with the rate r_p, which is recalculated every 50 ms according to:

$$r_p = \mu_p + \sigma \cdot \eta, \qquad \text{with} \qquad p \in [E1,\ E2] \tag{1}$$

where μ_p denotes the mean input rate of the excitatory populations, σ the noise intensity and η a normally distributed random variable. To induce a defined decision outcome, a small constant input bias is applied to one population. Concretely, the mean input rate μ_p to the selected of the two populations is slightly increased relative to the other, creating a systematic shift in evidence in favour of one decision alternative. This approach is necessary in order to evaluate the decision outcome and to obtain comparable results in multiple simulations. The decision outcome is evaluated within a predefined decision period $D_t = [t_0,\ t_1]$, which provides sufficient time for the network dynamics to evolve and form a stable decision state. The biased population is denoted by P, while the other is denoted by $\overline{P}$ (Fig. 2).

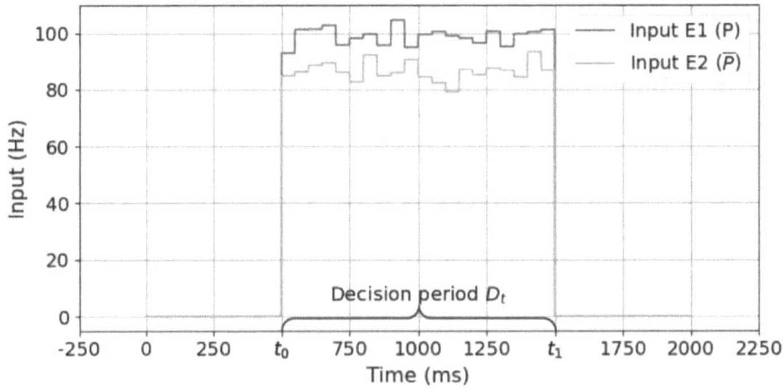

Fig. 2. An $E1$-biased stimulus is shown, with the decision period D_t highlighted. Population $E1$ (P) receives an input rate approximately 15 Hz higher than population $E2$ ($\overline{P}$). The input rates of both populations are recalculated every 50 ms.

In addition to the directed stimulus, all neurons receive independent Poisson-driven inputs that model background noise. All metrics are evaluated over n

independent simulation trials of the length T. The population activity in a time interval $[t_x,\ t_y]$ is defined as the mean firing rate of all neurons in the respective population [3]:

$$A[t_x;\ t_y] = \frac{1}{t_y - t_x} \int_{t_x}^{t_y} \frac{1}{N} \sum_{j=1}^{N} \sum_{f} \delta(t - t_j^{(f)}) dt \tag{2}$$

where N denotes the number of neurons in the population, $t_j^{(f)}$ the firing times of neuron j with f indexing all spike events in neuron j and the Dirac Delta function δ representing a spike event. To compare both population activities, the activity difference ΔA in a defined interval $[t_x;\ t_y]$ can be calculated through:

$$\Delta A[t_x;\ t_y] = A_P[t_x;\ t_y] - A_{\overline{P}}[t_x;\ t_y] \tag{3}$$

This quantity measures how strongly the biased population P outperforms the non-biased population $\overline{P}$ within the specified time interval. A positive value of ΔA indicates a correct decision in favor of the biased input, while values close to zero or negative values reflect ambiguous or incorrect outcomes. Hence, ΔA forms the basis for the novel metrics introduced here:

Decision Accuracy *Acc*: This metric evaluates whether the network converges to the attractor state corresponding to the stimulus bias. The metric therefore evaluates the activity difference $\Delta A[t_0,\ t_1]$ (Eq. 3) within the decision period D_t. Since small activity differences may arise purely from noise a threshold parameter $\theta > 0\ Hz$ is introduced to ensure that only sufficiently strong dominance of one population is counted as a decision. The specific value of θ is not universal but may be defined depending on the experimental context, the characteristics of the network activity, and the requirements of the particular decision-making task. A trial i is considered in the attractor state if

$$\Delta A_i[t_0;\ t_1] \geq \theta \tag{4}$$

i.e., if the biased population exhibits a sufficiently strong dominance during the decision window. Otherwise, the trial is classified as incorrect or undecided. The decision Accuracy is then defined as

$$Acc = \frac{1}{n} \sum_{i=1}^{n} a_i \qquad \text{with} \qquad a_i = \begin{cases} 1 & \text{if } \Delta A_i[t_0;\ t_1] \geq \theta \\ 0 & \text{else} \end{cases} \tag{5}$$

Here, n denotes the number of independent simulations with the same stimulus configuration. Thus, *Acc* quantifies how reliably the network converges to the attractor state within the predefined decision window in the presence of noise.

Clarity *C*: Another utilization of the activity difference (Eq. 3) is to calculate the clarity of the decisions in the model. Here, the absolute value of the activity difference is averaged across all n-simulations, regardless of the decision accuracy of the decision. Therefore the clarity is proposed as:

$$C = \frac{1}{n}\sum_{i=1}^{n} |\Delta A_i[t_0;\, t_1]| \tag{6}$$

Thus, this metric allows to estimate how clear the decision between the populations is: the higher C is, the sharper the decision.

Effective Inhibition I_{eff}: To quantify the strength of competition between the two excitatory populations, the Effective Inhibition metric is introduced. This metric measures how strongly the activity of the losing population L is suppressed during the decision period relative to its baseline activity outside the decision period. For each simulation trial i, the losing population is defined as the population with the lower mean activity (Eq. 2) during the decision period $D_t = [t_0;\, t_1]$:

$$L_i = \underset{p\in\{E1,E2\}}{\arg\min}\; A_i^p[t_0;\, t_1] \tag{7}$$

The corresponding activity during the decision period is denoted by $A_{L_i}[t_0;\, t_1]$ To quantify the degree of suppression, this activity is compared to a baseline activity representing the behavior outside the decision period. Since the network dynamics may differ before and after the decision period, both phases are taken into account. For each simulation trial i, the baseline activity B_{L_i} is therefore computed as the duration-weighted average of the activity before and after the decision period:

$$B_{L_i} = \frac{A_{L_i}[0;\, t_0]\cdot\, t_0 \;+\; A_{L_i}[t_1;\, T]\cdot\,(T - t_1)}{t_0 \;+\; T - t_1} \tag{8}$$

Using this baseline (Eq. 8), the Effective Inhibition I_{eff} is computed as:

$$I_{eff} = \frac{1}{n}\sum_{i=1}^{n} A_{L_i}[t_0;\, t_1] - B_{L_i} \tag{9}$$

where n denotes the number of independent simulation trials. Thus, lower values of I_{eff} indicate stronger suppression of the losing population and therefore more effective inhibition.

Stability S: To analyze the temporal robustness of the decision dynamics, the dominant population is examined whether it remains consistent throughout the decision period D_t. For this purpose, the decision period is divided into K equally sized time bins $\mathcal{B}_k$ of length Δt_B:

$$\mathcal{B}_k = [t_0 + (k-1)\Delta t_B;\, t_0 + k\Delta t_B] \qquad \text{with} \quad k \in [1,\, ...,\, K-1,\, K] \tag{10}$$

For each simulation trial i and bin k, the activity difference ΔA_i^k is computed according to (Eq. 3). Using (Eq. 5), the Decision Accuracy Acc_i^k of each simulation trial i and bin k can be determined through:

$$Acc_i^k = \begin{cases} 1 & \text{if } \Delta A_i^k \geq \theta \\ 0 & \text{else} \end{cases} \tag{11}$$

A decision in trial i is considered stable if the decision accuracy remains consistent across all bins of the decision window. This can be formalized as:

$$s_i = \begin{cases} 0 & \text{if } \exists k : Acc_i^k \neq Acc_i^{k-1} \\ 1 & \text{else} \end{cases} \tag{12}$$

Finally, the Stability metric S is defined as the mean proportion of trials that remain stable (Eq. 12) throughout the decision period:

$$S = \frac{1}{n} \sum_{i=1}^{n} s_i \tag{13}$$

Thus, S quantifies how consistently the network maintains the same dominant population during the decision period across all n simulations.

2.2 Model Architecture and Simulation Setup

A simplified leaky integrate-and-fire (LIF) neuron model with exponential membrane potential decay and discrete synaptic voltage updates is employed. The simulated network (Fig. 1) comprises a total of 220 neurons, organized into two excitatory populations, $E1$ and $E2$, each containing 100 neurons, and a shared inhibitory population I consisting of 20 neurons. The neurons are partially connected, with connections established probabilistically according to $p_{pre,\,post}$, where the indices denote the presynaptic and postsynaptic populations (e.g., p_{IE} specifies the probability of connections from inhibitory to excitatory neurons). The strength of these connections is determined by synaptic weights $w_{pre,\,post}$, such as w_{IE}. Each simulation runs for a total duration of $T = 2000\,\text{ms}$. The decision stimulus is presented between $t_0 = 500\,\text{ms}$ and $t_1 = 1500\,\text{ms}$, resulting in a simulation setup similar to the Wang model [10]. To generate the stimulus input, 20 independent Poisson spike trains are used, which are randomly projected onto neurons of the respective excitatory population with probability $p_{StE} = 5\%$. Consequently, individual neurons receive varying numbers of superimposed input streams.

2.3 Hyperparameter Optimization

The proposed metrics are explicitly used to guide the hyperparameter optimization through multiple grid searches, providing a structured and quantitative basis for evaluating the model behavior. The goal of this optimization is to identify

parameter configurations that yield reliable, distinct, and stable decisions. In this approach, predefined parameter ranges are combined into a grid of all possible value combinations. For each parameter combination, the model is simulated multiple times ($n = 20$) and evaluated using the above defined performance metrics. Each simulation represents an independent stochastic realization of the same stimulus configuration. Since the stimulus parameters remain identical across repetitions and only the stochastic input realizations differ, the model is expected to reach an accuracy of 100% through the hyperparameter optimization. The investigated parameter ranges and their discretization were determined empirically based on preliminary proof-of-concept experiments. These experiments were used to identify ranges in which meaningful competitive network dynamics emerged while avoiding configurations associated with inactive or unstable network behavior. Based on the same preliminary experiments, the threshold parameter θ was set to 30 Hz, and $K = 10$ was used for the Stability metric.
In order to limit the scope of hyperparameter optimization, only parameters that directly influence inhibitory control and, consequently, the competitive decision dynamics are taken into account. The focus here is primarily on the competitive aspect of the inhibitory population. In particular, the interaction between excitatory and inhibitory populations plays a central role in shaping the winner-takes-all behavior. The optimization therefore focuses on the following parameters:

$$w_{EI}, \quad w_{IE}, \quad p_{EI}, \quad p_{IE}, \quad \tau_I, \quad v_{th_I}$$

Here, w_{EI} and w_{IE} determine the synaptic weights between excitatory and inhibitory populations, while p_{EI} and p_{IE} determine the corresponding connection probabilities. These parameters directly control the strength of the feedback and thus the stability and selectivity of the competitive dynamics. The inhibitory spike threshold v_{th_I} is included in the optimization, as it controls the excitability of the inhibitory population. By determining how easily inhibitory neurons fire in response to incoming input, v_{th_I} directly affects the onset and strength of inhibitory feedback, and thus plays a crucial role in balancing excitation and inhibition. In addition, the inhibitory time constant τ_I is optimized, as it determines the integration time of the inhibitory population and thus the temporal dynamics of inhibition. In preliminary experiments, τ_I proved to be the most influential parameter for the emergence of stable competition between the excitatory populations. Consequently, it serves as a key baseline parameter in our grid searches and defines the basis around which the remaining parameters are systematically varied. All other parameters are kept constant, as they either have no direct influence on competitive interaction or primarily affect the general activity baseline, but not the selectivity of the decision. This targeted reduction significantly reduces the search space for hyperparameter optimization, while at the same time allowing the mechanisms central to decision-making behavior to be systematically optimized.
The results are visualized as heatmaps, each depicting the model's performance across parameter combinations for a specific metric. From these heatmaps, the

optimal parameter settings are identified by maximizing decision accuracy, clarity, and stability while minimizing the effective inhibition metric.

3 Results

The results focus on the grid search between the inhibitory time constant τ_I and the excitatory-to-inhibitory synaptic weight w_{EI} as representative example. Additional results can be found in the GitHub Repository specified in the Code Availability section.

To determine suitable parameter regions, the heatmaps shown in Fig. 3 are analyzed. Each heatmap visualizes one evaluation metric across the parameter space defined by τ_I and w_{EI}.

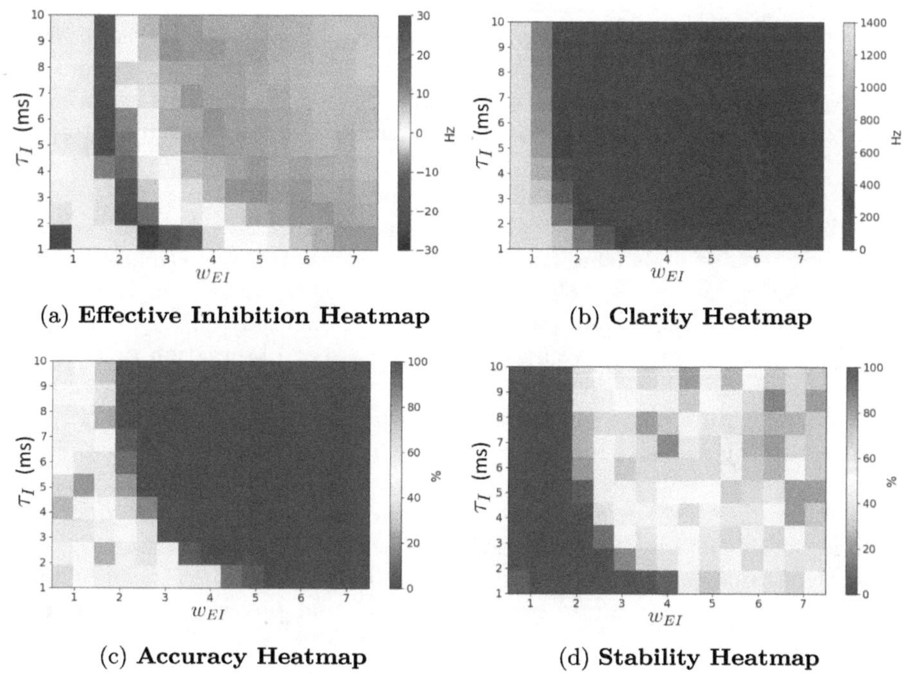

(a) **Effective Inhibition Heatmap** (b) **Clarity Heatmap**

(c) **Accuracy Heatmap** (d) **Stability Heatmap**

Fig. 3. Evaluation metrics across the parameter space defined by the inhibitory time constant τ_I and the excitatory-to-inhibitory synaptic weight w_{EI}. Each heatmap shows one metric: (a) decision accuracy, (b) effective inhibition, (c) decision clarity, and (d) stability. A similar pattern can be observed across all heatmaps, indicating that the interaction between τ_I and w_{EI} strongly determines the network dynamics.

The effective inhibition heatmap (Fig. 3a) reveals a narrow region of strong inhibition (blue area) for each value of τ_I. This region corresponds to parameter

combinations where the inhibitory population is able to suppress the losing excitatory population efficiently. Notably, the location of this region shifts slightly as τ_I increases, indicating that stronger synaptic weights are required to maintain effective inhibition when inhibitory integration becomes slower.
The clarity heatmap (Fig. 3b) provides additional insight into the quality of the decision. Very low values of w_{EI} lead to extremely high clarity values due to uncontrolled excitation, which results in unrealistically high firing rates. As inhibitory strength increases, clarity stabilizes at lower firing rates. This transition marks the boundary between unstable excitation and balanced competitive dynamics.
The accuracy heatmap (Fig. 3c) shows that high decision accuracy is achieved across a relatively large region of the parameter space. In particular, accuracy quickly approaches 100% once w_{EI} exceeds a certain threshold. This indicates that sufficiently strong inhibitory feedback reliably enforces a winner-take-all decision between the two excitatory populations.
Finally, the stability heatmap (Fig. 3d) illustrates how consistently the network maintains its decision over time. While large regions of the parameter space produce stable dynamics, certain combinations of small τ_I and intermediate w_{EI} lead to temporary instabilities caused by synchronous spiking activity within the excitatory populations.
Taken together, these heatmaps illustrate that the optimal parameter region cannot be determined from a single metric alone. Instead, the final parameter selection requires balancing trade-offs between several competing objectives: strong inhibition to suppress the losing population, sufficiently high clarity to produce distinct decisions, and stable network dynamics.
Based on this multi-metric evaluation and consistent observation of favorable behavior across the other grid searches, $\tau_I = 5\,ms$ and $w_{EI} = 2.3$ are selected as the central operating point of the model. Although this configuration does not correspond to the absolute minimum in the effective inhibition heatmap or the maximum in the stability heatmap, it represents a balanced compromise across all evaluation metrics. For example, on the one hand, τ_I values less than $5\,ms$ lead to less stable and less accurate decisions, while on the other hand, τ_I values greater than $5\,ms$ lead to less clear decisions. Similar observations can be made for w_{EI} values above (reduced decision clarity) and below (reduced decision accuracy) $2.3\,mV$. Furthermore, these parameter values also exhibited favorable decision-making behavior in other parameter combinations (see GitHub). The additional parameters mentioned before, such as the connectivity probabilities (p_{EI}, p_{IE}) were subsequently adjusted to further refine the competitive dynamics.
All grid search results ultimately led to the final parameter configuration summarized in Table 1. The corresponding metric scores averaged across 20 stochastic simulations are reported in Table 2.
With this final parameter setting, the model achieves a decision accuracy of 100% while maintaining a decision clarity of $112.08\,Hz$. The stability metric indicates that approximately 85% of simulations exhibit fully stable decision dynamics

Table 1. Final parameter setting

Parameter	Value
τ_I	$5ms$
w_{EI}	$2, 3$
w_{IE}	$2, 3$
p_{EI}	15%
p_{IE}	11%
v_{th_i}	$16mV$

Table 2. Metric scores of final model

Metric	Value
Decision Accuracy	100%
Clarity	$112.08Hz$
Stability	85%
Effective inhibition	$6.6Hz$

throughout the stimulus period. The effective inhibition value of $6.6\,Hz$ further confirms that the losing population is strongly suppressed during the decision process.

Fig. 4 illustrates the effect of the hyperparameter optimization on the network dynamics. Before optimization (Fig. 4A), the excitatory populations exhibit similar activity levels, and no clear decision emerges despite the presence of a biased stimulus. This behavior reflects insufficient inhibitory feedback within the network. The parameters of the model that had to be optimized were chosen arbitrary in the same value range as the heatmaps.

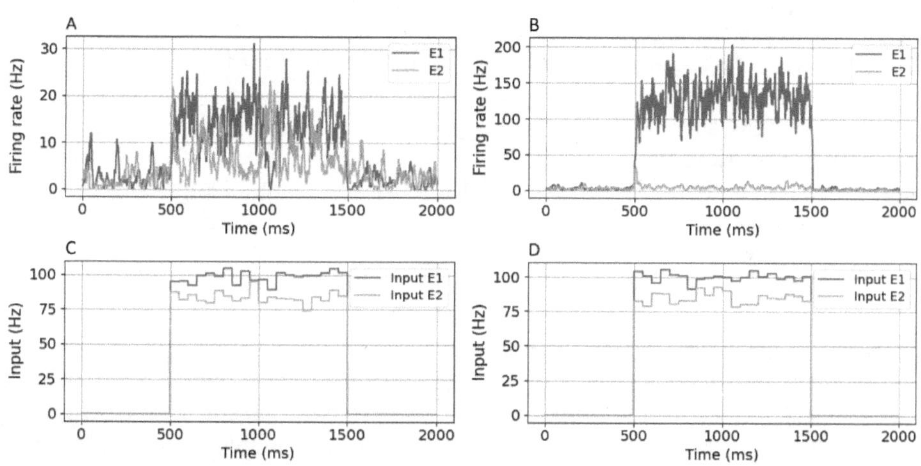

Fig. 4. Decision-making simulation before (A+C) and after (B+D) the Hyperparameter Optimization. **A**: Average population activity of $E1$ and $E2$. No clear decision is visible due to similar activity levels. Sometimes, $E1$ has a slightly higher population activity then $E2$, but this higher activity is not stable. **B**: Average population activity of $E1$ and $E2$. $E1$ develops almost instantly a high population activity due to static postsynaptic weights while suppressing $E2$. The decision remains stable over the decision period. **C** and **D**: The corresponding $E1$-biased stimuli to the two population activity figures with the same stimulus configuration but different noise.

After optimization (Fig. 4B), the network exhibits a clear winner-take-all dynamic. The stimulus-biased population rapidly increases its firing rate and maintains a high activity level throughout the decision period, while the competing population is strongly suppressed. The corresponding stimulus inputs (Fig. 4C and D) only vary in noise, demonstrating that the improved decision behavior arises solely from the optimized network parameters rather than changes in the stimulus configuration.

4 Discussion and Future Work

This work introduces a set of intuitive metrics for characterizing decision-making dynamics in Wang-type spiking neural networks, providing a basis for more systematic analysis and model refinement. The proposed metrics quantify complementary aspects of the decision process, namely decision accuracy, clarity, effective inhibition, and stability. Together, they provide a systematic framework for evaluating and comparing different parameter configurations.
Using these metrics in a grid search, it is shown that network parameters can be systematically optimized to achieve a desired decision-making behavior. In addition the analysis of the heatmaps showed that, different parameter configurations can produce similar decision outcomes while exhibiting substantially different underlying dynamics. This highlights the importance of multi-dimensional evaluation when tuning attractor-based spiking decision networks.
The present study focused on the Wang model as a proof of concept. However, the proposed metrics are based on general properties of competitive decision-making networks, such as differences in population activity, suppression of non-dominant populations, and temporal consistency of the selected state. Therefore, our framework (including the novel metrics) may also be applicable to other spiking neural network models [5] that implement decision-making through competing populations and winner-take-all dynamics.
At the same time, this broader applicability was not explicitly tested here and remains an important direction for future work. Further studies should examine the transferability of the metrics to other model classes and more complex stimulus settings, as well as to networks incorporating short-term synaptic plasticity [1]. In addition, combining the proposed evaluation framework with more advanced optimization methods may provide a scalable approach for tuning larger and more biologically detailed spiking decision models.

Acknowledgments. This work was partially funded by the Federal Ministry of Research, Technology and Space of Germany (BMFTR) in the program 'Center of Excellence for AI Research' through ScaDS.AI Dresden/Leipzig, and by BMFTR in DAAD project 57616814 (SECAI, https://secai.org/).

Disclosure of Interests. The authors have no competing interests to declare that are relevant to the content of this article.

Code Availability. The source code used in this study is available online in this GitHub repository (Code available at https://github.com/paulqu03/metrics).

References

1. Beltran-Velandia, F., Scherf, N., Bogdan, M.: Exploring the relationship between synaptic dynamics properties: gain-control and temporal filtering. In: Proceedings of the 34th European Symposium on Artificial Neural Networks, Computational Intelligence and Machine Learning, pp. 187–192 (2026). https://doi.org/10.14428/esann/2026.ES2026-236
2. Firmin, T., Boulet, P., Talbi, E.G.: Parallel hyperparameter optimization of spiking neural networks. Neurocomputing **609**, 128483 (2024). https://doi.org/10.1016/j.neucom.2024.128483, https://www.sciencedirect.com/science/article/pii/S0925231224012542
3. Gerstner, W., Kistler, W.M., Naud, R., Paninski, L.: Neuronal Dynamics: From Single Neurons to Networks and Models of Cognition. Cambridge University Press, Cambridge (2014)
4. Gold, J.I., Shadlen, M.N.: The neural basis of decision making. Ann. Rev. Neurosci. **30**, 535–574 (2007). https://doi.org/10.1146/annurev.neuro.29.051605.113038, https://www.annualreviews.org/content/journals/10.1146/annurev.neuro.29.051605.113038
5. Krenzer, D., Bogdan, M.: Reinforced liquid state machines—new training strategies for spiking neural networks based on reinforcements. Front. Comput. Neurosci. **19** (2025). https://doi.org/10.3389/fncom.2025.1569374, https://www.frontiersin.org/journals/computational-neuroscience/articles/10.3389/fncom.2025.1569374
6. Newsome, W., Pare, E.: A selective impairment of motion perception following lesions of the middle temporal visual area (mt). J. Neurosci. **8**(6), 2201–2211 (1988). https://doi.org/10.1523/JNEUROSCI.08-06-02201.1988, https://www.jneurosci.org/content/8/6/2201
7. Pfeiffer, M., Pfeil, T.: Deep learning with spiking neurons: opportunities and challenges. Front. Neurosci. **12** (2018). https://doi.org/10.3389/fnins.2018.00774, https://www.frontiersin.org/journals/neuroscience/articles/10.3389/fnins.2018.00774
8. Ratcliff, R., McKoon, G.: The diffusion decision model: Theory and data for two-choice decision tasks. Neural Comput. **20**(4), 873–922 (2008). https://doi.org/10.1162/neco.2008.12-06-420
9. Usher, M., Mcclelland, J.: The time course of perceptual choice: the leaky, competing accumulator model. Psychol. Rev. **108**, 550–592 (2001). https://doi.org/10.1037/0033-295x.108.3.550
10. Wang, X.J.: Probabilistic decision making by slow reverberation in cortical circuits. Neuron **36**(5), 955–968 (2002). https://doi.org/10.1016/S0896-6273(02)01092-9, https://www.sciencedirect.com/science/article/pii/S0896627302010929

Predictive Coding Light with Different Spike Timing-Dependent Learning Rules

Antony W. N'dri[1,3(✉)], Jiahui Fu[2], Céline Teulière[1], and Jochen Triesch[2]

[1] Université Clermont Auvergne, Clermont Auvergne INP, CNRS, Institut Pascal, Clermont-Ferrand 63000, France
antony.n-dri@orange.com, celine.teuliere@uca.fr
[2] Frankfurt Institute for Advanced Studies, Frankfurt am Main, Germany
{jfu,triesch}@fias.uni-frankfurt.de
[3] Orange Labs, Sophia Antipolis, France

Abstract. Learning in biological brains is thought to rely on diverse synaptic plasticity mechanisms. While many plasticity rules have been observed experimentally, it is still unclear how the combination of various plasticity rules at different synapse types supports learning at the network level. Here we systematically study the effects of combining three different spike timing-dependent plasticity (STDP) rules (causal, acausal and symmetric) in the recently proposed Predictive Coding Light (PCL) model, a recurrent spiking neural network model of unsupervised representation learning in visual cortex. We show that successful learning critically depends on the choice of STDP rule for excitatory synapses, with learning failing completely for the acausal STDP rule. Surprisingly, however, the network learns successfully with any choice of STDP rule for the inhibitory synapses. Overall, our work sheds new light on how combinations of plasticity rules shape learning at the network level.

Keywords: spiking neural network · spike timing-dependent plasticity · primary visual cortex · predictive coding · unsupervised learning

1 Introduction

Biological brains use a wide array of plasticity mechanisms to adapt and learn. A widely studied synaptic mechanism is spike timing-dependent plasticity (STDP) [2,8], where the change of synaptic efficacy depends on the relative timing of pre and post-synaptic spikes. STDP has been observed across many species and brain areas and has also been extensively studied as a learning mechanism in artificial spiking neural networks [1,5,13,17]. However, biological STDP rules come in many different flavors. Even for the same connection type, e.g., from an inhibitory neuron onto an excitatory neuron, vastly different STDP rules have been observed experimentally [4]. At present, it is unclear what the reasons for

L. Pasa et al. (Eds.): ICANN 2026, LNCS 17089, pp. 224–235, 2027.
https://doi.org/10.1007/978-3-032-38398-3_19

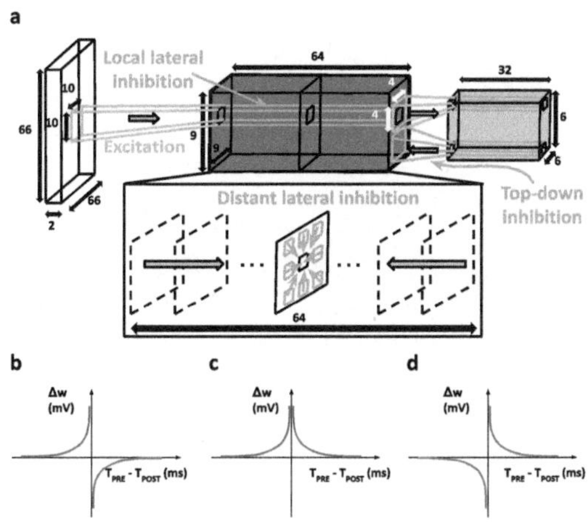

Fig. 1. The Predictive Coding Light network and different spike timing-dependent plasticity rules considered in this work. a, PCL's network architecture adapted from [12], with authorization. b, Causal STDP. c, Symmetric STDP. d, Acausal STDP.

this diversity are and how combinations of different plasticity rules at different connection types within a network support different forms of learning [7,19].

Here, we want to shed light on this issue by systematically investigating the effects of combining different kinds of STDP rules in an established model of unsupervised visual representation learning. Our starting point is the recently proposed Predictive Coding Light (PCL) model [9,12], a recurrent spiking neural network model for unsupervised learning of visual representations (see Fig. 1). We chose this model, because it reproduces many properties of primary visual cortex (V1), including simple and complex cell receptive fields and different non-classical receptive field effects (surround suppression, orientation-tuned suppression, cross-orientation suppression), which are considered a hallmark of predictive coding theories [15]. Furthermore, it only uses a single biologically plausible local STDP rule, establishing a link between STDP and classic theories of predictive coding [11].

We wondered how critical these achievements of PCL depend on the choice of STDP rule for the different connection types within the network. To answer this question, we considered three different STDP rules: the original "causal" rule where a pre-before-post pairing induces long-term potentiation (LTP), an "acausal" rule where the same kind of pairing induces long-term depression (LTD), and a "symmetric" STDP rule (see Fig. 1). We systematically studied various combinations of these rules for excitatory and inhibitory synapses in the network. As expected, we found that successful learning hinges on the type of STDP rule used for the excitatory synapses, with the acausal rule preventing learning. Surprisingly, however, we found learning to be very robust to changes in the STDP rule for inhibitory synapses, with all rules supporting successful learning. Overall, our work sheds light on how combinations of different STDP rules support learning at the network level.

2 Methods

2.1 Predictive Coding Light

Predictive Coding Light (PCL) is a recently proposed spiking neural network architecture for unsupervised learning of visual representations [12]. In this work, we use the same training protocol, hyperparameters, and datasets as in the original PCL publication [12]. For the sake of completeness, we summarize the central mechanisms of the PCL network below (for additional details see [12]).

Network Architecture. Like in the original PCL work, our network consists of 2 layers whose neurons come to resemble simple and complex cells of primary visual cortex when the network is trained on natural images (see Fig. 1). All feed-forward connectivity is excitatory and all recurrent and feedback connectivity is inhibitory. In PCL, there is no distinction between excitatory and inhibitory neurons, but a single neuron may excite some of its targets and inhibit others. The network is driven by input from an event-based vision sensor, which excites the simple cell layer, which in turn excites the complex cell layer. Both simple and complex cells make inhibitory synapses toward other neurons. In total, there are three kinds of inhibitory connections: local inhibitory connections among simple cells with identical receptive field location, long-ranging inhibitory connections among more distant simple cells, and top-down inhibitory connections from complex cells back to simple cells (see Fig. 1). The latter two connection types are essential for reproducing non-classical receptive field effects.

Leaky Integrate-and-Fire (LIF) Neuron Model. The PCL network uses LIF neurons with a refractory period. The simulation is event-based. Upon arrival of a synaptic input, the membrane potential is updated as follows:

$$\tilde{V}(t+\Delta t) = \max\{V_{\min}, V(t)e^{-\frac{\Delta t}{\tau_m}} + w_i(t) - \eta_{\mathrm{RP}} e^{-\frac{t+\Delta t - t_s}{\tau_{\mathrm{RP}}}}\} \tag{1}$$

$$V(t+\Delta t) = \begin{cases} \tilde{V}(t+\Delta t) & : \tilde{V}(t+\Delta t) < V_\theta \\ V_{\mathrm{reset}} & : \tilde{V}(t+\Delta t) \geq V_\theta \,, \end{cases} \tag{2}$$

where $V_{\min}$ is a minimum value of the membrane potential, Δt is the time since an input spike was last received by the neuron, τ_m is the membrane time constant, w_i is the weight of a synaptic excitatory or inhibitory input (in mV), η_{RP} is the amplitude of our relative refractory period, τ_{RP} is the time constant of this relative refractory period, t_s is the time when the neuron last spiked, V_{reset} is the reset value of the potential and V_θ is the membrane threshold. A spike is emitted by the LIF neuron whenever the potential exceeds this threshold.

STDP Rules. All synapses of the original PCL network are updated using a *causal STDP* (cSTDP) rule. For this rule, presynaptic spikes that occur shortly before a postsynaptic spike lead to synaptic weight increases modeling long term potentiation (LTP). However, presynaptic spikes that occur shortly after a postsynaptic spike lead to synaptic weight decreases modeling long term depression (LTD). Additionally, all plastic synapses of the PCL network use a softbound mechanism to limit weight growth. All synaptic updates of the PCL network follow:

$$\Delta w_i^{\mathrm{LTP}} = \eta_{\mathrm{LTP}} e^{\frac{t_i - t_\mathrm{s}}{\tau_{\mathrm{LTP}}}} \left(w_{\max} - w_i\right), \tag{3}$$

$$\Delta w_i^{\mathrm{LTD}} = -\eta_{\mathrm{LTD}} e^{\frac{t_{s-1} - t_i}{\tau_{\mathrm{LTD}}}} \left(w_i - w_{\min}\right), \tag{4}$$

where η_{LTP} and η_{LTD} are respectively the learning rates for potentiation and depression, t_i is the spike time of the last input received from synapse i, t_s is the spike time of the postsynaptic neuron, t_{s-1} is the spike time of the before-the-last spike of the postsynaptic neuron, τ_{LTP} and τ_{LTD} are respectively time constants that control the temporal window of potentiation and depression, $w_{\max}$ and $w_{\min}$ are respectively the maximum and minimum allowable synaptic weight values.

In this work, we consider two additional STDP rules: an acausal one and a symmetric one (see Fig. 1). In the *acausal STDP* (aSTDP), the effect of LTD and LTP are reversed such that the occurrence of presynaptic spikes shortly before a postsynaptic spike leads to weight decreases and the occurrence of presynaptic spikes shortly after a postsynaptic spike leads to weight increases:

$$\Delta w_i^{\mathrm{LTP}} = \eta_{\mathrm{LTP}} e^{\frac{t_{s-1} - t_\mathrm{i}}{\tau_{\mathrm{LTP}}}} \left(w_{\max} - w_i\right), \tag{5}$$

$$\Delta w_i^{\mathrm{LTD}} = -\eta_{\mathrm{LTD}} e^{\frac{t_i - t_s}{\tau_{\mathrm{LTD}}}} \left(w_i - w_{\min}\right), \tag{6}$$

Finally, we also consider a *symmetric STDP* (sSTDP) rule, for which the occurrence of presynaptic spikes shortly before or after a postsynaptic spike lead to potentiation:

$$\Delta w_i = \eta_{\mathrm{LTP}} \left(e^{\frac{t_{s-1} - t_\mathrm{i}}{\tau_{\mathrm{LTP}}}} + e^{\frac{t_i - t_\mathrm{s}}{\tau_{\mathrm{LTP}}}}\right) \left(w_{\max} - w_i\right). \tag{7}$$

Learning Rule Combinations. The PCL network uses STDP for excitatory and inhibitory synapses (eSTDP and iSTDP). To investigate how different forms of STDP affect learning at the network level, we combine the three different STDP rules introduced above for excitatory and inhibitory synapses. This leads to a total of nine combinations of STDP rules: **cc**, **cs**, **ca**, **sc**, **ss**, **sa**, **ac**, **as** and **aa**, where the first letter indicates the rule used for eSTDP, and the second letter that for iSTDP. We use these rule combinations to train nine different PCL networks and study their properties.

Weight Normalization. To model competition of synapses of the same type for a limited amount of a neuron's synaptic resources, we use an L1-normalization mechanism:

$$w_i \leftarrow \lambda \frac{w_i}{\sum_i w_i} \; \forall i, \tag{8}$$

where λ is the amount of available synaptic resources for incoming synapses received by a postsynaptic neuron.

Datasets. We use the same datasets as in [12]. The first is an event-based version of the Van Hateren natural image dataset [18]. It consists of 2000 samples that each have a duration of 800 ms. The second is a set of counterphase sinusoidal gratings of different orientations, frequencies and phases to test the tuning curves of simple and complex cells. The third is a set of drifting gratings that are used to test surround suppression, orientation-tuned suppression and cross-orientation suppression of simple and complex cells. All grating stimuli have a duration of 2 s.

2.2 Evaluation Metrics

Simple Cell Receptive Fields. Since the receptive fields (RFs) of simple cells in V1 are well described by Gabor functions, we fit the RFs of simple cells in the PCL network to Gabor filters and quantify the fit error.

Diversity of Tuning Properties. Biological simple and complex cells are tuned to diverse orientations and spatial frequencies. Table 1 details the expression of these metrics for each parameter and each cell type. To quantify the diversity of orientations and frequencies of the receptive fields, we measure their relative entropy to a uniform distribution. For this, we discretize the number of orientations into 8 bins that cover the 180° scope used in the testing set. Similarly, we also discretize frequencies into 18 bins also used in the testing set.

Magnitude of Suppression Effects. Simple cells in V1 have been shown to exhibit *surround suppression*, where the response of a neuron increases as its classical receptive field (cRF) gets excited but starts decreasing when both the cRF and the non-CRF (nCRF) are excited simultaneously. They also exhibit *orientation-tuned suppression*, where a neuron's response is minimal when its cRF and nCRF are excited with gratings of the same orientation, but the response increases as the orientation of the nCRF input deviates from the orientation of the cRF input. Lastly, they exhibit *cross-orientation suppression*, where a neuron's cRF is excited with the superposition of two gratings and the response is maximal when both gratings have the same orientation, but the response decreases as one of the superposed grating grows orthogonal towards the other one. The original PCL network showcases all of those suppressive effects. In Table 1, we define the three metrics M_s, M_o, M_c to quantify the magnitude of these three suppressive effects. In the definition, N is the number of gratings (with each grating varying only one parameter, e.g., orientation of the surround or bar length etc.), r is the normalized response of a cell to a stimulus, r_{cRF} is the response of a cell when only its classical receptive field is fully excited and r_θ is the response of the cell when surround and center are excited by the same orientation.

Table 1. Expression of the metrics of diversity of orientation, frequency and performance on suppression for simple and complex cells.

	Cell type	
	Simple	Complex
E_o (rel. entropy, orientation)	$\sum_i P_{\theta_i} \log\left(\frac{P_{\theta_i}}{P_{\theta,\text{uniform}}}\right)$	$\sum_i P_{\theta_i} \log\left(\frac{P_{\theta_i}}{P_{\theta,\text{uniform}}}\right)$
E_f (rel. entropy, frequency)	$\sum_i P_{f_i}\ log\left(\frac{P_{f_i}}{P_{f,\text{uniform}}}\right)$	$\sum_i P_{f_i}\ log\left(\frac{P_{f_i}}{P_{f,\text{uniform}}}\right)$
M_s (surround)	$\frac{1}{N}\sum_{i=0}^{N}(r_{\text{cRF}} - r_i)$	Not evaluated
M_o (orientation-tuned)	$\lvert 1 - \frac{1}{N}\sum_{i=0}^{N}(r_\theta - r_i)\rvert$	Not evaluated
M_c (cross-orientation)	$\frac{1}{N}\sum_{i=0}^{N}(r_\theta - r_i)$	Not evaluated

Synaptic Weight Distributions. To characterize learning across different combinations of STDP rules, we also calculate the variance and skewness of the resulting weight distributions.

3 Results

We trained PCL networks following the protocol of [12], but using 9 combinations of causal, symmetric and acausal STDP rules (see Methods). Here, we quantify their effect on the evaluation metrics defined above.

3.1 An Excitatory Acausal STDP Prevents the Learning of V1-Like Responses

While training the PCL networks, we observed that both causal and symmetric excitatory STDP led to Gabor filter-like receptive fields. However, using an excitatory aSTDP systematically impeded such learning, as the receptive fields remained similar to their random initializations and elicited a high Gabor fit error (average value 0.5 ± 0.13). However, excitatory cSTDP and excitatory sSTDP led to low fit errors (average value 0.03 ± 0.01 in both cases). The fit errors were averaged over 2 simulations and further averaged over networks trained with different learning rules, and the errors denoted one standard deviation across networks. Due to the poor results observed with excitatory aSTDP, in the following sections, we only consider networks trained with either causal or symmetric STDP for excitatory synapses.

3.2 The Pairing of Excitatory and Inhibitory Causal STDP Promotes the Diversity of Receptive Fields in both Simple and Complex Cells

We next wondered what effect had the different pairing of rules on the diversity of RFs as well as their tuning properties. Table 2 shows the diversity of orientations

Table 2. Diversity of receptive fields' orientation and frequencies for simple and complex cells for different STDP rule pairings. N_o and N_f indicate the discrete number of preferred orientations and frequencies observed in a network. In *ab*STDP, *a* designates the connectivity type (e for excitation and i for inhibition) and *b* designates the STDP rule (with c for causal, s for symmetric and a for acausal). Bold indicates results for complex cells and red color indicates the maximal diversity achieved by complex cells for all rules pairing. Results are averaged over 2 network simulations. Errors indicate one standard deviation.

		ecSTDP			esSTDP		
		icSTDP	isSTDP	iaSTDP	icSTDP	isSTDP	iaSTDP
Simple cells	$\mathbf{N_o}$ ↑	7 ± 1	7 ± 0	8 ± 1	8 ± 1	7 ± 0	8 ± 1
	$\mathbf{N_f}$ ↑	14 ± 1	13 ± 1	13 ± 1	11 ± 1	10 ± 0	16 ± 2
	$\mathbf{E_o}$ ↓	0.6 ± 0.03	0.58 ± 0.02	0.47 ± 0.01	0.56 ± 0.08	0.63 ± 0.03	0.43 ± 0.04
	$\mathbf{E_f}$ ↓	0.78 ± 0.07	0.84 ± 0.08	0.78 ± 0.08	1.04 ± 0.01	1.025 ± 0.03	0.55 ± 0.15
Complex cells	$\mathbf{N_o}$ ↑	8 ± 1	$\mathbf{8 \pm 2}$	$\mathbf{6 \pm 1}$	$\mathbf{5 \pm 1}$	$\mathbf{7 \pm 1}$	$\mathbf{5 \pm 1}$
	$\mathbf{N_f}$ ↑	12 ± 0	$\mathbf{10 \pm 2}$	$\mathbf{9 \pm 2}$	$\mathbf{8 \pm 1}$	$\mathbf{9 \pm 2}$	$\mathbf{9 \pm 1}$
	$\mathbf{E_o}$ ↓	0.3 ± 0.05	$\mathbf{0.42 \pm 0.22}$	$\mathbf{0.59 \pm 0.2}$	$\mathbf{0.68 \pm 0.07}$	$\mathbf{0.48 \pm 0.06}$	$\mathbf{0.94 \pm 0.09}$
	$\mathbf{E_f}$ ↓	0.63 ± 0.04	$\mathbf{0.86 \pm 0.22}$	$\mathbf{1.14 \pm 0.15}$	$\mathbf{1.14 \pm 0.05}$	$\mathbf{1.05 \pm 0.18}$	$\mathbf{0.99 \pm 0.09}$

Table 3. Performance of different excitatory/inhibitory pairings of causal, symmetric and acausal STDP rules on the emergence of surround, orientation-tuned and cross-orientation suppression for simple cells. In *ab*STDP, *a* designates the connectivity type (e for excitation and i for inhibition) and *b* designates the STDP rule (with c for causal, s for symmetric and a for acausal). Results are first separated by excitatory STDP (block of 3 columns), and further separated by inhibitory STDP (separated columns in a block). Bold indicates performance achieved on cross-orientation suppression and blue color indicates the minimal performance achieved by an inhibitory rule in a block of 3 columns. The performance is first normalized by the maximal performance achieved by the rules, and then averaged over 2 network simulations. Errors indicate one standard deviation.

	ecSTDP			esSTDP		
	icSTDP	isSTDP	iaSTDP	icSTDP	isSTDP	iaSTDP
Surround	0.94 ± 0.01	0.88 ± 0.01	$1. \pm 0$	0.95 ± 0	0.89 ± 0	0.98 ± 0
Orientation-tuned	0.9 ± 0.06	0.87 ± 0.07	0.75 ± 0.01	0.95 ± 0.05	0.95 ± 0.05	0.72 ± 0.04
Cross-orientation	$\mathbf{0.75 \pm 0}$	$\mathbf{0.15 \pm 0.05}$	$\mathbf{1 \pm 0}$	$\mathbf{0.66 \pm 0.11}$	$\mathbf{-0.11 \pm 0.06}$	$\mathbf{0.87 \pm 0.07}$

and frequencies of PCL's simple and complex cells for the different rule combinations. While simple cells exhibited similar degrees of diversity of orientations and frequencies, surprisingly, the pairing of an excitatory sSTDP with an inhibitory aSTDP led to the lowest divergence from the optimal distribution. When looking at complex cells' (bold row), we however observed significant differences between the different rules. First, this specific excitatory sSTDP/inhibitory aSTDP rule leads to high relative entropy in the RFs' orientations and frequencies, which rules it out as a generic rule for learning in V1. Our results showcase that the rules that lead to the lowest relative entropy in complex cells are excitatory cSTDP paired with inhibitory cSTDP. This would suggest that a fully causal STDP network performs best in learning diverse set of receptive fields in different layers of the network. Interestingly, we also tested the tuning properties of the different networks, and observed that all rules led to biological simple and complex cell-like tuning curves (results not shown due to space purposes).

3.3 An Inhibitory Symmetric STDP Prevents the Emergence of Cross-Orientation Suppression

We tested our different PCL networks on suppression. For simplicity, we focused on the emergence of suppression in simple cells as they were the ones that exhibited similar diversity in their RFs. In the original PCL work, while complex cells exhibited some suppressive effects, only simple cells exhibited all of them. Table 3 shows the performance metrics on all types of suppression for simple cells with the different combinations of rules. While all rules led to surround and orientation-tuned suppression, inhibitory sSTDP always prevented the emergence of cross-orientation suppression regardless of the excitatory STDP rule used. Importantly, an inhibitory aSTDP led to the strongest surround and cross-orientation suppression effects, while an inhibitory cSTDP led to the strongest orientation-tuned suppression. Interestingly, in the original PCL paper, cross-orientation suppression was primarily attributed to the local lateral inhibition of the network, while surround and orientation-tuned suppression were caused by lateral and top-down suppression.

Our results might thus indicate that different inhibitory STDP rules could be used for different inhibitory connection types to achieve higher suppression. Figure 2 shows suppression for all rules. Please, note that the patch size that corresponds to the classical RF (cRF) is 0.5.

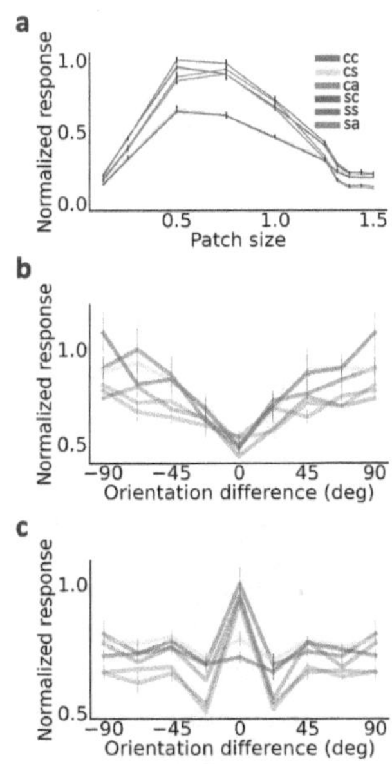

Fig. 2. Suppression effect of PCL's simple cells using different STDP rules. Results show average over 2 networks and error bars denote one standard deviation. Labels denote e/i STDP pairing a, Surround suppression. **b,** Orientation-tuned suppression. **c,** Cross-orientation suppression.

Interestingly, for surround suppression, we observe that the amplitude of the response at the cRF changes depending on both the excitatory and inhibitory rules used, with inhibitory sSTDP reducing the most the amplitude. Still for surround suppression, we observe that with an inhibitory aSTDP, simple cells' responses are not maximal yet when the cRF alone is excited as they still exhibit increases even when the non classical RF is excited. For orientation-tuned and cross-orientation suppression however, the amplitudes of responses for preferred orientations appear to be relatively the same across different learning rules (inhibitory sSTDP excluded for cross-orientation suppression).

3.4 Inhibitory STDP Leads to Skewed Weight Distributions

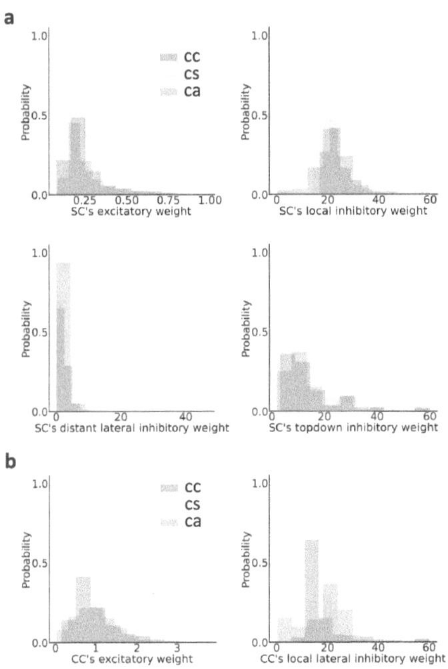

Fig. 3. Weight distributions of PCL networks trained with different learning rules. a, Simple cells' weight distributions. **b,** Complex cells' weight distributions.

To elucidate the differences between the different inhibitory STDP rules, we looked at the distribution of the weights after training. Due to space purposes, we consider only the case of a causal excitation, however similar results are observed with symmetric excitation. Figure 3 shows example distributions obtained. Interestingly, for simple cells, most rules appear to lead to right-skewed distributions, widely observed in brains [3,16] for all weights but local inhibition where aSTDP leads to no skewing compared to other rules. This might indicate why an aSTDP rule leads to poor results in the diversity of the receptive fields, as observed in Sect. 3.2. For complex cells, we note that the pairing of excitatory and inhibitory cSTDP also gives right-skewed distributions, while symmetric and acausal inhibitory STDP specifically give left-skewed distributions for inhibitory weights. As brains' synaptic strengths distributions have been observed to exhibit some level of skewness, STDP rules that lead to such distributions might be favored for the emergence of diverse receptive fields and suppressive effects. Table 4 details the distributions parameters and support our hypothesis with large inhibitory variance as well as non-null skewness.

Table 4. PCL's weight distribution parameters. Results are averaged over 2 network simulations. Errors indicate one standard deviation.

		ccSTDP	csSTDP	caSTDP
Variance				
Simple cells	Exc.	0.02 ± 0	0.02 ± 0	0.01 ± 0
	Loc. inh.	24.64 ± 0.16	50.71 ± 0	76.41 ± 0.34
	Dist. inh.	2.8 ± 0.01	2.82 ± 0.01	2.59 ± 0.02
	Td. inh.	123.03 ± 0.18	139.41 ± 0.08	115.92 ± 0.13
Complex cells	Exc.	0.27 ± 0.01	0.24 ± 0	0.26 ± 0.02
	Loc. inh.	93.54 ± 3.33	13.44 ± 0.46	99.71 ± 1.86
Skewness				
		ccSTDP	csSTDP	caSTDP
Simple cells	Exc.	1.96 ± 0.04	2.03 ± 0	1.81 ± 0.04
	Loc. inh.	−0.24 ± 0.11	0.78 ± 0.02	1.1 ± 0.07
	Dist. inh.	1.6 ± 0.01	1.59 ± 0.01	4.9 ± 0.04
	Td. inh.	2.08 ± 0.01	1.81 ± 0	1.81 ± 0
Complex cells	Exc.	1.44 ± 0.02	1.51 ± 0.02	0.56 ± 0.09
	Loc. inh.	2.58 ± 0.01	−2.19 ± 0.1	0.85 ± 0.16

4 Discussion

We have investigated the use of different STDP rules in the Predictive Coding Light (PCL) model, a recent recurrent spiking neural network model of unsupervised representation learning in visual cortex [12]. We compared different combinations of a causal, an acausal and a symmetric STDP rule for both excitatory and inhibitory synapses. Importantly, we stress that the testing of those different combinations was motivated by experimental observations of those different plasticity motifs in biological brains, and we aimed at uncovering the differences that these various STDP rules would cause during learning. We observed that successful learning of receptive fields strongly depends on the STDP rule for excitatory synapses, with the causal and symmetric STDP rules giving satisfactory results, while the acausal rule failed. Surprisingly, however, the results depended only weakly on the STDP rule used for inhibitory synapses. Various tuning properties, non-classical receptive field effects and long-tailed distributions of synaptic efficacies [3] emerged robustly for inhibitory synapses using any of the three STDP rules. We suspect that this is partly due to the weight normalization scheme in the PCL network, which ensures that the total amount of inhibition of a particular type that any neuron receives remains constant. We also stress again that this normalization scheme is used to mimic the competition over synaptic resources, that is thought to occur in biological synapses. At the same time, all STDP rules lead to similarly skewed distributions of inhibitory synaptic efficacies, i.e., the STDP rules only gradually affect the overall shape of the inhibitory weight distributions. This insensitivity to the precise STDP rule may help to explain the large diversity of inhibitory STDP rules observed in the nervous system [4]. While some differences were observed, these were generally weak and did not jeopardize the essential function of the network. Among all combinations of learning rules tested, the one that appeared to give the most brain-like results across all of our evaluations was using the causal STDP rule for both excitatory and inhibitory synapses. This is consistent with the frequent observation of this learning rule in biological experiments [2,8]. Nevertheless, the ability of various STDP rules to produce similar results in the PCL network is an interesting finding considering the large diversity of STDP rules observed in biology [4,17]. While in this work, we studied how predictive coding-like phenomena could emerge with a variety of STDP rules, other works that used rules based on predictive plasticity such as [6], inherently leads to classic STDP phenomena in the update of the synaptic weights. Studying the emergence of STDP rules in the minimization of the free-energy in spiking predictive coding models such as [14] might further help uncover how different rules can lead to a wide variety of phenomena. Furthermore, how such rules could alleviate runaway activity issues in recurrent excitatory spiking networks remains a critical question [10]. Overall, more research is needed to better understand the (dis-)advantages of combining specific learning rules in biological and artificial spiking neural networks.

Acknowledgments. This work benefited from a French government grant managed by the National Research Agency (ANR) under the "Investissements d'Avenir" program (PIA) with the following reference: ANR-20-SFRI-0003, and another grant with reference ANR-23-CE23-0034. This work was also supported by the Deutsche Forschungsgemeinschaft (DFG, German Research Foundation) through the Excellence Cluster EXC3066 "The Adaptive Mind' ' and the priority program SPP 2041, project numbers TR 881/7-2; TR 881/8-2 and priority program SPP 2411, project numbers 520617944; TR 881/11-1 and Research Unit FOR 5368. JT acknowledges support from the Johanna Quandt foundation.

Disclosure of Interests. The authors have no competing interests to declare that are relevant to the content of this article.

References

1. Barbier, T., Teulière, C., Triesch, J.: Spike timing-based unsupervised learning of orientation, disparity, and motion representations in a spiking neural network. In: Proceedings of the IEEE/CVF Conference on Computer Vision and Pattern Recognition Workshops, pp. 1377–1386 (2021)
2. Bi, G., Poo, M.: Synaptic modifications in cultured hippocampal neurons: dependence on spike timing, synaptic strength, and postsynaptic cell type. J. Neurosci. **18**(24), 10464–10472 (1998)
3. Buzsáki, G., Mizuseki, K.: The log-dynamic brain: how skewed distributions affect network operations. Nature Rev. Neurosci. **15**(4), 264–278 (2014)
4. Caporale, N., Dan, Y.: Spike timing–dependent plasticity: a Hebbian learning rule. Annu. Rev. Neurosci. **31**(1), 25–46 (2008)
5. Diehl, P.U., Cook, M.: Unsupervised learning of digit recognition using spike-timing-dependent plasticity. Front. Comput. Neurosci. **9**, 149773 (2015)
6. Halvagal, M.S., Zenke, F.: The combination of Hebbian and predictive plasticity learns invariant object representations in deep sensory networks. Nature Neurosci. **26**(11), 1906–1915 (2023)
7. Larisch, R., Hamker, F.H.: A systematic analysis of the joint effects of ganglion cells, lagged LGN cells, and intercortical inhibition on spatiotemporal processing and direction selectivity. Neural Netw. **186**,107273 (2025)
8. Rylan S Larsen et al. STDP in the developing sensory neocortex. Front. Synaptic Neurosci. **2**, 1456 (2010)

9. N'dri, A.W., et al.: Predictive coding light: learning compact visual codes by combining excitatory and inhibitory spike timing-dependent plasticity. In: Proceedings of the IEEE/CVF Conference on Computer Vision and Pattern Recognition. pp. 3997–4006
10. N'dri, A.W., et al.: Predictive Coding Light+: learning to predict visual sequences with spike timing-dependent plasticity and synaptic delays. arXiv preprint arXiv:2605.12732 (2026)
11. N'dri, A.W., et al.: Predictive coding with spiking neural networks: a survey. Neural Netw. 108371 (2025)
12. N'dri, A.W., et al.: Predictive coding light. Nature Commun. **16**(1), 8880 (2025)
13. Nessler, B., Pfeiffer, M., Maass, W.: STDP enables spiking neurons to detect hidden causes of their inputs. In: Advances in Neural Information Processing Systems, p. 22 (2009)
14. Ororbia, A.: Spiking neural predictive coding for continually learning from data streams. Neurocomputing,**544**,126292 (2023)
15. Rajesh, P.N., Rao, Dana, H.: Ballard: predictive coding in the visual cortex: a functional interpretation of some extra-classical receptive-field effects. Nat. Neurosci. **2**(1), 79–87 (1999)
16. Scheler, G.: Logarithmic distributions prove that intrinsic learning is Hebbian. F1000Research **6**, 1222 (2017)
17. Suvrathan, A.: Beyond STDP—towards diverse and functionally relevant plasticity rules. Curr. Opin. Neurobiol. **54**, 12–19 (2019)
18. Van Hateren, J.H., der Schaaf, A.V.: Independent component filters of natural images compared with simple cells in primary visual cortex. In: Proceedings of the Royal Society of London. Series B: Biological Sciences,vol. 265.1394, pp. 359–366 (1998)
19. Vogels, T.P., et al.: Inhibitory plasticity balances excitation and inhibition in sensory pathways and memory networks. Science **334**(6062), 1569–1573 (2011)

Test-Time Adaptation of Spiking Neural Networks for Intracortical Neural Decoding Using Membrane Potential Alignment

Guangzhi Tang(✉)

Department of Advanced Computing Sciences, Maastricht University, Maastricht, The Netherlands
guangzhi.tang@maastrichtuniversity.nl

Abstract. Intracortical braincomputer interfaces suffer from day-to-day neural signal shifts that degrade pretrained decoders. Existing unsupervised adaptation methods rely on deep recurrent or adversarial architectures that are too computationally expensive for implantable hardware. We propose Membrane Potential Alignment (MPA), a test-time adaptation method for spiking neural networks that realigns a pretrained decoder to shifted recordings by only matching membrane potential distributions via KL divergence. By restricting updates to low-rank (LoRA) weights, MPA adapts fewer than 9% of parameters. On a non-human primate reaching task spanning over one month, MPA achieves performance competitive with the state-of-the-art NoMAD method, while using a simpler architecture and finer temporal resolution (4 ms vs. 20 ms). These results show that efficient SNN-based test-time adaptation is a practical path toward long-term, recalibration-free braincomputer interfaces.

Keywords: spiking neural network · intracortical neural decoding · test-time adaptation

1 Introduction

Intracortical braincomputer interfaces (iBCIs) have become a critical technology for restoring communication and motor function in individuals with severe neurological disabilities [1,2]. Yet achieving reliable decoding in practice remains challenging, as decoders must handle low signal-to-noise ratios, high-dimensional neural variability, and strict real-time low-power constraints [3]. Once a decoder is trained and the device is chronically implanted, multiple factors can drive domain shifts between training and incoming data. For example, micromotion and hardware-related disruptions alter the recorded units [4], firing rates and spike amplitudes fluctuate over time [5], and progressive glial encapsulation reshapes the tissueelectrode interface [6]. These combined signal changes cause decoder performance to drop over time, yet standard recalibration is impractical

L. Pasa et al. (Eds.): ICANN 2026, LNCS 17089, pp. 236–247, 2027.
https://doi.org/10.1007/978-3-032-38398-3_20

for daily use as it requires the user to stop and repeat structured tasks, often with help from a technician to support supervised retraining [7].

Test-time adaptation (TTA) offers a promising way to address this challenge by updating a pretrained decoder at inference time using only unlabeled neural data, relying on unsupervised objectives such as distribution alignment or entropy minimization to close the gap between training and deployment conditions [8]. This idea has recently been applied to iBCI. For instance, NoMAD [9] first trains a variational autoencoder-based recurrent neural network (RNN) on an initial recording session to capture the latent dynamics of neural activities. At test time, it introduces a feedforward alignment network that maps each new day's spiking data back onto the original manifold. This is achieved by minimizing the KL divergence between latent state distributions alongside a self-supervised reconstruction loss, all without behavioral labels. The approach maintained stable decoding for over three months in non-human primate experiments. However, NoMAD, and other similar approaches [10] all rely on complex architectures, including gated RNNs and adversarial networks, that are computationally demanding, making them difficult to deploy on the power- and latency-constrained hardware typical of implantable iBCI systems.

Spiking neural networks (SNNs) offer a compelling alternative for neural decoding, as their event-driven, spike-based computation is inherently sparse and energy-efficient, well matched to the ultra-low power budgets of implantable devices [11]. Recent work has shown that SNN decoders can match or exceed the accuracy of RNNs on intracortical decoding tasks while consuming orders-of-magnitude less energy [12]. However, existing TTA methods designed for SNNs have focused almost exclusively on image classification benchmarks, where inputs are static [13]. These approaches do not account for the non-stationary nature of neural recordings, leaving a critical gap between the efficiency advantages of SNNs and the adaptation capabilities needed for long-term iBCI deployment.

In this paper, we propose Membrane Potential Alignment (MPA), a test-time adaptation method that aligns a pretrained SNN to shifted neural recordings by matching membrane potential distributions across sessions using KL divergence[1] By restricting adaptation to low-rank (LoRA) weights in the first hidden layer, MPA updates fewer than 9% of model parameters while keeping the pretrained decoder frozen. We benchmark MPA on a non-human primate reaching task spanning over one month and show that it achieves competitive decoding performance with the state-of-the-art NoMAD method, while using a substantially simpler architecture and operating at finer temporal resolution (4 ms vs. 20 ms). With only supervised pretraining on a single day-0 source session, MPA restores decoding on target sessions more than one month later using as few as a few seconds of data. The paper's code will be open-sourced after acceptance.

[1] https://github.com/ERNIS-LAB/snn-mpa-intracortical-neural-decoding-icann.

2 Methodology

2.1 Test-Time Adaptation for Non-human Primate Reaching

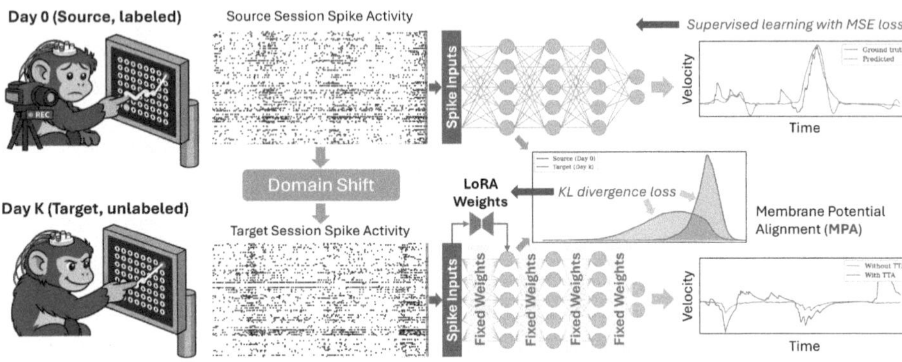

Fig. 1. Overview of the proposed cross-day SNN test-time adaptation pipeline. An SNN pretrained on day 0 is applied to a later target session, day k, where cross-session drift causes a mismatch in neural inputs and labels are unavailable. We therefore keep the decoder fixed and adapt the SNN's LoRA weights by aligning source and target membrane potential distributions through membrane potential alignment (MPA), improving target-session decoding without supervised behavior labels.

We propose a TTA method for unsupervised SNN adaptation on a target session using a model pretrained on the day-0 source session, as illustrated in Fig. 1. During day-0 pretraining, the model is trained in a supervised manner with ground-truth targets, allowing the SNN to learn the mapping from neural activity to the intended motor output. This produces a decoder that performs well on the source session, but its performance can degrade on later days because neural activity statistics change across sessions. In a realistic deployment setting at day k on the same subject, target labels are not available; instead, the model only has access to unlabeled spike recordings and the corresponding SNN internal dynamics from the source and target sessions. The goal of day-k alignment is therefore not to retrain the decoder from scratch, but to adapt it to the shifted input distribution while preserving the decoding function learned on day 0. To achieve this, we align the membrane potential distributions between source and target sessions using the MPA method proposed in this work, which will be described in detail in Sect. 2.3.

2.2 Spiking Neural Network for Behavior Decoding

We adapt SNN training in [12] for day-0 pretraining. The SNN processes binary spike trains from the neural recording and predicts continuous hand velocities at each timestep. We use the leaky integrate-and-fire (LIF) neuron model for the SNN. For a neuron i in the network, the membrane potential u_i evolves as:

$$u_i(t) = \beta\, u_i(t-1) + I_i(t) - S_i(t-1)V_{\text{th}} \tag{1}$$

where β is the membrane decay, $I_i(t)$ is the input current at timestep t, V_{th} is the firing threshold, and $S_i(t-1)$ is the spike output at the previous timestep.

Unlike the hard reset used for LIF neurons in [12], we use the soft reset that preserves the residual membrane potential after a spike. This produces more continuous membrane potential distributions, which provide more informative KL-divergence gradients during MPA training for target day k sessions. In contrast, a hard reset forces all suprathreshold states back to zero, making the distribution less expressive and reducing the sensitivity of the KL objective to distribution shift.

2.3 Membrane Potential Alignment Using LoRA Weights

Let $f_{\boldsymbol{\theta}}$ denote the pretrained SNN decoder with parameters $\boldsymbol{\theta}$, trained on day 0 source session. When applied to day k target session, the distribution shift in the input spike patterns propagates through the network, altering the statistics of the internal membrane potentials. We seek an adapted model $f_{\boldsymbol{\theta}+\Delta\boldsymbol{\theta}}$ using MPA such that the membrane potential distributions of the SNN on day k match those of the original model on day 0, thereby restoring the learned input-output mapping without requiring new labels.

The key insight of our MPA approach is that the membrane potentials of spiking neurons provide a rich, continuous-valued representation of the network's internal state that is well-suited for distribution matching. For a given hidden layer l, let $\mathbf{u}^{(l)}(t) \in \mathbb{R}^H$ denote the vector of membrane potentials at timestep t. Over a batch of B input samples with timestep T, we collect the membrane potential vectors across all samples and timesteps, yielding a set of $N = B \times T$ observations $\{\mathbf{u}_n^{(l)}\}_{n=1}^N$ in $\mathbb{R}^H$. We model the membrane potential distribution as a multivariate Gaussian:

$$p(\mathbf{u}^{(l)}) = \mathcal{N}\Big(\boldsymbol{\mu}^{(l)}, \boldsymbol{\Sigma}^{(l)}\Big) \tag{2}$$

where the mean and covariance are estimated from the collected samples:

$$\boldsymbol{\mu}^{(l)} = \frac{1}{N}\sum_{n=1}^{N} \mathbf{u}_n^{(l)}, \qquad \boldsymbol{\Sigma}^{(l)} = \frac{1}{N-1}\sum_{n=1}^{N} \Big(\mathbf{u}_n^{(l)} - \boldsymbol{\mu}^{(l)}\Big)\Big(\mathbf{u}_n^{(l)} - \boldsymbol{\mu}^{(l)}\Big)^{\top} \tag{3}$$

To align the adapted model's internal representations with the pretrained model's, we minimize the Kullback–Leibler (KL) divergence between the day 0 reference distribution and the day k adapted distribution. Let $p_0^{(l)} = \mathcal{N}(\boldsymbol{\mu}_0^{(l)}, \boldsymbol{\Sigma}_0^{(l)})$ denote the membrane potential distribution obtained by running the frozen pretrained model on day 0 data, and $p_k^{(l)} = \mathcal{N}(\boldsymbol{\mu}_k^{(l)}, \boldsymbol{\Sigma}_k^{(l)})$ denote the distribution obtained by running the adapted model on day k. The MPA adaptation loss is:

$$\mathcal{L}_{\mathrm{MPA}} = D_{\mathrm{KL}}\left(p_0^{(l)} \,\|\, p_k^{(l)}\right) \tag{4}$$

Directly fine-tuning all hidden weights with MPA risks overfitting to the limited unlabeled adaptation data and catastrophically forgetting the pretrained representations. We instead adopt Low-Rank Adaptation (LoRA) [14], which parameterizes the weight update as a low-rank matrix. For a pretrained weight matrix $\mathbf{W} \in \mathbb{R}^{d_{\mathrm{out}} \times d_{\mathrm{in}}}$, the adapted weight is:

$$\mathbf{W}' = \mathbf{W} + \mathbf{AB} \tag{5}$$

where $\mathbf{A} \in \mathbb{R}^{d_{\mathrm{out}} \times r}$ and $\mathbf{B} \in \mathbb{R}^{r \times d_{\mathrm{in}}}$ are the trainable low-rank matrices with rank $r \ll \min(d_{\mathrm{in}}, d_{\mathrm{out}})$, and the original weight $\mathbf{W}$ remains frozen. Matrix $\mathbf{A}$ is initialized to zero so that $\mathbf{AB} = \mathbf{0}$ at initialization, ensuring the adapted model starts with identical behavior to the pretrained model. This reduces the number of trainable parameters from $d_{\mathrm{out}} \times d_{\mathrm{in}}$ to $r \times (d_{\mathrm{out}} + d_{\mathrm{in}})$, providing an effective regularization that constrains the adaptation to a low-dimensional subspace.

As shown in Fig. 1, we apply LoRA to the first hidden layer of the network. Since the first layer directly receives the neural spike input, it is most affected by the domain shifts. The remaining layers, which learn more abstract behavior-level representations, are kept frozen to preserve the pretrained decoding capability. This yields a highly parameter-efficient adaptation, where only a small fraction of the total model parameters are updated.

2.4 Source-Target Session Compatibility

TTA assumes that the source model trained on the day-0 session captures enough of the underlying neural structure to serve as a useful starting point for the target session. However, if the recording array has shifted or the neural population has reorganized, TTA can fail silently, producing confident but incorrect velocity predictions. We therefore need a way to assess, before adaptation, whether a given day-0 (source) session is a reasonable match for the day-k (target) session. We adopt two complementary metrics for this purpose: representational similarity [15], which checks whether channel-level relationships are preserved, and Grassmann distance [16], which measures how much the low-dimensional neural subspace has rotated between sessions.

To capture whether the pairwise statistical relationships among channels remain stable across sessions, we compute a representational similarity score. For each session we first construct the $C \times C$ Pearson correlation matrix $\mathbf{R}$ from the binned spike data, where C is the number of recorded channels and each entry r_{ij} gives the correlation between the spike trains of channels i and j over the full recording. We then extract the upper-triangular entries (excluding the diagonal) into a vector $\mathbf{u} \in \mathbb{R}^{C(C-1)/2}$ to represent each session. The representational similarity between the source session and a target session is simply the Pearson correlation between $\mathbf{u}^{(0)}$ and $\mathbf{u}^{(k)}$:

$$S_{\mathrm{rep}} = \mathrm{corr}(\mathbf{u}^{(0)},\ \mathbf{u}^{(k)}). \tag{6}$$

A value close to 1 indicates that the channels that were correlated (or uncorrelated) on day 0 remain so on day k. Therefore, the population's internal relational structure is intact. A substantially lower value suggests that the functional relationships among channels have reorganized, which undermines the feature representations learned by the source decoder.

To go beyond pairwise statistics and ask whether the population-level manifold has rotated, we turn to subspace alignment via the Grassmann distance. Raw spike trains are binned into non-overlapping 1-second windows to form a spike-count matrix $\mathbf{X} \in \mathbb{R}^{N \times C}$, where N is the number of bins and C is the number of channels. PCA is fit separately to the source session and the target session, and the top $K = 10$ principal components are retained, giving orthonormal basis matrices $\mathbf{U}_0, \mathbf{U}_k \in \mathbb{R}^{C \times K}$ for the corresponding subspaces $\mathcal{U}_0$ and $\mathcal{U}_k$. The principal angles $\theta_1, \ldots, \theta_K$ between the two subspaces are computed from the singular values of $\mathbf{U}_0^\top \mathbf{U}_k$. The Grassmann distance is

$$d_G(\mathcal{U}_0, \mathcal{U}_k) = \sqrt{\sum_{i=1}^{K} \theta_i^2}. \tag{7}$$

This quantity is zero when the two subspaces are identical and increases as they rotate apart. A small Grassmann distance means that neural activity on day k still lives in roughly the same low-dimensional manifold as day 0, so the decoder's learned input-output mapping remains geometrically valid. A large distance signals that the dominant directions of variability have shifted, and the source decoder is effectively projecting onto the wrong axes. We use this metric alongside representational similarity to flag sessions where TTA is unlikely to recover acceptable decoding performance from the source model alone.

3 Experiments and Results

3.1 Non-human Primate Reaching Dataset and TTA Baseline

All experiments are conducted on the non-human primate reaching dataset from [17]. This dataset contains simultaneously recorded sensorimotor cortex spike trains and reaching kinematics from monkeys performing self-paced reaching movements. In this study, we used 23 sessions from Monkey 1 (Indy), each recorded with 96 channels, collected between 2016-09-15 and 2017-01-24. Following the NeuroBench [18], we encoded the spike data into 4-ms bins and used velocity labels at the same 4-ms resolution. For each session, we included only movement-active recordings and split the data into 50% training, 25% validation, and 25% testing sets. To train the SNN, we used 50-timestep windows for both day-0 pretraining and TTA. During evaluation, testing was performed in a streaming manner, with the SNN processing the full test set continuously.

We use NoMAD [9] as the state-of-the-art TTA baseline in our comparisons. NoMAD builds on LFADS [19] and stabilizes cross-session decoding by aligning neural activity to a shared latent dynamical manifold through unsupervised distribution alignment and spike-rate reconstruction. The model is based on a variational autoencoder with multiple GRU layers. We implemented NoMAD in PyTorch and validated our reproduction on the original dataset from the paper, obtaining similar pretraining and adaptation results. We then transferred the method to our dataset using the same hyperparameters. Spike recordings are binned at 20 ms, and each training sample contained 30 timebins. The training and validation sets are generated with a sliding window and 120 ms overlap. We did not normalize the spike inputs because it had little effect on performance.

3.2 Experimental Setup

Table 1. Comparison of test-time adaptation performance across 11 recording sessions from the same non-human primate over a one-month period (Day 0 to Day 22). Models were supervisedly pretrained on the source session *indy_20161005_06* (Day 0). Reported values denote R^2 scores as mean ± standard deviation.

Sessions	LFADS[a] [9]	NoMAD[a] (TTA) [9]	SNN[b] [12]	MPA (ours) (TTA)
Day 0 **(Source)**	0.53 ± 0.05	-	$\mathbf{0.59 \pm 0.01}$	-
Day 1	-0.25 ± 0.10	0.33 ± 0.07	0.10 ± 0.05	$\mathbf{0.36 \pm 0.03}$
Day 2	0.21 ± 0.30	$\mathbf{0.46 \pm 0.03}$	0.20 ± 0.03	0.43 ± 0.04
Day 6	-0.12 ± 0.19	0.18 ± 0.08	0.27 ± 0.05	$\mathbf{0.37 \pm 0.05}$
Day 8	-0.61 ± 0.15	-0.57 ± 0.21	-0.03 ± 0.01	-0.04 ± 0.02
Day 9	-0.10 ± 0.23	$\mathbf{0.32 \pm 0.09}$	0.19 ± 0.04	0.28 ± 0.07
Day 12	0.03 ± 0.23	$\mathbf{0.49 \pm 0.08}$	0.11 ± 0.02	0.46 ± 0.01
Day 19	-0.21 ± 0.27	$\mathbf{0.37 \pm 0.10}$	-0.26 ± 0.17	0.32 ± 0.02
Day 20	0.07 ± 0.20	$\mathbf{0.39 \pm 0.06}$	0.20 ± 0.03	0.38 ± 0.01
Day 21	-0.19 ± 0.24	$\mathbf{0.40 \pm 0.06}$	0.03 ± 0.04	0.37 ± 0.02
Day 22	-0.20 ± 0.17	0.28 ± 0.08	0.12 ± 0.03	$\mathbf{0.40 \pm 0.03}$

[a] LFADS and NoMAD denote our own implementations in PyTorch.
[b] SNN denotes our own implementation using soft-reset LIF neurons.

For all experiments, we trained three-layer fully connected SNNs. We adapted the training code from [12] for pretraining. Each hidden layer consisted of 64 LIF neurons with membrane decay $\beta = 0.96$, threshold $V_{\text{th}} = 1.0$, and a fast-sigmoid surrogate gradient with slope 20. Dropout with $p = 0.2$ was applied after each linear operation. All models were trained in PyTorch using AdamW. For day-0 pretraining, we used a learning rate of 0.001 and a batch size of 256. Training was run for 10 epochs, and the model with the best validation decoding loss was selected. We used a temporally weighted MSE loss, where the weights increase linearly across the 50-timestep window to place more emphasis on later timesteps. For TTA, we used a learning rate of 0.0005, a batch size of 1024, and a cosine annealing learning-rate scheduler, to train LoRA weights of $r = 8$ for the first SNN layer. Training was also run for 10 epochs, and the model with the lowest validation KL-divergence loss was selected. To account for training variability, day-0 pretraining was repeated with five random seeds, and TTA was performed across sessions using the pretrained model from each seed.

3.3 Month-Long Adaptation of SNN TTA Using MPA

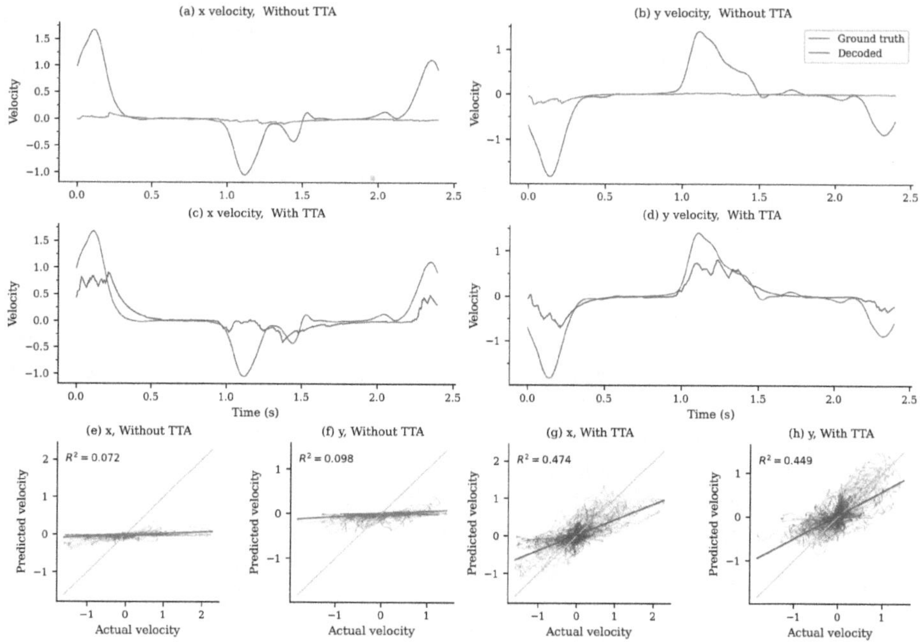

Fig. 2. Decoded velocities on session *indy_20161017_02* using an SNN pretrained on *indy_20161005_06*, with and without MPA. (ad) A representative 2.4 s segment of the testset selected to highlight the improvement from MPA TTA. Ground truth is shown in gray; decoded velocity in rose (without TTA) or blue (with TTA). (eh) Predicted vs. actual velocity across all test samples. Each panel shows individual samples (scatter), the identity line (light gray), and an ordinary least-squares fit (colored line). (Color figure online)

To evaluate long-term adaptation, we used the first October 2016 session as Day 0 for supervised pretraining and applied TTA to the remaining October sessions. As shown in Table 1, MPA achieves strong performance throughout this evaluation period. Compared with the non-adaptive SNN baseline, MPA improves performance on 9 of the 10 target sessions and increases the mean R^2 from 0.09 to 0.33. MPA is also competitive with the state-of-the-art NoMAD

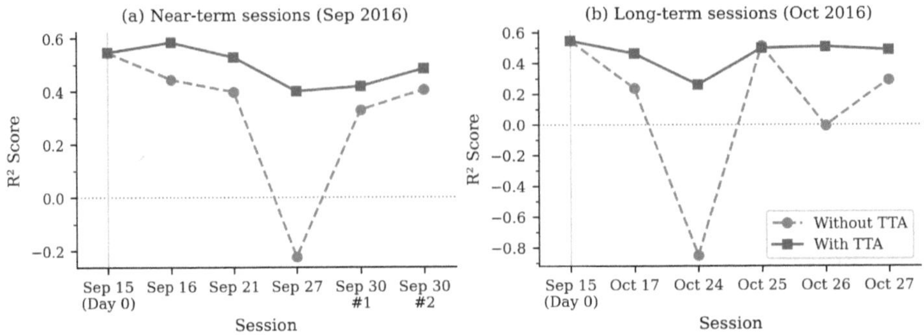

Fig. 3. Decoding performance across sessions for an SNN trained on a different Day 0 session (*indy_20160915_01*). (a) Near-term adaptation with MPA to five sessions in September 2016. (b) Long-term adaptation with MPA to five sessions in October 2016.

method. It outperforms NoMAD on three positively decoded sessions (Days 1, 6, and 22) and remains close on most of the others. The lower LFADS and NoMAD performance on our benchmark, compared with the results reported in the original paper, likely reflects the greater difficulty of our dataset. Unlike the structured 2D center-out task used in the original paper, our dataset involves self-paced grid reaching with random targets, which is less constrained and generally more challenging. In addition, MPA and the SNN baseline operate at 4 ms resolution, whereas LFADS and NoMAD use 20 ms bins. Both MPA and NoMAD fail on Day 8, indicating a session shift for which unsupervised alignment alone is insufficient. We analyze this failure case further in Sect. 3.4.

To highlight MPA's decoding gains, Fig. 2 compares SNN ($seed = 1$) behavior predictions for the October 17 target session (Day 12) with and without MPA. MPA improves performance from $R^2 \approx 0.08$ to $R^2 \approx 0.46$ for both x- and y-velocity, showing that unsupervised KL-divergence alignment can compensate for cross-session neural drift without movement labels from the new session.

We further provide a complementary view of the TTA behavior when the SNN is pretrained on a different Day 0 session, shown in Fig. 3. Without adaptation, decoding performance degrades markedly as the recording session changes, and in some cases even drops below zero, indicating that the pretrained model no longer generalizes reliably under session drift. In contrast, applying TTA with MPA consistently restores performance for both near-term and long-term sessions, showing that the unsupervised KL-divergence alignment objective can effectively compensate for cross-session neural drift without requiring movement labels from the new session.

Notably, the performance of the non-adapted LFADS and SNN does not show a monotonic decrease with increasing session interval. The absence of a monotonic performance drop with increasing session interval suggests that elapsed time alone is not the main driver of decoding difficulty. Instead, performance is likely dominated by session-specific instabilities in the recorded signals [4–6].

3.4 Target Session Compatibility to Day-0 Session

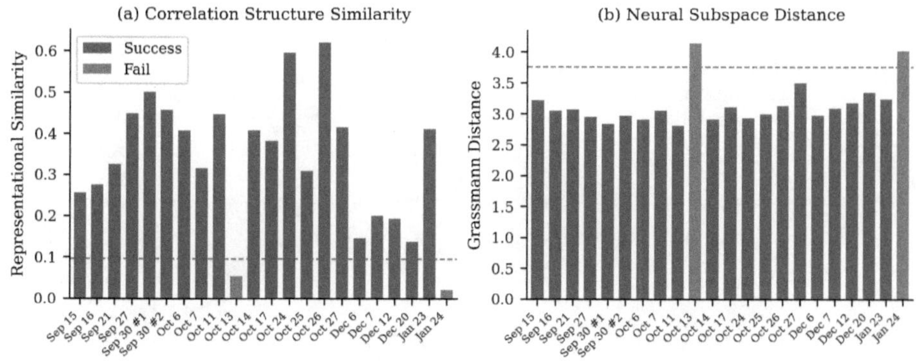

Fig. 4. Sourcetarget session compatibility between *indy_20161005_06* source session and all 22 target sessions measured by two metrics. (a) Representational similarity for correlation structure similarity. (b) Grassmann distance for neural subspace. Dashed lines indicate empirical thresholds separating sessions where TTA succeeded (blue) from those where it failed (red). (Color figure online)

Although our TTA method is effective for most target sessions, Table 1 shows that both NoMAD and MPA fail on the Day 8 session. We therefore examined source-target compatibility between the day-0 source session and this failed session, and asked whether our compatibility metrics could identify other incompatible sessions. To do so, we computed representational similarity and Grassmann distance between the October 5 day-0 source session and the other 22 sessions studied. As shown in Fig. 4, the failed session on October 13 (Day 8) has the second-lowest representational similarity and the highest Grassmann distance, indicating substantial mismatch in both channel-level structure and low-dimensional subspace. This suggests that TTA fails because the source session is too poorly matched to support reliable unsupervised alignment.

We further examined sessions with low representational similarity and high Grassmann distance. The January 24, 2017 session showed the lowest representational similarity and one of the highest Grassmann distances. For this session, both NoMAD and MPA failed to recover positive decoding performance. In contrast, both methods still worked on December 6 and December 20, despite their relatively low representational similarity. This suggests that TTA fails only when the mismatch is large in both metrics. These metrics may therefore help identify incompatible target sessions and choose the most suitable source session.

3.5 Effectiveness of Small TTA Training Data

MPA requires both source and target session data for alignment, so in practice, we aim to use as little training data as possible. Smaller datasets reduce memory

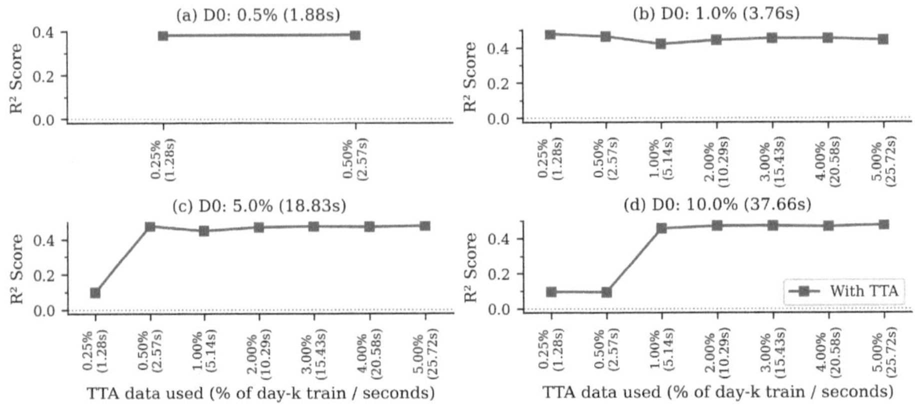

Fig. 5. TTA R^2 score on the target session (*indy_20161017_02*) as a function of TTA alignment data size, for four source day-0 training data sizes. Each panel shows the adapted model R^2 after TTA (blue) across seven alignment data fractions (0.25%5% of the target session training set, 1.2825.72 s). Panel titles indicate the source session data fraction and duration used for TTA alignment. (Color figure online)

demands, support edge-device deployment, and enable faster adaptation. To evaluate the data requirements of MPA, we tested different source–target data combinations, with results shown in Fig. 5. TTA still yields a substantial R^2 improvement even with very limited data, using as little as 1.88 s of source data and 1.28 s of alignment data. In these experiments, MPA used smaller batch sizes (128 in Fig. 5(a) and 256 in Fig. 5(b-d)), and the number of epochs was adjusted to keep the total number of gradient updates constant.

4 Conclusion

We presented MPA, a test-time adaptation method that enables pretrained SNNs to maintain decoding performance across weeks of neural recording drift without behavioral labels. On a month-long non-human primate reaching task, effective adaptation is possible with just seconds of unlabeled data, and the proposed compatibility metrics help identify when adaptation is likely to fail. The lightweight nature of LoRA-based SNN learning makes MPA well-suited for deployment on neuromorphic hardware, where on-device adaptation could eliminate the need for external computing. Performance may be further improved by large-scale pretraining across multiple subjects and sessions to learn more robust initial representations. Extending MPA to continual adaptation and validating it in additional iBCI tasks remain important future directions.

References

1. Wairagkar, M., et al.: An instantaneous voice-synthesis neuroprosthesis. Nature **644**(8075), 145–152 (2025)

2. Willsey, M.S., et al.: A high-performance brain-computer interface for finger decoding and quadcopter game control in an individual with paralysis. Nat. Med. **31**(1), 96–104 (2025)
3. Patrick-Krueger, K.M., Burkhart, I., Contreras-Vidal, J.L.: The state of clinical trials of implantable brain-computer interfaces. Nature Rev. Bioeng. **3**(1), 50–67 (2025)
4. Dunlap, C.F., Colachis, S.C., IV., Meyers, E.C., Bockbrader, M.A., Friedenberg, D.A.: Classifying intracortical brain-machine interface signal disruptions based on system performance and applicable compensatory strategies: a review. Front. Neurorobot. **14**, 558987 (2020)
5. Perge, J.A., et al.: Intra-day signal instabilities affect decoding performance in an intracortical neural interface system. J. Neural Eng. **10**(3), 036004 (2013)
6. Salatino, J.W., Ludwig, K.A., Kozai, T.D., Purcell, E.K.: Glial responses to implanted electrodes in the brain. Nature Biomed. Eng. **1**(11), 862–877 (2017)
7. Jarosiewicz, B., et al.: Virtual typing by people with tetraplegia using a self-calibrating intracortical brain-computer interface. Science Trans. Med. **7**(313), 313ra179–313ra179 (2015)
8. Liang, J., He, R., Tan, T.: A comprehensive survey on test-time adaptation under distribution shifts. Int. J. Comput. Vision **133**(1), 31–64 (2025)
9. Karpowicz, B.M., et al.: Stabilizing brain-computer interfaces through alignment of latent dynamics. Nat. Commun. **16**(1), 4662 (2025)
10. Ma, X., Rizzoglio, F., Bodkin, K.L., Perreault, E., Miller, L.E., Kennedy, A.: Using adversarial networks to extend brain computer interface decoding accuracy over time. elife **12**, e84296 (2023)
11. Rivelli, F., Popov, M., Kouzinopoulos, C.S., Tang, G.: Adaptively pruned spiking neural networks for energy-efficient intracortical neural decoding. In: 2025 47th Annual International Conference of the IEEE Engineering in Medicine and Biology Society (EMBC), pp. 1–7. IEEE (2025)
12. Hueber, P., et al.: Benchmarking of hardware-efficient real-time neural decoding in brain-computer interfaces. Neuromorphic Comput. Eng. **4**(2), 024008 (2024)
13. Luo, X., Chen, K., Sun, P.S.V., TIAN, C.X., Basu, A., Li, H.: Space: spike-aware consistency enhancement for test-time adaptation in spiking neural networks. In: The Thirty-ninth Annual Conference on Neural Information Processing Systems (2025)
14. Hu, E.J., et al.: Lora: low-rank adaptation of large language models. Iclr **1**(2), 3 (2022)
15. Diedrichsen, J., Kriegeskorte, N.: Representational models: a common framework for understanding encoding, pattern-component, and representational-similarity analysis. PLoS Comput. Biol. **13**(4), e1005508 (2017)
16. Hamm, J., Lee, D.D.: Grassmann discriminant analysis: a unifying view on subspace-based learning. In: Proceedings of the 25th International Conference on Machine Learning, pp. 376–383 (2008)
17. Makin, J.G., O'Doherty, J.E., Cardoso, M.M., Sabes, P.N.: Superior arm-movement decoding from cortex with a new, unsupervised-learning algorithm. J. Neural Eng. **15**(2), 026010 (2018)
18. Yik, J., et al.: The neurobench framework for benchmarking neuromorphic computing algorithms and systems. Nat. Commun. **16**(1), 1545 (2025)
19. Pandarinath, C., et al.: Inferring single-trial neural population dynamics using sequential auto-encoders. Nat. Methods **15**(10), 805–815 (2018)

Data Transfer Limitations and Accuracy Trade-Offs in Neuromorphic Real-Time Systems

Carmen Weigelt[1,2](✉), Sirine Arfa[3](✉), Pascal Gerhards[1], Fabian Kress[4], Mathias Haberjahn[1], Christian Mayr[3,5], and Juergen Becker[2,6]

[1] Infineon Technologies Dresden AG and Co. KG, Dresden, Germany
{carmen.weigelt,pascal.gerhards,mathias.haberjahn}@infineon.com
[2] Karlsruhe Institute of Technology, Karlsruhe, Germany
becker@kit.edu
[3] Technische Universitaet Dresden, Dresden, Germany
{sirine.arfa,christian.mayr}@tu-dresden.de
[4] Infineon Technologies Semiconductor GmbH, Langen, Germany
fabian.kress@infineon.com
[5] Centre for Tactile Internet with Human-in-the-Loop (CeTI), Dresden, Germany
[6] FZI Research Center for Information Technology, Karlsruhe, Germany

Abstract. Neuromorphic computing is an event-driven and sparse, biologically inspired computing paradigm, offering strong potential for resource-efficient real-time inference. However, event-based sensors produce highly variable event rates, which can challenge predictable deployment on performance constrained systems. In this work, we analyze data-transfer limitations and accuracy trade-offs for DVS128 Gesture classification on SpiNNaker2 using a passive input buffer and buffer-aware training. We show that a buffer could constrain the number of input events to mitigate event fluctuations and still preserves performance. On SpiNNaker2 (High PL, 1000 timesteps, 1 ms tick), buffered models achieve 89.39–91.29% on-chip accuracy, compared to 91.67% for the no-buffer baseline. At the same time, average on-chip energy is reduced from 0.800 J without buffering to 0.510 J with buffering. These results demonstrate that buffer-aware deployment can substantially improve energy efficiency while maintaining competitive accuracy, enabling more predictable and robust neuromorphic real-time systems.

Keywords: Spiking Neural Networks · DVS128 Gesture · Hardware-aware Training · Resource Efficiency · Neuromorphic Computing · SpiNNaker2

C. Weigelt and S. Arfa—Contributed equally.

L. Pasa et al. (Eds.): ICANN 2026, LNCS 17089, pp. 248–259, 2027.
https://doi.org/10.1007/978-3-032-38398-3_21

1 Introduction

Neuromorphic computing promises low-latency and energy-efficient processing by leveraging event-driven computation, sparse activity, and local memory. Unlike conventional synchronous systems, computation is triggered only by relevant events, making neuromorphic architectures well suited for processing temporally sparse signals from event-based sensors [16]

Dynamic Vision Sensors (DVS) encode visual information as asynchronous events generated by local intensity changes, achieving microsecond temporal resolution and high dynamic range [4,12]. This representation reduces redundant data and is particularly effective for tasks such as gesture recognition. However, event streams are inherently non-stationary: in datasets such as DVS128 Gesture [1], event rates can vary significantly across samples and over time.

While such variability is manageable in offline settings, it becomes a critical challenge in real-time neuromorphic systems. On platforms such as SpiNNaker2 [5], performance is constrained not only by computation but also by communication and memory bandwidth. Bursts of high event activity can saturate interconnects, introduce latency variability, and violate real-time constraints. Limiting the number of processed events improves predictability and efficiency, but may remove task-relevant information and degrade accuracy.

Existing methods for reducing event streams include spatial downsampling [6], filtering with a spiking neural network (SNN) [15], or cutting event-streams when a time limit is reached [7]. It has also been investigated how different buffering schemes impact the event selection limiting the data rate through a system [18]. However, these approaches are typically applied as preprocessing steps and are not explicitly aligned with hardware constraints. In particular, the relationship between data transfer limitations, real-time execution, and model accuracy remains insufficiently explored. More broadly, data movement is known to dominate system cost in modern computing architectures [8,17], making it a critical factor in neuromorphic deployments.

In this work, we investigate how constraining input event streams affects inference accuracy in a real-time neuromorphic setting. Using the DVS128 Gesture benchmark on SpiNNaker2, we add a buffer to the system that limits the amount of transferred data and evaluate the impact on the task and hardware performance. In addition, we propose buffer-aware training to adapt the model to constrained input statistics and improve robustness under deployment conditions.

Our contributions are as follows:

- Analysis of event-rate variability as a bottleneck in real-time neuromorphic systems
- Investigation of suitable points where the buffer could be integrated into the processing chain
- Buffer-aware training to improve robustness under constrained inputs
- Evaluation on SpiNNaker2 demonstrating reduced data transfer with minimal accuracy loss

2 Background

2.1 Neuromorphic Computing and Spiking Neural Networks

Neuromorphic systems implement computation using spiking neural networks (SNNs), where information is encoded in discrete spike events and processed asynchronously. In contrast to conventional artificial neural networks, SNNs operate in a time-continuous manner and exploit sparse activity, making them well suited for energy-efficient and low-latency inference [3,10].

A commonly used neuron model in SNNs, which is also used for our model, is the Leaky Integrate-and-Fire (LIF) neuron. The membrane potential $V(t)$ evolves according to:

$$\tau_m \frac{dV(t)}{dt} = -V(t) + RI(t), \tag{1}$$

where τ_m is the membrane time constant, R is the membrane resistance, and $I(t)$ is the synaptic input current. When the membrane potential reaches a threshold V_{th}, the neuron emits a spike and its potential is reset:

$$\text{if } V(t) \geq V_{\text{th}} \Rightarrow \text{spike and } V(t) \leftarrow V_{\text{reset}}. \tag{2}$$

In discrete-time simulations, which are commonly used for deployment on digital neuromorphic hardware, the dynamics can be approximated using Euler integration:

$$V[t+1] = \alpha V[t] + (1-\alpha)I[t], \tag{3}$$

with $\alpha = \exp(-\Delta t/\tau_m)$. This formulation enables efficient implementation on hardware platforms such as SpiNNaker2.

2.2 DVS128 Gesture Inference on Different Hardware Platforms

The DVS128 Gesture dataset, recorded with a 128 × 128 pixel DVS camera, contains 10 gestures and a background class. Since its introduction by IBM [1], it has become a prominent benchmark for evaluating neuromorphic and spiking neural network implementations across various platforms. IBM initially solved this task with a convolutional SNN on their TrueNorth neuromorphic chip, achieving 94.59 % classification accuracy [1]. Similarly, Arfa et al. [2] reported 94.69 % accuracy with a quantization-aware trained convolutional SNN deployed on SpiNNaker2. Intel's Loihi chip demonstrated competitive performance on gesture classification, reaching 89.64 % accuracy through a deep neural network (DNN) to SNN conversion approach [13]. On GPU platforms, state-of-the-art performance was achieved with 94.76 % accuracy on an Nvidia RTX 3080 [11].

2.3 SpiNNaker2 Hardware Constraints

Neuromorphic hardware platforms such as SpiNNaker2 are designed to efficiently execute SNN workloads through massively parallel processing and distributed memory [5]. Each processing element operates with limited local memory and communicates via a packet-based interconnect. Real-time SNN inference on SpiNNaker2 has been demonstrated for various sensor modalities, such as radar-based gesture classification [9], further motivating the study of hardware-aware deployment strategies under strict timing constraints.

In such systems, performance is not only determined by computational workload but also by communication overhead. Event-based processing requires the transmission of spikes between neurons and cores, making system behavior sensitive to input event rates. High event bursts can saturate communication links, increase latency, and disrupt real-time guarantees.

More generally, data movement is a dominant contributor to energy consumption and system cost in modern computing architectures [8,17]. This is particularly relevant for neuromorphic systems, where fine-grained spike communication amplifies the impact of input variability.

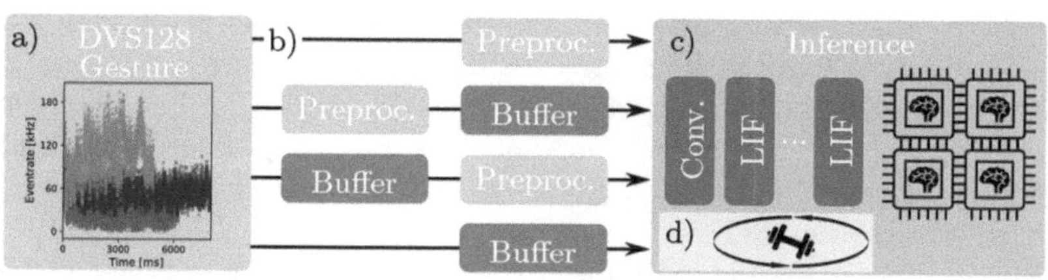

Fig. 1. Overview of the performed experiments with a) three samples from the DVS128 Gesture dataset being processed by b) a different order of preprocessing (denoising and downsampling) and buffering steps. The data samples were then c) inferred in simulation and on neuromorphic hardware. For further experiments the basemodel was d) fine-tuned with buffer-aware training.

3 Methods

As a baseline the publicly available, quantization-aware trained network from Arfa et al. was used [2]. The network consists of three convolutional blocks (Conv2D with 5×5 and 3×3 kernels, each followed by LIF neurons and SumPooling layers), succeeded by two fully connected layers with 256 and 11 units respectively. It solves the DVS128 Gesture classification task with an accuracy of 94.69 %. For training and testing, the DVS128 Gesture dataset was preprocessed with a denoising step that ensures that events are not spatio-temporally separated from other events. In this way background events can be detected and removed. Then the data gets downsized to 32×32 pixel 1 ms frames, which does not affect the overall number of events. For the deployment on SpiNNaker2, the event frames are going to be converted into event lists that contain only relevant events for processing. The SNN models used for the on-chip inference are NIR graphs [14].

Figure 1a) presents the event rates of three different samples from the DVS128 Gesture dataset, having different sample lengths and fluctuating event rates. In order to mitigate this effect, the number of incoming events was constrained by buffers of different sizes. To simulate the buffering behavior, a temporal window of 1 ms was applied. When the number of events within this window exceeded the buffer size, only the first events up to the buffer capacity were retained for inference in order of their appearance, while the remaining events were discarded. We also tested the skip-override buffering scheme introduced in [18]; however, as it did not yield measurable differences in classification accuracy compared to the first-in buffering strategy, these results are not presented in this work.

3.1 Integrating the Buffer Into the Processing Pipeline

In order to integrate the buffer into the processing pipeline, multiple positions are possible. Figure 1b) provides an overview on the performed experiments with the buffer placement. The top case represents the baseline. Then the buffer could be either placed before or after the preprocessing steps (denoise and downsampling). Since the downsampling operation reduces spatial resolution but preserves the temporal event count, the buffer's position relative to this step does not impact the number of buffered events. That is why the buffer position was not evaluated relative to this processing step.

Since both, denoise step and buffering, filter out events, the buffer's effectiveness as a standalone filtering solution was also tested. In those experiments, the denoising step was entirely omitted from the preprocessing pipeline and the raw event streams were fed directly into the buffer, followed by the spatial-temporal downsampling step. Since the buffer cannot distinguish between events from a gesture or from background noise, these experiments can further provide insights into the robustness of the SNN.

3.2 Buffer-Aware Training

To mitigate the gap between training and deployment conditions, we fine-tuned the baseline model on data that replicated real-world buffering and preprocessing operations. While the initial model was trained on complete, denoised datasets, deployment scenarios with varying buffer sizes introduce substantial input variability that degrades classification performance. We therefore augmented the dataset to incorporate the buffering behavior and trained the baseline model for 50 epochs. This adaptation aligns the model with actual deployment constraints, enhancing its robustness to buffer-induced input variations. To get even closer to the deployment scenario, the classification accuracy is also tracked for a subset of a gesture that is described in the following section.

Table 1. Impact on event statistics and classification accuracy of integrating buffer into the processing chain.

Buffer Size	Denoise → Buffer			Buffer → Denoise			Buffer Only		
	Mean±Std [kHz]	Total [M]	Acc. [%]	Mean±Std [kHz]	Total [M]	Acc. [%]	Mean±Std [kHz]	Total [M]	Acc. [%]
10	9.2±1.4	16.7	82.20	3.4±1.7	6.2	51.14	9.9±0.4	17.8	53.79
20	17.0±3.7	30.9	94.32	9.2±3.2	16.7	75.00	19.2±1.6	34.6	70.08
30	23.1±6.4	42.2	95.08	15.1±4.9	27.5	78.41	27.5±3.5	49.8	**75.38**
50	31.2±10.8	57.1	**95.45**	24.9±8.5	45.4	**78.79**	40.3±8.4	73.3	74.24
100	39.2±16.1	71.8	95.08	36.5±14.6	67.0	75.00	54.7±16.3	100.0	71.97
none	41.2±18.3	75.4	94.69	–	–	–	60.2±20.8	109.7	70.45

3.3 Deployment on SpiNNaker2

The baseline for this paper is exactly the QAT model from [2] that was deployed on SpiNNaker2 with 600 simulation timesteps and 3.5 ms system ticks. This deployment baseline has been adjusted to the real-time processing setting, where we run the experiments for 1000 timesteps and with 1 ms system tick. For each sample, the effective timestep count is defined as

$$T_{\text{eff}} = \min(T_{\text{cfg}}, t_{\max} + 1)\,, \tag{4}$$

where $T_{\text{cfg}} = 1000$ and $t_{\max}$ is the maximum input spike timestamp. Input spikes are clipped to $t < T_{\text{eff}}$ before deployment to guarantee strict alignment between injected input and simulated time horizon. The power management strategy employed is the High Performance Level (High PL), in which all cores operate at 0.8 V and 300 MHz. At this clock frequency, all neuron and synapse processing completes well within the 1 ms system tick, enabling the system to deliver a gesture classification result every second, evaluated on the first second of each sample. This is significantly closer to a real-time processing regime than the original dataset recordings, where gestures are often repeated across multiple seconds. An idle-reference measurement $E_{\text{idle,ref}}$ is recorded once per run using empty input. The per-sample dynamic component E_{dyn} is then estimated using the total measured energy under load E_{meas}:

$$E_{\text{dyn}} = E_{\text{meas}} - \frac{E_{\text{idle,ref}}}{T_{\text{cfg}}} T_{\text{eff}}. \tag{5}$$

4 Results

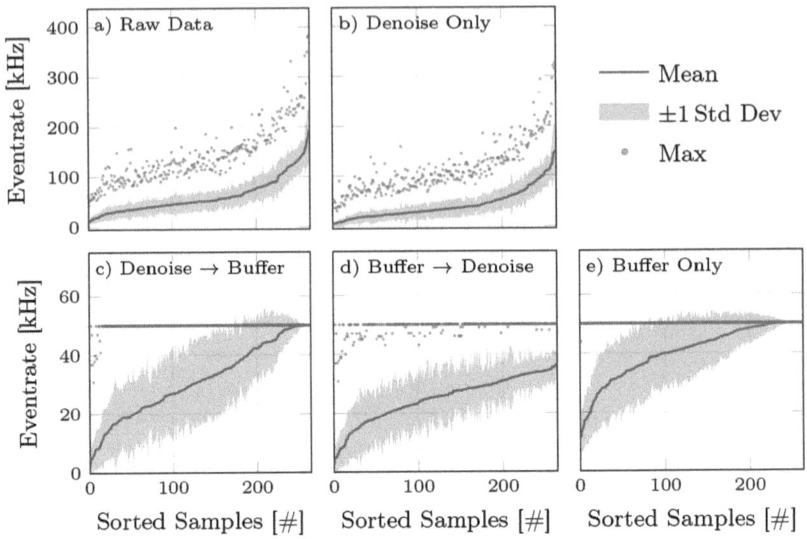

Fig. 2. Event statistics of DVS128 Gesture test set samples. The mean, standard deviation, and maximum number of events is based on event counts of 1 ms time windows in each sample after different preprocessing steps were applied: a) none, showing the raw data, b) denoising as it was used in the baseline model, adding a buffer with the size 50 c) after and d) before the denoise step, and e) only using a buffer of size 50. For better visualization, all samples have been sorted by their average event rate.

4.1 Including the Buffer in the Processing Chain

The effect of introducing a buffer at different stages of the processing chain is visualized in Fig. 2 with the average event rates, its fluctuations and maximum number of events within a 1 ms time window for each sample of the DVS128 Gesture test dataset. The measured accuracies and input event statistics for different buffer sizes across all test samples can be found in Table 1.

Figure 2a shows the raw samples without any processing, exhibiting an average event rate of 60.2±20.8 kHz (109.7 M total event count) with substantial variability across samples, resulting in an inference accuracy of 70.45%. When denoising is applied without buffering, the overall event rate is reduced but still has similar fluctuations. Introducing a buffer sets a hard limit on the maximum number of input events for a 1 ms time window. When the buffer is placed after the denoising step, the baseline accuracy could be maintained for most buffer sizes, while less events are processed. In contrast, applying the buffer before the denoising step further decreases the event rates significantly, which is expected

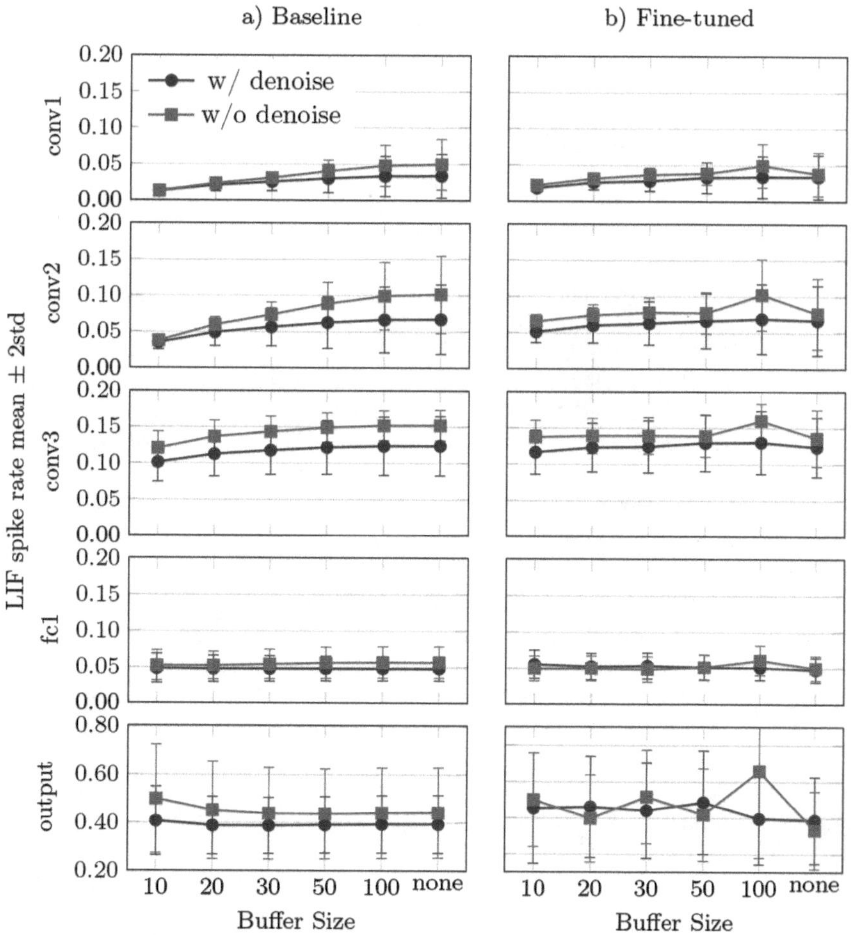

Fig. 3. Spike rates of the LIF neuron layer after convolutional (conv), fully connected (fc) and at the output layer for different input buffer sizes. The spike activity was measured for networks a) without and b) with extra buffer-aware training to fine-tune the networks. When denoising is used, the buffer is placed after the denoising step.

since the buffering mechanism may remove sufficient events such that the remaining events appear temporally separated and are subsequently labeled as background noise by the denoising algorithm. This effect is particularly evident in Table 1: at the smallest buffer size, the event rate drops to only 3.4±1.7 kHz (a 91.7% reduction compared to raw data), with inference accuracy degrading to 51.14%. When only the buffer is applied without denoising, a larger overall number of events is retained compared to the combined approaches and accuracies similar to the case where the buffer is placed before the denoise step, are measured. Overall, smaller buffer sizes lead to reduced input event rate with smaller standard deviations and number of total events. When the buffer is placed after the denoising step, the accuracy can be maintained for most configurations. The impact on the SNNs' internal spike rates can be observed in Fig. 3a where the impact of low fluctuations due to the constrained input remain present until the LIF neurons of the last convolution layer. When denoising is used, the average spike rates are lower and do not differ a lot for different buffer sizes. Still, smaller buffer size leads to lowest spike rates and spike rate deviations except for the output LIF layer.

4.2 Hardware-Aware Training

Since constraining the input event rate through buffer size limitations showed promising results but impacts the accuracy, we incorporated this constraint during training. The retrained models achieved accuracies of 94.5±0.5% with the denoising step and 93.8±0.5% without denoising for all five buffer sizes. This small difference demonstrates that the buffer can effectively replace the denoising step while recovering the observed accuracy losses when applying buffering to the baseline model. These results indicate that the SNNs can be successfully fine-tuned to handle reduced input event rates. Regarding the internal spiking behavior as shown in Fig. 3b, the spike rate fluctuations dependent on buffer size remain similar to those of the baseline model. However, the difference in spike rates between configurations with and without denoising is less pronounced in the retrained models compared to the baseline. Furthermore, for inference on SpiNNaker2, only a 1 s sample length is used, as this would be representative of a practical deployment scenario. As expected, classifying gestures using only 1 s windows decreases the accuracy compared to using full-length samples, since less temporal information is available. However, this configuration represents a more realistic real-world deployment use case where continuous, low-latency gesture recognition is required.

4.3 Real-Time Processing on SpiNNaker2

To evaluate deployment behavior under real-time constraints, all on-chip measurements were performed with a fixed simulation budget of 1000 timesteps and 1 ms system tick duration in High PL mode. In addition to measured total energy, we report an idle-reference-adjusted dynamic energy estimate to separate fixed execution costs from activity-dependent costs. Table 2 shows that buffered configurations substantially reduce on-chip energy compared with the no-buffer baseline. Total energy decreases from 0.800 J (none) to 0.510–0.517 J (buffers 10–100). The dynamic component decreases from 0.304 J to 0.016–0.022 J, confirming that limiting input event traffic effectively reduces activity-dependent hardware cost. Across buffered settings, the energy trend is consistent with buffer aggressiveness: lower buffer sizes yield lower energy. At the same time, on-chip accuracy remains competitive ranging from 89.39% up to 91.29% relative to the no-buffer baseline with 91.67%, highlighting a favorable trade-off between constrained input traffic and classification performance.

Table 2. Buffer-size sweep on SpiNNaker2 under an unified measurement protocol (High PL, 1000 timesteps, adaptive timestep evaluation, input clipping, idle-reference measurement).

	Off-Chip		On-Chip		
Buffer Size	Acc. [%]	Acc.* [%]	Acc.* [%]	Energy [J]	Dynamic Energy [J]
10	93.94	92.42	90.53	0.510	0.016
20	94.32	90.15	89.39	0.514	0.019
30	94.70	91.67	89.77	0.515	0.019
50	94.70	92.42	90.15	0.517	0.021
100	95.45	91.67	91.29	0.517	0.022
none	94.69	92.05	91.67	0.800	0.304

*1-second inference time window

5 Discussion

As a consequence for an end-to-end system, even though the DVS camera produces a fluctuating event stream, the data transfer rate can be reduced substantially when this reduction is incorporated in the training step, so that events can directly be filtered out at the sensor or after an initial preprocessing step such as denoising. Our experiments demonstrate that buffer sizes as small as 10 events per 1 ms time window remain viable, corresponding to roughly 1% of the input pixels after downsampling to 32 x 32 pixels. This translates to a maximum data rate of 10 000 events per second—a predictable bandwidth budget.

An important aspect of our deployment setup is that classification is performed on 1 s samples in real-time, reflecting an application scenario where gestures should be recognized as soon as possible. This contrasts with prior work that typically waits to classify the full gesture sequence [1,13]. Consequently, the observed accuracy drops due to reduced input data are acceptable given the priority for low-latency responsiveness.

In terms of predictability for neuromorphic real-time systems, the buffer serves a dual purpose: it limits the input data rate and, as a consequence, constrains internal spike rates on DVS128 Gesture as well. This cascading effect directly impacts power consumption. Aggressive input buffering not only reduces event transmission cost but also minimizes the activity-dependent dynamic energy within the SNN. The measurements confirm this relationship: the smallest buffer size (10 events per 1 ms) yields the lowest dynamic energy (0.016 J), demonstrating that input-rate control propagates through the entire processing chain to deliver predictable and efficient neuromorphic execution.

6 Conclusion

This work demonstrates that a simple buffer mechanism to limit event rates substantially improves the predictability and energy efficiency of neuromorphic gesture recognition systems. When applied after denoising, the buffer maintains baseline accuracy of 95.45% while reducing input event rates from 60.2±20.8 kHz to 31.2±10.8 kHz and stabilizing input variability.

Hardware-aware training enables the buffer to effectively replace computationally expensive preprocessing. Models trained with buffer constraints achieve 94.5±0.5% accuracy with denoising and 93.8±0.5% without, demonstrating that robustness can shift from preprocessing to learned network weights. On-chip deployment on SpiNNaker2 validates these benefits: buffered configurations reduce dynamic energy by over 90% (from 0.304 J to 0.016–0.022 J) while maintaining competitive accuracy (89.39–91.29% vs. 91.67% baseline) under realistic 1-second inference windows.

These findings demonstrate how a neuromorphic real-world gesture recognition task can be co-designed to handle fluctuating event streams through a constraining buffering mechanism, leading to reduced internal variability and improved predictability in spike activity and resource consumption. By mitigating event fluctuations at the source, the inherent energy efficiency and lightweight nature of neuromorphic systems can be preserved with better predictable power budgets and data transfer rates for practical deployment.

Acknowledgments. The authors thank Bernhard Vogginger for his valuable feedback on the manuscript. This work has been supported by the RIGOLETTO project, funded by the Chips Joint Undertaking and its members, including top-up funding from CERN and its member states, under grant agreement No. 101194371. This work was also funded by the European Union within the programme Horizon Europe under grant agreement no. 101120727 (PRIMI). The authors gratefully acknowledge the computing time made available to them on the high-performance computer at the NHR Center of TU Dresden. This center is jointly supported by the Federal Ministry of Education and Research and the state governments participating in the NHR (www.nhr-verein.de/unsere-partner).

References

1. Amir, A.: A low power, fully event-based gesture recognition system. In: Proceedings of the IEEE Conference on Computer Vision and Pattern Recognition, pp. 7243–7252 (2017)
2. Arfa, S., Vogginger, B., Liu, C., Partzsch, J., Schöne, M., Mayr, C.: Efficient deployment of spiking neural networks on spinnaker2 for DVS gesture recognition using neuromorphic intermediate representation. In: 2025 Neuro Inspired Computational Elements (NICE) (2025). https://doi.org/10.1109/NICE65350.2025.11065119
3. Davies, M.: Loihi: a neuromorphic manycore processor with on-chip learning. IEEE Micro (2018)
4. Gallego, G.: Event-based vision: a survey. IEEE TPAMI (2020)
5. Gonzalez, H.A.: Spinnaker2: a large-scale neuromorphic system for event-based and asynchronous machine learning (2024). arXiv:2401.04491 arXiv preprint
6. Gruel, A., Martinet, J., Linares-Barranco, B., Serrano-Gotarredona, T.: Performance comparison of DVS data spatial downscaling methods using SNNs, WACV (2023)

7. Hamara, A., Kilpatrick, B., Baratta, A., Kofink, B., Freeman, A.C.: Low-latency scalable streaming for event-based vision (2024). arXiv:2412.07889 arXiv preprint
8. Horowitz, M.: 1.1 computing's energy problem (and what we can do about it). ISSCC (2014)
9. Huang, J., et al.: Real-time radar gesture classification with spiking neural network on spinnaker 2 prototype. In: 2022 IEEE 4th International Conference on Artificial Intelligence Circuits and Systems (AICAS), pp. 362–365 (2022). https://doi.org/10.1109/AICAS54282.2022.9869987
10. Indiveri, G., Liu, S.C.: Memory and information processing in neuromorphic systems. Proc. IEEE (2015)
11. Kosta, A.K., Apolinario, M.P.E., Roy, K.: Live demonstration: ANN vs SNN vs hybrid architectures for event-based real-time gesture recognition and optical flow estimation. In: 2023 IEEE/CVF Conference on Computer Vision and Pattern Recognition Workshops (CVPRW), pp. 4148–4149. IEEE, Vancouver, BC, Canada (2023). https://doi.org/10.1109/CVPRW59228.2023.00436
12. Lichtsteiner, P., Posch, C., Delbruck, T.: A 128x128 120 DB 15 μs latency asynchronous temporal contrast vision sensor. IEEE J. Solid-State Cir. (2008)
13. Massa, R., Marchisio, A., Martina, M., Shafique, M.: An efficient spiking neural network for recognizing gestures with a DVS camera on the loihi neuromorphic processor. In: 2020 International Joint Conference on Neural Networks (IJCNN), pp. 1–9. IEEE (2020)
14. Pedersen, J.E., et al.: Neuromorphic intermediate representation: a unified instruction set for interoperable brain-inspired computing. Nat. Commun. **15**(1), 8122 (2024)
15. Rizzo, C.P., Schuman, C.D., Plank, J.S.: Neuromorphic downsampling of event-based camera output. In: Neuro-Inspired Computational Elements (2023)
16. Schuman, C.D., Kulkarni, S.R., Parsa, M., Mitchell, J.P., Date, P., Kay, B.: Opportunities for neuromorphic computing algorithms and applications. Nature Comput. Sci. **2**(1), 10–19 (2022). https://doi.org/10.1038/s43588-021-00184-y
17. Sze, V., Chen, Y.H., Yang, T.J., Emer, J.S.: Efficient processing of deep neural networks. Morgan and Claypool (2020)
18. Weigelt, C., et al.: Mitigating event fluctuations with adaptive buffering for robust neuromorphic systems. In: Koprinska, I., Mendes-Moreira, J., Branco, P. (eds.) Machine Learning and Principles and Practice of Knowledge Discovery in Databases, pp. 174–185. Springer Nature Switzerland, Cham (2026). https://doi.org/10.1007/978-3-032-19099-4_13

Working Memory in Recurrent Spiking Neural Networks With Heterogeneous Synaptic Delays

Laurent U Perrinet(✉)

Institut de Neurosciences de la Timone (UMR7289), Aix Marseille Univ,CNRS, 27 Bd Moulin, 13005 Marseille, France
laurent.perrinet@univ-amu.fr
https://laurentperrinet.github.io/publication/perrinet-26-icann/

Abstract. Working memory—the ability to store and recall precise temporal patterns of neural activity—remains an open challenge for spiking neural networks (SNNs). We propose a recurrent SNN of $N = 1024$ neurons in which each synapse is equipped with $D = 41$ heterogeneous delays, parameterised as a weight tensor $\mathbf{W} \in \mathbb{R}^{N \times N \times D}$ and trained end-to-end with surrogate-gradient backpropagation through time. Each stored pattern is represented as a sequential chain of overlapping Spiking Motifs: contiguous context windows of length D that uniquely predict the activity at the next time step. A closed-form Hebbian initialisation, derived by deconvolving the LIF membrane response and targeting a sub-threshold membrane potential value superior to the threshold ϑ, achieves an accuracy as measured by the F_1-score relative close to that expected when noise is present and before any gradient step on a benchmark of $M = 16$ patterns of duration $T = 1000\,\text{ms}$. With learning, the network tolerates up to 25% bit-flip noise, and reaches F_1 scores closer to the value expected from the noise level. These results demonstrate attractor-like retrieval dynamics consistent with hippocampal pattern completion.

Keywords: Spiking Neural Networks · Heterogeneous synaptic delays · Working memory

1 Introduction

A century ago, Adrian and Zotterman [2] established that neural communication relies on all-or-none action potentials, *spikes*. Whether information is encoded in precise spike timing or in the coarser measure of instantaneous firing rate remains debated [4,11,22]—a question that becomes especially sharp for working memory, where the brain must maintain and manipulate information across delays of seconds. In traditional artificial neural networks, long-range temporal dependencies are addressed by gated recurrent unit (GRU) architectures or Transformers [24], which decouple temporal binding range from the timescale of

L. Pasa et al. (Eds.): ICANN 2026, LNCS 17089, pp. 260–271, 2027.
https://doi.org/10.1007/978-3-032-38398-3_22

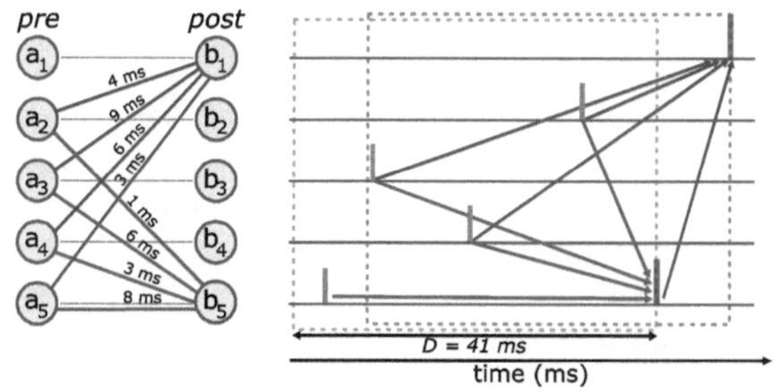

Fig. 1. Working memory as a chained sequential spike prediction via heterogeneous delays. *Left:* A recurrent HD-SNN of five neurons numbered 1–5, connected by two subsets of recurrent connections that define two Spiking Motifs from the pre-synaptic side (neurons a) to the post-synaptic side (neurons b): dark-red connections from a_2, a_3, a_4, and a_5 target b_1 with delays 4, 9, 6, 3 ms; green connections from a_1, a_2, a_3, a_4 target b_5 with delays 1, 6, 3, 8 ms. *Right:* Raster plot illustrating how the network predicts successive spikes from a past context window, see text for details.

individual units. The analogous problem for spiking neural networks (SNNs)—binding activity patterns separated by intervals far exceeding a single neuron's time constant—lacks a general solution and remains a key challenge for biologically plausible sequence processing [27].

One principled approach relies on the *precise relative timing* of spikes across a population. Abeles [1] was a precursor in proposing that cortical neurons act as coincidence detectors; Bienenstock [3] formalised this into *synfire chains*, feed-forward pools in which synchronous volleys propagate reliably. Experimental support came from Bruno and Sakmann [5], who showed that thalamocortical inputs are individually weak but effective through precise synchrony; theoretical work further showed such synchrony can be learned in an unsupervised manner [20]. Ikegaya et al. [14] then directly observed recurring spatiotemporal spike sequences in neocortex at both sub- and supra-second timescales, and complementary hippocampal evidence shows that such sequences self-organise spontaneously as internal population templates [25]. The recent study by Xie et al. [26] demonstrates that in humans, information is encoded using stereotypical spatiotemporal spike sequences within population bursts, complementing traditional rate- and latency-based neural codes.

Izhikevich [15] extended the synfire framework by introducing heterogeneous axonal delays in a recurrent population of spiking neurons trained with spike-timing-dependent plasticity (STDP). The resulting *polychronous neuronal groups* (PNGs)—time-locked, non-synchronous firing patterns at millisecond precision—vastly outnumber the neurons in the network, yielding a large short-term memory capacity exploited for visual feature binding [7] and sequence classification [19]. Szatmáry and Izhikevich [23] showed that short-term plasticity can cue a PNG's reactivation and sustain a working memory. However, storing spike patterns of arbitrary length is inherently difficult: the sequential structure of the task means each context window is not an independent query but is itself the output of the network at the previous step, so small errors at early time steps propagate and amplify towards the end of the sequence—a compounding difficulty absent from static associative memory like Hopfield networks.

The limitation of the original model from Izhikevich [15] is that delays are fixed rather than learned. Recent work addresses this:Gltz et al. [9] demonstrated fast, energy-efficient inference on neuromorphic hardware using first-spike-time coding, while Hammouamri et al. [13] introduced gradient-based delay learning via dilated convolutions with learnable spacings (DCLS), achieving sparse solutions in which only ~20% of delay channels remain active. Applied to event-based vision, heterogeneous learned delays in a single spiking layer suffice for fast motion detection [12], scalable to always-on object recognition [10]. Building on this, *Spiking Motifs* (SMs) generalise PNGs to a gradient-trainable, hardware-compatible framework by replacing fixed STDP rules with end-to-end learned delays, here referred to as Heterogeneous-Delay SNNs (HD-SNNs) [21]. However these motifs have a short time span equal to the ranbge of possible delays, while it has been shown in mice stimulations that sequences of the order of seconds influence neural activity [6]. Kronland-Martinet et al. [16] recently extended this to motifs of arbitrary length by chaining sub-motif detectors with bounded synaptic delays, achieving robust encoding of complete Spiking Heidelberg Digits patterns.

The present work extends the chain-of-motifs framework of Kronland-Martinet et al. [16] to a fully recurrent topology for working memory. Rather than relying on fixed delays and unsupervised STDP as in the original polychronisation model, the network is trained end-to-end with surrogate gradients [18]: once primed by a short trigger window of D steps, it autonomously recalls arbitrary-length spike sequences, bridging polychronisation theory with gradient-based machine learning.

2 Methods

2.1 Recurrent HD-SNN Architecture

In a previous work, we have demonstrated that a feedforward HD-SNN can learn to detect Spiking Motifs [21]. However, this architecture is limited to motifs of length D, the maximum synaptic delay. To store and recall longer sequences, it is possible to extend this architecture to a fully recurrent topology, where each neuron receives inputs from all other neurons across all delays. Our core concept is to retrieve a memory progressively, in which each motif of length D would allow to predict the next spikes in the memory to retrieve. This allows the network to generate a chain of overlapping Spiking Motifs, where each motif provides the context for predicting the next spikes in the sequence (see Fig. 1).

In practice, we will study a recurrent Heterogeneous-Delay SNN (HD-SNN) which consists of a population $\mathcal{A}$ of $N = 1024$ leaky integrate-and-fire (LIF) neurons implemented in SNNTORCH [8] and simulated over $T = 1000$ steps of 1 ms each (total duration: 1 s). Extending the feedforward Spiking Motif Detector [21], our architecture will include full recurrence: each neuron $j \in \mathcal{A}$ integrates inputs from *all* other neurons across *all* preceding $D = 41$ time steps. Such network is the most general for learning purposes, and at inference, pruning of the synapses may be added to gain from the sparseness of synapses on the energy budget. Moreover, the choice for the maximum delay value D is justified from biological observations and similar to the value used in the original polychronisation study [15]. The membrane potential $u_j(t)$ evolves as

$$u_j(t) = \beta \cdot u_j(t-1) \cdot (1 - s_j(t-1)) + \sum_{i=1}^{N} (\sum_{d=1}^{D} \mathbf{W}_{j,i,d} \cdot s_i(t-d)), \quad (1)$$

where $\beta \in (0, 1)$ is the membrane decay constant ($\beta = 0.7$ in our simulations, that is, a time constant of $\tau = 1/\ln(\frac{1}{\beta}) \approx 2.8$ steps), $\mathbf{W}_{j,i,d}$ is the synaptic weight to neuron j from neuron i at delay d, and $s_j(t) \in \{0, 1\}$ is the spike indicator $s_j(t) = \mathbb{1}[u_j(t) \geq \vartheta]$, that is, $s_j(t) = 0$ if $u_j(t) < \vartheta$ and a spike is emitted $s_j(t) = 1$ when $u_j(t)$ exceeds the threshold ϑ. After spiking, the membrane potential that is accumulated in Eq. 1 is reset to zero.

In our implementation, the spike history of all N neurons over the past D steps is concatenated into a flattened context vector of dimension $N \cdot D$ at each time step, so that $\mathbf{W}$ is reshaped as $\bar{\mathbf{W}} \in \mathbb{R}^{N \times (N \cdot D)}$. This is equivalent to the temporal convolution that we used previously in a feedforward setting [21], here extended to a fully recurrent topology. The HD-SNN thus maps onto a standard static computational graph that can be trained with surrogate gradient descent as classical SNNs (see Fig. 2, left panel).

2.2 Synthetic Dataset

To assess working-memory capacity in all generality, we generated a dataset of $M = 16$ distinct target patterns $\mathbf{S}^{(\mu)} \in \{0, 1\}^{N \times T}$. Following previous work [21], each pattern is generated using a stochastic spiking pattern generator with both a structured and background components. Specifically, for each pattern we define a *frozen* pattern as $P^{(\mu)} \in \{0, 1\}^{N \times T}$ with $P^{(\mu)} \sim \text{Bernoulli}(p_A)$ where we set $p_A = 0.16 \times 10^{-4}$ spike per neuron per time step (0.16 Hz if we consider 1 ms time bins) as the average spike probability. This follows the estimation of the average spiking frequency in the human brain [17].

To account for the stochasticity inherent to neural activity, we further include a balanced bit flipping mechanism with probability $p_{\text{flip}} = 0.05$. This is defined by the binary mask $M^{(\mu)} \in \{0, 1\}^{N \times T}$ of bit flips locations : $M^{(\mu)} \sim \text{Bernoulli}(p_{\text{flip}})$. Values are left untouched where $M^{(\mu)} = 0$ and will be drawn with probability p_A where $M^{(\mu)} = 1$. Finally, the spike pattern $S^{(\mu)} \in \{0, 1\}^{N \times T}$ are those of the frozen pattern $P^{(\mu)}$ except where $M^{(\mu)} = 1$ for which they are obtained by drawing with probability p_A. Formally, this can be expressed as:

$$\mathbf{S}^{(\mu)} \sim P^{(\mu)} \cdot (1 - M^{(\mu)}) + \text{Bernoulli}(p_A) \cdot M^{(\mu)} \quad (2)$$

Note that the bit flipping mechanism introduces a stochasticity into the structure of the spike patterns, however, it does not change the average firing rate. Indeed, since the generation of $\mathbf{P}^{(\mu)}$ and $\mathbf{M}^{(\mu)}$ are independent, the expected firing rate $E(\mathbf{S}^{(\mu)})$ is equal to $p_A \cdot (1 - p_{\text{flip}}) + p_A \cdot p_{\text{flip}} = p_A$.

2.3 Working Memory Task: Memory Cuing

Given a set of target patterns as defined above, a generic working memory task consists of training the network to reproduce the target pattern that follow a short *trigger* window. Formally, given $\mathbf{S}^{(\mu)}$, each contiguous window of length D, i.e. the slice $\mathbf{S}^{(\mu)}[\cdot, t{-}D : t{-}1]$, constitutes a unique context that may be used to predict the spiking activity $s_j(t)$ of all neurons j at time t (see Fig. 1). The network is therefore trained to generate a sequential chain of overlapping Spiking Motifs [16,21] that faithfully reproduces the target. Memory retrieval is cued by clamping all neuron activities to the target pattern during the trigger window which corresponds to the initial D time steps ($0 \leq t < D$), that is, corresponding to one elementary Spiking Motif (see Fig. 2, right panel); the goal of the network is then to generate the remainder of the sequence autonomously.

As an accuracy metric, we will use the mean F_1 score between the predicted and target spike trains, averaged over time (in the prediction window), neurons and patterns. The F_1 score is the harmonic mean of precision (proportion of predicted activity that are correct) and recall (proportion of predicted activity that is successfully reproduced). Knowing a priori that average neural activity is sparse, maximising $\mathcal{L}$ simultaneously penalises over-active networks (low precision) and silent networks (low recall).

2.4 Analytical Weight Initialisation

From Eq. 1, by normaizing the potential to arbitrary units, we wish the membrane potential to reach a sub-threshold value close to one whenever a spiking motif from a target pattern occurs. Before training, the weight tensor $\mathbf{W}$ can be initialised analytically by treating this task as a linear regression problem. For each target pattern $\mathbf{S}^{(\mu)} \in \{0,1\}^{N \times T}$, the context window at time t is the flattened vector $\mathbf{c}^{(\mu)}(t) \in \{0,1\}^{N \cdot D}$ containing the spike history of all neurons over steps $t - D$ to $t - 1$ and the corresponding desired output $\mathbf{I}(t) \in \{0,1\}^N$. Stacking all $K = M \cdot (T - D)$ context–output pairs across patterns and time steps yields

$$\mathbf{C} \in \{0,1\}^{(N \cdot D) \times K}, \qquad \mathbf{I}^* \in \mathbb{R}^{N \times K},$$

and the initialisation problem becomes $\mathbf{W}\,\mathbf{C} \approx \mathbf{I}$.

Since the system is strongly underdetermined ($K \ll ND$), the minimum-ℓ_2-norm solution is the Moore–Penrose pseudo-inverse $\mathbf{W}^* = \mathbf{I}^* \cdot \mathbf{C}^+$. With sparse neural activity at rate $p_A = .16 \cdot 10^{-3}$, consecutive context windows are nearly orthogonal, so the Gram matrix satisfies $\mathbf{C}\mathbf{C}^\top \approx N \cdot D \cdot p_A \cdot \mathbf{I}$, giving the closed-form initialisation

$$w_{i,j,d} = \frac{1}{N \cdot D \cdot p_A \cdot M} \sum_{\mu=1}^{M} \sum_{t=D+1}^{T} s_i^{*(\mu)}(t-d) \cdot \left(s_j^{*(\mu)}(t) - \beta \cdot s_j^{*(\mu)}(t-1)\right). \quad (3)$$

Each weight $w_{i,j,d}$ is proportional to the cross-correlation between the deconvolved spike train of neuron j (scaled by $\vartheta_0 = 0.8$) and the raw spike train of neuron i at delay d, averaged over all M patterns.

2.5 Training

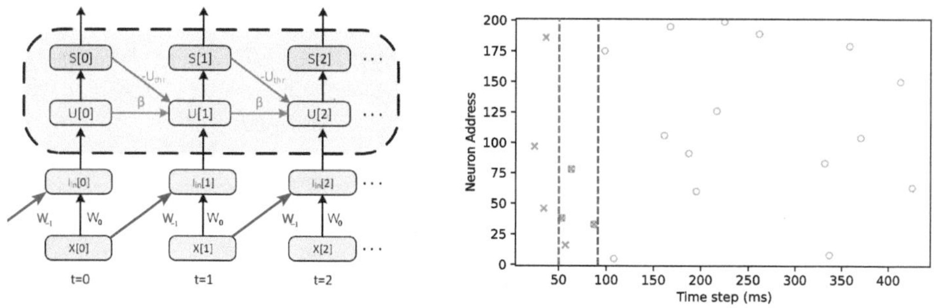

Fig. 2. Network architecture and a sample target pattern. *Left:* Unrolled view of the recurrent HD-SNN over time (one column representing one time step), illustrating delayed connections between neurons: Blue arrows indicate synaptic connections that integrate past activity across heterogeneous delays; for clarity only one delay is shown. *Right:* One representative target spike pattern ($N = 1024$ neurons, $T = 1000$ time steps, of which we show 256 neurons and 400 time steps); each red circle marks one target spike. The goal of the network is to memorise $M = 16$ such patterns. Each trial is flanked by 50 time steps of spontaneous background activity (random spikes at rate p_A) before and after the stimulus. The trigger window (green crosses between the dashed lines) consists of the first $D = 41$ time steps of the target pattern clamped to the network input; the recurrent network then autonomously predicts the remaining spikes. (Color figure online)

The loss is defined as $\mathcal{L} = 1 - F_1$, such that minizing this loss (with a maximum of 1 and minimum of 0) maximises the mean F_1 score across all M target patterns. Learning is carried out with surrogate-gradient BPTT [18], using a fast-sigmoid surrogate with sharpness parameter $\alpha = 15$. Weights are optimised with AdamW (learning rate $\eta = 10^{-3}$, momentum $\mu = 0.99$, weight decay $\lambda = 0$). Dropout regularization with probability 0.37 is applied. To mitigate overfitting and stabilise delay learning, the learning rate is annealed with a cosine schedule with warmup, as used for instance in [13]. The validation loss is monitored for detecting overfitting.

Our implementation is available at https://github.com/laurentperrinet/MNESIS. The recurrent HD-SNN was implemented using the snnTorch [8] library with custom extensions for heterogeneous delay handling. The implementation leverages PyTorch with the CUDA backend on the Jean Zay supercluster at GENCI. The MPS and CUDA backends enabled efficient training and simulation of the large recurrent networks required for working memory tasks, leveraging the parallel processing capabilities of the GPU cores while maintaining low-latency access to the substantial memory resources necessary for storing all tensors' parameters. Training typically converged within 128 gradient steps, requiring approximately 5 minutes of computation time per training.

3 Results

3.1 Training Dynamics and Results

Training the recurrent HD-SNN with surrogate-gradient BPTT [18] consistently maximised the mean F_1 score across all $M = 16$ target patterns to a perfect score of 1.0, with $N = 1024$ neurons, $D = 41$ delays, $T = 1000$ time steps, $p_A = 10^{-3}$, and $\beta = 0.7$ (time constant $\tau \approx 2.8\,\text{ms}$). Without the analytical initialisation, two characteristic failure modes arise during early training: episodes of complete silence, in which the network emits no spikes, and over-activation, in which nearly all neurons fire at every time step. Both are well-known pathological attractors of recurrent SNNs [15]. The F_1 score is the harmonic mean of precision and recall, and maximising it therefore penalises both failure modes symmetrically: low precision corresponds to an over-active network and low recall to a silent one. The cosine learning-rate schedule [13] helped guide the network away from these extremes towards a balanced, sparse-firing regime (Fig. 3).

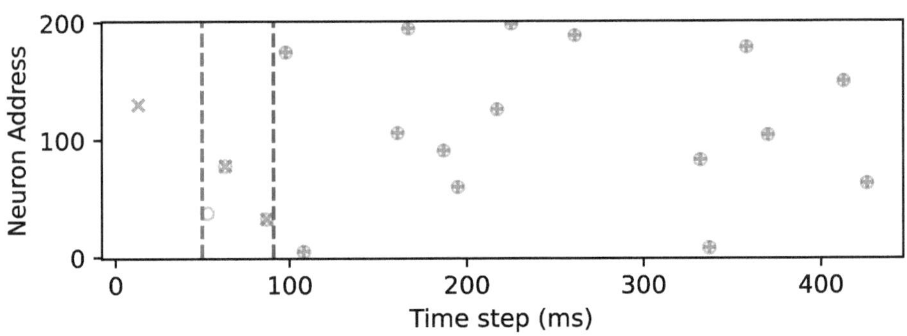

Fig. 3. Pattern retrieval from the recurrent HD-SNN. After clamping the network with input spikes (green crosses) with 50 steps of spontaneous activity (left of the red dashed line) and a trigger window consisting of the initial sequence of one of the $M = 16$ target patterns ($0 \leq t < D = 41$, delimited by the dashed red and blue line), we simulate the network after training output the duration the full sequence duration ($T = 1000$ steps). We observe that the raster plot shows network activity (blue crosses, $N = 1024$ neurons) which perfectly matches the target pattern which was cued during the trigger window. Note that the target spikes after the trigger window (red circles right of the blue dashed line) are hidden from the network. (Color figure online)

3.2 Working Memory and Robustness to Perturbations

A hallmark of biological working memory is graceful degradation: recall should remain possible under incomplete or corrupted retrieval cues, mirroring the tolerance of hippocampal pattern completion to partial input [25]. We therefore evaluated three distinct perturbations of the trigger window, each probing a different dimension of robustness (Fig. 4).

Robustness to Spike Noise (Figure 4, Left). We corrupted the trigger window by independently flipping each spike with probability $p_{\text{flip}} \in [0, 1]$ (bit-flip noise). At $p_{\text{flip}} = 0.25$, the network still achieved precision $= 1.000$, recall $= 0.937$, and $F_1 = 0.967$, demonstrating high tolerance to noise in the trigger window. The F_1 score decreases monotonically as p_{flip} increases, reaching chance level only near $p_{\text{flip}} = 0.5$ where the trigger window is entirely randomised. This behaviour is consistent with the network operating as an attractor: small perturbations of the initial state are corrected by the recurrent dynamics, which converge back to the stored pattern.

Robustness to Trigger-Window Duration (Figure 4, Middle). We varied the duration of the trigger window from 0 to $D - 1 = 40$ steps, scanning from no cue at all to a trigger that is one step short of the full motif. At 75% of the full trigger duration (≈ 30 steps out of $D = 41$), the network achieved precision $= 1.000$, recall $= 0.757$, and $F_1 = 0.862$. The F_1 score increases monotonically with trigger duration, rising steeply once the trigger exceeds roughly $D/2$, consistent with the requirement that the context window must contain enough active synapses to uniquely identify the stored pattern above the interference floor.

Robustness to Partial Neuron Coverage (Figure 4, Right). We varied the number of neurons whose activity was silenced during the trigger window, from 0 (full trigger) to $N = 1024$ (no trigger). With $N/8 = 128$ neurons silenced—meaning 87.5% of the population remained active—the network still achieved perfect recall ($F_1 = 1.000$). The F_1 score degrades smoothly as more neurons are silenced and drops sharply only when the majority of the trigger population is inactive, again consistent with an attractor interpretation in which a sufficient fraction of the pattern suffices to pull the network dynamics towards the stored memory.

3.3 Effect of Delay Depth and Pattern Duration

To characterise how the architectural parameters govern working-memory performance, we conducted a systematic one-at-a-time scan over the maximum delay D, the pattern duration T, and the background firing rate p_A, holding all other hyperparameters at their default values (Sect. 2). Each condition was evaluated with $N_{\text{cv}} = 10$ independent random seeds and results are reported as mean loss $\mathcal{L} = 1 - F_1$ (Fig. 5).

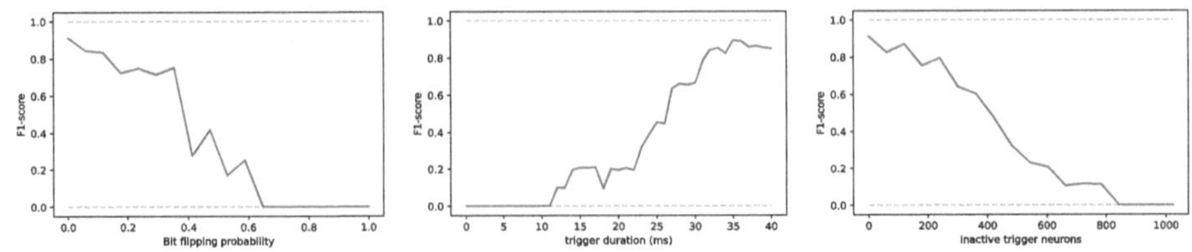

Fig. 4. Robustness of working-memory retrieval to three types of trigger-window perturbation (F_1 score averaged over patterns). *Left:* Bit-flip noise applied to all spikes in the trigger window ($p_{\text{flip}} \in [0, 1]$); the network maintains high F_1 up to $p_{\text{flip}} \approx 0.4$. *Middle:* Truncated trigger window (trigger duration from 0 to $D - 1 = 40$ steps); retrieval remains reliable for durations above roughly $D/2$. *Right:* Partial neuron coverage (number of silenced neurons from 0 to $N = 1024$); perfect recall is maintained until the majority of the trigger population is inactive. Orange dashed lines indicate $F_1 \in \{0, 1\}$. (Color figure online)

Delay Depth D (Figure 5*, Left).* The loss decreases monotonically as D grows from 3 to 127 ms, dropping from $\mathcal{L} \approx 1.$ at $D = 3$ to near zero at $D = 60$. This confirms the capacity argument: the parameter budget $N^2 \cdot D$ and the context-window size $N \cdot D$ both scale linearly with D. A wider delay window provides more temporally diverse synaptic pathways, increasing the number of simultaneously representable Spiking Motifs and reducing interference between overlapping context windows. At $D = 3$ the context vector has only $N \cdot D = 3072$ dimensions with expected activity $N \cdot D \cdot p_A \approx 3.07$—a regime where contexts are nearly all-zero and indistinguishable. As a comparison, at $D = 127$ the context has $N \cdot D = 129{,}024$ dimensions with expected activity ≈ 129, and near-orthogonality is strongly satisfied.

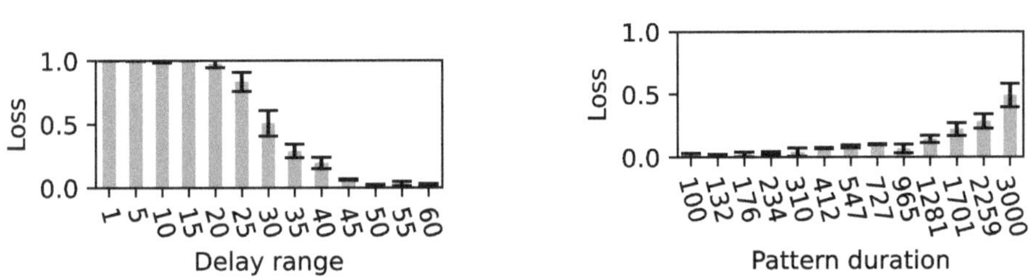

Fig. 5. Effect of some key parameters on working-memory performance (loss $\mathcal{L} = 1 - F_1$, mean $\pm$ s.e. over $N_{\text{cv}} = 10$ cross-validation folds). *Left:* Increasing the maximum delay D consistently reduces the loss, confirming that deeper delay windows provide a larger, more orthogonal context and that longer maximum delays directly increase the capacity of the recurrent HD-SNN. *Right:* Loss grows slowly and monotonically with pattern duration T, consistent with the compounding-error and gradient-vanishing arguments: longer sequences require the network to maintain coherent predictions further from the trigger window.

Pattern Duration T (Figure 5, Middle). The loss grows monotonically with T, from $\mathcal{L} \approx 0.004$ at $T = 100$ to $\mathcal{L} \approx .49$ at $T = 3000$. This is the empirical signature of the two compounding difficulties identified in the capacity analysis: longer sequences impose a larger open-loop rollout, and the error propagation recurrence accumulates over $T - D$ steps. Even a sub-critical propagation coefficient $\lambda < 1/D$ eventually allows errors to reach a non-negligible fraction of N spikes.

4 Discussion

4.1 Memory Capacity and the Role of Heterogeneous Delays

We have demonstrated that a recurrent SNN with heterogeneous synaptic delays can store and reproduce arbitrary target spike patterns, with the analytical weight initialisation alone achieving perfect recall ($F_1 = 1.0$) on a benchmark of $M = 16$ patterns of $T = 1000$ steps. The result is noteworthy from a capacity standpoint: the entire network is parameterised by a single weight tensor $\mathbf{W} \in \mathbb{R}^{N \times N \times D}$ of dimension $N^2 \cdot D = 1024^2 \cdot 41 \approx 43 \times 10^6$ parameters, yet it faithfully stores M patterns each of $N \cdot T$ binary values. This compression is made possible precisely by the temporal structure introduced by the D delay channels: rather than requiring one parameter per stored bit, the network exploits the sequential dependencies between consecutive context windows of length D to represent the entire pattern as a chain of overlapping Spiking Motifs [16,21].

The robustness experiments further strengthen this interpretation. The network tolerates up to 25% bit-flip noise in the trigger window, requires only 75% of the full trigger duration for reliable recall, and maintains perfect retrieval with up to 87.5% of trigger neurons silenced. This progressive degradation profile is consistent with the network operating as an attractor: the stored patterns correspond to stable fixed points of the recurrent dynamics, and the trigger window provides a partial cue that the network completes autonomously—the same mechanism proposed for hippocampal pattern completion [25].

4.2 Limitations and Directions for Improvement

A systematic evaluation of memory capacity—how many patterns of what duration can be reliably stored and retrieved—remains to be carried out beyond the present $M = 16$ benchmark. The theoretical analysis predicts that capacity scales as $N^2 \cdot D/(T - D)$, giving approximately 5,700 patterns for the current architecture, but whether gradient-based training reliably reaches this bound at scale is an open empirical question, as the implementation does not scale well with respect to the size of the GPU's memory. Interference between stored patterns at high memory load may degrade precision in ways that the F1 training objective does not fully penalise, and the sensitivity of recall to perturbations of the trigger window is expected to decrease as M grows.

We used the simplest LIF model with a fixed membrane decay β. Introducing adaptive threshold dynamics or more complex single-neuron models, as studied in Szatmáry and Izhikevich [23], should increase the diversity of intrinsic timescales available to the network and improve both capacity and retrieval fidelity. In particular, the deconvolution analysis (Sect. 2) shows that the optimal weight profile depends critically on β: larger values of β require exponentially amplified weights at long delays, suggesting that adaptive β per neuron could allow the network to exploit the full delay range more efficiently.

The current architecture connects every neuron directly to every other neuron across all delays. Partitioning the population into input-output neurons and a larger pool of latent "reservoir" neurons, as in delay-based reservoir computing [19], would enrich the internal spiking dynamics without increasing the number of trained parameters proportionally. Such a hybrid architecture could combine the representational power of random recurrent networks with the precision of gradient-trained delay weights [18], potentially closing the gap between biologically inspired models and sparse selective-update architectures [27].

The present model requires a fixed set of target patterns known in advance. A natural extension is an online learning rule that updates the weights continuously as new patterns are presented. The closed-form Hebbian initialisation (Eq. 3) suggests a natural online update: each new spike $s_j(t)$ incrementally adjusts $w_{i,j,d}$ in proportion to $s_i(t-d)$, recovering a local STDP-like rule [15]. Whether such online updates are stable under recurrent feedback and can support continual working memory remains to be established.

References

1. Abeles, M.: Role of the cortical neuron: integrator or coincidence detector? Isr. J. Med. Sci. **18**(1), 83–92 (1982)
2. Adrian, E.D., Zotterman, Y.: The impulses produced by sensory nerve endings. J. Physiol. **61**(4), 465–483 (1926)
3. Bienenstock, E.: A model of neocortex. Netw. Comput. Neural Syst. **6**(2), 179–224 (1995)
4. Bohte, S.M.: The evidence for neural information processing with precise spike-times: a survey. Nat. Comput. **3**(2), 195–206 (2004)
5. Bruno, R.M., Sakmann, B.: Cortex Is driven by weak but synchronously active thalamocortical synapses. Science **312**(5780), 1622–1627 (2006)
6. Deveau, C.E., et al.: Recurrent cortical networks encode natural sensory statistics via sequence filtering. Neuron, p. S0896627325009808 (2026)
7. Eguchi, A., Isbister, J.B., Ahmad, N., Stringer, S.: The emergence of polychronization and feature binding in a spiking neural network model of the primate ventral visual system. Psychol. Rev. **125**(4), 545–571 (2018)
8. Eshraghian, J.K., et al.: Training spiking neural networks using lessons from deep learning. Proc of the IEEE **111**(9), 1016–1054 (2023)
9. Göltz, J., et al.: Fast and energy-efficient neuromorphic deep learning with first-spike times. Nature Mach. Intell. **3**(9), 823–835 (2021)
10. Grimaldi, A., Boutin, V., Ieng, S.H., Benosman, R., Perrinet, L.U.: A robust event-driven approach to always-on object recognition. Neural Netw. **178**, 106415 (2024)
11. Grimaldi, A., Gruel, A., Besnainou, C., Jérémie, J.N., Martinet, J., Perrinet, L.U.: Precise spiking motifs in neurobiological and neuromorphic data. Brain Sci. **13**(1), 68 (2023)

12. Grimaldi, A., Perrinet, L.U.: Learning heterogeneous delays in a layer of spiking neurons for fast motion detection. Biol Cyb **117**, 373–387 (2023)
13. Hammouamri, I., Khalfaoui, I., Masquelier, T.: Learning delays in spiking neural networks using dilated convolutions with learnable spacings (2024)
14. Ikegaya, Y., et al.: Synfire chains and cortical songs: temporal modules of cortical activity. Science **304**(5670), 559–564 (2004)
15. Izhikevich, E.M.: Polychronization: computation with Spikes. Neural Comput. **18**(2), 245–282 (2006)
16. Kronland-Martinet, T., Viollet, S., Perrinet, L.U.: Detection of spiking motifs of arbitrary length in neural activity using bounded synaptic delays (2025)
17. Lennie, P.: The cost of cortical computation. Curr. Biol. **13**(6), 493–497 (2003)
18. Neftci, E.O., Mostafa, H., Zenke, F.: Surrogate gradient learning in spiking neural networks: bringing the power of gradient-based optimization to spiking neural networks. IEEE Signal Process. **36**(6), 51–63 (2019)
19. Paugam-Moisy, H., Martinez, R., Bengio, S.: Delay learning and polychronization for reservoir computing. Neurocomputing **71**, 7–9 (2008)
20. Perrinet, L., Samuelides, M.: Coherence detection in a spiking neuron via Hebbian learning. Neurocomputing **44–46**, 133–139 (2002)
21. Perrinet, L.U.: Accurate detection of spiking motifs in multi-unit raster plots. In: Iliadis, L., Papaleonidas, A., Angelov, P., Jayne, C. (eds.) Artificial Neural Networks and Machine Learning — ICANN 2023, pp. 369–380 (2023)
22. Softky, W.R., Koch, C.: The highly irregular firing of cortical cells is inconsistent with temporal integration of random EPSPs. J Neur 13(1) (1993)
23. Szatmáry, B., Izhikevich, E.M.: Spike-timing theory of working memory. PLoS Comput. Biol. **6**(8), e1000879 (2010)
24. Vaswani, A., et al.: Attention is all you need. In: Advances in Neural Information Processing Systems, vol. 30, pp. 5998–6008, arXiv (2017)
25. Villette, V., Malvache, A., Tressard, T., Dupuy, N., Cossart, R.: Internally recurring hippocampal sequences as a population template of spatiotemporal information. Neuron **88**(2), 357–366 (2015)
26. Xie, W., Wittig, J.H., Chapeton, J.I., El-Kalliny, M., Jackson, S.N., Inati, S.K., Zaghloul, K.A.: Neuronal sequences in population bursts encode information in human cortex. Nature **635**(8040), 935–942 (2024)
27. Yin, B., Wang, S., Tan, H., Bohte, S., Corradi, F., Li, G.: Efficient sparse selective-update RNNs for long-range sequence modeling (2026)

Spike-Transmission Delays Improve the Accuracy–Efficiency Trade-Off for Linear Readouts of Spiking Populations

Alina Lytovchenko(✉), Nicolas Martalog, and Jeff Orchard

Cheriton School of Computer Science, University of Waterloo, Waterloo, Canada
alytovch@uwaterloo.ca

Abstract. Linear decoders for spiking populations typically use only the instantaneous filtered activity of each neuron and therefore do not fully exploit the temporal structure present in neural responses. We study a delay-augmented linear readout in which each neuron contributes several time-shifted copies of its activity, giving the decoder access to recent response history without introducing recurrence or nonlinear processing. To ensure a fair comparison, we keep the total number of decoder inputs fixed across conditions: when more delay taps are added per neuron, the number of neurons is reduced accordingly. On one-dimensional reconstruction of band-limited white-noise signals, we compare this approach with a baseline readout that uses current activity alone in populations of leaky integrate-and-fire (LIF) and adaptive LIF (ALIF) neurons. Delay-augmented decoding improves accuracy under matched decoder dimensionality, with gains that depend on signal timescale and firing regime. The largest improvements occur for ALIF populations and reach nearly 40% relative MSE reduction in the best settings, while LIF populations show smaller and more regime-dependent benefits. These results show that fixed spike-transmission delays can improve decoding efficiency without recurrence or learned temporal dynamics.

Keywords: Spiking Neural Networks · Temporal Coding · Neural Adaptation · Energy-Efficient Computing

1 Introduction

Spiking neural networks (SNNs) are widely explored as brain-inspired models that communicate through discrete spike events, enabling sparse, event-driven computation suitable for neuromorphic hardware [4,10,15,16]. However, achieving good task performance often requires a large number of neurons or high firing rates, reducing overall efficiency. This brings the main question into focus: how efficiently can signals be reconstructed from spiking activity?

A common approach is linear decoding from filtered population activity [6, 18,25]. While simple and interpretable, standard feedforward decoders typically use only instantaneous neural activity. The temporal filtering introduced by

L. Pasa et al. (Eds.): ICANN 2026, LNCS 17089, pp. 272–283, 2027.
https://doi.org/10.1007/978-3-032-38398-3_23

neural and synaptic dynamics can make the desired signal difficult to recover from instantaneous activity alone.

Previous work addresses this limitation by embedding temporal structure directly into the network. Recurrent and reservoir models transform recent history into richer internal states [11,12]. Statistical decoders explicitly incorporate spike history or latent temporal structure [14,21,26], while other approaches introduce learnable delays inside the network itself [5,9,19,20,24,27]. Although effective, these methods modify the encoding network or add complexity to optimization, leaving open a narrower question: how much can decoding accuracy improve if explicit temporal context is provided only through fixed delays between the neurons and the readout?

This paper explores that question using per-neuron spike-transmission delays applied to the spike trains of a fixed feedforward spiking population. To ensure a fair comparison, we match the total number of inputs to the linear decoder across conditions. Because delay taps yield multiple inputs to the readout node per neuron, the delayed setting uses fewer neurons so that the overall decoder size remains the same. This lets us test whether temporal context helps beyond simply increasing the number of decoder features. We perform this matched-dimensionality comparison on continuous signal reconstruction tasks with leaky integrate-and-fire (LIF) and adaptive LIF (ALIF) neuron populations. Adaptation in ALIF neurons creates temporal dependencies in spike responses [3,13,17], potentially increasing the benefits of explicit spike-transmission delays.

Thus, the primary contribution of this paper is a controlled evaluation of fixed per-neuron delays inserted between a spiking population and a linear readout. Specifically, we introduce a matched-dimensionality comparison to isolate the effect of these delays and show that using delayed copies of neural activity in the readout can improve decoding accuracy under matched decoder dimensionality, especially for adaptive neuron populations.

2 Related Work

2.1 Linear Decoding from Spiking Populations

Linear decoding from neural populations is a long-established idea. Classical work on population codes showed that continuous variables can often be reconstructed accurately by linear estimators applied to nonlinear neural responses [18,25]. In computational neuroscience, this perspective was developed further in the Neural Engineering Framework (NEF), where low-dimensional variables are represented by spiking populations and recovered through linear decoders [6]. Our baseline readout belongs to this family: a feedforward spiking population encodes a low-dimensional signal, and a linear decoder reconstructs it from filtered population activity.

2.2 Temporal Decoding from Spike Trains

Another approach is to decode signals directly from spike trains using statistical or Bayesian models. Point-process and generalized linear model (GLM) methods include spike history, population history, and external variables [14,21]. State-space methods such as Kalman filters are also widely used for continuous decoding, especially in motor tasks [26]. For leaky integrate-and-fire populations, Bayesian decoders have also been derived directly from spiking neuron models [7,8]. These methods are often more flexible than a simple linear readout, but they also depend more strongly on the choice of model. Our goal is simpler: we ask how much temporal information can be recovered by adding fixed delay taps to a linear decoder.

2.3 Temporal Processing Through Recurrence and Adaptation

Temporal processing in spiking networks is more commonly achieved through recurrence. Liquid state machines and related reservoir approaches use recurrent spiking dynamics to map recent input history into a high-dimensional state from which a linear readout is trained [11,12,23]. In contrast, our encoder remains feedforward, as we do not rely on recurrent state, learned recurrent dynamics, or memory stored in the network connectivity itself.

Another relevant line of work studies intrinsic neuronal mechanisms for temporal computation. Adapting spiking neurons have been shown to support temporally extended computations more effectively than non-adapting neurons, both in recurrent spiking models and in population-level analyses of coding and decoding [3,13,17]. This literature motivates our comparison between LIF and ALIF populations. If adaptation makes the neural response depend on recent spike history, then a decoding strategy with access to delayed copies of that activity should be better positioned to extract the encoded signal.

2.4 Delays in Spiking Neural Networks

The closest related work involves temporal processing using explicit delays. Voelker et al. [22] showed that accurate delays and richer synaptic dynamics can support delay-line-like computations in spiking networks. In the broader SNN literature, many methods focus on learning delays inside the network. For example, SLAYER introduced gradient-based training of SNNs with axonal delays [19], and later work learned delays using differentiable spike representations [24] or joint delay-weight plasticity [27]. More recent approaches learn axonal delays for temporal tasks [20], use dilated temporal structures [9], or jointly learn delays, weights, and adaptation [5].

In contrast, we do not learn delays or use recurrence. Instead, we study a fixed feedforward spiking population with a readout-side fixed delay bank and a linear decoder. Our goal is to test whether this simple design improves decoding under a fixed decoder size. Overall, our work provides a controlled study of how adding a small amount of temporal context at the readout affects the accuracy–efficiency trade-off of linear decoding.

3 Methods

3.1 Overview

We encode a time-varying scalar signal $x(t)$ into a population of N spiking neurons and attempt to recover it with a linear readout. Each neuron n receives the input signal as a driving current and produces a spike train $S_n(t)$. In the baseline setting, the readout is computed from the instantaneous filtered spike trains of the population. In the delayed setting, each neuron contributes multiple time-shifted copies of its spike train, forming a delay bank. The spikes synapse onto the readout node, where a synaptic weight and filter are applied. Finally, all inputs to the readout node are added together to form $\hat{x}(t)$, an estimate of the desired output. The connection weights of this linear decoder are learned on a separate training segment.

3.2 Neural Models

Leaky Integrate-and-Fire (LIF). Each LIF neuron maintains a membrane potential $v_n(t)$ that integrates its input current $J_n(t)$ according to

$$\tau_m \, \dot{v}_n = -v_n + J_n(t), \tag{1}$$

where τ_m is the membrane time constant.

Whenever v_n reaches a threshold v_{th}, a spike is emitted and v_n is reset to zero. The neuron resumes integrating after a refractory period τ_{ref}.

Adaptive LIF (ALIF). The ALIF neuron extends the LIF dynamics with a spike-triggered adaptation current $q_n(t)$:

$$\tau_m \, \dot{v}_n = -v_n - q_n + J_n(t), \tag{2}$$

$$\tau_a \, \dot{q}_n = -q_n + \beta \, S_n(t) \tag{3}$$

where τ_a is the adaptation time constant and β is the adaptation strength. Because q_n builds up after each spike and decays slowly, it suppresses neuronal excitability over time and reduces the firing rate even under constant input. As a result, the relationship between the neuron's input and spike response becomes history-dependent, so fixed decoders trained based on stationary assumptions may not work well unless adaptation is explicitly accounted for.

3.3 Spike Trains and Delay Taps

The spike train, $S_n(t)$, of neuron n is the sum of Dirac delta functions at its spike times $\{t_{n,k}\}$,

$$S_n(t) = \sum_k \delta(t - t_{n,k}), \tag{4}$$

where $t_{n,k}$ is the time of its kth spike. In the baseline setting, each neuron contributes a single temporally filtered activity trace obtained by convolving its spike train with a causal exponential synaptic filter,

$$a_n(t) = (h_{\tau_s} * S_n)(t), \tag{5}$$

where $h_{\tau_s}(t)$ is a causal exponential synaptic filter with time constant τ_s.

In the delay-augmented setting, each neuron contributes M tapped copies of its activity with delays $\{d_0, d_1, \ldots, d_{M-1}\}$, where $d_0 = 0$ corresponds to the instantaneous channel. The mth feature of neuron n is

$$a_n^{(m)}(t) = (h_{\tau_s} * S_n)(t - d_m), \tag{6}$$

with $a_n^{(m)}(t) = 0$ for $t < d_m$. In implementation, these taps are realized as explicit per-neuron spike-transmission delays applied before a shared readout synapse. For a linear time-invariant synaptic filter, this gives the feature representation above.

For each neuron n, we collect the M delayed activity channels into the row vector

$$\mathbf{a}_n(t) = \left[a_n^{(0)}(t), \ldots, a_n^{(M-1)}(t)\right] \in \mathbb{R}^{1\times M}. \tag{7}$$

The complete delayed feature vector from the entire population is then

$$\mathbf{z}(t) = \left[\mathbf{a}_1(t)|\ldots|\mathbf{a}_N(t)\right] \in \mathbb{R}^{1\times NM}. \tag{8}$$

Thus, the instantaneous decoder uses one feature per neuron, whereas the delayed decoder uses M features per neuron. The delay bank introduces temporal context into the readout while keeping the decoder itself linear and feedforward.

3.4 Linear Readout and Decoder Training

Although the readout is linear in the feature vector $\mathbf{z}(t)$, it can be trained to approximate nonlinear functions of the input because the neural population encoding is itself nonlinear. The reconstructed signal is obtained from a linear decoding of the delayed population features,

$$\hat{x}(t) = \mathbf{z}(t)\mathbf{w} \ . \tag{9}$$

where $\mathbf{w} \in \mathbb{R}^{NM\times 1}$ is the vector of decoding weights.

Suppose we sample the input signal at times $t_1, \ldots, t_P$, store the target values in the vector $\mathbf{x} \in \mathbb{R}^{P\times 1}$, and store the corresponding feature vectors in the rows of $\mathbf{Z} \in \mathbb{R}^{P\times NM}$. The optimal linear decoding weights are computed by ridge-regularized least squares,

$$\mathbf{w}^* = \arg\min_{\mathbf{w}} \|\mathbf{Z}\mathbf{w} - \mathbf{x}\|_2^2 + \lambda\|\mathbf{w}\|_2^2 \ . \tag{10}$$

We set $\lambda = 0.01\,\sigma_{\max}^2$, where $\sigma_{\max}$ is the largest singular value of $\mathbf{Z}$.

3.5 Interpretation as an LTI Readout

The delay-augmented decoder can be interpreted as a linear time-invariant (LTI) readout acting on the population's spike trains. This is because each decoder channel applies only three operations to the spike trains: a fixed time delay, convolution with a fixed synaptic filter h_{τ_s}, and a weighted sum across channels. Each of these operations is linear and time-invariant, and their composition is therefore also LTI. In particular,

$$\hat{x}(t) = \sum_{n=1}^{N} \sum_{m=0}^{M-1} w_{n,m}(h_{\tau_s} * S_n)(t - d_m), \tag{11}$$

which can be rewritten as

$$\hat{x}(t) = \sum_{n=1}^{N} (g_n * S_n)(t), \tag{12}$$

where

$$g_n(t) = \sum_{m=0}^{M-1} w_{n,m} h_{\tau_s}(t - d_m) \tag{13}$$

is the effective temporal filter associated with neuron n. Thus, instead of assigning each neuron a single instantaneous weight, the delayed decoder assigns each neuron a short temporal filter. This allows the readout to use recent spike history, not just current activity, which is helpful when useful information is distributed over time.

3.6 Computational Cost and Practical Choice of Delays

Adding delays introduces several practical costs compared to the instantaneous baseline.

First, delayed decoding requires buffering spikes so that they can be delivered after the prescribed offsets. In contrast, the instantaneous decoder needs no delay buffer. The required buffer length is determined by the maximum delay $d_{\max}$, so longer delays increase the amount of stored spike history.

Second, under matched dimensionality, the number of decoder inputs and hence the number of multiply-accumulate operations in the linear readout is the same in both settings. The additional implementation cost comes from maintaining the delay buffer and routing each spike to the corresponding delayed channels.

Third, the delayed decoder introduces latency: although the instantaneous tap preserves an immediate decoding path, the full delayed feature vector is not available until the longest delay has elapsed. Thus, larger delays provide a longer temporal window to the decoder, but also increase the effective response lag.

Finally, increasing the number of taps improves the temporal resolution available to the decoder, but under matched dimensionality it also reduces the number of neurons in the population and therefore the diversity of tuning curves. In this work, we use $M = 3$ taps as a simple trade-off between temporal coverage and population size.

4 Experiments

4.1 Overview

We compared baseline readouts (which use non-delayed activity alone) with delay-based readouts in LIF and ALIF populations under matched dimensionality across several one-dimensional decoding benchmarks. Each trial consisted of two independent 10-second simulations, one used for training the decoder and one for evaluating test performance.

All simulations were implemented in Nengo [2], which provides the spiking neuron dynamics, synaptic filtering, and connection infrastructure. The per-neuron delay mechanism was implemented as a custom Nengo node that buffers spikes and replays them at the appropriate offsets, as described in Sect. 3.3.

4.2 Setup

Benchmark Signals and Targets. We evaluated the decoder on a continuous one-dimensional reconstruction task in which the input signal $x(t)$ was band-limited white noise and the decoding target was the input signal itself. To probe how temporal scale affects decoding, we varied the cutoff frequency of the input across conditions (e.g. 1 Hz, 5 Hz, 10 Hz), producing signals with different characteristic timescales. Training and test signals were generated independently for each run and bandwidth condition, using different random seeds.

Network Parameters. All experiments used a synaptic filter time constant $\tau_s = 5\,\text{ms}$ and a simulation timestep $dt = 0.1\,\text{ms}$. For LIF neurons, we used the default Nengo parameters ($\tau_m = 20\,\text{ms}$, $\tau_{\text{ref}} = 2\,\text{ms}$). ALIF neurons shared the same base parameters with adaptation strength $\beta = 0.03$ and adaptation time constant $\tau_a = 1.5\,\text{s}$. In the delay-augmented condition, each neuron contributed $M = 3$ taps: an instantaneous channel at $d_0 = 0$ and two delayed channels at offsets d_1 and d_2. The nonzero delays were matched to the temporal scale of the input signal, with larger delays for slower signals and shorter delays for faster ones. In the primary experiments, the delayed population comprised $N_{\text{delay}} =$ 100 neurons and the baseline used $N_{\text{inst}} = 300$ neurons, giving the same total number of decoder inputs in both conditions. We also repeated the comparison at larger proportional population sizes (for example, $N_{\text{delay}} = 300$ and $N_{\text{inst}} = 900$) and observed the same qualitative pattern.

Matched Dimensionality. To compare delay-based and instantaneous decoding fairly, we matched the total number of inputs to the linear decoder across conditions. In the instantaneous setting, each of the N_{inst} neurons contributes one decoder input, so the decoder has N_{inst} input features. In the delayed setting, each of the N_{delay} neurons contributes M features, one for each delay tap, so the decoder has MN_{delay} input features in total, where M is the number of delay taps per neuron. In all matched-dimensionality experiments, population sizes were chosen such that

$$N_{\text{inst}} = MN_{\text{delay}}. \tag{14}$$

For example, with $M = 3$, a delayed population of 100 neurons was compared with an instantaneous population of 300 neurons.

Effect of Neuron Model and Delay Scaling. The usefulness of delay taps depends on the temporal scale of the signal. For slower-changing input signals, larger delays produced better results, whereas higher-frequency inputs required shorter delays to remain informative.

Specifically, the delay taps are scaled to the signal's characteristic timescale

$$\theta = \frac{1}{2\pi f_c}, \tag{15}$$

where f_c is the bandwidth of the input. The two taps are then defined as $d_1 = \eta_1\theta$ and $d_2 = \eta_2\,\theta$, with $\eta_1 \approx 0.42$ and $\eta_2 \approx 0.71$. These coefficients were selected using Optuna [1], a Bayesian hyperparameter optimization framework, by minimizing reconstruction MSE on a held-out validation segment. The resulting values performed well across the conditions tested here, though they may not be globally optimal and could benefit from further tuning for different network configurations or input statistics.

Evaluation Metric. Decoder performance was evaluated using mean squared error (MSE) between the target signal $x(t)$ and the reconstructed output $\hat{x}(t)$. To compare delay-augmented decoding with the no-delay baseline, we report the relative percentage improvement in MSE:

$$\Delta\text{MSE}(\%) = 100 \cdot \frac{\text{MSE}_{\text{no-delay}} - \text{MSE}_{\text{delay}}}{\text{MSE}_{\text{no-delay}}}. \tag{16}$$

Here, $\text{MSE}_{\text{delay}}$ denotes the reconstruction error obtained with delayed decoding, and $\text{MSE}_{\text{no-delay}}$ denotes the error obtained with instantaneous decoding. Positive values therefore indicate that adding delays improves reconstruction accuracy, while negative values indicate that the no-delay decoder performs better.

4.3 Results

Each reported result is averaged over 10 different runs with different train/test seed pairs.

Figure 1 summarizes the relative MSE improvement of delay-augmented decoding over the no-delay baseline across white-noise bandwidths and firing-rate regimes. The strongest and most consistent gains occur for the 5 Hz condition; for LIF neurons, the improvement ranges from 11.7% to 33.3%, while for ALIF neurons it ranges from 4.1% to 39.3%. For the 10 Hz condition, the gains are smaller and concentrated in intermediate firing regimes, peaking at 18.7% for LIF and 20.3% for ALIF. The 1 Hz condition reveals the clearest difference between neuron models. ALIF remains strongly positive across all firing-rate ranges, with improvements from 7.4% to 39.2%, whereas LIF becomes negative at moderate to high firing rates, dropping to -10.8% in the highest range.

In additional experiments, we also decoded nonlinear target functions of the input, such as $|x|$ and $\sin(2\pi x)$. The qualitative pattern was again the same. Overall, these results show that delay taps are most beneficial when useful information is distributed over recent spike history, and that this effect can be enhanced by adaptive dynamics.

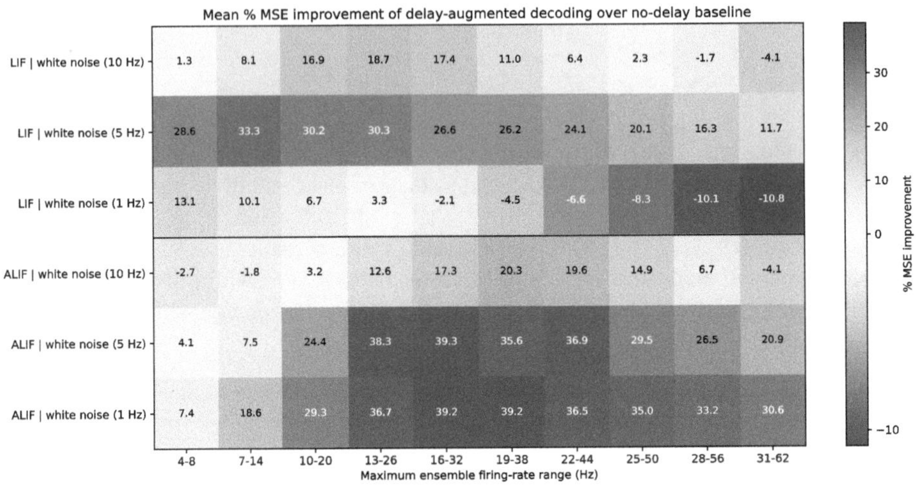

Fig. 1. Mean percentage MSE improvement of delay-augmented decoding over the no-delay baseline for LIF and ALIF populations under white-noise inputs with different frequencies. The no-delay population uses 300 neurons, while the delay-augmented population uses 100 neurons with $M = 3$ taps per neuron (one instantaneous, two delayed), yielding matched decoder dimensionality.

A complementary view is provided by the spike-count/error curves in Fig. 2, which plot test MSE against mean spikes per neuron per second. For LIF neurons, the delayed decoder consistently lies below the no-delay baseline across the plotted range, indicating lower reconstruction error despite using one third as many neurons. For ALIF neurons, the delayed decoder offers the clearest advantage at low and intermediate firing rates, where it reaches substantially lower MSE than the 300-neuron no-delay baseline, while the gap narrows at the highest firing rates. This pattern supports the interpretation that delay augmentation is most useful in regimes where spikes are informative but not yet abundant.

These trends support the interpretation in Sect. 3.5. Delay augmentation gives each neuron a short temporal filter instead of just one instantaneous weight. For ALIF neurons, whose responses depend on history because adaptation builds up from recent spikes, this temporal filter aligns better with the structure of the encoded signal. The weaker and more regime-dependent gains for LIF suggest that delayed readout features are most useful when the neural response depends strongly on recent history, as in ALIF populations with adaptation.

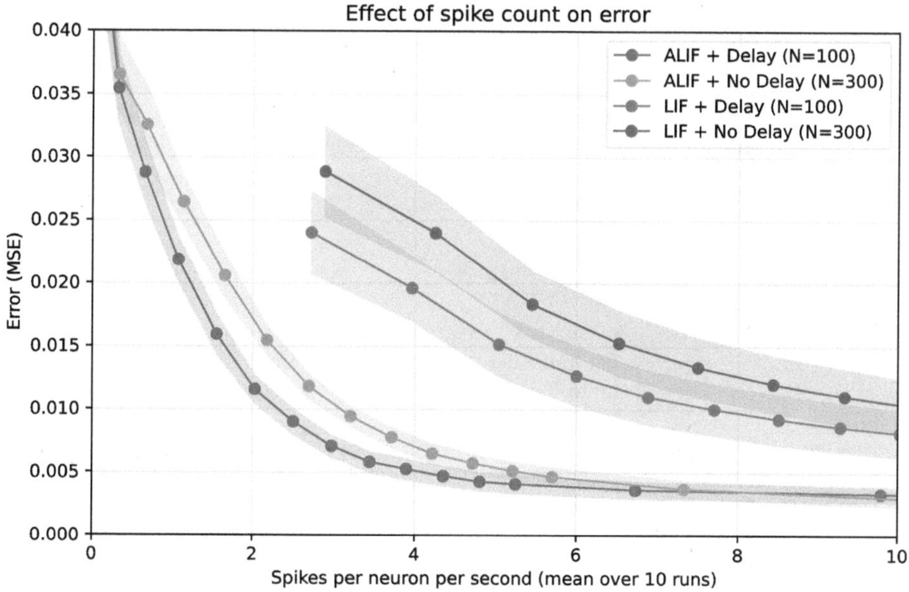

Fig. 2. Reconstruction error vs. spike rate for LIF and ALIF populations with and without delay augmentation under 5 Hz white-noise input, using matched decoder dimensionality. Here, N denotes the number of neurons in the population. Lines show the mean over 10 runs, and shaded bands indicate ± 1 standard deviation across runs.

5 Conclusion and Future Work

We studied a simple way to add temporal context to linear decoding in feed-forward spiking networks: fixed per-neuron delay taps between the population and the readout. Under matched decoder dimensionality, delay augmentation frequently improved reconstruction accuracy, compared to an instantaneous decoder, and the largest gains appeared for ALIF populations, consistent with the fact that adaptation makes their responses depend on recent history. The results therefore suggest that even without recurrence, learned delays, nonlinear readouts, or increased population size, a small delay bank can substantially improve the accuracy–efficiency trade-off of spiking population decoding.

These findings are also consistent with the hypothesis that the brain may use adaptive neurons as a strategy to simultaneously maximize decoding accuracy and minimize energy expenditure, at least for rate-based readouts.

Several directions remain open. First, the present results suggest that delay spacing should be matched to the signal timescale, so a more systematic procedure for selecting the number and placement of delays would be useful. Second, future experiments should compare models under explicitly matched total spike budgets and hardware cost, since buffering delayed spikes introduces additional memory and routing overhead. Third, it would be valuable to evaluate the same readout on richer temporal benchmarks, multidimensional signals, and downstream control tasks.

Acknowledgments. We gratefully acknowledge NSERC, Intel, and the University of Waterloo for financial and in-kind support of this work.

Disclosure of Interests. Authors declare no competing interests.

References

1. Akiba, T., Sano, S., Yanase, T., Ohta, T., Koyama, M.: Optuna: a next-generation hyperparameter optimization framework. In: Proceedings of the 25th ACM SIGKDD International Conference on Knowledge Discovery & Data Mining (2019)
2. Bekolay, T.: Nengo: a Python tool for building large-scale functional brain models. Front. Neuroinform. **7**, 48 (2014). https://doi.org/10.3389/fninf.2013.00048
3. Bellec, G., Salaj, D., Subramoney, A., Legenstein, R., Maass, W.: Long short-term memory and learning-to-learn in networks of spiking neurons. In: Advances in Neural Information Processing Systems 31, Curran Associates, Inc (2018)
4. Davies, M., et al.: Loihi: a neuromorphic manycore processor with on-chip learning. IEEE Micro **38**(1), 82–99 (2018). https://doi.org/10.1109/MM.2018.112130359
5. Deckers, L., Van Damme, L., Van Leekwijck, W., Tsang, I.J., Latré, S.: Co-learning synaptic delays, weights and adaptation in spiking neural networks. Front. Neurosci. **18**, 1360300 (2024). https://doi.org/10.3389/fnins.2024.1360300
6. Eliasmith, C., Anderson, C.H.: Neural Engineering: Computation, Representation, and Dynamics in Neurobiological Systems. MIT Press, Cambridge, MA (2003)
7. Gerwinn, S., Macke, J.H., Bethge, M.: Bayesian population decoding of spiking neurons. Front. Comput. Neurosci. **3**, 21 (2009). https://doi.org/10.3389/neuro.10.021.2009
8. Gerwinn, S., Macke, J.H., Bethge, M.: Reconstructing stimuli from the spike times of leaky integrate-and-fire neurons. Front. Neurosci. **5**, 1 (2011). https://doi.org/10.3389/fnins.2011.00001
9. Hammouamri, I., Khalfaoui-Hassani, I., Masquelier, T.: Learning delays in spiking neural networks using dilated convolutions with learnable spacings. In: The Twelfth International Conference on Learning Representations (2024)
10. Maass, W.: Networks of spiking neurons: the third generation of neural network models. Neural Netw. **10**(9), 1659–1671 (1997). https://doi.org/10.1016/S0893-6080(97)00011-7
11. Maass, W., Natschläger, T., Markram, H.: A model for real-time computation in generic neural microcircuits. In: Advances in Neural Information Processing Systems 15, MIT Press (2002)
12. Natschläger, T., Maass, W.: Information dynamics and emergent computation in recurrent circuits of spiking neurons. In: Advances in Neural Information Processing Systems 16, MIT Press (2003)
13. Naud, R., Gerstner, W.: Coding and decoding with adapting neurons: a population approach to the peri-stimulus time histogram. PLoS Comput. Biol. **8**(10), e1002711 (2012). https://doi.org/10.1371/journal.pcbi.1002711
14. Paninski, L., Pillow, J., Lewi, J.: Statistical models for neural encoding, decoding, and optimal stimulus design. In: Progress in Brain Research, vol. 165, pp. 493–507. Elsevier (2007). https://doi.org/10.1016/S0079-6123(06)65031-0

15. Pfeiffer, M., Pfeil, T.: Deep learning with spiking neurons: opportunities and challenges. Front. Neurosci. **12** (2018). https://doi.org/10.3389/fnins.2018.00774
16. Roy, K., Jaiswal, A., Panda, P.: Towards spike-based machine intelligence with neuromorphic computing. Nature **575**(7784), 607–617 (2019). https://doi.org/10.1038/s41586-019-1677-2
17. Salaj, D., Subramoney, A., Kraisnikovic, C., Bellec, G., Legenstein, R., Maass, W.: Spike frequency adaptation supports network computations on temporally dispersed information. eLife **10**, e65459 (2021). https://doi.org/10.7554/eLife.65459
18. Salinas, E., Abbott, L.F.: Vector reconstruction from firing rates. J. Comput. Neurosci. **1**(1–2), 89–107 (1994). https://doi.org/10.1007/BF00962720
19. Shrestha, S.B., Orchard, G.: SLAYER: Spike layer error reassignment in time. In: Advances in Neural Information Processing Systems, p. 31 (2018)
20. Sun, P., Chua, Y., Devos, P., Botteldooren, D.: Learnable axonal delay in spiking neural networks improves spoken word recognition. Front. Neurosci. **17**, 1275944 (2023). https://doi.org/10.3389/fnins.2023.1275944
21. Truccolo, W., Eden, U.T., Fellows, M.R., Donoghue, J.P., Brown, E.N.: A point process framework for relating neural spiking activity to spiking history, neural ensemble, and extrinsic covariate effects. J. Neurophysiol. **93**(2), 1074–1089 (2005). https://doi.org/10.1152/jn.00697.2004
22. Voelker, A.R., Eliasmith, C.: Improving spiking dynamical networks: accurate delays, higher-order synapses, and time cells. Neural Computation **30**(3), 569–609 (2018). https://doi.org/10.1162/neco_a_01046
23. Voelker, A.R., Kajic, I., Eliasmith, C.: Legendre memory units: continuous-time representation in recurrent neural networks. In: Advances in Neural Information Processing Systems, vol. 32 (2019)
24. Wang, X., Lin, X., Dang, X.: A delay learning algorithm based on spike train kernels for spiking neurons. Front. Neurosci. **13**, 252 (2019). https://doi.org/10.3389/fnins.2019.00252
25. Westover, M.B., Eliasmith, C., Anderson, C.H.: Linearly decodable functions from neural population codes. Neurocomputing **44–46**, 691–696 (2002). https://doi.org/10.1016/S0925-2312(02)00459-9
26. Wu, W., Gao, Y., Bienenstock, E., Donoghue, J.P., Black, M.J.: Bayesian population decoding of motor cortical activity using a Kalman filter. Neural Comput. **18**(1), 80–118 (2006). https://doi.org/10.1162/089976606774841585
27. Zhang, M., et al.: Supervised learning in spiking neural networks with synaptic delay-weight plasticity. Neurocomputing **409** (2020). https://doi.org/10.1016/j.neucom.2020.03.079

Design of a Hardware-Efficient Digital Inference Circuit for SNN-Based Reservoir Computing Systems

Satoshi Moriya(✉), Tetsuta Sakai, Yuki Majima, Hideaki Yamamoto, and Shigeo Sato

Research Institute of Electrical Communication, Tohoku University, Sendai, Japan
satoshi.moriya.e6@tohoku.ac.jp

Abstract. With the rapid proliferation of IoT devices and the increasing demand for high-precision, large-volume data processing, the growth in communication traffic is highlighting the limitations of conventional cloud-centric information processing. Edge computing has hence attracted significant attention as a paradigm that enables data processing close to data sources, thereby reducing communication latency and bandwidth requirements. In this study, we designed a hardware-efficient digital inference circuit based on the reservoir computing framework to perform on-chip inference using spike outputs from an ultra-low-power analog spiking neural network (SNN) circuit. The proposed inference circuit employs a bit-shift-based filtering scheme to generate reservoir states without multipliers, and it supports adjustable decay intervals to adapt the effective time constant to task-specific temporal dynamics. The functionality of the inference logic was verified through RTL simulation. Furthermore, the proposed circuit was applied to a speech classification task, and the effects of the internal-state bit width and decay interval were evaluated. The simulation results demonstrate that the proposed circuit achieves a maximum classification accuracy of 96%, comparable to that obtained with the first-order exponential filter, while reducing power consumption by approximately 30%.

Keywords: Spiking neuron · spiking neural network · reservoir computing · time-series classification · edge computing

1 Introduction

Spiking neural networks (SNNs), which emulate the dynamics of spiking neurons and synapses to process information as time-series spike trains, are well suited for information processing at the edge, i.e., in close physical proximity to data sources owing to extremely low power consumption and high computational efficiency [1–5]. Edge computing is a distributed processing paradigm that addresses key limitations of conventional cloud computing, including latency and bandwidth congestion caused by centralized data aggregation in the cloud, as well

L. Pasa et al. (Eds.): ICANN 2026, LNCS 17089, pp. 284–294, 2027.
https://doi.org/10.1007/978-3-032-38398-3_24

as the difficulty of guaranteeing real-time processing [6,7]. However, hardware deployment at the edge inevitably entails stringent resource constraints, such as limited computational capacity and power budgets. In this context, spike-based communication and information representation, together with event-driven operation provide a key technological foundation for realizing energy-efficient information processing.

SNNs have been implemented in both digital and analog circuits [8–23]. Representative digital implementations include TrueNorth [9] and Loihi [10], which realize large-scale networks while maintaining high scalability of the network architecture. In contrast, analog implementations typically employ smaller networks but achieve superior energy efficiency, because computation can be performed by exploiting device-level physical properties, such as transistor operation in the subthreshold region.

Reservoir computing has attracted attention as a model well suited for processing spike signals in SNNs [24,25]. In reservoir computing, an input signal is projected to a high-dimensional feature space by a reservoir layer, making it linearly separable by a readout layer. Training is performed by updating only the weights between the reservoir and the readout layer. Consequently, computationally expensive algorithms such as backpropagation through time (BPTT) [26] are not required, enabling low-power and fast training and inference on edge devices. Moreover, because the computational cost of training is low, reservoir computing facilitates retraining in analog SNN circuits, where device and circuit variations can be substantial.

In SNN-based reservoir computing, spike signals generated in the reservoir layer are first converted into continuous reservoir state values by a filter, and the inference results at each time step are obtained by calculating a weighted sum of the reservoir state values and pre-trained readout weights. The filter is typically implemented as a first-order exponential function [24,27], but the hardware implementation costs are high due to the requirement for multipliers. To enable efficient inference, hardware-friendly filters are required; however, research on hardware-oriented filter design remains limited, with only a few prior works such as [28] et al.

In this study, we designed a dedicated digital inference circuit that performs on-chip inference based on spike trains output from an analog SNN circuit. To effectively utilize the limited hardware resources available on edge devices, the proposed architecture reduces the use of computationally expensive multipliers and instead relies primarily on bit-shift and addition operations, thereby minimizing circuit area and complexity.

This paper is organized as follows. Section 2 describes the overall architecture of the reservoir computing system, with a particular focus on the design of the proposed digital inference circuit. Section 3 demonstrates that speech signals can be classified using spike trains generated by the analog SNN circuit through circuit-level simulations. In addition, we discuss experimental results evaluating the effects of the filter time constant and the bit width on inference performance.

Section 4 analyzes the obtained results from the viewpoint of hardware resource usage. Finally, Sect. 5 concludes the study.

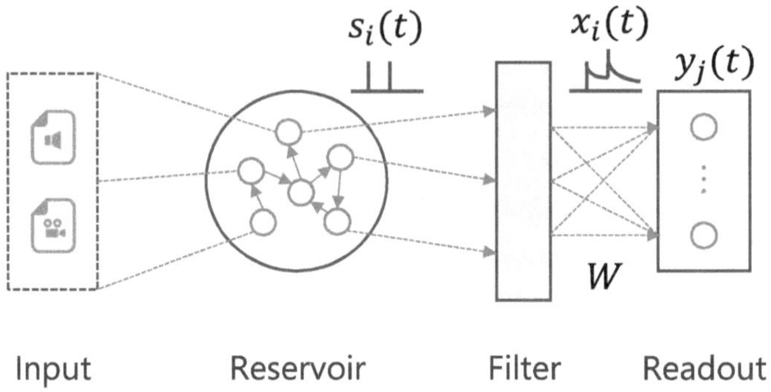

Fig. 1. Overview of the reservoir computing system.

2 Architecture of Reservoir Computing Systems

Figure 1 illustrates an overview of the reservoir computing system used in this study. The system consists of four layers: an input layer, a reservoir layer, a filter layer, and a readout layer. In the input layer, input signals are transmitted to the reservoir layer through the input connections. The reservoir layer maps the input into a high-dimensional feature space. We employ an analog SNN circuit comprising $N = 256$ analog spiking neuron circuits as the reservoir to minimize power consumption for edge-device applications. The filter layer converts binary spike signals into continuous-valued signals over a temporal window. In the readout layer, multiply–accumulate (MAC) operations are performed between the filtered continuous-valued signals and the synaptic weights to produce the inference results.

The asynchronous spike trains generated by the analog SNN circuit are directly captured by a two-stage flip-flop-based circuit, in which the incoming spike and the flip-flop output are combined by an OR gate and fed back to the flip-flop input. The captured spike-event signal is then supplied to the subsequent digital inference circuit. This ADC-less interface reduces power consumption and circuit area because only spike-event information is required. The two-stage flip-flop configuration mitigates metastability caused by the asynchronous analog spikes.

The digital inference circuit, corresponding to the filter and readout layers, processes the spike-event signals to perform the classification task. Figure 2 shows the overall architecture of the inference circuit. The circuit comprises two functional blocks: a spike filter and a MAC operation unit. At each time step, the circuit receives a binary vector $S(t) = [s_1(t), s_2(t), \ldots, s_N(t)]$ indicating the presence or absence of spikes from the analog SNN circuit. To incorporate not only the instantaneous spike events but also past spike history (e.g., firing frequency and timing) into inference, the circuit generates reservoir states. To optimize the filter time constant according to the temporal variation of the input signal, the circuit allows adjustment of the update frequency of the right-shift operation that decays the reservoir state. The reservoir state $x_i(t)$ is updated as follows:

$$x_i(t+1) = s_i(t) + \beta x_i(t), \beta = \begin{cases} 0.5 & t \equiv 0 \pmod{k}, \\ 1 & \text{otherwise}, \end{cases} \tag{1}$$

where i is the neuron index and t denotes the current time step. Each time step corresponds to a real-time duration of 2 µs, determined based on the maximum firing frequency of each neuron in the analog SNN circuit. At the shift timing (i.e., when t is a multiple of k), the state from the previous time step is attenuated by a factor of 1/2, whereas the previous state is not attenuated at other time steps. Since both the shift timing detection and the 1/2 scaling are implemented using a k-bit counter and a right bit-shift operation, the proposed design realizes the filtering function in a hardware-efficient manner without requiring a multiplier. As a result, the spike trains can be converted into a reservoir state vector $X(t) = [x_1(t), x_2(t), \ldots, x_N(t)]^\top$ with an appropriate time constant for the target task.

The on-chip memory (SRAM) stores a pre-trained synaptic weight matrix $W \in \mathbb{R}^{M \times N}$ for $M = 10$ classes, where w_{ji} denotes the synaptic weight from the i-th reservoir state to the j-th output neuron. At each time step, the class-wise inference vector $Y(t) = [y_1(t), y_2(t), \ldots, y_{10}(t)]$ is given by $Y(t) = WX(t)$. This operation yields an instantaneous inference result at time step t. The final inference result is obtained by accumulating $Y(t)$ over all time steps to form Y^{sum}, and the predicted label is determined as the class with the maximum value in Y^{sum}.

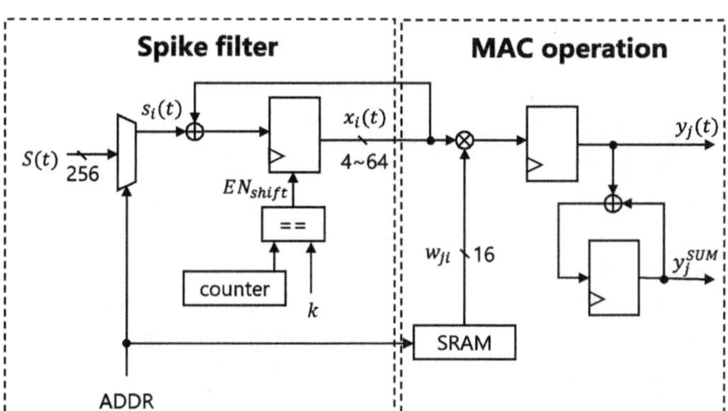

Fig. 2. Diagram of digital inference circuit.

3 Simulation Results

3.1 Design of the Inference Circuit

To investigate the hardware efficiency of the proposed bit-shift-based inference circuit, the inference performance was evaluated by increasing the bit width of the reservoir states and varying k. We examined whether the proposed method can achieve inference performance comparable to that of the exponential filter approach while reducing hardware resource usage.

Table 1 summarizes the bit configurations of $s_i(t)$, $x_i(t)$, and w_{ji}, together with their representable ranges. The spike signal $s_i(t)$ is a 1-bit binary vector, which can be directly processed from the analog SNN circuit outputs without additional encoding. The reservoir state variable $x_i(t)$ is represented as an unsigned fixed-point number with a 4-bit integer part and a fractional part of 0–60 bits. The weight w_{ji} is represented as a signed fixed-point number consisting of a 1-bit sign, a 1-bit integer part, and a 14-bit fractional part.

Table 1. Representation of Variables

Variable	Sign bit	Integer bits	Fractional bits	Value
$s_i(t)$	0	0	1	0, 1
$x_i(t)$	0	4	$0 \sim 60$	$0 \sim 8$
w_{ji}	1	1	14	$-1 \sim +1$

Figure 3 shows the temporal evolution of $x_i(t)$. The dots in the figure correspond to spike inputs, i.e., $s_i(t) = 1$. Figure 3 (a) and (b) show the results for bit widths of 4 and 64, respectively, with $k = 1$. In the 4-bit case, the value rapidly decays to zero due to the absence of fractional bits, whereas in the 64-bit case, the result closely matches that of a first-order exponential filter with a decay factor of 0.5. Figure 3 (c) and (d) show the results for a bit width of 8 with shift periods of $k = 4$ and $k = 8$, respectively. A larger k allows spike information to be retained for a longer duration. However, the smooth decay characteristic of the original exponential filter is degraded, and this effect becomes more pronounced when spikes are densely clustered.

To confirm the logical correctness of the circuit, we developed a numerical simulation environment on a PC. This software model was designed to precisely replicate the hardware specifications, employing fixed-point arithmetic with 8-bit reservoir states and 16-bit weights. We provided identical 256×1024 spike-train outputs from the analog SNN circuit (Fig. 4) to both the RTL design and the software model, and then conducted a comparative analysis of the intermediate inference processes and the final classification results.

Figure 5(a) shows the temporal evolution of the inference output $y_j(t)$ for the correct class "Zero", and Fig. 5(b) presents the accumulated outputs Y^{sum} for each class after all time steps. As shown in these results, the hardware

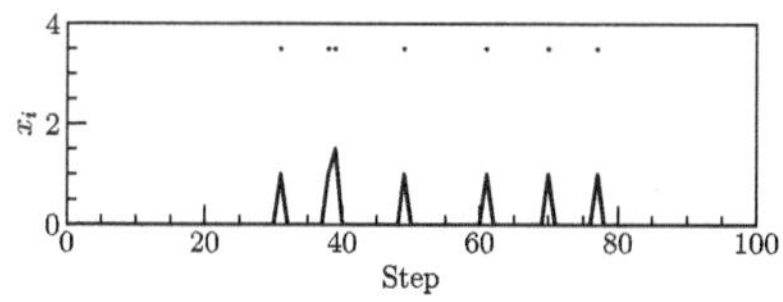

(a) Bit width of $x_i(t)$: 4, $k = 1$.

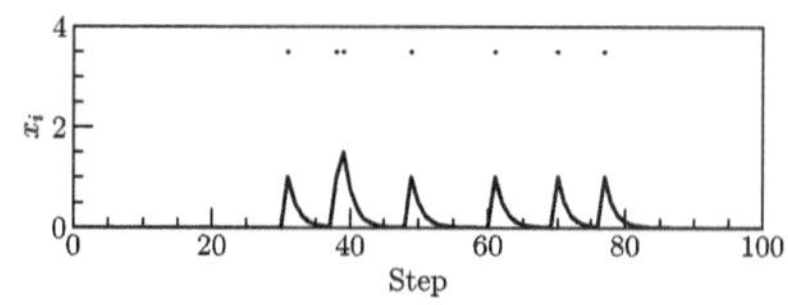

(b) Bit width of $x_i(t)$: 64, $k = 1$.

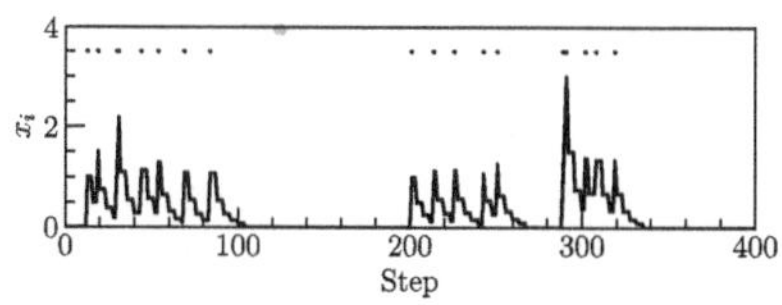

(c) Bit width of $x_i(t)$: 8, $k = 4$.

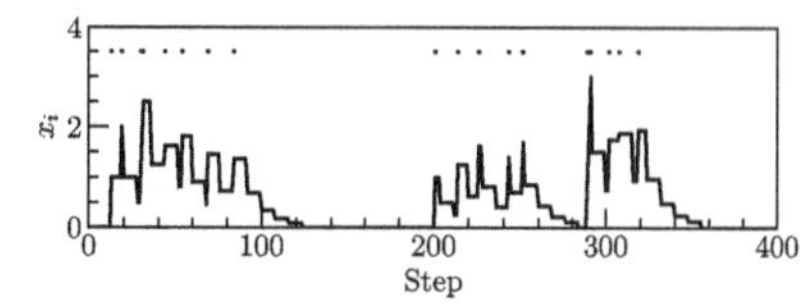

(d) Bit width of $x_i(t)$: 8, $k = 8$.

Fig. 3. Temporal evolution of $x_i(t)$ for different bit widths and shift periods.

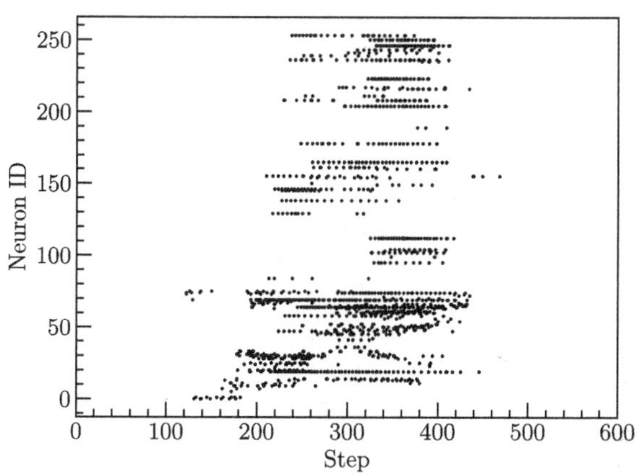

Fig. 4. Raster plot of the analog SNN circuit for digit "Zero" input.

output values obtained from the RTL simulation are in exact agreement with those obtained from the numerical simulation model. This demonstrates that the internal-state update using bit-shift operations and MAC operations are correctly implemented according to the design specifications.

3.2 Information Processing Performance of the Inference Circuit

To investigate the efficacy of the spike filtering method using bit-shift operations, inference performance was evaluated under different internal-state bit widths and shift periods. It was examined whether the proposed method could achieve or even exceed the inference performance of the conventional approach. The evaluation was conducted by numerical simulation on a PC while varying the

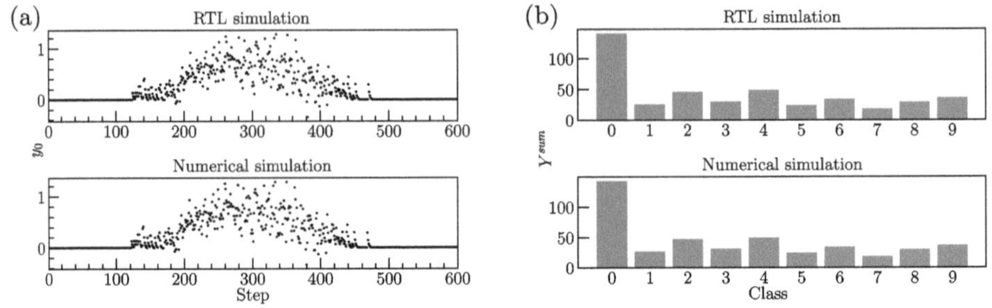

Fig. 5. (a) The temporal evolution of the inference output $y_0(t)$ for the correct class, (b) The accumulated outputs y_j^{sum} for each class after all time steps, where the correct class is "Zero".

bit width of the reservoir states from 4 to 64 bits and the shift period k from 1 to 8. The spoken-digit recognition task was evaluated using the TI-46 dataset, a speech corpus consisting of ten digit classes (0–9). The speech signals were converted into discrete binary signals and then fed into the analog SNN circuit. A total of 100 training samples and 100 test samples were used to evaluate the 10-class classification performance for a single female speaker based on the output spikes. The reservoir states were represented as unsigned fixed-point numbers, consisting of a 4-bit integer part and a fractional part comprising the remaining bits. For comparison, the classification accuracy was also evaluated using an 8-bit exponential filter with the same fixed-point representation, i.e., $x_i(t+1) = s_i(t) + \alpha x_i(t)$, where the decay factor $\alpha = 0.9$ was optimized for the target task.

Figure 6(a) shows the relationship between bit width and classification accuracy. The results indicate that the accuracy saturates at 8 bits and does not reach the accuracy achieved by the exponential filter. This suggests that the impact of high-precision computation of decay values on the overall inference accuracy is limited.

Figure 6(b) illustrates the relationship between the shift period k and classification accuracy when the bit width of $x_i(t)$ is set to 8 bits. By appropriately setting the shift period k and effectively adjusting the information retention time according to the task, reservoir states suitable for the temporal features of the input data can be generated. This approach achieves classification accuracy comparable to that of a conventional exponential filter. The reduced accuracy of the exponential filter is attributed to the restriction of the bit width of $x_i(t)$ to 8 bits; in contrast, a 64-bit implementation achieves an accuracy of 96%. These findings demonstrate that the proposed shift-interval control method is effective for achieving flexible and high-accuracy inference for specific tasks while reducing hardware costs.

3.3 Hardware Resources and Energy Consumption of the Circuits

The hardware area and power consumption of an exponential filter implemented with multipliers and those of the proposed shift-based filter were estimated.

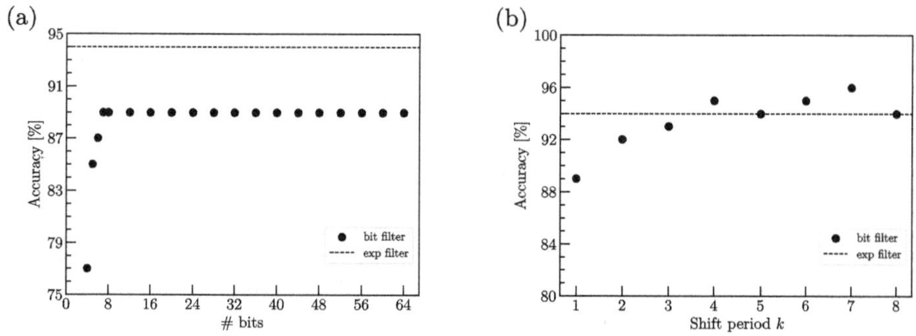

Fig. 6. Classification accuracy versus (a) bit width at a fixed shift period of $k = 1$, and (b) shift period k at a fixed bit width of 8 bits. The baseline corresponds to an 8-bit exponential filter ($\alpha = 0.9$).

Both circuits were implemented using the TSMC 65 nm CMOS process. The circuit areas were obtained after logic synthesis using Synopsys Design Compiler, while power consumption was estimated using Synopsys PrimeTime PX based on simulations performed with Synopsys VCS on the post-synthesis netlist.

Table 2 summarizes the circuit area, power consumption, and classification accuracy for each filter. In addition to the proposed circuit (Proposed), comparisons were made with a first-order exponential filter (Exponential, $\alpha = 0.9$) and a filter with a fixed shift interval of 1 (Shift). Fixed-point arithmetic with 8-bit reservoir states and 16-bit weights was employed in all conditions. Compared with the exponential filter, the proposed circuit achieves less than one-fifth of the circuit area. This reduction is attributed to the elimination of area-intensive multipliers, allowing the circuit to be composed solely of shift operations and counters. Furthermore, since the counter for the shift interval is shared among the reservoir state registers, the area overhead is less than 1% compared with the Shift condition. The power consumption of the proposed circuit is reduced by approximately 30% compared to the exponential filter.

Although the proposed and bit-shift designs reduce the multiplier-related combinational logic, the total number of registers remains almost unchanged among the three implementations. As a result, the contribution of sequential elements and their clocking remains significant, which explains why the power reduction is more moderate than the area reduction.

Table 2. Area and power consumption

Filter	Area [mm^2]	Power [mW]	Accuracy [%]	J/Infer.
Proposed	3.10×10^{-2}	2.10	96	2.05×10^{-6}
Exponential	1.71×10^{-1}	3.03	94	2.96×10^{-6}
Shift	3.08×10^{-2}	2.10	89	2.05×10^{-6}

4 Discussion

Experimental results show that the proposed inference circuit employing the bit-shift-based filter achieves inference accuracy comparable to exponential filter. This outcome is likely attributable to the nature of the information processed by SNNs. In SNNs, spike events and their firing rates are more dominant information carriers than precise analog values. Consequently, even with the stepwise attenuation inherent in bit-shift operations, the proposed system retains sufficient temporal dynamics for feature extraction. These results suggest that the proposed hardware approximation, which eliminates computationally expensive multipliers, is effective for resource-constrained implementations.

A key feature of this design is the flexibility to modify the execution frequency of the right-shift operation responsible for decay during the state update of $X(t)$. Adjusting k corresponds to tuning the effective filter time constant. Since time-series inputs such as speech have inherent temporal scales, selecting k to match the decay rate of the reservoir state to the input dynamics enables accurate feature extraction.

The proposed circuit is a hardware-efficient approximation of temporal smoothing and state retention, rather than a task-specific feature extraction method. Therefore, similar performance trends are expected for temporal classification tasks whose input dynamics have comparable time scales to those evaluated in this work. However, the optimal bit width and shift period may vary depending on the temporal statistics of the task, including the required memory length and the complexity of the class boundaries. Evaluation with larger speaker sets and different tasks is therefore left for future work.

To further reduce hardware resource usage, an early-termination scheme can be considered [29], where inference is stopped once Y^{sum} for a certain class exceeds the others by more than a predefined margin. This approach is expected to be particularly effective for inference tasks such as N-MNIST, where spatio-temporal patterns exhibit redundancy.

5 Conclusion

In this study, we designed a dedicated digital inference circuit for edge environments, enabling on-chip inference while preserving the low-power advantages of analog SNN circuits. The proposed digital inference circuit employs a hardware-efficient architecture based on bit-shift and addition operations for temporal processing of input spikes, thereby eliminating the need for multipliers, which typically incur significant area and power overhead. In particular, by introducing a mechanism to variably control the execution interval of the right-shift operation for state decay, the proposed design achieves both hardware simplification and adaptability to task-specific time scales.

Simulation results demonstrated that the circuit satisfies timing constraints under physical delay conditions and produces outputs consistent with a software-based simulation model. Furthermore, evaluations on speech classification tasks confirmed that optimizing the shift period to adjust the effective time constant achieves inference accuracy comparable to conventional exponential-filter-based methods, while reducing the circuit area to less than one-fifth and reducing power consumption by approximately 30% compared to the exponential filter. Device-level evaluation of an SNN chip integrated with the proposed inference circuit and extending the evaluation to a broader range of tasks is future work.

Acknowledgments. This study was supported by the Cooperative Research Project Program of the Research Institute of Electrical Communication of Tohoku University, MEXT Initiative to Establish Next-generation Novel Integrated Circuits Centers (X-NICS) (Grant No. JPJ011438), and JSPS KAKENHI (23K28179, 24H02332, 25H00447, 25K24740, 26K02992 and 26H02512), Japan. This work was also supported through the activities of VDEC, d.lab, The University of Tokyo, in collaboration with NIHON SYNOPSYS G.K.

References

1. Maass, W.: Networks of spiking neurons: the third generation of neural network models. Neural Netw. **10**(9), 1659–1671 (1997)
2. Verstraeten, D., Schrauwen, B., Stroobandt, D., Van Campenhout, J.: Isolated word recognition with the liquid state machine: a case study. Inf. Process. Lett. **95**(6), 521–528 (2005)
3. Paugam-Moisy, H., Bohte, S. M.: Computing with spiking neuron networks. In: Rozenberg, G., Bäck, T., Kok, J.N. (eds) Handbook of Natural Computing, pp. 335–376, Springer, Berlin, Heidelberg (2012). https://doi.org/10.1007/978-3-540-92910-9_10
4. Furber, S.: Large-scale neuromorphic computing systems. J. Neural Eng. **13**(5), 051001 (2016)
5. Roy, K., Jaiswal, A., Panda, P.: Towards spike-based machine intelligence with neuromorphic computing. Nature **575**(7784), 607–617 (2019)
6. Shi, W., Cao, J., Zhang, Q., Li, Y., Xu, L.: Edge computing: vision and challenges. IEEE Internet Things J. **3**(5), 637–646 (2016)
7. Cao, K., Liu, Y., Meng, G., Sun, Q.: An overview on edge computing research. IEEE Access **8**, 85714–85728 (2020)
8. Furber, S.B., Galluppi, F., Temple, S., Plana, L.A.: The SpiNNaker project. Proc. IEEE **102**(5), 652–665 (2014)
9. Akopyan, F., et al.: TrueNorth: design and tool flow of a 65 mW1 million neuron programmable neurosynaptic chip. IEEE Trans. Comput.-Aided Design Integr. Circ. Syst. **34**(10), 1537–1557 (2015)
10. Davies, M., et al.: Loihi: a neuromorphic manycore processor with on-chip learning. IEEE Micro **38**(1), 82–99 (2018)
11. Frenkel, C., Lefebvre, M., Legat, J.-D., Bol, D.: A 0.086-mm^2 12.7-pJ/SOP 64k-synapse 256-neuron online-learning digital spiking neuromorphic processor in 28-nm CMOS. IEEE Trans. Biomed. Circ. Syst. **13**(1), 145–158 (2018)
12. Frenkel, C., Legat, J.-D., Bol, D.: MorphIC: a 65-nm 738k-Synapse/mm^2 quad-core binary-weight digital neuromorphic processor with stochastic spike-driven online learning. IEEE Trans. Biomed. Circuits Syst. **13**(5), 999–1010 (2019)

13. Stuijt, J., Sifalakis, M., Yousefzadeh, A., Corradi, F.: μBrain: an event-driven and fully synthesizable architecture for spiking neural networks. Front. Neurosci., 538 (2021)
14. Qiao, N., et al.: A reconfigurable on-line learning spiking neuromorphic processor comprising 256 neurons and 128K synapses. Front. Neurosci. **9**, 141 (2015)
15. Moradi, S., Qiao, N., Stefanini, F., Indiveri, G.: A scalable multicore architecture with heterogeneous memory structures for dynamic neuromorphic asynchronous processors (DYNAPs). IEEE Trans. Biomed. Circuits Syst. **12**(1), 106–122 (2017)
16. Neckar, A., et al.: Braindrop: a mixed-signal neuromorphic architecture with a dynamical systems-based programming model. Proc. IEEE **107**(1), 144–164 (2018)
17. Wijekoon, J.H., Dudek, P.: Compact silicon neuron circuit with spiking and bursting behaviour. Neural Netw. **21**(2–3), 524–534 (2008)
18. Rangan, V., Ghosh, A., Aparin, V., Cauwenberghs, G.: A subthreshold aVLSI implementation of the Izhikevich simple neuron model. In: Proceeding Annual International Conference of the IEEE Engineering in Medicine and Biology, pp. 4164–4167 (2010)
19. Mizoguchi, N., Nagamatsu, Y., Aihara, K., Kohno, T.: A two-variable silicon neuron circuit based on the Izhikevich model. Artif. Life Rob. **16**(3), 383–388 (2011)
20. Sato, S., et al.: A subthreshold spiking neuron circuit based on the Izhikevich model. In: Proceedings International Conference on Artificial Neural Networks (ICANN), pp. 177–181 (2021)
21. Moriya, S., Yamamoto, H., Sato, S., Yuminaka, Y., Horio, Y., Madrenas, J.: A fully analog CMOS implementation of a two-variable spiking neuron in the subthreshold region and its network operation. In: Proceeding International Joint Conference Neural Networks (IJCNN), pp. 1–7 (2022)
22. Moriya, S., et al.: Analog VLSI implementation of subthreshold spiking neural networks and its application to reservoir computing. IEEE Trans. Circuits Syst. I: Reg. Papers **72**(10), 5571–5582 (2025)
23. Sato, S., Moriya, S., Ishikawa, M., Yamamoto, H.: Analog CMOS spiking neural network for time-series signal recognition. In: Proceeding IEEE International Symposium Multiple-Valued Logic (ISMVL), p. 69 (2025)
24. Maass, W., Natschläger, T., Markram, H.: Real-time computing without stable states: a new framework for neural computation based on perturbations. Neural Comput. **14**(11), 2531–2560 (2002)
25. Jaeger, H.: The "echo state" approach to analysing and training recurrent neural networks-with an erratum note. Ger. National Res. Cent. Inf. Technol. GMD Tech. Rep. **148** (2001)
26. Williams, R.J., Peng, J.: An efficient gradient-based algorithm for on-line training of recurrent network trajectories. Neural Comput. **2**(4), 490–501 (1990)
27. Legenstein, R., Maass, W.: What makes a dynamical system computationally powerful?. In: New Directions in Statistical Signal Processing: From Systems to Brain, pp. 127–154 (2007)
28. Saraswat, V., Gorad, A., Naik, A., Patil, A., Ganguly, U.: Hardware-friendly synaptic orders and timescales in liquid state machines for speech classification. In: Proceeding International Joint Conference on Neural Networks (IJCNN), pp. 1–8. IEEE (2021)
29. Li, Y., Geller, T., Kim, Y., Panda, P.: SEENN: towards temporal spiking early exit neural networks. Adv. Neural. Inf. Process. Syst. **36**, 63327–63342 (2023)

OverNeuron: a Spiking Recurrent Cell as Neuromorphic LSTM by Higher-Order Connectivity

Aurora Gruber[1(✉)], Walter Gallego Gomez[1], Yulia Sandamirskaya[2], Enrico Macii[1], Gianvito Urgese[1], and Vittorio Fra[1,2]

[1] Politecnico di Torino, Turin, Italy
{aurora.gruber,walter.gallego,enrico.macii, gianvito.urgese,vittorio.fra}@polito.it
[2] ZHAW, Wädenswil, Switzerland
yulia.sandamirskaya@zhaw.ch

Abstract. With the convergence of neuroscience and Artificial Intelligence (AI) into the highly interdisciplinary field of Neuro-AI, a significant influence of Deep Learning (DL) can be observed in the domain of neuromorphic computing. As a result, established DL architectures have been adapted to more bio-inspired counterparts in the attempt to obtain the advantages offered by event-based computation. In this work, we adopt a deeper approach than just converting activation functions to binarized alternatives, and we look at neuroscience to draw inspiration for the neuromorphic redesign of the Long Short-Term Memory (LSTM) cell. Specifically, we employ the idea of higher-order connectivity in the brain to build a spiking recurrent cell that implements the same structure as the LSTM but only relies on populations of Leaky Integrate-and-Fire (LIF) neurons. For benchmarking purposes, we use the Spiking Heidelberg Digits (SHD) and the event-based Braille letter datasets, and we report on effective execution of our architecture, named *OverNeuron*, on Intel Loihi 2. With a best on-hardware test accuracy of 88.75% for SHD and 85.93% for Braille, we show that the OverNeuron can outperform the LSTM cell on a temporal and a spatio-temporal event-based task.

Keywords: Neuromorphic computing · Spiking neural networks · LSTM · Audio classification · Spatio-temporal tactile patterns · Neuromorphic HW

1 Introduction

In the understanding of how to employ the mechanisms underlying natural intelligence to equip artificial systems, the convergence of different disciplines in recent development of Artificial Intelligence (AI) is shaping the new and inherently multidisciplinary domain of Neuro-AI [5,14,31]. In this context, neuromorphic computing is often closely associated with Deep Learning (DL) as a

L. Pasa et al. (Eds.): ICANN 2026, LNCS 17089, pp. 295–306, 2027.
https://doi.org/10.1007/978-3-032-38398-3_25

suitable pathway for improved efficiency [18]. Although such an association only partially corresponds to what neuromorphic computing and engineering is [15], huge efforts have been made along the line of brain-inspired DL, either at the software level [3,28,43] or from the hardware perspective [26,36,40]. As one of the direct consequences of these research interests, some established DL architectures have been explored through the lens of Spiking Neural Networks (SNNs) to embed the brain-inspired principle of spike-based computation, as in the case, for instance, of the Long Short-Term Memory (LSTM) cell [21], the Legendre Memory Unit (LMU) [7], ResNet [4], YOLO [17] and Mamba [12]. The conversion process from Artificial Neural Networks (ANNs) to SNNs can be handled with two main strategies: on the one hand, a direct conversion of specific components is used to emulate spike emission through binary responses, neglecting the neuronal dynamics; on the other hand, neuro-inspired models rely on simplified neuronal dynamics as the source of spiking activity. In this work, we adopt the latter by drawing inspiration from the brain's higher-order connectivity [23,33,37] to redesign the LSTM cell as a spiking recurrent cell suitable for deployment on neuromorphic hardware. Higher-order connectivity denotes more complex interactions compared to pairwise neural correlations, involving groups of neurons interacting with each other [34,42], and, similar to [19], forms the basis for our neuromorphic redesign based on neural populations. We name the proposed architecture OverNeuron, as its operating principle is rooted in ensembles of Leaky Integrate-and-Fire (LIF) neurons that interplay to build a cell having an internal state with the characteristics of a membrane potential, and a spiking output [8].

The two main contributions we propose in this paper are: (i) the OverNeuron, a new LSTM-inspired recurrent spiking architecture based on populations of LIF neurons; (ii) the deployment of the OverNeuron on Loihi 2 [16] with hardware profiling for effective execution.

2 Methods

At their very inner core, SNNs are characterized by spiking computational units that inherently exploit temporal dynamics [22]. On the other hand, in the ANN domain, the capability to effectively work with time-dependent signals is obtained through specific recurrent architectures offering memory states, a prominent example being the LSTM cell [11].

The OverNeuron aims to enhance the LSTM architecture by redesigning it in the neuromorphic domain, to leverage the native suitability of SNNs to work with time as a computational resource. However, unlike previous works focusing on ANN-to-SNN conversion mainly through changes in the activation functions [2, 13,30], our approach is characterized by the use of LIF neuron populations as fundamental building blocks.

2.1 OverNeuron Design

The LSTM-based structure of the OverNeuron architecture is illustrated in Fig. 1, where the different populations of LIF neurons replacing the original gates [11] are highlighted. Unlike previous spiking variants of the LSTM [20,21, 27,44], which directly replace the gates with spiking activations, the architecture we propose exploits richer dynamics by relying on the membrane potential of neuron populations too. With this conceptualization, all populations process the same input and contribute to an internal state of the OverNeuron which rules the output spiking activity of the whole cell in analogy to what the membrane potential does in the operation of a single neuron. The role of the population-based gates can be summarized as follows:

- **Forget gate**: at each time t, it weighs the internal state of the cell from the previous time step;
- **Input gate**: it weighs the actual contribution to the internal state by the candidate gate;
- **Candidate gate**: it produces an update of the internal state;
- **Output gate**: it produces the output spiking activity h_t of the cell depending on the gated internal state, on the input x_t, and on the spikes emitted at the previous time step h_{t-1}.

To fairly describe the underlying mechanisms of the LSTM cell [11,32] by relying on neuron populations only, specific attention is given to the role of the hyperbolic tangent. The *tanh* function is crucial, as it allows both positive and negative values, while LIF neurons are typically not intended to provide a negative output. Therefore, to implement a similar behavior, we relied on an interplay between excitatory and inhibitory neural activities, using two distinct populations of neurons and computing the difference between their output.

From [32], the equations defining the various gates are reformulated in terms of neural interactions:

$$\mathbf{f}_t = (\mathbf{W}_{hf}\mathbf{h}_{t-1,spk} + \mathbf{W}_{if}\mathbf{x}_{t,spk})_{mem} \tag{1}$$

$$\mathbf{i}_t = (\mathbf{W}_{hi}\mathbf{h}_{t-1,spk} + \mathbf{W}_{ii}\mathbf{x}_{t,spk})_{mem} \tag{2}$$

$$\begin{aligned} \tilde{\boldsymbol{c}}_t &= (\tilde{\boldsymbol{c}}^+{}_t)_{mem} - (\tilde{\boldsymbol{c}}^-{}_t)_{mem} = \\ &= (\mathbf{W}_{h\tilde{c}^+}\mathbf{h}_{t-1,spk} + \mathbf{W}_{i\tilde{c}^+}\mathbf{x}_{t,spk})_{mem} + \qquad (3)\\ &\quad - (\mathbf{W}_{h\tilde{c}^-}\mathbf{h}_{t-1,spk} + \mathbf{W}_{i\tilde{c}^-}\mathbf{x}_{t,spk})_{mem} \qquad (4) \end{aligned}$$

$$\mathbf{mem}_t = \mathbf{f}_t \odot \mathbf{c}_{t-1} + \mathbf{i}_t \odot \tilde{\boldsymbol{c}}_t \tag{5}$$

$$\mathbf{c}_t = (\mathbf{o}_t)_{mem} = (\mathbf{W}_{ho}\mathbf{h}_{t-1,spk} + \mathbf{W}_{io}\mathbf{x}_{t,spk} \mid \mathbf{mem}_t)_{mem} \tag{6}$$

$$\mathbf{h}_{t,spk} = (\mathbf{o}_t)_{spk} = (\mathbf{W}_{ho}\mathbf{h}_{t-1,spk} + \mathbf{W}_{io}\mathbf{x}_{t,spk} \mid \mathbf{mem}_t)_{spk} \tag{7}$$

where $\mathbf{f}_t$, $\mathbf{i}_t$, $\tilde{\boldsymbol{c}}^+{}_t$, $\tilde{\boldsymbol{c}}^-{}_t$ and $\mathbf{h}_t$ are populations of LIF neurons, $\mathbf{W}$ represents the synaptic weight matrices for the different interactions among such populations, and the subscripts *mem* and *spk* specify whether the membrane potential or the spiking activity is used as means of interaction with other populations.

The internal dynamics produced through the specific functional interactions among the different neuron populations, described by Eq. (1)–(6), encompass the higher-order connectivity that ultimately rules the emission of output spikes by the OverNeuron according to Eq. (7). The latter collects and summarizes all the distinguishing elements of the OverNeuron: it is a spiking cell because its output $h_{t,spk}$ is in the form of spike trains; it is a recurrent cell because the past activity $h_{t-1,spk}$ is fed back at the cell's input; it is an expansion of the single-neuron mechanism because the binarized output in the form of spiking activity depends on weighted spikes received as input.

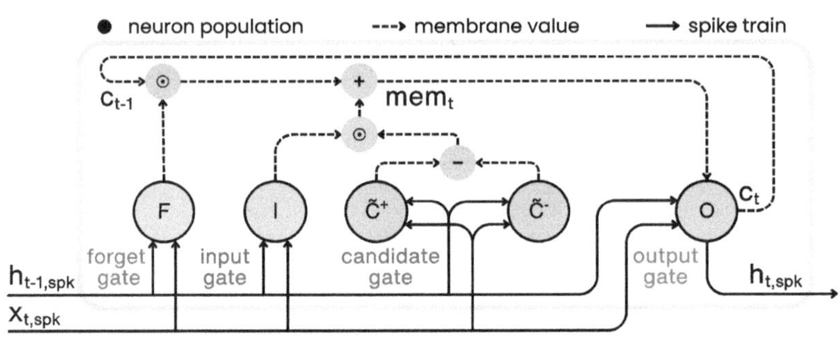

Fig. 1. Representation of the OverNeuron mechanism. Starting from the LSTM architecture, all gates and activation functions are replaced by populations of LIF neurons. To preserve the zero-crossing codomain of the hyperbolic tangent, an inhibition mechanism between two populations is used for the candidate gate, resulting in a subtraction.

2.2 Leaky Integrate-and-Fire (LIF) Neuron Model

In snnTorch [3], the simulation framework we adopted, the LIF neuron is implemented by the `Leaky`[1] class, with the temporal dynamics described by Eq. (8):

$$U_t = \beta U_{t-1} + W X_t - R_\theta|_{S_{t-1}} \tag{8}$$

with X_t being the neuron's input at time t, W the synaptic weight matrix, U_t the membrane potential at time t, β its decay factor, and $R_\theta|_{S_{t-1}}$ the reset mechanism occurring after spike emission, where S_{t-1} is the spiking activity of the neuron at the previous time step, and θ the threshold voltage [3]. The process of spike emission is then modeled with a Heaviside function by Eq. (9):

$$S_t = H(U_t - \theta) \quad \text{with} \quad H(0) = 0 \tag{9}$$

[1] https://snntorch.readthedocs.io/en/latest/snn.neurons_leaky.html.

2.3 snnTorch Implementation

We implemented and trained our architecture within the snnTorch framework, relying on the `Leaky` model for the LIF neurons and on the `Linear` layer for synaptic weights. Additionally, to adapt OverNeuron to classification problems, we introduced an input projection layer and a readout layer. The former is intended to match the input data dimension with the hidden state, while the latter features as many neurons as the number of classes needed.

Data Preparation. Following the analysis in [39], we evaluated the OverNeuron on two classification datasets with different spatial and temporal hierarchies, to test its ability to capture patterns: the Spiking Heidelberg Digits (SHD) dataset [1] and the event-based Braille letter reading dataset [25]. For the SHD dataset, to preserve the intrinsic temporal structure of the data, we employed a preprocessing strategy that retains all input channels and all the original 1000 timesteps, avoiding any form of temporal compression or channel selection. To enable batch-wise processing, samples shorter than the maximum were padded to ensure consistent sequence lengths across inputs. As the SHD dataset does not include a predefined validation split, a subset of the training data, equal in size to the test set, was reserved for validation (approximately 30% of the training data). For the Braille dataset, the same preprocessing as in [19] was adopted: raw pressure signals were converted into event-based data via a delta-sigma modulator ($\theta = 1$) with a 5 ms binning step. The resulting sequences have a length of 256 time steps.

Training Setup. We trained the OverNeuron by means of Backpropagation Through Time (BPTT) with the Adam optimizer and the `ce_count_loss` loss function. The latter was chosen among snnTorch's built-in loss functions to primarily focus the training of the network on the output spiking activity. This spike-count-based cross-entropy loss indeed accumulates the spikes produced by the output neurons, so that class evidence is computed based on the spiking activity over the full sequence.

Table 1. Search space for hyperparameter optimization.

Hyperparameter	Values / Range
Population size	32, 64, 128, 256, 512
β	$[0.05, 1]$ with step 0.05
θ	$[0.05, 1]$ with step 0.05
Batch size	64, 128, 256
Learning rate	0.0001, 0.0002, 0.0005, 0.001, 0.002, 0.005, 0.01

Hyperparameter Optimization. To identify optimal hyperparameter configurations for the OverNeuron, we performed Hyperparameter Optimization (HPO) Optimization (HP for each dataset, using the Neural Network Intelligence (NNI) toolkit[2] by adapting the procedure described in [6]. The search space included both neuron- and architecture-related parameters, together with learning settings, as reported in Table 1. As objective metric, we used the classification accuracy on the validation set. Each model was trained for a maximum of 300 epochs with early stopping enabled. The models were then evaluated on the test set using the checkpoint corresponding to the best validation accuracy.

2.4 Loihi 2 Deployment

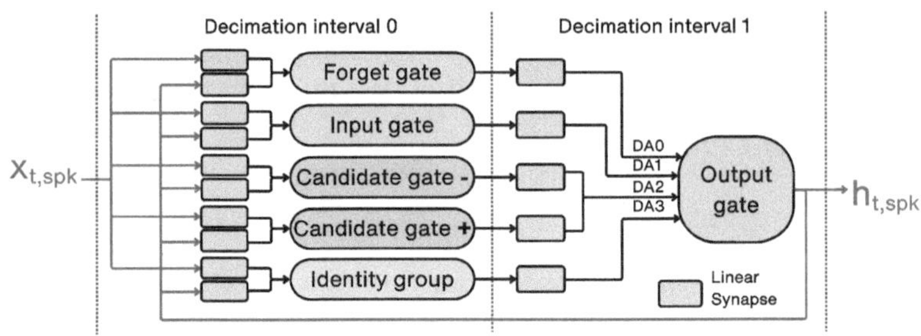

Fig. 2. NxKernel implementation of the OverNeuron. Gates are described as NxKernel groups, and interconnected using NxKernel linear synapses. The execution is alternated: The output gate is executed on odd timesteps, while the rest of the gates are executed on even ones. The identity group simply propagates its input to its output.

We chose Intel Loihi 2 as our target hardware, considering its high level of flexibility in neuron models and the possibility of detailed execution profiling [16].
NxKernel Implementation. Figure 2 depicts the OverNeuron implementation in Intel's NxKernel, a neuromorphic programming interface for Loihi 2 provided to Intel Neuromorphic Research Community members [24]. Gates are written as separate NxKernel groups, interconnected using NxKernel linear synapses. The output gate required a custom model written in Loihi 2's specific assembly language, as it receives four different inputs that are combined as described in Sect. 2.1 to produce the cell's spiking output. To guarantee equivalent dynamics between the snnTorch and NxKernel, we exploited the decimation capabilities of Loihi 2 to alternate the execution of the gates, as the default (pipelined) execution mode introduces undesired delays in the recurrent connections of the OverNeuron. Moreover, a supporting *identity group* was used to ensure all inputs to the output gate were properly synchronized.
Quantization. Neuron states were quantized during training to 24 bits using the `state_quantize` option of the `Leaky` model. Betas, thresholds, and weights were quantized after training with bit-widths set to 12, 14, and 8 for the SHD dataset, and 12, 15, and 9 for the Braille dataset, respectively. The chosen bit resolutions

[2] https://github.com/microsoft/nni.

are compatible with efficient execution on Loihi 2 while providing enough precision to avoid high accuracy drop.
Partitioning and Mapping. Efficient execution on neuromorphic hardware requires knowledge of the system's bottlenecks to inform the model deployment process. Works that have studied Loihi 2 to uncover these bottlenecks include [9, 38, 41]. We followed the partitioning and mapping heuristic approach described in [41], which uses execution profiling information for each group in the network to determine the potentially slowest ones, which are then partitioned into more cores, and evaluates different mapping strategies to mitigate Network-on-Chip (NoC) congestion.
IO Mode. Execution was performed in IO–unthrottled mode, one test sample at a time, as done in previous works [29, 35], by storing the sample in the local memory of dedicated neuron cores, before starting the network execution. This enabled a precise profiling of the network using Loihi 2's and NxKernel's rich measuring features, unaffected by slow IO operations.
Evaluation. Accuracy for each model was calculated on the test set. Additionally, each model was profiled in terms of execution time and network activity using a subset of it, composed of ten randomly chosen samples for each class.

3 Results and Discussion

The results we obtained with our OverNeuron on both the SHD and the Braille dataset show that the architecture successfully captures information with different spatio-temporal hierarchies. To demonstrate the hardware compatibility of the OverNeuron, we deployed and profiled it on a N3C1-revision single-chip Loihi 2 system (Oheo Gulch) available locally, using Intel's proprietary framework NxKernel.

3.1 Classification Task

On SHD, our model achieves a test accuracy of 91.91% in snnTorch, placing the OverNeuron in the top-10 of the official leaderboard[3]. Limited performance degradation is observed when the OverNeuron is deployed on Loihi 2, with a best accuracy of 88.75%. Both in simulation and on Loihi 2, the OverNeuron outperforms the classical LSTM cell evaluated on the same dataset [1], while introducing event-based benefits.

Before deploying the model on Loihi 2, we evaluated it across different stages, as summarized in Table 2. We first evaluated the model in simulation using snnTorch, with the results obtained during the HPO process. The same model was then evaluated on snnTorch with quantized parameters, i.e., weights, betas, and thresholds, to assess the impact of reduced numerical precision on accuracy and to approximate the expected hardware performance. When the same parameters were transferred to Loihi 2, a modest accuracy drop was observed, depending on the model. This gap can be attributed to hardware-specific constraints not fully captured in the snnTorch simulation, such as overflows or saturations.

[3] https://zenkelab.org/resources/spiking-heidelberg-datasets-shd/.

Table 2. Comparison of test accuracies across snnTorch, quantized snnTorch, and Loihi 2, ordered by decreasing snnTorch accuracy. FLOPs values are shown per time step, and consider synaptic and neuron operations. Dense FLOPS in snnTorch were estimated using NeuroBench [40]. Effective FLOPs in Loihi 2 were calculated from NxKernel profiling metrics. Highlighted rows are Pareto-optimal points from Loihi 2.

Test accuracy (%)			Params. $(\times 10^6)$	Dense FLOPs snnTorch $(\times 10^5)$	Effective FLOPs Loihi 2 $(\times 10^5)$
snnTorch	Q. snnTorch	Loihi 2			
91.91	84.32	85.13	1.35	27.02	0.93
90.76	83.52	84.16	1.35	27.02	0.87
90.28	**89.58**	**88.75**	**1.35**	**27.02**	**0.86**
88.03	**87.37**	**85.69**	**0.46**	**9.26**	**0.60**
87.32	**85.51**	**84.16**	**0.28**	**5.69**	**0.19**
87.32	86.66	86.11	1.35	27.02	1.05

While higher accuracies are generally achieved by larger models, Table 2 also highlights that competitive performances can be obtained with smaller configurations. On the other hand, as shown in Fig. 3a, models with a high total number of neurons do not always achieve better results than models with fewer neurons. In the same plot, the models are additionally characterized by their neuronal activity, which further supports this observation. In particular, no clear correlation emerges between activity levels and classification accuracy: low-activity models can still achieve strong performances, while highly active configurations may be less accurate.

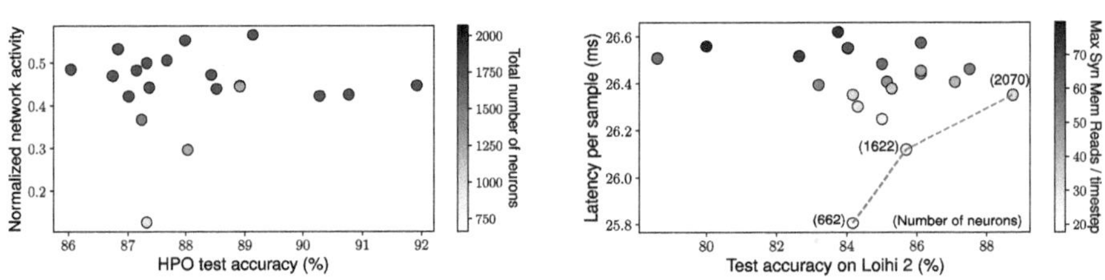

(a) Accuracy vs activity on snnTorch. Neuronal activity was estimated by aggregating synaptic operations across all gates and timesteps, taking inspiration from the metrics proposed in NeuroBench [40]. In grayscale, the total number of neurons.

(b) Accuracy vs latency on Loihi 2. In grayscale, the synaptic memory reads per timestep of the neuron core with the highest value. The dashed line connects the Pareto points, and for each one, the number of neurons is reported.

Fig. 3. Profiling results of different models on (a) snnTorch and (b) Loihi 2 [40].

On the event-based Braille dataset, the same experimental methodology as for SHD was adopted, leading to a best test accuracy of 85.93% on Loihi 2 and thus overcoming the result reported in [25] when using LSTM with event-based data (eLSTM). To the best of our knowledge, the previous state-of-the-art accuracy was reported to be 85.6% in [19].

3.2 Hardware Execution

For the SHD dataset, the best partitioning configuration we found following the introduced heuristic approach consists of 8 cores for the output gate and 4 cores for each of the other gates. This is expected as the output gate is the most complex in terms of synaptic operations and neuron updates, followed by the other gates, all with comparable complexity. Random mapping prevented the execution slowdown caused by traffic congestion in the NoC.

Figure 3 shows the results of execution on Loihi 2. Firstly, we note that latency is strongly correlated with the maximum synaptic memory reads by any core in the network, suggesting the model was in the memory-bound state reported by [41]. Secondly, larger and more active models are, in general, slower, due to the synaptic memory accesses bottleneck. However, through the HPO process, we were able to obtain moderately active models that matched or surpassed the accuracy of the most active ones, while also reducing latency. This results in three Pareto models as reported in the figure, whose accuracies are detailed in Table 2. Additionally, the computational cost of these models is also reported, through the theoretical dense FLOPs in snnTorch and the effective FLOPs executed on Loihi 2. Particularly, the latter turns out to be lower than the classical LSTM's baseline (8.53×10^5) [1].

An analogous process on the Braille dataset resulted in a partitioning configuration with 20 cores for the output gate and 10 for the other gates. The most accurate model on Loihi 2, with 1.9×10^6 parameters, 10.2×10^5 dense FLOPs, and 1.57×10^5 effective FLOPs, achieved a latency per sample of 2.58 ms.

4 Conclusion

In this work, we presented the OverNeuron, a spiking recurrent cell, built on the principles of higher-order connectivity of neurons, that brings the computational flow of the classical LSTM cell into the neuromorphic domain through LIF neuron populations. We showed that it achieves a test accuracy of 91.91% on the SHD dataset, improving on the 85.7% obtained by the classical LSTM cell. Moreover, the OverNeuron is natively compatible with customizable neuromorphic architectures such as Loihi 2 and SpiNNaker2 [10], and we demonstrated its deployment and hardware execution achieving a test accuracy of 88.75%. To further demonstrate the model's capability to capture spatio-temporal characteristics, we evaluated it on the Braille dataset, surpassing the current state-of-the-art (85.6%) with an on-hardware accuracy of 85.93%. Although we took into consideration Loihi 2's fixed-point constraints and execution bottlenecks, and followed a deployment optimization process, there is still room for improvement on the hardware execution. Tailored quantization approaches such as QuantizationAware Training (QAT), applied to all the neuron and synaptic parameters, could help to further mitigate accuracy losses. Network activity reduction techniques such as weight pruning and Sparsity-Aware Training (SAT) could allow for the exploitation of structured and unstructured sparsity, supported by neuromorphic hardware to reduce latency and energy consumption.

Our work showcases the potential benefits of deep brain-inspired approaches for ANN-to-SNN architecture conversion, their suitability for hardware deployment, and the possible improvement with respect to DL solutions.

Acknowledgments. This research is funded by the European Union - NextGenerationEU Project 3A-ITALY (PE0000004, CUP E13C22001900001). We acknowledge a contribution from the Italian National Recovery and Resilience Plan (NRRP), M4C2, funded by the European Union - NextGenerationEU (Project IR0000011, CUP B51E22000150006, "EBRAINS-Italy").

Disclosure of Interests. The authors have no competing interests to declare that are relevant to the content of this article.

References

1. Cramer, B., Stradmann, Y., Schemmel, J., Zenke, F.: The Heidelberg spiking data sets for the systematic evaluation of spiking neural networks. IEEE Trans. Neural Netw. Learn. Syst. **33**(7), 2744–2757 (2022)
2. Deng, S., Gu, S.: Optimal conversion of conventional artificial neural networks to spiking neural networks. arXiv preprint arXiv:2103.00476 (2021)
3. Eshraghian, J.K., et al.: Training spiking neural networks using lessons from deep learning. Proc. IEEE **111**(9), 1016–1054 (2023)
4. Fang, W., Yu, Z., Chen, Y., Huang, T., Masquelier, T., Tian, Y.: Deep residual learning in spiking neural networks. Adv. Neural. Inf. Process. Syst. **34**, 21056–21069 (2021)
5. Fellous, J.M., Thomas, P., Tiesinga, P., Lindner, B.: Beyond the Nobel prizes: towards new synergies between computational neuroscience and artificial intelligence. Biol. Cybern. **119**(1), 1 (2024). https://doi.org/10.1007/s00422-024-01002-0
6. Fra, V.: Application-oriented automatic hyperparameter optimization for spiking neural network prototyping. arXiv preprint arXiv:2502.12172 (2025)
7. Gaurav, R., Agarwal, S., Stewart, T.C., Yi, Y.: Benchmarking deep legendre-SNN for time series classification-analysis and enhancements. IEEE Trans. Emerg. Top. Comput. Intell. (2025)
8. Gerstner, W., Kistler, W.M.: Spiking Neuron Models: Single Neurons, Populations, Plasticity. Cambridge University Press, Cambridge, U.K. (2002)
9. Gomez, W.G., Pignata, A., Pignari, R., Fra, V., Macii, E., Urgese, G.: First steps towards micro-benchmarking the lava-Loihi neuromorphic ecosystem. In: 2023 IEEE 16th International Symposium on Embedded Multicore/Many-core Systems-on-Chip (MCSoC), pp. 462–469 (2023)
10. Gonzalez, H.A., et al.: SpiNNaker2: a large-scale neuromorphic system for event-based and asynchronous machine learning (2024)
11. Hochreiter, S., Schmidhuber, J.: Long short-term memory. Neural Comput. **9**(8), 1735–1780 (1997)
12. Huang, Y., et al.: SpikingMamba: towards energy-efficient large language models via knowledge distillation from mamba. Trans. Mach. Learn. Res. (2026)
13. Huang, Z., et al.: Differential coding for training-free ANN-to-SNN conversion. In: Forty-second International Conference on Machine Learning (2025)

14. Indiveri, G.: Biological fidelity: the engine driving the neuromorphic renaissance. Proc. Natl. Acad. Sci. **122**(50), e2525734122 (2025)
15. Indiveri, G.: Neuromorphic is dead. Long live neuromorphic. Neuron **113**(20), 3311–3314 (2025)
16. Intel: taking neuromorphic computing to the next level with Loihi 2 (2021). https://www.intel.com/content/www/us/en/research/neuromorphic-computing-loihi-2-technology-brief.html
17. Kim, S., Park, S., Na, B., Yoon, S.: Spiking-YOLO: spiking neural network for energy-efficient object detection. In: Proceedings of the AAAI Conference on Artificial Intelligence, vol. 34, pp. 11270–11277(2020)
18. Kudithipudi, D., et al.: Neuromorphic computing at scale. Nature **637**(8047), 801–812 (2025)
19. Leto, B., Urgese, G., Macii, E., Fra, V.: A LIF-based legendre memory unit as neuromorphic state space model benchmarked on a second-long Spatio-temporal task. In: 2025 Neuro Inspired Computational Elements (NICE), pp. 1–9. IEEE (2025)
20. Li, Y., Wang, X., Chen, F., Zhao, B., Fu, Q.: Online learning behavior analysis and prediction based on spiking neural networks. J. Soc. Comput. **5**(2), 180–193 (2024)
21. Lotfi Rezaabad, A., Vishwanath, S.: Long short-term memory spiking networks and their applications. In: International Conference on Neuromorphic Systems 2020, pp. 1–9 (2020)
22. Maass, W.: Networks of spiking neurons: the third generation of neural network models. Neural Netw. **10**(9), 1659–1671 (1997)
23. Majhi, S., et al.: Patterns of neuronal synchrony in higher-order networks. Phys. Life Rev. **52**, 144–170 (2025)
24. Meszaros, B., Knight, J.C., Timcheck, J., Nowotny, T.: A complete pipeline for deploying SNNS with synaptic delays on Loihi 2. In: 2025 International Conference on Neuromorphic Systems (ICONS), pp. 225–228. IEEE (2025)
25. Muller-Cleve, S.F., et al.: Braille letter reading: a benchmark for spatio-temporal pattern recognition on neuromorphic hardware. In: 2022 IEEE Biomedical Circuits and Systems Conference (BioCAS), pp. 424–428. IEEE (2022). https://doi.org/10.1109/BioCAS54905.2022.9948630
26. Ottati, F., et al.: To spike or not to spike: a digital hardware perspective on deep learning acceleration. IEEE J. Emerg. Sel. Top. Circ. Syst. **13**(4), 1015–1025 (2023)
27. Pang, J., Xia, T., Zhang, P., Shi, Z.: Application of the LSTM-SNN hybrid temporal modeling approach in complex gait activity recognition. In: 2025 5th International Conference on Neural Networks, Information and Communication Engineering (NNICE) (2025)
28. Pedersen, J.E., et al.: Neuromorphic intermediate representation: a unified instruction set for interoperable brain-inspired computing. Nat. Commun. **15**(1), 8122 (2024)
29. Pierro, A., Abreu, S., Timcheck, J., Stratmann, P., Wild, A., Shrestha, S.B.: Accelerating linear recurrent neural networks for the edge with unstructured sparsity (2025)
30. Rueckauer, B., Lungu, I.A., Hu, Y., Pfeiffer, M., Liu, S.C.: Conversion of continuous-valued deep networks to efficient event-driven networks for image classification. Front. Neurosci. **11**, 682 (2017)
31. Sadeh, S., Clopath, C.: The emergence of neuroAI: bridging neuroscience and artificial intelligence. Nat. Rev. Neurosci. **26**(10), 583–584 (2025)

32. Sak, H., Senior, A., Beaufays, F.: Long short-term memory based recurrent neural network architectures for large vocabulary speech recognition. arXiv preprint arXiv:1402.1128 (2014)
33. Santoro, A., Battiston, F., Lucas, M., Petri, G., Amico, E.: Higher-order connectomics of human brain function reveals local topological signatures of task decoding, individual identification, and behavior. Nat. Commun. **15**(1), 10244 (2024)
34. Schneidman, E., Berry, M.J., Segev, R., Bialek, W.: Weak pairwise correlations imply strongly correlated network states in a neural population. Nature **440**(7087), 1007–1012 (2006)
35. Shrestha, S.B., Timcheck, J., Frady, P., Campos-Macias, L., Davies, M.: Efficient video and audio processing with Loihi 2. In: ICASSP 2024 - 2024 IEEE International Conference on Acoustics, Speech and Signal Processing (ICASSP), pp. 13481–13485. IEEE (2024)
36. Sun, P.S.V., et al.: Exploiting deep learning accelerators for neuromorphic workloads. Neuromorphic Comput. Eng. **4**(1), 014004 (2024)
37. Swanson, L.W.: Brain Architecture: Understanding the Basic Plan. Oxford University Press, Oxford (2012)
38. Timcheck, J., Pierro, A., Shrestha, S.B.: A compute and communication runtime model for Loihi 2 (2026)
39. Walters, B., et al.: A temporally structured dataset for neuromorphic computing. Neuromorphic Comput. Eng. **5**(2), 027001 (2025). https://doi.org/10.1088/2634-4386/add36c
40. Yik, J., et al.: The NeuroBench framework for benchmarking neuromorphic computing algorithms and systems. Nat. Commun. **16**(1), 1545 (2025)
41. Yik, J., et al.: Modeling and optimizing performance bottlenecks for neuromorphic accelerators (2025)
42. Yu, S., Yang, H., Nakahara, H., Santos, G.S., Nikolić, D., Plenz, D.: Higher-order interactions characterized in cortical activity. J. Neurosci. **31**(48), 17514–17526 (2011)
43. Zhan, Q., Liu, G., Xie, X., Zhang, M., Sun, G.: Bio-inspired active learning method in spiking neural network. Knowl.-Based Syst. **261**, 110193 (2023)
44. Zhu, N.: Inductive conformal prediction enhanced LSTM-SNN network: applications to birds and UAVs recognition. IEEE Geosci. Remote Sens. Lett. **21**, 3502705 (2024)

Temporal Evolution of Synaptic Weights in Spatiotemporal Learning Spiking Neural Networks

Takemori Orima(✉) and Yoshihiko Horio

ACRO, Teikyo University, Itabashi-ku, Tokyo, Japan
orima.takemori.zu@teikyo-u.ac.jp

Abstract. Spiking neural networks (SNNs) are promising candidates for next-generation artificial intelligence that is highly energy-efficient and enables biologically plausible information processing. Although various methods have been proposed as learning rules for SNNs, most employ statistical methods that average information. Therefore, it is difficult for SNNs to learn subtle differences in contextual information. While a spatiotemporal learning rule (STLR) based on local coincidence is effective for solving the challenge, conventional STLR models have typically relied on synchronous global clocks. This discrete-time approach artificially quantizes spike timings, thereby destroying the precise temporal context and diminishing the inherent advantages of SNNs. To overcome the limitation, this study employs an asynchronous, continuous-time SNN model using an event-driven STLR. By eliminating synchronous clocks, the event-driven STLR strictly preserves the exact temporal context of input sequences. The performance of the model was evaluated using a multistage template-matching method based on F1-scores, specifically testing its discriminatory power for similar spatiotemporal patterns generated from spatial vectors with small Hamming distances. Numerical simulations demonstrated that the model achieves effective pattern separation. Furthermore, an analysis using multi-dimensional scaling revealed that distinct synaptic weight structures emerged for different spatiotemporal patterns. This approach confirms that the network can learn contextual structures and successfully memorize the sequence context.

Keywords: Neuromorphic computing · Spatiotemporal learning rule · Spiking neural network · Hippocampus

1 Introduction

Spiking neural networks (SNNs), inspired by the physiological information processing of the brain, have emerged as highly promising paradigms for next-generation artificial intelligence (AI) [1]. By utilizing the continuous-time, event-driven, and asynchronous nature of spike signals, SNNs offer ultra-low power consumption and inherently efficient temporal processing, making them highly

L. Pasa et al. (Eds.): ICANN 2026, LNCS 17089, pp. 307–317, 2027.
https://doi.org/10.1007/978-3-032-38398-3_26

suitable for edge computing devices [2]. However, realizing the full potential of brain-inspired computing presents fundamental challenges, particularly regarding how spatiotemporal context is spike-encoded and learned.

While substantial progress has been made, mainstream SNNs increasingly rely on deep learning methodologies, utilizing synchronous, discrete-time updates driven by a global clock to facilitate backpropagation through time and surrogate gradient methods [3]. Although this discrete-time approach achieves high accuracy on standard benchmarks, it fundamentally compromises biological plausibility. It enforces redundant computations even in the absence of spikes, diminishing the inherent energy efficiency of neuromorphic hardware, and artificially quantizes spike timings into fixed time bins [4]. This temporal quantization destroys the fine-grained contextual structures that are essential for cognitive tasks such as sequence identification.

Beyond the issue of temporal quantization, a fundamental challenge is learning contextual structures without heavily relying on the statistical averaging of large datasets [5]. Conventional statistical learning, while robust to noise, cannot learn the temporal context embedded in individual events. To address these challenges, it is essential to use biologically plausible local learning rules that can be learned instantaneously. Therefore, a spatiotemporal learning rule (STLR) [6,7] is a good candidate for preserving precise temporal context without relying on massive data averaging and global error signals.

An asynchronous, continuous-time, event-driven neural network using STLR synapses has been proposed [8–12]. The proposed model utilizes a circuit model employing two-variable spiking neurons (TSNs) [13], which are designed to achieve low power consumption and miniaturization. The TSN is based on the Izhikevich neuron model, and it has been demonstrated that various spiking behaviors, including chaotic dynamics, can be reproduced by adjusting the circuit's voltage parameter values [14].

A method has been proposed to evaluate the learning performance of STLR in synchronous, discrete-time, clock-driven models [15,16]. Specifically, a multistage template matching method was proposed [17,18], in which the ideal characteristics of the network are represented by multiple templates, and the errors relative to these templates are evaluated. Analyses of the network structure within a specific metric space have confirmed that learned spatiotemporal information is stored in the synaptic weight space with a fractal structure [19]. Additionally, the network demonstrated effectiveness in a spatiotemporal pattern separation task using handwritten character datasets, which cannot be learned by the Hebbian learning rule [20]. However, a detailed analysis of the synaptic weight space structure in asynchronous, continuous-time, event-driven models has not yet been conducted. Therefore, in this study, a detailed analysis of the temporal evolution of the synaptic weight space structure in the asynchronous, continuous-time, event-driven model was conducted using multistage template matching and multi-dimensional scaling (MDS).

2 Event-Driven Asynchronous Continuous-Time Model

A single-layer feedforward SNN is employed. The network structure, synapse model, and neuron model are illustrated in Fig. 1. Detailed definitions are provided below.

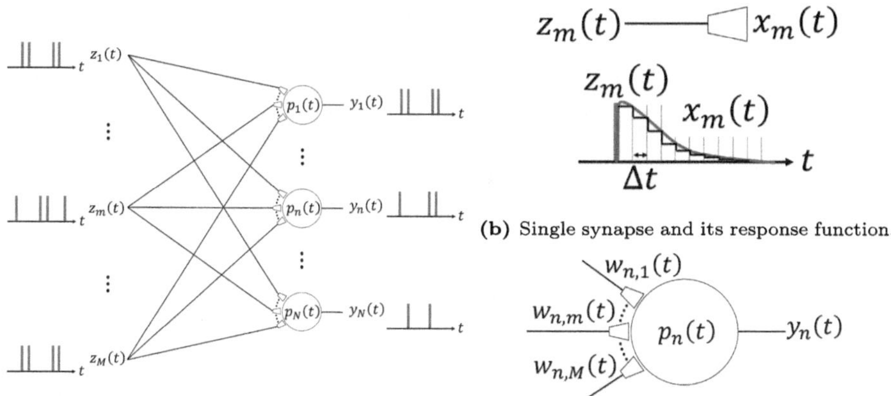

(a) Single-layer feedforward spiking neural network with STLR synapses.

(b) Single synapse and its response function.

(c) Single neuron with STLR synapses.

Fig. 1. Network architecture: (a) Single-layer feed-forward spiking neural network with STLR synapses. (b) Output of the mth synapse in response to the mth spike input. (c) The nth neuron has M synapses.

2.1 Synapse Model

In biological synapses, conductance typically decays over time following the arrival of an input spike. For computational efficiency and simplicity in our numerical simulations, a synaptic model with rapid decay kinetics, analogous to those of AMPA receptors [21], was employed. Consequently, a first-order linear filter model was used, in which the synaptic output rises instantaneously upon spike arrival and subsequently decays exponentially over time [22].

As illustrated in Fig. 1a, input spikes are asynchronously and continuously fed into the synapses. The incoming spike train is processed by the synaptic response function as shown in Fig. 1b to generate the continuous-time synaptic output for the post-synaptic neuron. The dynamics of the mth synapse are described by the following differential equation:

$$\tau_{\rm syn}\frac{\mathrm{d}x_m(t)}{\mathrm{d}t} = -x_m(t) + z_m(t), \tag{1}$$

where t and $\tau_{\rm syn}(> 0)$ denote the continuous time and the synaptic time constant, respectively. $z_m(t)$ and $x_m(t)$ denote the synaptic input and output of the mth synapse at time t, respectively.

2.2 Neuron Model

Although the proposed architecture is compatible with various continuous-time neuron models, the standard leaky integrate-and-fire (LIF) model [22] is used in this study. The LIF model provides a mathematically tractable yet biologically relevant foundation that facilitates efficient numerical simulations.

As illustrated in Fig. 1c, the nth neuron receives the weighted sum of the continuous-time outputs from M synapses as

$$s_n(t) = \sum_{m=1}^{M} w_{n,m}(t)x_m(t), \tag{2}$$

where $w_{n,m}(t)$ represents the synaptic weight connecting the mth synapse to the nth neuron. The internal state of the neuron $p_n(t)$ is updated based on the total input $s_n(t)$ according to:

$$\tau_{\text{neu}} \frac{\mathrm{d}p_n(t)}{\mathrm{d}t} = -p_n(t) + s_n(t), \tag{3}$$

where $\tau_{\text{neu}}(> 0)$ denotes the neuronal time constant. The output of the neuron $y_n(t)$, is determined by the threshold function:

$$y_n(t) = \begin{cases} 1 \ (p_n(t) \geq \theta), \\ 0 \ (p_n(t) < \theta), \end{cases} \tag{4}$$

where θ is the firing threshold.

2.3 Spatiotemporal Learning Rule

The synapses are updated using the STLR based on input coincidence as

$$c_{n,m}(t) = w_{n,m}(t)x_m(t) \left(p_n(t) - w_{n,m}(t)x_m(t)\right), \tag{5}$$

$$\lambda \frac{\mathrm{d}q_{n,m}(t)}{\mathrm{d}t} = -q_{n,m}(t) + c_{n,m}(t), \tag{6}$$

where $c_{n,m}(t)$ represents the input coincidence for the mth synapse of the nth neuron, $q_{n,m}(t)$ tracks the time history of $c_{n,m}(t)$, and $\lambda(> 0)$ is a time constant.

The initial synaptic weights $w_{n,m}(0)$ are set by a uniform distribution over $[-1, 1]$. Synapses with $w_{n,m}(0) > 0$ are defined as excitatory, $w^{\mathrm{E}}_{n,m}(t)$, whereas those with $w_{n,m}(0) < 0$ are defined as inhibitory, $w^{\mathrm{I}}_{n,m}(t)$. This balanced random initialization ensures a diverse range of initial synaptic dynamics and network stability. The synaptic weight update rules are given by

$$\frac{\mathrm{d}w^{\mathrm{E}}_{n,m}(t)}{\mathrm{d}t} = \eta h(q_{n,m}(t)), \tag{7}$$

$$\frac{\mathrm{d}w^{\mathrm{I}}_{n,m}(t)}{\mathrm{d}t} = -\eta h(q_{n,m}(t)), \tag{8}$$

where $\eta(\geq 0)$ is the learning coefficient. The learning function $h(\cdot)$ is defined as

$$\begin{aligned} h(x) &= h_{\mathrm{LTP}}(x) + h_{\mathrm{LTD}}(x) \\ &= \frac{1}{1+\exp\left(-(x-\theta_{\mathrm{LTP}})/\varepsilon_{\mathrm{LTP}}\right)} \\ &+ \frac{1}{1+\exp\left(-(x-\theta_{\mathrm{LTD}})/\varepsilon_{\mathrm{LTD}}\right)} - 1, \end{aligned} \tag{9}$$

where θ_{LTP} and θ_{LTD} are thresholds for long-term potentiation (LTP) and long-term depression (LTD), respectively. To prevent LTP and LTD from occurring simultaneously, the learning function is set such that $\theta_{\mathrm{LTP}} > \theta_{\mathrm{LTD}}$.

3 Numerical Experiments

In the simulation, Eqs. (1), (3), and (6)–(8) were discretized using the fourth-order Runge-Kutta method with time step $\Delta t = 10^{-3}$ that was chosen as it provides sufficient accuracy to qualitatively observe the branching structures of the synaptic weight space while maintaining computational efficiency. The simulation parameters were set as follows: $M = 120$, $N = 120$, $\eta = 0.5$, $\tau_{\mathrm{syn}} = 5 \times 10^{-3}$, $\tau_{\mathrm{neu}} = 0.02$, and $\lambda = 0.223$. Additionally, Eq. (9) is changed to a step function as $\varepsilon_{\mathrm{LTP}} = \varepsilon_{\mathrm{LTD}} \to \infty$.

3.1 Generation of Spatiotemporal Patterns

We designed a task to generate input patterns that enabled a comprehensive evaluation of the network's learning capability and performance in separating spatiotemporal information. First, we generated $K = 2$ spatial binary column vectors of dimension $M = 120$ such that the Hamming distance is $HD = 4$. These vectors are defined as $\mathbf{A}_1$ and $\mathbf{A}_2$. Next, $I = 5$ spatial vectors were used to create $L(= K^I = 2^5 = 32)$ different spatiotemporal patterns. These patterns are highly similar, both spatially and temporally, making their separation and learning challenging.

In this study, the L spatiotemporal patterns are labeled in ascending order according to the indices of the spatial vectors. The lth spatiotemporal pattern is shown on the left side of Fig. 2. In the figure, $a_m^{k_i^l}$ and $a_{m'}^{k_i^l}$ are the mth and m'th element of the spatioal vector $\mathbf{A}_{k_i^l}$, respectively, where $k_i^l \in \{1, 2\}$ is the index of the ith input spatial vector in the lth spatiotemporal pattern.

3.2 Rate Coding

Each spatial binary column vector is presented for a duration of $T = 1$ as shown in Fig. 2. During this interval, the vector elements were converted into spike firing rates using the number of spikes (FR). Specifically, $a_m^{k_i^l} = 0$ correspond to a firing rate of 10 spikes per T, whereas $a_m^{k_i^l} = 1$ corresponds to 100 spikes per T. These distinct firing rates were chosen to provide sufficient contrast in the synaptic output $x_m(t)$ within the given synaptic time constant τ_{syn}. Specific spike timings were randomly generated according to a Poisson distribution.

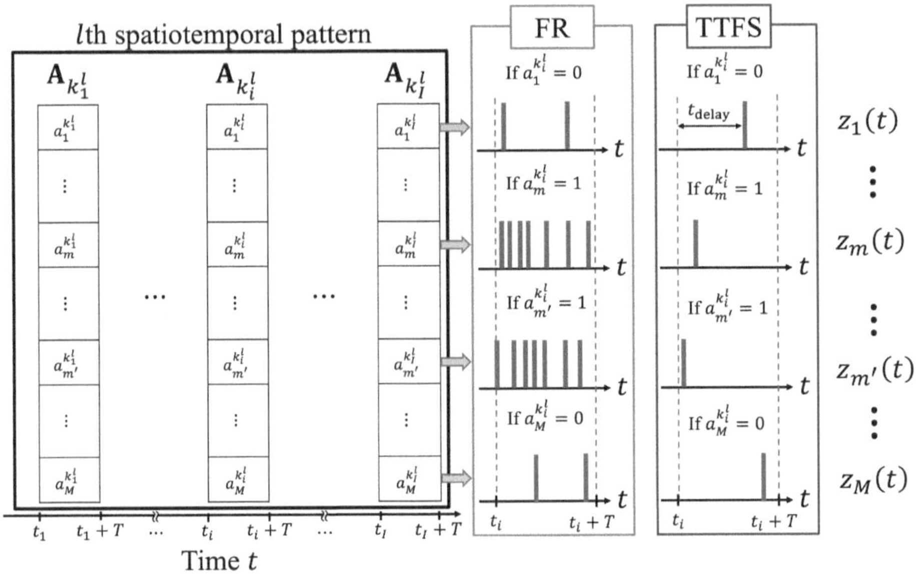

Fig. 2. Input spatiotemporal pattern and spike coding. The spatial vector consists of bit elements converted into spike trains through rate coding (FR) and temporal coding (TTFS).

3.3 Temporal Coding

To explicitly evaluate the capability of the STLR to decode precise temporal contexts, a core principle of neuromorphic computing, a temporal coding scheme is introduced, specifically time-to-first-spike (TTFS), for comparison with the rate coding. In the TTFS scheme, as shown in Fig. 2, the spatiotemporal pattern is encoded by the exact timing of a single spike per a spatial vector, not by the statistical firing rate. The precise spike time t_i for the mth synapse is defined as

$$t_i^{\text{fire}} = t_i + t_{\text{delay}}(1 - a_m^{k_i^l}) + \xi_i, \tag{10}$$

where t_i and t_{delay} are the input onset time and a delay, respectively. To ensure biological plausibility and prevent artificial perfect synchrony, a small temporal jitter ξ_i drawn from a uniform distribution [0, 0.002] was added to each spike. By computing this temporally precise, highly sparse coding with the conventional rate coding, the temporal discrimination power of the asynchronous continuous-time network is robustly assessed.

3.4 Results

The evolution of synaptic weights was evaluated, where in accordance with the number of spikes representing spatial vectors, $\theta_{\text{LTP}} = 0.04$ and $\theta_{\text{LTD}} = -0.02$

for rate coding, and $\theta_{\mathrm{LTP}} = 0.04 \times 10^{-3}$ and $\theta_{\mathrm{LTD}} = -0.02 \times 10^{-3}$ for TTFS, respectively.

Figure 3 shows the results of evaluating the network's learning performance when the spike coding is changed, using a multi-stage template matching method based on the F1-score. The vertical and horizontal axes in the figure represent the F1-score and the template number j, respectively.

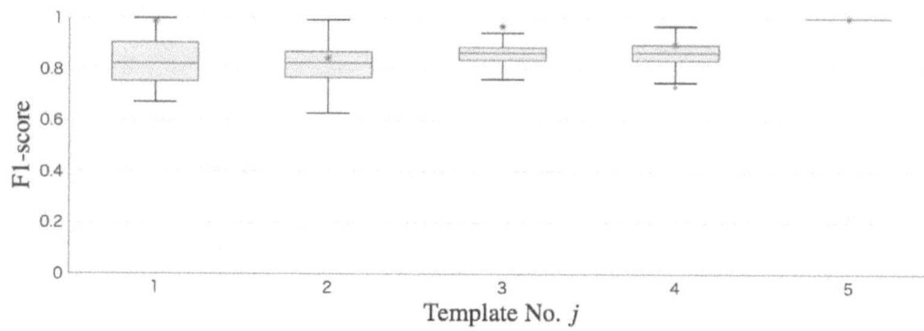

(a) Firing rate.

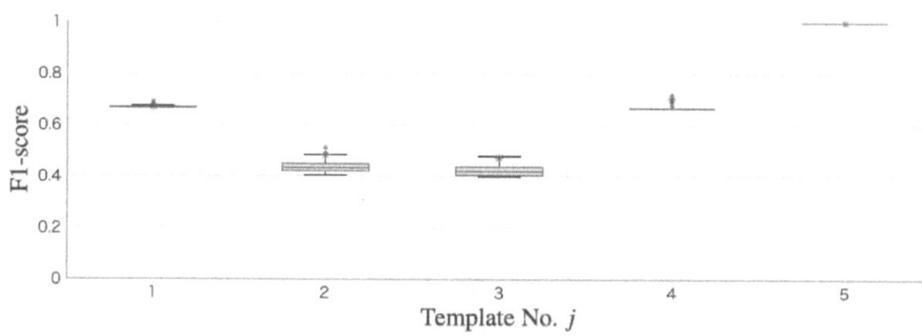

(b) Time-to-first-spike.

Fig. 3. Distribution of F1-scores for each template over 100 trials. The asterisk indicates the best result. (a) Spike coding through FR. (b) Spike coding through TTFS.

The method enables quantitative evaluation of a network's ability to separate spatiotemporal patterns and its memory performance. In Figs. 3a and 3b, the F1-scores are 1 at $j = 5$. This indicates that the network has successfully learned to distinguish all of the input spatiotemporal patterns. From $j = 1$ to 4, we can verify the extent to which the network learns the temporal structure of the input spatiotemporal patterns. Figure 3a shows that the network learns the temporal structure of almost all spatiotemporal patterns. On the other hand, Fig. 3b shows that the information of the spatial vectors input at $j = 2$ and 3 has been forgotten. The reason the F1-score for the spatial vector information input at the beginning ($j = 1$) is higher than for $j = 2$ and 3 is that the STLR learns the spatial vector at the beginning more effectively [19,20].

Furthermore, Figs. 4a and 4b show the results of the MDS analysis in the distance space using the initial synaptic weight values that achieved the highest average F1-score when using FR and TTFS, respectively.

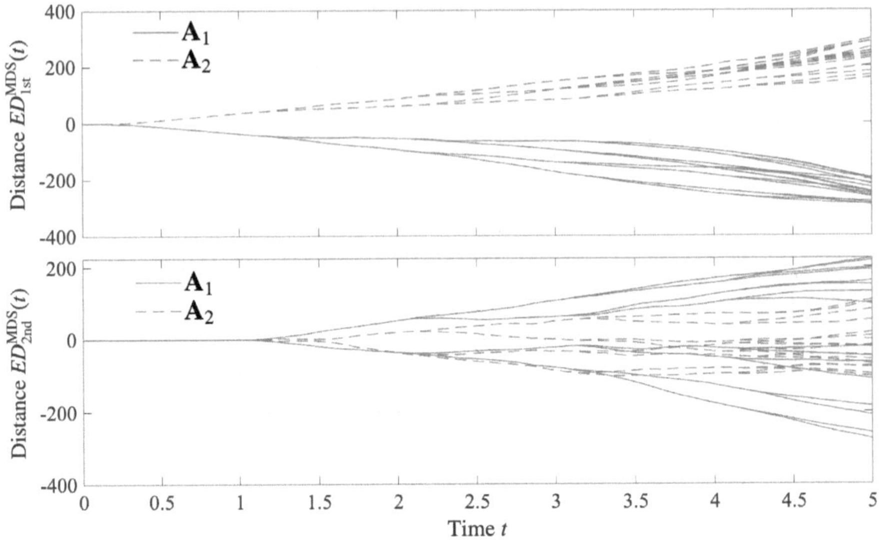

(a) Firing rate.

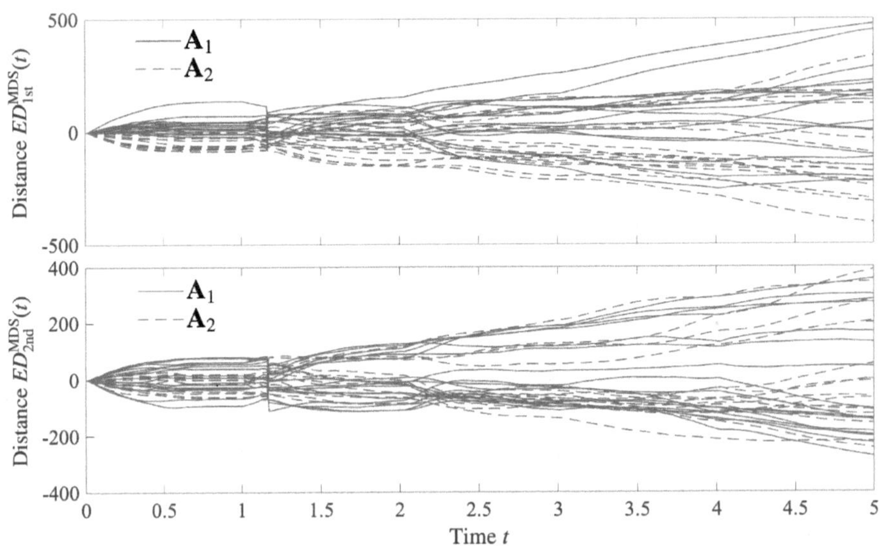

(b) Time-to-first-spike.

Fig. 4. Analysis of Euclidean distances between synaptic weights using the MDS. The solid and dashed lines represent when $\mathbf{A}_1$ and $\mathbf{A}_2$ are input at t_1, respectively. (a) Spike coding through FR. (b) Spike coding through TTFS.

To analyze the structure of the synaptic weights, let $\mathbf{W}_l(t)$ and $\mathbf{W}_{l'}(t)$ denote the synaptic weight matrices after learning the lth and l'th spatiotemporal pattern, respectively. We quantified the difference between the learned structures using the Euclidean distance:

$$ED_{l,l'}(t) = \|\mathbf{W}_l(t) - \mathbf{W}_{l'}(t)\|_2, \tag{11}$$

$$\mathbf{ED}(t) = \begin{bmatrix} ED_{1,1}(t) & \cdots & ED_{1,l'}(t) & \cdots & ED_{1,L}(t) \\ \vdots & \ddots & & & \vdots \\ ED_{l,1}(t) & & ED_{l,l'}(t) & & ED_{l,L}(t) \\ \vdots & & & \ddots & \vdots \\ ED_{L,1}(t) & \cdots & ED_{L,l'}(t) & \cdots & ED_{L,L}(t) \end{bmatrix}. \tag{12}$$

The trajectory of the weight evolution using the MDS (Figs. 4a and 4b) was visualized. In the figure, $ED^{\mathrm{MDS}}_{\mathrm{1st}}(t)$ and $ED^{\mathrm{MDS}}_{\mathrm{2nd}}(t)$ represent the Euclidean distances of the first and second principal components, respectively, obtained from the MDS analysis of $\mathbf{ED}(t)$. The branching patterns indicate that the network successfully differentiated spatial patterns over time.

In Figs. 4a and 4b, trajectories are drawn as solid and dashed lines based on the initially input spatial vector at t_1. From this result, it is possible to determine whether the synaptic weight space has acquired a contextual structure. For example, in the top of Fig. 4a, when $\mathbf{A}_1$ is input first, the final synaptic weights can be distinguished from those of $\mathbf{A}_2$ in the first principal component. However, in Fig. 4b, they cannot be clearly distinguished in the first and second principal components.

4 Conclusions

This study employed an asynchronous, continuous-time, event-driven SNN model incorporating synapses governed by a biologically plausible STLR. We verified the model's operation and evaluated its learning performance using an interpretable, multistage template-matching method. The results demonstrate that the proposed model exhibits effective spatiotemporal pattern separation without relying on global error signals and synchronous clocks. We confirmed that distinct synaptic weight trajectories were formed, indicating that the input temporal context was embedded in the synaptic weight space.

Furthermore, we confirmed through the multi-stage template matching and the MDS analysis that TTFS exhibits inferior learning performance compared to FR. However, spike information representation through temporal coding is required to implement more energy-efficient SNN hardware. Therefore, it is necessary to investigate spike coding methods that can achieve performance equivalent to rate coding even with temporal coding.

In future studies, we plan to develop a readout mechanism to extract contextual information from the synaptic weight space for real-world cognitive tasks such as sequence identification. Moreover, investigating the scalability of the proposed STLR to larger, multi-layer networks remains a key objective for realizing highly efficient neuromorphic edge computing devices.

Acknowledgments. This work was supported by JSPS KAKENHI 25K21292, 25H00447, and 25K24740; the Cooperative Research Project of RIEC, Tohoku University; and ACRO Research Grants of Teikyo University.

References

1. Roy, K., Jaiswal, A., Panda, P.: Towards spike-based machine intelligence with neuromorphic computing. Nature **575**(7784), 607–617 (2019)
2. Davies, M., et al.: Loihi: a neuromorphic manycore processor with on-chip learning. IEEE Micro **38**(1), 82–99 (2018)
3. Neftci, E.O., Mostafa, H., Zenke, F.: Surrogate gradient learning in spiking neural networks: bringing the power of gradient-based optimization to spiking neural networks. IEEE Signal Process. Mag. **36**(6), 51–63 (2019)
4. Wunderlich, T., Pehle, C.: Event-based backpropagation can compute exact gradients for spiking neural networks. Sci. Rep. **11**(1), 12829 (2021)
5. LeCun, Y., Bengio, Y., Hinton, G.: Deep learning. Nature **521**(7553), 436–444 (2015)
6. Tsukada, M., Aihara, T., Saito, H., Kato, H.: Hippocampal LTP depends on the spatial and temporal correlation of inputs. Neural Netw. **9**(8), 1357–1365 (1996)
7. Tsukada, M., Pan, X.: The spatiotemporal learning rule and its efficiency in separating spatiotemporal patterns. Biol. Cybern. **92**(2), 139–146 (2005)
8. Orima, T., Tsuji, T., Horio, Y., Moriya, S., Sato, S.: Memory state evaluation of spatio-temporal contextual learning memory network based on output spike rate. In: Proceedings of the International Symposium on on NOLTA (2023)
9. Orima, T., Tsuji, T., Horio, Y.: An extended spatiotemporal contextual learning and memory network model for hardware implementation. Proc. Comput. Sci. **222**, 478–487 (2023)
10. Orima, T., Horio, Y., Tsuji, T.: Analysis of learning process of synaptic weights in spatio-temporal learning networks for hardware implementation. In: Proceedings of the IEEE WCCI (2024)
11. Tsuji, T., Orima, T., Horio, Y.: An event-driven mixed analog/digital spiking neural network circuit model for hippocampal spatiotemporal context learning and memory. In: Proceedings of the IJCNN (2024)
12. Orima, T., Horio, Y., Tsuji, T.: Spatiotemporal contextual learning network with excitatory and inhibitory synapses for spiking neural network hardware. IEICE NOLTA. **15**(4), 796–810 (2024)
13. Moriya, S., Yamamoto, H., Sato, S., Y., Yuminaka, Horio, Y., Madrenas, J.: A fully analog CMOS implementation of a two-variable spiking neuron in the subthreshold region and its network operation. In: Proceedings of the IJCNN (2022)
14. Orima, T., Horio, Y., Moriya, S., Sato, S.: Bifurcation phenomena observed from two-variable spiking neuron integrated circuit. In: Proceedings of the ISCAS (2024)
15. Kitahara, Y.: Analysis of learning rule based on spatial coincidence and time history. Master's Thesis for Grand. School of Engineering, Tokyo Denki Univ. (2000) (in Japanese)

16. Tsuji, T., Orima, T., Horio, Y.: Detailed evaluation of spatiotemporal learning rule based on hamming distances among output vectors. In: Proceedings of the International Symposium on NOLTA (2023)
17. Furuya,M., Orima, T., Horio, Y., Tsuji, T.: An evaluation method for spatiotemporal learning network outputs using multistage template matching. In: Proceedings of theIEICE General Conference. (2024) (in Japanese)
18. Peter, D., Furuya, M., Tsuji, T., Orima, T., Horio, Y.: Parameters search for spatiotemporal learning neural networks with PWM input sequences. In: Proceedings of the IEICE General Conference (2024)
19. Orima, T., Tsuda, I., Tsukada, M., Tsukada, H., Horio, Y.: Fractal memory structure in the spatiotemporal learning rule. Front. Comput. Neurosci. **19** (2025)
20. Orima, T., Horio, Y.: Effect of time history in coincidence among inputs on spatiotemporal learning rule. IEICE, NOLTA **17**(3) (2026)
21. Destexhe, A., Mainen, Z.F., Sejnowski, T.J.: Kinetic models of synaptic transmission. In: Koch, C., Segev, I. (eds.) Methods in Neuronal Modeling, 2nd edn., pp. 1–25. MIT Press, Cambridge (1998)
22. Gerstner, W., Kistler, W.M.: Spiking Neuron Models: Single Neurons, Populations. Plasticity. Cambridge University Press, Cambridge (2002)
23. Thorpe, S., Delorme, A., Van Rullen, R.: Spike-based strategies for rapid processing. In: Wermter, S., Austin, J., Willshaw, D. (eds.) Emergent Neural Computational Architectures Based on Neuroscience, pp. 1–9. Springer, Heidelberg (2001)
24. Shi, W., Cao, J., Zhang, Q., Li, Y., Xu, L.: Edge computing: vision and challenges. IEEE Int. Things J. **3**(5), 637–646 (2016)

Learning Algorithms

Spectral Feature Refinement: Enhancing Robustness to Image Corruptions

Savvas Karatsiolis[1(✉)] and Andreas Kamilaris[1,2]

[1] CYENS Centre of Excellence, Dimarchou Lellou Demetriadi 01, Nicosia, Cyprus
{s.karatsiolis,a.kamilaris}@cyens.org.cy
[2] University of Twente, Enschede, The Netherlands

Abstract. We propose Spectral Feature Refinement (SFR), a simple yet effective framework achieving robustness to corruptions and noise through unsupervised training. The proposed method synergistically combines three key components: adaptation of Batch Normalization statistics using unlabeled target data, feature distillation from a source-trained model and hypothesis transfer. By aligning feature distributions at the statistical and representational levels, SFR produces robust and domain-invariant feature spaces that generalize across similar domains without requiring target labels, complex architectures or sophisticated training objectives. Extensive experiments on benchmark datasets, including CIFAR10-C, CIFAR100-C, MNIST-C and PatchCamelyon-C, demonstrate that the proposed algorithm outperforms prior methods in robustness to corruption. Theoretical and empirical analyses further reveal that our method achieves improved feature alignment, increased mutual information between the domains and reduced sensitivity to input perturbations.

Keywords: Model Robustness · Feature Alignment · Distribution Shift

1 Introduction

Despite the remarkable results of Deep Learning (DL) models on benchmark datasets, they systematically show significant performance drops when encountering even minor distribution shifts or natural corruptions during testing [1]. Thus, the ability of Artificial Intelligence (AI) to operate reliably in dynamic environments hinges on its difficulty in processing and acting upon data that deviates from the controlled training conditions. This behavior poses a barrier to deploying AI models in the real world, especially in safety-critical domains like autonomous car driving and medical diagnosis. Unfortunately, unexpected variations in production environments are the norm, not the exception. Bridging the gap between development-time performance and real-world resilience is a fundamental challenge requiring novel approaches to advance model generalization and adaptation. Nevertheless, irreducible uncertainty inevitably arises

L. Pasa et al. (Eds.): ICANN 2026, LNCS 17089, pp. 321–332, 2027.
https://doi.org/10.1007/978-3-032-38398-3_27

from the inherent complexity of the data or from outliers. Existing techniques such as Deep Ensembles, Monte Carlo (MC) Dropout and temperature scaling offer only marginal improvements in mitigating the problem. Model calibration, in particular, fails to transfer effectively from the source to the target domain. While Deep Ensembles generally outperform other methods, their performance remains significantly below satisfactory levels. There is also often a trade-off between clean accuracy and robustness: many techniques enhance one while diminishing the other. The robustness and transferability of models, specifically Convolutional Neural Networks (CNNs), tend to be linked to scaling: increasing the training dataset size and the model's capacity enhances the CNN's robustness to distributional shifts, i.e., larger models trained on more extensive datasets demonstrate improved generalization to Out-Of-Distribution (OOD) data. More recently, the vulnerability of deep learning models to distribution shifts in vision has been addressed through a plethora of research strategies. These endeavors seek to align data from a deviating domain more closely with the feature space of the source domain. Such methods encompass strategically selected augmentations during training, adversarial domain adaptation techniques, self-supervised learning approaches, adaptation of test domain statistics [2] and pseudo-labeling. A more recent and significant development is Test-Time Adaptation (TTA) [3]. TTA methodologies are generally classified into two groups depending on the accessibility of the original training data during adaptation. One group comprises training data-free TTA, performing adaptation using only the test data. The other group, termed Test Time Training (TTT), integrates the training data (or source domain information) alongside the test data to direct the adaptation process. While offering substantial performance gains, many of these approaches often suffer from sensitivity to misleading patterns and hyperparameter tuning and exhibit training instability. In this paper, we propose a method for making models more robust to data corruptions based on synergistically training two models: one on labeled source data and another on unlabeled target data. This joint training paradigm enables effective information transfer from the source to the target model, resulting in excellent performance on the target domain without compromising source domain performance.

2 Spectral Feature Refinement

To ensure clarity and consistency throughout this paper, we define the key symbols and notations used. To address the challenge of domain shift, our objective is to learn a robust classification model for the target domain. We assume access to a source domain p_s with data instances $x \in X$ and their corresponding labels $y \in Y$. For the target domain p_t, we have a set of unlabeled data instances $z \in Z$. The primary task is to develop a model $\mathcal{F}_t : X \to Y$ and $\mathcal{F}_t : Z \to Y$ whose classification performance on the source data is comparable to that of a model $\mathcal{F}_s : X \to Y$ trained directly on the labeled source data. Crucially, $\mathcal{F}_t$ must be robust to the distribution shift between the source domain and the target domain, i.e., $\mathcal{F}_t : Z \to Y$ significantly outperforms $\mathcal{F}_s : Z \to Y$ when inferring on data from the target domain. Our approach relies on synergistically training two models: a source domain model and a target domain model. The source model is trained using labeled source domain data to classify input samples and serves as a knowledge distillation proxy for the target model. The target model, in contrast, is trained on unlabeled data from both domains and learns to approximate the internal feature representations of the source model. While the two models do not need to be architecturally identical, the source model must include BN layers, which are crucial for adapting to target domain statistics. Figure 1 illustrates the overall SFR workflow. Each training loop iteration consists of two sequential steps: supervised training of the source model which is updated using cross-entropy (CE) loss on labeled source domain data and unsupervised training of the target model via feature distillation. During the second step, both source and target domain data are passed through the (frozen) source model to obtain internal representations. The target model is then trained to minimize the MSE between its features and those of the source model separately for source and target domain inputs. No labels are used, and the source model is not updated during this step. However, its BN layers adapt their running statistics, thereby internalizing the target domain distribution. While it is possible to implement the algorithm's steps in alternative ways, such as updating BN statistics via additional forward passes during the supervised step, combining statistics adaptation and feature distillation in a single pass offers greater computational efficiency. These two training steps implement our dual feature refinement strategy: Target domain statistics adaptation and feature distillation. Together, they align the feature spaces of both domains without requiring target domain labels. During inference, we attach the source model's classifier head g_s to the target model's feature extractor. Because the feature spaces are now aligned, the source model's classifier can produce valid logits for target domain inputs. This constitutes the hypothesis transfer from the source to the target domain. As long as the two domains exhibit semantic similarity—despite differences in style, presentation, or depiction—SFR's feature alignment enables effective and label-free hypothesis transfer across domains. Algorithm 1 outlines the steps of the SFR training procedure. The primary advantages of this algorithm lie in its simplicity, straightforward implementation, and the absence of additional hyperparameters beyond those typically required for training, such as the learning rate and weight decay, which are task-dependent. Furthermore, SFR is very stable and does not suffer from mode collapsing like many domain adaptation algorithms. A key advantage of SFR is its ability to build a single model that performs well across both the source and target domains. This model not only achieves at least the same performance as a model trained with supervised learning on the source domain, but also demonstrates robustness to image corruptions. This dual-domain proficiency distinguishes SFR from other approaches that prioritize target domain performance at the expense of accuracy on the source domain. By excelling in both domains, the model intrinsically demonstrates its robust quality. The three mechanisms of the algorithm are essential in the domain adaptation process, and excluding any one of them leads to poor final results. While the step of adapting to the target domain's data distribution via BN statistics adaptation can yield performance improvements on its own, the other two steps are not impactful if used without it. The effects of feature distillation and hypothesis transfer heavily rely on the BN's statistics adaptation. Feature distillation, in particular, builds upon the adaptation of the source model to the principal components shared by the source and target domain data and amplifies the principal signals shared by both domains. Without the BN adaptation step, these signals are obscured by intricate correlations and are diffused across a vast, noisy feature space, rendering the subsequent steps ineffective.

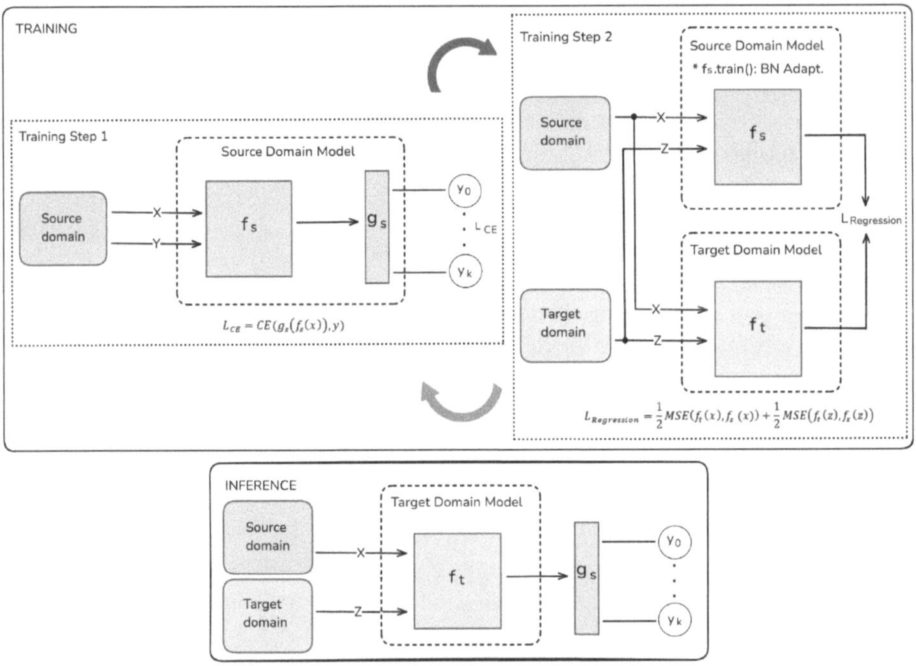

Fig. 1. The SFR framework comprises two steps: a) Learning the features of the source domain via supervised learning and b) Transferring this learning to the target domain model with feature distillation. During Step 2, the source model adapts its BN layers on both domains' data. At inference, both source and target domain data pass through the target domain model f_t, and subsequently, classification is conducted by the source domain classifier g_s.

2.1 Target Domain Statistics Adaptation

Batch Normalization (BN) [4] normalizes layer activations to zero mean and unit variance, mitigating internal covariate shift and accelerating training. A less

Algorithm 1. The SFR Algorithm

function TRAIN
  Source and target models in training mode:
    f_s.train(), g_s.train(), f_t.train()
  for epoch = 1 to *MaxEpoch* **do**
    Step 1:
    Sample $x \in \mathcal{X},\ y \in \mathcal{Y}$
    Compute $g_s(f_s(x))$
    Compute $\mathcal{L}_{CE} = \text{CE}(g_s(f_s(x)), y)$
    Back-propagate grads for $\mathcal{L}_{CE}$, update f_s, g_s
    Step 2:
    Sample $x \in \mathcal{X},\ z \in \mathcal{Z}$
    Compute $f_s(x),\ f_s(z)$ `// BN stats updates are still enabled`
    Compute $f_t(x),\ f_t(z)$
    Compute $\mathcal{L}_{\text{Regression}} = \frac{1}{2}\text{MSE}(f_t(x), f_s(x)) + \frac{1}{2}\text{MSE}(f_t(z), f_s(z))$
    Back-propagate grads for $\mathcal{L}_{\text{Regression}}$ and update f_t
  end for
end function

function INFERENCE(x)
  Put f_t and g_s in evaluation mode:
  f_t.eval(), g_s.eval()
  Predict $\hat{y} = g_s(f_t(x))$
end function

recognized but critical property is that BN *standardizes* each feature channel to unit variance, removing scale discrepancies across domains when its statistics are adapted. We formalize this effect and prove that adapting BN statistics to the target domain provably reduces the cross-domain feature discrepancy.

Given activations $X \in \mathbb{R}^{m \times d}$ with feature-wise mean μ and variance matrix $D = \text{diag}(\sigma_1^2, \ldots, \sigma_d^2)$, the covariance before BN is $\Sigma = \frac{1}{m}(X - \mathbf{1}\mu^\top)^\top (X - \mathbf{1}\mu^\top)$. BN produces $\hat{X} = (X - \mathbf{1}\mu^\top) D^{-1/2}$, yielding:

$$\Sigma_{\text{BN}} = D^{-1/2}\, \Sigma\, D^{-1/2}. \tag{1}$$

Proposition 1 (Effect of BN on Covariance Structure). *The post-BN covariance equals the Pearson correlation matrix R of the original features: $\Sigma_{\text{BN}} = R$, where $R_{ii} = 1$ and $R_{ij} = \rho_{ij} = \Sigma_{ij}/(\sigma_i \sigma_j)$ for $i \neq j$.*

Proof. From Eq. (1), $(\Sigma_{\text{BN}})_{ij} = \Sigma_{ij}/(\sigma_i \sigma_j) = \rho_{ij}$, confirming unit diagonal (unit variance per feature) and Pearson correlations off-diagonal.

BN thus performs standardization but not full whitening ($\Sigma_{\text{BN}} = I$ holds iff all features are uncorrelated). Nonetheless, enforcing unit variance per feature is the property that enables effective domain adaptation. Assuming Gaussian features $\mathcal{N}(\mu_s, \Sigma_s)$ on source and $\mathcal{N}(\mu_t, \Sigma_t)$ on target data, with diagonal variance matrices D_s, D_t and correlation matrices R_s, R_t:

Proposition 2 (BN Adaptation Reduces Cross-Domain Feature Discrepancy). *Define the cross-domain discrepancy* $\Delta = \Delta_{\text{mean}} + \Delta_{\text{var}} + \Delta_{\text{corr}}$*, where* $\Delta_{\text{mean}} = \|m_s - m_t\|^2$*,* $\Delta_{\text{var}} = \sum_i (v_{s,i} - v_{t,i})^2$*, and* $\Delta_{\text{corr}} = \sum_{i \neq j}(C_s^{ij} - C_t^{ij})^2$ *measure differences in the BN-normalized means, per-feature variances, and off-diagonal covariances.*
(a) Without BN Adaptation *(Source Statistics Used for Both Domains): the normalized target features have mean* $D_s^{-1/2}(\mu_t - \mu_s)$ *and covariance* $D_s^{-1/2}\Sigma_t D_s^{-1/2}$*, yielding:*

$$\Delta_{\text{mean}} = \textstyle\sum_i (\mu_{t,i} - \mu_{s,i})^2/\sigma_{s,i}^2, \tag{2}$$

$$\Delta_{\text{var}} = \textstyle\sum_i (1 - \sigma_{t,i}^2/\sigma_{s,i}^2)^2, \tag{3}$$

$$\Delta_{\text{corr}} = \textstyle\sum_{i \neq j} (\rho_{ij}^s - \rho_{ij}^t \sigma_{t,i}\sigma_{t,j}/(\sigma_{s,i}\sigma_{s,j}))^2. \tag{4}$$

(b) With BN Adaptation *(Each Domain Uses its Own Statistics): both domains are normalized to zero mean and unit variance, so:*

$$\Delta_{\text{mean}} = 0, \quad \Delta_{\text{var}} = 0, \quad \Delta_{\text{corr}} = \textstyle\sum_{i \neq j}(\rho_{ij}^s - \rho_{ij}^t)^2. \tag{5}$$

BN adaptation eliminates mean and variance discrepancies, reducing the gap to the difference in correlation structures alone.

Proof. **(a)** Normalizing target $z \sim \mathcal{N}(\mu_t, \Sigma_t)$ with source statistics gives $\hat{z} = D_s^{-1/2}(z - \mu_s)$, so $\mathbb{E}[\hat{z}] = D_s^{-1/2}(\mu_t - \mu_s)$ and $\text{Cov}(\hat{z}) = D_s^{-1/2}\Sigma_t D_s^{-1/2}$. Comparing with the source (mean $\mathbf{0}$, covariance R_s) yields Eqs. (2)–(4). **(b)** Normalizing with target statistics gives $\hat{z} = D_t^{-1/2}(z - \mu_t)$, so $\mathbb{E}[\hat{z}] = \mathbf{0}$ and $\text{Cov}(\hat{z}) = R_t$, yielding Eq. (5).

Corollary 1. *When the two domains share semantic content (e.g., clean vs. corrupted images of the same objects),* $\rho_{ij}^s \approx \rho_{ij}^t$ *and the residual* Δ_{corr} *is small, so BN adaptation eliminates the dominant sources of distribution shift.*

By adapting BN statistics using target domain samples during training, activations are normalized with domain-appropriate statistics, producing discriminative, domain-invariant features. While prior work [2] adapts BN layers heuristically, this is the first formal analysis showing how BN adaptation reduces cross-domain distributional discrepancy. BN adaptation forms a cornerstone of our approach and operates synergistically with feature distillation and hypothesis transfer.

2.2 Feature Distillation

Feature distillation trains a target model to replicate the internal representations of a source model [22]. We use MSE regression combined with weight decay regularization, which acts as a spectral low-pass filter retaining dominant, domain-invariant features while suppressing domain-specific noise. We decompose the source embedding as $f_s(x) = h(x) + \delta(x)$, where $h(x)$ is the domain-invariant signal and $\delta(x)$ is domain-specific noise. The target model f_t is trained via:

$$\mathcal{L}(f_t) = \mathbb{E}_{x\sim p}\big[\|f_t(x) - f_s(x)\|^2\big] + \lambda\,\Omega(f_t), \tag{6}$$

where $p = \frac{1}{2}p_s + \frac{1}{2}p_t$ and $\Omega(f_t) = \|\theta_t\|^2$ (L_2 weight decay). Without regularization, the optimum would replicate the full source model including noise; regularization is essential for denoising. We analyze this objective in the RKHS framework associated with a kernel κ and integral operator T_p with spectral decomposition $\{(\eta_k, \varphi_k)\}$, where η_k quantifies the importance of each mode under the data distribution.

Proposition 3 (Regularized Feature Distillation as Spectral Filtering). *Let $f_s = \sum_k c_k\varphi_k$ be the source embedding expanded in the eigenbasis of T_p. The solution to Eq.* (6) *is:*

$$f_t^*(x) = \sum_{k=1}^{\infty} \frac{\eta_k}{\eta_k + \lambda}\, c_k\, \varphi_k(x). \tag{7}$$

The factor $\eta_k/(\eta_k + \lambda)$ is a low-pass filter: modes with $\eta_k \gg \lambda$ are preserved; modes with $\eta_k \ll \lambda$ are suppressed.

Proof. Setting the functional derivative of Eq. (6) to zero and using the reproducing property yields the operator equation $(T_p + \lambda I)f_t^* = T_p f_s$. Expanding in the eigenbasis gives $c_k^* = \eta_k c_k/(\eta_k + \lambda)$.

The signal h, being shared across both domains, projects onto modes with large η_k (receiving eigenvalue contributions from both T_{p_s} and T_{p_t}) and is preserved. The noise δ, being domain-specific and typically high-frequency, concentrates in the spectral tail (small η_k) and is suppressed. Thus $f_t^* \approx h$, recovering the domain-invariant signal. In practice, this filtering is implemented through explicit weight decay, spectral bias of gradient-trained networks [23], finite model capacity, and the multi-domain tension that penalizes fitting domain-specific artifacts. The synergy with BN adaptation is critical: by eliminating mean and variance discrepancies (Proposition 2), BN adaptation reduces δ before distillation, widening the spectral gap and improving filtering quality.

2.3 Hypothesis Transfer and Integration-Synergy

Since the target domain is unlabeled, we reuse the source classifier g_s atop the target feature extractor: $\mathcal{F}_t = g_s(f_t(x))$. Because feature distillation aligns f_t with f_s, the source classifier's decision boundaries remain valid, effectively transferring the classification hypothesis from the labeled source to the unlabeled target domain. SFR's three components are mutually reinforcing. BN adaptation eliminates mean and variance discrepancies (Proposition 2), reducing δ and improving the spectral filtering of feature distillation (Proposition 3). Conversely, feature distillation addresses the residual correlation discrepancy Δ_{corr}

that BN cannot eliminate alone. Hypothesis transfer then capitalizes on the refined, aligned feature space for accurate classification. Together, statistical alignment and representational alignment constitute a dual feature refinement mechanism.

3 Experimental Results

We evaluated SFR on a suite of popular benchmark datasets for robustness, MNIST-C, CIFAR10-C, CIFAR100-C and on a real-world medical image dataset, PatchCamelyon-C [6]. The latter provides a critical real-world test case, as natural corruptions from the image acquisition process severely degrade the accuracy of models trained on clean data, causing them to perform only marginally better than random on this dataset. All datasets are corrupted versions of the original respective datasets, created to evaluate the robustness of machine learning models (especially deep neural networks) to common visual corruptions and perturbations. The *'-C'* suffix in these dataset names indicates that common corruptions (e.g., noise, blur, weather effects, digital distortions, etc.) have been artificially applied to the images of these datasets at various severity levels (5 severity levels in total – severity 1 being a random mild corruption and severity 5 being a heavy random corruption). We utilize a Residual Network of 18 layers (ResNet-18) [7] for all experiments. We opted to use the ResNet-18 model ($\approx$ 11.7 million parameters) rather than significantly larger models, as used in other works, because experiments with much larger architectures such as WideResNet-28-10 ($\approx$ 36.5 million parameters) yielded only marginal improvements at the cost of a substantial increase in training time. However, careful hyperparameter tuning and the use of an appropriate learning rate scheduler (cosine annealing [9]) enable ResNet-18 to achieve performance comparable to that of larger models. For feature distillation, we extract output from a linear layer appended after the average pooling layer of the ResNet-18 model, just before the final linear classifier. This added linear layer matches the size of the pooling layer's output. Aside from this addition, the models remain otherwise unmodified. We train the source models using the stochastic gradient descent optimizer, while the target models are optimized with AdamW [8].

CIFAR10-C/CIFAR100-C: SFR was evaluated across all 19 distinct corruption types (at 5 different severity levels) present within the dataset. As a baseline, we consider the classification success rate obtained when using the model trained on the original (clean) data, which comprises the source domain of the problem. Specifically, we train a ResNet-18 model in a supervised manner on clean data (CIFAR10/CIFAR100) and test its performance on the corrupted images (CIFAR10-C/CIFAR100-C). Our baseline results are on par with the baselines reported by previous works. SFR achieves a significantly improved success rate on the corrupted images compared to the baseline, while maintaining the performance on source domain. Especially for certain corruptions such as the Gaussian noise, the improvement is substantial. Figure 2 shows the classification success rate and the improvement achieved by SFR on the datasets per data corruption type, averaged over 5 runs. Table 1 provides a detailed comparison of SFR against other algorithms representing a diverse range of domain adaptation strategies.

MNIST-C: We applied SFR to the MNIST-C dataset and present the results in Fig. 2. Our baseline results were slightly lower than those reported by [11]. Specifically, they achieved an average classification success rate of 72.3%, whereas our baseline yielded 71.54%. This performance drop primarily stems from our baseline model's poor performance on brightness and fog corruptions when trained solely on source domain data. A possible explanation for this discrepancy is our use of a ResNet-18 model, which is significantly larger than the LeNet-5 architecture employed in the referenced work and may be more susceptible to overfitting specific data categories. Nevertheless, our baseline results on certain corruptions, like motion blur and translation, are significantly higher as shown in Table 1.

Table 1. Average error rates (%) on three corruption benchmarks. Each column reports the mean error across all corruption types and severities for the respective dataset. The suffix '-cont' denotes continual test-time adaptation. Best results are in **bold**. ↓ indicates lower is better.

Method	CIFAR-10-C Error ↓	CIFAR-100-C Error ↓	MNIST-C Error ↓
AdaBN [12]	–	–	5.80
Shot-IM [21]	–	43.47	2.80
FR [11]	–	–	3.30
BUFR [11]	–	–	3.60
TTA [16]	19.90	–	–
MEMO [13]	19.60	–	–
TTT-episodic [15]	21.50	–	–
TTT [15]	15.60	–	–
Tent [10]	14.00	36.91	–
Tent-cont [10]	–	37.50	–
ETA [14]	19.40	40.18	–
EATA [14]	19.70	39.76	–
RoTTA [17]	25.20	–	–
RMT [18]	12.50	–	–
TEA [19]	16.67	34.50	–
SAR [5]	20.23	37.05	–
SAR-cont [5]	–	35.00	–
CoTTa-cont [20]	–	38.20	–
SFR (ours)	**10.83**	**34.38**	**2.59**

PatchCamelyon-C: Besides famous benchmark datasets, we applied SFR on a dataset used for real-world medical imaging applications. The dataset focuses

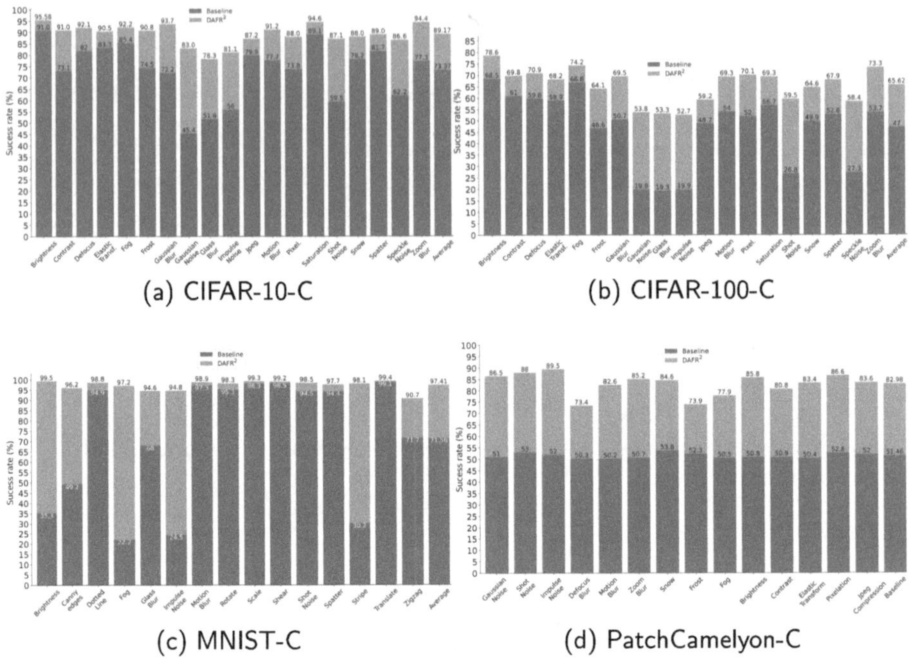

Fig. 2. Per-corruption accuracy of SFR across four benchmarks. SFR consistently improves over the clean-data baseline on all corruption types, with particularly large gains on corruptions that cause severe degradation in the baseline model (e.g., noise and blur corruptions on CIFAR-10-C and CIFAR-100-C).

on the task of detecting metastatic tissue in histopathological scans. Specifically, the PatchCamelyon dataset addresses the important clinical problem of identifying metastatic breast cancer in lymph node sections, which is crucial for cancer staging and treatment and contains patches extracted from histopathologic whole-slide images (WSIs) of sentinel lymph node sections. Each patch in the dataset is 96×96 pixels with 3 color channels (RGB) and is assigned a positive or negative label depending on whether at least one pixel of metastatic tumor tissue exists within the central 32×32 pixel region of the patch. Figure 2 illustrates that the baseline model's performance is only marginally better than random chance, whereas SFR demonstrates a substantial improvement.

4 Conclusions

This work introduces the SFR framework which comprises a robust and effective strategy for making models robust to corruption and noise by leveraging feature refinement through statistics alignment, feature distillation and hypothesis transfer. Unlike many existing approaches that rely heavily on complex objectives and require complex optimization schemes, SFR operates through conceptually simple yet powerful mechanisms. Its strong performance across multiple benchmark datasets underscores its effectiveness in bridging the distribution gap between source and target domain without relying on target labels or requiring complex model architectures.

Acknowledgments. This work received funding from the European Union's Horizon 2020 research and innovation programme under grant agreement No. 739578, as well as from the Government of the Republic of Cyprus through the Deputy Ministry of Research, Innovation and Digital Policy. Also received funding from CLIMAAX-NIC project. The latter is funded by the European Union as part of the cascade funding call of the CLIMAAX Climate Ready Regions project, which has received funding from the European Union's Horizon Europe – the Framework Programme for Research and Innovation (2021-2027) under grant agreement No. 101093864.

References

1. Yang, J., Zhou, K., Li, Y., Liu, Z.: Generalized out-of-distribution detection: a survey. Int. J. Comput. Vis. **132**(12), 5635–5662 (2024). https://doi.org/10.1007/s11263-024-02117-4
2. Nam, H., Kim, H.-E.: Batch-instance normalization for adaptively style-invariant neural networks. In: Proceedings of the 32nd International Conference on Neural Information Processing Systems, pp. 2563–2572. Curran Associates Inc., Red Hook, NY (2018)
3. Liang, J., He, R., Tan, T.: A comprehensive survey on test-time adaptation under distribution shifts. Int. J. Comput. Vis. **133**(1), 31–64 (2024)
4. Ioffe, S., Szegedy, C.: Batch normalization: accelerating deep network training by reducing internal covariate shift. In: Bach, F., Blei, D. (eds.) Proceedings of the 32nd International Conference on Machine Learning, Proceedings of Machine Learning Research, vol. 37, pp. 448–456. PMLR (2015)
5. Niu, S., et al.: Towards stable test-time adaptation in dynamic wild world. In: International Conference on Learning Representations (2023)
6. Zhang, Y., Sun, Y., Li, H., Zheng, S., Zhu, C., Yang, L.: Benchmarking the robustness of deep neural networks to common corruptions in digital pathology. In: Wang, L., Dou, Q., Fletcher, P.T., Speidel, S., Li, S. (eds.) Medical Image Computing and Computer Assisted Intervention - MICCAI 2022, pp. 242–252. Springer Nature Switzerland, Cham (2022)
7. He, K., Zhang, X., Ren, S., Sun, J.: Deep residual learning for image recognition. In: 2016 IEEE Conference on Computer Vision and Pattern Recognition, CVPR 2016, pp. 770–778. IEEE Computer Society (2016). https://doi.org/10.1109/CVPR.2016.90
8. Loshchilov, I., Hutter, F.: Decoupled weight decay regularization. In: 7th International Conference on Learning Representations, ICLR 2019. OpenReview.net, (2019)
9. Loshchilov, I., Hutter, F.: SGDR: stochastic gradient descent with warm restarts. In: 5th International Conference on Learning Representations, ICLR 2017. OpenReview.net (2017)
10. Wang, D., Shelhamer, E., Liu, S., Olshausen, B.A., Darrell, T.: Tent: fully test-time adaptation by entropy minimization. In: 9th International Conference on Learning Representations (2021) ICLR 2021. OpenReview.net

11. Eastwood, C., Mason, I., Williams, C.K.I., Scholkopf, B.: Source-free adaptation to measurement shift via bottom-up feature restoration. In: The Tenth International Conference on Learning Representations (2022). ICLR 2022. OpenReview.net
12. Li, Y., Wang, N., Shi, J., Liu, J., Hou, X.: Revisiting batch normalization for practical domain adaptation. In: 5th International Conference on Learning Representations, ICLR 2017. OpenReview.net (2017)
13. Zhang, M., Levine, S., Finn, C.: MEMO: test time robustness via adaptation and augmentation. In: Proceedings of the 36th International Conference on Neural Information Processing Systems. Curran Associates Inc., Red Hook, NY (2022)
14. Niu, S., Wu, J., Zhang, Y., Chen, Y., Zheng, S., Zhao, P., Tan, M.: Efficient test-time model adaptation without forgetting. In: Chaudhuri, K., Jegelka, S., Song, L., Szepesvari, C., Niu, G., Sabato, S. (eds.) International Conference on Machine Learning, ICML 2022, Proceedings of Machine Learning Research, vol. 162, pp. 16888–16905. PMLR (2022)
15. Sun, Y., Wang, X., Liu, Z., Miller, J., Efros, A.A., Hardt, M.: Test-time training with self-supervision for generalization under distribution shifts. In: Proceedings of the 37th International Conference on Machine Learning, ICML 2020, Proceedings of Machine Learning Research, vol. 119, pp. 9229–9248. PMLR (2020)
16. Ashukha, A., Lyzhov, A., Molchanov, D., Vetrov, D.P.: Pitfalls of in-domain uncertainty estimation and ensembling in deep learning. In: 8th International Conference on Learning Representations, ICLR 2020. OpenReview.net, (2020)
17. Yuan, L., Xie, B., Li, S.: Robust test-time adaptation in dynamic scenarios. In: IEEE/CVF Conference on Computer Vision and Pattern Recognition, CVPR 2023, pp. 15922–15932. IEEE (2023). https://doi.org/10.1109/CVPR52729.2023.01528
18. Dobler, M., Marsden, R.A., Yang, B.: Robust mean teacher for continual and gradual test-time adaptation. In: IEEE/CVF Conference on Computer Vision and Pattern Recognition, CVPR 2023, pp. 7704–7714. IEEE (2023). https://doi.org/10.1109/CVPR52729.2023.00744
19. Yuan, Y., Xu, B., Hou, L., Sun, F., Shen, H., Cheng, X.: TEA: test-time energy adaptation. In: IEEE/CVF Conference on Computer Vision and Pattern Recognition, CVPR 2024, pp. 23901–23911. IEEE (2024). https://doi.org/10.1109/CVPR52733.2024.02256
20. Wang, Q., Fink, O., Van Gool, L., Dai, D.: Continual test-time domain adaptation. In: IEEE/CVF Conference on Computer Vision and Pattern Recognition, CVPR 2022, pp. 7191–7201. IEEE (2022). https://doi.org/10.1109/CVPR52688.2022.00706
21. Liang, J., Hu, D., Wang, Y., He, R., Feng, J.: Source data-absent unsupervised domain adaptation through hypothesis transfer and labeling transfer. IEEE Trans. Pattern Anal. Mach. Intell. **44**(11), 8602–8617 (2022). https://doi.org/10.1109/TPAMI.2021.3103390
22. Hinton, G.E., Vinyals, O., Dean, J.: Distilling the knowledge in a neural network (2015). arXiv:1503.02531 arXiv preprint
23. Rahaman, N., et al.: On the spectral bias of neural networks. In: Chaudhuri, K., Salakhutdinov, R. (eds.) Proceedings of the 36th International Conference on Machine Learning, Proceedings of Machine Learning Research, vol. 97, pp. 5301–5310. PMLR (2019)

Minimum Distance Classification for Nonlinear Dynamical Systems

Dominique Martinez(✉)

Aix-Marseille Université, CNRS, ISM, 13009 Marseille, France
dominique.martinez@univ-amu.fr

Abstract. We address the problem of classifying trajectories or sequences generated by nonlinear dynamical systems, where each class corresponds to a distinct dynamical system. We propose Dynafit, a kernel-based method that learns a distance metric between training data and the underlying dynamics. New observations are assigned to the class whose dynamics best fit the observations according to the learned metric. The learning algorithm approximates the Koopman operator, which globally linearizes the dynamics in a (potentially infinite-dimensional) feature space associated with a kernel function. The distance metric is computed in the feature space independently of its dimensionality by exploiting the kernel trick commonly used in machine learning. The kernel function can be tailored to incorporate prior knowledge of the dynamics when available. We consider a classical test example, the logistic map as a discrete dynamical system, and derive analytically the kernel function from the polynomial Koopman basis that exactly linearizes the dynamics. Dynafit is applicable to a wide range of classification tasks involving nonlinear dynamical systems and sensors. We illustrate its effectiveness through three examples: chaos detection in the logistic map, recognition of handwritten dynamical patterns, and classification of visual dynamic textures.

Keywords: Nonlinear dynamics · Koopman operator · Classification, Machine Learning · Kernel machines · Kernel trick

1 Introduction

Nowadays, the multiplication of sensors and portable devices promote the development of lightweight machine learning (ML) or tiny ML capable of performing sensor data classification at the edge, thereby avoiding the need to transfer data to the cloud [22,23]. Deep learning [11] is not the ideal candidate for classification on embedded systems as it requires a large number of training data combined with lots of computational and memory resources. In contrast, traditional ML is less expensive in terms of memory and computation. This is particularly the case for distance-based classifiers, such as k-nearest neighbors [7], k-means [12] or Parzen windows [20], which are among the simplest classification algorithms. The classification rule is based on the (dis)similarity, as

L. Pasa et al. (Eds.): ICANN 2026, LNCS 17089, pp. 333–344, 2027.
https://doi.org/10.1007/978-3-032-38398-3_28

measured by a distance function, between test data and stored prototypes from the training set. Distance-based classifiers have been developed, for static data, using the Euclidean or Mahalanobis distance, and, for time series, using the DTW (Dynamic Time Warping) distance [1]. Kernel machines, such as SVMs (Support Vector Machines) [5], can also be seen as distance-based classifiers as the kernel function can be interpreted as a measure of similarity between data points in transformed feature space [25].

Distance-based classifiers evaluate the similarity between pairs of samples (observed data vs stored prototypes) either in the original or feature space. In most cases, the underlying dynamical system is not explicitly taken into account when dealing with sequential data. However, we argue that the generative process behind the data can carry meaningful information for classification. A representative example is pen dynamics in signature verification: unlike the static image of a signature, the motion data is much harder for a skilled forger to reproduce. Therefore, when data originates from a dynamical system, classification could leverage a similarity measure not between individual samples, but between the observed data and the dynamics of the underlying system itself (Fig. 1).

Sequential data are frequently modeled using state-space representations, for which various distance metrics have been proposed. The Martin distance, for instance, compares two state-space models based on the principal angles between their observability subspaces [6,13]. Alternatively, the parity space method enables the comparison of a state-space model with observed data by projecting the data onto the parity space—the subspace orthogonal to the observability subspace [21]. However, both approaches rely on prior knowledge of the system dynamics, specifically the state-space matrices. In the absence of an explicit model, data-driven methods have been developed to learn a suitable metric, typically inspired by parity space-based distances [3,14,15]. These methods, however, are currently limited to linear systems, and extending them to nonlinear dynamics remains an open challenge.

Nonlinear dynamics have traditionally been handled using Jacobian and first-order Taylor series approximation. However, such linearization techniques are inherently local and only valid in the vicinity of fixed points. In contrast, the Koopman operator theory [9,10] offers a global linear representation of nonlinear dynamics, albeit in an infinite-dimensional space. Koopman spectral analysis has recently gained renewed interest for the identification and control of nonlinear systems [4,16,17]. For comparing dynamical systems, Mezić and collaborators provide a theoretical framework based on the spectral properties of the Koopman operator [18,19]. In practice, data-driven methods such as the extended Dynamic Mode Decomposition [28] can approximate the Koopman operator as a set of eigenvalues, eigenfunctions, and modes that span a Koopman-invariant subspace [4]. However, the dimension of this subspace can become intractably large in practice.

Kernel methods, by expressing distances through inner products, circumvent the need for explicit mappings into high-dimensional feature spaces [25]. In this work, we propose Dynafit, a data-driven algorithm for classifying nonlinear dynamical trajectories that leverages kernel methods and Koopman operator approximations. Dynafit classifies trajectories by identifying the minimum distance to the underlying dynamics. It differs from [30] and [26], which compare subspace representations of dynamical systems and emphasize geometric similarity, and from [29], which analyzes the spectral properties of dynamics in a reproducing kernel Hilbert space.

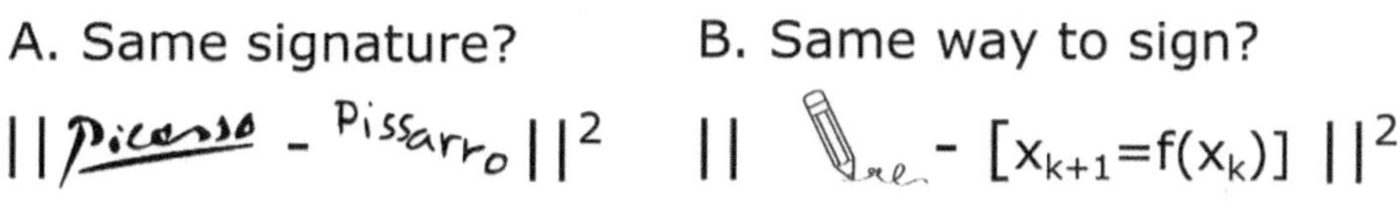

Fig. 1. Illustration of distance-based classification. A). In traditional classifiers, the distance is a measure of (dis)similarity between a test sample (here the image of Picasso's signature) and a stored prototype (Pissarro's signature). B). In Dynafit, the distance is between a trajectory sample (here the handwritten dynamics) and the dynamical system that has potentially generated the data.

2 Problem Statement

Let us consider a discrete-time system represented as

$$x_{k+1} = f(x_k) \tag{1}$$

with $x \in \mathbb{R}^n$ the n-dimensional state and f a nonlinear function. A trajectory vector $X \in \mathbb{R}^{nN}$ consists of a sequence of N states

$$X = \begin{pmatrix} x_0 \\ x_1 \\ \dots \\ x_{N-1} \end{pmatrix} \tag{2}$$

The Koopman operator theory [4,9,10] provides a linear but infinite-dimensional representation of nonlinear dynamics so that (1) can be rewritten as

$$\phi(x_{k+1}) = A\,\phi(x_k) \tag{3}$$

with $m \times m$ transition matrix A. The augmented state $\phi(x) \in \mathbb{R}^m$ is obtained through a nonlinear transformation $\phi(\cdot)$ onto a high-dimensional ($m >> n$), potentially infinite ($m = \infty$), feature space. Equation (3) does not represent a linearized version of the dynamics in the neighborhood of a fixed point that could be obtained when A is a Jacobian for example. Instead, it provides a global linear representation of the dynamics in a high-dimensional Hilbert space. In this feature space, observability matrix O_b maps initial state $\phi(x_0)$ into a l-dimensional trajectory vector $\phi(X) \in \mathbb{R}^l$ with $l = mN$:

$$\phi(X) = O_b\, \phi(x_0) \tag{4}$$

$$\text{with } \phi(X) = \begin{pmatrix} \phi(x_0) \\ \phi(x_1) \\ \cdots \\ \phi(x_{N-1}) \end{pmatrix} \text{ and } O_b = \begin{pmatrix} I_m \\ A \\ A^2 \\ \cdots \\ A^{N-1} \end{pmatrix} \tag{5}$$

where I_m is the $m \times m$ identity matrix. The initial state is given by $\phi(x_0) = O_b^+ \phi(X)$ in which $O_b^+ = (O_b^T O_b)^{-1} O_b^T$ is the pseudo-inverse of the observability matrix. This pseudo-inverse solution is the least-squares solution that minimizes the distance $d(X)$ in feature space between the collected data $\phi(X)$ and the underlying generating process $O_b \phi(x_0)$ and, such that :

$$d(X) = ||\phi(X) - O_b \phi(x_0)||^2 = ||\phi(X) - O_b O_b^+ \phi(X)||^2 \tag{6}$$

The problem is that the dynamical system is unknown so that $O_b O_b^+$ is unknown, and the dimension of $\phi(X)$ may be infinite. Thus, direct computation of the distance $d(X)$ is not tractable. In the next section, we propose a learning method that allows to estimate $O_b O_b^+$ based on training data and compute (6) independently of the (potentially infinite) dimensionality of the feature space.

3 Dynafit Learning

An estimation of $O_b O_b^+$ in (6) can be obtained by minimizing the distance in feature space for the training data $X^{(\mu)} \in \mathbb{R}^{nN}$, $\mu = 1 \cdots p$, which are particular trajectory samples (2) generated by system (1). Let

$$X_{train} = (X^{(1)}, X^{(2)}, \cdots, X^{(p)}) \in \mathbb{R}^{nN \times p}$$

and

$$\phi(X_{train}) = (\phi(X^{(1)}), \phi(X^{(2)}), \cdots, \phi(X^{(p)})) \in \mathbb{R}^{l \times p}$$

be the training set in original space and in feature space, respectively. The sum of individual distances for the training data in feature space writes as follows :

$$\begin{aligned} d(X_{train}) &= \sum_{\mu=1}^{p} d(X^{(\mu)}) \\ &= \sum_{\mu=1}^{p} ||\phi(X^{(\mu)}) - O_b O_b^+ \phi(X^{(\mu)})||^2 \\ &= ||\underbrace{\phi(X_{train}) - O_b O_b^+ \phi(X_{train})}_{\mathrm{E}}||_F^2 \end{aligned} \quad (7)$$

with $||E||_F^2 = tr(E^T E)$ is the Frobenieus norm and tr is the trace of the matrix.

Let us consider the SVD of $\phi(X_{train})$

$$\phi(X_{train}) = U\ \Sigma\ V^T \quad (8)$$

with U and V being the $l \times k$ and $p \times k$ matrices of the left and right singular vectors, respectively. The square $k \times k$ diagonal matrix Σ has non-zero singular values $\lambda_1, \cdots, \lambda_k$ in the diagonal in decreasing order, i.e. $\lambda_i \geq \lambda_{i+1}$.

Given that $U^T U = I_k$, it is easy to show that $d(X_{train}) = 0$ if $O_b O_b^+ = UU^T$. Thus, an estimation of $O_b O_b^+$ can be obtained as UU^T with U being the eigenvectors of $\phi(X_{train})\phi(X_{train})^T$ corresponding to non-zero singular values $\lambda_1, \cdots, \lambda_k$. In that case, one has

$$\begin{aligned} d(X_{train}) &= ||\phi(X_{train}) - U\ U^T\ \phi(X_{train})||_F^2 \\ &= tr\left(\phi(X_{train})^T \phi(X_{train}) - \phi(X_{train})^T U U^T \phi(X_{train})\right) \end{aligned} \quad (9)$$

as UU^T is an idempotent matrix, i.e. $UU^T UU^T = UU^T$. Yet, finding eigenvectors of $\phi(X_{train})\phi(X_{train})^T$, which is $l \times l$, is not tractable when $l = \infty$. To circumvent this problem, we consider a dual optimization. From (8) we have

$$U = \phi(X_{train})\ V\ \Sigma^{-1} \quad (10)$$

with Σ^{-1} being just the diagonal matrix with $1/\lambda_1, 1/\lambda_2, \cdots, 1/\lambda_k$ on the diagonal.

Now, V can be obtained as the eigenvectors corresponding to the non-zero eigenvalues of the $p \times p$ Gram matrix $K_{train} = \phi(X_{train})^T \phi(X_{train})$, with $p << l$ (number of training patterns much smaller than dimensionality of feature space). The eigenvalues of K_{train} are the squared singular values Σ^2 of $\phi(X_{train})$. Replacing (10) into (9) leads to

$$d(X_{train}) = tr\left(K_{train} - K_{train} H K_{train}\right) \quad (11)$$

$$\text{with } H = V\ \Sigma^{-2}\ V^T \quad (12)$$

K_{train} is the $p \times p$ kernel gram matrix with entries $k(X^{(i)}, X^{(j)}) = \phi(X^{(i)})^T \phi(X^{(j)})$. It is worth noting that the kernel function $k(\cdot)$ expresses an inner product in a high-dimensional feature space so that the kernel trick from kernel machines can be used to avoid the explicit mapping ϕ [5,25].

Thus, the learning process consists in the eigendecomposition of K_{train} to extract the eigenvectors V and eigenvalues Σ^2 and compute matrix H (Eq. 12). The total distance for the entire training set, i.e. $\sum_{\mu=1}^{p} d(X^{(\mu)})$, is obtained by Eq. (11). Individual distances $d(X^{(\mu)})$for each training data $X^{(\mu)}$ are obtained by replacing the trace operator $tr(\cdot)$ in Eq. (11) with the function $diag(\cdot)$ that returns the diagonal elements of the matrix.

4 Dynafit Inference

Let us consider a set of test data $X_{test} = (X'^{(1)}, X'^{(2)}, \cdots, X'^{(q)}) \in \mathbb{R}^{nN \times q}$. The total distance on the test set writes

$$d(X_{test}) = tr\left(K_{test} - K_{test,train} H K_{test,train}^T\right) \tag{13}$$

K_{test} is the $q \times q$ matrix with entries $k(X'^{(i)}, X'^{(j)}), i, j = 1 \cdots q$, and $K_{test,train}$ is the $q \times p$ matrix with entries $k(X'^{(i)}, X^{(j)}), i = 1 \cdots q, j = 1 \cdots p$.

Individual distances $d(X'^{(\mu)})$ for each test sample $X'^{(\mu)}$, $\mu = 1 \cdots q$, are obtained by replacing the trace operator $tr(\cdot)$ in Eq. (13) with the function $diag(\cdot)$ that returns the diagonal elements of the matrix. The class of a test sample $X'^{(\mu)}$ is obtained as follows :

- In one-class classification, the unique class represents the normal operating mode of the system. Unusual events (e.g. fault events) are detected whenever $d(X'^{(\mu)})$ exceeds a predetermined threshold (deviation from the normal operating mode).
- In multi-class classification, different classes represent different dynamics leading to one distance per class. The class of $X'^{(\mu)}$ is determined as the one with the smallest distance, that is, class i if $d_i(X'^{(\mu)}) < d_j(X'^{(\mu)}), \forall j \neq i$.

5 Kernel Engineering

Dynafit has been written in terms of kernel functions that correspond to inner products in some high dimensional (potentially infinite) feature space. When the dynamics is unknown, general kernel functions can be used such that the polynomial kernel $k(X, Y) = (1 + X^T Y)^d$ or the Gaussian kernel $k(X, Y) = \exp(-||X - Y||^2/\sigma^2)$. When the dynamics is partially known, specific kernels can be designed to incorporate prior knowledge. Here, we illustrate kernel engineering by considering the logistic map, whose state evolves according to the discrete dynamics :

$$x_{k+1} = r x_k (1 - x_k)$$

with $x \in [0,1)$. To enforce this domain, we clamp the dynamics to $x = 1 - \epsilon$ whenever $x = 1$, where ϵ is a very small positive quantity. The dynamics is partially known as the bifurcation parameter r is unknown. Depending on the value of r, the sequence may settle down to a regular regime (fixed point, periodic orbit) or may produce aperiodic chaos. As shown in [4], the Koopman operator for the logistic map can be represented in a polynomial basis with augmented state :

$$\phi(x) = \begin{pmatrix} x \\ x^2 \\ x^3 \\ \vdots \end{pmatrix} \tag{14}$$

In infinite-dimensional feature space, the state has linear dynamics $\phi(x_{k+1}) = A_\infty \phi(x_k)$ with :

$$A_\infty = \begin{pmatrix} r & -r & 0 & 0 & 0 & \cdots & \cdots & \cdots \\ 0 & r^2 & -2r^2 & r^2 & 0 & \cdots & \cdots & \cdots \\ 0 & 0 & r^3 & -3r^3 & 3r^3 & r^3 & 0 & \cdots \\ \vdots & \vdots & \vdots & \vdots & \vdots & \vdots & \vdots & \ddots \end{pmatrix}$$

As noted in [4] the rows of the transition matrix A_∞ are related to the rows of Pascal's triangle.

Let us consider truncating the state to degree ω (this constraint will be relaxed later) so that $\phi(x_{k+1}) \approx A_\omega \phi(x_k)$ with

$$\phi(x) = \begin{pmatrix} x \\ x^2 \\ \vdots \\ x^\omega \end{pmatrix} \tag{15}$$

The truncated system, with $\omega \times \omega$ transition matrix A_ω, approximates the true dynamics only for very large ω. Yet, for large ω, the feature space becomes intractably large which motivates the use of kernels.

Let us consider trajectory vectors $\phi(X)$ made of N consecutive observations (Eq 5). The kernel function expresses an inner product in feature space between two trajectories; that is :

$$k(X,Y) = \phi(X)^T \phi(Y) \tag{16}$$

Using (15), one has

$$k(X,Y) = \sum_{k=0}^{N-1} \left(\sum_{\xi=1}^{\omega} (x_k y_k)^\xi \right)$$

By identifying a geometric series for the summation between brackets, the kernel function simplifies to :

$$k(X,Y) = \sum_{k=0}^{N-1} \frac{x_k y_k (1 - (x_k y_k)^{\omega})}{1 - x_k y_k}$$

For $\omega \to \infty$, one has $(x_k y_k)^{\omega} \to 0$ as $x_k, y_k \in [0, 1)$. Thus, the logistic map kernel writes as follows

$$k(X,Y) = \sum_{k=0}^{N-1} \frac{x_k y_k}{1 - x_k y_k} \tag{17}$$

Evaluating the logistic map kernel (17) is much cheaper than calculating the mapping ϕ using (15) and the inner product (16) in feature space of high dimensionality. Also, the logistic map kernel (17) exactly represents the infinite-dimensional dynamics as $\omega = \infty$ and not the approximated dynamics of a truncated system.

6 Results

The Matlab code to reproduce the experiments can be found in the following Github repository : https://github.com/dynafit-sketch/dynafit.

Chaos Detection

In the past, chaos detection has been applied to neural data, meteorological data, financial and stock market, among others. Detecting chaotic regimes is critical for understanding and modeling a dynamical system, for predicting its future behavior and performing decision-making. With chaotic dynamics, one will never be able to predict the data beyond a brief interval determined by the Lyapunov exponent (example 4–5 days for weather prediction). Here, we take the logistic map as an example of chaotic system and we use Dynafit with the logistic map kernel (17) to discriminate between regular and chaotic regimes. To compare with deep learning we used the same dataset and parameter settings as in [2]. In this dataset[1], trajectories of $N = 1000$ samples are labeled as regular or chaotic depending of their Lyapunov exponent. We trained two metrics, d_1 for regular and d_2 for chaotic, using 2000 training data for each class. After training, Dynafit classified the test data as regular when $d_1 < d_2$ and chaotic otherwise. Table 1 summarizes the results obtained as compared to those reported in [2] for Multi-Layer Perceptron (MLP), Cellular Neural Network (CNN) and Long Short-Term Memory (LSTM). We note that Dynafit outperforms most of the

[1] https://doi.org/10.17632/k4x675k5dm.2.

deep learning classifiers. Training accuracy for Dynafit is higher than for MLP, CNN and LSTM. Test accuracy for Dynafit is better than for MLP and LSTM and slightly lower to that of CNN (99.2% vs 99.4%). In addition, Training is very fast for Dynafit, that is 37 sec on a standard laptop DELL latitude 7410 for computing the 2000×2000 kernel matrices and solving the eigenvalue problems for the two classes (regular and chaotic).

Table 1. Logistic map experiments. Train and test accuracy on the training set (2000 samples) and the test set (1000 samples) for Dynafit as compared to deep learning. Dynafit is trained with the logistic map kernel given in Eq. (17). Results for MLP, CNN and LSTM are taken from [2].

	Train acc. (in %)	Test acc. (in %)
Dynafit	99.75	99.2
MLP	99.302±0.161	97.865±0.098
CNN	99.383±0.335	99.410±0.239
LSTM	97.979±0.584	98.565±0.279

Character Recognition From Pen Trajectory

We consider a real dataset from the UCI Machine learning archive[2]. The dataset consists of 2858 handwriting characters of 20 classes collected using a WACOM pen tablet. The sensor data consists in pen trajectories with three components recorded at 200 Hz: pen velocity in x-position (mm/s), pen velocity in y-position (mm/s) and rate of the pen tip force (N/s). We used Dynafit with one distance metric per class using the first N=10, 20, 50 and 100 pen trajectory data. The training set consisted of 10% of the recordings, randomly sampled from the dataset, and the test set consisted of the remaining 90%. Training and test were repeated 100 times. Table 2 summarizes the results obtained with Dynafit as compared to SVM. Standard SVM classification is performed by following the one-vs-all strategy that splits the multiclass problem into one binary classification problem per class. We note that Dynafit outperforms SVM both in terms of test accuracy and training time with significant speedup. In SVM, we solve one quadratic programming problem per class that involves all the data. In Dynafit, we solve one eigenvalue problem per class that involves merely the data from one class. This simpler problem makes Dynafit up to 260 times faster than the standard SVM.

Dynamic Texture Classification

Dynamic textures (also called temporal textures) are time-varying visual patterns that are modeled with linear state-space representations [8]. Here, we use

[2] https://archive.ics.uci.edu/dataset/175/character+trajectories.

Table 2. Test accuracy on the test set with the first N =10, 20, 50 or 100 sampled points in the pen trajectory. The polynomial kernel has degree $d = 2$ for all N. The Gaussian kernel has width $\sigma = 2.9$, 4.2, 7.5 and 15 for N =10, 20, 50 and 100, respectively. Results are represented as means±S.D estimated over 100 trials. 'Train time' is the time in millisecond for training the classifier in each trial.

Kernel	N	Dynafit		SVM	
		Test Acc. (in %)	Train time (in ms)	Test Acc. (in %)	Train time (in ms)
polynomial	10	37.6±1.1	1.1±0.5	35.3 ±1.2	268.2±41.8
	20	50.9±1.1	1.2±0.5	51.5±1.4	319.9±57.3
	50	81.3±1.0	1.6±0.7	80.0±1.3	188.1±22.3
	100	94.2±0.7	1.8±0.8	93.9±0.8	174.3±17.5
gaussian	10	38.5±1.0	4.1±1.3	31.5 ±1.3	147.2±18.3
	20	52.5±1.1	5.0±1.2	49.3±1.4	156.3±18.9
	50	82.5±1.0	7.5±1.6	78.3±1.3	159.8±14.8
	100	93.7±0.7	8.9±1.3	91.3±0.9	175.5±16.4

Dynafit to classify dynamic textures. The UCLA dynamic texture dataset [24] consists of 50 classes of different textures, each class having 4 sequences of 75 frames of 48×48 pixels. For each class, we used three sequences for training and the remaining one for testing. Figure 2 shows the first images from the test sequence for each class. Using Dynafit, we obtain a test accuracy of 78% and 86% with a polynomial kernel of degree 1 and 3, respectively. These performance are well above the chance level of 2% and are comparable to those obtained in [24] with the Martin distance [6,13], i.e. 71% and 89.5% with features extracted from independent component analysis and principal component analysis, respectively.

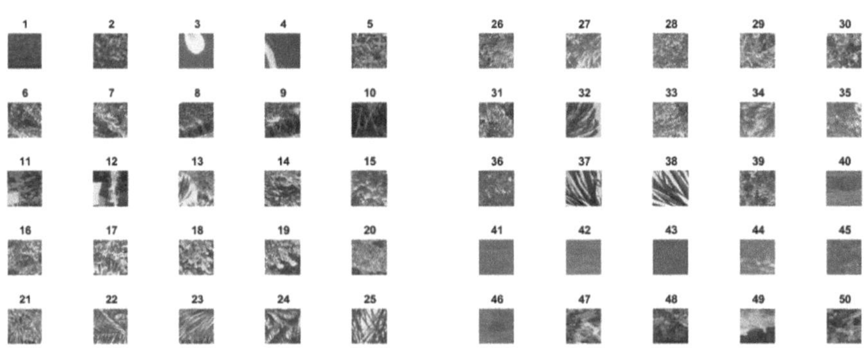

Fig. 2. Image samples from the UCLA dynamic texture dataset (50 classes of different dynamic textures). Each panel corresponds to the first image in the test sequence of a given class.

7 Conclusion

We introduced a data-driven distance metric for the classification of trajectories and sequences generated by nonlinear dynamical systems. The metric is based on a kernel-induced projection that approximates the Koopman operator, thereby enabling a global linearization of the dynamics in a potentially infinite-dimensional feature space. As a proof of concept, we analytically derived a kernel function from the polynomial Koopman basis that linearizes the dynamics of the logistic map. Using this kernel, the proposed metric successfully discriminates between regular and chaotic regimes, achieving performance comparable to—and in some cases exceeding—that of deep learning approaches. Unlike deep learning methods, the proposed framework does not require the end-to-end training of large parametric models. Instead, it relies on solving single-class eigenvalue problems, resulting in a computationally efficient learning procedure. This advantage was demonstrated on a character trajectory benchmark, where the proposed approach achieved a 260× speedup compared with SVM-based classification. It should be emphasized, however, that the objective of the experimental evaluation was not to provide an exhaustive comparison with state-of-the-art machine learning or deep learning techniques. Rather, the goal was to demonstrate that the proposed Koopman-based metric captures discriminative information about the underlying dynamics and can therefore serve as an effective tool for trajectory classification. Future work will focus on further reducing the computational requirements of the method to facilitate deployment in resource-constrained edge-learning environments. In particular, the Nyström approximation [27] will be investigated as a means of efficiently scaling the approach to large datasets and high-dimensional problems.

Acknowledgment. This work was funded by France 2030 under the investment program ANR-20-PCPA-0007 Pherosensor and by ANR-23-CE51-0037 Bucolyc.

References

1. Abanda, A., Mori, U., Lozano, J.: A review on distance based time series classification. Data Min. Knowl. Disc. **33**, 378–412 (2019)
2. Barrio, R., et al.: Deep learning for chaos detection. Chaos **33**(073146) (2023)
3. Battistelli, G., Tesi, P.: Classification for dynamical systems: model-based and data-driven approaches. IEEE Trans. Autom. Control **6**, 1741–1748 (2021)
4. Brunton, J., Brunton, B., Proctor, J., Kutz, J.: Koopman invariant subspaces and finite linear representations of nonlinear dynamical systems for control. PLoS One (2016)
5. Burges, C.: A tutorial on support vector machines for pattern recognition. Data Min. Knowl. Disc. **2**, 121–167 (1998)
6. Cock, K.D., Moor, B.D.: Subspace angles and distances between arma models. Syst. Control Lett. **46**(4), 265–270 (2002)
7. Cover, T., Hart, P.: Nearest neighbor pattern classification. IEEE Trans. Inf. Theory **13**(1), 21–27 (1967)

8. Doretto, G., Chiuso, A., Wu, Y., Soatto, S.: Dynamic textures. Intl. J. Comput. Vis. **51**(2), 91–109 (2003)
9. Koopman, B.: Hamiltonian systems and transformation in hilbert space. Proc. Natl. Acad. Sci. **17**(5), 315–318 (1931)
10. Koopman, B., Neumann, J.: Hamiltonian systems and transformation in hilbert space. Proc. Natl. Acad. Sci. **18**(3), 255 (1932)
11. LeCun, Y., Bengio, Y., Hinton, G.: Deep learning. Nature **521**, 436–444 (2015)
12. Lloyd, S.: Least squares quantization in PCM. IEEE Trans. Inf. Theory **28**(2), 129–137 (1982)
13. Martin, R.: A metric for arma processes. IEEE Trans. Signal Process. **48**(4), 1164–1170 (2000)
14. Martinez, D., Boutayeb, M.: Classification for dynamical systems: model-based and data-driven approaches. IEEE Sens. J. **23**(22), 28454–28461 (2023)
15. Martinez, D., Boutayeb, M.: Nullspace-based metric for classification of dynamical systems and sensors. In: LNCS, Springer, Cham, vol. 15016, no. 18–29 (2024)
16. Mezić, I.: Spectral properties of dynamical systems, model reduction and decompositions. Nonlinear Dyn. **41**(1), 309–325 (2005)
17. Mezić, I.: Analysis of fluid flows via spectral properties of the Koopman operator. Annual Rev. Fluid Mech. **45**, 357–378 (2013)
18. Mezić, I.: On comparison of dynamics of dissipative and finite-time systems using Koopman operator methods. IFAC-PapersOnLine **49**(18), 454–461 (2016), 10th IFAC Symposium on Nonlinear Control Systems NOLCOS 2016
19. Mezić, I., Banaszuk, A.: Comparison of systems with complex behavior. Physica D **197**(1), 101–133 (2004)
20. Parzen, E.: On estimation of a probability density function and mode. IAnn. Math. Stat **33**, 1065–1076 (1962)
21. Patton, R., Chen, J.: A review of parity space approaches to fault diagnosis. IFAC Proc. **24**(6), 65–81 (1991)
22. Ravindran, S.: Cutting AI down to size. Science **387**(6736), 818–821 (2025)
23. Ray, P.: A review on tinyML: state-of-the-art and prospects. J. King Saud Univ.-Comput. Info. Sci. **34**(4), 1595–1623 (2022)
24. Saisan, P., Doretto, G., Wu, Y., Soatto, S.: Dynamic texture recognition. IEEE Conf. CVPR **2**, 58–63 (2001)
25. Scholkopf, B.: The kernel trick for distances. In: Advances in Neural Information Processing Systems (NIPS), vol. 2 (2000)
26. Vishwanathan, S., Smola, A., Vidal, R.: Binet-cauchy kernels on dynamical systems and its application to the analysis of dynamic scenes. Int. J. Comput. Vis. **73**, 95–119 (2007)
27. Williams, C., Seeger, M.: Using the nyström method to speed up kernel machines. In: Advances in Neural Information Processing Systems (NIPS), p. 13 (2000)
28. Williams, M., Kevrekidis, I., Rowley, C.: A data-driven approximation of the Koopman operator: extending dynamic mode decomposition. J. Nonlinear Sci. **25**, 1307–1346 (2015)
29. Williams, M., Rowley, C., Kevrekidis, I.: A kernel-based method for data-driven Koopman spectral analysis. J. Computat. Dynam. **2**(2), 247–265 (2015)
30. Wolf, L., Shashua, A.: Learning over sets using kernel principal angles. J. Mach. Learn. Res. **4**, 913–931 (2003)

Fine-Tuning Pre-trained Vision-Language Models in a Human-Annotation-Free Manner

Qian-Wei Wang[1,2], Guanghao Meng[1,2], Ren Cai[1,3], Yaguang Song[1(✉)], and Shu-Tao Xia[1,2]

[1] Pengcheng Laboratory, Shenzhen, China
songyg01@pcl.ac.cn
{wanggw21,menggh22}@mails.tsinghua.edu.cn, 2101112278@stu.pku.edu.cn,
songyg01@pcl.ac.cn

[2] Tsinghua Shenzhen International Graduate School, Tsinghua University, Beijing, China
xiast@sz.tsinghua.edu.cn

[3] School of Electronic and Computer Engineering, Peking University Shenzhen Graduate School, Shenzhen, China

Abstract. Large-scale vision–language models (VLMs) such as CLIP exhibit strong zero-shot generalization, but adapting them to downstream tasks typically requires costly labeled data. Existing unsupervised self-training methods rely on pseudo-labeling, yet often suffer from unreliable confidence filtering, confirmation bias, and underutilization of low-confidence samples. We propose collaborative fine-tuning (CoFT), an unsupervised adaptation framework that leverages unlabeled data through a dual-model, cross-modal collaboration mechanism. CoFT introduces a dual-prompt learning strategy with positive and negative textual prompts to explicitly model pseudo-label cleanliness in a sample-dependent manner, removing the need for hand-crafted thresholds or noise assumptions. The negative prompt also regularizes lightweight visual adaptation modules, improving robustness under noisy supervision. CoFT employs a two-phase training scheme, transitioning from parameter-efficient fine-tuning on high-confidence samples to full fine-tuning guided by collaboratively filtered pseudo-labels. Extensive experiments demonstrate consistent gains over existing unsupervised methods and even few-shot supervised baselines.

Keywords: Pre-trained vision-language models · Unsupervised learning · Pseudo-labeling

1 Introduction

In recent years, large-scale pre-trained vision-language models (VLMs) [1,16,18] such as CLIP [25] have revolutionized cross-modal understanding by leveraging

Yaguang Song—Corresponding author, and this work is fully supported by Institute of Perceptual Intelligence, Pengcheng Laboratory.

L. Pasa et al. (Eds.): ICANN 2026, LNCS 17089, pp. 345–357, 2027.
https://doi.org/10.1007/978-3-032-38398-3_29

massive amounts of general-domain data to learn robust alignments between visual and textual representations. These models have demonstrated remarkable versatility in zero-shot and few-shot settings across diverse downstream tasks, including image classification, retrieval, and visual reasoning. Their ability to generalize from pre-trained knowledge to unseen tasks has made them a cornerstone of modern AI research, enabling applications in fields ranging from medical imaging to autonomous driving.

However, despite their strong zero-shot performance, fine-tuning remains a critical step to unlock their full potential on specific downstream tasks. Traditional fine-tuning paradigms, whether full-shot (using large labeled datasets) or few-shot (relying on limited annotations), share a common limitation: they depend on manually labeled task-specific data. This reliance on human annotations is not only labor-intensive and costly but also impractical for domains where labeled data is scarce (e.g., rare disease diagnosis) or rapidly evolving (e.g., emerging object categories in robotics).

To address this challenge, researchers have explored unsupervised learning strategies that leverage unlabeled data to enhance model performance. Self-training, a prominent unsupervised approach, involves generating pseudo-labels for unlabeled samples using a pre-trained model and iteratively refining the model with these labels. While promising, existing self-training methods for VLMs [3,11,21,22,33,35] often suffer from two key issues: (1) These methods lack effective mechanisms for exploiting low-confidence samples. Some approaches utilize only a small number of highest-confidence samples per class for fine-tuning, discarding the remaining unlabeled data; this severely limits the utilization of available data and leaves the model reliant on a narrow subset of potentially biased initial pseudo-labels. Other methods iteratively expand the coverage of pseudo-labels, but fail to incorporate effective safeguards against confirmation bias, making them prone to performance degradation. (2) Their fine-tuning processes usually rely on uni-model adjustments, lacking a collaborative mechanism between multiple models that could leverage cross-modal alignment to generate and validate pseudo-labels more robustly.

In this paper, we propose a novel **Co**llaborative **F**ine-**T**uning (CoFT) framework for effectively adapting vision–language models without any human annotations. The core of CoFT lies in a dual-prompt-based, dual-model cross-modal collaboration mechanism, which enables robust pseudo-label generation and validation without relying on manually designed confidence thresholds or prior assumptions about the label noise ratio. Specifically, we introduce a pair of positive and negative prompts to explicitly model the probability that a pseudo-label is clean. The negative prompt acts as a learnable, data-driven similarity reference that provides a sample-adaptive decision criterion, allowing the model to automatically distinguish clean pseudo-labels from noisy ones. In addition, our method adopts a two-phase training paradigm that progressively transitions from parameter-efficient fine-tuning to full fine-tuning. In the early phase, only a small set of high-confidence samples is used to update lightweight tunable parameters in both the textual and visual branches, effectively suppressing confirmation bias. In the later phase, the two models collaboratively perform cross-modal pseudo-label generation and validation, reliably expanding supervision to the majority of samples and enabling full fine-tuning of the visual encoder. This design establishes an effective bridge between conservative high-confidence learning and large-scale task-specific alignment, allowing CoFT to exploit unlabeled data more thoroughly and robustly.

Extensive experiments demonstrate that our methods effectively leverage unlabeled data to boost performance, even outperforming fine-tuning with few manually labeled samples. This work highlights the potential of collaborative self-training in VLMs, offering a cost-effective alternative to labeled data-dependent fine-tuning for task-specific adaptation.

2 Related Work

Large-scale vision–language models (VLMs), such as CLIP [25], learn aligned visual and textual representations from massive image–text pairs [1,16,18,32], enabling strong zero-shot generalization. To adapt these models to downstream tasks, fine-tuning strategies range from full fine-tuning, which updates all parameters but risks overfitting on small datasets, to parameter-efficient fine-tuning (PEFT) [7], which updates only a small subset of parameters while freezing the backbone. Representative PEFT methods include prompt learning [12,13,36,37], adapters [6,9,31], and low-rank adaptation (LoRA) [10,19,38], all of which aim to balance adaptation flexibility and parameter efficiency.

Self-training has been widely explored to leverage unlabeled data via pseudo-labels [26,28,30]. In the context of VLMs, recent works exploit cross-modal alignment to generate pseudo-labels without manual annotation. UPL [11] utilizes CLIP's zero-shot predictions to perform unsupervised prompt learning by selecting top-confidence samples, while DEFT [29] treats VLMs as noise detectors to identify clean samples under label noise, demonstrating the potential of self-guided training with pre-trained multimodal knowledge.

3 Methodology

Our CoFT framework (see Fig. 1) adopts a two-phase collaborative training paradigm, in which two sub-models are jointly optimized and mutually reinforce each other. In Phase I, we perform parameter-efficient fine-tuning (PEFT) on a small set of high-confidence pseudo-labeled samples to establish reliable task alignment while explicitly preventing early-stage noise amplification (see Sect. 3.1). In Phase II, pseudo-label generation is systematically expanded to the entire unlabeled dataset through bidirectional collaboration between the two task-aligned models. Leveraging their complementary predictions and learned cross-modal discrimination criteria, CoFT performs sample-dependent pseudo-label filtering, retaining only reliable supervision for subsequent training. The resulting high-quality pseudo-labels are then used to fully fine-tune the visual encoder, enabling robust large-scale task adaptation and further boosting task-specific performance (see Sect. 3.2).

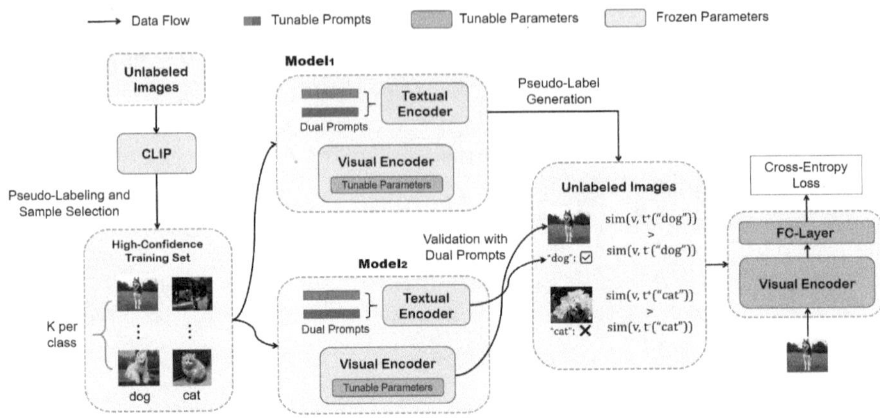

Fig. 1. Overall framework of the proposed collaborative fine-tuning (CoFT) method.

3.1 PEFT with High-Confidence Pseudo-labels

In the first phase, we initialize two CLIP models (denoted as $Model_1$ and $Model_2$) and introduce lightweight trainable parameters into both the textual and visual branches of the models, while keeping the pre-trained parameters frozen. The models are then fine-tuned using a set of highest-confidence initial pseudo-labels obtained from zero-shot inference.

Pseudo-label Generation and High-Confidence Selection. We start by generating initial pseudo-labels for the unlabeled downstream images using the pre-trained CLIP model via zero-shot inference. For each image $\boldsymbol{x}$ in the given unlabeled set $\mathcal{U}$, we compute the similarity between its visual embedding $v(\boldsymbol{x})$ encoded by CLIP's visual encoder and textual embeddings of natural language descriptions of all task categories $\{t(\boldsymbol{c}_k)\}_{k=1}^{C}$ encoded by CLIP's textual encoder, where $v(\cdot)$ and $t(\cdot)$ denote the visual encoder and textual encoder of CLIP respectively, $\boldsymbol{c}_k$ is the natural language description of the k-th task category and C is the total number of classes, yielding a probability distribution over classes:

$$p(\boldsymbol{c}_k \mid \boldsymbol{x}) = \frac{\exp(\mathrm{sim}(v(\boldsymbol{x}), t(\boldsymbol{c}_k))/\tau)}{\sum_{k'=1}^{C} \exp(\mathrm{sim}(v(\boldsymbol{x}), t(\boldsymbol{c}'_k))/\tau)}. \tag{1}$$

Here, τ is a temperature parameter, and $\mathrm{sim}(\boldsymbol{a}, \boldsymbol{b}) = \frac{\boldsymbol{a} \cdot \boldsymbol{b}}{\|\boldsymbol{a}\| \cdot \|\boldsymbol{b}\|}$ denotes cosine similarity. The predicted pseudo-label for each sample is computed as $y = \arg\max_k p(\boldsymbol{c}_k \mid \boldsymbol{x})$. Then, for each class, we select the top-K samples with the highest confidence scores. These selected samples form the high-confidence subset $\mathcal{S}^*$ with a size of $C \cdot K$, which serves as the training data for the first phase.

PEFT of Visual and Textual Encoders. To mitigate the accumulation of errors caused by confirmation bias, we restrict the early-stage optimization to a small set of high-confidence samples and lightweight trainable parameters.

For the textual encoder, we introduce a **dual prompt learning** mechanism to explicitly model clean and noisy pseudo-labels more robustly. Existing methods for pseudo-label filtering often rely on prior knowledge of the noise ratio or manually designed confidence thresholds, which limits their practical applicability. To overcome this limitation, we maintain a pair of learnable prompts: a positive prompt and a negative prompt.

Formally, for each class $k \in \{1 \ldots C\}$, we define:

$$\text{prompt}_k^+ = [V_1^+], [V_2^+], ..., [V_M^+], [\text{CLS}]_k, \tag{2}$$

$$\text{prompt}_k^- = [V_1^-], [V_2^-], ..., [V_M^-], [\text{CLS}]_k, \tag{3}$$

where $V_i^+, V_i^- \in \mathbb{R}^d$ are learnable context tokens, M is the total number of tunable textual context tokens and $[\text{CLS}]_k$ denotes the class token.

The positive prompt is trained to uncover discriminative features by maximizing the similarity between image features and their corresponding text features when paired with the true label. In contrast, the negative prompt is designed to behave oppositely: when paired with the true label, its textual representation should be dissimilar to the image features, whereas when paired with a non-true (complement) label, it should exhibit higher similarity. In this way, the negative prompt functions as a learnable, sample-dependent similarity threshold, enabling automatic identification of clean and noisy pseudo-labels without manual heuristics. Meanwhile, the learning of the negative prompt also supervises the lightweight trainable modules in the visual branch from a "negative discrimination" perspective, effectively acting as a form of regularization that improves the robustness of visual adaptation.

On the visual side, we insert lightweight trainable modules to modulate visual representations. In implementation, we adopt Visual Prompt Tuning (VPT) [12], where a small number of learnable visual prompt tokens are prepended to the input sequence of the visual transformer. These prompts are optimized to adapt CLIP's visual representations to task-specific characteristics without modifying the pre-trained transformer weights, thereby enabling efficient and stable adaptation.

Dual Loss Training for Dual Prompts. We jointly optimize $Model_1$ and $Model_2$ using two complementary loss functions that correspond to the distinct roles of the positive and negative prompts. This dual-loss design allows the model to both learn discriminative representations from high-confidence samples and identify noisy pseudo-labels in a principled manner.

- Loss for Positive Prompts and Visual Prompts ($\mathcal{L}_1$): This loss aligns visual embeddings with their corresponding positive textual embeddings using the high-confidence subset $\mathcal{S}^*$. For a sample $\boldsymbol{x}$ with pseudo-label y, we minimize the cross-entropy between the predicted distribution and the pseudo-label:

$$\mathcal{L}_1 = -\frac{1}{|\mathcal{S}^*|} \sum_{\boldsymbol{x} \in \mathcal{S}^*} \log \frac{\exp(\text{sim}(v_{\theta_v}(\boldsymbol{x}), t^+_{\theta_t}(\boldsymbol{c}_y))/\tau)}{\sum_{k=1}^{C} \exp(\text{sim}(v_{\theta_v}(\boldsymbol{x}), t^+_{\theta_t}(\boldsymbol{c}_k))/\tau)}, \tag{4}$$

where θ_v and θ_t denote the tuned parameters of visual and textual encoder, $v_{\theta_v}(\boldsymbol{x})$ denotes the visual embedding modulated by visual prompts, and $t^+_{\theta_t}(\boldsymbol{c}_k)$ represents the textual embedding of the positive prompt for class k.

- Loss for Negative Prompts ($\mathcal{L}_2$): This loss ensures that positive prompts are more discriminative than negative prompts for true labels, while the opposite holds for noisy labels. For each sample $\boldsymbol{x}$ with pseudo-label y, we randomly sample a complement label $\hat{y}$ from the remaining classes to simulate noise. We formulate the probability of a pseudo-label y being a true label as:

$$p_y^{clean} = \frac{\exp(\text{sim}(v_{\theta_v}(\boldsymbol{x}), t^+_{\theta_t}(\boldsymbol{c}_y))/\tau)}{\exp(\text{sim}(v_{\theta_v}(\boldsymbol{x}), t^+_{\theta_t}(\boldsymbol{c}_y))/\tau) + \exp(\text{sim}(v_{\theta_v}(\boldsymbol{x}), t^-_{\theta_t}(\boldsymbol{c}_y))/\tau)}. \tag{5}$$

We encourage high-confidence pseudo-labels to be classified as clean, while suppressing the probability that the complement label $\hat{y}$ is clean. The resulting loss is:

$$\mathcal{L}_2 = -\frac{1}{|\mathcal{S}^*|} \sum_{\boldsymbol{x} \in \mathcal{S}^*} \log p_y^{clean} + \log(1 - p_{\hat{y}}^{clean}). \tag{6}$$

This loss enforces that $\text{sim}(v_{\theta_v}(\boldsymbol{x}), t^+(\boldsymbol{c}_y)) > \text{sim}(v_{\theta_v}(\boldsymbol{x}), t^-(\boldsymbol{c}_y))$ for true labels and $\text{sim}(v_{\theta_v}(\boldsymbol{x}), t^+(\boldsymbol{c}_{\hat{y}})) < \text{sim}(v_{\theta_v}(\boldsymbol{x}), t^-(\boldsymbol{c}_{\hat{y}}))$ for noisy labels.

This objective enforces that, for clean samples, the similarity between the image features and the positive prompt exceeds that of the negative prompt, while the opposite holds for noisy labels. By jointly optimizing $\mathcal{L}_1$ and $\mathcal{L}_2$, the model learns to adapt safely under noisy supervision while preventing error accumulation in the early training stage. The total loss for this phase is $\mathcal{L}_{\text{Phase1}} = \mathcal{L}_1 + \lambda\mathcal{L}_2$, where λ balances the two losses. $Model_1$ and $Model_2$ are trained with the same loss but different initialization to encourage diversity.

3.2 Collaborative Fine-Tuning

In this phase, $Model_1$ and $Model_2$ collaborate to generate and filter reliable pseudo-labels from the entire unlabeled set $\mathcal{U}$, which are then used to fully fine-tune the visual encoder.

Collaborative Pseudo-label Generation and Validation. The collaboration between $Model_1$ and $Model_2$ is bidirectional, involving mutual pseudo-label generation and validation to improve pseudo-label quality. Without loss of generality, we describe the process where $Model_1$ performs pseudo-label generation and $Model_2$ conducts validation.

- **Pseudo-label generation by** $Model_1$**:** $Model_1$ generates pseudo-labels for all samples in $\mathcal{U}$ equipped with its fine-tuned textual and visual prompts. For each $\boldsymbol{x} \in \mathcal{U}$, the pseudo-label $y_1(\boldsymbol{x})$ is assigned as the class with the highest similarity to $v_{\theta_v^1}(\boldsymbol{x})$ under $t^+_{\theta_t^1}$, where θ_t^1, θ_v^1 and θ_t^2, θ_v^2 denotes the textual and visual attached parameters of $Model_1$ and $Model_2$, respectively.
- **Validation by** $Model_2$**:** $Model_2$ validates $y_1(\boldsymbol{x})$ using its dual prompts. A sample $\boldsymbol{x}$ with pseudo-label $y_1(\boldsymbol{x})$ is retained in the clean set $\mathcal{S}^1_{\text{clean}}$ if

$$\text{sim}(v_{\theta_v^2}(\boldsymbol{x}), t^+_{\theta_t^2}(\boldsymbol{c}_{y_1(\boldsymbol{x})})) > \text{sim}(v_{\theta_v^2}(\boldsymbol{x}), t^-_{\theta_t^2}(\boldsymbol{c}_{y_1(\boldsymbol{x})})).$$

Otherwise, it is assigned to the noise set $\mathcal{S}^1_{\text{noise}}$ and excluded from fully fine-tuning.

This collaborative procedure ensures that only samples exhibiting consistent cross-modal alignment, characterized by high positive similarity and low negative similarity, are selected for subsequent supervised training. Finally, we obtain two pairs of noisy and clean subset: $(\mathcal{S}^1_{\text{clean}}, \mathcal{S}^1_{\text{noise}})$ and $(\mathcal{S}^2_{\text{clean}}, \mathcal{S}^2_{\text{noise}})$.

Fully Fine-Tuning. We append a task-specific fully connected (FC) head to the visual encoder to map visual embeddings to class logits: $f(\boldsymbol{v}) = W_{\text{fc}}\boldsymbol{v} + \boldsymbol{b}_{\text{fc}}$, where $W_{\text{fc}} \in \mathbb{R}^{C\times d}$ and $\boldsymbol{b}_{\text{fc}} \in \mathbb{R}^C$. We then fine-tune the entire visual encoder together with the FC head using $\mathcal{S}^{1(2)}_{\text{clean}}$, optimized by the cross-entropy loss over the filtered pseudo-labels:

$$\mathcal{L}_{\text{FFT}} = -\frac{1}{|\mathcal{S}^{1(2)}_{\text{clean}}|} \sum_{\boldsymbol{x}\in\mathcal{S}^{1(2)}_{\text{clean}}} \log\left(\frac{\exp(f(v_{\theta_v}(\boldsymbol{x}))[y^{1(2)}(\boldsymbol{x})])}{\sum_{k=1}^{C}\exp(f(v_{\theta_v}(\boldsymbol{x}))[k])}\right). \tag{7}$$

4 Experiments

4.1 Datasets and Baselines

Following previous methods [11,37], we use ten publicly available image classification datasets including Caltech101 [5], CIFAR-100N [15], DTD [4], EuroSAT [8], FGVCAircraft [20], Flowers102 [23], Food101 [2], OxfordPets [24], StandfordCars [14], and UCF101 [27].

Except zero-shot inference baseline of CLIP, we compare our methods with five unsupervised fine-tuning approaches: **UPL** [11], **IFPL** [21], **GRIP** [21], **LaFTer** [22] and **CPL** [33] and three few-shot fine-tuning approaches: **CoOp** [37], **Tip-Adapter** [34], and **Tip-Adapter-F** [34]. Moreover, in the setting of learning from naturally noisy dataset, our methods are compared with nine learning from noisy labels approaches such as: **DEFT** [29]; and in the setting of distillation, our methods are compared with **PromptKD** [17]. To ensure a fair comparison, all algorithms were re-evaluated under the same backbone, i.e., ViT-B/16, unless otherwise specified.

Implementation Details. In implementation, we select the top $K = 16$ highest-confidence samples for each class for training in the first phase. For training in both phases, we use SGD as the optimizer, with initial learning rates set to 0.03 (for training a small number of tunable parameters) and 5e-4 (for training the entire visual encoder), respectively, along with a cosine decay scheduler. The weight decay is set to 5e-4 and momentum to 0.9. The number of training epochs is selected from $\{1, 10, 20\}$. For both positive and negative in dual-prompt learning, the number of tunable context tokens is $M = 16$, with the class token positioned at the end, and the tunable context tokens for visual branch is $M_v = 20$. We conduct all the experiments on one NVIDIA RTX 3090 GPU.

4.2 Main Results

Table 1. Performance comparison on various benchmark datasets. Best results are shown in **bold**. All results are reported in percentage (%).

Methods	CLIP	CoFT	UPL	IFPL	GRIP	LaFTer	CPL	CoOp	Tip-A	Tip-AF
Settings	ZSL	UL	UL	UL	UL	UL	UL	2-shot	2-shot	2-shot
Caltech101	89.27	**95.50**	94.56	93.47	92.13	93.43	82.80	94.40	93.75	94.44
DTD	44.73	**52.13**	44.68	48.24	50.90	51.77	50.11	50.18	52.42	**55.97**
EuroSAT	48.75	**86.02**	66.20	39.31	59.68	45.15	63.59	59.06	68.68	72.69
FGVCAircraft	25.11	**29.49**	23.67	23.10	10.14	20.91	22.86	27.48	31.59	**33.45**
Flowers102	71.73	**78.08**	73.37	68.86	68.25	65.25	65.73	83.96	87.29	**89.69**
Food101	84.83	**90.58**	84.71	86.55	89.16	83.72	88.64	84.65	86.09	86.28
OxfordPets	86.01	**92.40**	90.43	89.42	91.69	87.03	90.08	89.32	90.27	91.50
StanfordCars	66.52	**77.08**	66.19	50.55	60.84	60.76	58.23	66.30	68.55	70.87
UCF101	67.44	**78.35**	71.82	68.70	70.50	69.47	68.23	72.91	71.16	75.71
Average	64.93	**75.51**	68.40	63.13	65.92	64.17	65.59	69.81	72.20	74.51

Table 1 summarizes the classification accuracy under the fully unlabeled setting. It can be seen that our method consistently outperforms CLIP zero-shot and other unsupervised CLIP adaptation methods across all nine benchmarks. Our method achieves the best average accuracy of 75.51%, yielding a substantial improvement of 10.58% over the CLIP baseline. Compared with recent unsupervised approaches, our method demonstrates clear performance advantages on most datasets, surpassing the strongest competitor by an average margin of 7.11%. These results highlight the benefits of the proposed dual-prompt, dual-model collaborative pseudo-labeling mechanism, which explicitly models pseudo-label reliability through cross-model validation and effectively reduces confirmation bias while enabling more comprehensive exploitation of unlabeled data.

4.3 Results on Real-World Noisy Dataset

Table 2. Performance comparison on CIFAR-100N dataset, whose labels are derived from crowd-sourcing with a noisy rate $r \approx 0.4$.

Methods	CE	ELR	SCE	GMM	RoLT	UNIC	PMix	DEFT	CoFT
Perf.(%)	72.41	72.83	72.52	76.06	75.91	77.68	75.97	79.04	**79.40**

Table 2 reports results on real-world noisy dataset CIFAR-100N, where labels are obtained via crowd-sourcing with a high noise rate ($r! \approx!0.4$). We compare our methods with representative noisy-label learning approaches, including loss-based methods (CE, SCE, ELR, GMM), sample-selection techniques (RoLT, UNICON, ProMix) and recent CLIP incorporating method DEFT, all of which rely on the provided noisy annotations for training. In contrast, our method completely discard human labels and perform fully annotation-free adaptation. Despite this, CoFT achieves an accuracy of 79.40%, surpassing all competing methods, including the recent state-of-the-art DEFT (79.04%). These results demonstrate that collaborative pseudo-labeling with pre-trained vision–language models may generate more reliable supervision than noisy human annotations, offering a more effective and cost-efficient alternative.

4.4 Experiments of Distillation Scenario

Table 3. Performance comparison of distillation experiments from CLIP ViT-L/14 to CLIP ViT-B/16.

Methods	Caltech	DTD	ESAT	FGVC	Flowers	Food	Pets	Cars	UCF	Avg.
CLIP	89.27	44.73	48.75	25.11	71.73	84.83	86.01	66.52	67.44	64.93
PromptKD	93.61	55.08	63.74	26.24	75.33	88.84	91.59	73.93	76.39	71.33
CoFT	**95.88**	**57.97**	**87.10**	**36.86**	**81.86**	**92.59**	**95.50**	**83.38**	**80.84**	**79.11**

We further evaluate our method under distillation setting, transferring knowledge from a larger CLIP ViT-L/14 model to a compact CLIP ViT-B/16 model without any human annotations. Compared with PromptKD (see Table 3), which follows transductive zero-shot learning paradigm and relies on auxiliary data for teacher model pre-training, our methods achieve substantially better performance on all benchmarks. These results highlight the practicality of CoFT for efficient model compression while maintaining strong recognition performance.

Table 4. Stepwise performance analysis of CoFT across datasets.

Dataset	Phase I	Co-Anno.	PL Acc.	PL Ratio	Train Acc.	Phase II
Caltech101	93.51	94.06	95.22	93.34	95.13	95.50
DTD	47.81	48.62	53.17	85.64	52.48	52.13
EuroSAT	76.51	76.30	79.12	75.33	86.21	86.02
FGVCAircraft	26.04	25.98	26.47	92.77	30.46	29.49
Flowers102	71.78	71.90	74.21	95.97	75.62	78.08
Food101	87.51	84.24	85.49	96.10	85.92	90.58
OxfordPets	88.88	88.93	89.70	96.30	91.17	92.40
StanfordCars	70.56	71.53	71.89	99.14	77.72	77.08
UCF101	72.54	73.58	74.57	97.03	78.06	78.35

4.5 Ablation Studies

The stepwise results of our method are shown in Table 4, including the intermediate accuracy of Phase I (PEFT with high-confidence pseudo-labels), Co-Anno. (collaborative annotation accuracy), PL Acc. (filtered pseudo-label accuracy), PL Ratio (proportion of filtered pseudo-labels in total samples), Train Acc. (accuracy on unlabeled training set), and final Phase II performance. These results clearly illustrate the effectiveness of each component in our method.

4.6 Results Under Different Backbones

Table 5. Performance comparison under other CLIP backbones.

Backbone	Dataset	CLIP	CoFT	UPL	IFPL	GRIP	LaFTer	CPL
ViT-B/32	DTD	40.67	**50.22**	43.36	46.71	48.82	46.68	47.53
	EuroSAT	41.73	**80.55**	64.31	38.92	57.84	73.91	62.41
	StanfordCars	59.49	**70.21**	65.02	49.64	59.72	54.74	56.61
	UCF101	61.16	**74.80**	70.36	67.83	69.61	68.24	67.11
ViT-L/14	DTD	55.33	**62.30**	50.82	53.61	55.21	56.03	53.79
	EuroSAT	59.92	**88.18**	68.13	40.31	61.48	46.79	64.21
	StanfordCars	77.31	**87.39**	72.61	54.23	63.12	62.03	60.53
	UCF101	76.24	**85.33**	79.11	75.61	77.23	72.62	74.52

Table 5 reports the performance of our methods and representative baselines under different CLIP backbones, including ViT-B/32 and ViT-L/14. As shown, our method consistently achieves superior performance across all datasets and backbone configurations, demonstrating its strong robustness over backbones.

5 Conclusion

We propose CoFT, a collaborative fine-tuning framework for annotation-free adaptation of vision–language models that mitigates noise accumulation and confirmation bias via a two-phase training strategy. CoFT introduces a dual-prompt, dual-model collaboration mechanism to perform sample-adaptive pseudo-label filtering and extract reliable supervision from unlabeled data. Extensive experiments across unsupervised, noisy-label, and distillation settings demonstrate the effectiveness our method.

References

1. Bai, S., et al.: Qwen2. 5-VL Technical report. arXiv preprint arXiv:2502.13923 (2025)
2. Bossard, L., Guillaumin, M., Van Gool, L.: Food-101-mining discriminative components with random forests. In: European Conference on Computer Vision, pp. 446–461. Springer (2014)
3. Cao, A.Q., Jaritz, M., Guillaumin, M., de Charette, R., Bazzani, L.: LatteCLIP: unsupervised clip fine-tuning via LMM-synthetic texts. In: IEEE/CVF Winter Conference on Applications of Computer Vision, pp. 5030–5040 (2025)
4. Cimpoi, M., Maji, S., Kokkinos, I., Mohamed, S., Vedaldi, A.: Describing textures in the wild. In: IEEE Conference on Computer Vision and Pattern Recognition, pp. 3606–3613 (2014)
5. Fei-Fei, L., Fergus, R., Perona, P.: Learning generative visual models from few training examples: an incremental bayesian approach tested on 101 object categories. In: Conference on Computer Vision and Pattern Recognition Workshop, pp. 178–178 (2004)
6. Gao, P., et al.: CLIP-Adapter: better vision-language models with feature adapters. Int. J. Comput. Vision **132**(2), 581–595 (2024)
7. Han, Z., Gao, C., Liu, J., Zhang, J., Zhang, S.Q.: Parameter-efficient fine-tuning for large models: a comprehensive survey. arXiv preprint arXiv:2403.14608 (2024)
8. Helber, P., Bischke, B., Dengel, A., Borth, D.: EuroSAT: a novel dataset and deep learning benchmark for land use and land cover classification. IEEE J. Sel. Top. Appl. Earth Observations Remote Sens. **12**(7), 2217–2226 (2019)
9. Houlsby, N., et al.: Parameter-efficient transfer learning for NLP. In: International Conference on Machine Learning, pp. 2790–2799 (2019)
10. Hu, E.J., et al.: LoRA: low-rank adaptation of large language models. In: International Conference on Learning Representations, vol. 1, p. 3 (2022)
11. Huang, T., Chu, J., Wei, F.: Unsupervised prompt learning for vision-language model. arXiv preprint arXiv:2204.03649 (2022)
12. Jia, M., et al.: Visual prompt tuning. In: European Conference on Computer Vision, pp. 709–727. Springer (2022)
13. Khattak, M.U., Rasheed, H., Maaz, M., Khan, S., Khan, F.S.: MaPLE: multimodal prompt learning. In: IEEE/CVF Conference on Computer Vision and Pattern Recognition, pp. 19113–19122 (2023)
14. Krause, J., Stark, M., Deng, J., Fei-Fei, L.: 3D object representations for fine-grained categorization. In: IEEE International Conference on Computer Vision, pp. 554–561 (2013)

15. Krizhevsky, A., Hinton, G.: Learning multiple layers of features from tiny images (2009)
16. Li, J., Li, D., Xiong, C., Hoi, S.: BLIP: bootstrapping language-image pre-training for unified vision-language understanding and generation. In: International Conference on Machine Learning, pp. 12888–12900 (2022)
17. Li, Z., et al.: PromptKD: unsupervised prompt distillation for vision-language models. In: IEEE/CVF Conference on Computer Vision and Pattern Recognition, pp. 26617–26626 (2024)
18. Liu, H., Li, C., Wu, Q., Lee, Y.J.: Visual instruction tuning. In: Advances in Neural Information Processing Systems, vol. 36, pp. 34892–34916 (2023)
19. Luo, S., et al.: LCM-LoRA: a universal stable-diffusion acceleration module. arXiv preprint arXiv:2311.05556 (2023)
20. Maji, S., Rahtu, E., Kannala, J., Blaschko, M., Vedaldi, A.: Fine-grained visual classification of aircraft. arXiv preprint arXiv:1306.5151 (2013)
21. Menghini, C., Delworth, A., Bach, S.: Enhancing CLIP with CLIP: exploring pseudolabeling for limited-label prompt tuning. In: Advances in Neural Information Processing Systems, vol. 36, pp. 60984–61007 (2023)
22. Mirza, M.J., et al.: LaFTer: label-free tuning of zero-shot classifier using language and unlabeled image collections. In: Advances in Neural Information Processing Systems, pp. 5765–5777 (2023)
23. Nilsback, M.E., Zisserman, A.: Automated flower classification over a large number of classes. In: Indian Conference on Computer Vision, Graphics & Image Processing, pp. 722–729 (2008)
24. Parkhi, O.M., Vedaldi, A., Zisserman, A., Jawahar, C.: Cats and dogs. In: IEEE Conference on Computer Vision and Pattern Recognition, pp. 3498–3505 (2012)
25. Radford, A., et al.: Learning transferable visual models from natural language supervision. In: International Conference on Machine Learning, pp. 8748–8763 (2021)
26. Sohn, K., et al.: FixMatch: simplifying semi-supervised learning with consistency and confidence. In: Advances in Neural Information Processing Systems, pp. 596–608 (2020)
27. Soomro, K., Zamir, A.R., Shah, M.: UCF101: a dataset of 101 human actions classes from videos in the wild. arXiv preprint arXiv:1212.0402 (2012)
28. Tarvainen, A., Valpola, H.: Mean teachers are better role models: weight-averaged consistency targets improve semi-supervised deep learning results. In: Advances in Neural Information Processing Systems (2017)
29. Wei, T., Li, H.T., Li, C., Shi, J.X., Li, Y.F., Zhang, M.L.: Vision-language models are strong noisy label detectors. Adv. Neural. Inf. Process. Syst. **37**, 58154–58173 (2024)
30. Xie, Q., Luong, M.T., Hovy, E., Le, Q.V.: Self-training with noisy student improves imagenet classification. In: IEEE/CVF Conference on Computer Vision and Pattern Recognition, pp. 10687–10698 (2020)
31. Yang, L., Zhang, R.Y., Wang, Y., Xie, X.: MMA: multi-modal adapter for vision-language models. In: IEEE/CVF Conference on Computer Vision and Pattern Recognition, pp. 23826–23837 (2024)
32. Yang, Z., Li, L., Lin, K., Wang, J., Lin, C.C., Liu, Z., Wang, L.: The dawn of LMMs: Preliminary explorations with GPT-4V (ISION). arXiv preprint arXiv:2309.17421 **9**(1), 1 (2023)
33. Zhang, J., Wei, Q., Liu, F., Feng, L.: Candidate pseudolabel learning: enhancing vision-language models by prompt tuning with unlabeled data. In: International Conference on Machine Learning, pp. 60004–60020 (2024)

34. Zhang, R., et al.: Tip-Adapter: training-free clip-adapter for better vision-language modeling. arXiv preprint arXiv:2111.03930 (2021)
35. Zhang, Y., Zhang, C., Hu, X., He, Z.: Unsupervised prototype adapter for vision-language models. In: Chinese Conference on Pattern Recognition and Computer Vision, pp. 197–209 (2023)
36. Zhou, K., Yang, J., Loy, C.C., Liu, Z.: Conditional prompt learning for vision-language models. In: IEEE/CVF Conference on Computer Vision and Pattern Recognition, pp. 16816–16825 (2022)
37. Zhou, K., Yang, J., Loy, C.C., Liu, Z.: Learning to prompt for vision-language models. Int. J. Comput. Vision **130**(9), 2337–2348 (2022)
38. Zhu, C., Xiao, B., Shi, L., Xu, S., Zheng, X.: Customize segment anything model for multi-modal semantic segmentation with mixture of lora experts. arXiv preprint arXiv:2412.04220 (2024)

ARETS: Regularized Evolution Combined with Tabu Search for Efficient Auto-tuning in Deep Learning Compilers

Haibin Liang, Tianping Wu, Zihua Liu, Wenhui Chen, and Shaojian Qiu(✉)

College of Software Engineering, South China Agricultural University, Guangzhou, China
qiushaojian@scau.edu.cn

Abstract. Deep learning compilers typically adopt auto-tuning to generate optimized tensor programs. While current auto-tuning methods combine evolutionary search with learned cost models, their fitness-based elimination often leads to long-term dominance of elite schedules, reducing population diversity. Furthermore, prediction biases in early-stage cost models can prioritize pseudo-elite schedules, causing premature convergence to local optima. To address these issues, we propose ARETS, an auto-tuning algorithm based on regularized evolution and tabu search. ARETS introduces a regularized evolution mechanism to mitigate elite over-dominance and initial prediction biases. Additionally, it integrates a short-term memory mechanism from tabu search into the mutation stage to avoid redundant exploration, enhancing search diversity. We evaluate the performance of ARETS across multiple deep learning models, including ResNet, MobileNet, EfficientNet, and others. Experimental results demonstrate that ARETS achieves an average speedup of 1.58× in search efficiency and a speedup of 1.11× in the execution performance of the generated programs.

Keywords: Deep Learning Compiler · Auto-Tuning · Regularized Evolution · Tabu Search

1 Introduction

The rapid proliferation of deep learning across diverse domains, such as autonomous driving and face recognition, has rendered the inference efficiency of neural networks a critical bottleneck for large-scale deployment. Modern models, including ResNet [4] and EfficientNet [18], involve complex tensor computations whose performance depends heavily on the scheduling strategies of underlying operators on target hardware. To address this issue, researchers have developed deep learning compilers, such as TVM [2], Glow [14] and AKG [19], to generate optimized programs for various hardware architectures. Given the significant disparities in memory hierarchies and instruction sets across different hardware

L. Pasa et al. (Eds.): ICANN 2026, LNCS 17089, pp. 358–369, 2027.
https://doi.org/10.1007/978-3-032-38398-3_30

platforms, such as GPU and TPU, the internal optimization strategies of compilers constitute a vast and complex discrete search space. In this context, relying solely on predefined heuristics or template-based methods fails to fully exploit the potential of the underlying hardware.

In recent years, auto-tuning frameworks such as Ansor [20] have been widely adopted to automatically explore high-performance scheduling strategies. Their primary objective is to automatically explore optimal scheduling decisions, such as loop tiling and operator fusion, guided by a learned cost model. Although Ansor has made substantial progress, the traditional Genetic Algorithm [9] it employs exhibits intrinsic limitations in high-dimensional discrete spaces. Specifically, fitness-based elimination strategies allow a limited number of elite schedules to dominate the population for extended periods, which leads to population homogenization and inhibits the exploration of diverse candidates. Furthermore, the scarcity of training data during initial tuning phases amplifies prediction biases of the cost model under the roulette wheel selection mechanism. This enables pseudo-elite schedules, characterized by high predicted scores but poor actual performance, to prevail, thereby misleading the search trajectory and causing premature convergence.

In response to these challenges, this paper proposes ARETS, a synergistic evolutionary search algorithm based on Regularized Evolution [13] and Tabu Search [5]. At the population evolution level, ARETS introduces a First-In-First-Out (FIFO) aging mechanism to forcibly replace stale individuals, thereby mitigating the over-dominance of elite schedules. Simultaneously, the algorithm adopts tournament selection to replace the original roulette wheel mechanism, leveraging local competition to reduce the misleading effects of cost model biases during the initial stages of tuning. At the mutation level, ARETS incorporates a short-term memory mechanism from tabu search by constructing a tabu list to dynamically avoid redundant exploration of structurally similar schedules. By integrating macro-level population updates with micro-level mutation constraints, ARETS maintains multi-granularity search diversity and enhances both the robustness and efficiency of the auto-tuning process. The core contributions of this work are as follows:

- We propose ARETS, a synergistic evolutionary search method. By introducing a FIFO aging mechanism and Tournament Selection at the population level, we mitigate the long-term dominance of elite schedules and improve search stability during initial tuning phases.
- We implement a short-term memory mechanism at the mutation stage to prevent repetitive exploration of similar scheduling structures, thereby enhancing search diversity and reducing the risk of trapping in local optima.
- We implement ARETS in Ansor and evaluated across various representative deep learning models. Experimental results demonstrate that ARETS achieves improvements in both search efficiency and quality.

2 Related Work

Execution efficiency of deep learning models varies significantly across hardware platforms. High-level frameworks, such as TensorFlow [1] and PyTorch [12], often struggle to achieve comprehensive performance optimizations due to their high abstraction levels. To bridge this gap, deep learning compilers have been proposed to generate hardware-specific code using their own intermediate representations (IR). These compilers leverage IR-level graph optimizations and operator scheduling to minimize computational overhead and data transfer costs [10]. A critical enabling technique within this pipeline is auto-tuning, which automates the search for optimal program parameters (e.g., tiling sizes and loop transformations). Early methods like AutoTVM [3] rely on manually defined templates, which limits the search space to human expertise. In contrast, Ansor [20] introduces a template-free approach that automatically constructs a hierarchical search space through program sampling and sketches, significantly expanding the optimization potential.

As a high-performance Auto-Scheduler within Apache TVM [2], Ansor [20] employs cost-model-guided evolutionary search to explore the scheduling space. The auto-tuning process of Ansor primarily consists of the following steps. First, the computational graph is partitioned into subgraphs for independent optimization. For each subgraph, Ansor constructs a parameter space and generates a set of optimization templates called sketches. After random initialization, an evolutionary search is conducted to optimize the parameters within these sketches. Only a subset of candidate schedules, predicted by a cost model to have low execution latency, is selected for measurement on physical hardware. The cost model is then updated through online learning using these performance results. This iterative process continues until the predefined time budget is exhausted.

In recent years, auto-tuning methods for deep learning compilers have been extensively researched. For instance, FamilySeer [11] groups similar subgraphs and trains a separate cost model for each group, thereby accelerating cost model convergence and enhancing prediction performance. Jeong and Lee proposed Bayesian Code Diffusion [8], which is a method that introduces the concepts of Bayesian priors and posteriors into program optimization. By generating program code through iterative diffusion, this approach efficiently searches for optimal schedules within a narrowed search space, which further accelerates the auto-tuning process of deep learning compilers.

3 Method

3.1 Overview Design

The evolutionary search method of Ansor [20] is based on a standard Genetic Algorithm, which evolves schedules through iterative selection, crossover, and mutation. However, its fitness-based elimination mechanism and roulette wheel selection strategy are prone to population homogenization and exhibit high sensitivity to the prediction biases of cost models during the initial stages of tuning.

To address these issues and more effectively explore the search space of tensor programs, we proposes the ARETS algorithm. By organically fusing Regularized Evolution [13] and Tabu Search [5], this algorithm introduces two key improvements to the original evolutionary search framework.

First, the algorithm incorporates a regularized evolution mechanism. The original roulette wheel selection and fitness-driven elimination are replaced with tournament selection and a FIFO-based aging strategy. By limiting the maximum lifespan of individuals within the population, the algorithm suppresses the long-term dominance of pseudo-elite schedules while maintaining population diversity and convergence stability. Second, the algorithm integrates a short-term memory mechanism. A tabu search strategy is introduced during the mutation stage by maintaining a tabu list that records the feature signatures of the schedules. When the signature of a newly generated individual matches an entry in the tabu list, a retry mechanism is triggered to avoid redundant exploration. This approach reduces ineffective cycles in isomorphic or narrow search spaces during the later stages of evolution, further enhancing search diversity.

The workflow of ARETS is illustrated in Fig. 1. The algorithm first generates an initial population based on sketches and initializes the aging queue and the tabu list. In each evolutionary iteration, parent schedules are selected via tournament selection for mutation. The newly generated offspring must pass a tabu check to ensure structural uniqueness. After evaluation by the cost model, individuals enter the FIFO queue to begin the next iteration. This process continues until the termination criteria are met, ultimately outputting an optimal set of candidate schedules for physical hardware measurement and cost model refinement.

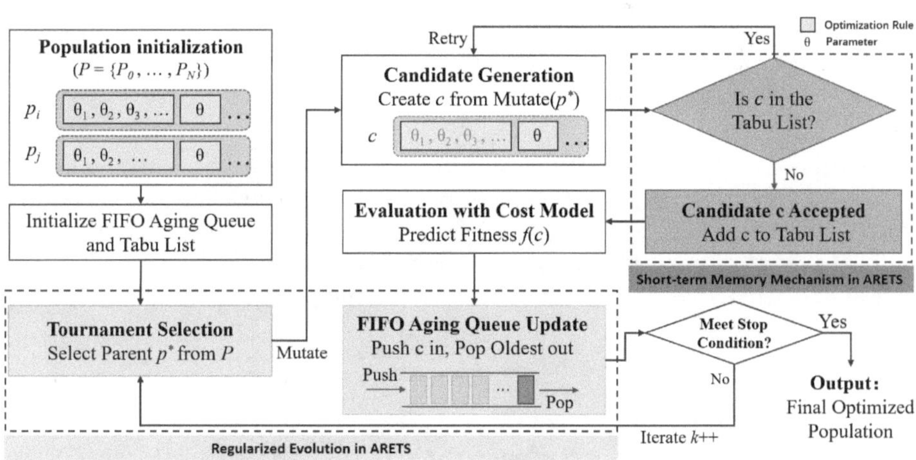

Fig. 1. Workflow of ARETS.

3.2 Regularized Evolution in ARETS

Regularized Evolution [13] was originally proposed by Real et al. in the field of neural architecture search. Its core principle lies in replacing traditional fitness-driven elimination strategies with an age-based regularization mechanism to maintain population diversity and mitigate the issue of premature convergence. In ARETS, we apply the principles of RE to the evolutionary search process of Ansor and implement regularized control over the population through two key mechanisms: a FIFO Aging Queue and Tournament Selection.

FIFO Aging Queue. The FIFO aging queue is designed to mitigate the issue where elite schedules dominate the population for excessive durations. ARETS replaces the conventional population structure with a FIFO aging queue Q with a capacity of N. Each element in the queue records a schedule and its corresponding fitness, which is represented by the execution latency predicted by the cost model. Elements are ordered by their insertion time, with the head of the queue representing the individual with the longest survival time. When a new offspring schedule is generated and added to the queue, the oldest individual at the head is removed if the queue has reached its capacity:

$$Q \leftarrow (Q \cup \{(c, \hat{f}(c))\}) \setminus \{\text{oldest}\}, \quad |Q| \leq N \tag{1}$$

where c denotes the newly generated offspring schedule and $\hat{f}(c)$ represents its predicted fitness from the cost model. This elimination strategy is independent of the fitness of the schedules and depends solely on their survival time within the queue. Under the FIFO aging mechanism of Regularized Evolution, any individual is inevitably eliminated after at most N iterations regardless of its fitness score. Consequently, this mechanism effectively prevents pseudo-elite schedules resulting from prediction biases from dominating the population over the long term and continuously introduces fresh diversity into the population.

Tournament Selection. ARETS employs tournament selection to determine the parent schedule before each generation of offspring. Specifically, a subset S containing T candidate schedules, where T represents the tournament size, is sampled uniformly at random from the aging queue Q. The schedule with the highest predicted fitness within S is selected as the parent:

$$p^* = \arg\max_{p \in S} \hat{f}(p), \quad S \sim \text{Uniform}(Q), \quad |S| = T \tag{2}$$

where $\hat{f}(p)$ denotes the predicted fitness of schedule p, and p^* is the selected parent with the highest fitness in S. Compared with traditional roulette wheel selection, tournament selection relies on the relative ranking of individuals within the candidate subset rather than absolute prediction scores, which provides enhanced robustness against noise in cost model predictions. The tournament size T serves as a critical parameter for controlling selection pressure. A larger T favors high-scoring individuals as parents, which accelerates population convergence at the expense of diversity. Conversely, a smaller T leads to a more randomized selection process, thereby enhancing global exploration capabilities.

3.3 Short-Term Memory Mechanism in ARETS

Tabu Search [5] employs a tabu list to record recently visited solutions, preventing redundant exploration and guiding the search to escape local optima. In ARETS, this short-term memory mechanism is integrated into the mutation stage to suppress the redundant generation of similar schedules, thereby enhancing search diversity.

Feature Representation of Schedule. In Ansor, a schedule is defined by a sequence of transformation steps. While stringifying the entire sequence enables exact equivalence checking, its high computational overhead and the rarity of identical schedules during evolution make it impractical for a tabu list. Instead, the search process primarily needs to avoid redundant sampling of structurally similar schedules.To this end, ARETS employs a loop structure representation G as a lightweight intermediate form to characterize the loop nest without full compilation. For a computation graph with M stages, each stage g_i consists of a set of loop iterators:

$$G = \{g_1, g_2, \ldots, g_M\}, \quad g_i = \{e_{i,1}, e_{i,2}, \ldots, e_{i,n_i}\} \tag{3}$$

where each iterator $e_{i,j}$ is defined by attributes including name, extent, type, and annotations. Based on the loop structure representation G, ARETS extracts the type and range of all iterators within each computation stage to construct a feature signature $\phi(s)$ for the schedule:

$$\phi(s) = \bigoplus_{i=1}^{|G|} \bigoplus_{j=1}^{|g_i|} [\mathrm{kind}_{i,j}, \mathrm{extent}_{i,j}] \tag{4}$$

where $\bigoplus$ denotes string concatenation. This feature signature provides a lightweight and efficient description of the loop structure, enabling the tabu list to rapidly identify and bypass structurally similar schedules, thereby enhancing search efficiency.

Tabu Check and Retry Mechanism. ARETS introduces a tabu check and retry mechanism to avoid redundant exploration of structurally similar schedules. The tabu list $\mathcal{T}$ is implemented as a FIFO queue with a capacity of L, storing the feature signatures of schedules visited during the mutation process. For each candidate offspring c generated via mutation, the algorithm first extracts its feature signature $\phi(c)$ and verifies whether it exists within the tabu list $\mathcal{T}$. The acceptance policy for offspring c is defined as follows:

$$\mathrm{accepted}(c) = \begin{cases} \text{accept}, & \phi(c) \notin \mathcal{T} \\ \text{accept}, & \phi(c) \in \mathcal{T} \wedge r = 0 \\ \text{reject and retry mutation}, & \phi(c) \in \mathcal{T} \wedge r > 0 \end{cases} \tag{5}$$

where r denotes the maximum number of retries allowed for each offspring. If the offspring is accepted, its feature signature $\phi(c)$ is appended to the tail of the tabu list; if the queue reaches its capacity, the oldest entry at the head is evicted. Unlike standard tabu search, ARETS does not implement aspiration criteria. Since $\phi(c)$ is an approximate feature representation rather than a complete state mapping, a tabu hit indicates a high degree of structural similarity rather than exact equivalence, the risk of missing a global optimum due to a tabu hit is negligible compared to the gain in diversity. Therefore, the offspring is still accepted once the retry limit is exhausted to ensure the efficiency and continuity of the search process.

3.4 Synergistic Evolutionary Search in ARETS

The complete evolutionary search process of ARETS, which integrates regularized evolution with a short-term memory mechanism during mutation, is presented in Algorithm 1. The procedure begins with the initialization of the population through random sampling and the prediction of fitness via the cost model. This is followed by the construction of a FIFO aging queue Q with capacity N and an empty tabu list $\mathcal{T}$. In the subsequent main evolutionary loop, ARETS employs tournament selection to sample T schedules from Q and identifies the individual p^* with the highest predicted fitness as the parent. Subsequently, p^* undergoes mutation to generate an offspring c, from which the feature signature $\phi(c)$ is extracted for a tabu check. If the offspring c matches an entry in the tabu list, a limited number of mutation retries is triggered. Once a new schedule is accepted, its fitness is evaluated by the cost model. This process strictly follows a steady-state update strategy, where the new schedule is appended to the tail of the FIFO aging queue while the oldest individual at the head is evicted, with the tabu list updated simultaneously. This iterative process continues until the convergence criteria are met, ultimately returning the set of schedules H with the best predicted performance.

4 Evaluation

4.1 Experimental Setup

The proposed ARETS is implemented on Ansor. Experiments are conducted on an NVIDIA RTX 3090 GPU platform, employing a diverse set of deep learning models for evaluation, including ResNet-18 [4], VGG-16,19 [16], MobileNet [6], MobileNet-V2 [15], SqueezeNet-V1.1 [7], Inception-V3 [17], and EfficientNet [18]. The search efficiency and quality of ARETS are compared against several baselines, specifically Ansor [20], FamilySeer [11], and BayesianCodeDiffusion [8]. ARETS maintains the same time budget as Ansor, performing $200 \times n$ physical measurements per DNN, where n represents the number of subgraphs. General parameters for the evolutionary search and cost model remain consistent across ARETS, Ansor, FamilySeer, and BayesianCodeDiffusion to ensure a fair comparison. Specifically, ARETS is configured with a tournament size of 10, a tabu list capacity of 100, and a maximum of 3 mutation retries.

Algorithm 1 ARETS

```
Input: Initial population P_0, population size N, tournament size T, tabu length
  L, retry limit r, mutation probability p_m, iterations K
Output: Best schedule set H
Q ← Initial aging queue from P_0; 𝒯 ← ∅
for k = 1 to K do
    C ← ∅
    repeat
        S ← UniformSample(Q, T)
        p* ← arg max_{p∈S} f̂(p); retry ← r
        repeat
            c ← Mutate(p*); is_tabu ← φ(c) ∈ 𝒯
            if ¬is_tabu ∨ retry = 0 then
                C ← C ∪ {c}; 𝒯 ← 𝒯 ∪ {φ(c)} (maintain size L)
                break
            end if
            retry ← retry − 1
        until false
    until |C| = N
    Update Q by appending C and evicting oldest elements to maintain size N
    Update H with the top-performing schedules from Q
end for
return H
```

4.2 Search Quality Improvement

Search quality is closely correlated with program execution latency; specifically, a lower execution latency indicates a higher quality of the corresponding schedule. Figure 2 illustrates the performance improvements of ARETS in terms of program execution latency compared to baseline methods. Experimental results demonstrate that ARETS effectively reduces overall program execution time, achieving an average speedup of 1.11× over Ansor, with a peak speedup reaching 1.22×. In contrast, BayesianCodeDiffusion achieves an average speedup of 1.04×, while FamilySeer succeeds in generating programs with comparable execution latency only for a limited number of models. These results indicate that ARETS consistently achieves lower execution latency for the generated programs compared to baseline methods.

4.3 Search Efficiency Improvement

Search efficiency is defined as the tuning time required to achieve a target execution latency for a given model; higher search efficiency implies less tuning time to reach the target performance. Figure 3 illustrates the search efficiency of ARETS compared to baseline methods on the GPU platform. Experimental results demonstrate that ARETS achieves an average search speedup of 1.58× compared to Ansor, with a peak speedup of 1.82×, exhibiting stable and consistent efficiency improvements across most models. In comparison, Bayesian-

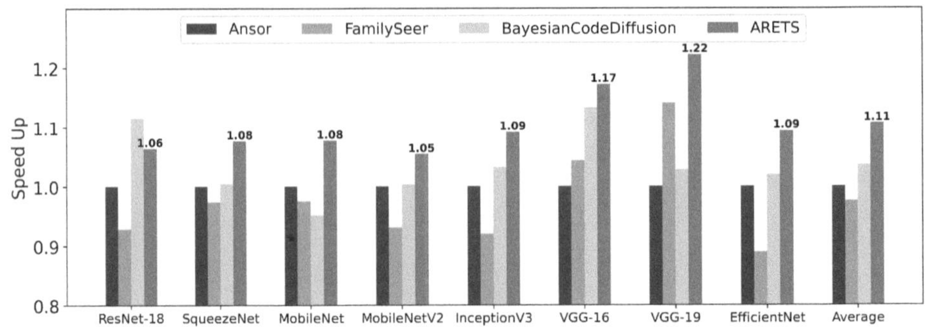

Fig. 2. Comparison of generated program execution latency among Ansor, FamilySeer, BayesianCodeDiffusion, and ARETS. Results are normalized to the Ansor baseline.

CodeDiffusion achieves an average speedup of 1.54×, while FamilySeer fails to generate programs with comparable execution latency for most models, as denoted by × in Fig. 3. Consequently, ARETS outperforms baseline methods in terms of search efficiency.

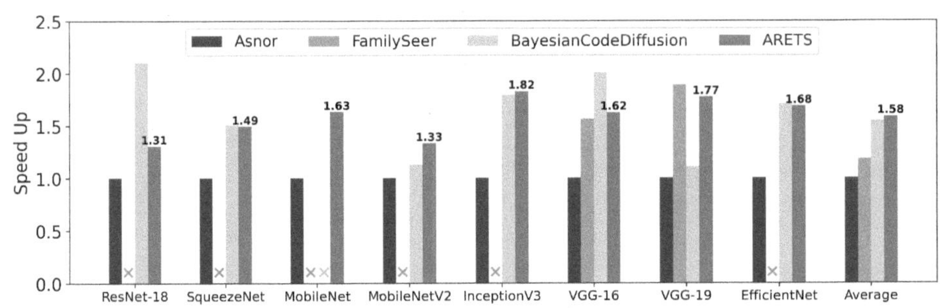

Fig. 3. Tuning time speedup to reach the equivalent optimal execution latency as Ansor. Results are normalized to the Ansor baseline. The symbol "×" indicates that the corresponding method failed to generate programs with execution latency comparable to Ansor's best results.

Furthermore, Fig. 4 presents the execution latency curves for ResNet-18 [4] and MobileNet-V2 [15] during tuning with ARETS and Ansor. The results indicate that ARETS surpasses Ansor in both execution latency and tuning time dimensions. Notably, the latency curves of Ansor plateau in the later stages, forming a pronounced horizontal tail that signifies a stagnation in the search process. In contrast, the ARETS curves display more inflection points with significant latency drops, indicating that the algorithm continuously discovers candidate programs with lower execution latency, thereby obtaining superior scheduling strategies.

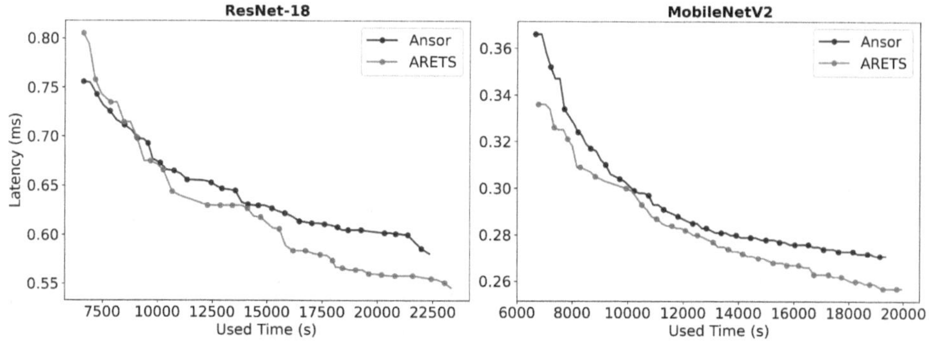

Fig. 4. End-to-end latency curve of ResNet-18 and MobileNetV2 during the tuning process of Ansor and ARETS.

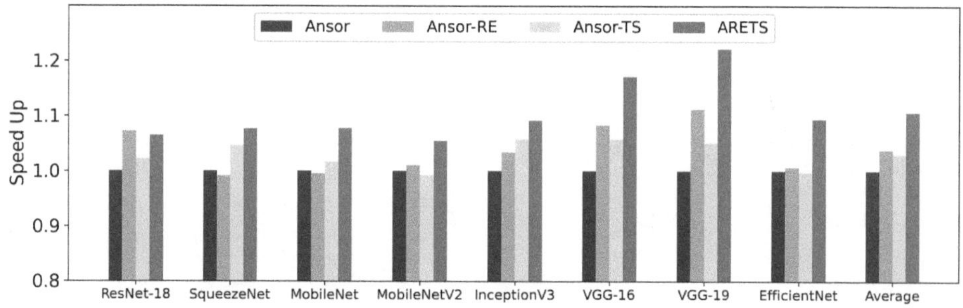

Fig. 5. Comparison of generated program execution latency among Ansor, Ansor-RE, Ansor-TS, and ARETS. Results are normalized to the Ansor baseline.

4.4 Ablation Study

To evaluate the individual and synergistic contributions of regularized evolution and the tabu-based short-term memory mechanism, we conduct an ablation study using two variants: Ansor-RE (incorporating only regularized evolution) and Ansor-TS (incorporating only tabu search). Figure 5 compares the program execution latency of Ansor, Ansor-RE, Ansor-TS, and ARETS across various models. The results indicate that Ansor-RE and Ansor-TS achieve average performance speedups of 1.04× and 1.03× over the baseline Ansor, respectively. Although the individual gains appear modest, their synergistic integration in ARETS provides a non-linear improvement in robustness, especially in the most complex models like VGG-19. Consequently, the complete ARETS achieves the best results in most cases, confirming that the two mechanisms effectively synergize at different levels to enhance overall search capability and discover programs with lower program execution latency.

5 Conclusion

We propose ARETS, an evolutionary search algorithm based on regularized evolution and tabu search, devised to accelerate the auto-tuning process of deep learning compilers. This method introduces a regularized evolution mechanism by combining a FIFO aging queue with a tournament selection strategy, effectively mitigating the long-term dominance of elite individuals and the adverse effects of cost model prediction biases during early search stages. These features enhance the ability of the algorithm to escape local optima. Additionally, a short-term memory mechanism from tabu search is integrated into the mutation stage to record and constrain the redundant generation of structurally similar schedules, thereby reducing ineffective exploration. Experimental results on a GPU platform demonstrate that ARETS outperforms Ansor in both search efficiency and quality, achieving average speedups of 1.58× and 1.11×, respectively.

Acknowledgments. The authors would like to thank the anonymous reviewers for their valuable comments. This work was supported in part by National Key Research and Development Program of China (2024YFC2814400), in part by the Guangdong Provincial Key Laboratory of Power System Cyber Security (GPKLPSNS-2024-KF-07).

References

1. Abadi, M., et al.: {TensorFlow}: a system for {Large-Scale} machine learning. In: 12th USENIX Symposium on Operating Systems Design and Implementation (OSDI 16), pp. 265–283 (2016)
2. Chen, T., et al.: {TVM}: an automated {End-to-End} optimizing compiler for deep learning. In: 13th USENIX Symposium on Operating Systems Design and Implementation (OSDI 18), pp. 578–594 (2018)
3. Chen, T., et al.: Learning to optimize tensor programs. Adv. Neural. Inf. Process. Syst. **31** (2018)
4. He, K., Zhang, X., Ren, S., Sun, J.: Deep residual learning for image recognition. In: Proceedings of the IEEE Conference on Computer Vision and Pattern Recognition, pp. 770–778 (2016)
5. Hertz, A., Taillard, E., de Werra, D.: Tabu search. In: Local Search In Combinatorial Optimization, pp. 121–136 (1997)
6. Howard, A.G., et al.: MobileNets: efficient convolutional neural networks for mobile vision applications. arXiv preprint arXiv:1704.04861 (2017)
7. Iandola, F.N., Han, S., Moskewicz, M.W., Ashraf, K., Dally, W.J., Keutzer, K.: SqueezeNet: alexnet-level accuracy with 50x fewer parameters and< 0.5 mb model size. arXiv preprint arXiv:1602.07360 (2016)
8. Jeong, I., Lee, S.: Bayesian code diffusion for efficient automatic deep learning program optimization. In: 19th USENIX Symposium on Operating Systems Design and Implementation (OSDI 25), pp. 295–311 (2025)
9. Katoch, S., Chauhan, S.S., Kumar, V.: A review on genetic algorithm: past, present, and future. Multimedia Tools Appl. **80**(5), 8091–8126 (2021)

10. Li, M., Liu, Y., Liu, X., Sun, Q., You, X., Yang, H., Luan, Z., Gan, L., Yang, G., Qian, D.: The deep learning compiler: a comprehensive survey. IEEE Trans. Parallel Distrib. Syst. **32**(3), 708–727 (2020)
11. Li, M., et al.: Exploiting subgraph similarities for efficient auto-tuning of tensor programs. In: Proceedings of the 52nd International Conference on Parallel Processing, pp. 786–796 (2023)
12. Paszke, A., et al.: PyTorch: an imperative style, high-performance deep learning library. Adv. Neural. Inf. Process. Syst. **32** (2019)
13. Real, E., Aggarwal, A., Huang, Y., Le, Q.V.: Regularized evolution for image classifier architecture search. In: Proceedings of the AAAI Conference on Artificial Intelligence, vol. 33, pp. 4780–4789 (2019)
14. Rotem, N., et al.: Glow: graph lowering compiler techniques for neural networks. arXiv preprint arXiv:1805.00907 (2018)
15. Sandler, M., Howard, A., Zhu, M., Zhmoginov, A., Chen, L.C.: MobileNetV2: inverted residuals and linear bottlenecks. In: Proceedings of the IEEE Conference on Computer Vision and Pattern Recognition, pp. 4510–4520 (2018)
16. Simonyan, K., Zisserman, A.: Very deep convolutional networks for large-scale image recognition. arXiv preprint arXiv:1409.1556 (2014)
17. Szegedy, C., Vanhoucke, V., Ioffe, S., Shlens, J., Wojna, Z.: Rethinking the inception architecture for computer vision. In: Proceedings of the IEEE Conference on Computer Vision and Pattern Recognition, pp. 2818–2826 (2016)
18. Tan, M., Le, Q.: EfficientNet: rethinking model scaling for convolutional neural networks. In: International Conference on Machine Learning, pp. 6105–6114. PMLR (2019)
19. Zhao, J., et al.: AKG: automatic kernel generation for neural processing units using polyhedral transformations. In: Proceedings of the 42nd ACM SIGPLAN International Conference on Programming Language Design and Implementation, pp. 1233–1248 (2021)
20. Zheng, L., et al.: Ansor: generating {High-Performance} tensor programs for deep learning. In: 14th USENIX Symposium on Operating Systems Design and Implementation (OSDI 20), pp. 863–879 (2020)

TaskFusion: Continual Anomaly Detection for Heterogeneous Tabular Data

Dayananda Herurkar[1,2](✉), Federico Raue[1], Joachim Folz[1], Jörn Hees[1,3], and Andreas Dengel[1,2]

[1] German Research Center for Artificial Intelligence (DFKI), Kaiserslautern, Germany
{dayananda.herurkar,federico.raue,joachim.folz, jorn.hees,andreas.dengel}@dfki.de
[2] RPTU Kaiserslautern-Landau, Kaiserslautern, Germany
[3] Hochschule Bonn-Rhein-Sieg (H-BRS), Bonn, Germany

Abstract. Continual anomaly detection in tabular data is challenging and remains largely underexplored, particularly in settings with heterogeneous feature schemas, distribution shifts, and severe class imbalance. In many real-world applications, data arrive sequentially from diverse domains, rendering conventional continual learning methods ineffective due to their reliance on a fixed input space. We propose a continual learning (CL) method, which can overcome these challenges and continually learn from different tasks. Our method consists of three main parts: our AGF model, Taskfusion augmentation, and outlier exposure. The *AGF-model* maps task-specific features into a shared space, then aligns distributions to reduce representation drift, and learns anomaly decision boundaries in the aligned space. To improve stability, we introduce *Taskfusion augmentation*, combining boundary-aware interpolation within tasks to refine the model anomaly boundaries and cross-task mixing to transfer anomaly structure across datasets. To handle class imbalance and memory constraints, we employ tabular dataset distillation to store compact synthetic replay samples, which are jointly used with augmented data in an *outlier exposure* objective for robust anomaly detection. We evaluate the approach on 21 heterogeneous datasets across multiple domains. Results show that our approach substantially improves continual anomaly detection performance over sequential fine-tuning and other CL baselines while reducing catastrophic forgetting and maintaining stable detection across heterogeneous datasets.

Keywords: Continual Learning · Lifelong Learning · Anomaly Detection · Tabular Data · Outlier Exposure · Data Augmentation

1 Introduction

Outlier detection in tabular data plays a critical role in many real-world applications, including fraud detection, system monitoring, healthcare analytics, and industrial inspection [9]. In practical deployments, however, data rarely arrive as a single static dataset. Instead, new tabular datasets emerge sequentially from different sources or domains, each characterized by distinct feature schemas, statistical distributions, and

L. Pasa et al. (Eds.): ICANN 2026, LNCS 17089, pp. 370–382, 2027.
https://doi.org/10.1007/978-3-032-38398-3_31

anomaly patterns. This setting naturally gives rise to a situation where a model must incrementally or continuously incorporate new knowledge while preserving previously learned behavior. A key opportunity in continual anomaly detection lies in aggregating outlier information across datasets. Anomalies observed in different domains often capture complementary failure modes, and combining such signals can lead to more robust and generalizable detectors. However, realizing this potential is particularly challenging for heterogeneous tabular data. As a result, joint training is often infeasible, preventing effective knowledge sharing across datasets. While effective in homogeneous settings, existing continual learning (CL) strategies do not address schema mismatch or cross-domain anomaly representation inconsistency. The challenge is further amplified in anomaly detection, where labeled anomalies are scarce, with large class imbalance, and decision boundaries are sparse and unstable. Consequently, naive sequential training rapidly forgets earlier anomaly patterns, while standard replay methods struggle with severe class imbalance and storage constraints.

Existing continual anomaly detection approaches typically assume homogeneous feature spaces or operate within a single domain, making them unsuitable for heterogeneous settings where feature schemas and distributions differ across tasks [4,6]. Moreover, standard replay mechanisms struggle under severe class imbalance and limited memory, limiting their ability to preserve anomaly knowledge over long sequences [14]. In this work, we address these limitations by shifting the focus from feature-level transfer to representation-level knowledge accumulation. We propose a continual learning framework called *AGF*, that first maps heterogeneous tabular inputs into a shared latent space and aligns task-specific distributions, enabling anomaly knowledge to be combined across datasets with different sets of feature spaces. Furthermore, we introduce TaskFusion augmentation, which improves robustness through boundary-aware interpolation to refine sparse anomaly decision regions within tasks and cross-task latent mixing to transfer anomaly structure across domains. To further stabilize learning under class imbalance and memory constraints, we integrate tabular dataset distillation [8] as a compact replay mechanism and leverage outlier exposure to reinforce anomaly discrimination. Together, these components enable scalable, continual anomaly detection without requiring access to raw historical data. This work makes the following contributions:

- We propose a continual anomaly detection approach called AGF for heterogeneous tabular data.
- We introduce TaskFusion augmentation, combining boundary-aware interpolation and cross-task mixing to improve robustness and anomaly generalization.
- We integrate tabular dataset distillation to enable efficient replay under limited memory and to overcome class imbalance constraints.

2 Related Work

Continual Learning: Continual learning (CL) studies how to learn from a sequence of tasks without catastrophic forgetting. A prominent direction constrains parameter drift, such as Elastic Weight Consolidation (EWC), which penalizes updates to parameters estimated to be important for previously learned tasks [11]. A complementary direction performs functional regularization through distillation, e.g., Learning without Forgetting (LwF), which preserves prior behavior by matching outputs of the previous model while training on new-task data [12]. Replay-based CL mitigates forgetting by maintaining a memory of past examples and interleaving them during training. iCaRL popularized exemplar-based rehearsal for incremental representation learning [14]. Furthermore, continual and lifelong anomaly detection have been explored through unlearning-based mechanisms [3], sequential one-class formulations such as ARCADe [6], and broader evaluations of lifelong anomaly learning protocols [4].

Anomaly Detection for Tabular Data: Recent advances in tabular anomaly detection have shifted toward deep learning and self-supervised representation learning to better capture complex feature interactions. Reconstruction-based methods remain widely used, but modern variants enhance them with retrieval or contextual mechanisms, such as retrieval-augmented models that leverage sample dependencies to improve reconstruction quality [15]. Self-supervised learning has become a prominent direction, with masked modeling approaches (e.g., masked cell modeling) capturing feature dependencies through reconstruction of partially observed inputs [19]. Transformer-based and context-conditioned models further improve performance by learning feature interactions and adapting to heterogeneous contexts within tabular data [10]. Additionally, recent works explore hybrid architectures that combine representation learning with classical principles, such as deep extensions of isolation-based methods [18] and influence-based anomaly scoring [16]. Despite these advances, most existing approaches operate in a static setting and assume a fixed feature schema, limiting their applicability to continual and heterogeneous scenarios where datasets evolve over time.

Outlier Exposure and Data Augmentation: Outlier Exposure (OE) enhances anomaly detection by training models against auxiliary outlier datasets, improving robustness to unseen anomalies [7]. Such a technique highlights the importance of exposing models to diverse anomaly patterns over time. Complementary to OE, data augmentation and interpolation strategies have emerged as powerful regularizers. MixUp [20] and Manifold Mixup [17] improve generalization by interpolating samples or hidden representations to smooth decision boundaries. For tabular domains, specialized augmentation mechanisms have been proposed, including pseudo-anomaly generation via nearest-neighbor Gaussian Mixup [2] and feature corruptionâĂŞbased contrastive learning (SCARF) [1]. While these approaches demonstrate that representation-space corruption and interpolation are effective for tabular learning, they are primarily designed for static or single-dataset settings. Our work integrates outlier exposure with augmentation within a continual learning framework, enabling progressive aggregation of anomaly knowledge across heterogeneous tabular datasets.

3 Approach

We consider a CL setting consisting of a sequence of tabular tasks $\mathcal{T}_1, \mathcal{T}_2, \ldots, \mathcal{T}_T$, where each task $\mathcal{T}_t = \{X_t, Y_t\}$ corresponds to a dataset with feature dimension d_t. Each sample $x \in \mathbb{R}^{d_t}$ is associated with a binary label $y \in \{0, 1\}$ indicating inlier (normal) or outlier (anomaly). Unlike continual learning in homogeneous settings, feature schemas can differ across tasks ($d_t \neq d_{t'}$). The objective is to learn a unified anomaly detector that incrementally integrates knowledge from heterogeneous datasets while preventing catastrophic forgetting. Formally, the model learns a function $f_\Theta : \mathbb{R}^{d_t} \rightarrow [0, 1]$, shared across all tasks, which predicts anomaly probability. The main challenge is therefore twofold: (1) aligning heterogeneous tabular schemas into a shared representation space, and (2) maintaining stable anomaly detection under continual distribution shifts.

3.1 AGF Model

The proposed framework follows a model architecture composed of three stages, referred to as AGF (see Fig. 1):

$$f_\Theta(X_t) = F_{\theta_f}(G_{\theta_g}(A^t_{\theta_a}(X_t))) \tag{1}$$

The parameters θ_g and θ_f are shared across tasks and updated sequentially while θ_a is updated only during the respective task. The task-specific feature spaces are mapped into a shared latent space via a learnable schema adapter. While raw input dimensionalities differ across tasks, the downstream modules operate in a fixed latent space.

Schema Adapter (A): The adapter is a multiplayer MLP that maps heterogeneous tabular inputs into the same latent dimensionality d_z i.e., $A_{\theta_a} : \mathbb{R}^{d_t} \rightarrow \mathbb{R}^{d_z}$. This module enables learning across datasets with different feature dimensions by performing schema-level harmonization.

General Distribution (G): The distribution alignment component is a shared encoder network that maps task-specific embeddings into a unified latent distribution $G_{\theta_g} : \mathbb{R}^{d_z} \rightarrow \mathbb{R}^{d_g}$. The alignment emerges implicitly through joint optimization across tasks, parameter sharing, and replay, where shared parameters are updated to minimize anomaly classification loss across heterogeneous domains. Shared latent representation module that encourages implicit alignment through parameter sharing and replay. By calibrating representations across tasks, the model reduces domain discrepancies and enables knowledge transfer between datasets.

Given an input x, its latent embedding becomes $g = G_{\theta_g}(A_{\theta_a}(x))$.

Anomaly Classifier (F): The final module performs binary anomaly classification based on aligned representations. It learns a robust decision boundary separating normal and anomalous samples while adapting continuously as new tasks arrive.

$F_{\theta_f} : \mathbb{R}^{d_g} \rightarrow \mathbb{R}^2$ produces logits used for binary anomaly classification. The anomaly probability is computed via softmax.

Together, these components allow the model to accumulate knowledge across heterogeneous domains without requiring identical feature schemas. The task loss is defined as standard cross-entropy: $L_{\text{task}} = \frac{1}{N}\sum_{i=1}^{N} \text{CE}\big(f_\Theta(x_i), y_i\big)$. The architecture enables combining outlier information from multiple datasets through a shared latent representation, which would otherwise be infeasible in raw feature space.

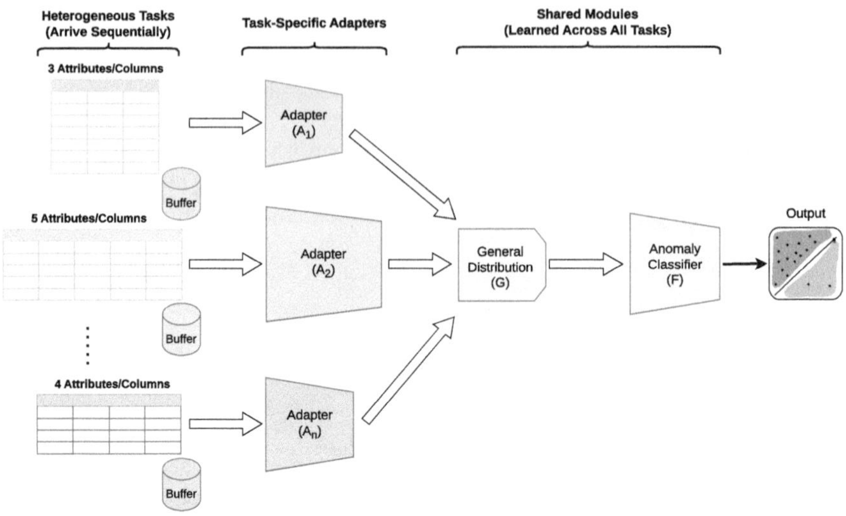

Fig. 1. AGF architecture for heterogeneous continual anomaly detection. Task-specific tabular datasets with different feature dimensions are learned using AGF to perform anomaly prediction by knowledge accumulation without shared raw samples. After each task, the buffer is created via tabular dataset distillation to store compact synthetic samples and added to the replay memory that is used in subsequent tasks.

3.2 Augmentation Techniques

To improve robustness and mitigate catastrophic forgetting, we introduce TaskFusion Augmentation, which operates in the learned latent space (G).

Within-Task Boundary-Aware Augmentation: Samples with uncertain predictions, typically located near the anomaly decision boundary, are identified within the current task. Interpolation among these samples generates additional training points that densify boundary regions. This encourages smoother decision surfaces and improves calibration in sparse anomaly regimes. Within each task, uncertain samples located near the anomaly boundary are identified using prediction confidence: $\mathcal{U}_t = \{x_i \mid \tau_1 < p_{\text{anom}}(x_i) < \tau_2\}$ where τ_1 and τ_2 are adaptive quantile-based thresholds over predicted scores.

Pairs (α_i, α_j) drawn from $\mathcal{U}_t$ are then combined using a MixUp with $\lambda \in (0, 1)$: $a\tilde{u}g_{ij} = \lambda\alpha_i + (1-\lambda)\alpha_j$. Mixing is restricted to samples with identical labels (anomalies with anomalies or normal with normal samples) to preserve semantic consistency. This augmentation densifies sparse boundary regions and encourages smooth decision surfaces.

Cross-Task Augmentation: To transfer knowledge between tasks, embeddings from the current and previous tasks are combined in latent space. Nearest-neighbor similarity is used to identify compatible samples, and convex interpolation produces hybrid representations. Pre- and post-consistency checks ensure that generated samples remain semantically valid and consistent with anomaly labels. To combine anomaly information across datasets, embeddings from the current task and previous tasks are mixed. For a current embedding $\alpha_i^{(t)}$, a compatible historical embedding is selected via nearest-neighbor search: $j = \arg\min_k \|\alpha_i^{(t)} - \alpha_k^{(t-1)}\|_2$. MixUp then generates $\tilde{aug}_{ij} = \lambda\alpha_i^{(t)} + (1-\lambda)\alpha_j^{(t-1)}$. Furthermore, compatibility constraints ensure safe mixing with $\cos(\alpha_i^{(t)}, \alpha_j^{(t-1)}) \geq \tau_{\text{cos}}, \quad \|\alpha_i^{(t)} - \alpha_j^{(t-1)}\|_2 \leq \epsilon$. where, the cosine similarity enforces directional alignment, while Euclidean proximity controls local density compatibility. Mixing is performed only when cosine similarity exceeds τ_{cos} and Euclidean distance is below ϵ. This mechanism enables aggregation of outlier characteristics across heterogeneous datasets, which is otherwise infeasible in raw tabular feature space.

3.3 Outlier Exposure Loss

Augmented samples are incorporated as additional training signals through an outlier exposure mechanism. Synthetic samples generated via augmentation, together with distilled samples of replay memory from previous tasks, provide diverse anomaly-related supervision during training. This strategy allows the model to reuse historical anomaly information without requiring raw data storage. As learning progresses, the detector benefits from accumulated exposure to heterogeneous anomaly patterns originating from multiple datasets. Augmented and replay synthetic samples are incorporated using a supervised outlier exposure (OE) objective. Let $z_i = F_{\theta_f}(aug_i)$ denote classifier logits and $y_i \in \{0, 1\}$ the anomaly label. The weighted cross-entropy loss is defined as $\ell_i = \text{CE}(z_i, y_i)$.

To emphasize anomaly samples, anomaly instances (y_i=1) are assigned a weight(w_i) 1.0, while normal instances (y_i=0) receive a weight of 0.9. The OE loss becomes $\mathcal{L}_{\text{OE}} = \frac{1}{N}\sum_{i=1}^{N} w_i\,\ell_i$. This formulation corresponds to supervised OE using weighted binary classification, where anomaly samples receive stronger optimization pressure. The final optimization objective combines task learning, augmentation, and outlier exposure:

$$\mathcal{L}_{\text{total}} = \mathcal{L}_{\text{task}} + \lambda_{\text{oe}}\mathcal{L}_{\text{OE}}. \quad (2)$$

This objective promotes stable continual learning by aligning heterogeneous representations, refining decision boundaries, and aggregating anomaly information across tasks.

4 Experimental Setup

4.1 Datasets

We evaluate the proposed framework on 21 heterogeneous tabular benchmark datasets spanning multiple domains, including finance, image-derived representations, astronautics and sensor monitoring, and document-based tabular features. The datasets vary substantially in feature dimensionality, sample size, and anomaly ratio. Feature dimensions range from low-dimensional structured attributes to high-dimensional derived representations, and anomaly rates vary across domains, creating realistic class imbalance scenarios. Crucially, feature schemas differ across datasets ($d_t \neq d_{t'}$), making joint training in raw feature space infeasible. This heterogeneous setting reflects practical deployment conditions where new tabular sources emerge over time with incompatible schemas. Detailed statistics for all datasets are provided in the Appendix.

4.2 Continual Learning Protocol

Datasets are presented sequentially as tasks $\mathcal{T}_1, \ldots, \mathcal{T}_T$. When a new dataset arrives, a new schema adapter is instantiated while the shared general distribution and anomaly classifier are retained. At each step, the model has access only to current task data and a fixed-size replay memory constructed via tabular dataset distillation [8]. After completing task $\mathcal{T}_t$, a distilled subset of size K (instances per class, IPC) is generated with the distribution matching [21] approach and stored in memory. Replay capacity remains fixed throughout training. Unless otherwise specified, results are averaged over multiple random task orderings to reduce ordering bias. After completing training on each task, the model is evaluated on all observed tasks to measure knowledge retention and forward transfer. This protocol measures adaptation to new heterogeneous datasets, knowledge retention, and cross-task anomaly accumulation.

4.3 Baseline Methods

We compare the proposed framework against several representative baselines:

Independent Training (w/o CL): Standard anomaly detection models are trained independently on each dataset without any continual learning. The final performance is reported as the average across all datasets/tasks. These methods represent conventional single-domain approaches.

Finetune: Sequential training or finetuning of the model on each task.

Multitask: A non-continual upper-bound setting where all datasets are jointly trained. This provides a reference performance assuming simultaneous access to all data.

CaSSLe [5]**:** A continual self-supervised learning baseline where representation drift is mitigated via a predictor-based distillation mechanism that aligns current representations with their past states.

EDSR [13]**:** We adopt Effective Data Selection and Replay (EDSR), a CL method that selects representative samples based on entropy and replays them with noise-enhanced knowledge distillation.

To analyze the contribution of individual components, we consider the following ablations of the proposed method:

AGF (No Augmentation): The proposed AGF architecture trained without TaskFusion augmentation.

AGF (No Outlier Exposure): A variant where the outlier exposure is disabled, isolating the impact of augmented and other task synthetic outliers across heterogeneous datasets.

All methods use identical architecture, capacity, and optimization settings for fair comparison. The complete implemetation details are described in Appendix. Performance is evaluated using Balanced Accuracy, ROC-AUC, and PR-AUC, along with task-wise performance evolution to assess how well models retain past knowledge while adapting to new tasks. We report average continual performance and forgetting, where high final performance and low forgetting indicate effective knowledge retention and accumulation across heterogeneous tasks. Details on evaluation metrics can also be found in Appendix.

5 Results

We evaluate our approach under two research questions:

Research Question 1: Can AGF effectively combine outlier information across heterogeneous tabular datasets for continual learning?

Research Question 2: Do TaskFusion and outlier exposure (OE) stabilize learning under tasks with long sequences and mitigate catastrophic forgetting?

5.1 Overall Performance on Heterogeneous Benchmarks

Table 1 summarizes the performance of continuous anomaly detection across 21 heterogeneous tabular datasets grouped by domain, with results averaged over three independent runs. Sequential finetuning exhibits severe degradation across all domains, confirming that naive sequential learning fails under schema mismatch and strong distribution shifts. Multitask training provides a strong upper bound but is impractical; notably, the proposed AGF framework with TaskFusion and outlier exposure (OE) substantially narrows this gap and even matches or exceeds multitask performance in some cases. Representation alignment via AGF already improves stability over finetuning, and replacing random replay with dataset distillation further enhances performance. Incorporating OE yields consistent gains across all metrics (e.g., Balanced Accuracy improves), highlighting the importance of exposing the model to diverse anomaly structures. TaskFusion augmentation provides additional improvements, with combined within-task and cross-task augmentation achieving the best results across nearly all settings, particularly in Image and Finance domains, where feature heterogeneity is high. Gains are more moderate in Astronautics and Document domains, where baseline performance is already strong or task diversity is limited. Overall, the results demonstrate that integrating representation alignment, augmentation, and OE enables effective knowledge accumulation and robust continual anomaly detection across heterogeneous tabular datasets.

Table 1. Continual anomaly detection performance across 21 heterogeneous tabular datasets. Best continual results are shown in bold. RS: Replay with Random Sampling, DD: Replay with Dataset Distillation, WT: WithinTask Augmentation, CT: CrossTask Augmentation.

Field	Tasks	w/o CL	Finetune	Multitask	**CaSSLe**	EDSR	AGF+ RS	AGF+ DD	AGF+OE +DD	AGF+OE +WT +DD	AGF+OE +CT +DD	AGF+OE +WT+CT +DD
						Balanced Accuracy						
All	21	78.74	51.12	79.73	63.52	67.43	72.97	75.09	80.37	82.99	83.34	**83.92**
Image	6	71.55	53.02	72.80	54.27	62.75	69.54	72.06	74.72	75.87	75.29	**76.91**
Finance	5	70.92	51.92	71.01	56.57	58.43	67.82	69.09	72.93	72.68	73.10	**74.75**
Astronautics	3	92.12	66.21	**94.37**	68.43	75.83	88.56	92.55	92.91	92.54	92.87	93.77
Document	2	81.02	56.65	**83.05**	62.70	64.70	73.32	76.20	76.99	78.64	78.39	78.43
						PR-AUC						
All	21	68.14	32.88	**70.09**	46.27	48.91	56.97	62.76	65.26	66.66	66.76	68.85
Image	6	54.74	32.96	**55.49**	27.01	40.26	46.48	50.32	51.14	52.44	52.12	52.83
Finance	5	54.56	35.84	**55.85**	35.91	40.71	46.14	51.98	52.60	53.85	53.85	55.10
Astronautics	3	95.61	66.53	**96.56**	47.31	41.23	85.84	94.87	94.78	95.30	95.89	96.36
Document	2	76.15	43.69	**77.67**	46.72	56.55	62.28	67.04	69.11	69.03	70.42	70.88
						ROC-AUC						
All	21	86.26	51.50	88.88	69.58	71.46	80.52	87.00	88.28	89.53	89.53	**90.54**
Image	6	84.00	59.43	85.06	65.94	74.91	80.86	84.60	84.72	84.65	84.13	**85.20**
Finance	5	81.78	53.47	82.30	68.69	72.83	74.47	81.35	82.14	81.34	82.02	**82.41**
Astronautics	3	97.43	71.19	98.63	62.85	60.46	90.25	97.95	98.02	98.66	98.24	**98.85**
Document	2	88.94	61.74	**90.95**	65.46	75.36	79.34	85.80	85.88	87.31	87.37	90.88

5.2 Ablation Study

Impact of Augmentation Components: We further analyze the contribution of individual components by progressively enabling augmentation mechanisms within the AGF framework (in Table 1). While representation alignment with tabular dataset distillation already improves over finetuning, it remains insufficient in long sequences (e.g., in the 21-task setting), indicating accumulated boundary drift. Introducing within-task boundary-aware augmentation significantly improves performance by densifying uncertain regions and stabilizing anomaly decision boundaries. Cross-task augmentation further enhances results, with notable gains in PR-AUC, suggesting improved detection of rare anomalies through shared representation exposure across tasks. The combination of both mechanisms achieves the best and most consistent performance across domains, demonstrating complementary effects: within-task augmentation refines local decision boundaries, while cross-task mixing facilitates knowledge transfer across heterogeneous datasets. Overall, these results confirm that augmentation acts as an effective structural regularizer, mitigating representation drift and improving robustness in long continual learning sequences.

Boundary Selection Analysis: We analyze the influence of the boundary-quality (BQ) threshold used for selecting samples in task-wise augmentation. Table 2 reports the resulting Balanced Accuracy, True Positive Rate (TPR), and True Negative Rate (TNR). The results reveal a clear trade-off between anomaly sensitivity and normal-class preservation. Lower BQ thresholds favor conservative augmentation, yielding very high TNR values (e.g., 95.27 at $BQ = 0.6$) but comparatively lower TPR, indicating limited improvement in detecting difficult anomalies. As the threshold increases, augmentation focuses more strongly on uncertain boundary samples, leading to substantial gains in TPR, which steadily improves and reaches its maximum at $BQ = 0.95$. Balanced Accuracy peaks at a moderate threshold ($BQ = 0.8$), suggesting that selecting samples near but not exclusively at the decision boundary provides the best overall balance between detecting anomalies and maintaining stable inlier classification. Beyond this point, further increasing the threshold continues to improve anomaly sensitivity but causes a noticeable reduction in TNR, indicating overemphasis on ambiguous regions. Overall, these results confirm that boundary-aware augmentation is most effective when applied to moderately uncertain samples. Rather than uniformly augmenting the dataset, focusing on informative boundary regions enables improved anomaly detection while preserving overall classification stability.

Table 2. Effect of boundary-quality (BQ) threshold in task-wise augmentation. Moderate thresholds provide the best balance between anomaly detection (TPR) and inlier preservation (TNR), yielding the highest Balanced Accuracy at BQ $= 0.8$.

Metric (BQ)	0.6	0.7	0.8	0.9	0.95
Bal Acc	75.68	77.95	**78.92**	77.18	78.25
TPR	56.09	62.95	67.94	68.44	**74.21**
TNR	**95.27**	92.95	89.91	85.92	82.31

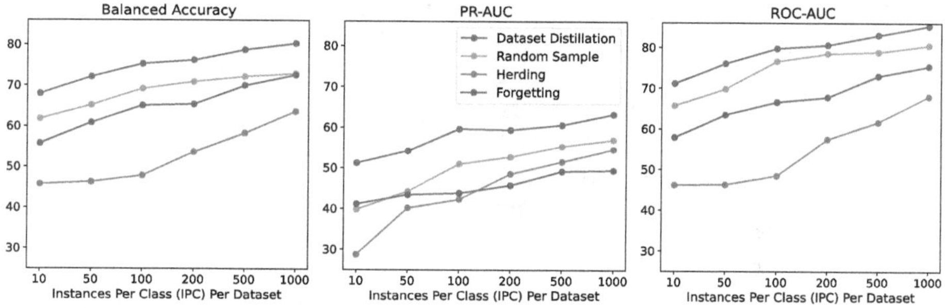

Fig. 2. Influence of replay capacity measured by instances per class (IPC). Increasing IPC consistently improves performance across Balanced Accuracy, PR-AUC, and ROC-AUC. Dataset distillation achieves the strongest results across all capacities, while herding is highly sensitive to memory size and random sampling shows moderate gains.

5.3 Influence of Replay Method and Capacity

Figure 2 evaluates replay capacity by varying instances per class (IPC) and under different replay strategies. Across all metrics, increasing IPC consistently improves performance. Dataset distillation achieves the strongest results at every capacity level, indicating that replay quality is as important as replay quantity [8]. Random sampling improves steadily with IPC, but remains below distillation. Herding is highly sensitive to small memory budgets, performing poorly at low IPC and improving only when sufficient capacity is available. In contrast, dataset distillation maintains strong performance even under moderate memory constraints. These findings demonstrate that structured data representations in replay memory enable efficient knowledge retention and improved anomaly discrimination.

6 Conclusion

We addressed continual anomaly detection on heterogeneous tabular data, where datasets differ in feature dimensionality, distribution, and anomaly structure. Unlike continual learning in homogeneous settings, raw feature spaces are incompatible, making direct data reuse or joint optimization infeasible. We introduced a continual learning framework that enables progressive aggregation of outlier knowledge across heterogeneous domains. The approach integrates three complementary components: (i) the AGF architecture for schema adaptation and distribution alignment in a shared latent space, (ii) TaskFusion augmentation combining boundary-aware within-task interpolation and cross-task latent mixing, and (iii) outlier exposure with distilled replay representations and augmented samples to refine anomaly boundaries without storing raw historical data. Experiments on 21 benchmark datasets demonstrate improved stability and detection performance over sequential fine-tuning and non-continual baselines. The framework achieves reduced forgetting, knowledge accumulation across domains, and robust behavior under long heterogeneous task sequences. More broadly, our results suggest that heterogeneous continual anomaly detection is best formulated as progressive outlier exposure in representation space rather than parameter-constrained preservation in raw feature space. By combining alignment, augmentation, and structured replay, our approach provides a practical and scalable solution for continual anomaly detection across diverse structured data sources.

While the proposed framework improves continual anomaly detection on heterogeneous tabular data, challenges remain in handling extremely long task sequences. Additionally, the current approach relies on implicit feature alignment and fixed augmentation strategies, which may limit robustness under diverse or highly dissimilar domains. Future work will focus on adaptive augmentation, semantic-aware alignment, and extending the framework to fully online or streaming settings for scalable real-world deployment.

Acknowledgments. This work was supported by the BMBF project Albatross (Grant 01IW24002).

References

1. Bahri, D., Jiang, H., Tay, Y., Metzler, D.: SCARF: self-supervised contrastive learning using random feature corruption (2022). https://arxiv.org/abs/2106.15147
2. Dong, H., Frusque, G., Zhao, Y., Chatzi, E., Fink, O.: NNG-Mix: improving semi-supervised anomaly detection with pseudo-anomaly generation. IEEE Trans. Neural Netw. Learn. Syst. **36**(6), 10635–10647 (2025). https://doi.org/10.1109/TNNLS.2024.3497801

3. Du, M., Chen, Z., Liu, C., Oak, R., Song, D.: Lifelong anomaly detection through unlearning. In: Proceedings of the 2019 ACM SIGSAC Conference on Computer and Communications Security, pp. 1283–1297. CCS '19, Association for Computing Machinery, New York, NY, USA (2019). https://doi.org/10.1145/3319535.3363226. https://doi.org/10.1145/3319535.3363226
4. Faber, K., Corizzo, R., Sniezynski, B., Japkowicz, N.: Lifelong continual learning for anomaly detection: New challenges, perspectives, and insights. IEEE Access **12**, 41364–41380 (2024). https://doi.org/10.1109/ACCESS.2024.3377690
5. Fini, E., Da Costa, V.G.T., Alameda-Pineda, X., Ricci, E., Alahari, K., Mairal, J.: Self-supervised models are continual learners. In: 2022 IEEE/CVF Conference on Computer Vision and Pattern Recognition (CVPR), pp. 9611–9620 (2022). https://doi.org/10.1109/CVPR52688.2022.00940
6. Frikha, A., Krompass, D., Tresp, V.: ARCADe: a Rapid Continual Anomaly Detector. In: 2020 25th International Conference on Pattern Recognition (ICPR), pp. 10449–10456. IEEE Computer Society, Los Alamitos, CA, USA (2021). https://doi.org/10.1109/ICPR48806.2021.9412627. https://doi.ieeecomputersociety.org/10.1109/ICPR48806.2021.9412627
7. Hendrycks, D., Mazeika, M., Dietterich, T.: Deep anomaly detection with outlier exposure (2019). https://arxiv.org/abs/1812.04606
8. Herurkar, D., Raue, F., Dengel, A.: Tab-Distillation: impacts of dataset distillation on tabular data for outlier detection. In: Proceedings of the 5th ACM International Conference on AI in Finance, pp. 804–812. ICAIF '24, Association for Computing Machinery, New York, NY, USA (2024).https://doi.org/10.1145/3677052.3698660, https://doi.org/10.1145/3677052.3698660
9. Hilal, W., Gadsden, S.A., Yawney, J.: Financial Fraud: a review of anomaly detection techniques and recent advances. Expert Syst. Appl. **193**, 116429 (2022)
10. King, S., Zhang, Z., Yu, R., Coskun, B., Ding, W., Cui, Q.: Contextual learning for anomaly detection in tabular data (2025). https://arxiv.org/abs/2509.09030
11. Kirkpatrick, J., et al.: Overcoming catastrophic forgetting in neural networks. Proc. Natl. Acad. Sci. **114**(13), 3521–3526 (2017)
12. Li, Z., Hoiem, D.: Learning without forgetting (2017). https://arxiv.org/abs/1606.09282
13. Liu, H., Di, S., Li, H., Li, S., Chen, L., Zhou, X.: Effective data selection and replay for unsupervised continual learning. In: 2024 IEEE 40th International Conference on Data Engineering (ICDE), pp. 1449–1463 (2024).https://doi.org/10.1109/ICDE60146.2024.00119
14. Rebuffi, S.A., Kolesnikov, A., Sperl, G., Lampert, C.H.: iCaRL: incremental classifier and representation learning. In: 2017 IEEE Conference on Computer Vision and Pattern Recognition (CVPR), pp. 5533–5542 (2017).https://doi.org/10.1109/CVPR.2017.587
15. Thimonier, H., Popineau, F., Rimmel, A., Doan, B.L.: Retrieval augmented deep anomaly detection for tabular data. In: Proceedings of the 33rd ACM International Conference on Information and Knowledge Management, pp. 2250–2259. CIKM '24, Association for Computing Machinery, New York, NY, USA (2024). https://doi.org/10.1145/3627673.3679559, https://doi.org/10.1145/3627673.3679559
16. Thimonier, H., Popineau, F., Rimmel, A., Doan, B.L., Daniel, F.: TracinAD: measuring influence for anomaly detection. In: 2022 International Joint Conference on Neural Networks (IJCNN), pp. 1–6 (2022). https://api.semanticscholar.org/CorpusID:248505883
17. Verma, V., et al.: Manifold Mixup: better representations by interpolating hidden states. In: International Conference on Machine Learning (2018). https://api.semanticscholar.org/CorpusID:59604501
18. Xu, H., Pang, G., Wang, Y., Wang, Y.: Deep isolation forest for anomaly detection. IEEE Trans. Knowl. Data Eng. **35**(12), 12591–12604 (2023). https://doi.org/10.1109/TKDE.2023.3270293

19. Yin, J., Qiao, Y., Zhou, Z., Wang, X., Yang, J.: MCM: masked cell modeling for anomaly detection in tabular data. In: The Twelfth International Conference on Learning Representations (2024). https://openreview.net/forum?id=lNZJyEDxy4
20. Zhang, H., Cissé, M., Dauphin, Y.N., Lopez-Paz, D.: Mixup: beyond empirical risk minimization. In: 6th International Conference on Learning Representations, ICLR 2018, Vancouver, BC, Canada, April 30 - May 3, 2018, Conference Track Proceedings, OpenReview.net (2018). https://openreview.net/forum?id=r1Ddp1-Rb
21. Zhao, B., Bilen, H.: Dataset condensation with distribution matching. In: Proceedings of the IEEE/CVF Winter Conference on Applications of Computer Vision, pp. 6514–6523 (2023)

On the Numerical Stability of Mixed-Precision Mutual Information Feature Selection

Simón Noya-Domínguez, Samuel Suárez-Marcote(✉), Laura Morán-Fernández, and Verónica Bolón-Canedo

Universidade da Coruña, CITIC, A Coruña, Spain
s.suarez.marcote@udc.es

Abstract. Feature selection methods based on information-theoretic criteria are widely adopted due to their efficiency and ability to capture non-linear dependencies. However, their reliance on double-precision arithmetic limits their deployment on resource-constrained edge devices, where memory and energy budgets are severely limited. Despite the growing interest in mixed-precision computation, the numerical stability of information-based feature selection under precision reduction remains largely unexplored. This work presents a systematic study of mixed-precision fixed-point arithmetic applied to three representative feature selection algorithms: Mutual Information Maximization (MIM), Joint Mutual Information (JMI), and minimum Redundancy Maximum Relevance (mRMR). We analyze how precision reduction affects ranking stability and downstream classification performance across both low- and high-dimensional datasets. Our results show that numerical robustness strongly depends on the algorithmic structure. While univariate approaches remain highly stable under precision reduction, multivariate methods, particularly those involving cumulative joint information terms, are significantly more sensitive to error accumulation in high-dimensional settings. A mixed configuration combining 8-bit and 16-bit achieves performance comparable to double precision for MIM and mRMR, although this does not hold for JMI across all high-dimensional settings. These findings provide preliminary, simulation-based guidance for the design of precision-aware feature selection, motivating further validation on embedded hardware.

Keywords: Feature Selection · Mutual Information · Mixed Precision · Fixed-Point Arithmetic · Edge Computing

1 Introduction

Feature selection reduces dimensionality by identifying relevant variables while discarding irrelevant ones [1,2]. This mitigates the curse of dimensionality and

L. Pasa et al. (Eds.): ICANN 2026, LNCS 17089, pp. 383–394, 2027.
https://doi.org/10.1007/978-3-032-38398-3_32

reduces computational overhead. Among existing approaches, filter-based methods grounded in Mutual Information (MI) are attractive due to their efficiency, classifier independence, and ability to capture non-linear relationships [3].

MI-based feature selection typically relies on double-precision (64-bit) floating-point arithmetic for numerical stability, at the cost of increased memory usage and energy consumption. Mixed-precision computation, where different numerical precisions are assigned to distinct computational components [4], has emerged as an effective strategy to alleviate these costs while preserving performance. Recent work has further demonstrated the effectiveness of mixed-precision strategies for neural network quantization and deployment on resource-constrained platforms [5]. Although widely adopted in deep learning, its impact on the numerical behavior of information-theoretic methods remains largely unexplored.

In the context of feature selection, preliminary work has considered reduced-precision implementations for resource-constrained environments [6], showing promising results for univariate methods. However, there is currently no systematic understanding of how precision reduction affects the numerical stability of MI-based feature selection algorithms, particularly for multivariate methods in high-dimensional settings, where error accumulation may play a critical role.

This paper addresses this gap by presenting a systematic analysis of mixed-precision arithmetic in MI-based feature selection. We study three representative algorithms, Mutual Information Maximization (MIM) [7], Joint Mutual Information (JMI) [8], and minimum Redundancy Maximum Relevance (mRMR) [9], and analyze how precision reduction impacts both feature ranking stability and downstream classification performance. To this end, the MI computation is decomposed into three stages with distinct numerical sensitivities, namely marginal probabilities, intermediate operations, and critical arithmetic. This decomposition allows different precision levels to be assigned to each component and facilitates the analysis of numerical error propagation.

The experimental results show that numerical robustness strongly depends on the algorithmic structure. While univariate methods such as MIM remain highly stable under precision reduction, multivariate approaches exhibit significantly higher sensitivity, particularly in high-dimensional settings where error accumulation becomes dominant. Among the evaluated configurations, a mixed-precision scheme combining 8-bit and 16-bit representations provides a favorable trade-off between efficiency and stability for MIM and mRMR, though JMI remains sensitive in high-dimensional scenarios.

The remainder of the paper is organized as follows. Section 2 introduces the background on MI and the three feature selection methods under study. Section 3 details the experimental methodology, including datasets, evaluation metrics, and the evaluation procedure. Section 4 presents the mixed-precision methodology, including fixed-point arithmetic, the decomposition of MI computation, bit allocation, and configuration selection. Section 5 reports and analyzes the experimental results. Finally, Sect. 6 concludes the paper.

2 Background on Mutual Information-Based Feature Selection

Mutual information is a well-established criterion for feature selection due to its ability to quantify non-linear dependencies between variables without being restricted to a specific predictive model. This makes it particularly suitable for filter-based methods, where the selection process is independent of the subsequent classifier.

For a random variable U representing a feature (or set of features) and a random variable V representing the class label, mutual information is formally defined as:

$$I(U;V) = \sum_{u \in U} \sum_{v \in V} p(u,v) \log \frac{p(u,v)}{p(u)p(v)}, \tag{1}$$

where $p(u,v)$ denotes the joint probability mass function and $p(u)$ and $p(v)$ represent marginal distributions.

This study analyzes three filter-based feature selection algorithms that rely on mutual information: MIM, JMI, and mRMR. These methods were chosen because they represent different approaches to feature selection with varying levels of computational complexity and sensitivity to numerical precision. In the context of feature selection, we denote by K the number of features to be selected. Each algorithm will generate a ranking or scoring of all candidate features, from which the top-K features are chosen for further analysis.

2.1 MIM

MIM [7] is a univariate ranking method. It computes $I(X_i;Y)$ independently for each feature X_i and generates a ranking of all features according to descending MI values. The top-K features are then selected based on this ranking.

MIM was selected for its conceptual simplicity and computational efficiency. As a univariate method, it evaluates each feature in isolation without considering interactions among features. This independent evaluation mechanism makes MIM particularly interesting for studying the effects of precision reduction, as errors introduced in the MI calculation for one feature do not propagate to others. However, its inability to address feature redundancy means that the selected subset may contain highly correlated variables.

2.2 JMI

JMI [8] is a multivariate selection strategy designed to promote complementarity among features. The selection process is iterative: starting with an empty set, at each step the feature that maximizes the JMI criterion is added to the selected set S until the desired number of features K is reached. For a candidate feature X_i, the evaluation criterion given the current selected set S is:

$$\mathrm{JMI}(X_i) = \sum_{X_j \in S} I(X_i, X_j; Y). \tag{2}$$

JMI was chosen for its demonstrated accuracy and stability in feature selection tasks. By considering the joint information between candidate features and previously selected ones, it favors features that contribute new information beyond what is already captured. However, the summation of multiple joint MI terms increases computational complexity. We hypothesize that this aggregation may amplify numerical errors when computation is performed under reduced precision, and we empirically investigate this claim in Sect. 5.

2.3 mRMR

The mRMR method [9] explicitly balances two competing objectives: maximizing relevance to the target class while minimizing redundancy among selected features. Its selection criterion for a candidate feature X_i is defined as:

$$\mathrm{mRMR}(X_i) = I(X_i; Y) - \frac{1}{|S|} \sum_{X_j \in S} I(X_i; X_j). \tag{3}$$

mRMR was selected as a representative multivariate method that approaches feature selection from a different perspective than JMI. While JMI sums joint MI terms, mRMR subtracts pairwise redundancy terms. This distinction in mathematical formulation may lead to different numerical behavior under precision constraints. We hypothesize that the subtraction operation may exhibit different error propagation characteristics compared to cumulative summations. This hypothesis is examined through the experiments reported in Sect. 5.

Since mutual information computation involves non-linear operations such as divisions, logarithms, and cumulative sums, its numerical behavior under reduced precision is expected to vary across algorithms, which is the central question investigated in this work.

3 Experimental Methodology

3.1 Dataset Selection and Dimensionality Criteria

For this study, six datasets are selected and divided into two distinct groups based on a threshold of 500 features. The first group comprises low-dimensional datasets with fewer than 500 features: Splice, Waveform, and Congress. The second group comprises high-dimensional datasets with more than 500 features: Human Activity, TOX-171, and DLBCL. Within this group, two microarray datasets are deliberately included as stress tests for numerical stability. For low-dimensional datasets, thresholds of 5, 10, and 15 features are used; for high-dimensional datasets, thresholds of 25, 50, and 75 features are established. Table 1 summarizes the principal characteristics of these datasets.

Table 1. Datasets used in the study. Star (*) identifies microarray-type datasets.

Dataset	Samples	Features	Classes	Reference
splice	3,175	60	3	[10]
waveform	5,000	40	3	[10]
congress	435	16	2	[10]
humanActivity	7,352	561	6	[11]
TOX171*	171	5,748	4	[12]
DLBCL*	47	4,026	2	[13]

Continuous features were discretized inside each cross-validation fold to avoid information leakage from the test partition into the training process. A equal-width discretization scheme was used, with 5 bins per feature. The same discretization procedure was applied consistently across all precision configurations so that any observed differences could be attributed to numerical precision rather than to changes in preprocessing.

3.2 Evaluation Metrics

The evaluation is conducted in two complementary phases using 64-bit double precision as the reference baseline.

In the first phase, the True Positive Rate (TPR) measures the agreement between feature rankings obtained under reduced precision and those obtained using the double-precision baseline [14]. For K selected features, TPR is defined as the number of common features between both rankings divided by K.

In the second phase, classification accuracy is used to assess the practical impact of precision reduction. A k-nearest neighbors (k-NN) classifier with $k = 3$ is employed and evaluated using a 3×5-fold cross-validation scheme [15], where both feature selection and classification are performed within each fold.

4 Applying Mixed-Precision Arithmetic to MI-Based Feature Selection

The central hypothesis of this work is that different computational stages within MI-based feature selection algorithms exhibit varying sensitivity to numerical precision. Rather than applying a uniform precision reduction to all computations, a mixed-precision strategy is adopted: higher precision is allocated to operations that are more sensitive to numerical errors, while lower precision is used for more stable stages. This approach aims to maximize computational efficiency, reducing memory footprint and energy consumption, while preserving the quality of the feature selection process [6].

Implementing this mixed-precision strategy requires a numerical representation that allows fine-grained control over bit width. Fixed-point arithmetic provides this capability, as it enables independent specification of total bits and their distribution between integer and fractional parts.

4.1 Fixed-Point Arithmetic

Fixed-point arithmetic represents numbers using integers combined with a shared scaling factor. A real value is represented as an integer multiplied by 2^{-f}, where f denotes the number of fractional bits. Compared to floating-point arithmetic, this representation reduces hardware complexity and energy consumption, making it suitable for devices without floating-point units.

The distribution between integer and fractional bits determines numerical range and precision. For probabilities in the interval $[0, 1]$, an unsigned representation without integer bits is employed to maximize resolution. The fixed-point arithmetic operations described above were implemented using the `fxpmath` library [16], a Python tool for fixed-point arithmetic emulation. Unless otherwise stated, fixed-point values were rounded using the nearest rule, and values outside the representable range were handled by saturation. These settings were applied consistently across all mixed-precision configurations to ensure that differences among results could be attributed to the precision budget rather than to implementation artifacts.

4.2 Decomposition of MI Computation

The MI calculation involves several mathematical operations that are not equally susceptible to numerical errors when precision is reduced. To design an effective mixed-precision strategy, it is necessary to identify which parts of the computation can tolerate lower precision and which require higher fidelity. Through analysis of the data flow in Eq. (1), three stages with distinct numerical characteristics were identified:

1. **Marginal probabilities**: The computation of probabilities for each possible feature value and class label. These are obtained by summing contingency tables, operations that involve only addition of integer counts followed by normalization. Additions are numerically stable and do not amplify existing errors; moreover, probabilities are bounded between zero and one, which prevents overflow. This stage can therefore tolerate significant precision reduction without compromising the final result.
2. **Intermediate operations**: Linear transformations such as outer products and matrix reshaping that combine values from the first stage. While they do not introduce non-linear error amplification, they propagate input inaccuracies to subsequent steps. Their stability depends on the precision of their inputs, making them a middle ground in terms of sensitivity.
3. **Critical arithmetic**: Operations that determine the final MI value, including the division of joint probabilities by the product of marginals, the logarithm of the resulting ratio, and the final summation. These steps are particularly sensitive to reduced precision, as division may amplify denominator errors,

the logarithm is unstable near zero, and repeated summation can accumulate rounding errors. Therefore, this stage requires the highest numerical fidelity.

This decomposition, illustrated in Algorithm 1, allows different precisions to be assigned to each stage according to its numerical sensitivity, enabling a fine-grained analysis of how precision reduction affects the overall calculation.

Algorithm 1. Mutual Information Calculation with Mixed Precision

Require: X, Y ▷ feature and label vectors
Require: $p_{\text{pro}}, p_{\text{int}}, p_{\text{cal}}$ ▷ precision per stage
Ensure: MI
table $\leftarrow$ ContingencyTable(Discretize(X), Y, p_{int}) ▷ joint counts $n(x, y)$
    Stage 1 — Marginal probabilities (precision p_{pro})
$p_X \leftarrow$ Quantize(RowSums(*table*)$/N$, p_{pro})
$p_Y \leftarrow$ Quantize(ColSums(*table*)$/N$, p_{pro})
    Stage 2 — Intermediate operations (precision p_{int})
$P_{XY} \leftarrow$ Quantize(*table*$/N$, p_{int}) ▷ joint probabilities
$P_X \cdot P_Y \leftarrow$ Quantize($p_X \otimes p_Y$, p_{int}) ▷ outer product of marginals
    Stage 3 — Critical arithmetic (precision p_{cal})
MI $\leftarrow 0$
for each (i, j) with $P_{XY}[i,j] > 0$ and $(P_X \cdot P_Y)[i,j] > 0$ **do**
    $r \leftarrow \textsc{Quantize}\left(\frac{P_{XY}[i,j]}{(P_X \cdot P_Y)[i,j]},\ p_{\text{cal}}\right)$
    MI $\leftarrow$ MI + Quantize($P_{XY}[i,j] \cdot \log(r)$, p_{cal})
end for
return MI

4.3 Bit Allocation Strategy

The choice of how many bits to allocate to the fractional part is critical for maintaining numerical accuracy. To determine the optimal allocation, a preliminary experiment was conducted using all datasets and the three feature selection methods (MIM, JMI, and mRMR). For each combination of total bits and fractional bits, the mean ranking agreement TPR was computed, averaging across all datasets, methods, and selection thresholds. Table 2 presents the results.

The configuration allocating approximately 75% of bits to the fractional part yields the best performance across all total bit widths and is therefore adopted as the default throughout the experiments.

4.4 Configuration Selection

Based on the precision requirements of each component, several mixed-precision configurations were established for experimental evaluation. These configurations are designated by a three-number notation, where the first number represents

Table 2. Mean TPR and standard deviation for fixed-point bit allocations, averaged over all datasets, methods, and selection thresholds.

Total Bits	Fractional Bits	Mean	Std. Dev.
32	24	1.000	0.000
	16	0.979	0.073
	8	0.867	0.183
16	12	0.940	0.173
	8	0.867	0.183
	4	0.630	0.227
8	6	0.696	0.260
	4	0.630	0.227
	2	0.352	0.326
4	3	0.333	0.349
	2	0.352	0.326
	1	0.000	0.000

the bit width assigned to marginal probabilities, the second to intermediate operations, and the third to arithmetic operations.

The first configuration uses 8-bit precision for marginal probabilities, while employing 16-bit precision for both intermediate values and arithmetic operations, codified as 8–16–16. This configuration maximizes computational fidelity while achieving significant memory reduction compared to double precision.

The 4–8–16 configuration represents a balanced approach, reducing marginal probabilities to 4 bits while maintaining intermediate operations at 8 bits for stability and preserving arithmetic operations at 16 bits as a precision anchor.

The most aggressive configuration, 4–8–8, extends lower precision to the final arithmetic operations, using 8 bits for the critical division and logarithm steps.

5 Experimental Results and Analysis

The results are analyzed in terms of ranking similarity to the double-precision baseline and downstream classification performance. To focus on the most challenging scenarios, classification results are reported for the highest feature thresholds in each dimensionality group (15 features for low-dimensional datasets and 75 for high-dimensional datasets).

5.1 Ranking Similarity Analysis

The TPR metric quantifies the agreement between feature rankings from mixed-precision configurations and the double-precision baseline.

For low-dimensional datasets, distinct patterns emerge. Figure 1 shows that MIM with 8–16–16 achieves near-perfect TPR, while 4–8–16 and 4–8–8 show acceptable degradation due to MIM's univariate nature, which limits error propagation. JMI's 4–8–16 and 4–8–8 configurations remain close to the baseline, indicating that precision reduction does not significantly impact ranking quality in low dimensions. For mRMR, the 8–16–16 configuration performs well, but the aggressive configurations show a more noticeable decline.

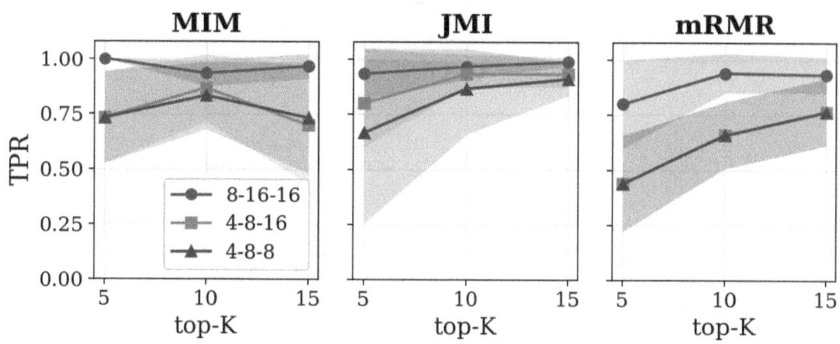

Fig. 1. Mean and standard deviation of TPR over low-dimensional datasets for MIM, JMI and mRMR methods under different mixed-precision configurations.

For high-dimensional datasets, the behavior presented on Fig. 2 changes dramatically. MIM with 8–16–16 remains stable, but 4–8–16 and 4–8–8 degrade significantly. With more features, MI values become densely distributed, making rankings more sensitive to small errors. JMI exhibits severe sensitivity: even 8–16–16 yields low TPR for larger subsets, and aggressive configurations perform poorly. This degradation is consistent with the hypothesis that JMI's cumulative summation of joint MI terms over previously selected features, where errors accumulate across iterations, potentially corrupting the selected set and causing cascade failures. mRMR fares better: 8–16–16 maintains high TPR, and aggressive configurations degrade less than JMI. The subtraction operation in mRMR averages redundancy terms, which may average errors and enable partial cancellation.

5.2 Classification Performance

Since different feature subsets may encode similar amounts of relevant information, classification performance is used to assess whether the selected features preserve discriminative capability. Accuracy is evaluated using a k-NN classifier, which provides a direct measure of the information retained by the selected features without introducing additional model-specific learning effects that could obscure the impact of feature selection. For brevity and to focus on the most challenging scenarios, results are presented for the highest threshold values: 15 features for low-dimensional datasets and 75 features for high-dimensional datasets, as shown in Tables 3, 4 and 5.

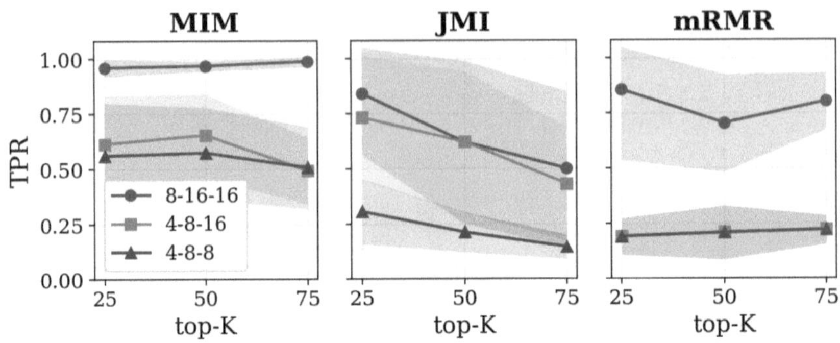

Fig. 2. Mean and standard deviation of TPR over high-dimensional datasets for MIM, JMI and mRMR methods under different mixed-precision configurations.

Table 3. Classification accuracy (%) for MIM method with top-15 features (low-dimensional) and top-75 features (high-dimensional) across different mixed precision configurations and datasets.

Config.	Low-dimension			High-dimension		
	splice	wave.	congress	human.	TOX171	DLBCL
4–8–8	83.24±2.00	76.99±1.78	91.88±2.67	85.58±1.33	75.05±5.66	91.41±9.02
4–8–16	82.64±1.41	76.33±1.63	92.03±2.86	86.63±0.76	74.64±7.65	92.15±8.30
8–16–16	84.47±1.58	79.85±1.11	92.34±2.71	86.17±0.61	71.19±6.88	94.96±6.75
64	84.84±1.57	79.85±1.11	92.34±2.71	86.08±0.62	72.16±5.92	94.96±5.40

For low-dimensional datasets, the trends observed in the ranking analysis are confirmed. For all three methods, the 8–16–16 configuration delivers accuracy very close to the double-precision baseline. For MIM and mRMR, the 4–8–16 and 4–8–8 configurations show some divergence on certain datasets. For JMI, the 4–8–16 and 4–8–8 configurations remain close to the baseline values.

For high-dimensional datasets, the classification results reinforce the patterns observed in the ranking analysis. For MIM, the 8–16–16 configuration remains very close to the baseline across all datasets. For JMI, the sensitivity to precision is evident. The 8–16–16 configuration remains reasonably close to the baseline on some datasets but shows significant divergence on others. The 4–8–16 and 4–8–8 configurations deviate substantially from the baseline in most cases. For mRMR, the 8–16–16 configuration stays close to the baseline across all datasets. The 4–8–16 and 4–8–8 configurations show moderate degradation, with performance remaining within a few percentage points of the baseline on some datasets but diverging more significantly on others.

Table 4. Classification accuracy (%) for JMI method with top-15 features (low-dimensional) and top-75 features (high-dimensional) across different mixed precision configurations and datasets.

Config.	Low-dimension			High-dimension		
	splice	wave.	congress	human.	TOX171	DLBCL
4–8–8	84.05±0.78	78.24±1.14	92.26±2.70	94.10±0.57	76.07±6.78	84.96±10.52
4–8–16	83.87±1.10	78.55±1.38	91.88±2.24	93.62±0.52	74.90±7.40	86.37±9.31
8–16–16	84.45±1.64	79.85±1.11	92.26±2.70	94.33±0.52	79.75±6.25	92.81±7.68
64	84.63±1.34	79.85±1.11	92.26±2.70	85.18±7.46	74.10±7.13	96.52±6.14

Table 5. Classification accuracy (%) for mRMR method with top-15 features (low-dimensional) and top-75 features (high-dimensional) across different mixed precision configurations and datasets.

Config.	Low-dimension			High-dimension		
	splice	wave.	congress	human.	TOX171	DLBCL
4–8–8	72.90±3.12	64.45±4.19	94.64±2.74	80.01±5.21	65.69±6.32	89.41±7.53
4–8–16	72.90±3.12	64.40±4.23	94.64±2.74	80.01±5.21	65.50±6.30	89.41±7.53
8–16–16	89.32±0.80	72.30±1.78	95.86±2.05	74.16±1.11	65.74±8.69	88.00±8.53
64	89.50±0.98	73.88±1.70	95.63±1.89	75.31±1.16	66.13±7.85	88.52±10.21

6 Conclusions

This work has analyzed the impact of reduced numerical precision on MI-based feature selection, highlighting how algorithmic structure influences robustness under mixed-precision arithmetic.

The results show that univariate methods such as MIM are highly stable even under aggressive precision reduction, particularly in high-dimensional settings. In contrast, multivariate approaches exhibit different sensitivity patterns. JMI is the most affected in high-dimensional datasets consistent with the accumulation of joint mutual information terms across iterations, which amplifies numerical errors. mRMR presents an intermediate behavior, offering improved robustness compared to JMI which we can attribute to its redundancy-subtraction formulation, although performance depends on the selected precision configuration.

Overall, the experiments suggest that 8–16–16 is the most stable configuration among those evaluated. It provides behavior close to double precision for MIM and mRMR and remains the most robust configuration under reduced precision.

These findings provide preliminary, simulation-based guidance for the design of information-theoretic feature selection methods in resource-constrained environments. However, the observed behavior depends on the discretization strategy and the fixed-point simulation. Moreover, the experiments were conducted in software simulations, so the results may differ when deployed on actual embedded hardware due to hardware-specific limitations in fixed-point arithmetic and memory management. Future work will extend the evaluation to additional datasets, alternative numerical schemes, and real embedded hardware platforms.

Acknowledgments. This work was supported by the Ministry of Science and Innovation of Spain (MCIN/AEI/10.13039/501100011033 under Grant PID2023-147404OB-I00) and together with "NextGenerationEU"/PRTR by the Ministry for Digital Transformation and Civil Service under grant TSI-100925–2023-1 and by Xunta de Galicia (Grants ED431G 2023/01 and ED431C 2026/38).

Disclosure of Interests. The authors have no competing interests to declare that are relevant to the content of this article.

References

1. Guyon, I., Elisseeff, A.: An introduction to variable and feature selection. J. Mach. Learn. Res. **3**, 1157–1182 (2003)
2. Blum, A., Langley, P.: Selection of relevant features and examples in machine learning. Artif. Intell. **97**(1–2), 245–271 (1997)
3. Venkatesh, B., Anuradha, J.: A review of feature selection and its methods. Cybern. Inf. Technol. **19**(1), 3–26 (2019)
4. Micikevicius, P., et al.: Mixed precision training. In: International Conference on Learning Representations (ICLR) (2018)
5. Wu, D., Wang, Y., Fei, Y., Gao, G.: A novel mixed-precision quantization approach for CNNs. IEEE, IEEE Access (2025)
6. Suárez-Marcote, S., Morán-Fernández, L., Bolón-Canedo, V.: Less is more: low-precision feature selection for wearables. In: International Joint Conference on Neural Networks (IJCNN), pp. 1–8. IEEE (2022)
7. Lewis, D.D.: Feature selection and feature extraction for text categorization. In: Proceedings of the Workshop on Speech and Natural Language, pp. 212–217. Association for Computational Linguistics (1992)
8. Yang, H.H., Moody, J.: Data visualization and feature selection: new algorithms for non-gaussian data. In: Advances in Neural Information Processing Systems, pp. 687–693 (2000)
9. Peng, H., Long, F., Ding, C.: Feature selection based on mutual information criteria of max-dependency, max-relevance, and min-redundancy. IEEE Trans. Pattern Anal. Mach. Intell. **27**(8), 1226–1238 (2005)
10. School of Information and Computer Sciences: UCI machine learning repository. University of California, Irvine () (2021). Accessed Mar 2026
11. Google LLC Kaggle: Kaggle datasets (2021). [Online; Accessed Mar 2026]
12. Arizona State University: Feature selection datasets (2021). [Online; Accessed Mar 2026]
13. Kent Ridge, Biomedical: Data Set Repository: elvira biomedical data set repository (2001). Accessed Mar 2026
14. Bolón-Canedo, V., Sánchez-Maroño, N., Alonso-Betanzos, A.: A review of feature selection methods on synthetic data. Knowl. Inf. Syst. **34**(3), 483–519 (2013)
15. Kohavi, R.: A study of cross-validation and bootstrap for accuracy estimation and model selection. In: International Joint Conference on Artificial Intelligence (IJCAI), vol. 14, pp. 1137–1145 (1995)
16. Alcaraz, F.: fxpmath: fixed-point arithmetic emulation in Python (2020). https://github.com/francof2a/fxpmath

Contextual Bandits with Randomized Neural Networks

M. Angioli[1(✉)], A. Rosato[1], M. Olivieri[1], A. Loutfi[2,3], and D. Kleyko[2,4]

[1] Sapienza University of Rome, Rome 00184, Italy
marco.angioli@uniroma1.it
[2] Örebro University, Örebro 70281, Sweden
[3] Linköping University, Linköping 58183, Sweden
[4] RISE Research Institutes of Sweden, Kista 16440, Sweden

Abstract. Contextual bandits are sequential decision-making problems under partial feedback that arise in many adaptive systems. Hyperdimensional Contextual Bandits (HD-CB) have been proposed as a lightweight alternative to classical linear methods, replacing matrix inversions with high-dimensional nonlinear projection and prototype-based learning, offering scalability, faster convergence, and reduced memory footprint. The HD-CB formulation, however, relies on abstractions for hyperdimensional computing that may not be interpretable by the broader machine learning community. Furthermore, its evaluation has been limited to linear reward settings with a single transformation. In this paper, we show that the HD-CB pipeline: transformation, action selection, and learning admits a one-to-one correspondence with randomized feed-forward neural networks, where the transformation of context vectors is done by a fixed random feature map and the per-action prototype vectors are the sole learnable weights of the output layer. Building on this equivalence, we derive an equivalent matrix reformulation of the algorithm that exposes data parallelism for BLAS backends, and we conduct a systematic ablation of six transformations across four different environments, showing that the optimal feature map is task-dependent and that HD-CB is not limited to settings where the reward depends linearly on the context. These findings link HD-CB with the area of randomized neural networks that include extreme learning machines, random vector functional link networks, and approximation of kernel methods, providing a principled basis for deriving approximation-theoretic guarantees under partial feedback and facilitating deployment on emerging hardware.

Keywords: Random Vector Functional Link Networks · Hyperdimensional Computing · Extreme Learning Machines · Contextual Bandits

1 Introduction

Sequential decision-making under uncertainty has many applications, e.g., robotic control [5], recommendation systems [1,2], clinical trial design [9]. Con-

L. Pasa et al. (Eds.): ICANN 2026, LNCS 17089, pp. 395–407, 2027.
https://doi.org/10.1007/978-3-032-38398-3_33

textual bandit (CB) algorithms provide a principled abstraction for these settings: at each round, an agent observes a context, selects an action from a finite set, and observes a reward for this action, learning to maximize cumulative reward over time [26]. Unlike general reinforcement learning, the selected action does not influence the next context, and the agent maximizes immediate returns.

Standard linear approaches [18] for modelling and solving CB problems require per-action matrix inversions and quadratic memory in the context dimensionality [4], making them impractical for resource-constrained and latency-sensitive deployments. To address these limitations, recent work has introduced Hyperdimensional Contextual Bandits (HD-CB) [3], which replace the dense linear algebra of standard methods with the lightweight, parallel operations of hyperdimensional computing, also known as vector symbolic architectures (HD/VSA) [14,16]. HD-CB achieves linear time complexity, faster convergence, improved scalability, and a reduced memory footprint, making it particularly suitable for hardware acceleration and deployment on resource-constrained devices. Yet, despite these advantages, HD-CB has primarily been developed and interpreted through the lens of hyperdimensional computing, which may make it less immediately accessible to the broader machine learning community. To the best of our knowledge, this is the first work to show that HD-CB also admits a formal equivalence within the framework of randomized feed-forward neural networks: the transformation stage constitutes a fixed random projection with nonlinearities (hidden layer), the per-action prototype matrix is the sole learnable readout matrix, and the reward-modulated Hebbian update rule is an online learning rule adapted to the partial-feedback bandit setting.

This connection places HD-CB within the well-established literature on randomized neural networks [23] including random vector functional link networks [13,19], extreme learning machines [11,12], and random features [22], and motivates two further contributions. First, we derive equivalent matrix-form algorithms for HD-CB and its variants, exposing data parallelism for BLAS backends and yielding significant speedups over baseline implementations. Second, it reveals that the transformation design is equivalent to choosing the random feature map for the network, motivating us to perform an ablation study of alternative transformations, showing how the choice of feature map affects decision quality.

The paper is organized as follows. Section 2 reviews CB, HD/VSA, and randomized neural networks. Section 3 presents the formal equivalence and derives the matrix reformulation. Section 4 quantifies the resulting speedup. Section 5 explores several transformations through systematic ablations. Section 6 discusses broader implications and concludes the paper.

2 Background

This section reviews bodies of work that converge in this paper: the CB problems (Sect. 2.1), HD/VSA (Sect. 2.2), HD-CB (Sect. 2.3), and randomized feed-forward neural networks (Sect. 2.4). Table 1 summarizes the notation used across these areas and highlights their corresponding elements.

Table 1. Notation used in HD-CB and its RVFL/ELM correspondence.

Notation	Meaning
d, D, N	Input dim., hypervector/hidden dim., and num of actions/outputs
$\mathbf{x}_t(\mathbf{x}_{t,a})$	Input context at round t (shared or action-dependent)
$\mathcal{X}_t(\mathcal{X}_{t,a}) \leftrightarrow \sigma(\mathbf{W}_{\text{in}}\mathbf{x} + \mathbf{b})$	Context hypervector / hidden-layer activation
$\mathbf{\Phi} \leftrightarrow \mathbf{W}_{\text{in}}$	Fixed transformation parameters / frozen input weights
$\mathcal{A}_a, \mathbf{A} \leftrightarrow \mathbf{W}_{\text{out}}$	Action prototype, stacked prototypes / readout matrix
$\mathcal{B}_a, \mathbf{B}$	Confidence hypervector, stacked confidence matrix
$a_t, e_{t,a}, c_{t,a}$	Selected action, estimated payoff, and confidence score
$r_{t,a_t}, \mathcal{R}_{t,a_t}$	Observed reward, reward hypervector

2.1 Contextual Bandits

Contextual multi-armed bandits are a class of sequential decision-making problems in which an agent interacts with an environment through a reward-driven process. Given a finite set of N possible actions (or "arms"), at each round $t = 1, \ldots, T$ the agent observes a context $\mathbf{x}_{t,a} \in \mathbb{R}^d$ for each action a, selects an action a_t, and observes a scalar reward r_{t,a_t}. Crucially, only the reward for the selected action is observed, while the rewards for the actions not taken remain unobserved. This partial-feedback structure, known as bandit feedback, distinguishes the setting from full-information supervised learning and is what makes the problem both practically relevant and theoretically challenging [6,26].

The agent's objective is to learn a policy π that maps contexts to actions so as to maximize the cumulative reward $\sum_{t=1}^{T} r_{t,a_t}$ over a horizon of T rounds. Unlike general reinforcement learning, in the CB setting the selected action does not influence the next context, and the agent aims to maximize immediate rather than long-term returns [17,25].

2.2 Hyperdimensional Computing

HD/VSA is a neuro-inspired computing framework that represents information using high-dimensional distributed vectors (hypervectors) of fixed dimensionality $D \gg d$ [14,16,24]. Thanks to the concentration of measure in high-dimensional spaces, independently drawn random hypervectors are quasi-orthogonal with high probability, providing an exponentially large number of such hypervectors within a fixed-dimensional space [14].

These hypervectors are combined using three lightweight operations: i) *superposition*, to combine multiple hypervectors together; ii) *binding*, which associates hypervectors; and iii) *permutation*, to encode positional or role information; and are compared using a similarity measure. Together, these operations form the complete arithmetic of HD/VSA and can be used to encode compositional structures (e.g., sets, records, sequences) and solve learning tasks without increasing dimensionality. The specific mathematical definitions for these operations vary across different HD/VSA models [16,24]. In this paper, we utilize the MAP [10] model, where the binding is a component-wise multiplication, superposition is addition, permutation is a cyclic shift, and similarity is cosine similarity.

2.3 HD-CB: Hyperdimensional Contextual Bandits

HD-CB [3] moves the entire CB pipeline into a high-dimensional space. In the baseline variant, each action $a \in \{1, \ldots, N\}$ is represented by a *prototype hypervector* $\mathcal{A}_a \in \mathbb{R}^D$, initialized to a zero vector.

At each round t, the agent first transforms the observed context $\mathbf{x}_t$ into a hypervector $\mathcal{X}_t \in \mathbb{R}^D$ via a fixed nonlinear transformation. This approach is *transformation-agnostic* [3] and the HD/VSA literature offers a rich catalog of transformations [8,15,16,21], ranging from dense or sparse random projections that may include component-wise nonlinearities to compositional schemes that exploit HD/VSA operations. Once the context is encoded, the agent computes the cosine similarity between each action prototype and the context hypervector,

$$e_{t,a} = \frac{\mathcal{A}_a^\top \mathcal{X}_t}{\|\mathcal{A}_a\|_2 \|\mathcal{X}_t\|_2}, \tag{1}$$

and selects an action according to an exploration strategy (e.g., ε-greedy or confidence-based) [3]. Since prototypes are initialized to zero, the cosine similarity in Eq. (1) is zero for actions that have not yet been selected. After observing the reward r_{t,a_t} (transformed into a reward hypervector $\mathcal{R}_{t,a_t}$ via a thermometer code [20,21]), the prototype hypervector of the selected action is updated as:

$$\mathcal{A}_{a_t} \leftarrow \mathcal{A}_{a_t} + \big(\mathcal{X}_t \odot \mathcal{R}_{t,a_t}\big). \tag{2}$$

This update rule adjusts the selected action prototype closer to or further from the context hypervector, depending on the magnitude of the reward.

The uncertainty-aware variant (HD-CB$_{\text{UNC}}$) introduces a *confidence hypervector* $\mathcal{B}_a$ that tracks how frequently each action a has been chosen in similar contexts, enabling a confidence-like exploration mechanism. For each action a, a confidence score is computed as the similarity between $\mathcal{B}_a$ and $\mathcal{X}_t$, $c_{t,a} = \frac{\mathcal{B}_a^\top \mathcal{X}_t}{\|\mathcal{B}_a\|_2 \|\mathcal{X}_t\|_2}$, and the action selection score combines the estimated payoff with an exploration bonus controlled by a tunable exploration factor $\alpha > 0$: $p_{t,a} = e_{t,a} + \alpha(1 - c_{t,a})$. After selecting an action a_t, the confidence hypervector $\mathcal{B}_{a_t}$ is updated as:

$$\mathcal{B}_{a_t} \leftarrow \mathcal{B}_{a_t} + \boldsymbol{m}_t \odot \mathcal{X}_{t,a_t}, \tag{3}$$

where $\boldsymbol{m}_t \in \{0,1\}^D$ is a binary mask whose components are set to 1 with probability $\min(1, \max(0, \frac{1+c_{t,a_t}}{2}))$, i.e., the cosine confidence mapped from $[-1,1]$ to $[0,1]$; for an all-zero $\mathcal{B}_{a_t}$ the score is taken as zero, yielding probability 0.5.

HD-CB achieves $O(ND)$ time complexity per round for action selection and $O(D)$ for the update, with an $O(ND)$ memory footprint: linear in both the number of actions and the dimensionality of hypervectors. This contrasts sharply with the $O(Nd^2)$ memory and per-round computation of standard linear CB, where the operations involve linear algebra operations that cannot be trivially parallelized. Empirically, HD-CB has demonstrated orders-of-magnitude speedups over linear CB on both general-purpose and custom hardware [3].

2.4 Randomized Feed-Forward Neural Networks

Randomized feed-forward neural networks that include Random Vector Functional Link (RVFL) networks [13,19] also known as Extreme Learning Machines (ELM) [11,12], are single-hidden-layer networks characterized by a distinct approach to learning within the network: the weights between input and hidden layers are randomly initialized and held fixed; only the output-layer weights are updated. Formally, an RVFL/ELM network with d input neurons, D hidden neurons, and N output neurons computes:

$$f(\mathbf{x}) = \mathbf{W}_{\texttt{out}}\,\sigma\big(\mathbf{W}_{\texttt{in}}\,\mathbf{x} + \mathbf{b}\big), \tag{4}$$

where $\mathbf{W}_{\texttt{in}} \in \mathbb{R}^{D\times d}$ and $b \in \mathbb{R}^{D}$ are frozen random parameters, $\sigma(\cdot)$ is a nonlinear activation function, and $\mathbf{W}_{\texttt{out}} \in \mathbb{R}^{N\times D}$ is the learnable readout matrix.

The essential principle is a separation between a fixed nonlinear transformation $(\mathbf{x} \mapsto \sigma(\mathbf{W}_{\texttt{in}}\,\mathbf{x}+\mathbf{b})) \in \mathbb{R}^{D}$ and the learnable readout $\mathbf{W}_{\texttt{out}}$. When the activation function is bounded and non-constant, RVFL/ELM networks are universal approximators whose mean-square error decays as $O(C/\sqrt{D})$ with hidden-layer width [13]. Rahimi and Recht [22] further showed that the hidden-layer output can be interpreted as a random approximation of a kernel function, converging to the exact kernel as $D \to \infty$ and connecting RVFL/ELM networks to the reproducing kernel Hilbert spaces.

3 HD-CB as RVFL/ELM Networks

This section presents the central contribution of this paper. We first establish a formal equivalence between HD-CB and RVFL/ELM networks (Sect. 3.1), then leverage it to derive equivalent matrix-form algorithms (Sect. 3.2).

3.1 Formal RVFL/ELM Equivalence

The three stages of the HD-CB pipeline map one-to-one onto the components of an RVFL/ELM network with a single hidden-layer, see Fig. 1.

The Transformation as a Fixed Random Feature Map: The HD-CB transforms the context $\mathbf{x}_t \in \mathbb{R}^{d}$ into a hypervector $\mathcal{X}_t \in \mathbb{R}^{D}$, $D \gg d$. This transformation is determined by parameters that are drawn at random during the initialization and are never modified during learning, whether it takes the form of a random projection, a record-based transformation, or any other nonlinear transformation. These are precisely the defining characteristics of the RVFL/ELM hidden layer. For nonlinear random projections the correspondence is immediate:

$$\mathcal{X}_t = \mathrm{sign}(\mathbf{\Phi}\,\mathbf{x}_t) = \sigma(\mathbf{W}_{\texttt{in}}\,\mathbf{x}_t + \mathbf{b}), \tag{5}$$

with $\mathbf{W}_{\texttt{in}} = \mathbf{\Phi}$, $\mathbf{b} = \mathbf{0}$, and $\sigma(\cdot) = \mathrm{sign}(\cdot)$. For the record-based transformation used in the original formulation [3], the functional form is more complicated than a single matrix-vector product, but the main elements are the same: all parameters (the random hypervectors assigned to each input feature and the structured ones assigned to each discretized value level) are drawn during initialization and held fixed throughout learning. This observation was formalized in [15], showing that the record-based transformation, despite being composed of two sub-layers: a density-based representation layer, where each input feature is independently mapped to its thermometer code, and a hidden layer that binds thermometer codes with columns of a random weight matrix and superimposes with sign binarization, fulfills the role of a single RVFL/ELM hidden layer, since all parameters before the output layer remain either deterministic or fixed at initialization. Regardless of the specific transformation variant adopted, crucially, the transformation stage is set once and never updated: no gradients flow through it, and no learning signal modifies its parameters.

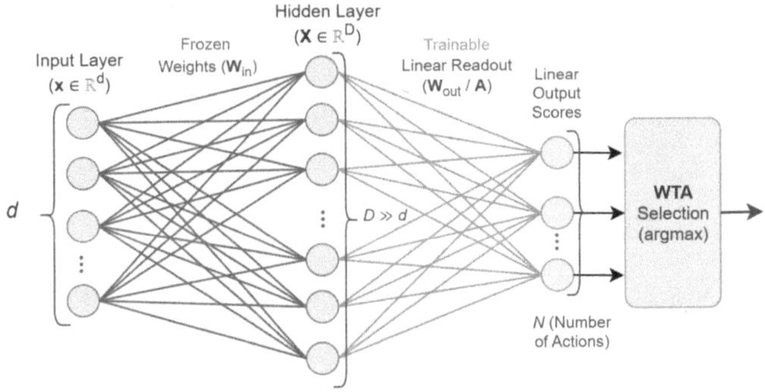

Fig. 1. Graphical equivalence between RVFL/ELM networks and HD-CB.

The Action Prototypes as Learnable Output Weights: Once the context has been transformed, HD-CB performs action selection via a linear model. As described in Sect. 2.3, each action a is associated with a prototype hypervector $\mathcal{A}_a \in \mathbb{R}^D$, and the estimated payoff is the cosine similarity, Eq. 1. When the transformation produces bipolar vectors, $\mathcal{X}_t \in \{-1,+1\}^D$, the norm $\|\mathcal{X}_t\|_2 = \sqrt{D}$ is constant and can be folded into a global scaling factor, so that normalization requires only the per-action prototype norms. In either case, the score for each action is a *linear function* of the hidden-layer activation $\mathcal{X}_t$, weighted by the prototype $\mathcal{A}_a$. This is exactly the RVFL/ELM output layer, where each action prototype corresponds to a row of the readout matrix $\mathbf{W}_{\mathtt{out}}$. The collection of action prototypes $\{\mathcal{A}_a\}_{a=1}^N$ constitutes the only set of parameters modified during learning. Action selection, $a_t = \arg\max_a e_{t,a}$ with probability $1-\varepsilon$ (and a random action otherwise), can be viewed as a stochastic Winner-Take-All output nonlinearity.

The Update Rule as Online RVFL/ELM Learning: After selecting a_t and observing r_{t,a_t}, HD-CB$_{\mathtt{EPS}}$ updates the selected action prototype as in Eq. (2). In the thermometer code of the reward, $\mathcal{R}_{t,a_t}$, the number of $+1$ components increases monotonically with the reward magnitude, so that a high reward produces a hypervector largely aligned with $\mathcal{X}_t$ (pulling the action prototype toward the context hypervector) while a low reward produces one largely anti-aligned (pushing it away). This is a reward-modulated Hebbian rule applied to the output weights: only the prototype corresponding to the selected action is modified, in the direction of the hidden-layer activation $\mathcal{X}_t$ modulated by the reward hypervector $\mathcal{R}_{t,a_t}$. This is the standard form of learning the output weights in RVFL/ELM networks, with the sole deviation being the partial-feedback structure in which only the selected action's prototype is updated.

3.2 Matrix Reformulation

The RVFL/ELM equivalence makes the matrix reformulation immediate: once the per-action prototypes are recognized as rows of the readout matrix, expressing the HD-CB pipeline in a matrix form requires only standard linear algebra. In the original formulation [3], the estimated payoffs are computed via a per-action loop. From the RVFL/ELM perspective, this loop is simply an unrolled matrix-vector product. By stacking N action prototypes as rows of a matrix,

$$\mathbf{A} = [\mathcal{A}_1, \ldots, \mathcal{A}_N]^\top \in \mathbb{R}^{N \times D}, \tag{6}$$

the normalized vector of estimated payoffs for all actions is obtained directly. When the context is shared across actions:

$$\mathbf{e}_t = \frac{\mathbf{A}\,\mathcal{X}_t}{\sqrt{\operatorname{diag}(\mathbf{A}\mathbf{A}^\top)}\|\mathcal{X}_t\|_2} \in \mathbb{R}^N. \tag{7}$$

When contexts are action-dependent, stacking the per-action transformations as rows of $\mathbf{X}_t \in \mathbb{R}^{N \times D}$ yields:

$$\mathbf{e}_t = \frac{\operatorname{diag}(\mathbf{A}\,\mathbf{X}_t^\top)}{\sqrt{\operatorname{diag}(\mathbf{A}\mathbf{A}^\top)}\,\sqrt{\operatorname{diag}(\mathbf{X}_t\mathbf{X}_t^\top)}}. \tag{8}$$

For bipolar hypervectors ($\mathcal{X}_t \in \{-1, +1\}^D$), the norm is constant $\|\mathcal{X}_t\|_2 = \sqrt{D}$ and can be absorbed into a global scaling factor. The update remains as in Eq. (2), and the action prototypes' norms can be modified incrementally.

The same principle extends to HD-CB$_{\text{UNC}}$, which maintains the second matrix $\mathbf{B} = [\mathcal{B}_1, \ldots, \mathcal{B}_N]^\top \in \mathbb{R}^{N \times D}$: the second set of RVFL/ELM output weights operating on the same hidden-layer activations. The estimated payoff and confidence vectors are computed in parallel:

$$\mathbf{e}_t = \frac{\operatorname{diag}(\mathbf{A}\,\mathbf{X}_t^\top)}{\sqrt{\operatorname{diag}(\mathbf{A}\mathbf{A}^\top)}\sqrt{\operatorname{diag}(\mathbf{X}_t\mathbf{X}_t^\top)}}, \qquad \mathbf{c}_t = \frac{\operatorname{diag}(\mathbf{B}\,\mathbf{X}_t^\top)}{\sqrt{\operatorname{diag}(\mathbf{B}\mathbf{B}^\top)}\sqrt{\operatorname{diag}(\mathbf{X}_t\mathbf{X}_t^\top)}}, \tag{9}$$

and the updates follow the same component-wise form as Eqs. (2) and (3).

Algorithm 1. HD-CB$_{\text{EPS}}$, Matrix Form

Require: Exploration rate $\varepsilon \in [0, 1]$

$\mathbf{A} \leftarrow \mathbf{0}_{N \times D}$
for $t = 1, \ldots, T$ **do**
  Transform context(s) $\mathbf{x}_t$ ($\mathbf{x}_{t,a}$)
  $\mathbf{e}_t \leftarrow \dfrac{\mathbf{A}\,\mathcal{X}_t}{\sqrt{\operatorname{diag}(\mathbf{A}\mathbf{A}^\top)}\,\|\mathcal{X}_t\|_2}$ ▷ matrix scoring + cosine normalization
  Draw $u \sim \mathcal{U}(0, 1)$
  if $u < \varepsilon$ **then**
    $a_t \leftarrow$ RandomAction()
  else
    $a_t \leftarrow \arg\max_a e_{t,a}$
  end if
  Observe r_{t,a_t}; encode $\mathcal{R}_{t,a_t} \in \{-1, +1\}^D$
  $\mathbf{A}_{a_t} \leftarrow \mathbf{A}_{a_t} + \mathcal{X}_{t,a_t} \odot \mathcal{R}_{t,a_t}$ ▷ rank-one row update
end for

Algorithm 1 summarizes the complete matrix-form HD-CB$_{\text{EPS}}$; the HD-CB$_{\text{UNC}}$ extension adds a parallel $\boldsymbol{c}_t$ computation via Eq. (9) and a rank-one update to $\mathbf{B}$. Both formulations produce identical decisions to their loop-based counterparts; their benefit is purely computational: the matrix-vector products and norm computations are dense linear algebra operations dispatchable to optimized BLAS kernels, replacing the sequential per-action loop.

4 Execution Time Analysis Results

To quantify the practical benefit of the matrix reformulation, we implemented both HD-CB$_{\text{EPS}}$ and HD-CB$_{\text{UNC}}$ in two versions: the original per-action loop and the matrix form derived in Sect. 3.2. All implementations use record-based transformation with bipolar hypervectors of dimensionality $D = 1000$. We benchmark on two problem sizes ($N \in \{10, 20\}$ actions, $d = 10$ context features) across a CPU (Intel Core i7-11800H, NumPy/MKL) and a GPU (NVIDIA RTX 3050, CuPy/cuBLAS). Per-round execution times cover the complete round (context encoding, scoring, action selection, and the prototype update), averaged over 5 independent runs of 200 rounds with the first 20 discarded as warm-up; because per-round action selection forces a device-to-host transfer, the GPU measurements include synchronization. Figure 2 reports the results.

On CPU, the matrix reformulation yields a 1.4×–2.1× speedup for HD-CB$_{\text{EPS}}$ and 1.1×–2.4× for HD-CB$_{\text{UNC}}$ across the evaluated configurations. On GPU, the gains are substantially larger: 2.9×–3.4× for HD-CB$_{\text{EPS}}$ and up to 4.8× for HD-CB$_{\text{UNC}}$ at $N = 20$. The GPU benefits more because each Python-level iteration in the loop-based version incurs a CUDA kernel launch; the matrix reformulation replaces N individual inner products with a single matrix–vector product, eliminating $N - 1$ launches per scoring step. The uncertainty-aware variant benefits most because its loop requires *two* per-action similarity computations, doubling the number of kernel launches that the matrix form removes.

In absolute terms the GPU is slower at this scale: the per-round workload is small enough that kernel-launch and synchronization overhead exceeds the actual compute time, consistent with GPU backends on low-arithmetic-intensity workloads [4]. We, therefore, expect the GPU to overtake the CPU only once arithmetic intensity rises, at $D \sim 10^4$ or N in the hundreds, so at the per-round granularity typical of CB, CPU deployment with BLAS-optimized matrix operations is the preferred target, supporting resource-constrained deployments.

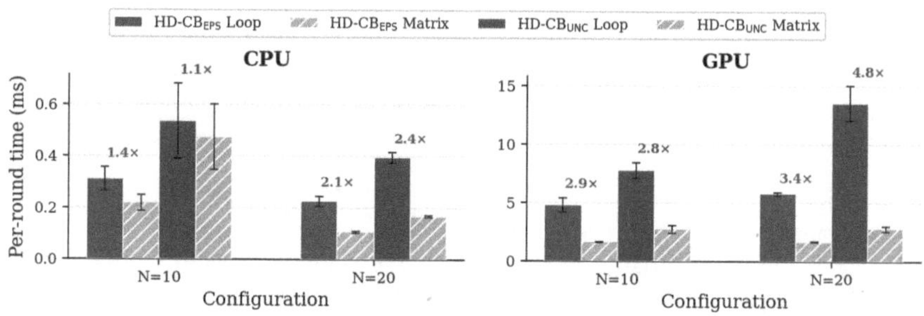

Fig. 2. Per-round execution time of loop-based vs. matrix-form for HD-CB_{EPS} and HD-CB_{UNC} ($D = 1000$, $d = 10$). Left: CPU (Intel i7-11800H); right: GPU (NVIDIA RTX 3050). Annotations report speedups of the matrix form over the loop one.

5 Transformation Ablation Study

The original HD-CB evaluation [3] assessed performance exclusively on environments with linear rewards using the record-based transformation. However, unlike traditional linear CB algorithms [18], whose correctness relies on the linear rewards, HD-CB imposes no structural constraint on the reward function: the transformation–prototype–learning pipeline can, in principle, approximate arbitrary reward structures. The RVFL/ELM equivalence established in Sect. 3 makes this point precise: if the transformation is a random feature map, then the quality of the policy depends on how well that map represents the reward function. This section tests this observation by comparing six transformations: `Record-based` [27] (record-based transformation with best number of quantization levels $L \in \{10, 20, 50\}$, two-sub-layer RVFL/ELM hidden layers [15]); `Linear RP` (dense Gaussian random projection (RP)); `Sign RP` (dense Gaussian RP and sign(·)); `RFF` [22] (random Fourier features, $\sqrt{2/D}\cos(\mathbf{\Phi x}+\mathbf{b})$, where γ is the kernel bandwidth, tuned per environment); `FlyHash` [8] (sparse RP with connection probability s followed by k-Winners-Take-All retaining a fraction k/D of active neurons, both tuned per environment); and `Sparse Binary RP` (sparse $\{-1, 0, +1\}$ matrix + sign(·), sparsity swept over $\{0.05, 0.10, 0.20\}$), across four synthetic environments of increasing reward nonlinearity, where $q_a(\mathbf{x}_t)$ denotes the expected reward for action a given context $\mathbf{x}_t$, and each observed reward is a Bernoulli draw $r_{t,a} \sim \text{Bernoulli}\big(q_a(\mathbf{x}_t)\big)$: (i) `Linear-Small` ($N = 5$, $d = 5$): $q_a = \sigma(\mathbf{x}_t^\top \mathbf{w}_a)$; (ii) `Linear-Large` ($N = 20$, $d = 30$): same functional form, larger scale; (iii) `Quadratic` ($N = 10$, $d = 15$): $q_a = \sigma(\mathbf{x}_t^\top \mathbf{M}_a \mathbf{x}_t)$ with a full-rank symmetric $\mathbf{M}_a = \frac{1}{2}(\mathbf{A}_a + \mathbf{A}_a^\top)$, $\mathbf{A}_a$ i.i.d. $\mathcal{N}(0,1)$ (second-order interactions); (iv) `Cosinusoidal` ($N = 10$, $d = 10$): $q_a = \sigma\big(\sum_{j \in S_a} \varsigma_{a,j} \cos(\omega_{a,j} x_{t,j})\big)$ over a random three-feature subset S_a, with signs $\varsigma_{a,j} \in \{-1, +1\}$ and frequencies $\omega_{a,j} \in [8, 16]$ (periodic, linearly inseparable). All experiments use HD-CB$_{\text{EPS}}$ with $\varepsilon = 0.05$, $T = 5000$ rounds, averaged over 50 simulation runs. In all configurations, the observed reward is encoded via thermometer code as in the original study [3]; the six transformations differ only in how the context is transformed. As a reference we include LinEPS [18], a standard linear CB that estimates per-action payoffs by ridge regression and explores with the same ε-greedy strategy, isolating the effect of the transformation.

Table 2. Transformation ablation: average reward (mean $\pm$ std), with $D = 1000$, $\varepsilon = 0.05$, $T = 5000$, and four environments, each averaged over five random instances $\times$ 10 simulations (50 runs). The best policy per environment is in bold.

	`Lin.-Small`	`Lin.-Large`	`Quadratic`	`Cosinusoidal`
	N=5, d=5	N=20, d=30	N=10, d=15	N=10, d=10
HD-CB Record	0.738±.025	0.767±.002	0.568±.003	**0.622±.010**
HD-CB FlyHash	0.676±.024	0.763±.005	**0.670±.001**	0.553±.011
HD-CB Sign RP	**0.751±.027**	**0.792±.006**	0.639±.008	0.549±.011
HD-CB Linear RP	**0.751±.026**	0.790±.006	0.640±.004	0.545±.010
HD-CB RFF	0.739±.027	0.758±.003	0.668±.003	0.558±.010
HD-CB Sparse Bin.	0.747±.023	0.792±.006	0.637±.005	0.550±.014
LinEPS *(baseline)*	0.729±.025	**0.803±.007**	0.620±.005	0.542±.007
Random policy	0.500±.001	0.500±.000	0.498±.003	0.502±.001

Table 2 summarizes the results. No single transformation dominates: the best strategy changes across environments, confirming that transformation choice is a consequential feature-map design decision. Dense random projections (`Sign RP`, `Linear RP`, `Sparse Binary RP`) lead on the linear environments, matching the linear baseline; the kernel-approximating transformations (`FlyHash`, `RFF`) prevail on `Quadratic`, whose full-rank form induces feature interactions; and `Record-based` prevails on the additive-periodic `Cosinusoidal`.

6 Discussion and Conclusion

This paper established a formal one-to-one correspondence between hyperdimensional contextual bandits (HD-CB) and randomized feed-forward neural networks (RVFL/ELM), showing that the complete HD-CB pipeline is a single-hidden-layer neural network with a fixed random feature map and learnable output weights. This equivalence yields two concrete outcomes. First, it provides a matrix reformulation that replaces sequential per-action loops with dense linear algebra primitives, delivering measurable speedups on both CPU and GPU backends. Second, it reframes transformation design as random feature map selection, connecting HD-CB to the established literature on randomized neural networks and random features. The ablations across four reward environments of varying complexity demonstrate that the choice of the feature map has a significant, task-dependent impact on decision quality and highlights that HD-CB is not inherently limited to the linear reward settings where it was originally evaluated.

Beyond these direct contributions, the RVFL/ELM equivalence suggests several theoretical connections that we frame as directions for future work rather than as results established under partial feedback. Since HD-CB shares the RVFL/ELM network, the universal approximation results recalled in Sect. 2.4 offer a quantitative characterization of the observed empirical trend that decision quality improves with D [3]. A formal extension of the $O(C/\sqrt{D})$ bound to partial feedback remains an open question. The connection to kernel methods [22] carries over directly: inner products between transformed contexts converge to a kernel function as $D \to \infty$ arc-cosine kernel for random projections with $\text{sign}(\cdot)$ function [7] or Gaussian kernel when the transformation follows one of the random Fourier features constructs, so that HD-CB can be understood as online linear learning in an implicitly defined reproducing kernel Hilbert space without ever computing or storing a kernel matrix. The neural network interpretation places HD-CB within the broader family of neural CB methods [28], as the special case in which the first layer is frozen and only the output layer is learned. As a result, while this paper focuses on the original Hebbian update of HD-CB, the RVFL/ELM equivalence admits other standard learning rules on the output weights. Finally, the RVFL/ELM network also maps naturally onto spiking neural network substrates: the encoder becomes a fixed spiking feature extractor, the action prototypes correspond to learnable synaptic weights, and action selection is implemented via a Winner-Take-All circuit.

Acknowledgments. The work of AL and DK was supported by Knut and Alice Wallenberg Foundation under the Wallenberg Scholars program (Grant No. KAW2023.0327). DK acknowledges funding from the Swedish Strategic Research Foundation under the Future Research Leaders program (Grant No. FFL24-0111), the Swedish Research Council under the Starting Grant program (Grant No. 2025–05421), and the Swedish Knowledge Foundation (Contract No. 20210016). This work was supported in part by the AFOSR under award number FA8655-25-1-7007.

Disclosure of Interests. The authors have no competing interests to declare that are relevant to the content of this paper.

References

1. Agarwal, A.: Multiworld testing decision service: a system for experimentation, learning, and decision-making. Whitepaper Microsoft, 1–40 (2016)
2. Amat, F., Chandrashekar, A., Jebara, T., Basilico, J.: Artwork personalization at Netflix. In: ACM Conference on Recommender Systems (RecSys), pp. 487–488 (2018)
3. Angioli, M., Rosato, A., Barbirotta, M., Martino, R., Menichelli, F., Olivieri, M.: HD-CB: the first exploration of hyperdimensional computing for contextual bandits problems. IEEE Open J. Comput. Soc. **7**, 105–116 (2026)
4. Angioli, M., et al.: Efficient implementation of LinUCB through algorithmic improvements and vector computing acceleration for embedded learning systems. ACM Trans. Embed. Comput. Syst. **24**(4), 1–23 (2025)
5. Banerjee, R.: To ask or not to ask: human-in-the-loop contextual bandits with applications in robot-assisted feeding. In: IEEE International Conference on Robotics and Automation (ICRA), pp. 1378–1384 (2025)
6. Bouneffouf, D., Rish, I., Aggarwal, C.: Survey on applications of multi-armed and contextual bandits. IEEE Congress on Evolutionary Computation (CEC), pp. 1–8 (2020)
7. Cho, Y., Saul, L.: Kernel methods for deep learning. In: Advances in Neural Information Processing Systems (NeurIPS), pp. 342–350 (2009)
8. Dasgupta, S., Stevens, C.F., Navlakha, S.: A neural algorithm for a fundamental computing problem. Science **358**(6364), 793–796 (2017)
9. Durand, A., Achilleos, C., Iacovides, D., Strati, K., Mitsis, G.D., Pineau, J.: Contextual bandits for adapting treatment in a mouse model of de novo carcinogenesis. In: Machine Learning for Healthcare (MLHC), pp. 67–82 (2018)
10. Gayler, R.W.: Multiplicative binding, representation operators & analogy. In: Advances in Analogy Research: Integration of Theory and Data from the Cognitive, Computational, and Neural Sciences, pp. 1–4 (1998)
11. Huang, G.B., Zhu, Q.Y., Siew, C.K.: Extreme learning machine: theory and applications. Neurocomputing **70**(1–3), 489–501 (2006)
12. Huang, G.B.: Extreme learning machine for regression and multiclass classification. IEEE Trans. Syst. Man Cybern. B Cybern. **42**(2), 513–529 (2012)
13. Igelnik, B., Pao, Y.H.: Stochastic choice of basis functions in adaptive function approximation and the functional-link net. IEEE Trans. Neural Netw. **6**(6), 1320–1329 (1995)
14. Kanerva, P.: Hyperdimensional computing: an introduction to computing in distributed representation with high-dimensional random vectors. Cogn. Comput. **1**(2), 139–159 (2009)
15. Kleyko, D., Kheffache, M., Frady, E.P., Wiklund, U., Osipov, E.: Density encoding enables resource-efficient randomly connected neural networks. IEEE Trans. Neural Netw. Learn. Syst. **32**(8), 3777–3783 (2021)
16. Kleyko, D., Rachkovskij, D.A., Osipov, E., Rahimi, A.: A survey on hyperdimensional computing aka vector symbolic architectures, Part I: models and data transformations. ACM Comput. Surv. **55**(6), 1–40 (2022)
17. Lattimore, T., Szepesvári, C.: Bandit Algorithms. Cambridge University Press (2020)
18. Li, L.: A contextual-bandit approach to personalized news article recommendation. In: International Conference on World Wide Web (WWW), pp. 661–670 (2010)

19. Pao, Y.H., Park, G.H., Sobajic, D.J.: Learning and generalization characteristics of the random vector functional-link net. Neurocomputing **6**(2), 163–180 (1994)
20. Penz, P.A.: The closeness code: an input integer to binary vector transformation suitable for neural network algorithms. In: IEEE First Annual International Conference on Neural Networks (ICNN), pp. 515–522 (1987)
21. Rachkovskij, D.A., Slipchenko, S.V., Kussul, E.M., Baidyk, T.N.: Sparse binary distributed encoding of scalars. J. Autom. Inf. Sci. **37**(6), 12–23 (2005)
22. Rahimi, A., Recht, B.: Random features for large-scale kernel machines. In: Advances in Neural Information Processing Systems (NeurIPS), pp. 1–8 (2007)
23. Scardapane, S., Wang, D.: Randomness in neural networks: an overview. Data Min. Knowl. Disc. **7**(2), 1–18 (2017)
24. Schlegel, K., Neubert, P., Protzel, P.: A comparison of vector symbolic architectures. Artif. Intell. Rev. **55**(6), 4523–4555 (2022)
25. Slivkins, A.: Introduction to multi-armed bandits. Found. Trends Mach. Learn. **12**(1–2), 1–286 (2019)
26. Sutton, R.S., Barto, A.G.: Reinforcement Learning: An Introduction. MIT Press (1998)
27. Vergés, P., et al.: Classification using hyperdimensional computing: a review with comparative analysis. Artif. Intell. Rev. **58**(6), 1–41 (2025)
28. Zhang, W., Zhou, D., Li, L., Gu, Q.: Neural Thompson sampling. In: International Conference on Learning Representations (ICLR) (2021)

Adaptive Weighted LS-SVM for Multi-view Classification

Farnaz Faramarzi Lighvan(✉), Mehrdad Asadi, and Lynn Houthuys

AI Lab, Vrije Universiteit Brussel, Brussels, Belgium
farnaz.faramarzi.lighvan@vub.be

Abstract. Multi-view learning integrates diverse representations of the same instances and can improve performance when interactions across views are effectively exploited. Most existing kernel-based multi-view learning methods either rely on fusion techniques without explicitly enforcing a consensus or complementary collaboration across views, or use co-regularization-based loss functions that impose only pairwise interactions, thereby limiting global collaboration. We propose AW-LSSVM, an adaptive weighted LS-SVM that explicitly enforces complementary learning across all views through an iterative global coupling mechanism. At each iteration, each view not only learns from its own data but is also guided to compensate for samples misclassified by other views in previous iterations by assigning adaptive sample weights. We introduce two strategies for computing these weights: (1) based on averaging misclassification errors across other views and, (2) based on a dissimilarity-aware error aggregation that puts more emphasis on errors from more dissimilar views. Experiments demonstrate that AW-LSSVM outperforms existing multi-view methods on most benchmark datasets.

Keywords: Multi-view Learning · Kernel Methods · LS-SVM

1 Introduction

Multi-view learning integrates multiple representations of the same data to improve performance. Views may come from different data types or feature groups and provide correlative or complementary information. Correlative views enhance generalization, while complementary ones lead to more robust models. A core component of multi-view learning is the fusion strategy [16]. Early fusion merges input-level features for joint modeling, but can be vulnerable to noise and prone to issues arising from high dimensionality. Late fusion combines predictions from separate models but may miss inter-view interactions. Mid fusion aims to balance both approaches.

Kernel-based methods are commonly used in multi-view classification tasks due to their capacity for nonlinear modeling, strong generalization performance on small datasets, and support for flexible regularization. Multiple kernel learning [1] aggregates kernels from each view to learn a shared decision function

L. Pasa et al. (Eds.): ICANN 2026, LNCS 17089, pp. 408–419, 2027.
https://doi.org/10.1007/978-3-032-38398-3_34

without explicitly enforcing a specific type of collaboration among views. Co-regularized methods impose pairwise inter-view collaborations. SVM-2K [8] encourages pairwise alignment, while coupling-based MV-LSSVM [9] incorporates pairwise products of the errors into Least-Squares SVM (LS-SVM) leading to complementary views that can offset each otherâĂŹs errors. Co-regularization methods fail to capture global interactions among multiple views. To address this limitation, ϱTMV-RKM [10] uses tensor-based Restricted Kernel Machines with a joint optimization to create a shared latent space that aligns all views at once. Mumbo [12] is a boosting scheme that combines weak learners from each view through sample-weight updates. It promotes a limited form of complementarity by encouraging weaker views to focus on samples that are difficult for the major view. Consequently, its global collaboration mechanism is mainly centered around the dominant view.

We propose Adaptive Weighted LS-SVM (AW-LSSVM) for multi-view classification, a method that explicitly enforces complementary collaboration across views. Unlike MV-LSSVM and Mumbo, it takes into account the weaknesses of all views when guiding each viewâĂŹs learning, rather than restricting the interaction to pairwise or dominant-view-centered compensation, and unlike ϱTMV-RKM, which aligns all views in a joint latent space, it enforces global complementarity directly in the original feature spaces through adaptive error-based sample weighting, retaining the distinctiveness of each view. Experiments show that AW-LSSVM achieves strong classification performance on a variety of benchmark multi-view datasets. An added advantage of the proposed method is that raw features remain separated across views, making it naturally suitable for distributed settings such as Vertical Federated Learning (VFL) [14], though formal privacy-preserving realization and experiments are left for future work.

2 Proposed Model

Multi-view data often exhibit differences in distribution across views, and this diversity allows them to provide complementary information that can be leveraged to enhance model performance. Motivated by this, we propose Adaptive Weighted LS-SVM (AW-LSSVM), an iterative approach for multi-view classification that captures global collaboration and promotes compensation of misclassification errors across views. For each view, this is achieved by adaptively assigning higher weights to samples misclassified by other views. Unlike SVM [5] where only margin-violating samples contribute to the loss, in LS-SVM [19], each sample has an error variable that measures the deviation from the ideal margin constraint and contributes to the loss. This property is particularly suitable for our adaptive approach. Given a dataset $\{(\mathbf{x}_k^{(v)}, y_k)\}_{k=1,v=1}^{N,V}$ with N samples and V views, at each iteration $t > 0$, AW-LSSVM solves the following **primal** problem for each view v:

$$\min_{\mathbf{w}^{(v,t)},\, b^{(v,t)},\, e^{(v,t)}} \frac{1}{2}\|\mathbf{w}^{(v,t)}\|^2 + \frac{\gamma^{(v,t)}}{2}\sum_{k=1}^{N}(e_k^{(v,t)})^2 + \frac{\rho^{(v,t)}}{2}\sum_{k=1}^{N} s_k^{(v,t-1)}(e_k^{(v,t)})^2$$
$$\text{s.t.} \quad y_k\left(\mathbf{w}^{(v,t)\mathrm{T}}\varphi^{(v,t)}(\mathbf{x}_k^{(v)}) + b^{(v,t)}\right) = 1 - e_k^{(v,t)}, \quad \forall k \tag{1}$$

where, $\mathbf{w}^{(v,t)}$ is the weight vector, $b^{(v,t)}$ is the bias value, $\varphi^{(v,t)}(\cdot)$ is the feature map, and $\gamma^{(v,t)} > 0$ is the regularization parameter. While the standard LS-SVM regularization term $\sum_{k=1}^{N}(e_k^{(v,t)})^2$ penalizes all sample errors uniformly, the proposed global coupling term $\sum_{k=1}^{N} s_k^{(v,t-1)}(e_k^{(v,t)})^2$, controlled by the hyperparameter $\rho^{(v,t)}$, links the optimization of each view to the misclassification behavior of all other views through $s_k^{(v,t-1)}$. This additional term encourages each view-specific model not only to generalize on its own data but also to compensate for the weaknesses of other views by assigning higher weights to samples that were misclassified by them. To induce this error-compensation mechanism, the weights are computed from the aggregated errors of misclassified samples across views. Specifically, the error vectors from the previous iteration are first modified as $\tilde{\mathbf{e}}^{(v,t-1)}$, retaining only the errors of misclassified samples while setting the rest ($e_k \leq 1$) to zero. The resulting modified errors from other views are then aggregated and incrementally updated to maintain stability across iterations. For this aggregation, we consider two strategies; in addition to simple averaging (Eq. (2)), we introduce a novel dissimilarity-aware weighted aggregation using Eq. (3). This strategy leverages the normalized pairwise Euclidean distance between modified error vectors to assign higher importance to views with more dissimilar error patterns.

$$s_{k,\text{avg}}^{(v,t-1)} = \beta^{t-2}\frac{1}{V-1}\sum_{\substack{u=1\\u\neq v}}^{V}(\tilde{e}_k^{(u,t-1)})^2 + s_k^{(v,t-2)} \tag{2}$$

$$s_{k,dissim}^{(v,t-1)} = \beta^{t-2}\sum_{u=1}^{V}\frac{\left\|(\tilde{\mathbf{e}}^{(v,t-1)})^{\circ 2} - (\tilde{\mathbf{e}}^{(u,t-1)})^{\circ 2}\right\|_2}{\sum_{u'}\left\|(\tilde{\mathbf{e}}^{(v,t-1)})^{\circ 2} - (\tilde{\mathbf{e}}^{(u',t-1)})^{\circ 2}\right\|_2}(\tilde{e}_k^{(u,t-1)})^2 + s_k^{(v,t-2)} \tag{3}$$

with $s_k^{(v,0)} = 0$ at the initial round $t = 1$, $\circ$ denotes element-wise operation, and $\beta^{(t-2)} \in (0, 1)$ is the decay factor that gradually reduces the increments across iterations, serving as an informal stabilizer. Notably, this stabilization is enforced by β independently of ρ, the kernel, or the cross-view error signals. In practice, the choice of aggregation strategy depends on the structure of inter-view diversity. The average-based variant is preferable when the views are relatively consistent, as it provides a more stable coupling mechanism. The dissimilarity-aware variant is more suitable when the views are less redundant and exhibit complementary information, as it emphasizes informative differences and promotes stronger cross-view compensation. Such inter-view diversity can be assessed using metrics such as Centered Kernel Alignment (CKA) [4].

By taking the Lagrangian of the primal problem, deriving the KKT optimality conditions, and eliminating the primal variables $\mathbf{w}$ and e, we obtain the following **dual** problem:

$$\left[\begin{array}{c|c} 0 & \mathbf{y}^{(v,t)^T} \\ \hline \mathbf{y}^{(v,t)} & \boldsymbol{\Omega}^{(v,t)} + \Lambda^{(v,t)} \end{array}\right] \left[\begin{array}{c} b^{(v,t)} \\ \hline \boldsymbol{\alpha}^{(v,t)} \end{array}\right] = \left[\begin{array}{c} 0 \\ \hline \mathbf{1}_N \end{array}\right] \tag{4}$$

where, $\boldsymbol{\Lambda}^{(v,t)} = \text{diag}\big((\gamma^{(v,t)} + \rho^{(v,t)}\, s_1^{(v,t-1)})^{-1}, \ldots, (\gamma^{(v,t)} + \rho^{(v,t)}\, s_N^{(v,t-1)})^{-1}\big) \in \mathbb{R}^{N \times N}$ is a sample-specific inverse weighting matrix, $\boldsymbol{\Omega}^{(v,t)}$ is a labeled kernel matrix with elements $\Omega_{ij} = y_i y_j K(\mathbf{x}_i^{(v)}, \mathbf{x}_j^{(v)})$, using a kernel function $K : \mathbb{R}^{d_v} \times \mathbb{R}^{d_v} \to \mathbb{R}$, where d_v denotes the dimensionality of view v. Solving Eq. (4) yields dual variables $\boldsymbol{\alpha}^{(v,t)} \in \mathbb{R}^N$ and a bias variable $b^{(v,t)} \in \mathbb{R}$ to compute the error vector of training samples for the next iteration as follows:

$$\mathbf{e}^{(v,t)} = \mathbf{y} - (\boldsymbol{\Omega}^{(v,t)} \boldsymbol{\alpha}^{(v,t)} + b^{(v,t)} \mathbf{1}) \tag{5}$$

For multiclass datasets, one-vs-all label encoding produces multiple binary AW-LSSVM subproblems for each view. After the optimal number of iterations T, the prediction for a new test sample $\mathbf{x}^* = \{\mathbf{x}^{*(v)}\}_{v=1}^V$ is obtained using Eq. (7), which aggregates the soft decision values of all view-specific classifiers, each computed via the decision function in Eq. (6):

$$f^{(v,T)}(\mathbf{x}^{*(v)}) = \sum_{k=1}^{N} \alpha_k^{(v,T)} y_k K\left(\mathbf{x}^{*(v)}, \mathbf{x}_k^{(v)}\right) + b^{(v,T)} \tag{6}$$

$$\hat{y}(\mathbf{x}^*) = \text{sign}\left(\frac{1}{V} \sum_{v=1}^{V} f^{(v,T)}\left(\mathbf{x}^{*(v)}\right)\right) \tag{7}$$

Figure 1 illustrates the overall framework of the proposed method. As demonstrated, this algorithm's ability to keep raw features isolated across views suggests potential applicability in VFL, though privacy guarantees remain outside the scope of this paper. The detailed training and inference procedures are summarized in Algorithm 1.

3 Experiments

In this section, we apply the proposed AW-LSSVM model to several benchmark multi-view datasets and compare its performance with other kernel-based multi-view classification methods. The code will be made publicly available upon publication.

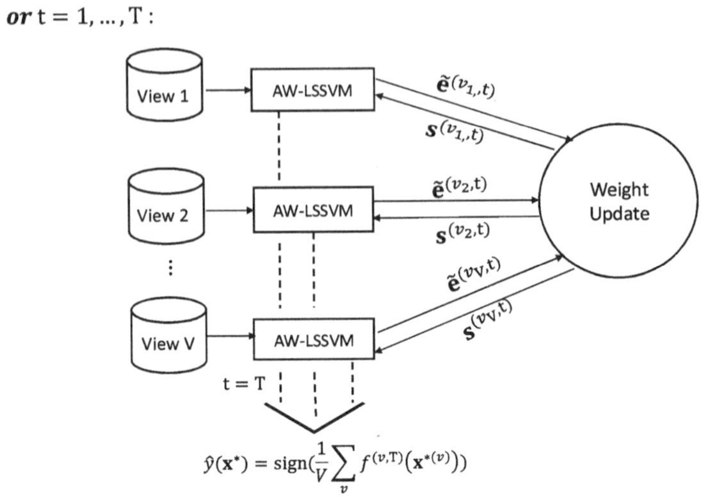

Fig. 1. Illustration of the proposed algorithm. At each iteration, view-specific misclassified error vectors $\tilde{\mathbf{e}}^{(v,t)}$ are aggregated to update sample weights for the subsequent iterations.

3.1 Datasets and Baseline Models

We conducted experiments on ten benchmark multi-view datasets[1] [2,3,7,11,13,15,17,18,20], whose statistical properties are summarized in Table 1. Due to the large size of the YT (YouTube video games) dataset [15], stratified downsampling was applied to this dataset to subsample 500 instances from seven most frequent categories. From its multiple high-dimensional views, two auditory views (Spectrogram and MFCC) and one visual view (HOG) were chosen.

We compare AW-LSSVM with the baseline models **BSV**, **Early Fusion**, and **Late Fusion** as well as the state-of-the-art kernel-based classifiers **MV-LSSVM** [9], **EasyMKL** [1], **Mumbo** [12], and ϱ**TMV-RKM** [10]. **BSV** applies LS-SVM on each view separately, reporting the best-performing view. **Early Fusion** concatenates all view features before training a single LSSVM, while **Late Fusion** combines predictions of view-specific LS-SVM classifiers via majority voting.

3.2 Model Selection

For robust hyperparameter tuning, each dataset is divided into three distinct train-test splits, and for each training split, hyperparameter optimization is carried out using dual annealing combined with three-fold cross-validation.

In order to reduce computational complexity, for all one-vs-all binary subproblems of a given view, the same hyperparameters are shared. In AW-LSSVM,

[1] Data are available at https://github.com/ChuanbinZhang/Multi-view-datasets.

Algorithm 1: Multi-View Classification by AW-LSSVM

Training

Input: $\mathcal{D} = \{(\mathbf{x}_k^{(v)}, y_k)\}_{k=1,v=1}^{N,V}$, $K(\cdot,\cdot)$, γ, ρ, β, T, $\mathcal{A} \in \{\text{avg}, \text{dissim}\}$

Initialize $s_k^{(v,0)} = 0$

for $t = 1$ **to** T **do**

  foreach $v = 1, \ldots, V$ **do**

    Solve AW-LSSVM using $s_k^{(v,t-1)}$ // Eq. (4)

    Compute error vector $\mathbf{e}^{(v,t)}$ // Eq. (5)

    $\tilde{\mathbf{e}}^{(v,t)} \leftarrow \mathbf{e}^{(v,t)}$ with entries $\tilde{e}_k^{(v,t)} = 0$ whenever $e_k^{(v,t)} \leq 1$

  foreach $v = 1, \ldots, V$ **do**

    if $\mathcal{A} = \text{avg}$ **then**

      Update $s_{k,\text{avg}}^{(v,t)}$ // Eq. (2)

    if $\mathcal{A} = \text{dissim}$ **then**

      Update $s_{k,\text{dissim}}^{(v,t)}$ // Eq. (3)

Output: trained model $\{(\boldsymbol{\alpha}^{(v,T)}, b^{(v,T)})\}_{v=1}^{V}$

Inference

Input: Test sample $\mathbf{x}^* = \{\mathbf{x}^{*(v)}\}_{v=1}^{V}$, trained model $\{(\boldsymbol{\alpha}^{(v,T)}, b^{(v,T)})\}_{v=1}^{V}$

foreach $v = 1, \ldots, V$ **do**

  Compute $f^{(v,T)}(\mathbf{x}^{*(v)})$ // Eq. (6)

Compute $\hat{y}(\mathbf{x}^*)$ // Eq. (7)

Output: $\hat{y}(\mathbf{x}^*)$

Algorithm 2: Model Selection

Input: $\mathcal{D} = \{(\mathbf{x}_k^{(v)}, y_k)\}_{k=1,v=1}^{N,V}$, $K(\cdot,\cdot)$, search spaces for γ, σ, $\rho^{(v)}$, β, T

Split $\mathcal{D} \rightarrow \{\mathcal{D}_i^{\text{train}}, \mathcal{D}_i^{\text{test}}\}_{i=1}^{3}$

for $i = 1, 2, 3$ **do**

  $(\gamma_i^*, \sigma_i^*, \rho_i^{(v)*}, \beta_i^*, T_i^*) \leftarrow$ DA & CV_3, Alg. 1 on $\mathcal{D}_i^{\text{train}}$

  $\hat{\mathbf{y}}_i^{test} \leftarrow$ Alg. $1\big(\mathcal{D}_i^{\text{test}}, K, \gamma_i^*, \sigma_i^*, \rho_i^{(v)*}, \beta_i^*, T_i^*\big)$

Output: $\hat{\mathbf{y}}_1^{test}, \hat{\mathbf{y}}_2^{test}, \hat{\mathbf{y}}_3^{test}$

the same hyperparameter values are also used across all iterations. Although separate kernel widths and regularization parameters could in principle be tuned for each view, we use a shared RBF kernel width σ and a common regularization parameter γ across views for all methods except BSV. The remaining method-specific hyperparameters of each baseline, as well as those of AW-LSSVM ($\rho^{(v)}$, T with a maximum of five, and β), are tuned accordingly. Algorithm 2 describes the complete model selection procedure used in the experiments.

Table 1. Properties of the multi-view datasets used in experiments.

Dataset	Samples	Views	Classes	Dataset	Samples	Views	Classes
3Sources	169	3	6	Caltech	1474	6	7
YT	500	3	7	Flower	1360	7	17
NUSWIDE	2400	5	12	ACM	3025	5	3
Prokaryotic	551	3	4	Cora	2708	4	7
ProteinFold	694	12	27	OutdoorScene	2688	4	8

3.3 Results

The final results in Table 2 report the balanced accuracies on three hold-out test splits using the corresponding optimal hyperparameters for each benchmark multi-view dataset. The last two rows correspond to the proposed AW-LSSVM model with two aggregation strategies: AW-LSSVM$_{avg}$ and AW-LSSVM$_{dissim}$, which use average-based and dissimilarity-aware error aggregation, respectively, as defined in Eq. (2) and Eq. (3). The results show that AW-LSSVM performs strongly across the benchmark datasets and achieves the highest mean balanced accuracy on seven out of ten datasets, highlighting the benefit of effective sample-weighted global error compensation among views. Figure 2 shows the mean and standard deviation of the test balanced accuracy of AW-LSSVM$_{\text{avg}}$ and AW-LSSVM$_{\text{dissim}}$ over five iterations across three different test splits[2]. As shown, three iterations (with $t = 1$ corresponding to the initial LS-SVM round) generally yield the highest test performance. This suggests that the method has limited sensitivity to the choice of T, and that for datasets with similar sizes and properties, a small fixed value such as $T = 3$ may already be sufficient in practice without requiring extensive tuning. This early convergence can improve communication efficiency in FL settings, where secure data exchange among parties holding different views is essential but costly.

Examining the individual performances of the LS-SVM view-specific classifiers in Fig. 3 for ProteinFold and Cora, where AW-LSSVM does not achieve the best overall result, shows that the majority of views in ProteinFold and half of the views in Cora are weak, with balanced accuracy below 50%. These weak views are either insufficiently informative or noisy. Although AW-LSSVM benefits from cross-view interaction and therefore outperforms most baselines on these datasets, its error-coupling mechanism may become less effective when several views are weak. In such cases, strong views may be guided by unreliable error signals from weak views, while weak views may be encouraged to focus on hard samples that they cannot model well. This can degrade some view-specific classifiers and limit the overall gain under uniform score aggregation (Fig. 3).

[2] Note that, although the results in Table 2 are reported using the tuned iterations T, the models in Fig. 2 were trained for five iterations on all datasets and splits to get a consistent visualization.

Table 2. Mean and std. of balanced accuracy on three hold-out test splits. **Bold** values indicate the highest score. In the Flower dataset, since the kernels have already been precomputed, the Mumbo model is not applicable.

Method	Datasets				
	3Sources	YT	NUSWIDE	Prokaryotic	ProteinFold
BSV	72.50(±12.31)	79.66(±2.18)	38.89(±2.41)	66.88(±5.90)	59.76(±0.53)
Early Fusion	67.13(±4.24)	73.17(±0.16)	49.03(±3.46)	66.56(±6.85)	58.22(±1.93)
Late Fusion	63.89(±2.41)	68.83(±2.15)	43.54(±1.16)	66.31 (±8.73)	47.96(±2.99)
MV-LSSVM	68.98(±9.85)	77.14(±0.14)	47.64(±2.04)	62.17(±2.25)	57.37(±3.45)
EasyMKL	81.48(±9.25)	77.82(±1.52)	**49.58**(±3.15)	67.08(±2.66)	**68.09** (±3.72)
Mumbo	55.56(±8.67)	76.21(±4.40)	39.31(±1.39)	63.39(±4.39)	46.44(±1.20)
ϱTMV-RKM	89.04(±2.10)	78.73(±4.03)	48.26(±1.70)	79.58(±2.11)	67.74(±5.19)
AW-LSSVM$_{avg}$	89.73(±1.68)	78.62(±2.10)	47.01(±1.58)	83.31 (±0.71)	65.79(±1.49)
AW-LSSVM$_{dissim}$	**91.07**(±0.71)	**80.57**(±2.22)	48.33(±1.78)	**83.77**(±1.38)	64.99(±1.40)
	Caltech	Flower	ACM	Cora	OutdoorScene
BSV	93.14(±3.20)	75.74(±2.41)	91.91(±0.56)	76.22(±0.76)	88.04 (±0.38)
Early Fusion	94.37(±1.68)	88.36(±1.29)	93.04(±0.77)	80.95(±1.83)	89.69(±0.61)
Late Fusion	88.81(±4.90)	83.95(±1.29)	89.78(±1.00)	39.19(±2.37)	89.55(±0.22)
MV-LSSVM	91.91(±0.75)	89.22(±1.49)	92.72(±1.59)	77.46(±3.04)	89.91 (±1.09)
EasyMKL	92.96(±2.06)	88.60(±1.10)	93.60(±1.44)	79.76(±5.51)	89.58(±0.58)
Mumbo	93.03(±0.43)	–	89.37(±1.95)	71.52(±2.72)	85.52 (±0.45)
ϱTMV-RKM	95.12(±2.70)	88.48(±1.29)	93.22(±1.21)	**85.02**(±0.14)	90.24(±0.99)
AW-LSSVM$_{avg}$	**96.23**(±3.84)	**90.20** (±1.29)	94.90(±0.74)	84.26 (±2.77)	**90.31**(±0.42)
AW-LSSVM$_{dissim}$	95.43(±4.24)	89.95(±1.12)	**95.04**(±0.80)	82.80(±1.73)	90.02(±0.99)

This also helps explain the relative behavior of the two AW-LSSVM variants. Overall, AW-LSSVM$_{dissim}$ performs only slightly better than AW-LSSVM$_{avg}$, with mean balanced accuracies of 82.20 and 82.04 over all datasets, respectively, corresponding to a gain of 0.16. However, this advantage becomes more noticeable when excluding ProteinFold and Cora, where the gain increases to 0.48. This is consistent with the discussion above, since the dissimilarity-aware variant places greater emphasis on views with more distinct error patterns, which in datasets containing both strong and weak views may undesirably amplify the influence of weak or noisy ones. In general, the dissimilarity-aware variant can therefore be used as the default strategy, except for datasets with a combination of strong and weak views in which a large proportion of the views are weak.

CKA can further support the choice between the two variants as an indicator of inter-view similarity. Figure 4 shows pairwise CKA heatmaps for the Caltech and YT datasets, which exhibit the largest gains of AW-LSSVM$_{avg}$ over AW-LSSVM$_{dissim}$ and of AW-LSSVM$_{dissim}$ over AW-LSSVM$_{avg}$, respectively. The CKA values are computed using RBF kernels with the default kernel width. The

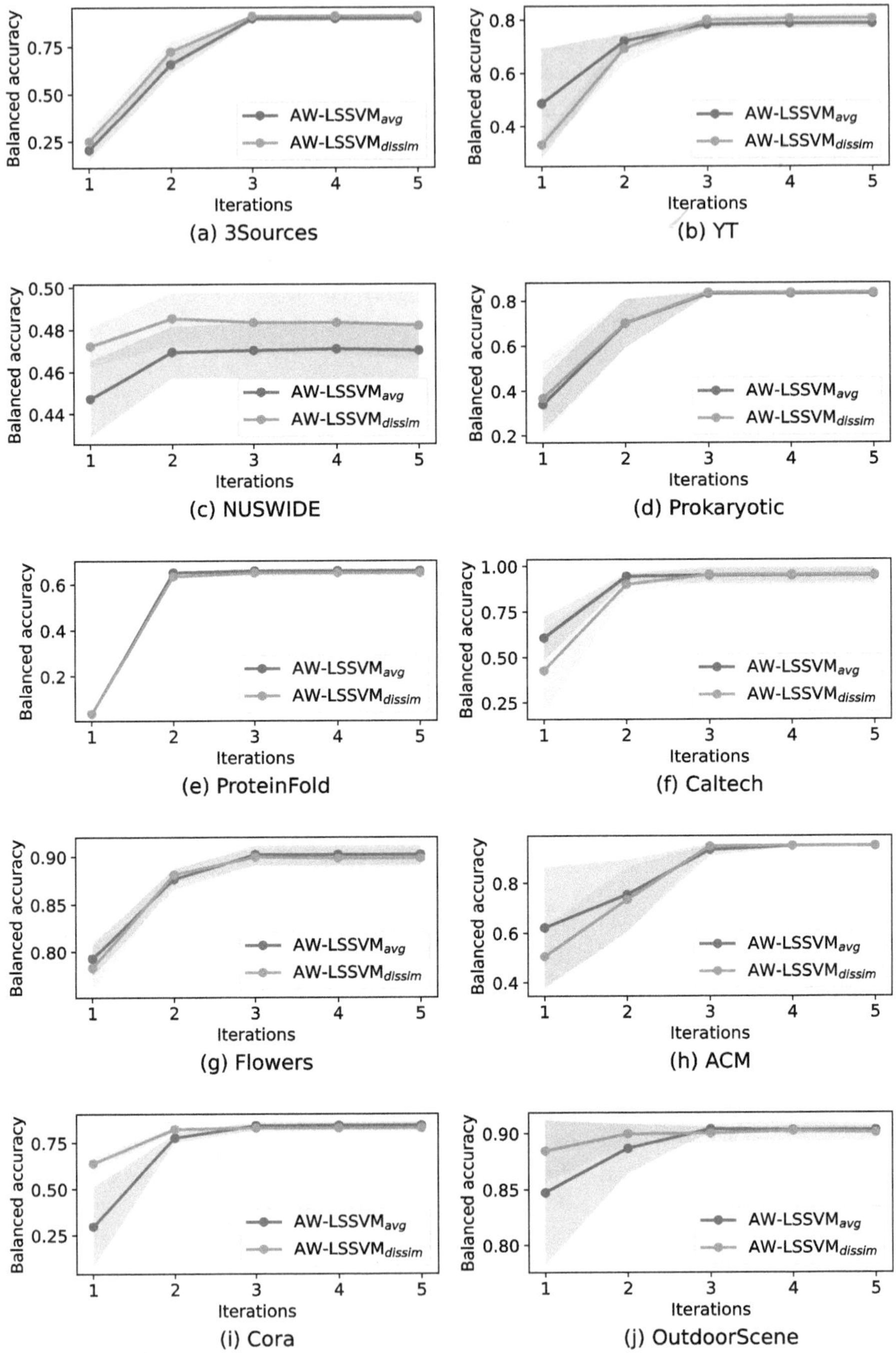

Fig. 2. Mean and standard deviation of balanced accuracy for AW-LSSVM$_{\text{avg}}$ and AW-LSSVM$_{\text{dissim}}$ over three test splits with respect to iterations t across different datasets. For clarity, note that the y-axis scales are not consistent across subplots.

relatively high average pairwise CKA in Caltech (0.58) indicates that the views are more similar and aligned, which is consistent with the stronger performance of the average-based variant. The lower average pairwise CKA in YT (0.17) indicates a more diverse view structure, which is consistent with the stronger performance of the dissimilarity-aware variant.

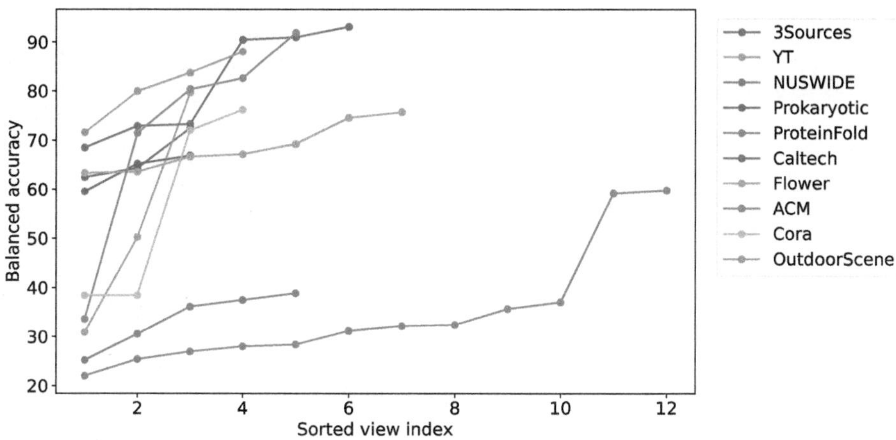

Fig. 3. Sorted single-view LS-SVM balanced accuracy scores for each dataset. Each curve shows the ordered performance of the individual views, highlighting the distribution of weak and strong views within each dataset.

EasyMKL, which performs weighted kernel averaging across views, and ϱTMV-RKM, which projects the views into a shared subspace, may be more effective on datasets with a large proportion of weak views (e.g., NUSWIDE, ProteinFold, and Cora), as they can reduce the impact of view-specific noise or overall lack of informativeness. Nevertheless, replacing the uniform aggregation in the decision rule of Eq. (7) with a weighted averaging scheme based on measures such as the error covariance matrix may further improve AW-LSSVM. In addition, kernel-level aggregation is compatible with our framework, allowing earlier fusion strategies to be incorporated when beneficial, for example on NUSWIDE, where the strong Early Fusion result suggests that the views can interact effectively at an earlier stage.

3.4 Statistical Analysis

For statistical analysis of the performance improvement of AW-LSSVM over baseline methods, we conducted pairwise Wilcoxon signed-rank [6] tests across all datasets. In particular, when compared against BSV, Early Fusion, Late Fusion, MV-LSSVM, EasyMKL, and Mumbo, the Wilcoxon statistics indicate that our method significantly outperforms these methods, with p-values below 0.05 in all cases. The comparison with ϱTMV-RKM produces $T = 11.0$ and p-value

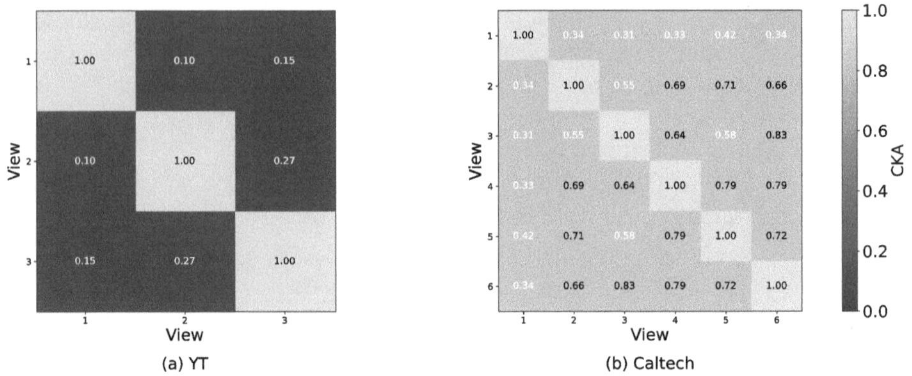

Fig. 4. Pairwise centered kernel alignment (CKA) between views for the Caltech and YT datasets, computed using RBF kernels with the default kernel width.

= 0.1, indicating no significant difference between the two methods. Overall, these results show that AW-LSSVM delivers statistically significant gains over the considered baseline methods, while performing comparably to the strongest competing method, ϱTMV-RKM.

4 Conclusion and Future Work

This paper introduced AW-LSSVM, a novel kernel-based method for multi-view classification that explicitly promotes complementary learning across all views through a global coupling term. This term adaptively assigns additional sample weights to each view, guiding it to compensate for samples misclassified by the other views. The weights are computed from the aggregated misclassification errors in previous iterations. For this aggregation, besides a uniform view-error aggregation strategy, we also proposed a dissimilarity-aware variant for more coordinated error compensation, in which each view is encouraged to place greater emphasis on errors from views with which it is less redundant. Experimental results on ten benchmark datasets show that AW-LSSVM improves performance over state-of-the-art multi-view methods that either do not explicitly enforce any particular form of collaboration across views or model collaboration only in a pairwise manner. By relying solely on view-level error and decision information, the proposed method keeps raw features separated across views and is therefore architecturally suitable for distributed multi-view settings such as VFL. Nevertheless, formal privacy guarantees are not established in this work. Future work will investigate federated implementations of AW-LSSVM and evaluate their communication, security, and privacy properties. Extending AW-LSSVM to a semi-supervised setting is another promising research direction.

Acknowledgments. We would like to thank the Flemish Government under the Onderzoeksprogramma Artificiele Intelligentie (AI) Vlaanderen programme for funding this research.

Disclosure of Interests. The authors have no competing interests to declare.

References

1. Aiolli, F., Donini, M.: EasyMKL: a scalable multiple kernel learning algorithm. Neurocomputing **169**, 215–224 (2015)
2. Brbić, M., Piškorec, M., Vidulin, V., Kriško, A., Šmuc, T., Supek, F.: The landscape of microbial phenotypic traits and associated genes. NAR **44**, 10074–10090 (2016)
3. Chua, T.S., Tang, J., Hong, R., Li, H., Luo, Y., Zheng, Y.: NUS-WIDE: a real-world web image database from national university of singapore. In: ACM International Conference on MIR, pp. 1–9 (2009)
4. Cortes, C., Mohri, M., Rostamizadeh, A.: Algorithms for learning kernels based on centered alignment. J. Mach. Learn. Res. **13**, 795–828 (2012)
5. Cortes, C., Vapnik, V.: Support-vector networks. Mach. Learn.**20**(3), 273–297 (1995)
6. Demšar, J.: Statistical comparisons of classifiers over multiple data sets. JMLR **7**, 1–30 (2006)
7. Ding, C.H., Dubchak, I.: Multi-class protein fold recognition using support vector machines and neural networks. Bioinformatics **17**(4), 349–358 (2001)
8. Farquhar, J., Hardoon, D., Meng, H., Shawe-Taylor, J., Szedmak, S.: Two view learning: SVM-2K, theory and practice. NIPS, pp. 1–8 (2005)
9. Houthuys, L., Langone, R., Suykens, J.A.K.: Multi-view least squares support vector machines classification. Neurocomputing **282**, 78–88 (2018)
10. Houthuys, L., Suykens, J.A.K.: Tensor-based restricted kernel machines for multi-view classification. Inf. Fusion **68**, 54–66 (2021)
11. Hu, Z., Nie, F., Wang, R., Li, X.: Multi-view spectral clustering via integrating nonnegative embedding and spectral embedding. Inf. Fusion **55**, 251–259 (2020)
12. Koço, S., Capponi, C.: A boosting approach to multiview classification with cooperation. In: ECML PKDD, pp. 209–228 (2011)
13. Liu, J., Wang, C., Gao, J., Han, J.: Multi-view clustering via joint nonnegative matrix factorization. In: ICDM, pp. 252–260 (2013)
14. Liu, Y., et al.: Vertical federated learning: concepts, advances, and challenges. IEEE TKDE **36**(7), 3615–3634 (2024)
15. Madani, O.: YouTube Multiview Video Games Dataset. UCI Machine Learning Repository (2013). https://doi.org/10.24432/C5MC7X
16. Mena, F., Arenas, D., Nuske, M., Dengel, A.: Common practices and taxonomy in deep multiview fusion for remote sensing applications. IEEE J-STARS **17**, 4797–4818 (2024). https://doi.org/10.1109/JSTARS.2024.3361556
17. Nilsback, M.E., Zisserman, A.: A visual vocabulary for flower classification. In: IEEE CVPR, vol. 2, pp. 1447–1454 (2006)
18. Sen, P., Namata, G., Bilgic, M., Getoor, L., Gallagher, B., Eliassi-Rad, T.: Collective classification in network data. In: Workshop on MLG at ICKDDM, pp. 1–6 (2008)
19. Suykens, J.A.K., Van Gestel, T., De Brabanter, J., De Moor, B., Vandewalle, J.P.: Least squares support vector machines, World scientific (2002)
20. Wang, X., Ji, H., Shi, C., Wang, B., Ye, Y., Cui, P., Yu, P.S.: Heterogeneous graph attention network. In: WWW Conference, pp. 2022–2032 (2019)

Tackling Catastrophic Forgetting in Split Learning via Orthogonal Projection

Thoop-hom Supannopaj[1], Nandu Voore[1], Andrew Taggart[1], Taeyeong Choi[2], and Chao Huang[1](✉)

[1] Montclair State University, Montclair, NJ, USA
{supannopajt1,vooren1,taggarta2,huangch}@montclair.edu
[2] Kennesaw State University, Kennesaw, GA, USA
tchoi3@kennesaw.edu

Abstract. Split Learning (SL) enables distributed training by partitioning a neural network between clients and a server. However, sequential client participation under heterogeneous data distributions can lead to catastrophic forgetting, where the model overfits to recent clients while degrading on earlier ones. Orthogonal projection is a widely used technique in continual learning, yet existing orthogonal projection methods are not directly applicable to SL because they require projecting updates in the full parameter space. In SL, the server does not have access to client-side parameters and interacts with clients only through cut-layer activations and the associated gradients. To address this limitation, this paper introduces a cut-layer orthogonal projection mechanism to mitigate forgetting in SL. The proposed approach maintains a reservoir of recent cut-layer activations and uses principal component analysis to construct a low-rank subspace that captures their dominant directions. Server-returned gradients are then projected to remove components aligned with this subspace, which helps preserve previously learned knowledge while still allowing adaptation to new clients. The proposed approach is also applicable to other SL variants, such as split federated learning. Experiments on CIFAR-10 and CIFAR-100 show that, under strong data heterogeneity, the proposed approach improves test accuracy by up to 3.36% over the corresponding non-projected split learning baselines.

Keywords: Split Learning · Catastrophic Forgetting · Continual Learning · Orthogonal Projection · Split Federated Learning

1 Introduction

Modern deep learning increasingly requires training on distributed, heterogeneous data. In healthcare, finance, and mobile sensing, data are privacy-sensitive and distributed across institutions or devices. Federated Learning (FL) addresses this setting by keeping data local and aggregating model updates [14], but

Supplementary Information The online version contains supplementary material available at https://doi.org/10.1007/978-3-032-38398-3_35.

L. Pasa et al. (Eds.): ICANN 2026, LNCS 17089, pp. 420–431, 2027.
https://doi.org/10.1007/978-3-032-38398-3_35

training full models locally can remain challenging under non-IID and resource-constrained settings [13].

Split Learning (SL) partitions a neural network between clients and a server [9,20]. Clients compute forward passes up to a cut layer and transmit intermediate activations to the server, which completes training and returns gradients. This reduces client-side computational burden while preserving data privacy. However, when clients participate sequentially under non-IID (non-independent and identically distributed) data, the model can overfit to recent clients while degrading on earlier ones. This phenomenon is known as **catastrophic forgetting** [3] and is the focus of this paper.

Orthogonal projection methods address forgetting in centralized continual learning by projecting gradients onto subspaces that preserve previously acquired knowledge [7,17]. In federated settings, methods such as Federated Orthogonal Training (FOT) [1] and Federated Weighted Inter-client Transfer (FedWeIT) [22] extend continual learning ideas to distributed training. **However, both centralized and federated orthogonal projection methods assume full access to model parameters across all layers**. In SL, the server interacts only through cut-layer activations and gradients, without access to raw data or client-side parameters. As a result, methods designed for centralized or FL settings cannot be directly applied to SL.

To address this limitation, this paper develops an orthogonal projection approach tailored to SL. The method projects server-returned gradients away from a low-rank subspace estimated from cut-layer activations using principal component analysis (PCA) [12]. It is worth noting that cut-layer representations evolve during training and depend on the split location, which affects how effectively projection can preserve task-relevant knowledge. Therefore, the method maintains a reservoir of recent activations and periodically rebuilds the protected subspace. It also uses warm-up scheduling, linear ramping of projection strength, and per-sample norm preservation to balance retention of past knowledge with adaptation to new clients.

The contributions of this paper are summarized as follows.

- Developing an approach for SL to mitigate catastrophic forgetting under data heterogeneity. The approach operates exclusively at the cut layer through orthogonal gradient projection without accessing the full model.
- To handle the evolving structure of cut-layer representations, the approach maintains a reservoir of recent cut-layer activations and periodically reconstructs a low-rank projection subspace via PCA, ensuring the protected subspace adapts to shifting data distributions across clients.
- To balance preserving past knowledge against learning from new data, a warm-up schedule is introduced with linear ramping of projection strength and per-sample gradient norm preservation, preventing disruption of early training dynamics while gradually introducing forgetting mitigation.
- The proposed method is also applicable to other SL variants such as split federated learning (SFL). Experiments on CIFAR-10 and CIFAR-100 under non-IID partitions show that the method improves performance in several split learning settings relative to their non-projected counterparts.

2 Related Work

Split Learning. Split learning partitions a neural network between clients and a server, enabling collaborative training without sharing raw data [9,20]. SplitFed extends this framework with federated aggregation [18].

Related work has studied practical aspects of split and split-federated learning, including incentive mechanisms [11], cut-layer placement [2], and convergence under heterogeneous data [10]. Orthogonal projection has also been explored at the split interface for machine unlearning, where projected gradient caches are used to remove the influence of targeted data points [6]. Despite these advances, split learning remains vulnerable to catastrophic forgetting when clients participate sequentially under non-IID data. Recent work has started to study this issue more directly. For example, [19] analyzes forgetting in SFL-V2 under data heterogeneity, while [21] mitigates forgetting in a two-server SFL framework using knowledge replay from assistant clients.

Forgetting Mitigation Strategies. Catastrophic forgetting has been addressed through several families of continual-learning methods, including rehearsal-based approaches that retain exemplar data [15], architectural expansion methods that allocate new model capacity for new tasks [16], and gradient-projection approaches that restrict update directions to reduce interference with previously learned knowledge [17]. In federated continual learning, additional challenges arise from distributed data and client heterogeneity. FOT constructs per-layer principal subspaces from prior tasks and projects aggregated updates to reduce interference [1]. Federated Weighted Inter-client Transfer (FedWeIT) decomposes weights into global and sparse task-specific parts to enable selective inter-client transfer [22], and Federated Class-Incremental Learning (GLFC) mitigates class-incremental forgetting using local distillation with a global generator of past-class representations [4]. However, these methods assume access to full-model parameter updates, whereas in SL the server interacts only through cut-layer activations and cut-layer gradients, requiring redesign to apply them in SL.

3 Problem Formulation

Consider N clients, where each client n holds a private local dataset $\mathcal{D}_n$ of size $D_n = |\mathcal{D}_n|$. Each data sample can be represented as a tuple (x_i, y_i), where x_i represents the input and y_i represents the label. The full model is $f(x) = f_s(f_c(x))$, where f_c represents the client-side network mapping raw inputs to intermediate representations, and f_s represents the server-side network mapping representations to final predictions. For input x, the client computes cut-layer activations $z = f_c(x)$ and transmits them to the server, which then computes logits $f_s(z)$.

This work studies two variants of split learning, illustrated in Fig. 1: Sequential Split Learning (SL) [9,20] and Split Federated Learning (SFL) [18]. For SFL, the focus is on SFL-V2, since SFL-V1 behaves similarly to FedAvg.

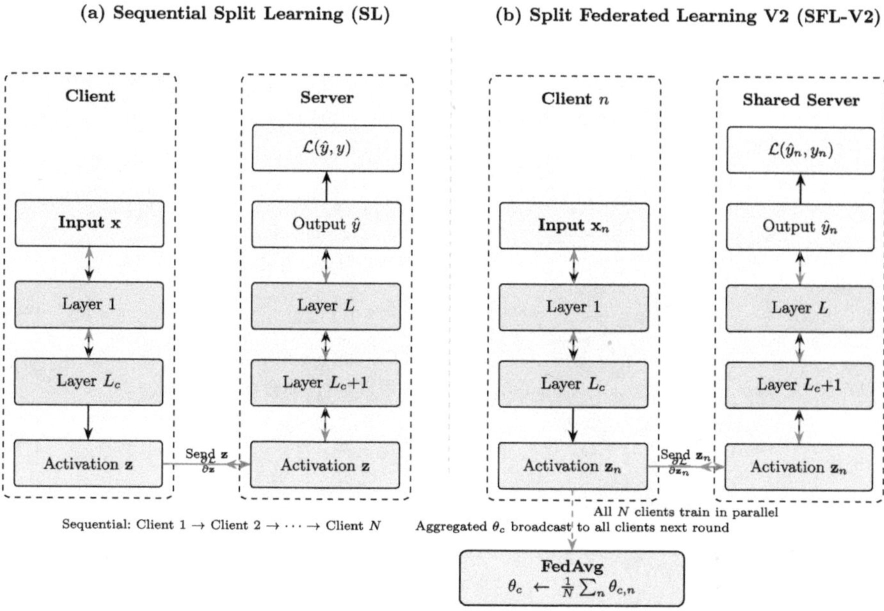

Fig. 1. Conceptual overview of the two split learning variants studied in this work. **(a) Sequential Split Learning (SL).** A single shared model is updated sequentially as clients participate one at a time. The client transmits cut-layer activations $\mathbf{z}$ to the server in orange, which returns gradients $\partial\mathcal{L}/\partial\mathbf{z}$ (red dashed). **(b) Split Federated Learning V2 (SFL-V2):** all N clients train in parallel against a shared server-side model. After local training, client-side models are aggregated via FedAvg. Both variants suffer from catastrophic forgetting under heterogeneous data, but their distinct aggregation mechanisms lead to different forgetting dynamics.

Sequential Split Learning (SL). The framework maintains a single shared parameter pair $\theta = (\theta_c, \theta_s)$ updated sequentially across clients. In each round t, clients are visited sequentially in a randomized manner. For each client n, the training procedure consists of:

- *Client forward pass:* Compute cut-layer activations $z_i = f_c(x_i; \theta_c)$ for each sample $(x_i, y_i) \in \mathcal{D}_n$ and transmit z_i to the server.
- *Server forward pass:* Compute predictions $\hat{y}_i = f_s(z_i; \theta_s)$ and evaluate loss $\ell(\hat{y}_i, y_i)$.
- *Server backward pass:* Compute the gradient $\nabla_{\theta_s}\ell$ and update the server parameters $\theta_s \leftarrow \theta_s - \eta\nabla_{\theta_s}\ell$, where η denotes the learning rate. Then compute the cut-layer gradient $g_i = \nabla_{z_i}\ell$ and return it to the client.
- *Client backward pass:* Backpropagate g_i through f_c to obtain $\nabla_{\theta_c}\ell$ and update client parameters $\theta_c \leftarrow \theta_c - \eta\nabla_{\theta_c}\ell$.

After completing local training, the updated client-side model θ_c is passed to the next client, which initializes its local training from this shared state. This

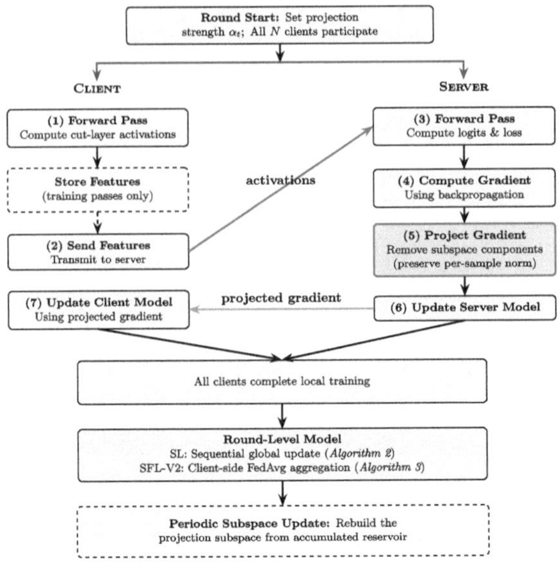

Fig. 2. Workflow of the proposed projection-based training in SL and SFL-V2.

sequential handoff continues until all clients in the round have trained, and the process repeats across rounds until convergence.

Split Federated Learning V2 (SFL-V2). Each client n maintains its own local copy $\theta_{c,n}$ of the client-side model. In round t, the training procedure is as follows:

- *Client initialization:* Download the global client-side model $\theta_{c,n} \leftarrow \theta_c^{(t)}$.
- *Client forward pass:* Compute $z_i = f_c(x_i; \theta_{c,n})$ for $(x_i, y_i) \in \mathcal{D}_n$ and transmit z_i to the server.
- *Server forward pass:* Compute $\hat{y}_i = f_s(z_i; \theta_s)$ and evaluate the loss $\ell(\hat{y}_i, y_i)$.
- *Server backward pass (sequential w.r.t. clients):* Using only client n's mini-batch, update the shared server parameters $\theta_s \leftarrow \theta_s - \eta \nabla_{\theta_s} \ell$ and return the cut-layer gradient $g_i = \nabla_{z_i} \ell$ to client n.
- *Client backward pass:* Update $\theta_{c,n} \leftarrow \theta_{c,n} - \eta \nabla_{\theta_{c,n}} \ell$.
- *Aggregation:* After all clients complete local training, the client-side models are averaged to produce the updated global model: $\theta_c^{(t+1)} \leftarrow \frac{1}{N} \sum_{n=1}^{N} \theta_{c,n}^{(t+1)}$.

4 Approach: Orthogonal Projection at the Cut Layer

The approach maintains a low-rank orthonormal basis $U \in \mathbb{R}^{d \times r}$ spanning the principal directions of past cut-layer features, where d is the cut-feature dimension and r is the number of principal components. The basis U is obtained via PCA [12] on a reservoir of recent train-side cut activations.

The key idea is that the leading principal directions of past cut-layer activations span the representation subspace that earlier clients rely on for prediction. A gradient component aligned with this subspace is likely to overwrite shared features as the server adapts to a new client. Removing it preserves directions tied to past knowledge, while leaving the orthogonal complement for client-specific adaptation. This transfers the rationale of gradient-projection methods in centralized continual learning [17] to the quantities available to the server, namely cut-layer activations and their gradients. The projection filters direction, not magnitude, since per-sample norm preservation in Eq. (2) restores each gradient's original ℓ_2 norm.

For a minibatch (x, y), the client computes cut activations $F = f_c(x; \theta_c) \in \mathbb{R}^{B \times d}$, where B is the batch size. The client then detaches F from its local computation graph and transmits $\tilde{F} = \text{detach}(F)$ to the server, so that the server-side backward pass does not propagate gradients through the client's forward computation. The server computes predictions $\hat{y} = f_s(\tilde{F}; \theta_s)$, evaluates the loss $\ell(\hat{y}, y)$, and backpropagates to obtain the cut-layer gradient $G = \nabla_{\tilde{F}} \ell \in \mathbb{R}^{B \times d}$.

Before returning the gradient, the server projects away components along the subspace spanned by U and applies norm-preserving rescaling with a warm-up/ramp schedule:

- **Cut-gradient projection:**

$$\tilde{G} = G - \alpha_t \, GUU^\top, \tag{1}$$

  where $\alpha_t \in [0, \alpha_{\max}]$ controls the projection strength.
- **Per-sample norm preservation:** Let $\tilde{g}_i$ and g_i denote the i-th row of $\tilde{G}$ and G, respectively, that is, the projected and original cut-layer gradient vectors for sample i. To maintain gradient magnitudes after projection, the server rescales each projected gradient to preserve its original ℓ_2 norm:

$$\tilde{g}_i \leftarrow \tilde{g}_i \cdot \frac{\|g_i\|_2}{\|\tilde{g}_i\|_2 + \varepsilon}, \tag{2}$$

  where $\varepsilon > 0$ is a small constant introduced for numerical stability to prevent division by zero.
- **Warm-up and linear ramp schedule:** The server applies a linearly ramped schedule for projection strength [8]:

$$\alpha_t = \min\left\{\alpha_{\max},\ \max\left(0,\ \alpha_{\min} + (\alpha_{\max} - \alpha_{\min})\frac{t - T_{\text{warm}}}{T_{\text{ramp}}}\right)\right\}, \tag{3}$$

  where T_{warm} defines an initial warm-up period during which no projection is applied, allowing the model to establish stable initial representations.

After warm-up, α_t ramps linearly from $\alpha_{\min}$ to $\alpha_{\max}$ over T_{ramp} rounds, a simple and predictable transition controlled by the single hyperparameter T_{ramp} that gradually introduces forgetting mitigation without disrupting early training dynamics [8]. The subspace U is periodically rebuilt by the server every R_{rebuild}

Algorithm 1 Orthogonal Projection Module

Require: T_{warm}, T_{ramp}, $\alpha_{\min}$, $\alpha_{\max}$, R_{rebuild}, r, C

Initialization: $\mathcal{R} \leftarrow [\,]$, $n_{\text{seen}} \leftarrow 0$, $U \leftarrow$ None

function gate(t)
if $t < T_{\text{warm}}$ **then**
  return 0
end if
$\tau \leftarrow \min\left(1, \frac{t - T_{\text{warm}}}{\max(1, T_{\text{ramp}})}\right)$
return $\alpha_{\min} + \tau(\alpha_{\max} - \alpha_{\min})$
end function

function project(g, U, α)
if $U =$ None **or** $\alpha \leq 0$ **then**
  return g
end if
$n_0 \leftarrow \|g\|_2$
$g' \leftarrow g - \alpha(gU)U^{\top}$
$n_1 \leftarrow \|g'\|_2$
$g' \leftarrow g' \odot (n_0 / \max(n_1, \epsilon))$
return g'
end function

function add_to_reservoir(z)
for each sample z_i **do**
  $n_{\text{seen}} \leftarrow n_{\text{seen}} + 1$
  if $|\mathcal{R}| < C$ **then**
    Append z_i to $\mathcal{R}$
  else
    Sample $j \sim \text{Uniform}\{0, \ldots, n_{\text{seen}} - 1\}$
    if $j < C$ **then**
      $\mathcal{R}[j] \leftarrow z_i$
    end if
  end if
end for
end function

function rebuild_subspace()
if $|\mathcal{R}| = 0$ **then**
  return None
end if
$X \leftarrow \text{stack}(\mathcal{R})$
Center X and compute top-r principal components
$U \leftarrow$ resulting basis
$\mathcal{R} \leftarrow [\,]$, $n_{\text{seen}} \leftarrow 0$
return U
end function

Algorithm 2 Sequential Split Learning with Projection

Require: N, T, E, projection module

for $t = 0$ to $T - 1$ **do**
  $\alpha \leftarrow \texttt{gate}(t)$
  $\mathcal{S}_t \leftarrow \{1, \ldots, N\}$ { Full participation}
  for each client $i \in \mathcal{S}_t$ sequentially **do**
    for $e = 1$ to E **do**
      for each minibatch (x, y) **do**
        Client computes $f_x \leftarrow f_c(x; \theta_c)$
        $\texttt{add\_to\_reservoir}(f_x)$
        Server computes cut-layer gradient g
        $g' \leftarrow \texttt{project}(g, U, \alpha)$
        Update server parameters θ_s
        Backpropagate g' and update client parameters θ_c
      end for
    end for
  end for
  if $(t + 1) \bmod R_{\text{rebuild}} = 0$ **then**
    $U \leftarrow \texttt{rebuild\_subspace}()$
  end if
end for

Algorithm 3 SFL-V2 with Projection

Require: N, T, E, projection module

for $t = 0$ to $T - 1$ **do**
  $\alpha \leftarrow \texttt{gate}(t)$
  $\mathcal{S}_t \leftarrow \{1, \ldots, N\}$ { Full participation}
  for each client $i \in \mathcal{S}_t$ **do**
    $\theta_{c,i} \leftarrow \theta_c$
    for $e = 1$ to E **do**
      for each minibatch (x, y) **do**
        Client computes $f_x \leftarrow f_c(x; \theta_{c,i})$
        $\texttt{add\_to\_reservoir}(f_x)$
        Server computes cut-layer gradient g
        $g' \leftarrow \texttt{project}(g, U, \alpha)$
        Update server parameters θ_s
        Backpropagate g' and update $\theta_{c,i}$
      end for
    end for
  end for
  $\theta_c \leftarrow \frac{1}{|\mathcal{S}_t|} \sum_{i \in \mathcal{S}_t} \theta_{c,i}$
  if $(t + 1) \bmod R_{\text{rebuild}} = 0$ **then**
    $U \leftarrow \texttt{rebuild\_subspace}()$
  end if
end for

rounds from a reservoir of recent training cut-layer activations, ensuring that the protected subspace adapts to evolving data distributions.

Figure 2 summarizes the training workflow. The client stores cut-layer activations in the reservoir and transmits them to the server (steps 1–2). The server computes the loss and cut-layer gradients (steps 3–4), applies orthogonal projection to suppress components aligned with the maintained subspace while preserving pre-sample norms (step 5), updates its parameters (step 6), and returned the projected gradient for the client update (step 7). The projection step is the key intervention that distinguishes the proposed approach from vanilla split learning. See Algorithms 1–3 for detailed pseudocode.

5 Experiments

The evaluation uses CIFAR-10 and CIFAR-100 split across $N = 10$ clients with Dirichlet label distributions, using $\alpha \in \{0.1, 0.5\}$, where smaller α indicates stronger heterogeneity. The base model is ResNet-18. It includes SL, SFL-V1, and SFL-V2, each with and without orthogonal projection, and centralized training as a non-distributed reference rather than a strict upper bound, since a strong split-federated baseline can occasionally exceed it.

Unless otherwise stated, experiments use the *early cut* after the third residual block of ResNet-18. Training runs for 200 rounds with 5 local epochs per client, full participation, SGD with momentum 0.9, learning rate 10^{-3}, batch size 256, and weight decay 5×10^{-4}. Projected variants use a reservoir of 4,096 cut-layer activations, rebuild the subspace every $R_{\text{rebuild}} = 5$ rounds with $r = 8$ principal components, warm up for $T_{\text{warm}} = 10$ rounds, and ramp projection strength over $T_{\text{ramp}} = 10$ rounds from $\alpha_{\text{min}} = 0$ to $\alpha_{\text{max}} = 0.1$. RandAugment is used for all methods, and projected variants are compared with their non-projected baselines and centralized training.

Performance is evaluated using test accuracy at round 200 and a window-based mean forgetting score adapted from continual federated learning metrics [5]. The test set is partitioned into $\mathcal{T}$ disjoint windows using indices $\{0, 1, 2, \ldots, \mathcal{T} - 1\}$, and forgetting at round t is

$$\text{mean\_FgT}(t) = \frac{1}{t} \sum_{w=0}^{t-1} \big(L_t(w) - L_{\text{ref}}(w)\big), \tag{4}$$

where $L_t(w)$ denotes the loss at round t on window w, and $L_{\text{ref}}(w)$ denotes the lowest loss recorded on that window in any earlier round. Each table reports the value at the final round ($t = 200$). The metric is therefore signed. A *positive* value means the current per-window loss exceeds the best previously achieved, indicating forgetting, whereas a *negative* value means the model is at or below its previous best on those windows, indicating no net forgetting. Lower (more negative) values are thus better, and the projected variants are expected to shift the score downward relative to their non-projected baselines.

5.1 Results

Figures 3 show the fixed early-cut comparison under RandAugment for CIFAR-10 and CIFAR-100. The augmented results follow the same pattern as the main comparison: projection helps SL and SFL-V2 the most, while SFL-V1 already performs well without it.

Tables 1 and 2 report final test accuracy and mean forgetting scores for the evaluated methods on CIFAR-10 and CIFAR-100 under two heterogeneity levels, $\alpha \in \{0.1, 0.5\}$.

Methods Comparison. Among the split-learning variants, SFL-V1 is the strongest baseline across nearly all settings, matching or exceeding the centralized references (e.g., 90.47% vs. 88.34% on CIFAR-10 at $\alpha = 0.5$) and degrading only modestly under stronger heterogeneity. SFL-V2 and SL degrade more

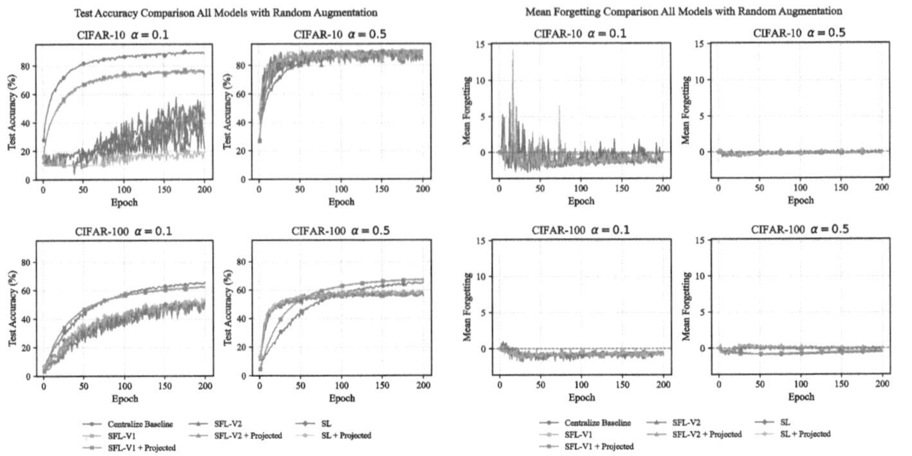

Fig. 3. Test accuracy and Mean forgetting comparisons under fixed early-cut evaluation with RandAugment across CIFAR-10 and CIFAR-100.

sharply as heterogeneity increases. On CIFAR-10 with $\alpha = 0.1$, SFL-V2 drops to 42.79% and SL to 21.68%, while at $\alpha = 0.5$ both recover to above 83%. The gap to SFL-V1 narrows at $\alpha = 0.5$ but remains large at $\alpha = 0.1$, so SFL-V2 and SL are clearly more sensitive to non-IID data distributions.

Impact of Projection. The effect of orthogonal projection varies with the method and heterogeneity level. On CIFAR-10 with $\alpha = 0.5$, projection gives large gains for SFL-V2 (+6.16 points, to 90.05%) and SL (+6.48 points, to 90.67%), while SFL-V1 stays flat at around 90%. At $\alpha = 0.1$, it still helps SFL-V2 (+3.36 points) and SFL-V1 (+1.68 points), but hurts SL, which drops from 21.68% to 18.99%. On CIFAR-100, the same pattern holds with smaller gains for SL and SFL-V2, roughly 0.4 to 1.6 points, while SFL-V1 decreases slightly (-0.75 at $\alpha = 0.1$, -0.80 at $\alpha = 0.5$).

These results suggest that projection works best when the baseline forgets but still learns a usable cut-layer representation, as with SL and SFL-V2 when performance remains above chance. It helps little when the baseline is already strong, as with SFL-V1, and can hurt when the baseline is too weak to provide a reliable subspace, as with SL on CIFAR-10 at $\alpha = 0.1$. The benefit therefore depends on protocol strength, heterogeneity, and where the model is split.

Forgetting Scores. The forgetting scores in Tables 1 and 2 generally follow the accuracy results, though not monotonically. Where projection produces the largest accuracy gains, they shift in the expected direction. On CIFAR-10 with $\alpha = 0.5$, SFL-V2 moves from 0.08 to -0.27 and SL from 0.13 to -0.16, and on CIFAR-100 with $\alpha = 0.1$ SL improves from -0.78 to -0.94. In other cases, forgetting changes only slightly despite noticeable accuracy improvements. This suggests that final performance depends on more than retention, including optimization dynamics and the quality of the cut-layer representation.

Table 1. CIFAR-10

Method	α	Acc (%)	mean_FgT_loss
Centralized	–	88.34	-0.14
SFL-V1	0.1	74.17	-0.15
SFL-V1 + Proj	0.1	75.85	-0.17
SFL-V2	0.1	42.79	-1.04
SFL-V2 + Proj	0.1	46.15	-0.67
SL	0.1	21.68	-0.01
SL + Proj	0.1	18.99	-0.64
SFL-V1	0.5	90.47	-0.15
SFL-V1 + Proj	0.5	90.39	-0.15
SFL-V2	0.5	83.89	0.08
SFL-V2 + Proj	0.5	90.05	-0.27
SL	0.5	84.19	0.13
SL + Proj	0.5	90.67	-0.16

Table 2. CIFAR-100

Method	α	Acc (%)	mean_FgT_loss
Centralized	–	65.52	-0.55
SFL-V1	0.1	62.75	-0.47
SFL-V1 + Proj	0.1	62.00	-0.47
SFL-V2	0.1	49.90	-0.97
SFL-V2 + Proj	0.1	51.10	-1.01
SL	0.1	52.63	-0.78
SL + Proj	0.1	54.27	-0.94
SFL-V1	0.5	67.59	-0.39
SFL-V1 + Proj	0.5	66.79	-0.37
SFL-V2	0.5	55.89	0.03
SFL-V2 + Proj	0.5	56.29	0.09
SL	0.5	56.34	0.16
SL + Proj	0.5	57.54	0.12

Note: Reported values correspond to a single representative run (seed 1234).

Table 3. Ablation study: Effect of cut layer placement on SFL-V2 + Projected performance (CIFAR-100, $\alpha = 0.1$, 200 rounds).

Cut Location	Final Acc (%)	mean_FgT_loss
Early Cut (after block 3)	51.10	-1.01
Mid Cut (after block 5)	51.46	-0.54
Late Cut (after GAP)	60.07	-0.46

Ablation Study on Cut-Layer Placement.

Table 3 reports the effect of cut-layer placement on SFL-V2 + Projected for CIFAR-100 with $\alpha = 0.1$, with all other hyperparameters fixed. Accuracy improves as the cut moves deeper, from 51.10% for the early cut after block 3, to 51.46% for the mid cut after block 5, and 60.07% for the late cut after GAP. Forgetting follows the same trend, with the late cut giving the least negative score (-0.46 vs. -1.01 for the early cut). The roughly 9% point gap between early and late cuts shows that cut placement is an important design choice for projection-based methods. Later layers provide more class-discriminative activations and a better-conditioned projection subspace, while early cuts expose low-level features where projection may interfere with representation learning. Prior work on cut-layer selection in SFL reports a similar dependence [2].

6 Conclusion

This paper introduced an orthogonal projection mechanism at the split interface to mitigate catastrophic forgetting in split learning. The method operates entirely on cut-layer gradients, requires no raw-data storage, and stays within the standard split-learning protocol.

Experiments on CIFAR-10 and CIFAR-100 show that projection improves accuracy for SFL-V2 in all tested settings and for SL in all except the unstable

CIFAR-10 case at $\alpha = 0.1$, with gains up to 6.48 points on CIFAR-10. SFL-V1 is already a strong baseline and benefits little from projection. The ablation study shows that cut placement also matters, with later cuts yielding substantially better performance when combined with projection. Together, these results indicate that the benefit of projection depends on the training protocol, data heterogeneity, and where the model is split.

Several limitations point to future work. This paper only evaluates CIFAR-10/100, ResNet-18, and full client participation. Future studies could test more architectures, more clients, and partial participation, while also comparing against continual-learning and replay-based baselines, reporting results across multiple random seeds, and analyzing sensitivity to projection hyperparameters such as r, rebuild interval, reservoir size, warm-up length, and ramp length. Another useful direction is to adjust the cut layer and projection strength α during training so the method can better respond when client data changes.

Acknowledgments. This work is supported partially by NSF under grant ERI-2552997.

Disclosure of Interests. The authors have no competing interests to declare that are relevant to the content of this article.

References

1. Bakman, Y.F., Yaldiz, D.N., Ezzeldin, Y.H., Avestimehr, S.: Federated orthogonal training: Mitigating global catastrophic forgetting in continual federated learning. In: International Conference on Learning Representations (2024). https://openreview.net/forum?id=nAs4LdaP9Y
2. Dachille, J., Huang, C., Liu, X.: The impact of cut layer selection in split federated learning. In: AAAI Workshop on FLUID (2025)
3. Delange, M., et al.: A continual learning survey: defying forgetting in classification tasks. IEEE Trans. Pattern Anal. Mach. Intell., 1–1 (2021). https://doi.org/10.1109/TPAMI.2021.3057446
4. Dong, J., et al.: Federated class-incremental learning. In: Proceedings of the IEEE/CVF Conference on Computer Vision and Pattern Recognition, pp. 10164–10173 (2022)
5. Dupuy, C., et al.: Quantifying catastrophic forgetting in continual federated learning. In: ICASSP 2023 - IEEE International Conference on Acoustics, Speech and Signal Processing, pp. 1–5 (2023). https://doi.org/10.1109/ICASSP49357.2023.10097140
6. Etheridge, A., Zhu, M., Liu, X., Huang, C.: Tackling sequential entanglement in split unlearning. In: IEEE International Conference on Multimedia Information Processing and Retrieval, pp. 290–296 (2025). https://doi.org/10.1109/MIPR67560.2025.00053
7. Farajtabar, M., Azizan, N., Mott, A., Li, A.: Orthogonal gradient descent for continual learning. In: Chiappa, S., Calandra, R. (eds.) Proceedings of the Twenty Third International Conference on Artificial Intelligence and Statistics. Proceedings of Machine Learning Research, vol. 108, pp. 3762–3773. PMLR (2020). https://proceedings.mlr.press/v108/farajtabar20a.html

8. Goyal, P., et al.: Accurate, large minibatch SGD: training ImageNet in 1 hour (2018)
9. Gupta, O., Raskar, R.: Distributed learning of deep neural network over multiple agents. J. Netw. Comput. Appl. **116**, 1–8 (2018)
10. Han, P., Huang, C., Tian, G., Tang, M., Liu, X.: Convergence analysis of split federated learning on heterogeneous data. In: Advances in Neural Information Processing Systems, vol. 37, pp. 103476–103544 (2024)
11. Huang, C., Dachille, J., Liu, X.: An accuracy-shaping mechanism for competitive distributed learning. In: Wand, M., Malinovská, K., Schmidhuber, J., Tetko, I.V. (eds.) Artificial Neural Networks and Machine Learning – ICANN 2024. Lecture Notes in Computer Science, vol. 15026, pp. 143–158. Springer Nature Switzerland (2024)
12. Jolliffe, I.T., Cadima, J.: Principal component analysis: a review and recent developments. Phil. Trans. R. Soc. A **374**(2065), 20150202 (2016). https://doi.org/10.1098/rsta.2015.0202
13. Li, Q., Diao, Y., Chen, Q., He, B.: Federated learning on non-IID data silos: an experimental study. In: IEEE 38th International Conference on Data Engineering, pp. 965–978 (2022)
14. McMahan, H.B., Moore, E., Ramage, D., Hampson, S., y Arcas, B.A.: Communication-efficient learning of deep networks from decentralized data. In: Artificial Intelligence and Statistics, pp. 1273–1282. PMLR (2017)
15. Rebuffi, S.A., Kolesnikov, A., Sperl, G., Lampert, C.H.: iCaRL: incremental classifier and representation learning. In: Proceedings of the IEEE Conference on Computer Vision and Pattern Recognition, pp. 5533–5542 (2017)
16. Rusu, A.A., et al.: Progressive Neural Networks (2016)
17. Saha, G., Garg, I., Roy, K.: Gradient projection memory for continual learning. In: International Conference on Learning Representations (2021). https://openreview.net/forum?id=3AOj0RCNC2
18. Thapa, C., Arachchige, P.C.M., Camtepe, S., Sun, L.: SplitFed: when federated learning meets split learning. In: Proceedings of the AAAI Conference on Artificial Intelligence, vol. 36, pp. 8485–8493 (2022)
19. Tirana, J., Tsigkari, D., Noguero, D.S., Kourtellis, N.: Data heterogeneity and forgotten labels in split federated learning. In: Proceedings of the AAAI Conference on Artificial Intelligence (2026)
20. Vepakomma, P., Gupta, O., Swedish, T., Raskar, R.: Split learning for health: distributed deep learning without sharing raw patient data (2018). https://arxiv.org/abs/1812.00564
21. Xia, Z., et al.: MultiSFL: towards accurate split federated learning via multi-model aggregation and knowledge replay. Proc. AAAI Conf. Artif. Intell. **39**(1), 914–922 (2025). https://doi.org/10.1609/aaai.v39i1.32076
22. Yoon, J., Jeong, W., Lee, G., Yang, E., Hwang, S.J.: Federated continual learning with weighted inter-client transfer. In: Proceedings of the 38th International Conference on Machine Learning. Proceedings of Machine Learning Research, vol. 139, pp. 12073–12086. PMLR (2021)

ResonatorLM: Causal Resonant Field Mixing for Efficient Long-Context Language Modeling

Archie Chaudhury(✉)

Axionic Labs, San Francisco, USA
archie@axioniclabs.ai

Abstract. Contemporary language models are dominated by the transformer architecture, which leverages self-attention mechanisms to enable more efficient, parallelized training across a wide set of documents and corpora. This has allowed transformers to effectively model data across a wide range of modalities and contexts. However, transformers, along with their conventional counterparts such as recurrent neural networks (RNNs) and convolutional neural networks (CNNs), often struggle to maintain efficiency when processing long contexts. We introduce ResonatorLM, a new mechanism that replaces attention with a physics-derived alternative. ResonatorLM treats token sequences as a single, driven one-dimensional latent field and replaces attention dot products with causal functions of damped resonators. We implement ResonatorLM on a traditional network architecture and test it on standard long-context modeling tasks. We find that in a small, 6M matched setting, training and prefill speedups increase with sequence length, decode speed reaches 6.47x compared to that of a standard, optimized transformer at 32K tokens, and accuracy reaches 61.31 percent (compared to 55.32 percent) on WikiText.

Keywords: long-context language modeling · efficient decoding · sequence modeling · alternative to attention

1 Introduction

Consistent, efficient long-range sequence modeling has been the primary goal of generative model architectures. Transformers have emerged as the current, most widely adopted state-of-the-art approach for modeling multimodal tasks with long contexts [14]. The adoption of transformers has led to the creation of generalized language models (LMs), which have shown themselves to be capable of modeling text and language across a wide range of use cases. Transformers, and by extension the majority of modern LMs, depend on the attention mechanism to efficiently parallelize training on large sets of data [3,15,23]. Despite the widespread adoption of transformers, their core computational profile remains inefficient and expensive at long context lengths [20].

L. Pasa et al. (Eds.): ICANN 2026, LNCS 17089, pp. 432–444, 2027.
https://doi.org/10.1007/978-3-032-38398-3_36

Recent alternatives have tried to achieve stronger results, mostly by modifying the attention mechanism rather than departing from it entirely. Linear and kernelized methods approximate attention with feature maps or low-rank structure [6,16]. Other long-context architectures use implicit convolutional filters or state-space recurrences to obtain linear-time sequence updates, including Hyena, S4, and Mamba [11,13,19]. These models show that explicit quadratic attention is not necessarily needed for strong sequence modeling; yet, they are still inherently derived from the existing attention-based architecture.

We introduce ResonatorLM, a language model architecture that replaces self-attention with causal resonant field mixing. ResonatorLM models a token sequence as a driven latent field and uses learned resonant kernels to transmit context across positions. This design preserves parallel computation during training and uses compact recurrent states during decoding for efficient long-context inference. This allows ResonatorLM to be significantly more efficient at language modeling, especially in long-context scenarios. Implementing ResonatorLM with a standardized language modeling architecture and benchmarking against a matched transformer, we find that it is both faster and more accurate. In a 6M-parameter WikiText-2 character setting, it reaches a decode speedup of 6.47x at 32K tokens while outperforming the baseline on both perplexity and accuracy. In a separate experiment focusing solely on a kernel reference benchmark, ResonatorLM reaches a 440.29x speedup at 8K tokens and 575.86x at 32K tokens.

2 Related Work

Long-context language modeling has been primarily shaped by the transformer and its algorithmic offshoots. These models leverage self-attention to give each token access to all preceding tokens within individualized layers, allowing for large-scale modeling regimes that remain efficient across various architectures and scenarios [3,14,15,23]. Other structured attention variants include the sparse transformer, reformer, BigBird, and Longformer, as well as projection-based and kernel-based formulations such as Linformer, linear attention, and Performer [2,5,6,16,17,24,25]. Linear and kernelized attention replace explicit similarity matrices with feature-map or low-rank formulations.

Current alternatives to the transformer architecture often replace self-attention regimes, specifically by replacing token-to-token dot products with efficient sequence operators. Early state-space formulations established a principled route from control-theoretic structure to deep sequence modeling via HiPPO and S4 [12,13]. Structured state-space models provide linear-time sequence updates with learned recurrences: recent architectures have formalized this as state-space models (SSMs), demonstrating that recurrence and implicit convolution can remain viable for sequence modeling while reducing the compute required during both training and decoding [11,19].

Physics and continuous-time algorithms have long been emphasized in machine learning and broader sequence modeling. Neural ODEs and Hamiltonian neural networks show how differential-equation structure and conserved dynamics can improve modeling priors, stability, and interpretability in learned systems [4,10]. ResonatorLM is directly inspired by these methods, with the motivation that embedding physical dynamics in a sequence operator can allow for both efficiency and the maintenance of computational structure.

Systems work remains essential for fair long-context claims. FlashAttention-style kernels reduce the practical overhead of exact attention on modern hardware, and decoding studies characterize the cache-driven memory bottlenecks that dominate long-context inference [7,8,20,21]. For this reason, our evaluation separates practical block benchmarks from kernel-only scaling benchmarks. This separation is critical for interpreting where gains come from: full-block wall-clock behavior versus asymptotic mixer behavior.

We implement a basic kernel-only scaling benchmark to measure pure algorithmic speedup. While this is significantly less practical than a wall-clock benchmark, it does allow us to see the algorithmic advantage of our method compared to the native baseline.[1]

3 Model Design

ResonatorLM replaces self-attention with a causal mixer built from damped resonant modes. The central design goal is to keep a parallel full-sequence path for training and prefill, while using fixed-size recurrent state for autoregressive decoding. Each block combines this resonant mixer with a local lexical path and a standard feed-forward sublayer.

We create a generalized model that leverages a causal resonant field mixer with per-head damped oscillatory dynamics as an alternative to self-attention. Each head is parameterized, and the resulting kernel is executed in two modes: causal FFT convolution for training and prefill, and fixed-size recurrent state updates for decoding. The same block includes a local lexical path and optional cross-head coupling, so token-level precision and long-range propagation remain in a single architecture.

At a high level, a token sequence is treated as a driven latent field rather than a set of token-to-token dot products. Learned resonant kernels transport information across positions; the local branch preserves short-range lexical detail, and optional head coupling shares information across resonant channels. The block therefore allows for two key algorithmic improvements: full-sequence training and prefill run through causal FFT convolution with $O(n \log n)$ complexity, while autoregressive decoding uses fixed-size recurrent state per head instead of an ever-growing cache. The local lexical path and cross-head coupling preserve short-range lexical detail alongside long-range propagation.

Figure 1 summarizes the full block and the two execution modes used by the same kernel family: FFT-based convolution for full-sequence passes and recurrent state updates for one-step decoding.

[1] An independent proposal on the Hugging Face forums introduced a closely related wave-field sequence mixer using damped oscillatory kernels and FFT-based causal

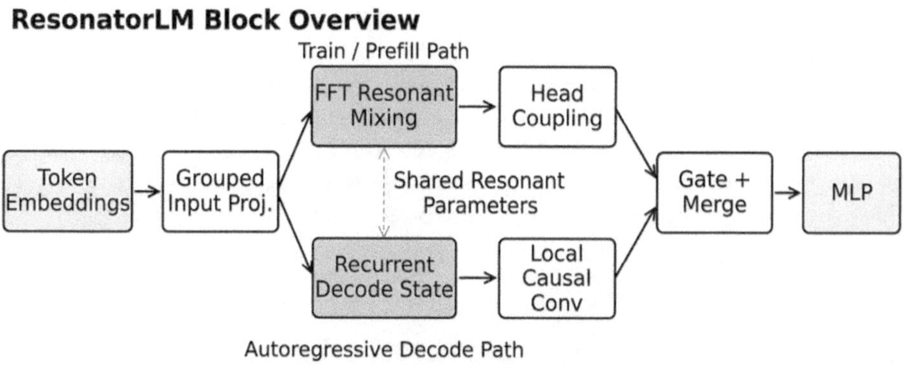

Fig. 1. ResonatorLM block overview. A grouped input projection feeds a resonant mixing path that uses FFT convolution during training and prefill and fixed-size recurrent state during autoregressive decoding. A local causal convolution and head coupling complement the global resonant field before the output is passed to the MLP.

3.1 Resonant Kernel Parameterization

For head h and discrete lag $t \geq 0$, the mixer kernel is

$$k_h[t] = \exp(-\alpha_h t) \cos(\omega_h t + \phi_h), \tag{1}$$

where $\alpha_h > 0$ controls damping, $\omega_h \in (0, \pi)$ is angular frequency, and ϕ_h is a phase offset. In implementation, unconstrained parameters are mapped to valid ranges with softplus/sigmoid transforms (with $\alpha_{\min} = 10^{-4}$), then initialized with log-spaced half-lives and frequencies so different heads begin at different timescales.

The input sequence $x \in \mathbb{R}^{B \times T \times d}$ is projected by a grouped linear layer into a drive tensor u and a gate tensor g. The gate $\gamma = \sigma(g)$ is applied after resonant mixing, which keeps a content-dependent modulation signal without constructing query-key similarity matrices.

3.2 Full-Sequence Training and Prefill

For training and other full-sequence passes without an existing cache, each head is mixed by causal convolution with its resonant kernel. Let $u_h \in \mathbb{R}^{T \times d_h}$ be the drive for head h. The mixer computes

$$y_h = k_h * u_h, \tag{2}$$

where $*$ denotes causal convolution over the sequence dimension. The implementation evaluates this operator with FFTs:

$$y_h = \mathcal{F}^{-1}(\mathcal{F}(u_h) \odot \mathcal{F}(k_h)), \tag{3}$$

convolution, while also using physics-based diagnostics, reinforcing the broader viability of physics-derived alternatives to attention [1].

using zero-padding to the next power of two. This gives an $O(T \log T)$ mixing path for training and prefill.

When the model is asked to build a cache from a prompt, it uses the same FFT path for outputs and then computes the terminal recurrent state analytically. For each head, the prompt state is the weighted sum of past drives under the complex decay factor $e^{(-\alpha_h + i\omega_h)t}$, stored as real and imaginary components. This keeps prefill and decode on the same underlying kernel family.

3.3 Recurrent Decoding

Autoregressive decoding uses a fixed-size per-head recurrent state instead of a growing key-value cache. For one token step, each head maintains a complex state $s_t = a_t + ib_t$ and updates it as

$$s_{t+1} = \lambda_h s_t + u_t, \qquad \hat{y}_{t+1} = \Re\big(e^{-i\phi_h} s_{t+1}\big), \tag{4}$$

$$\lambda_h = \exp(-\alpha_h)\,(\cos\omega_h + i\sin\omega_h)\,. \tag{5}$$

In code, the state is represented as two real tensors for numerical stability and efficient batching. Its size is $B \times H \times d_h \times 2$ and does not grow with decoded length. This split between FFT convolution for full-sequence passes and recurrence for one-step decoding is central to the architecture. It preserves the same damped resonant dynamics in both regimes while avoiding the quadratic prefill cost and the linear-in-context decode memory of standard attention.

3.4 Head Coupling and Local Lexical Path

The resonant field alone is complemented by two additional mechanisms. First, the model can mix information across heads with a learned coupling matrix

$$C = I + \frac{\lambda_c}{\sqrt{H}} \tanh(\widetilde{C}), \tag{6}$$

where λ_c is the configured coupling strength. After resonant mixing, the head tensor is transformed by

$$\bar{y}_{g,t,:} = \sum_{h=1}^{H} C_{gh} y_{h,t,:}. \tag{7}$$

Setting $\lambda_c = 0$ recovers uncoupled heads. This setting appears in the ablation study.

Second, the model includes an optional causal depthwise convolution on the token stream. This local path is intended to preserve short-range lexical precision that can be weak in small global mixers. If the kernel size is $K > 0$, the local branch produces l_t by causal depthwise convolution, scales it channel-wise with a learned vector s, and adds it to the gated resonant field:

$$z_t = \gamma_t \odot \text{flatten}(\bar{y}_t) + s \odot l_t. \tag{8}$$

The combined output is then projected back to model width by a grouped output projection and passed through dropout.

At the block level, ResonatorLM uses root mean square normalization before the mixer and before the feed-forward network, residual connections around both sublayers, and a SwiGLU MLP after mixing [22,26]. The model therefore changes the sequence-mixing operator while keeping the surrounding language-model stack conventional.

3.5 Physics Verification

Rather than utilizing attention maps to diagnose and verify the structure of the model, our implementation leverages learned parameters from the resonant field. These parameters are listed below:

- impulse response per head, computed from the learned kernel;
- head half-lives, defined as $\log 2/\alpha_h$;
- spectral decomposition from the FFT of the impulse response;
- state-energy decay, given by $\exp(-2\alpha_h t)$;
- a causality test that perturbs the suffix of an input sequence and measures the maximum change in the output prefix.

These quantities serve two roles. They verify that the mixer respects causal structure, and they provide an interpretable view of what timescales the learned modes occupy. In the reported configuration, the measured half-life range spans approximately 2 to 2048 tokens and the causality check passes with maximum prefix error below 10^{-6}.

4 Experimental Protocol

We implement ResonatorLM as a matched-scale 6M-parameter language model and compare it against a matched transformer in a study spanning quality, robustness, breadth, and efficiency. Our primary hypothesis is that a model using a sequence mixer derived from resonant fields can improve long-context efficiency while maintaining quality.

Across all experimental runs, we find that ResonatorLM improves both perplexity and accuracy. In long-context block-based benchmarks, it becomes substantially faster, and in the kernel-tail benchmark, it shows a strong theoretical scaling speedup against the baseline.

Table 1. Matched 6M WikiText-2 character results. Means and standard deviations are computed over six seeds.

Model	Runs	Test PPL	Test Acc. (%)	Train tok/s
Transformer	6	4.617 ± 0.070	55.32 ± 0.44	262610.5
ResonatorLM	6	3.764 ± 0.010	61.31 ± 0.09	172890.1

The primary quality evaluation uses WikiText-2 raw with a character tokenizer [18] and a matched parameter budget near 6M. The transformer baseline uses $d_{\text{model}} = 248$, 6 layers, and 8 heads. ResonatorLM uses $d_{\text{model}} = 256$, 6 layers, 8 heads, the full resonator preset, a local lexical path, and cross-head coupling. Both models are trained for 10,000 steps with the same split, optimizer family, schedule, and evaluation protocol. The main matched comparison used six seeds per model. We report test perplexity, test top-1 accuracy, and training throughput. Parity, transfer, long-context, and scale studies use three seeds per setting, and stress ablations use three seeds per family.

Our pipeline remained consistent across all experiments. Each seed was run with the same configuration. For any row in the matched, transfer, long-context, or scale studies, compared models share train/validation/test split, tokenizer, maximum sequence length, optimizer family, learning-rate schedule, batch size, evaluation batch size, and number of optimization steps. The only architectural change is the sequence mixer plus the small width adjustment needed to maintain the matched parameter budget. The same compute settings were used for all reported results.

We keep speed claims protocol-specific. The practical block benchmark measures full block behavior against optimized attention, including projections, cache/state construction, and decode-step cost. The kernel-tail benchmark isolates asymptotic mixer scaling against a quadratic causal reference. This separation allows us to report both pure wall-clock speedups in a practical setting and kernel-tail results that reflect theoretical algorithmic speedup. Figure 2 provides a compact visual summary of the practical block speedup trends used in this evaluation. Auxiliary studies extend the main setup across WikiText-103 and TinyStories [9,18], and test longer contexts than the 256-token matched-budget baseline.

5 Results

5.1 Matched Quality and Core Architecture Tests

Table 1 gives the central matched-budget result. ResonatorLM improves both test perplexity and test accuracy over the transformer baseline, while the transformer retains higher training throughput. Across six seeds, the quality gap is large and stable: transformer test perplexity has a 95% confidence interval of [4.561, 4.673] versus [3.757, 3.772] for ResonatorLM, and transformer accuracy has [54.968, 55.674] versus [61.239, 61.390] for ResonatorLM.

In mean terms, ResonatorLM reduces test perplexity from 4.617 to 3.764 (about an 18.5% relative reduction) and increases test accuracy from 55.32% to 61.31% (+5.99 percentage points). The throughput tradeoff is explicit: ResonatorLM trains at 172,890 tok/s versus 262,610 tok/s for the matched transformer, so the quality gain is not coming from an advantage in raw training throughput.

Table 2 tests whether this gain depends on one brittle configuration. Across six-seed core ablations, removing coupling or the local path changes full-model quality only modestly, and all variants remain tightly grouped. The spread in perplexity across the three variants is only 0.021 (3.771 to 3.750), and the spread in accuracy is only 0.171 points (61.275 to 61.446). This indicates the matched-quality result is not tied to a single fragile hyperparameter setting.

The same table reports physics diagnostics from the matched resonator checkpoint. The measured max prefix error of 7.75×10^{-7} confirms numerical causality at strict tolerance. The learned half-life range spans 2.0 to 2048.0 tokens, showing that the model occupies both short and long timescales rather than collapsing to a narrow temporal regime.

Table 2. Core ablations and physics diagnostics. Left: six-seed core ablations on WikiText-2 raw. Right: diagnostics from the matched resonator checkpoint.

Variant	Runs	Test PPL	Test Acc. (%)
Balanced preset	6	3.771 ± 0.013	61.275 ± 0.094
No coupling	6	3.765 ± 0.011	61.316 ± 0.109
No local path	6	3.750 ± 0.012	61.446 ± 0.099

Metric	Value
Max prefix error	7.75×10^{-7}
Min half-life	2.0
Max half-life	2048.0
Coupling radius	1.0

5.2 Breadth Across Data and Context

The breadth studies maintain the same trend after moving to three seeds per setting. Table 3 shows ResonatorLM ahead across transfer settings (WikiText-2 char/byte, WikiText-103 char, TinyStories char/byte) and long-context settings (WikiText-2 at 512 and 1024, plus TinyStories long-context).

In the transfer rows, the gap is consistent across datasets and tokenization choices. On WikiText-2, ResonatorLM remains around 1.0 perplexity lower and about 6.9–7.4 accuracy points higher under both char and byte tokenization. On WikiText-103 char, the pattern remains similar (5.086 vs 3.856 perplexity, 52.511% vs 60.306% accuracy). On TinyStories, ResonatorLM keeps a clear lead under both tokenizers, with the largest margin in byte-tokenized accuracy (73.653% vs 66.527%).

In the long-context rows, the quality difference widens as tasks become harder. At WikiText-2 length 512 and 1024, transformer perplexity rises sharply to 8.696 and 11.022, while ResonatorLM stays near 4.339 and 4.502. The same trend appears in accuracy. The TinyStories long-context task shows the same direction: 8.940 vs 2.950 perplexity and 33.778% vs 66.505% accuracy.

The shared parity sweep also preserves ordering: at learning rate 5×10^{-4} and zero dropout, the transformer reaches 4.058 ± 0.036 perplexity and $59.20 \pm 0.31\%$ accuracy, while ResonatorLM reaches 3.715 ± 0.013 and $61.86 \pm 0.14\%$. The short-run three-seed scale sweep keeps ResonatorLM ahead at 6M, 15M, and 30M.

Table 3. Transfer and long-context quality. All rows report mean and standard deviation over three seeds per setting.

Setting	Trans. PPL	Trans. Acc.	Res. PPL	Res. Acc.
WikiText-2 char	5.062 ± 0.050	52.696 ± 0.214	3.925 ± 0.008	60.116 ± 0.048
WikiText-2 byte	4.972 ± 0.064	53.176 ± 0.354	3.918 ± 0.015	60.084 ± 0.091
WikiText-103 char	5.086 ± 0.083	52.511 ± 0.515	3.856 ± 0.009	60.306 ± 0.065
TinyStories char	2.743 ± 0.017	68.452 ± 0.160	2.304 ± 0.004	73.566 ± 0.040
TinyStories byte	2.922 ± 0.072	66.527 ± 0.699	2.298 ± 0.001	73.653 ± 0.003
WikiText-2 512	8.696 ± 0.323	36.727 ± 1.001	4.339 ± 0.010	57.200 ± 0.124
WikiText-2 1024	11.022 ± 0.071	29.762 ± 0.123	4.502 ± 0.025	56.166 ± 0.182
TinyStories long-context	8.940 ± 0.029	33.778 ± 0.084	2.950 ± 0.013	66.505 ± 0.113

5.3 Practical Efficiency and Kernel-Tail Scaling

Table 4 reports practical long-context block behavior. Train and prefill speedups increase monotonically with sequence length. Decode is slower than attention at 2048 tokens, crosses over between 4K and 8K, and then becomes decisively faster, reaching 6.47x at 32K. This indicates that ResonatorLM achieves significant efficiency gains specifically in long-context modeling scenarios.

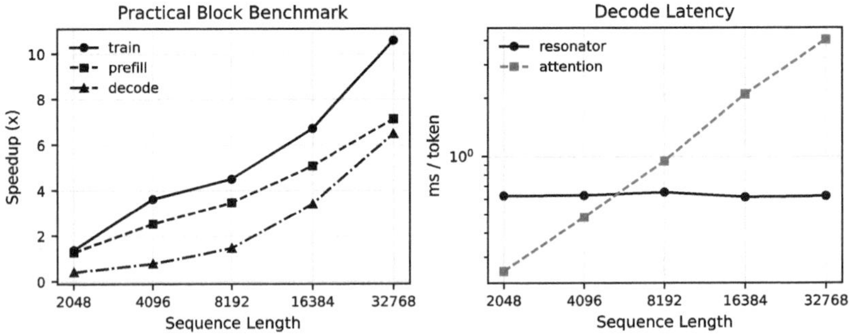

Fig. 2. Practical block speedup trends versus sequence length. Train and prefill speedups increase with context length, while decode transitions from slower at short context to faster at long context.

The decode crossover is visible directly in ms/token values. At a context size of 2048 tokens, ResonatorLM decode is 0.6254 ms/tok versus 0.2532 for attention. At 8192, the ordering flips (0.6511 vs 0.9423), and the advantage grows at 16384 (0.6138 vs 2.0897) and 32768 (0.6228 vs 4.0296). The prefill and train columns show the same context-length trend: practical gains strengthen as sequence length increases. All speed tables were measured on a single NVIDIA L4 GPU using PyTorch 2.10.0+cu128. Training uses CUDA automatic mixed precision with bfloat16. The kernel-tail benchmark uses float32 and reports the winning chunk size by sequence length in Table 5. The practical block benchmark uses batch size 1, decode length 128, five timed iterations after two warm-up iterations, and synchronized wall-clock timing with explicit device synchronization around timed regions. The baseline uses PyTorch scaled-dot-product attention, while ResonatorLM uses cuFFT-backed `torch.fft` for full-sequence passes and fixed-size recurrent state for decode.

Table 4. Practical block benchmark against an optimized attention block. This protocol includes projections, state or cache construction, and decode cost.

Seq. Len.	Train	Prefill	Decode	Res. ms/tok	Attn. ms/tok
2048	1.36x	1.27x	0.40x	0.6254	0.2532
4096	3.61x	2.52x	0.77x	0.6262	0.4829
8192	4.49x	3.46x	1.45x	0.6511	0.9423
16384	6.71x	5.07x	3.40x	0.6138	2.0897
32768	10.58x	7.12x	6.47x	0.6228	4.0296

We implement a separate kernel-tail benchmark to measure pure long-context speedup against a baseline asymptotic mixer, without overhead from projections or multiple layers. Table 5 shows that under the disclosed reference setting (float32, chunk size 32), the resonator kernel reaches 440.29x at 8K tokens and 575.86x at 32K tokens.

Table 5. Kernel-tail benchmark against a quadratic causal reference. This protocol is separate from the practical block benchmark and should not be interpreted as an end-to-end model speedup.

Seq. Len.	Dtype	Chunk	Best Speedup
8192	torch.float32	32	440.29x
32768	torch.float32	32	575.86x

6 Discussion

Our results show that a non-attention mixer with a physics-motivated foundation can outperform a transformer on language quality while also improving long-context efficiency. Across multiple runs, ResonatorLM consistently outperforms or matches a transformer baseline and shows a clear decode-speed advantage in long contexts. To test whether these gains come from the resonant mixer rather than auxiliary architectural choices, we conducted a multi-step ablation study that removed coupling and the local path; these changes produced only modest performance shifts. We also evaluated both practical long-context regimes and a separate kernel benchmark, showing that the method remains strong across settings and can realize much larger asymptotic speedups.

Limitations. This work is intended as a foundational demonstration that a physics-derived alternative to attention can remain competitive in language modeling while improving long-context algorithmic efficiency. Despite statistically significant results across multiple seeds, our experiments cover a limited benchmark set and small model sizes. Future work should test ResonatorLM at larger scales and on more complex long-context tasks, as well as on broader sequence-modeling domains, including multimodal settings. We also compare primarily against a matched transformer baseline; evaluating against newer architectures and stronger system-level optimizations is an important next step. Furthermore, we note that the practical block benchmark is not an actual production benchmark, where other factors may impact the actual speedup that is observed. As such, our work is meant to be foundational in nature, rather than serving as a clear empirical proof that Resonator has significant advantages over a transformer or other similar architectures.

7 Conclusion

In this work, we implemented ResonatorLM, a language model architecture that replaces attention with causal resonant field mixing while preserving consistent dynamics across FFT-based sequence modeling and recurrent decoding. We first formalized ResonatorLM and introduced our design. We then benchmarked it across multiple language-modeling tasks, including WikiText-2, WikiText-103, and TinyStories. Across three-seed breadth studies, the same trend holds under tokenizer changes, longer contexts, and short-run scaling. In the practical block benchmark, long-context efficiency grows with sequence length and reaches a 6.47x decode speedup at 32K. In the separate kernel-tail benchmark, the resonator kernel substantially exceeds reported long-context targets.

Disclosure of Interests. The authors have no competing interests to declare that are relevant to the content of this article.

Usage of LLMs. We utilized AI assistants to help draft, proofread, and formalize some of the key figures in our paper. All core ideas were implemented by the authors.

References

1. Badaramoni, A.: Wave field LLM — o(n log n) attention via wave equation dynamics, within 5% of standard transformer, hugging Face Forums post (2026). https://discuss.huggingface.co/t/wave-field-llm-o-n-log-n-attention-via-wave-equation-dynamics-within-5-of-standard-transformer/173625. Accessed 31 Mar 2026

2. Beltagy, I., Peters, M.E., Cohan, A.: Longformer: the long-document transformer. arXiv preprint arXiv:2004.05150 (2020)
3. Brown, T.B., et al.: Language models are few-shot learners. In: Advances in Neural Information Processing Systems (2020)
4. Chen, R.T.Q., Rubanova, Y., Bettencourt, J., Duvenaud, D.: Neural ordinary differential equations. In: Advances in Neural Information Processing Systems (2018)
5. Child, R., Gray, S., Radford, A., Sutskever, I.: Generating long sequences with sparse transformers. arXiv preprint arXiv:1904.10509 (2019)
6. Choromanski, K., et al.: Rethinking attention with performers. In: International Conference on Learning Representations (2021)
7. Dao, T.: FlashAttention-2: faster attention with better parallelism and work partitioning. arXiv preprint arXiv:2307.08691 (2023)
8. Dao, T., Fu, D.Y., Ermon, S., Rudra, A., Re, C.: FlashAttention: fast and memory-efficient exact attention with io-awareness. In: Advances in Neural Information Processing Systems (2022)
9. Eldan, R., Li, Y.: TinyStories: how small can language models be and still speak coherent english?. arXiv preprint arXiv:2305.07759 (2023)
10. Greydanus, S., Dzamba, M., Yosinski, J.: Hamiltonian neural networks. In: Advances in Neural Information Processing Systems (2019)
11. Gu, A., Dao, T.: Mamba: linear-time sequence modeling with selective state spaces. arXiv preprint arXiv:2312.00752 (2023)
12. Gu, A., Dao, T., Ermon, S., Rudra, A., Re, C.: HiPPO: recurrent memory with optimal polynomial projections. In: Advances in Neural Information Processing Systems (2020)
13. Gu, A., Goel, K., Re, C.: Efficiently modeling long sequences with structured state spaces. In: International Conference on Learning Representations (2022)
14. Hoffmann, J., et al.: Training compute-optimal large language models. arXiv preprint arXiv:2203.15556 (2022)
15. Kaplan, J., et al.: Scaling laws for neural language models. arXiv preprint arXiv:2001.08361 (2020)
16. Katharopoulos, A., Vyas, A., Pappas, N., Fleuret, F.: Transformers are RNNS: fast autoregressive transformers with linear attention. In: International Conference on Machine Learning (2020)
17. Kitaev, N., Kaiser, L., Levskaya, A.: Reformer: the efficient transformer. In: International Conference on Learning Representations (2020)
18. Merity, S., Xiong, C., Bradbury, J., Socher, R.: Pointer sentinel mixture models. In: International Conference on Learning Representations (2017)
19. Poli, M., et al.: Hyena hierarchy: towards larger convolutional language models. Int. Conf. Mach. Learn. (2023)
20. Pope, R., et al.: Efficiently scaling transformer inference. arXiv preprint arXiv:2211.05102 (2022)
21. Shazeer, N.: Fast transformer decoding: one write-head is all you need. arXiv preprint arXiv:1911.02150 (2019)
22. Shazeer, N.: GLU variants improve transformer. arXiv preprint arXiv:2002.05202 (2020)
23. Vaswani, A., et al.: Attention is all you need. In: Advances in Neural Information Processing Systems (2017)

24. Wang, S., Li, B.Z., Khabsa, M., Fang, H., Ma, H.: Linformer: self-attention with linear complexity. arXiv preprint arXiv:2006.04768 (2020)
25. Zaheer, M., et al.: Big bird: transformers for longer sequences. arXiv preprint arXiv:2007.14062 (2020)
26. Zhang, B., Sennrich, R.: Root mean square layer normalization. In: Advances in Neural Information Processing Systems (2019)

ADP: Activation-Driven Neuron Pruning

Dario Lazzaro[1,2], Antonio Emanuele Cinà[1(✉)], Maura Pintor[3], Ambra Demontis[3], Battista Biggio[3], Fabio Roli[1,3], Marcello Pelillo[4], and Luca Oneto[1]

[1] University of Genoa, Genoa, Italy
antonio.cina@unige.it
[2] Sapienza University of Rome, Rome, Italy
[3] University of Cagliari, Cagliari, Italy
[4] Ca' Foscari University of Venice, Venice, Italy

Abstract. Pruning enables the deployment of deep neural networks on resource-constrained hardware by substantially reducing model size and inference latency. Among existing approaches, sparsity-based pruning shows promise for removing neurons that contribute minimally, but insufficient control can lead to significant performance degradation. To address this limitation, we introduce ADP, an Activation-Driven Neuron Pruning approach that controls hidden-layer activation patterns to enable reliable neuron removal while maintaining stable performance. ADP operates directly on neuron activations rather than on weight statistics, shaping inactivity patterns through activation-aware fine-tuning. Building on ideas from energy-aware training and activation-manipulation techniques, ADP leverages a differentiable regularization mechanism that can both suppress unnecessary activations and re-stimulate remaining neurons. As a result, ADP induces structured sparsity that can be effectively exploited for pruning, while preserving the expressive capacity of the pruned network. Empirical evaluations on ResNet18 and VGG16 across three datasets show that ADP achieves competitive pruning performance with strong accuracy-compression trade-offs, often matching or surpassing baseline methods in FLOPs reduction or accuracy retention.

Keywords: Sparsity Pruning · Network Level Pruning · Adjustment

1 Introduction

Deep Neural Networks (DNNs) have become core components of modern systems, enabling advances in application domains such as medical imaging [17], autonomous driving [7], and virtual assistants [20]. Despite their strong predictive performance, the high computational and energy costs of such models limit deployment in resource-constrained environments. As a result, reducing inference-time computational cost, in terms of both FLOPs and energy consumption, has become a critical requirement for practical deployment. Pruning techniques address this challenge by simplifying over-parameterized networks while preserving predictive performance [1]. However, aggressive pruning may

L. Pasa et al. (Eds.): ICANN 2026, LNCS 17089, pp. 445–457, 2027.
https://doi.org/10.1007/978-3-032-38398-3_37

cause irrecoverable accuracy degradation, making the compression-performance trade-off particularly delicate.

In this work, we show that neuron activation patterns can be explicitly controlled during training and leveraged as a principled signal for structured pruning. Our study focuses on vision DNNs, where structured pruning is especially effective, while alternative compression strategies such as quantization are commonly preferred for large language models. Recent studies support this perspective by demonstrating that neuron activation patterns can be deliberately manipulated during optimization [4,13]. However, these approaches primarily targeted energy reduction or activation-density control, rather than leveraging activation manipulation as a direct signal for structured pruning.

Motivated by these results, we propose `ADP`, an `Activation-Driven` Neuron `Pruning` framework that explicitly controls hidden-layer activation patterns to enable reliable and structured neuron removal in DNNs. Rather than operating on weight statistics, `ADP` acts directly on neuron activations through dedicated fine-tuning. By shaping activation patterns during training, `ADP` induces neuron inactivity, making structural removal safe and reliable. To translate activation sparsity into effective model compression, `ADP` integrates an iterative pruning scheme that progressively removes inactive neurons. Within this framework, we also propose a structured pruning algorithm, termed *Thresholded APoZ*, built upon the Average Percentage of Zeros (APoZ) criterion [11]. By removing neurons whose inactivity exceeds a predefined threshold, the method provides explicit control over the trade-off between model size, computational cost, and predictive performance, while activation-aware adjustment re-stimulates and stabilizes the pruned architecture.

We empirically evaluate our approach on two CNN architectures, VGG16 [18] and ResNet18 [9], across three datasets: GTSRB [10], CIFAR-10 [12], and SVHN [16]. Results show that `ADP` achieves comparable or better performance than two state-of-the-art pruning methods, i.e., Network Slimming [15] and LAMP [14], in terms of accuracy retention and FLOPs reduction.

2 Related Work

Several approaches have addressed the redundancy of DNNs through pruning strategies that remove parameters or neurons with limited impact on the final prediction [5]. Among existing techniques, magnitude-based approaches rely on local importance measures defined on weights or neurons to guide pruning [2,3,19], yielding smaller architectures with limited performance degradation when starting from over-parameterized networks. Han et al. [8] propose a pruning framework in which weights below a given threshold are eliminated, using ℓ_1 and ℓ_2 norms as regularization terms to encourage sparsity. Hu et al. [11] introduce the Average Percentage of Zeros (APoZ), which measures the fraction of zero activations in feature maps as an indicator of neuron redundancy. Network Slimming [15] further enforces channel-level sparsity by applying ℓ_1 regularization to the scaling factors of batch normalization (BN) layers. While effective and easy to deploy, this approach is limited to architectures that include BN layers. Along this line of work, Lee et al. [14] propose Layer-Adaptive Magnitude-based Pruning (LAMP), an importance score that rescales weight magnitudes by accounting for their relative importance within each layer. By normalizing weight saliency across layers, LAMP enables more balanced pruning decisions and improves compression effectiveness. Complementary to these approaches, the Lottery Ticket Hypothesis [6] shows that dense networks contain smaller subnetworks capable of matching or exceeding the accuracy of the original model, but does not provide a concrete mechanism for identifying them.

3 Methodology

In this section, we introduce ADP, a structured Activation-Driven Neuron Pruning framework designed to reduce the computational cost of DNNs by shaping hidden-layer activation patterns. The idea behind ADP is to make redundancy observable by acting directly on neuron activations, inducing systematic inactivity that identifies neurons contributing marginally to the forward computation and can therefore be safely removed. Crucially, the same activation-aware mechanism can also be applied in the opposite direction, encouraging the remaining neurons to actively participate in the computation after pruning. This bidirectional control allows ADP to expose redundancy while stabilizing the pruned architecture. Overall, ADP combines three tightly coupled components: (i) a structured pruning strategy based on neuron activation patterns, (ii) an activation-aware fine-tuning mechanism supporting both pruning and stabilization, and (iii) an iterative framework that integrates these components to progressively compress the network while preserving performance.

Structured Pruning Strategy. ADP performs structured neuron removal based on activation patterns induced by activation-aware training. To this end, we propose a structured pruning algorithm, called *Thresholded APoZ*, which identifies and removes redundant neurons according to their degree of inactivity. The method builds upon the Average Percentage of Zeros (APoZ) criterion introduced by Hu et al. [11], which measures how often a neuron outputs zero after the ReLU non-linearity. Neurons with high APoZ values exhibit low activation frequency and can therefore be safely removed. For each ReLU layer, neurons whose APoZ exceeds a predefined threshold τ are structurally pruned, translating activation sparsity into a reduced model with fewer parameters and FLOPs. The corresponding pruning procedure is reported in Algorithm 1.

Activation-Aware Pruning Step. ADP builds on the observation that hidden-layer activation patterns can be explicitly manipulated to control redundancy and guide structured pruning. Early evidence of this behavior was provided by Ciná et al. [4], who introduced *sponge training* and showed that a differentiable approximation of neuron activity can be optimized to deliberately increase the number of firing neurons. Subsequent work, such as Energy-Aware Training (EAT) [13], demonstrated that the same mechanism can be leveraged in the

Algorithm 1: prune function (Thresholded APoZ).

Input: $\mathbf{w}$, model parameters; $\mathcal{V}$, reference set; τ, APoZ threshold.
Output: $\mathbf{w}_{\text{pr}}$, pruned parameters

```
w_pr ← w
for l ∈ Layers(w) do
    if l is a ReLU layer then
        u ← APoZ(l, V)
        for each neuron j in l do
            if u_j > τ then
                mark neuron j for removal
        w_pr ← trim(w_pr)
return w_pr
```

opposite direction to promote sparse representations and reduce inference-time energy consumption. In ADP, pruning and activation-aware fine-tuning are unified within a single optimization step. Rather than treating pruning and training as separate phases, each step first applies structured pruning and then performs activation-aware fine-tuning on the resulting network. As a result, both sparsification and adjustment operate on an already pruned architecture, while differing only in how neuron activations are regulated during fine-tuning. Formally, let $\mathcal{D}$ denote the training set, $\mathcal{L}$ the empirical risk minimization loss (e.g., cross-entropy), and f a neural network parameterized by $\mathbf{w} \in \mathbb{R}^m$. Following [4], we introduce an auxiliary differentiable term E that provides a smooth surrogate of the ℓ_0 norm of hidden activations, enabling gradient-based control over activation density. Let $\phi_k(\mathbf{x}, \mathbf{w}) \in \mathbb{R}^{d_k}$ denote the activation vector of the k-th hidden layer, for $k = 1, \ldots, H$. Accordingly:

$$E(\mathbf{x}, f, \mathbf{w}, \sigma) = \sum_{k=1}^{H} \sum_{j=1}^{d_k} \frac{\phi_{k,j}(\mathbf{x}, \mathbf{w})^2}{\phi_{k,j}(\mathbf{x}, \mathbf{w})^2 + \sigma}, \tag{1}$$

where H denotes the number of hidden layers and σ controls the smoothness of the approximation, trading off approximation accuracy and optimization stability. The resulting activation-aware objective is defined as:

$$\min_{\mathbf{w}} \ \textstyle\sum_{(\mathbf{x},y)\in\mathcal{D}} \mathcal{L}(\mathbf{x}, y, \mathbf{w}) \ - \ \lambda \textstyle\sum_{(\mathbf{x},y)\in\mathcal{D}} E(\mathbf{x}, f, \mathbf{w}, \sigma), \tag{2}$$

where $\lambda \in \mathbb{R}$ controls both the strength and the direction of the activation regularization. Depending on the sign of λ, the same objective induces two distinct regimes:

- `Sparsify` ($\lambda < 0$): neuron activity is penalized, promoting sparse hidden representations and exposing redundancy for pruning;
- `Adjust` ($\lambda > 0$): neuron activity is encouraged, stabilizing the pruned network and restoring expressive capacity.

When $\lambda = 0$, Eq. (2) reduces to standard fine-tuning driven solely by the task loss. The combined activation-aware pruning step is summarized in Algorithm 2.

Algorithm 2: `activation_prune` step.

Input: $\mathcal{D}$, training set; $\mathcal{V}$, reference set; $\mathbf{w}$, model parameters; τ, APoZ threshold; λ, activation coefficient; K, fine-tuning steps; σ, approximation parameter.

Output: $\mathbf{w}_{\text{best}}$

$\mathbf{w}_{\text{pr}} \leftarrow \texttt{prune}(\mathbf{w}, \mathcal{V}, \tau)$

$\mathbf{w}_{\text{ft}}^{(0)} \leftarrow \mathbf{w}_{\text{pr}}$

for $i = 1, \ldots, K$ **do**

  for $(\mathbf{x}, y) \in \mathcal{D}$ **do**

    $\mathbf{g} \leftarrow \nabla_{\mathbf{w}} \mathcal{L}(\mathbf{x}, y, \mathbf{w}_{\text{ft}}^{(i-1)}) - \lambda \nabla_{\mathbf{w}} E(\mathbf{x}, f, \mathbf{w}_{\text{ft}}^{(i-1)}, \sigma)$

    $\mathbf{w}_{\text{ft}}^{(i)} \leftarrow \mathbf{w}_{\text{ft}}^{(i-1)} - \alpha \mathbf{g}$

return $best(\mathbf{w}_{\text{ft}}^{(0)}, \ldots, \mathbf{w}_{\text{ft}}^{(K)})$

Iterative Scheme. Inspired by the approach in [1], `ADP` applies the activation-aware pruning step over N iterations, progressively compressing and adapting the network. Each iteration consists of two consecutive calls to `activation_prune` with opposite activation regimes: a `Sparsify` step followed by an `Adjust` step. The sparsification step employs an activation coefficient $\lambda_t^{\mathtt{S}} < 0$, which suppresses neuron activity and exposes redundancy for structured pruning. It is immediately followed by an adjustment step using a coefficient $\lambda_t^{\mathtt{A}} > 0$, which re-stimulates the remaining neurons and stabilizes the pruned architecture. In both cases, structured pruning is executed at the beginning of the call (Algorithm 2), while the sign of λ determines whether the subsequent fine-tuning suppresses or enhances hidden-layer activations. To control the strength of this bidirectional activation regularization over time, we introduce a γ-BidirectionalSchedule. We index the schedule at the iteration level, where each iteration t comprises exactly one sparsification step and one adjustment step. The geometric factor γ^{t-1} therefore represents a shared decay applied to both activation regimes at iteration t. Let $\lambda_0 < 0$ denote the initial sparsification strength and $\gamma \in (0, 1)$ a geometric damping factor. Under this schedule, the activation coefficients at iteration t are computed as:

$$\lambda_t^{\mathtt{S}} = \lambda_0 \, \gamma^{t-1}, \qquad \lambda_t^{\mathtt{A}} = -\left(1 - \gamma^2\right) \lambda_t^{\mathtt{S}}, \tag{3}$$

where $\lambda_t^{\mathtt{S}} < 0$ induces sparsification, while $\lambda_t^{\mathtt{A}} > 0$ provides a paired adjustment step whose magnitude decays consistently with the same factor γ. As t increases, the magnitude of both coefficients progressively decreases, shifting the optimization from aggressive sparsification toward mild adjustment and, eventually, standard fine-tuning. When the magnitude of an activation coefficient falls below a threshold $\varepsilon > 0$ (i.e., $|\lambda| < \varepsilon$), it is set to zero, transitioning the optimization to standard fine-tuning without activation regularization. The complete iterative pruning scheme driven by the γ-BidirectionalSchedule is summarized in Algorithm 3.

4 Experiments

In this section, we evaluate the effectiveness of `ADP` in terms of accuracy, FLOPs, and parameter count, and compare it with state-of-the-art methods.

Algorithm 3: Activation-Driven Neuron Pruning (ADP).

Input: $\mathbf{w}$, model parameters; $\mathcal{D}$, training set; $\mathcal{V}$, reference set; τ, APoZ threshold; N, iterations; K, fine-tuning steps; σ, approximation parameter; $\lambda_0 < 0$, initial coefficient; $\gamma \in (0, 1)$, damping factor; $\varepsilon > 0$, stop threshold.

Output: $\mathbf{w}$, pruned parameters

Function $\mathtt{Thr}(x, \varepsilon)$
  if $|x| < \varepsilon$ **then**
    return 0
  else
    return x

for $t = 1, \ldots, N$ **do**
  $\lambda_t^{\mathtt{S}} \leftarrow \mathtt{Thr}(\lambda_0\, \gamma^{t-1}, \varepsilon)$
  $\mathbf{w} \leftarrow \mathtt{activation\_prune}(\mathcal{D}, \mathcal{V}, \mathbf{w}, \tau, \lambda_t^{\mathtt{S}}, K, \sigma)$ // Sparsify ($\lambda_t^{\mathtt{S}} \leq 0$)
  $\lambda_t^{\mathtt{A}} \leftarrow \mathtt{Thr}(-\,(1 - \gamma^2)\, \lambda_t^{\mathtt{S}}, \varepsilon)$
  $\mathbf{w} \leftarrow \mathtt{activation\_prune}(\mathcal{D}, \mathcal{V}, \mathbf{w}, \tau, \lambda_t^{\mathtt{A}}, K, \sigma)$ // Adjust ($\lambda_t^{\mathtt{A}} \geq 0$)

return $\mathbf{w}$

4.1 Experimental Setup

Datasets. We evaluate our method on three datasets with different characteristics to assess its effectiveness across diverse contexts. CIFAR-10 consists of 60,000 RGB images of size 32×32 evenly distributed across 10 classes, with 50,000 training and 10,000 test samples. SVHN contains 99,289 RGB digit images of size 32×32 over 10 classes, split into 73,257 training and 26,032 test samples. GTSRB comprises 60,000 RGB traffic sign images across 43 classes and is highly class-imbalanced. Following [10], we use 39,209 samples for training and 12,630 for testing, with images resized to 32×32 pixels and augmented via random cropping and rotation.

Models and Training Phase. We evaluate ADP on VGG16 [18] and ResNet18 [9], trained on the datasets mentioned above. All models are trained from scratch for 100 epochs using stochastic gradient descent (SGD) with momentum 0.9, weight decay 5×10^{-4}, and a batch size of 512, together with an exponential learning rate scheduler (initial LR 0.1, decay 0.95). An exception is ResNet18 on CIFAR-10, for which we adopt a two-phase training strategy with 10 warm-up epochs using OneCycleLR (max LR 0.1), followed by 200 epochs with a CosineAnnealingLR scheduler. These trained models serve as the starting point for all pruning methods considered in our evaluation.

Pruning Setup. We compare ADP with two representative pruning methods implemented in Torch-Pruning[1], namely Network Slimming (NS) [15] and LAMP [14], using their official implementations and default or recommended configurations. All methods are evaluated under an iterative pruning protocol with $N = 25$ iterations. At each iteration, the model is fine-tuned to induce or reinforce sparsity, pruned, and then fine-tuned again to recover performance. For

[1] https://github.com/VainF/Torch-Pruning/tree/v0.2.8.

ADP, each iteration consists of a sparsify step followed by an adjust step, both trained for $K = 50$ epochs, with the γ-BidirectionalSchedule controlling activation regularization across different networks and datasets. For Torch-Pruning baselines requiring an explicit sparsification phase, we follow the framework recommendations by fine-tuning for one third of the pruning-iteration budget. Accordingly, the model is trained for $K/3 = 16$ epochs to induce sparsity, then pruned, and finally fine-tuned for $K = 50$ epochs to recover accuracy. We refer to this configuration as NS. To ensure a budget-matched comparison with ADP, we also introduce an extended version of Network Slimming (ENS), where the pre-pruning sparsification phase is increased to $K = 50$ epochs, matching the sparsification budget of ADP, while keeping the post-pruning recovery fixed to $K = 50$ epochs. For all Torch-Pruning baselines, we adopt structured channel pruning with a global target sparsity of 0.5, which corresponds to the default setting of the framework and provides a good trade-off between compression and accuracy. Across all experiments, we employ the same optimization strategy and training hyperparameters, using SGD with momentum, a fixed learning rate of 0.01, batch size 512, and exponential learning-rate scheduling during fine-tuning.

Evaluation Metrics. To evaluate the effectiveness of pruning across iterations, we consider three performance measures: classification accuracy, the number of floating-point operations (FLOPs) and the number of parameters. Accuracy is used to assess predictive performance, while FLOPs and parameter count quantify the computational and structural complexity of the pruned networks. FLOPs and parameters are computed using Microsoft's NNI toolkit[2]. As a selection criterion for the quantitative results reported, we retain, for each method, the pruned model achieving the largest or comparable FLOPs/parameter reduction while limiting the accuracy drop to at most 10% with respect to the unpruned baseline. This criterion reflects a practical trade-off between compression and predictive performance. Its validity is further supported by the pruning curves reported in the plots, which illustrate the evolution of accuracy and FLOPs across pruning iterations.

4.2 Experimental Results

In this section, we first provide a quantitative comparison of ADP against representative pruning baselines across datasets and architectures. We then analyze the role of the `Sparsify` and `Adjust` phases, highlighting their impact on activation patterns and iterative-pruning stability.

Comparisons Between Methods. Table 1 reports a comparison of the considered pruning methods across datasets and network architectures. For each setting, InitNet denotes the unpruned model trained from scratch and used as a reference. For each pruning method, we report two configurations: (i) a one-shot setting, corresponding to a single pruning step, and (ii) an iterative setting

[2] https://github.com/microsoft/nni.

Table 1. Pruning comparison across datasets on ResNet18 and VGG16. For each method, we report the selected pruned model obtained within 25 pruning iterations (It.). Accuracy (Acc.) is reported in %; FLOPs/Params are in millions.

ResNet18

	GTSRB				CIFAR-10				SVHN			
	It.	Acc.	FLOPs	Params	It.	Acc.	FLOPs	Params	It.	Acc.	FLOPs	Params
InitNet	-	93.35	555.44	11.18	-	91.75	555.42	11.16	-	95.86	555.42	11.16
NS	1	94.85	128.33	2.68	1	90.62	119.95	2.88	1	**95.67**	106.38	3.21
ENS	1	94.23	**128.25**	2.68	1	90.65	120.06	2.89	1	**95.67**	**106.25**	3.21
LAMP	1	94.36	240.67	**1.58**	1	91.07	247.08	**1.61**	1	95.62	244.58	**1.75**
OURS	1	**95.11**	293.04	7.75	1	**91.12**	273.42	6.57	1	95.36	190.56	6.46
NS	4	83.55	2.85	**0.02**	3	86.45	13.63	0.10	7	87.22	0.91	**0.01**
ENS	4	85.33	3.94	**0.02**	3	86.19	13.09	0.11	4	**93.50**	4.39	0.03
LAMP	4	90.83	4.71	**0.02**	3	85.05	19.49	**0.08**	4	91.71	4.78	0.02
OURS	24	**91.09**	**2.34**	0.12	23	**89.05**	**12.07**	0.20	20	90.81	**0.76**	0.04

VGG16

	GTSRB				CIFAR-10				SVHN			
	It.	Acc.	FLOPs	Params	It.	Acc.	FLOPs	Params	It.	Acc.	FLOPs	Params
InitNet	-	89.69	313.22	14.74	-	89.95	313.20	14.72	-	94.89	313.20	14.72
NS	1	91.73	79.44	3.96	1	89.40	76.78	4.18	1	**95.59**	75.49	3.88
ENS	1	91.01	79.47	3.96	1	89.45	76.90	4.18	1	95.53	**75.36**	3.87
LAMP	1	91.46	153.17	**3.22**	1	**89.59**	140.06	3.27	1	95.31	156.14	**3.16**
OURS	1	**93.10**	**69.43**	3.90	1	87.83	**60.81**	**3.20**	1	94.89	77.73	4.23
NS	4	87.24	5.50	0.04	3	87.46	15.62	0.17	4	86.97	1.42	**0.01**
ENS	4	82.23	5.45	0.05	3	**87.96**	15.78	0.17	4	86.39	1.35	0.02
LAMP	5	86.56	**1.84**	**0.01**	4	83.02	**6.16**	0.05	5	**90.68**	1.68	**0.01**
OURS	25	**90.08**	2.08	0.03	25	87.14	10.77	**0.04**	24	88.72	**0.58**	**0.01**

with up to $N = 25$ pruning iterations. In the one-shot configuration, all methods preserve high predictive performance, confirming that a single structured pruning step can already remove a significant amount of redundancy. In this regime, ADP consistently achieves competitive accuracy across all datasets while providing meaningful FLOPs reductions. This behavior is particularly evident for VGG16 on GTSRB and CIFAR-10, where ADP attains strong compression with negligible accuracy loss. In the iterative setting, differences among pruning strategies become evident. Different methods may select models at different pruning iterations, according to the criterion defined in Sect. 4.1. This behavior is empirically supported by the pruning curves in Figs. 1 and 2. In this scenario, ADP achieves competitive FLOPs reductions while attaining higher final accuracy than competing methods. Notably, ADP sustains a large number of pruning iterations before a significant accuracy degradation occurs, highlighting its robustness under iterative pruning. This behavior can be attributed to the scheduled control of the activation regularization through the γ-BidirectionalSchedule (see Sect. 3), which progressively attenuates the magnitude of the activation coefficient λ across pruning iterations. As λ decays to zero, the optimization naturally

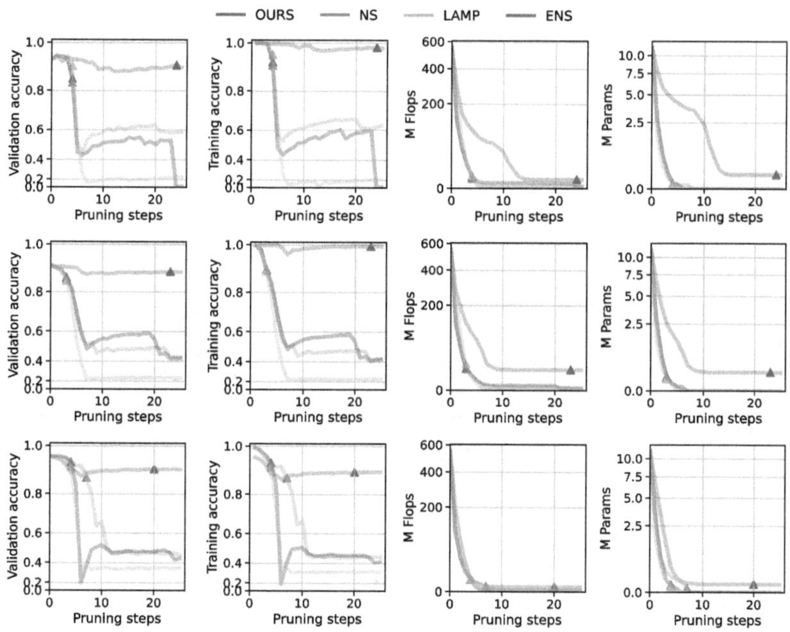

Fig. 1. Comparison of the pruning results on ResNet18 between ADP (OURS), Network Slimming (NS), Extended Network Slimming (ENS) and LAMP. GTSRB (first row), CIFAR-10 (second row) and SVHN (third row).

transitions to standard fine-tuning, enabling the network to stabilize without further perturbations.

Sparsification. As described in Sect. 3, ADP induces sparsity via the `Sparsify` phase of its iterative scheme (Algorithm 3), where a negative activation coefficient ($\lambda_t^{\mathsf{S}} < 0$) discourages hidden-layer activations while optimizing the task loss. Figure 3 illustrates the effect of this mechanism on a ResNet18 trained on CIFAR-10, showing a substantial reduction in the fraction of active units across ReLU and pooling layers. In particular, ReLU layers exhibit a markedly higher proportion of zero activations compared to the unregularized baseline, indicating that predictive capacity is progressively concentrated in a smaller subset of neurons. Importantly, the amount of induced sparsity is not fixed a priori but is adaptively determined by the interaction between activation regularization and the training objective. As a result, different layers naturally converge to different sparsity levels, reflecting their relative importance within the network. This activation-level sparsification directly facilitates subsequent structured pruning. Neurons with consistently low activation are effectively identified by the APoZ criterion and can be safely removed.

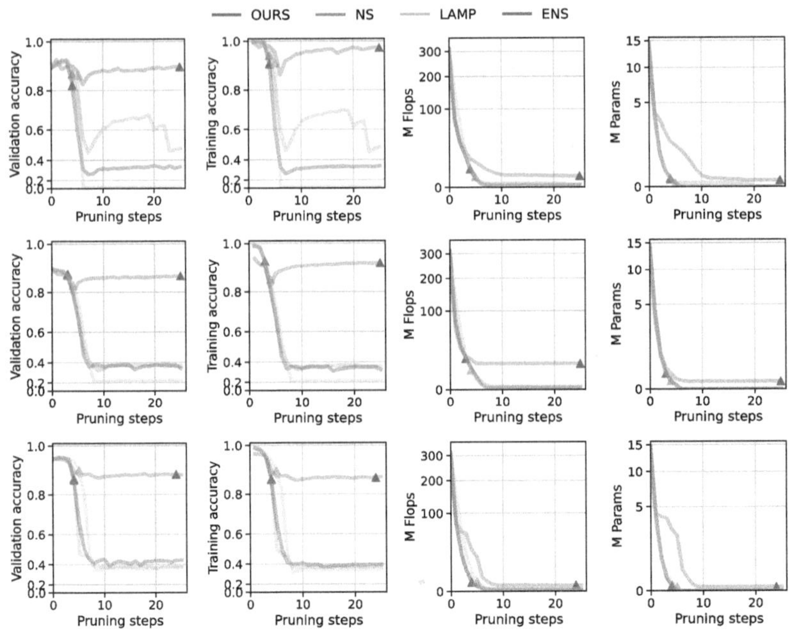

Fig. 2. Comparison of the pruning results on VGG16 between `ADP` (OURS), Network Slimming (NS), Extended Network Slimming (ENS) and LAMP. GTSRB (first row), CIFAR-10 (second row) and SVHN (third row).

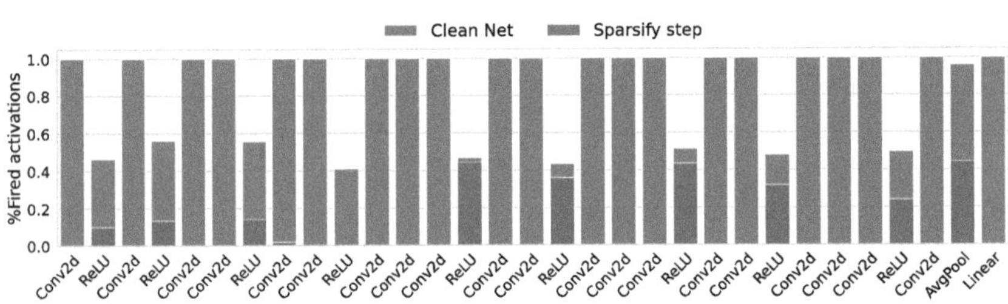

Fig. 3. Example of the `Sparsify` step applied to ResNet18 on CIFAR-10.

Adjustment. Complementary to sparsification, `ADP` counterbalances the induced inactivity through the `Adjust` phase of the iterative scheme (Algorithm 3), where a positive activation coefficient ($\lambda_t^{\mathtt{A}} > 0$) encourages the reactivation of the remaining neurons after pruning. Figure 4 illustrates the effect of this step on a pruned ResNet18 trained on CIFAR-10. Compared to the sparsified model, the fraction of firing activations is restored across the network, particularly in the ReLU layers, indicating that the remaining neurons are effectively re-engaged in the forward computation. Rather than negating sparsification, the adjust step consolidates it by ensuring that the reduced architecture operates in a stable and well-conditioned activation regime.

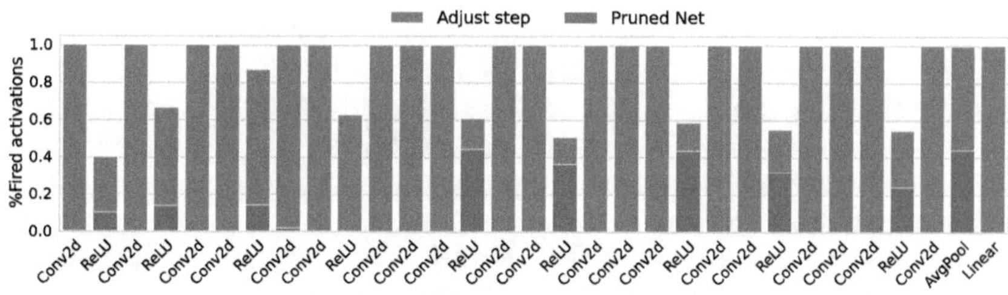

Fig. 4. Example of the `Adjust` step applied to ResNet18 on CIFAR-10.

Crucially, the adjust step improves the robustness of the pruning process under multiple iterations. Figure 5 compares a CIFAR-10 VGG16 model pruned using the complete `ADP` scheme against variants in which the adjust step is replaced by standard fine-tuning.

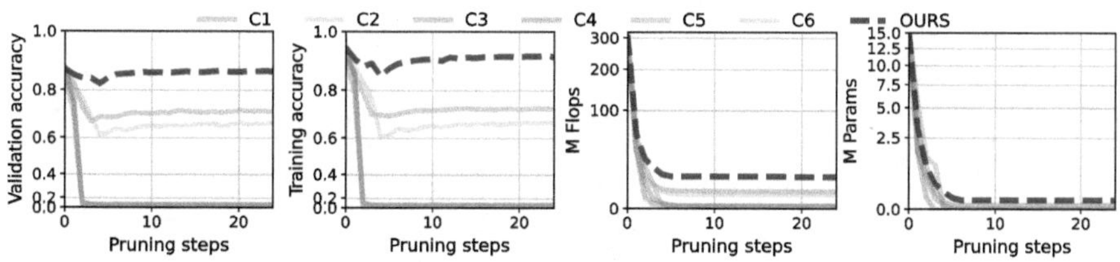

Fig. 5. Effect of the `Adjust` step on training stability. Dashed black line indicates the model with the adjust step, while solid lines correspond to standard fine-tuning. `ADP` with Thresholded APoZ ($\tau = 0.9$) on CIFAR-10 using VGG16.

In this experiment, the compared models (denoted as C1–C6 in Fig. 5) are obtained by varying the sparsification schedule, with initial activation coefficients $\lambda_0 \in \{-6, -9\}$ and geometric damping factors $\gamma \in [0.7, 0.8]$. While models trained without adjustment reach lower FLOPs, they exhibit unstable behavior and rapid accuracy degradation as pruning progresses. In contrast, `ADP` maintains consistent accuracy across iterations, preventing network collapse even under more aggressive sparsification schedules. This confirms the role of the adjust step as a stabilization mechanism that enables safe iterative pruning.

5 Conclusions and Limitations

In this work, we introduced `ADP`, an activation-driven framework for neuron pruning that exposes redundancy by shaping hidden-layer activation patterns during fine-tuning. By combining activation-aware regularization with Thresholded APoZ and an iterative `Sparsify`/`Adjust` scheme, `ADP` leverages controlled activation sparsity to achieve reliable structural compression while maintaining stability. Across datasets and architectures, our method achieves strong accuracy-compression trade-offs, while the γ-BidirectionalSchedule relaxes activation regularization to ensure a smooth transition to standard fine-tuning.

Despite these advantages, ADP introduces additional hyperparameters for activation regularization and scheduling. Although the proposed hyperparameters enhance stability across iterations, their selection may affect the compression-performance trade-off. Developing adaptive or data-driven strategies for automatic tuning remains future work. Iterative pruning increases computational cost over one-shot approaches, and reducing this overhead remains challenging. Finally, our evaluation relies on empirical accuracy-compression thresholds, and more principled criteria for pruned-model assessment could strengthen the comparison among methods.

Acknowledgments and Disclosure of Funding. This work is partially supported by (i) project ELSA - European Lighthouse on Secure and Safe AI funded by the European Union's Horizon Europe under the grant agreement No. 101070617, (ii) projects SERICS (PE00000014) and FAIR (PE00000013) under the NRRP MUR program funded by the EU - NGEU, and (iii) project FISA-2023–00128 funded by the MUR program "Fondo italiano per le scienze applicate". Lastly, this work was carried out while Dario Lazzaro was enrolled in the Italian National Doctorate on Artificial Intelligence run by the Sapienza University of Rome in collaboration with the University of Genoa, funded by "Unione europea-Next Generation EU, Missione 4 Componente 1 CUP B53C23003580004".

References

1. Castellano, G., Fanelli, A.M., Pelillo, M.: An iterative pruning algorithm for feedforward neural networks. IEEE Trans. Neural Netw. **8**, 519–531 (1997)
2. Chauvin, Y.: Back-propagation, weight-elimination and time series prediction. In: Connectionist Models (1991)
3. Chauvin, Y.: A back-propagation algorithm with optimal use of hidden units. In: Advances in Neural Information Processing Systems (1988)
4. Cinà, A.E., Demontis, A., Biggio, B., Roli, F., Pelillo, M.: Energy-latency attacks via sponge poisoning. Inf. Sci. **702**, 121905 (2025)
5. Denil, M., Shakibi, B., Dinh, L., Ranzato, M.A., de Freitas, N.: Predicting parameters in deep learning. In: Advances in Neural Information Processing Systems (2013)
6. Frankle, J., Carbin, M.: The lottery ticket hypothesis: finding sparse, trainable neural networks. arXiv preprint arXiv:1803.03635 (2018)
7. Grigorescu, S., Trasnea, B., Cocias, T., Macesanu, G.: A survey of deep learning techniques for autonomous driving. J. Field Robot. **37**(3), 362–386 (2020)
8. Han, S., Pool, J., Tran, J., Dally, W.: Learning both weights and connections for efficient neural network. In: Advances in Neural Information Processing Systems (2015)
9. He, K., Zhang, X., Ren, S., Sun, J.: Deep residual learning for image recognition. In: IEEE Conference on Computer Vision and Pattern Recognition (2016)
10. Houben, S., Stallkamp, J., Salmen, J., Schlipsing, M., Igel, C.: Detection of traffic signs in real-world images: the german traffic sign detection benchmark. In: International Joint Conference on Neural Networks (IJCNN) (2013)

11. Hu, H., Peng, R., Tai, Y.W., Tang, C.K.: Network trimming: a data-driven neuron pruning approach towards efficient deep architectures. arXiv preprint arXiv:1607.03250 (2016)
12. Krizhevsky, A., Hinton, G.: Learning multiple layers of features from tiny images (2009)
13. Lazzaro, D., Cinà, A.E., Pintor, M., Demontis, A., Biggio, B., Roli, F., Pelillo, M.: Minimizing energy consumption of deep learning models by energy-aware training. In: International Conference on Image Analysis and Processing (2023)
14. Lee, J., Park, S., Mo, S., Ahn, S., Shin, J.: Layer-adaptive sparsity for the magnitude-based pruning. In: International Conference on Learning Representations (2021)
15. Liu, Z., Li, J., Shen, Z., Huang, G., Yan, S., Zhang, C.: Learning efficient convolutional networks through network slimming. In: IEEE International Conference on Computer Vision (2017)
16. Netzer, Y., Wang, T., Coates, A., Bissacco, A., Wu, B., Ng, A.Y.: Reading digits in natural images with unsupervised feature learning. In: NIPS Workshop on Deep Learning and Unsupervised Feature Learning (2011)
17. Shen, D., Wu, G., Suk, H.I.: Deep learning in medical image analysis. Annu. Rev. Biomed. Eng. **19**, 221–248 (2017)
18. Simonyan, K., Zisserman, A.: Very deep convolutional networks for large-scale image recognition. arXiv preprint arXiv:1409.1556 (2014)
19. Weigend, A.S., Rumelhart, D.E., Huberman, B.A.: Generalization by weight-elimination applied to currency exchange rate prediction. In: IEEE International Joint Conference on Neural Networks (1991)
20. Xu, A., Liu, Z., Guo, Y., Sinha, V., Akkiraju, R.: A new chatbot for customer service on social media. In: Conference on Human Factors in Computing Systems (2017)

Theoretical Framework for Distance Metrics in k-Nearest Neighbor Imputation

Tim Kochs[1,2](✉), Andrea Castellani[2], and Barbara Hammer[1]

[1] University Bielefeld, Inspiration 1, 33619 Bielefeld, Germany
{tkochs,bhammer}@techfak.uni-bielefeld.de
[2] Honda Research Institute Europe, Carl-Legien-Straße 30, 63073 Offenbach am Main, Germany
andrea.castellani@honda-ri.de

Abstract. Missing data poses a pervasive challenge in data analysis, with significant implications for inference and decision-making. While k-Nearest-Neighbor (kNN) imputation is widely used, neighborhood selection under missingness remains underexplored. Existing methods rely on heuristic or naive assumptions that can bias neighbor selection and degraded imputation quality. In this work, we develop a theoretical framework for distance metrics in kNN imputation by analyzing the expected distances between samples under missing data. We develop a theoretical framework for distance metrics in kNN imputation under missing data, and introduce a data-centric extension of the classical missingness taxonomy that more precisely characterizes how missingness affects distance estimation in a machine learning context. Our results demonstrate an average improvement of 8–16% in NMAE and 5–11% in NRMSE over state-of-the-art imputation methods across multiple methodological domains, including statistical, classical machine learning, and deep learning approaches.

Keywords: Machine Learning · Missing Data · Imputation · Missing Not At Random · MNAR · kNN

1 Introduction

Missing data is a pervasive challenge across scientific and industrial domains, including healthcare, recommender systems, and the social sciences [10,22], often amplifying bias and impairing data analysis [11]. A standard remedy is imputation, where unobserved entries are inferred values prior to downstream analysis or learning [7]. Following [13], we assume an underlying ground-truth complete data and adopt the following missingness mechanisms, i.e., *Missing Completely at Random (MCAR)*, *Missing at Random (MAR)*, and *Missing Not at Random (MNAR)*. MCAR implies that the missingness is independent of the data, MAR assumes the missingness is dependent on the observed data, and MNAR assumes

L. Pasa et al. (Eds.): ICANN 2026, LNCS 17089, pp. 458–470, 2027.
https://doi.org/10.1007/978-3-032-38398-3_38

the missingness is dependent on unobserved data. We argue that these definitions fall short in a *Machine learning (ML)* context, where one wants to impute data for subsequent downstream learning tasks rather than predict missingness. While MNAR is traditionally considered most problematic [13, 24], we argue this hierarchy inadequately captures consequences for ML tasks.

Consider two nurses, Alice and Bob, recording blood pressure measurements. Alice is meticulous and records all measurements, Bob is less diligent, frequently resulting in missing entries. If Bob's omissions are purely random, no distribution shift occurs. If Bob omits certain measurements based on their magnitude, the distribution of observed values is altered. Here, the missingness mechanism can be either MAR, if nurse is an observed variable or MNAR if nurse is unobserved; albeit the impact on downstream ML models relies on the characteristic of omissions by Bob only. This illustrates that the conventional hierarchy of MCAR, MAR, and MNAR conflates the missigness indicator with its effect on the data distribution.

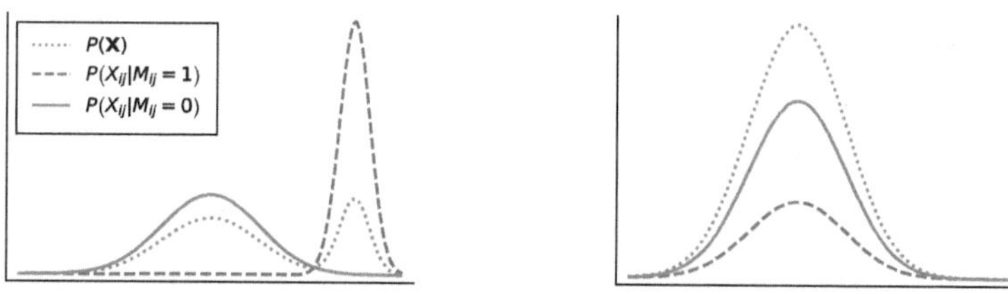

(a) Probability density under Disruptive Missingness (DM).

(b) Probability density under Non - Disruptive Missingness (NDM) (scaled for visual purposes).

Fig. 1. DM implies differing distributions of missing and observed values; NDM assumes approximate alignment. X denotes the total data distribution, M the binary missingness mask.

In this work, we propose a complementary definition of missingness: *Disruptive Missingness (DM)* where missing values are sampled from a distribution different from the observed values, and *Non - Disruptive Missingness (NDM)* where missing values approximate the observed distribution (see Fig. 1). For DM, these are not aligned, presenting challenges for subsequent ML algorithms, as typically associated with MNAR. This data-centric view cuts across the classical taxonomy—MAR, for instance, may be either DM or NDM depending on the omission pattern—and directly reflects the difficulty posed for downstream learning. Building on this framework, we derive a unifying theoretical explanation for the effectiveness of imputation methods, tracing it to implicit neighborhood restrictions that align with the data-generating process, and use it to motivate an improved neighbor-selection heuristic for kNN imputation [10, 12, 17, 18].
The key contributions of this research are: (i) A novel definition of missingness in

line with the effect of the missingness mechanism; (ii) An analysis of kNN neighbor selection in the presence of missing data; (iii) A variant of kNN imputation *Expected Distance - kNN imputation (ED-kNN)* which is better tailored to non-random missingness; and (iv) A novel algorithm to generate MNAR data, for evaluation. We evaluate on multiple benchmark datasets, across different missing data schemes, with the aim of demonstrating a data-centric perspective can improve imputation algorithm design.

2 Related Work

Missingness Mechanisms. A prominent framework for missing data is given by Rubin et al. [13]. While widely adopted, this taxonomy has been challenged in the ML context: [15] argues that *structured missingness* constitutes a practically relevant regime not adequately captured by Rubin's framework, since the taxonomy focuses on the missingness indicator rather than its effect on the observable data distribution In [9], an extension of Rubins taxonomy was proposed, yet practical guidance for algorithm design remains limited. Using DM we aim to make this problem more explicit and provide this guidance.

Imputation. A broad range of imputation techniques exists, including mean/-mode imputation [7], MICE [1], MissForest [20], kNN imputation [21], GAIN [23] and mDAE [5]. The majority of these methods are designed and evaluated under MCAR or MAR assumptions, where the observed data is representative of the full distribution. Methods targeting more realistic mechanisms tend to be highly domain-specific [12,18,24], relying on strong modeling assumptions or substantial prior knowledge about the missingness process, which limits their general applicability. Our proposed ED-kNN offers a solution, performing well even under realistic missingness mechanisms.

kNN Imputation and Neighbor Selection. We focus on kNN imputation as it is both widely used in practice [10,12,17,18] and admits a transparent geometric interpretation that connects naturally to missingness mechanisms. The standard neighbor-selection heuristic traces back to [3] and is based on intuition rather than formal analysis. [6] explored Euclidean distance metrics in the presence of missing values; we build directly on this work and show how distance-based neighbor selection relates to the underlying missingness mechanism. Recent work [17] introduced *kNNSampler*, providing theoretical guarantees for weight selection under MCAR and MAR, but explicitly sidestepping the MNAR setting. Some kNN variants do target MNAR [12,18], but remain domain-specific. Our proposed method bridges this gap by deriving principled neighbor-selection criteria that remain effective under non-random missingness, without requiring domain-specific assumptions.

3 Theoretical Background

We will use the following notation: $\mathbf{X} \in \mathbb{R}^{n \times m}$ is a Dataset with binary missing mask $\mathbf{M} \in [0, 1]^{n \times m}$, where $\mathbf{X}_{obs} = \{\mathbf{X}_{ij} : \mathbf{M}_{ij} = 0\}$ and $\mathbf{X}_{miss} = \{\mathbf{X}_{ij} : \mathbf{M}_{ij} =$

1} denote observed and missing entries, respectively. A single row is written $\mathbf{x} + \mathbf{X}_i \in \mathbb{R}^m$ with corresponding mask $\mathbf{m} \in \{0,1\}^m = \mathbf{M}_i$, and $m_x \in \{0,1\} : x \in \mathbf{x}$ denotes the mask value of $x \in \mathbf{x}$.

3.1 Missingness Mechanisms

Missingness mechanisms are defined as probability distributions the mask $\mathbf{M}$:

Missing Completely at Random (MCAR) MCAR holds if the probability of missingness is independent of the data: $P(\mathbf{M}|\mathbf{X}) = P(\mathbf{M})$.

Missing at Random (MAR) MAR holds if missingness depends on the observed values: $P(\mathbf{M}|\mathbf{X}) = P(\mathbf{M}|\mathbf{X_{obs}})$.

Missing Not at Random (MNAR) MNAR holds if the probability of missingness depends solely on the non-observed data $\mathbf{X_{miss}}$: $P(\mathbf{M}|\mathbf{X}) = P(\mathbf{M}|\mathbf{X_{miss}})$.

The completeness of the above taxonomy has been challenged by [15]. Jackson et al. [9] introduced an extension to *structured missingness (SM)*, the missingness of a variable depends on the missingness of other variables. A missingness mechanism can be *weakly structured* (probabilistic) or *strongly structured* (deterministic). However, SM does not sufficiently cover the impact of missingness on downstream tasks. Consider a feature $\mathbf{x} \in \mathbb{R}^n$ with missing values $\mathbf{m} \in [0,1]^n$, with $m_i = 1$ iff $x_i < a$ for some $a \in \mathbb{R}$. This missingness mechanism causes a disparity in the distribution of observed data and the missing data. While there is a clear structure [9], classifies this as unstructured.

Therefore we introduce *Disruptive Missingness* (DM), defined per feature: a mechanism is DM if missingness induces a distributional shift between observed and missing values, and NDM otherwise.

Definition 1 (Disruptive Missingness). *Let $\mathbf{X} \in \mathbb{R}^{n\times m}$ be a data matrix and $\mathbf{M} \in \{0,1\}^{n\times m}$ a missingness mask, where $M_{ij} = 1$ indicates that X_{ij} is missing. Fix a feature $j \in \{1,\dots,m\}$ and let*

$$S_j := \{X_{ij} : i = 1,\dots,n\}, \qquad S_j^{\text{miss}} := \{X_{ij} : M_{ij} = 1\}.$$

Let $\mathcal{F}$ be a class of bounded real-valued test functions on $\mathbb{R}$, and let $\varepsilon > 0$ be a tolerance parameter. Missingness in feature j is called ε-disruptive *if*

$$\sup_{f\in\mathcal{F}} \left| \frac{1}{|S_j^{\text{miss}}|} \sum_{x\in S_j^{\text{miss}}} f(x) - \frac{1}{|S_j|} \sum_{x\in S_j} f(x) \right| \geq \varepsilon.$$

Otherwise, missingness in feature j is called ε-non-disruptive.

Here $\mathcal{F}$ is a class of bounded test functions which are introduced to test possible changes of the distribution due to missing values in an efficient way. Each $f \in \mathcal{F}$ can probe a particular aspect of the feature distribution (e.g., location, scale, or tail behavior). For instance, choosing $\mathcal{F} = \{\mathbf{1}\{x \leq t\} : t \in \mathbb{R}\}$ recovers the Kolmogorov–Smirnov distance between the two empirical distributions. The parameter ε controls the minimum distributional deviation required for missingness to be considered disruptive.

Most ML methods are designed and evaluated under NDM [7]. DM parallels data drift in continual learning [8], and while MCAR is inherently NDM, both MAR and MNAR may be DM depending on the setting. Crucially, DM affects expected distances between samples, directly motivating a revised kNN imputation scheme based on principled distance bounds derived below.

3.2 Distance Estimation with Missing Values

Under DM, domain knowledge is required to derive bounds for imputation algorithms. We assume a probability P on $[a, b]^n : a, b \in \mathbb{R}$ and observations $\mathbf{x}$ following P. For any concrete observation with feature values x, and mask $\mathbf{m}$, we refer to $\mathbf{x}_m$ as the distribution which is peaked at x if $m_x = 1$ and following the marginal if $m_x = 0$. The uncertainty between two random variables gets maximized if we assume $x, y \sim Unif[a, b]$ [19]. This choice of $P = Unif[a, b]$ is reasonable, in absence of domain knowledge, as assuming any other distribution for P would lead to an increase in bias.

Theorem 1. *For probability P on $[a, b]^n$ with $\mathbf{x}$ following P and mask $\mathbf{m} \in \{0, 1\}^n$ the expected Euclidean distance d in case of $\mathbf{x}_m \sim Unif[a, b]$ is bounded by:*

$$\mathbb{E}[d(\mathbf{x}_m, \mathbf{y}_m)] \leq \begin{cases} |x - y| & \text{if } m_{\{x,y\}} = \mathbf{0}, \\ max\{x - a, b - x\} & \text{if } m_x = 0 \wedge m_y = 1, \\ max\{y - a, b - y\} & \text{if } m_x = 1 \wedge m_y = 0, \\ (b - a)/3 & \text{otherwise.} \end{cases}$$

Proof. Given $\mathbf{x} : x, y \in \mathbf{x}$ and $\mathbf{m} \in \{0, 1\}^n$, W.l.o.g. $\mathbf{x} \in [0, 1]^n$, in case $\mathbf{x} \in [a, b]^n$, $\mathbf{x} = (\mathbf{x} - a)/(b - a)$ scale $\mathbf{x}$ in the assumed range. And $\mathbf{x}_m \sim Unif[0, 1]$.

Case i: Both x and y are observed, we can calculate the exact distance:

$$\mathbb{E}\left[\sqrt{(\mathbf{x}_m - \mathbf{y}_m)^2} | \mathbf{x}_m = x, \mathbf{y}_m = y\right] = \sqrt{(x - y)^2}$$

Case ii: One variable is missing: w.l.o.g. $m_x = 1, m_y = 0$

$$|a - b| = \begin{cases} a - b \leq 1 - b & \text{if } b \leq a, \\ b - a \leq b & \text{else.} \end{cases}$$

Hence for all $x \in [0, 1]$:

$$\sqrt{(x - y)^2} \leq |x - y| \leq \max\{y, 1 - y\}$$

Case iii: $m_{\{x,y\}} = \mathbf{1}$

$$\begin{aligned}\mathbb{E}\left[\sqrt{(\mathbf{x}_m - \mathbf{y}_m)^2}\right] &= \int_0^1 \int_0^1 \sqrt{(\mathbf{x}_m - \mathbf{y}_m)^2}\, d\mathbf{x}_m\, d\mathbf{y}_m \\ &\leq \int_0^1 \int_0^1 |\mathbf{x}_m - \mathbf{y}_m|\, d\mathbf{x}_m\, d\mathbf{y}_m \\ &= 2\int_0^1 \frac{\mathbf{y}_m^2}{2}\, d\mathbf{y}_m = \frac{1}{3}\end{aligned}$$

Reversing scale derives the final bounds. □

The estimations in Theorem 1 lead to the least biased bounds, stronger assumptions might lead to better estimations without domain knowledge they introduce unjustified bias. Further electing for looser bounds in the observed-missing case is in line with injecting the least amount of bias into the missing values, without loosing the information provided. Assuming NDM we know that for any P over $\mathbb{R}^n$ the observed values are an approximation of the missing values, therefore we can show the following.

Theorem 2. *Given i.i.d. observations* $\mathbf{x} \in \mathbb{R} : x, y \in \mathbf{x}$ *according to a distribution* P *over* $\mathbb{R}^n$ *with missing values* $\mathbf{m} \in \{0,1\}^n$*, which are subject to NDM, then the expected squared-euclidean-distance is given by:*

$$\mathbb{E}\left[(\mathbf{x}_m, \mathbf{y}_m)^2\right] = \begin{cases} (x-y)^2 & \text{if } \mathbf{m}_{\{x,y\}} = \mathbf{0}, \\ Var(\mathbf{x}_m) + \mathbb{E}\left[\mathbf{x}_m\right]^2 - 2\mathbb{E}\left[\mathbf{x}_m\right] y + y^2 & \text{if } m_x = 1 \wedge m_y = 0 \\ Var(\mathbf{y}_m) + \mathbb{E}\left[\mathbf{y}_m\right]^2 - 2\mathbb{E}\left[\mathbf{y}_m\right] x + x^2 & \text{if } m_x = 0 \wedge m_y = 1 \\ 2 \cdot Var(\mathbf{x}_m) & \text{otherwise.} \end{cases}$$

Proof. We can distinguish 3 cases.

Case i: $\mathbf{m}_{\{x,y\}} = \mathbf{0}$

$$\mathbb{E}\left[(\mathbf{x_m} - \mathbf{y_m})^2 | \mathbf{x_m} = x, \mathbf{y_m} = y\right] = (x-y)^2$$

Case ii: w.l.o.g. $m_x = 1, m_y = 0$

$$\mathbb{E}\left[(\mathbf{x}_m - \mathbf{y}_m)^2 | \mathbf{y}_m = y\right] = Var(\mathbf{x}_m) + \mathbb{E}\left[\mathbf{x}_m\right]^2 - 2\mathbb{E}\left[\mathbf{x}_m\right] y + y^2$$

Case iii: $\mathbf{m}_{\{x,y\}} = \mathbf{1}$

$$\mathbb{E}\left[(\mathbf{x}_m - \mathbf{y}_m)^2\right] = 2 \cdot Var(\mathbf{x}_m)$$

Because of the DM assumption, expectation and variance of x and y coincide.

□

4 Expected Distance - kNN imputation

The expected distances derived for DM and NDM can be directly incorporated into kNN imputation, yielding Algorithm 1. Treating missing values as random variables, the expected distance must account for both their mean and variance; when the same feature is missing in both samples, this reduces to the variance of the observed distribution (Fig. 2b).

For the observed-missing case, the expected distance is bounded by $\max\{x, 1-x\}$; since DM assumes missing and observed values are fundamentally dissimilar, overestimating this distance is preferable. When both values are missing, the uniform assumption yields an expected distance of $1/3$, and underestimating here is intuitively acceptable as no further information is available.

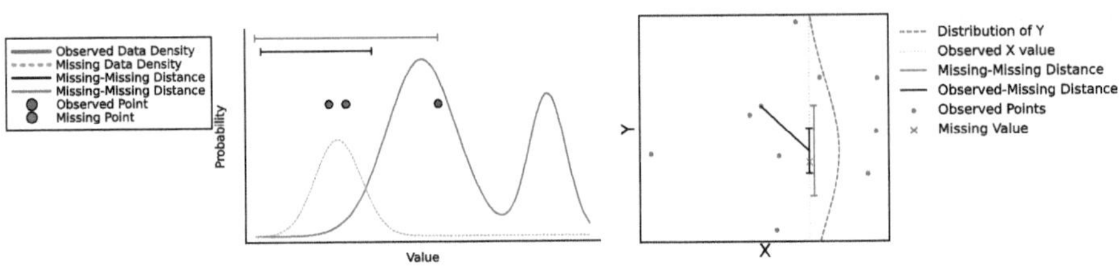

Fig. 2. (a) Under DM, a missing value increases the expected distance to observed values but not to other missing values (see Theorem 1). (b) The Non - Disruptive Missingness metric estimates distance to the distributional mean, weighted by variance.

Algorithm 1. Expected Distance - kNN imputation

Require: Dataset $\mathbf{X}$ with missing entries, number of neighbors k
Ensure: Imputed dataset $\hat{\mathbf{X}}$

Initialize $\hat{\mathbf{X}} \leftarrow \mathbf{X}$
Scale $\hat{\mathbf{X}}$ to $[0, 1]$
for each $\mathbf{x}_i \in \hat{\mathbf{X}}$ with missing entries **do**
  for each missing feature j in sample $\mathbf{x}_i$ **do**
    for each $x \in \mathbf{x}_i \setminus j$ **do**
      $y \leftarrow x$ of candidate neighbor
      Indentify candidate neighbors using δ

$$\delta_j(x, y) = \begin{cases} \sqrt{(x-y)^2} & \text{if } m_x = 0 \wedge m_y = 0 \\ \max\{x, 1-x\} & \text{if } m_x = 0 \wedge m_y = 1 \\ \max\{y, 1-y\} & \text{if } m_x = 1 \wedge m_y = 0 \\ 1/3 & \text{otherwise.} \end{cases}$$

    end for
  end for
  Select k nearest neighbors according to $\delta = \sum_j \delta_j$
  Impute j as the weighted average of neighbors
end for
return $\hat{\mathbf{X}}$

Using the metric derived in Theorem 2 we can define a custom algorithm for the DM case given assumptions about the distribution of missing values. Assuming the missing values are sampled from $\mathcal{N}(\mu, \sigma^2)$ we get:

$$\delta(x,y) = \begin{cases} (x-y)^2 & \text{if } m_x = 0 \wedge m_y = 0, \\ \sigma^2 + \mu^2 - 2\mu y + y^2 & \text{if } m_x = 1 \wedge m_y = 0, \\ \sigma^2 + \mu^2 - 2\mu x + x^2 & \text{if } m_x = 0 \wedge m_y = 1, \\ 2\sigma^2 & \text{otherwise.} \end{cases}$$

This allows any distributional assumption about missing values to be seamlessly integrated into a custom kNN variant (see Fig. 2b).

5 Experimental Setup

We evaluate on ten datasets from UCIML [4](Table 1), each dataset was cleaned of missing values. As baseline we use mean (simple) imputation, mice, MissForest, kNN and Gain; which cover imputation methods from all domains. We are limited to numerical datasets, since common libraries implementing these baselines do not work with categorical data. However, it is known that kNN does perform well when using categorical data (see [14]). For all variants of kNN imputation we fixed $k = 5$. We consider five different missingness mechanisms:

Table 1. Datasets sourced from UCIML.

Dataset	#Instances	#Features
Iris	150	4
Wine	178	13
Wholesale customers	440	7
Connectionist Bench (Sonar, Mines vs. Rocks)	208	60
Parkinsons Telemonitoring	5875	19
Rice (Cammeo and Osmancik)	3810	7
Glass Identification	214	9
Banknote Authentication	1372	4
Iranian Churn	3150	13
Cardiotocography	2126	21

NDM/MCAR Values are missing completely at random; hence, this can be categorized as Non - Disruptive Missingness.

Missing Tails Removed values from the tails of the distribution; assuming a normally distributed feature results in Disruptive Missingness.

GM Lower Quantile Missing values are sampled from a Gaussian distribution centered at the lower quantiles of each feature.

GM Upper Quantile Missing values are sampled from a Gaussian distribution centered at the upper quantiles of each feature.

GM Median Missing values are sampled from a Gaussian distribution centered at the median of each feature.

Algorithm 2 is used for the latter three. It draws samples from a Gaussian, then matches the sample to the closest value for given feature and sets this to missing. Assuming data is approximately Gaussian [16], this provides a principled way to introduce non-random missingness without domain knowledge; a requirement highlighted in [15,24]. This is used for all experiments labeled GM, further all experiments labeled GM use 10% of the features range as variance. While Algorithm 2 is expected to introduce DM, its actual disruption varies per dataset and feature. We selected two commonly used metrics for evaluating the quality of the imputation, *mean average error (MAE)* and *root mean square error (RMSE)* [7,24]. In order to make these comparable across datasets with different scales, we normalize them by each features range. We will refer to these as *normalized mean absolute error (NMAE)* and *normalized root mean square error NRMSE*, which are normalized by range.

6 Results and Discussion

Figure 3 shows MAE on the Wine dataset across experiments and imputation algorithms. When missing values align with the observed distribution, traditional algorithms perform well; ED-kNN however maintains competitive performance even when this is not the case. Figure 5 shows that most Wine features are left-skewed or normally distributed, meaning missing values at the upper quantile constitute DM, confirming our theoretical analysis empirically. Figure 4 shows NMAE and NRMSE averaged across all datasets at 30% missingness, further reinforcing this trend.

Algorithm 2. Gaussian Missingness (GM)

Require: Dataset $\mathbf{X} \in \mathbb{R}^{n \times m}$ without missing values, $\alpha \in (0, 1)$ percentage of missing values, $\tilde{\mu} \in [0, 1]$ relative mean, $\tilde{\sigma} \in [0, 1]$ relative variance.
Ensure: Missing dataset $\hat{\mathbf{X}}$ with $\approx \alpha \cdot nm$ missing values

```
Initialize X̂ ← X
μ, σ ← Quantiles(X, μ̃), σ̃·Range(X)
g ← N(μ, σ) // g has a Gaussian per feature in X
while CountMissingValues(X̂) ≤ α · n · m do
    s ← Sample(g)
// MatchClosest returns closest observed value in X̂ to s
    X̂[MatchClosest(X̂, s)] ← nan
end while
return X̂
```

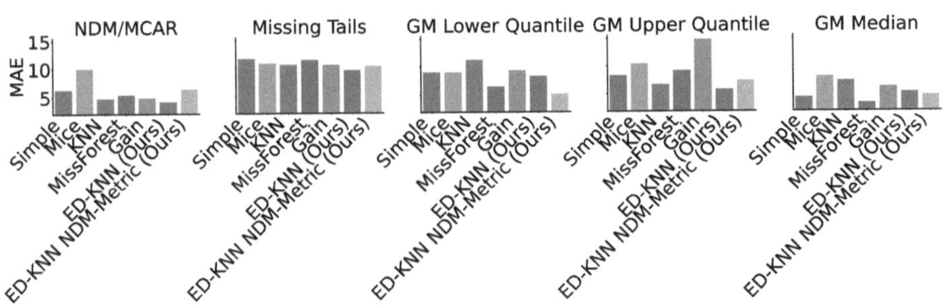

Fig. 3. Mean Absolute Error for the Wine dataset at 30% missing values.

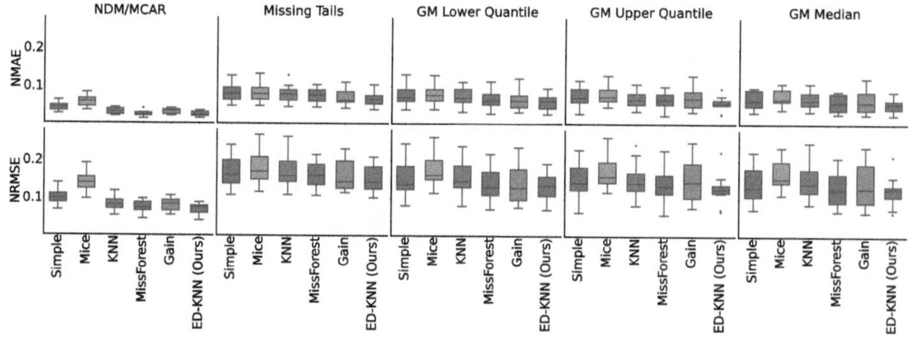

Fig. 4. Average NMAE and NRMSE over datasets mentioned in Table 1 at 30% Missingness.

Most imputation algorithms have a decent performance under NDM. The empirical results confirm our theoretical analysis: using the distances derived in Theorem 2 in the case of NDM show better performance than the distances from Theorem 1, as shown in Fig. 3 where Expected Distance - kNN imputation uses the bounds derived in Theorem 1 and ED NDM-kNN uses the sample expectation and variance. Using Theorem 1 we get an improvement in the DM case. This supports that given domain knowledge and stronger assumptions, increased performance can be achieved. Notably, all non-NDM/MCAR experiments involve MNAR, yet algorithms designed for MCAR or MAR—including ED NDM-kNN—still perform well, confirming that the mere presence of MNAR is insufficient to meaningfully disrupt the data distribution without accompanying DM.

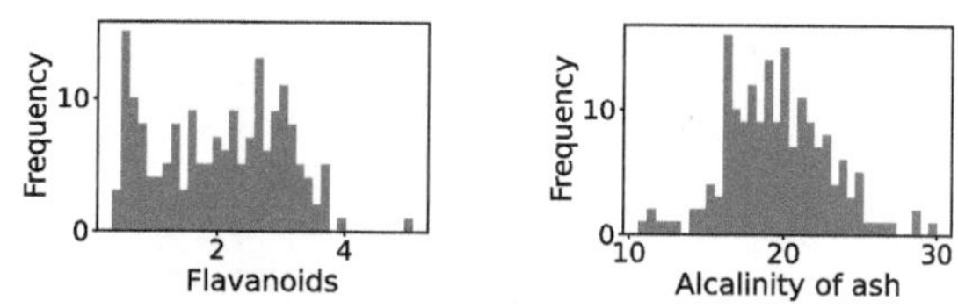

Fig. 5. Wine features are predominantly left-skewed or normally distributed, while upper quantiles are sparse.

Using ED-kNN with the distance estimations from Theorem 1, consistent results regardless of data distribution and across all experiments can be achieved. Hence, in the following, we only use the bounds derived in Theorem 1, as this is less dependent on choices of dataset and experiment type. Figure 4 shows the *normalized mean absolute error (NMAE)* and *normalized root mean square error (NRMSE)* averaged across datasets. This demonstrates the capability of ED-kNN to handle DM, whenever a large portion of missing values overlaps with observed values ED-kNN decays in performance.

During the runs of GM Lower Quantile, GM Median and GM Upper Quantile we observe that if for the majority of features their mass is centered around the lower quantile, median or upper quantile MissForest has superior performance. However if the missing values are distributed further away from the center of mass, its performance degrades.

ED-kNN shows more consistent performance, particularly when missing values lie far from the center of mass (Fig. 4 and Table 2). Our method accounts for an average improvement of 8–16% in NMAE and 5–11% in NRMSE, compared to the basic kNN. We summarize all experiments using the Critical Distance plot in Fig. 6, derived from a Friedman test followed by the Nemenyi post-hoc analysis ($\alpha = 0.01$). It shows that ED-kNN attains the best overall average rank. However, it lies within the critical distance of MissForest and GAIN, meaning that no statistically significant differences are detected between these methods.

Table 2. Average NMAE and NRMSE across experiments.

Algorithm	Missing %	NMAE	NRMSE
Simple	0.1	0.026 ± 0.011	0.090 ± 0.035
Mice		0.028 ± 0.010	0.108 ± 0.031
KNN		0.020 ± 0.011	0.077 ± 0.038
MissForest		0.020 ± 0.011	0.074 ± 0.036
Gain		0.021 ± 0.010	0.078 ± 0.033
ED-KNN (Ours)		**0.017 ± 0.008**	**0.065 ± 0.027**
Simple	0.3	0.070 ± 0.027	0.143 ± 0.049
Mice		0.075 ± 0.024	0.168 ± 0.044
KNN		0.066 ± 0.027	0.142 ± 0.053
MissForest		0.060 ± 0.027	0.129 ± 0.049
Gain		0.063 ± 0.029	0.136 ± 0.056
ED-KNN (Ours)		**0.055 ± 0.023**	**0.122 ± 0.045**

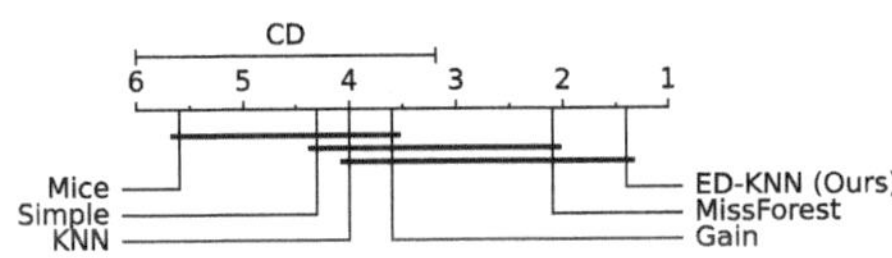

Fig. 6. Critical distance of the average NMAE at 30% Missingness across Datasets.

7 Conclusion

While Rubin's taxonomy [13] remains foundational, its sufficiency in ML contexts has been questioned [15]. We formalized an alternative, distribution-centric classification of missingness and showed that incorporating it into kNN imputation yields improved performance under complex missingness scenarios. Based on these improvements, further research into richer and more realistic models of missingness is possible. A key practical limitation is that Disruptive Missingness is difficult to detect, making targeted solutions an open problem. Due to the scaling operation in ED-kNN, outliers may pose a potential issue. Although this was not observed in our experiments, the presence of large outliers could still adversely affect the method's performance. A promising direction is the integration of representation learning with missing data modeling (see [2]). Such approaches embed missingness directly into the learning process, which may prove beneficial when applied to imputation-focused methods. More broadly, we argue that a data-centric perspective on missingness, as embodied by DM, is a productive lens for developing more effective imputation methods.

References

1. Buuren, S.V., Groothuis-Oudshoorn, K.: Journal of statistical software mice: Multivariate imputation by chained equations in r. Technical report (2011). http://www.jstatsoft.org/
2. Chang, L.W., Li, C.T., Yang, C.P., Lin, S.D.: Learning on missing tabular data: attention with self-supervision, not imputation, is all you need. ACM Trans. Intell. Syst. Technol. **16** (2025). https://doi.org/10.1145/3729241
3. Dixon, J.K., Member, A.: Pattern recognition with partly missing data. Technical report (1979)
4. Dua, D., Graff, C.: UCI machine learning repository (2017). http://archive.ics.uci.edu/ml
5. Dupuy, M., Chavent, M., Dubois, R.: MDAE: modified denoising autoencoder for missing data imputation (2024). http://arxiv.org/abs/2411.12847
6. Eirola, E., Lendasse, A., Vandewalle, V., Biernacki, C.: Mixture of gaussians for distance estimation with missing data. Neurocomputing **131**(5), 32–42 (2014). https://doi.org/10.1016/j.neucom.2013.07.050
7. Emmanuel, T., Maupong, T., Mpoeleng, D., Semong, T., Mphago, B., Tabona, O.: A survey on missing data in machine learning. J. Big Data **8** (2021). https://doi.org/10.1186/s40537-021-00516-9
8. Hinder, F., Vaquet, V., Hammer, B.: One or two things we know about concept drift - a survey on monitoring in evolving environments. Part A: detecting concept drift. Front. Artif. Intell. **7** (2024). https://doi.org/10.3389/FRAI.2024.1330257
9. Jackson, J., Mitra, R., Hagenbuch, N., McGough, S., Harbron, C.: A complete characterisation of structured missingness (2023). http://arxiv.org/abs/2307.02650
10. Joel, L.O., Doorsamy, W., Paul, B.S.: On the performance of imputation techniques for missing values on healthcare datasets (2024). http://arxiv.org/abs/2403.14687
11. Kochs, T., Castellani, A., Lanfermann, F., Hammer, B.: Evaluating and mitigating fairness risks in machine learning under structured missing data. Technical Report (2025)
12. Lee, J.Y., Styczynski, M.P.: NS-kNN: a modified k-nearest neighbors approach for imputing metabolomics data. Metabolomics **14**(12), 1–12 (2018). https://doi.org/10.1007/s11306-018-1451-8
13. Little, R.J., Rubin, D.B.: Statistical analysis with missing data (2014). https://doi.org/10.1002/9781119013563
14. Memon, S.M., Wamala, R., Kabano, I.H.: A comparison of imputation methods for categorical data. Inf. Med. Unlocked **42**, 101382 (2023). https://doi.org/10.1016/j.imu.2023.101382. https://linkinghub.elsevier.com/retrieve/pii/S2352914823002289
15. Mitra, R., et al.: Learning from data with structured missingness (2023). https://doi.org/10.1038/s42256-022-00596-z
16. Murphy, K.P.: Machine learning - a probabilistic perspective. In: Adaptive Computation and Machine Learning Series (2012). https://api.semanticscholar.org/CorpusID:17793133
17. Pashmchi, P., Benoit, J., Kanagawa, M.: knnsampler: Stochastic imputations for recovering missing value distributions (2025). http://arxiv.org/abs/2509.08366
18. Shah, J.S., Rai, S.N., DeFilippis, A.P., Hill, B.G., Bhatnagar, A., Brock, G.N.: Distribution based nearest neighbor imputation for truncated high dimensional data with applications to pre-clinical and clinical metabolomics studies. BMC Bioinf. **18** (2017). https://doi.org/10.1186/s12859-017-1547-6

19. Soch, J., et al.: Amvosk, maxgrozo: Statproofbook/statproofbook.github.io: Statproofbook 2025 (2025). https://doi.org/10.5281/zenodo.18096132
20. Stekhoven, D.J., Bühlmann, P.: Missforest-non-parametric missing value imputation for mixed-type data. Bioinformatics **28**(1), 112–118 (2012). https://doi.org/10.1093/bioinformatics/btr597
21. Troyanskaya, O., et al.: Missing value estimation methods for dna microarrays. Technical report (2001). http://smi-web.stanford.edu/projects/helix/pubs/impute/
22. Wei, R., et al.: Missing value imputation approach for mass spectrometry-based metabolomics data. Sci. Rep. **8** (2018). https://doi.org/10.1038/s41598-017-19120-0
23. Yoon, J., Jordon, J., Schaar, M.V.D.: Gain: missing data imputation using generative adversarial nets. Technical Report (2018)
24. Zhou, Y., Aryal, S., Bouadjenek, M.R.: Review for handling missing data with special missing mechanism (2024). http://arxiv.org/abs/2404.04905

Revisiting Gradient Centralization: A Unified View of Mean Subtraction and Singular Spectrum Flattening

Koji Kamma(✉)

Wakayama University, Wakayama-shi, Wakayama 640-8510, Japan
kkamma@wakayama-u.ac.jp

Abstract. We revisit *Gradient Centralization (GC)* from the perspective of singular spectrum flattening of gradient matrices and propose *Scaled Gradient Mean Subtraction (SGMS)* as a unified formulation. In mini-batch training, gradients from individual samples are expected to span diverse directions; however, in practice they are often highly correlated, concentrating updates into a low-rank subspace. As a result, many directions in the weight space remain underutilized during optimization. SGMS addresses this issue by subtracting a scaled batch mean from the gradient, which weakens the dominant singular direction and implicitly flattens the singular spectrum. With common ReLU-like activations (e.g., ReLU, GeLU, SiLU), this simple operation promotes complementary directions without relying on explicit covariance estimation or matrix decompositions. Compared to second-order or whitening-based approaches, SGMS achieves a similar rebalancing effect with minimal overhead and no architectural modifications. Formally, SGMS generalizes Gradient Centralization as a special case. By partially rather than fully suppressing the mean direction, it preserves useful gradient components that GC eliminates. Experiments on vision and language tasks across multiple architectures show that SGMS consistently improves accuracy for Convolutional Neural Networks (CNNs) and Transformers with negligible overhead, demonstrating its practicality as a lightweight framework for singular spectrum flattening of gradient matrices.

Keywords: Neural networks · Optimization · Gradient centralization

1 Introduction

Deep neural networks (DNNs) have achieved remarkable success across a wide range of tasks, including image recognition, natural language processing, and speech recognition. However, these advances are largely driven by increasing model scale, which makes both training and deployment computationally demanding. This has motivated research on improving training efficiency.

One of the fundamental reasons why DNN training can be inefficient lies in the structure of gradient updates. In mini-batch stochastic gradient descent, gradients from different samples are ideally expected to point in diverse directions.

L. Pasa et al. (Eds.): ICANN 2026, LNCS 17089, pp. 471–483, 2027.
https://doi.org/10.1007/978-3-032-38398-3_39

In practice, however, they are often strongly correlated and span only a few dominant directions in parameter space. This phenomenon has also been noted in prior work [6,17], which exploits low-rank structures in gradients. This *low-rank subspace concentration* manifests as a highly skewed singular spectrum of gradient matrices, leaving many potentially useful directions underutilized. *Flattening this singular spectrum*, thereby rebalancing gradient directions, is therefore a natural step toward more efficient training.

We propose *Scaled Gradient Mean Subtraction (SGMS)*, a lightweight method that improves training performance by amplifying underutilized gradient directions. SGMS simply subtracts a scaled mean from the gradients. When combined with common nonnegative activations (e.g., ReLU), this operation directly weakens the most dominant direction of gradients, allowing complementary directions to play a greater role in optimization. Typically, achieving such *singular spectrum flattening* requires costly matrix decompositions such as singular value decomposition (SVD). Notably, SGMS attains a similar effect through a single mean subtraction, adding minimal (often negligible) overhead and requiring no architectural changes. Moreover, SGMS is orthogonal to optimizer design: it can be used as a plug-in for optimizers such as SGD and Adam, and is neither a competitor nor a replacement for them.

Importantly, SGMS provides a unified reinterpretation of *Gradient Centralization (GC)* [15] from the perspective of singular spectrum flattening. Originally, GC was introduced as a gradient normalization technique that restricts weight updates to a zero-mean subspace, thereby improving optimization stability and generalization. From the viewpoint of singular spectrum flattening, fully subtracting the mean can be interpreted as suppressing a specific gradient direction. With common activation functions such as ReLU, this direction often aligns with the leading singular vector. As a result, GC effectively removes the dominant component of the gradient. While this may accelerate convergence, it can also discard useful signal contained in that direction. By contrast, SGMS allows partial suppression, preserving valuable gradient components while still amplifying weaker directions, thereby offering a more flexible middle ground.

Related approaches pursue the same goal of balancing gradient directions using preconditioning techniques that decorrelate gradients through matrix decompositions. AdaBK [16] is a representative example: it computes layer-wise second-moment estimates and constructs approximate inverse-root preconditioners, effectively decorrelating gradient updates and expanding the range of directions that contribute to learning. While effective against low-rank subspace concentration, such methods incur nontrivial computation and memory overhead in training. SGMS offers a lightweight alternative, achieving a similar singular spectrum flattening effect with only a single mean subtraction and minimal (often negligible) overhead.

Our contributions are:

- We revisit subspace concentration of gradient updates as a key inefficiency factor and interpret it as singular spectrum skewness in gradient matrices.

- We propose SGMS, a lightweight method that flattens the singular spectrum of gradients via a mean subtraction without costly matrix decompositions.
- We clarify the connection to GC and show that SGMS occupies a practical position between GC and covariance-based preconditioners such as AdaBK in terms of cost-benefit trade-offs.
- We demonstrate on vision and language tasks that SGMS improves accuracy while adding minimal overhead.

2 Related Work

We review areas of research most relevant to SGMS: (i) optimization methods for DNN training, (ii) methods that modify gradient directions to mitigate subspace concentration.

Optimization Methods. Stochastic gradient descent (SGD) [10] and its variants, including momentum [12] and adaptive methods such as Adam [7], form the foundation of DNN training. While adaptive optimizers rescale updates based on historical gradient statistics, they operate in an element-wise manner and do not address correlations among gradient directions. SGMS is orthogonal to optimizer design and can be combined with standard optimizers as a plug-in to alleviate subspace concentration.

Gradient Centralization and Preconditioners. Gradient Centralization (GC) [15] was originally introduced as a gradient normalization technique that subtracts the mean along feature dimensions, thereby projecting updates onto a zero-mean subspace and improving optimization stability and generalization. With common activation functions such as ReLU, this mean direction often aligns with the dominant singular direction of the gradients. From the perspective of singular spectrum flattening, fully subtracting the mean therefore suppresses the dominant component of the gradient. Other prior work has also discussed gradient centering [9,11], but has not explicitly connected mean subtraction to dominant singular-direction suppression. SGMS makes this connection explicit and further generalizes GC by enabling partial suppression of this direction.

Covariance-based preconditioners, such as AdaBK [16] and related methods [3,4], explicitly decorrelate gradient directions using second-order statistics. These methods also flatten the singular spectrum of gradient matrices. However, maintaining covariance matrices and computing their inverse square roots incur substantial computational and memory overhead. In contrast, SGMS achieves a similar spectrum flattening effect using only a single mean subtraction.

Overall, prior work has discussed gradient centralization and singular spectrum flattening separately. Our analysis shows that under modern ReLU-like activations, mean subtraction closely aligns with suppressing the dominant singular direction, unifying these perspectives into a single view. This connection directly motivates SGMS as a lightweight and principled alternative.

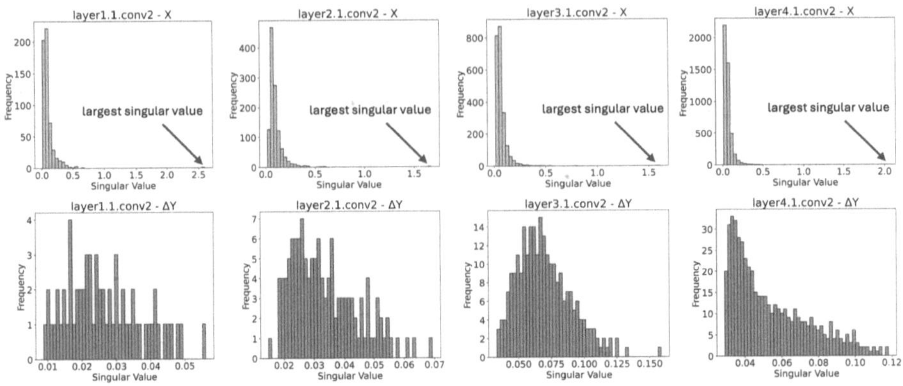

Fig. 1. Histograms of singular values of X (top) and ΔY (bottom) in several layers of ResNet-18 at random initialization.

3 Methodology

In this section, we first analyze the phenomenon that gradients concentrate in a low-rank subspace. Gradient Centralization (GC) has traditionally been understood as subtracting the mean to restrict updates to a zero-mean subspace. Based on our analysis, we reinterpret this mean-subtraction operation as suppressing the dominant singular direction of the gradient matrix. We then introduce SGMS as a generalized formulation that enables partial suppression of this direction.

3.1 Gradient Analysis and Covariance-Based Spectrum Flattening

We now analyze why weight updates in standard training naturally concentrate in a low-rank subspace.

Consider a linear transformation

$$Y = XW, \tag{1}$$

with input $X \in \mathbb{R}^{n\times k}$, weights $W \in \mathbb{R}^{k\times d}$, and output $Y \in \mathbb{R}^{n\times d}$. The gradient of the loss with respect to W is

$$\Delta W = \tfrac{1}{n} X^\top \Delta Y, \tag{2}$$

where $\Delta Y \in \mathbb{R}^{n\times d}$ is the matrix of backpropagated gradients. The prefactor $1/n$ results from averaging over the input samples and is omitted hereafter, as it does not affect the following rank analysis. From basic linear algebra it follows that

$$\operatorname{rank}(\Delta W) \le \min(\operatorname{rank}(X), \operatorname{rank}(\Delta Y)), \tag{3}$$

so the rank of ΔW is always bounded by those of X and ΔY. In particular, if either X or ΔY is degenerate, the weight updates are necessarily confined to a lower-dimensional subspace.

To examine this phenomenon empirically, we computed the singular values of X and ΔY at different layers of ResNet-18. The data were collected at random initialization using 10,240 training images from ImageNet-1K (40 mini-batches of size 256). As shown in Fig. 1, the singular value distributions of both X and ΔY are skewed, but the skewness is far more pronounced for X. In particular, the largest singular value of X is substantially larger than the rest, whereas the singular values of ΔY are more evenly distributed. We observed the same behavior across other architectures under nonnegative activations such as ReLU. These results indicate that X primarily drives the low-rank structure of ΔW, suggesting that mitigating the skewness of X is crucial for flattening the singular spectrum of ΔW.

When X lies in such a low-rank subspace, ΔW is effectively restricted to the span of X, and directions associated with small singular values contribute negligibly. A straightforward countermeasure is to precondition ΔW to cancel this skewness. Let the SVD of X be

$$X = U\Sigma V^\top, \quad \Sigma = \mathrm{diag}(\sigma_1, \ldots, \sigma_m), \tag{4}$$

where m is the number of nonzero singular values. Preconditioning with the inverse square root of the Gram matrix $X^\top X$ yields

$$\Delta \bar{W} = \left(X^\top X\right)^{-\frac{1}{2}} \Delta W = VU^\top \Delta Y, \tag{5}$$

which equalizes the singular directions of X up to rotation. This procedure corresponds to one component of the full preconditioning performed in AdaBK [16], which additionally accounts for ΔY. While effective, such covariance-based preconditioners require computing SVDs (or iterative alternatives) and matrix multiplications during training, leading to significant computational overhead. This observation motivates the search for lightweight approximations that retain the benefit of singular spectrum flattening without the cost of explicit SVD.

3.2 Scaled Gradient Mean Subtraction

Before introducing our method, we emphasize that we do not solve Eq. (5). Instead, our goal is instead to obtain a similar spectrum flattening effect—reducing the dominance of the largest singular direction—without the heavy computation required by Eq. (5). This becomes feasible when X is nonnegative, where the leading singular direction can often be well-approximated by the uniform vector.

Our proposed method, Scaled Gradient Mean Subtraction (SGMS), leverages this property. It smooths the singular value spectrum by subtracting a scaled column mean from ΔW, requiring minimal additional computation. Although it does not fully match the effectiveness of the SVD-based approach that directly solves Eq. (5), this trade-off results in a method that is extremely lightweight while still effective in practice.

As observed in Fig. 1, the largest singular value of X dominates the gradient updates. Flattening the entire spectrum would counter this imbalance, but requires costly SVD. In practice, suppressing only the dominant direction should already be effective, which simultaneously amplifies the relative contribution of other directions and keeps the method lightweight. This efficiency-effectiveness trade-off motivates the design of SGMS.

We can rewrite Eq. (4) as

$$X = \sum_{i=1}^{m} \sigma_i \boldsymbol{u}_i \boldsymbol{v}_i^\top, \tag{6}$$

with singular values ordered as $\sigma_1 \geq \sigma_2 \geq \cdots \geq \sigma_m \geq 0$. Although X is in principle full rank ($m = k$), many singular values are close to zero, so it is effectively degenerate. Rather than flattening the entire spectrum, we attenuate only the dominant singular direction associated with σ_1. Specifically, we introduce

$$P = I - \beta \boldsymbol{v}_1 \boldsymbol{v}_1^\top, \tag{7}$$

where I is the identity and $\beta \in [0, 1]$ controls the suppression. Preconditioning ΔW with P yields

$$\tilde{\Delta W} = P \Delta W = \left(X - \beta X \boldsymbol{v}_1 \boldsymbol{v}_1^\top\right)^\top \Delta Y. \tag{8}$$

This rank-one correction selectively suppresses the dominant direction without fully flattening the spectrum, thereby increasing the relative contribution of complementary directions. Hence, it suffices to approximate the leading right singular vector $\boldsymbol{v}_1$, which could in principle be obtained by power iteration on $X^\top X$. However, performing such iterations during training would incur significant computational overhead, which is undesirable in practice.

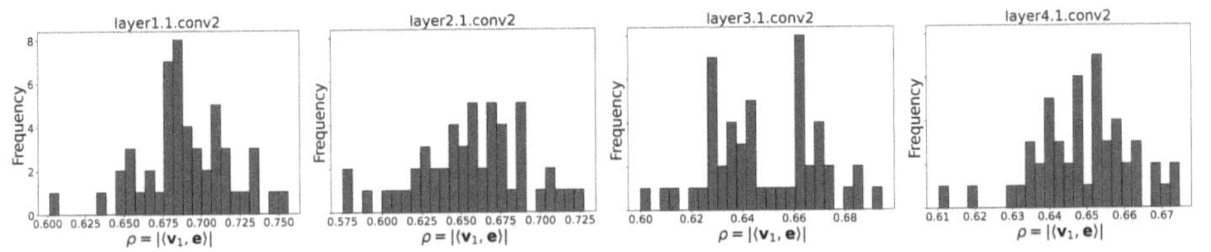

Fig. 2. Histogram of $\rho = |\langle \boldsymbol{v}_1, \boldsymbol{e} \rangle|$ across layers, using ResNet-18.

Fortunately, there is a simple and practically effective approximation for $\boldsymbol{v}_1$ that can be written as

$$\boldsymbol{e} = \frac{1}{\sqrt{k}}[1, 1, \ldots, 1]^\top \in \mathbb{R}^k. \tag{9}$$

When the elements of X are nonnegative, $X^\top X$ becomes a nonnegative symmetric matrix, and by the Perron-Frobenius theorem its leading eigenvector has nonnegative entries. Since the right singular vectors of X are precisely the eigenvectors of $X^\top X$, this leading eigenvector coincides with $\boldsymbol{v}_1$, the first right singular vector of X. (Although singular vectors are defined up to sign, we may choose the nonnegative representative without loss of generality.) Although this does not imply that $\boldsymbol{v}_1$ is exactly the uniform direction $\boldsymbol{e}$, it suggests that $\boldsymbol{v}_1$ is positively correlated with $\boldsymbol{e}$. In the ideal case where the columns of X exhibit identical activation statistics, $\boldsymbol{v}_1$ coincides with $\boldsymbol{e}$. These observations motivate using $\boldsymbol{e}$ as a lightweight approximation of the dominant direction $\boldsymbol{v}_1$.

Formally, we define

$$\rho = \left|\langle \boldsymbol{v}_1, \boldsymbol{e}\rangle\right|, \tag{10}$$

where $\langle \cdot, \cdot \rangle$ denotes the inner product, and ρ quantifies how closely $\boldsymbol{v}_1$ aligns with $\boldsymbol{e}$. Figure 2 shows the empirical distribution of ρ in ResNet-18. The distribution indicates strong alignment between $\boldsymbol{v}_1$ and $\boldsymbol{e}$. Even values around 0.6, which may appear moderate at first glance, imply strong correlation, since in high-dimensional spaces random vectors are nearly orthogonal. This observation suggests that weakening the projection onto $\boldsymbol{e}$, instead of $\boldsymbol{v}_1$, effectively suppresses the dominant direction associated with the largest singular value.

With this approximation, we can now summarize our method. The preconditioner P that originally required $\boldsymbol{v}_1$ can be replaced by a computationally lightweight form using $\boldsymbol{e}$:

$$P' = I - \beta \boldsymbol{e}\boldsymbol{e}^\top. \tag{11}$$

Accordingly, the updates become

$$\Delta\tilde{W}' = P'\Delta W = \Delta W - \beta \boldsymbol{e}\boldsymbol{e}^\top \Delta W. \tag{12}$$

Since $\boldsymbol{e}\boldsymbol{e}^\top$ simply averages the columns of a matrix, this operation amounts to subtracting the scaled column mean from each column of ΔW.

Ideally, the coefficient β would be chosen in a principled way, for example by selecting the value that maximizes the entropy of the singular spectrum of X after applying the scaled mean subtraction. However, computing such a value would introduce an overhead comparable to that of second-order preconditioners. To keep SGMS lightweight, we simply treat β as a hyperparameter.

Note that SGMS also applies to convolution and attention layers, as their core operations can be expressed as matrix multiplications, allowing the same mean-subtraction mechanism to be applied without modification.

3.3 Limitation

SGMS relies on the assumption that X is nonnegative (e.g., under ReLU activations). Empirically, SGMS also remains effective for ReLU-like activations with a positive bias, such as SiLU/Swish and GeLU, even though their activations are not strictly nonnegative. For zero-centered activations (e.g., tanh or linear), the effect of SGMS is weaker but neutral, as subtracting a near-zero mean has little impact on the updates.

4 Experiments

We evaluate SGMS across vision and language tasks. For image classification, we consider CIFAR-100 [8] as a lightweight benchmark and ImageNet-1K [1] as a large-scale benchmark, using ResNets [5] and DeiT [13] as representative architectures. For language task, we conduct experiments on a compact GPT-style model trained from scratch on a 1M-token subset of OpenWebText [2]. Unless otherwise noted, the SGMS scaling factor is set to $\beta = 0.9$.

4.1 Results on CIFAR-100

Table 1 reports CIFAR-100 results for two typical mid-sized ResNets (32 and 56). SGMS with $\beta = 0.9$ consistently improves accuracy over Vanilla, with only minor runtime overhead and no additional memory cost. SGMS with $\beta = 1$ (GC) also improves upon Vanilla, but is generally slightly worse than $\beta = 0.9$. This matches our interpretation: fully removing the mean direction can discard useful information in that direction, whereas partial suppression ($0 < \beta < 1$) preserves it while reducing its dominance.

Table 1. Top-1 accuracy (%, mean ± std over 5 runs) on CIFAR-100. "Acc. gain" indicates improvement relative to the Vanilla baseline. Time is per epoch (s, measured on an RTX 4060 GPU), Mem is peak GPU memory (GiB). Values in parentheses for Time/Mem are normalized to Vanilla (=1.00).

Model	Method	Top-1 Acc. (%)	Acc. gain	Time (s)	Mem (GiB)
ResNet-32	Vanilla	69.21 ± 0.38	–	4.39	0.35
	SGMS, $\beta = 0.9$	69.67 ± 0.26	+0.46	4.46 (×1.02)	0.35 (×1.00)
	SGMS, $\beta = 1$ (GC)	69.50 ± 0.27	+0.29	4.42 (×1.01)	0.35 (×1.00)
	AdaBK	70.57 ± 0.21	+1.36	4.76 (×1.08)	0.83 (×2.32)
ResNet-56	Vanilla	69.87 ± 0.37	–	7.25	0.55
	SGMS, $\beta = 0.9$	70.69 ± 0.42	+0.82	7.31 (×1.01)	0.55 (×1.00)
	SGMS, $\beta = 1$ (GC)	70.42 ± 0.23	+0.55	7.26 (×1.00)	0.55 (×1.00)
	AdaBK	72.41 ± 0.20	+2.54	7.90 (×1.09)	1.05 (×1.93)

AdaBK yields the largest improvements in accuracy (e.g., +2.54 points on ResNet-56), but incurs nearly 2× higher memory usage and noticeable runtime overhead. SGMS therefore provides a more favorable accuracy–efficiency trade-off as a lightweight alternative to costly preconditioners.

The observed runtime increase for SGMS (up to 2%) is larger than the theoretical estimate (less than 0.003% for ResNet-32/56). This discrepancy likely arises from implementation details: SGMS is implemented as a custom operation, whereas the rest of the training loop relies on highly optimized native kernels.

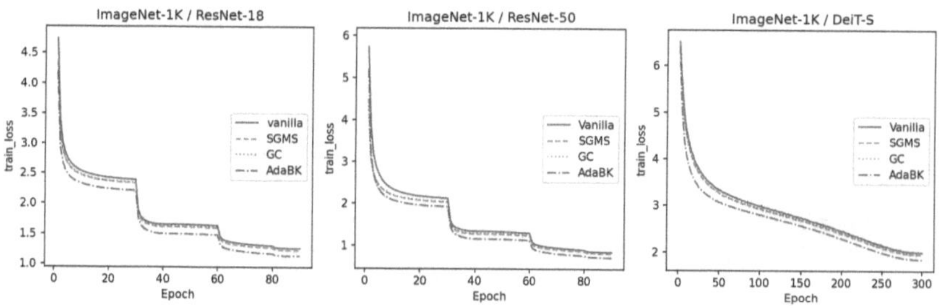

Fig. 3. Training loss vs. epoch on ImageNet (mean ± std. over 3 runs). Each panel compares Vanilla, SGMS, GC, and AdaBK, with shaded regions indicating the standard deviation across runs. *AdaBK (RB)* denotes AdaBK trained with a reduced batch size due to memory constraints.

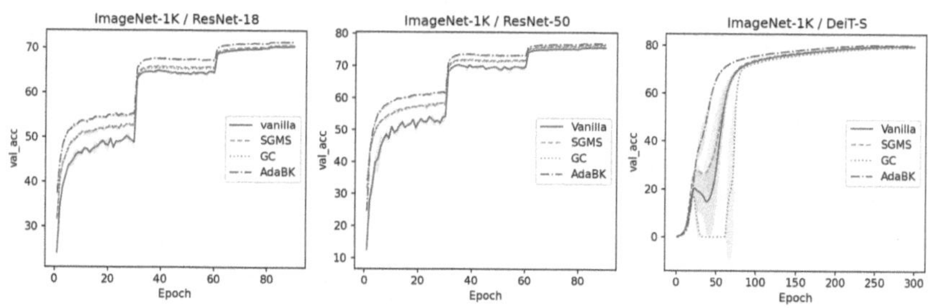

Fig. 4. Validation accuracy vs. epoch on ImageNet (mean ± std. over 3 runs). Each panel compares Vanilla, SGMS, GC, and AdaBK, with shaded regions indicating the standard deviation across runs. *AdaBK (RB)* denotes AdaBK trained with a reduced batch size due to memory constraints.

4.2 Results on ImageNet-1K

We next evaluate SGMS on ImageNet-1K dataset. For ResNet-18 and ResNet-50 [5], we follow the conventional 90-epoch training recipe. For DeiT-Small [13], we adopt the standard 300-epoch training.

Table 2 and Figs. 3 and 4 summarize the ImageNet-1K results. Across ResNet-18, ResNet-50, and DeiT-Small, SGMS consistently improves over the Vanilla baseline while incurring negligible overhead in both runtime and memory. For example, SGMS yields gains of +0.23 points on ResNet-18 and +0.83 points on ResNet-50 with essentially identical training cost. On DeiT-Small, SGMS also produces a clear improvement (+0.24 points), demonstrating that the method transfers to Transformer architectures as well. GC exhibits partially similar behavior, but SGMS consistently outperforms GC. This suggests that partial suppression ($\beta < 1$) is more effective than fully removing the mean direction. In contrast, AdaBK achieves larger gains (e.g., +1.01 on ResNet-18 and +0.37 on DeiT-Small) at substantially higher memory cost.

We further analyze the optimization dynamics using the learning curves shown in Fig. 3 and 4. Across all architectures, SGMS reduces the training loss

Table 2. Top-1 accuracy (%, mean ± std over 3 runs) on ImageNet-1K. "Acc. gain" indicates the accuracy improvement relative to the Vanilla baseline. Time is per epoch (measured on an RTX 3090 GPU), Mem is peak GPU memory. Values in parentheses for Time/Mem are normalized to Vanilla (=1.00); N.A. indicates cases where batch size differs. "RB" denotes results obtained with a reduced batch size due to memory constraints. GC has the same computational form as SGMS (per-layer mean subtraction) and therefore the same time and memory cost; we omit these redundant numbers for clarity. ResNets are trained with SGD, and DeiT-Small is trained with AdamW.

Model	Method	Acc. (%)	Acc. gain	Time (s/epoch)	Mem (GiB)
ResNet-18	Vanilla	70.35 ± 0.01	–	544.8	3.25
	SGMS	70.58 ± 0.15	+0.23	542.9 (×1.00)	3.25 (×1.00)
	GC	70.34 ± 0.08	-0.01	–	–
	AdaBK	71.36 ± 0.11	+1.01	590.9 (×1.08)	10.11 (×3.11)
ResNet-50	Vanilla	75.71 ± 0.01	–	1413.4	11.68
	SGMS	76.54 ± 0.03	+0.83	1416.3 (×1.00)	11.68 (×1.00)
	GC	76.30 ± 0.11	+0.59	–	–
	AdaBK	Out of memory			
	AdaBK (RB)	76.89 ± 0.11	+1.18	1599.8 (N.A.)	9.34 (N.A.)
DeiT-Small	Vanilla	79.58 ± 0.25	–	1160.6	8.81
	SGMS	79.82 ± 0.15	+0.24	1162.3 (×1.00)	8.81 (×1.00)
	GC	79.48 ± 0.07	-0.10	–	–
	AdaBK	79.95 ± 0.08	+0.37	1177.4 (×1.01)	15.23 (×1.73)

and increases validation accuracy more rapidly than the Vanilla baseline, while maintaining stable convergence. These results confirm that the method improves the underlying optimization process. GC also accelerates training compared to Vanilla, but consistently lags behind SGMS. AdaBK achieves faster loss reduction at the cost of a substantially higher memory footprint and training time.

4.3 Results on Language Models

We conducted experiments using nanoGPT, a minimal GPT-style model based on the implementation from [14]. Apart from adding SGMS, the only modification to the original code was to introduce an explicit split into train, validation, and test sets. We used a subset of OpenWebText containing 1M tokens, which we partitioned into training, validation, and test splits in an 8:1:1 ratio.

We adopt a compact GPT-style architecture (4 layers, 6 heads, 384-dim embeddings, context length 1024), closely following the original implementation. We use gradient accumulation with 4 steps, which corresponds to an effective batch size of roughly 250K tokens. Accordingly, we scale the learning rate to 1.2×10^{-3} (from 6×10^{-4} in the original script) following the standard $\sqrt{\text{batch\_size}}$ rule.

We vary the SGMS strength via $\beta \in \{0.0\ , 0.1, 0.3, 0.5, 0.7, 0.9, 1.0\}$. The test losses are summarized in Table 3. Although the improvements are modest, SGMS shows a consistent trend of lower loss compared to the vanilla baseline, with $\beta = 0.7$ achieving the best overall performance. The full training and validation loss trajectories are shown in Fig. 5, exhibiting the same pattern of steady but reliable improvement across the training process. These results indicate that SGMS provides a stable benefit in GPT-style language models.

Table 3. Test-set loss of nanoGPT for different SGMS strengths β.

β	0.0 (vanilla)	0.1	0.3	0.5	0.7	0.9	1.0 (GC)
Test loss	1.0423	1.0394	1.0415	1.0355	**1.0321**	<u>1.0326</u>	1.0374

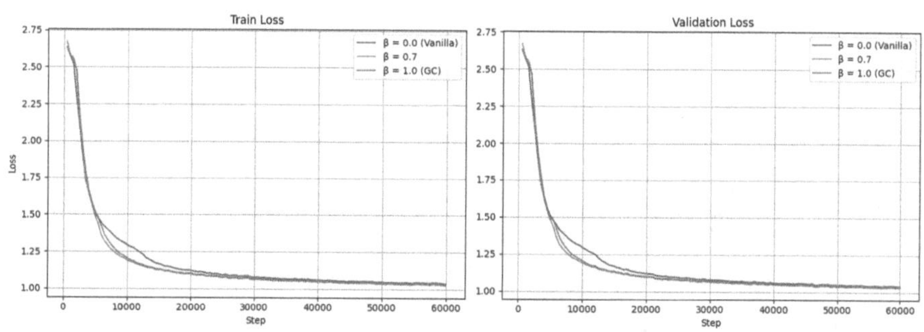

Fig. 5. Training (left) and validation (right) loss curves for nanoGPT with varying SGMS strengths β.

5 Conclusion

We introduced *Scaled Gradient Mean Subtraction (SGMS)*, a lightweight method that implicitly flattens the singular spectrum of gradient updates by partially suppressing the dominant direction through scaled mean subtraction. SGMS generalizes Gradient Centralization (GC), which corresponds to the special case $\beta = 1$, while preserving useful gradient components through partial suppression ($\beta < 1$). Experiments on CNNs and Transformers demonstrate consistent accuracy improvements with negligible computational overhead.

Acknowledgements. This work was supported by JSPS KAKENHI Grant Number 24K20796.

References

1. Deng, J., Dong, W., Socher, R., Li, L.J., Li, K., Fei-Fei, L.: Imagenet: a large-scale hierarchical image database. In: Proceedings of the IEEE Conference on Computer Vision and Pattern Recognition (CVPR), pp. 248–255 (2009)
2. Gokaslan, A., Cohen, V.: Openwebtext corpus (2019). http://Skylion007.github.io/OpenWebTextCorpus
3. Grosse, R., Martens, J.: A kronecker-factored approximate fisher matrix for convolution layers. In: Proceedings of the 33rd International Conference on International Conference on Machine Learning, ICML'16, vol. 48. p. 573–582. JMLR.org (2016)
4. Gupta, V., Koren, T., Singer, Y.: Shampoo: preconditioned stochastic tensor optimization. In: Dy, J., Krause, A. (eds.) Proceedings of the 35th International Conference on Machine Learning. Proceedings of Machine Learning Research, vol. 80, pp. 1842–1850. PMLR (2018). https://proceedings.mlr.press/v80/gupta18a.html
5. He, K., Zhang, X., Ren, S., Sun, J.: Deep residual learning for image recognition. In: Proceedings of the IEEE Conference on Computer Vision and Pattern Recognition (CVPR) (2016)
6. Hu, E.J., et al.: LoRA: low-rank adaptation of large language models. In: International Conference on Learning Representations (2022). https://openreview.net/forum?id=nZeVKeeFYf9
7. Kingma, D.P., Ba, J.: Adam: a method for stochastic optimization. In: Bengio, Y., LeCun, Y. (eds.) 3rd International Conference on Learning Representations, ICLR 2015, San Diego, CA, USA, 7–9 May 2015, Conference Track Proceedings (2015). http://arxiv.org/abs/1412.6980
8. Krizhevsky, A.: Learning multiple layers of features from tiny images, University of Toronto. cIFAR-10/100 dataset Technical Report (2009)
9. LeCun, Y., Kanter, I., Solla, S.A.: Eigenvalues of covariance matrices: application to neural-network learning. Phys. Rev. Lett. **66**, 2396–2399 (1991)
10. Robbins, H., Monro, S.: A stochastic approximation method. Ann. Math. Stat. **22**(3), 400–407 (1951). http://www.jstor.org/stable/2236626
11. Schraudolph, N.N.: Centering neural network gradient factors. In: Montavon, G., Orr, G.B., Müller, K.-R. (eds.) Neural Networks: Tricks of the Trade. LNCS, vol. 7700, pp. 205–223. Springer, Heidelberg (2012). https://doi.org/10.1007/978-3-642-35289-8_14
12. Sutskever, I., Martens, J., Dahl, G., Hinton, G.: On the importance of initialization and momentum in deep learning. In: Dasgupta, S., McAllester, D. (eds.) Proceedings of the 30th International Conference on Machine Learning. Proceedings of Machine Learning Research, vol. 28, pp. 1139–1147. PMLR, Atlanta (2013). https://proceedings.mlr.press/v28/sutskever13.html
13. Touvron, H., Cord, M., Douze, M., Massa, F., Sablayrolles, A., Jégou, H.: Training data-efficient image transformers & distillation through attention. In: Proceedings of the 38th International Conference on Machine Learning (ICML). Proceedings of Machine Learning Research, vol. 139, pp. 10347–10357. PMLR (2021)
14. whatdhack: Pre-training and generation of small-scale llms (2024). https://github.com/whatdhack/mini_llms/tree/main. Accessed 01 Dec 2025
15. Yong, H., Huang, J., Hua, X., Zhang, L.: Gradient centralization: a new optimization technique for deep neural networks. In: Vedaldi, A., Bischof, H., Brox, T., Frahm, J.-M. (eds.) ECCV 2020. LNCS, vol. 12346, pp. 635–652. Springer, Cham (2020). https://doi.org/10.1007/978-3-030-58452-8_37

16. Yong, H., Sun, Y., Zhang, L.: A general regret bound of preconditioned gradient method for dnn training. In: Proceedings of the IEEE/CVF Conference on Computer Vision and Pattern Recognition (CVPR), pp. 7866–7875 (2023)
17. Zhao, J., Zhang, Z., Chen, B., Wang, Z., Anandkumar, A., Tian, Y.: Galore: memory-efficient llm training by gradient low-rank projection. In: Proceedings of the 41st International Conference on Machine Learning, ICML'24, JMLR.org (2024)

Variance-Covariance Regularization for Unsupervised Domain Adaptation

Junchi Liu[1,2], Xiang Zhang[1,2](✉), and Yonggang Che[1,2]

[1] College of Computing, National University of Defense Technology, Changsha 410073, HuNan, China
{liujunchi12,zhangxiang08,ygche}@nudt.edu.cn

[2] Laboratory of Digitizing Software for Frontier Equipment, National University of Defense Technology, Changsha 410073, China

Abstract. In unsupervised domain adaptation (UDA), there exists an inherent conflict between transferability and discriminative ability, due to domain shift. Nevertheless, most UDA methods are biased towards transferability rather than discriminative ability. To this end, we introduce the newly identified phenomenon of neural collapse (NC) to analyze the model separation capacity. As a result, we find that the unlabeled target domain definitely disrupts the NC structure of the classifier trained on source domain. To alleviate this issue, this paper proposes a class-aware variance and covariance regularization for feature discriminant. In detail, the class-aware variance regularization prevents all features from collapsing into a single point, thereby enhancing feature discriminability, while the covariance regularization eliminates correlations among feature dimensions to promote feature diversity. Experiments on four benchmark datasets demonstrate the effectiveness of the proposed regularization.

Keywords: Variance Regularization · Covariance Regularization · Neural Collapse

1 Introduction

Unsupervised domain adaptation (UDA) [1] is well-known as making the model transferable from source domain to target domain. Despite its potential, UDA itself faces the distribution discrepancy challenge, namely, domain shift between the source and target domains. To address this issue, substantial research efforts have been devoted to improving the transferability of models across domains. However, recent studies [2] find that transferability often conflicts with the goal of improving discriminability. In UDA, enhancing feature discriminability while preserving transferability is crucial to avoid negative transfer and improve overall performance. The key to improving discriminability [3] lies in reducing intra-class feature distances and increasing inter-class feature distances. SHOT [4] uses pseudo-labels predicted by the model on the target domain to supervise training. CDAN [5] incorporates class-aware discriminative features into adversarial

L. Pasa et al. (Eds.): ICANN 2026, LNCS 17089, pp. 484–493, 2027.
https://doi.org/10.1007/978-3-032-38398-3_40

learning for feature discrimination. ProCA [6] integrates inter-class information into class-wise prototypes, simultaneously modeling both intra-class and inter-class distributions. TVT [7] employs a discriminative clustering module based on mutual information maximization to enhance feature diversity and separation.

Unlike such arts above, we will use the newly discovered phenomenon of neural collapse (NC) [8] to achieve this goal. In particular, this NC phenomenon comprises the following components: in the final layer of training, all feature representations within the same class collapse to their class mean (NC1); these class means converge to the vertices of a simplex equiangular tight frame (ETF) (NC2); the classifier's weight aligns with the class mean and likewise converges to the same simplex ETF structure (NC3/NC4). A few pioneering works also integrate this NC1-4 insight into different techniques, such as knowledge distillation [9], contrastive continual learning [10], imbalanced classification [11], out-of-distribution (OOD) detection [12], etcetera. Nonetheless, existing research has not addressed the question: *Can desirable UDA models benefit from the geometric structure of NC?*

In this paper, we use the Fisher's ratio as the metric to measure the discriminability of UDA models. Figure 1 provides the effect of neural collapse-inspired loss and general loss on feature discriminability and classification accuracy of individual domains, respectively. In the source domain, supervised training enables the features to converge to an elegant NC structure. However, in the target domain, the absence of label supervision will destroy such a NC structure. Therefore, this paper focuses on exploring neural collapse-based solutions to enhance model adaptation performance on the target domain in unsupervised domain adaptation scenarios.

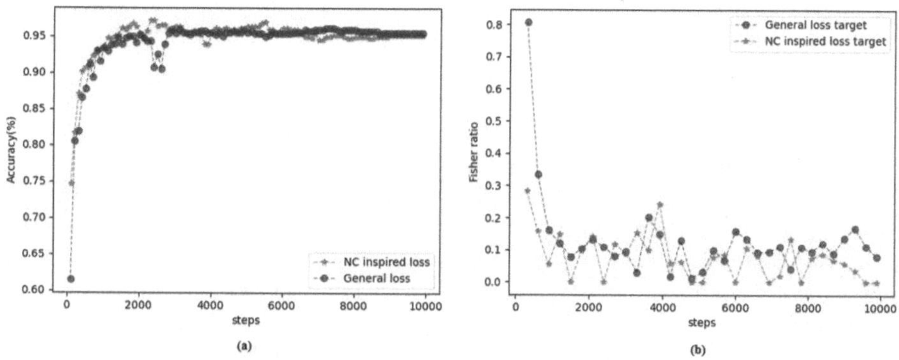

Fig. 1. Comparisons on Fisher ratio and accuracy of the general loss and the NC inspired loss on target domain.

Inspired by the above analysis, we adopt two regularization strategies to balance transferability and discriminability. On the one hand, class-aware variance regularization is applied to reduce within-class variance, thereby preventing within-class collapse (NC1) and enhancing feature discriminability; on the other hand, covariance regularization is used to minimize off-diagonal elements in the covariance matrix, eliminating redundancy among feature dimensions and promoting orthogonality between class means (NC2). Figure 1(a) shows that the neural collapse-inspired loss with the regularization strategy (red curve) improves the classification accuracy on target domain, as compared to the general loss (blue curve). In Fig. 1(b), the proposed loss (red curve) achieves more stable Fisher ratio than the general loss (blue curve). This implies that the neural collapse-inspired loss, combined with the regularization strategy, not only enhances model performance but also effectively guides the target domain features to maintain the neural collapse structure.

In summary, the contributions of this paper are summarized as follows.

- We propose a novel end-to-end domain adaptation model, namely VCRUDA, which incorporates class-aware variance regularization, it effectively prevents neural collapse in target domain features and enhances their discriminability.
- To further improve the performance of the model, the covariance regularization is introduced to constrain correlations among feature dimensions in the target domain, thereby improving feature representation and preventing neural collapse caused by feature redundancy.
- VCRUDA achieves well-behaved performance by a large margin compared to the state-of-the-art on Office-31, ImageCLEF-DA, and Office-Home datasets.

2 The Proposed Method

2.1 The Overall Architecture

VCRUDA comprises four main components: a Transformer Encoder for extracting image features, a class-aware regularization module and a covariance regularization module addressing the NC1 and NC2 issues in the target domain, respectively, and a classifier module for feature classification. The overall framework is illustrated in Fig. 2. The key contribution of this work lies in the first attempt to integrate NC into UDA. Moreover, we adopt class-aware regularization and covariance regularization to improve the neural collapse (NC) structure in the target domain, thereby enhancing both the discriminability and diversity of the target domain features.

As Fig. 2 shows, images from both source and target domains are split into patches, which are then rearranged and fed into Transformer Encoder. The "CLS-Token" feature is used for classification. The cross-entropy ($\mathcal{L}_{cls}$) is adopted to source branch and the unsupervised discriminative clustering ($-\mathcal{I}$) [13,14] is adopted to target branch, while minimizing adversarial loss ($\mathcal{L}_{adv}$) is to reduce discrepancy between cross-domain distributions. Under this adversarial domain adaptation framework, the target branch incorporates class-aware variance regularization ($\mathcal{L}_{var}$) and covariance regularization ($\mathcal{L}_{cov}$) to optimize the NC structure in the target domain, thereby achieving more effective cross-domain feature alignment. In summary, the overall objective function of VCRUDA is formulated as follows:

$$\mathcal{L}_{cls} + \alpha \cdot \mathcal{L}_{adv} - \beta \cdot \mathcal{I} + \gamma \cdot \mathcal{L}_{var} + \delta \cdot \mathcal{L}_{cov}. \tag{1}$$

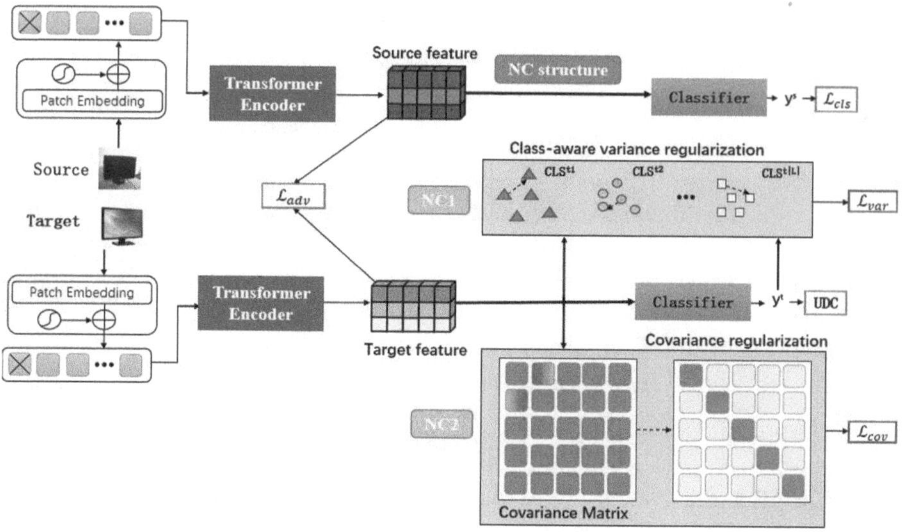

Fig. 2. The overall structure of VCRUDA.

where α, β, γ and δ are four hyper-parameters.

2.2 Regularization

To avoid the neural collapse, the class-aware variance regularization [15] is used to address NC1. The CLS-Token features output by the Transformer encoder in the target domain form a batch denoted as $C = [c_1, ..., c_n]$. This regularization is activated only when two or more samples in the batch share the same pseudo-label. It enforces a standard deviation of θ along each feature dimension, ensuring that embeddings of same-class samples are appropriately dispersed within the batch to prevent intra-class collapse into a single mean point. The class-aware variance regularization can be defined as

$$\mathcal{L}_{var} = \frac{1}{K} \sum_{i=1}^{K} \max(0, \theta - std(c_i | \hat{y}_i = k)) \tag{2}$$

where the vector $\hat{y}_i$ is the pseudo-label, the value of θ is fixed at 1.

To address NC2, covariance regularization [16] is introduced. Specifically, the covariance matrix of the target domain's CLS-Token feature after mean removal can be expressed as:

$$Cov(C) = \frac{1}{n-1} \sum_{i=1}^{n} (c_i - \frac{1}{n} \sum_{i=1}^{n} c_i)(c_i - \frac{1}{n} \sum_{i=1}^{n} c_i)^T \tag{3}$$

By minimizing the sum of squared off-diagonal elements in the covariance matrix, this term penalizes inter-dimensional covariances, thereby decorrelating feature dimensions and promoting feature orthogonality. The covariance regularization can be formally expressed as:

$$\mathcal{L}_{cov} = \frac{1}{d} \sum_{i \neq j} [Cov(C)_{ij}]^2 \tag{4}$$

where the d is the dimensionality of the feature vector CLS, i and j represent the row and column indices of the covariance matrix, respectively.

3 Experiments

3.1 Datasets and Implementation

To verify the effectiveness of VCRUDA, we evaluate the model on four object recognition benchmarks: Office-31 [17], Office-Home [18], ImageCLEF-DA and VisDA-2017 [19] datasets. VIT-B/16 model pre-trained on ImageNet 21k is used as backbone network to extract image features. The Transformer Encoder of VIT-B/16 consists of 12 Transformer Encoder layers. The input image size in experiment is 256 × 256, and the size of each patch is 16 × 16. We train the model using minibatch Stochastic Gradient Descent (SGD) optimizer with the momentum of 0.9, and initialize the learning rate as 0. We linearly increase its learning rate to 0.05 after 500 training steps, and then decrease it by the cosine decay strategy. For datasets Office-31, ImageCLEF-DA and Office-Home, the training step is set to 10,000. For large-scale datasets VisDA-2017, the training step is set to 20,000. Experiments are conducted on a single card 3090Ti with 24G memory. The hyperparameters α is set to 0.1, β is set to 0.01, γ is set to 0.1, δ is set to 0.005.

3.2 Results of Object Recognition

We evaluate object recognition performance of the compared methods on Office-31, ImageCLEF-DA, Office-Home, and VisDA-2017 datasets. The results are shown in Tables 1- 4. VCRUDA achieves comparable results to FFTAT on Office-31 and demonstrates significant improvements over previous best-performing UDA methods on both ImageCLEF-DA and Office-Home (95.0% vs 90.7%, 92.6% vs 91.4%). But VCRUDA is slightly inferior to FFTAT on VisDA-2017. This is because our loss is aware of class imbalance. We leave the topic on bypassing the class imbalance in NC inspired loss to the future work. Besides, as in Table 5, VCRUDA surpasses the Baseline method (96.03% vs 95.82%). This is also evidenced by the t-SNE visualization of learned features as shown in Fig. 3. Figure 3(a) shows that the source and target domains are not aligned very well where some points are hard to classify. In contrast, Fig. 3(b) shows that the source and target domains are totally aligned very well. Experiments show that the proposed class-aware variance and covariance regularization is effective in alleviating the problem of unsupervised domain adaptation (Tables 2 and 3).

Table 1. Performance comparison on Office-31.

Method	A→W	D→W	W→D	A→D	D→A	W→A	Avg
ResNet-50	68.4	96.7	99.3	68.9	62.5	60.7	76.1
CDAN+E [5]	94.1	98.6	**100.0**	92.9	71.0	69.3	87.7
SHOT [4]	90.1	98.4	99.9	94.0	74.7	74.3	88.6
ALSDA	95.2	99.2	**100.0**	95.8	78.1	77.5	91.0
ViT [20]	91.2	99.2	**100.0**	93.6	80.7	80.7	91.1
TVT [7]	96.4	**99.4**	**100.0**	96.4	84.9	86.1	93.9
CDTrans [21]	96.7	99.0	**100.0**	97.0	81.1	81.9	92.6
PMTrans [22]	**99.5**	**99.4**	**100.0**	**99.8**	86.7	86.5	95.3
FFTAT [23]	97.6	99.2	**100.0**	97.8	90.2	**91.0**	**96.0**
VCRUDA (Ours)	97.7	99.1	**100.0**	97.2	**91.4**	90.8	**96.0**

Table 2. Performance comparison on ImageCLEF-DA.

Method	I→P	P→I	I→C	C→I	C→P	P→C	Avg
ResNet-50	74.8	83.9	91.5	78.0	65.5	91.2	80.7
CDAN+E	77.7	90.7	97.7	91.3	74.2	94.3	87.7
DSAN	80.2	93.3	97.2	93.8	80.8	95.9	90.2
DALN [24]	80.5	93.8	97.5	92.8	78.3	95.0	89.7
MCC+NWD [24]	79.8	94.5	98.0	94.2	80.0	97.5	90.7
VCRUDA (Ours)	**86.2**	**98.8**	**99.0**	**99.3**	**86.7**	**99.7**	**95.0**

3.3 Parameter Sensitivity Analysis

In our model, the hyperparameter α controls the $\mathcal{L}_{avd}$. the hyperparameter β controls the $\mathcal{I}$, the hyperparameter γ controls the $\mathcal{L}_{var}$, the hyperparameter δ controls the $\mathcal{L}_{cov}$. This article fixes $\alpha = 0.1$ and $\beta = 0.01$. To better understand the effects of γ and δ, we report the sensitivities of γ in Fig. 4(a), and the sensitivities of δ in Fig. 4(b). The values in Fig. 4(a) represent the average results of the A→D and D→W tasks from the Office-31 dataset. The values in Fig. 4(b) represent the average results of the A→D and A→W tasks from the Office-31 dataset. It can be seen that our VCRUDA achieves the best results when $\gamma = 0.1$ and $\delta = 0.005$. Therefore, $\gamma = 0.1$ and $\delta = 0.005$ are used throughout experiments. Luckily, we find that the accuracy drops sharply when γ exceeds 1.0. From a geometric perspective, an excessively large γ forces the source domain features of each class to collapse towards their class centers thus inducing intra-class compactness, but this blurs the inter-class boundaries that should be discriminative across classes. So the small γ is favorable.

Table 3. Performance comparison on Office-Home.

Method	Ar→Cl	Ar→Pr	Ar→Rw	Cl→Ar	Cl→Pr	Cl→Rw	Pr→Ar	Pr→Cl	Pr→Rw	Rw→Ar	Rw→Cl	Rw→Pr	Avg
ResNet-50	44.9	66.3	74.3	51.8	61.9	63.6	52.4	39.1	71.2	63.8	45.9	77.2	59.4
SHOT	57.1	78.1	81.5	68.0	78.2	78.1	67.4	54.9	82.2	73.3	58.8	84.3	71.8
CDAN+E	54.6	74.1	78.1	63.0	72.2	74.1	61.6	52.3	79.1	72.3	57.3	82.8	68.5
ALSDA	60.4	78.9	81.9	67.8	77.4	77.5	68.6	57.9	83.0	80.1	62.8	85.0	73.4
ViT	54.7	83.0	87.2	77.3	83.4	85.6	74.4	50.9	87.2	79.6	54.8	88.8	75.5
TVT	74.9	86.8	89.5	82.8	88.0	88.3	79.8	71.9	90.1	85.5	74.6	90.6	83.6
CDTrans	68.8	85.0	86.9	81.5	87.1	87.3	79.6	63.3	88.2	82.0	66.0	90.6	80.5
PMTrans	81.3	**92.9**	92.8	88.4	93.4	93.2	87.9	80.4	93.0	89.0	80.9	84.8	89.0
FFTAT	83.2	**92.9**	95.2	91.1	93.5	95.2	89.7	85.0	94.9	93.0	87.5	95.9	91.4
VCRUDA (Ours)	**88.7**	92.6	**96.4**	**92.0**	**95.1**	**95.8**	**89.9**	**86.3**	**95.1**	**93.5**	**88.8**	**97.3**	**92.6**

Table 4. Performance comparison on VisDA-2017.

Method	plane	bcycl	bus	car	horse	knife	mcycl	person	plant	sktbrd	train	truck	Avg
ResNet-50	55.1	53.3	61.9	59.1	80.6	17.9	79.7	31.2	81.0	26.5	73.5	8.5	52.4
CDAN+E	85.2	66.9	83.0	50.8	84.2	74.9	88.1	74.5	83.4	76.0	81.9	38.0	73.9
SHOT	94.3	88.5	80.1	57.3	93.1	93.1	80.7	80.3	91.5	89.1	86.3	58.2	82.9
ALSDA	93.8	72.8	81.0	49.0	82.9	90.5	89.3	80.8	88.5	86.6	87.3	43.9	78.9
ViT	97.7	48.1	86.6	61.6	78.1	63.4	94.7	10.3	87.7	47.7	94.4	35.5	67.1
TVT	92.9	85.6	77.5	60.5	93.6	98.2	89.4	76.4	93.6	92.0	91.7	55.7	83.9
CDTrans	97.1	90.5	82.4	77.5	96.6	96.1	93.6	88.6	97.9	86.9	90.3	62.8	88.4
PMTrans	99.4	88.3	88.1	78.9	98.8	98.3	95.8	70.3	94.6	98.3	96.3	48.5	88.0
FFTAT	99.7	**98.5**	93.1	**81.1**	**99.8**	**99.5**	**97.8**	**89.6**	95.7	**99.8**	98.7	**72.4**	**93.8**
VCRUDA (Ours)	**100.0**	**98.5**	**93.3**	80.4	97.9	90.4	94.5	85.1	**98.1**	99.0	**98.9**	61.5	91.4

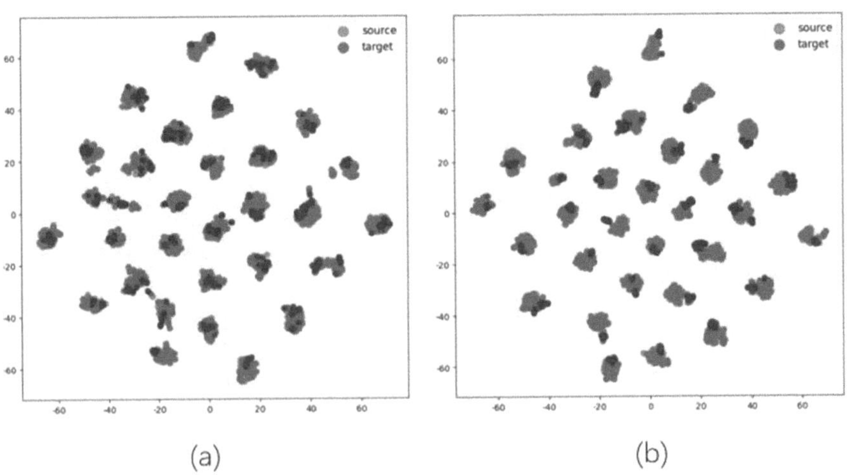

Fig. 3. Feature visualization of (a) the baseline and (b) VCRUDA using t-SNE on the task A→D of Office-31, where red and blue points indicate the source and the target domain, respectively. (Color figure online)

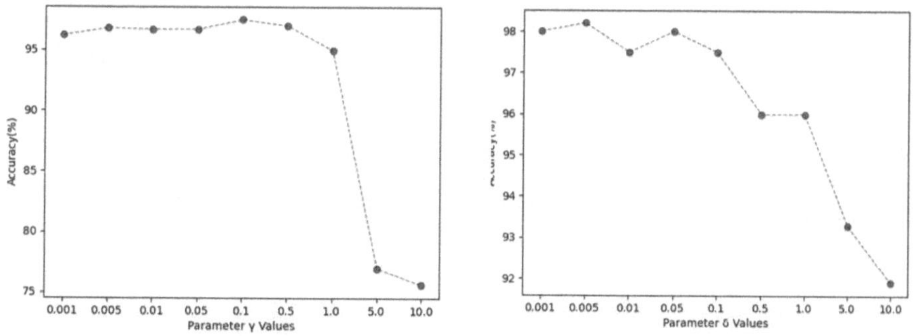

(a) (b)

Fig. 4. The sensitivity of (a) γ and (b) δ on Office-31.

3.4 Ablation Study

To illustrate the impact of class-aware variance and covariance regularization on the performance of VCRUDA, we conduct an ablation study. As in Table 5, without the class-aware variance regularization, the classification accuracy decreases by 0.08%, indicating that this regularization could enhance feature discriminability. Besides, without the covariance regularization, the classification accuracy is reduced by 0.3%, which indicates that the covariance regularization could enhance feature diversity. Besides, the pure baseline model is reduced by 0.21%. This hints the effect of such two components on model performance.

Table 5. Ablation Study on Office-31.

Method	A→W	D→W	W→D	A→D	D→A	W→A	Avg
Baseline	97.5	**99.4**	**100.0**	96.0	91.2	90.8	95.82
w/o $\mathcal{L}_{var}$	**97.7**	98.9	**100.0**	97.0	91.2	**90.9**	95.95
w/o $\mathcal{L}_{cov}$	97.4	**99.4**	**100.0**	95.6	91.3	90.7	95.73
VCRUDA	**97.7**	99.1	**100.0**	**97.2**	**91.4**	90.8	**96.03**

4 Conclusion

This work tackles neural collapse in unsupervised domain adaptation via the introduction of class-aware variance and covariance regularization. We propose a novel network architecture termed VCRUDA, an end-to-end framework based on the Transformer encoder. The class-aware variance and covariance regularization components can be flexibly integrated into existing unsupervised domain adaptation frameworks, enhancing both the discriminability and diversity of target domain features. The experimental results on widely used benchmarks demonstrate the effectiveness of the VCRUDA model in unsupervised domain adaptation.

Acknowledgments. This work was supported by National Key Research and Development Program of China (2025YFB3003605) and National Natural Science Foundation of China (U2530207).

References

1. Wang, M., Deng, W.: Deep visual domain adaptation: a survey. Neurocomputing **312**, 135–153 (2018)
2. Zhou, Y., et al.: Learning transferable and discriminative representations for 2D image-based 3D model retrieval. IEEE Trans. Circuits Syst. Video Technol. **32**(10), 7147–7159 (2022)
3. Cai, H., et al.: Discriminative object tracking by domain contrast. Comput. Vis. Image Underst. **239**, 103891 (2024)
4. Liang, J., Hu, D., Feng, J.: Do we really need to access the source data? source hypothesis transfer for unsupervised domain adaptation. In: Proceedings of the 37th International Conference on Machine Learning, pp. 6028–6039 (2020)
5. Long, M., Cao, Z., Wang, J., Jordan, M.I.: Conditional adversarial domain adaptation. In: Proceedings of the 32nd International Conference on Neural Information Processing Systems, pp. 1647–1657 (2018)
6. Jiang, Z., Li, Y., Yang, C., Gao, P., Wang, Y., Tai, Y., Wang, C.: Prototypical contrast adaptation for domain adaptive semantic segmentation. In: Proceedings of the European Conference on Computer Vision, pp. 36-54 (2022)
7. Yang, J., Liu, J., Xu, N., Huang, J.: TVT: Transferable vision transformer for unsupervised domain adaptation. In: Proceedings of the IEEE Winter Conference on Applications of Computer Vision, pp. 520–530 (2023)
8. Papyan, V., Han, X.Y., Donoho, D.L.: Prevalence of neural collapse during the terminal phase of deep learning training. In: Proceedings of the National Academy of Sciences, vol. 117, no. 40, pp. 24652-24663 (2020)
9. Zhang, S., Song, Z., He, K.: Neural collapse inspired knowledge distillation. In: Proceedings of the AAAI Conference on Artificial Intelligence, vol. 39, no. 21, pp. 22542–22550 (2025)
10. Montmaur, A., Larue, N., Vu, N.-S.: Neural collapse inspired contrastive continual learning. In: Proceedings of the 35th British Machine Vision Conference (2024).
11. Zhang, E., Li, C., Geng, C., Chen, S.: All-around neural collapse for imbalanced classification. IEEE Trans. Knowl. Data Eng. **37**(8), 4460–4470 (2025)
12. Ben Ammar, M., Belkhir, N., Popescu, S., Manzanera, A., Franchi, G.: NECO: Neural collapse based out-of-distribution detection. In: Proceedings of the 12th International Conference on Learning Representations, pp. 9539–9568 (2024)
13. Wang, S., et al.: Efficient multiple kernel k-means clustering with late fusion. IEEE Access **7**, 61109–61120 (2019)
14. Liu, X., et al.: Late fusion incomplete multi-view clustering. IEEE Trans. Pattern Anal. Mach. Intell. **41**(10), 2410–2423 (2018)
15. Bardes, A., Ponce, J., LeCun, Y.: VICReg: Variance-invariance-covariance regularization for self-supervised learning, In: Proceedings of the International Conference on Learning Representations, pp. 8941–8963 (2022)

16. Zbontar, J., Li J., Misra, I., LeCun, Y., Deny, S.: Barlow twins: self-supervised learning via redundancy reduction. In: Proceedings of the 38th International Conference on Machine Learning, vol. 139, pp. 12310–12320 (2021)
17. Saenko, K., Kulis, B., Fritz, M., Darrell, T.: Adapting visual category models to new domains. In: Proceedings of the European Conference on Computer Vision, pp. 213–226 (2010)
18. Venkateswara, H., Eusebio, J., Chakraborty, S., Panchanathan, S.: Deep hashing network for unsupervised domain adaptation. In: Proceedings of the IEEE Conference on Computer Vision and Pattern Recognition, pp. 5018–5027 (2017)
19. Peng, X., Usman, B., Kaushik, N., Hoffman, J., Wang, D., Saenko, K.: Visda: The visual domain adaptation challenge. arXiv:1710.06924 (2017)
20. Dosovitskiy, A., et al.: An image is worth 16 × 16 words: transformers for image recognition at scale. In: Proceedings of the International Conference on Learning Representations, pp. 611–631 (2021)
21. Xu, T., Chen, W., Wang, P., Wang, F., Li, H., Jin, R.: CDTrans: cross-domain transformer for unsupervised domain adaptation, In: Proceedings of the International Conference on Learning Representations, pp. 832-845 (2022)
22. Zhu, J., Bai, H., Wang, L.: Patch-mix transformer for unsupervised domain adaptation: a game perspective. In: Proceedings of the IEEE Conference on Computer Vision and Pattern Recognition, pp. 3561–3571 (2023)
23. Yu, X., Huang, Z.: Zhang: Z. Feature fusion transferability aware transformer for unsupervised domain adaptation. In: Proceedings of the IEEE Winter Conference on Applications of Computer Vision, pp. 6752–6761 (2025)
24. Chen, L., et al.: Reusing the task-specific classifier as a discriminator: Discriminator-free adversarial domain adaptation. In: Proceedings of the IEEE Conference on Computer Vision and Pattern Recognition, pp. 7181–7190 (2022)

Alignment with Ground Truth (GTA): An Optimization-Free Hyperparameter-Free Metric for Evaluating Learned Representations

Ahmad Mustapha[1(✉)] and Ali Chehab[2]

[1] Chehab Lab, American University of Beirut, Beirut, Lebanon
amm90@mail.aub.edu

[2] Department of Electrical and Computer Engineering, American University of Beirut, Beirut, Lebanon
chehab@aub.edu.lb

Abstract. Evaluating learned representations remains a central challenge in modern machine learning. Linear probing is widely adopted as a practical proxy for representation quality; however, it requires supervised optimization, hyperparameter tuning, and classifier training, introducing variability that is external to the representation itself.

We propose Ground-Truth Alignment (GTA), a training-free and hyperparameter-free metric for evaluating learned representations. GTA measures the degree of alignment between embeddings and ground-truth labels by integrating normalized conditional label entropy across multiple clustering scales of the embedding space. By tracking how label uncertainty evolves as cluster granularity increases, GTA captures the intrinsic semantic organization of the representations without requiring classifier training or hyperparameter tuning.

Across 13 pretrained encoders and 9 datasets, GTA demonstrates strong and consistent correlation with linear probe test accuracy, and it also aligns with retrieval performance. Sensitivity analyses further indicate that the metric remains stable with respect to sampling strategy and evaluation sample size.

These findings suggest that multiscale entropy integration provides a principled and optimization-independent signal of representation quality. As such, GTA offers a complementary diagnostic to probe-based evaluation and contributes toward more reproducible and robust assessment of learned representations.

Keywords: Embedding · Evaluation · Vision · Image Encoders · Image Representation · Representation Learning

1 Introduction

Transfer learning has been the dominant approach in AI applications since the early days of modern deep learning. AI practitioners typically adapted pretrained

L. Pasa et al. (Eds.): ICANN 2026, LNCS 17089, pp. 494–505, 2027.
https://doi.org/10.1007/978-3-032-38398-3_41

models to their own datasets and tasks. For a long time, the primary large-scale pretraining dataset was ImageNet. Researchers would train their models, whether Vision Transformers (ViTs), Convolutional Neural Networks (CNNs), or other architectures, on ImageNet in a supervised fashion and then release the resulting weights, which practitioners would in turn reuse.

Over time, the focus of researchers shifted from supervised methods to self-supervised algorithms. In these approaches, models are pretrained on large amounts of unlabeled data, extracting a training signal directly from the data itself. To compare the performance of different self-supervised methods, researchers commonly relied on linear probing on selected downstream tasks as a proxy for evaluating the quality of learned representations. Linear probing involves placing a single linear layer on top of the frozen representations and training only this added layer while keeping the underlying embedding fixed.

However, linear probing is known to be sensitive to hyperparameter choices, such as learning rate, normalization, and training configuration. Consequently, reported performance can vary substantially across experimental settings, potentially leading to inconsistent or misleading assessments of representation quality. To address this issue, we introduce the Ground-Truth Alignment (GTA) metric, a stable and consistent evaluation measure that requires neither hyperparameter tuning nor additional training and directly quantifies how well an embedding aligns with the ground-truth structure of a given dataset.

Our contributions can be summarized as follows:

- We introduce **Ground-Truth Alignment (GTA)**[1], a training free and hyperparameter free metric for evaluating learned representations based on multiscale conditional entropy.
- We formalize GTA as the area under a normalized label-entropy curve computed across clustering scales, yielding a measure of representationâĂŞlabel alignment.
- We demonstrate that GTA strongly correlates with downstream linear probe accuracy across 13 pretrained encoders and 9 datasets. And it correlates with retrieval performance on the GPR1200 dataset.
- We provide a systematic sensitivity analysis showing that GTA is stable with respect to sampling strategy, and sample size.

The remainder of the paper is organized as follows. Section 2 reviews related work. Section 3 introduces the proposed metric. Section 4 analyzes the effect of the metric parameters and studies their influence on the metric values. Section 5 presents theoretical and empirical validation of the metric. Section 6 examines the relationship between the proposed metric and linear probing. Section 7 examines the relationship between the proposed metric and retrieval performance. Section 8 discusses using one clustering scale. Section 9 concludes the paper.

[1] Code on Github: Chehab-Lab/alignment-with-ground-truth-gta-metric.

2 Related Work

Linear probing, as a formal evaluation methodology, was introduced by [1], although linear classifiers had previously been employed to assess learned representations. Its first application in the context of self-supervised learning—without an explicit designation was reported in [9]. Subsequently, this protocol has become a standard evaluation technique, widely adopted across numerous self-supervised learning frameworks [2–5,10,13]. Despite its widespread use, several studies have identified inherent limitations of linear probing, including sensitivity to hyperparameter choices and its restriction to linearly separable information [6,7]. Nevertheless, it remains a prevalent evaluation tool in contemporary research [11,12].

In this work, we propose a novel metric for evaluating learned embeddings: the evaluation is performed with respect to a specific dataset, reflecting the impracticality of optimizing a representation for all potential downstream tasks, particularly classification tasks that require high-level semantic features. For the past decade, the predominant evaluation approaches have been linear probing and k-nearest neighbors (kNN), both of which exhibit substantial sensitivity to hyperparameter configurations. [8] provides a comprehensive comparative analysis of these metrics across 26 pre-trained models.

3 Alignment with Ground Truth (GTA) Metric

We propose the *Alignment With Ground Truth (GTA)* metric for evaluating the semantic alignment of embedding spaces with respect to known dataset labels. GTA measures how consistently ground-truth semantics are preserved by unsupervised clustering of the embedding space across multiple levels of granularity.

Given a set of embeddings, we perform hierarchical clustering for a range of cluster counts $k \in \mathcal{K}$, where the number of clusters is normalized with respect to the sample size. Specifically, we define clustering scales as proportions of the number of samples, ensuring that GTA is comparable across datasets and evaluation settings. For each clustering scale, we compute the conditional entropy of the labels given the resulting cluster assignments. This yields an alignment curve that captures the amount of label information retained at different clustering resolutions. For a fixed k, the alignment score is based on the normalized conditional entropy and defined as:

$$A(Y; k) = 1 - \frac{H(Y \mid C_k)}{H(Y)}, \tag{1}$$

where $H(Y)$ denotes the entropy of the ground-truth label distribution and $H(Y \mid C_k)$ is the expected conditional entropy of labels within clusters induced by clustering with k clusters. By construction, $A(Y; k) \in [0, 1]$, with larger values indicates a stronger semantic alignment.

The GTA metric is defined as the area under the alignment curve across clustering scales:

Table 1. Summary of pretrained image encoders used in our experiments

Model	Output	Params	Architecture	Training	Notes
ResNet-50	2048	25.6M	CNN	Supervised	Standard residual backbone
EfficienNet-B0	1280	5.3M	CNN	Supervised	Parameter-efficient design
ConvNeXt-Base	1024	89M	CNN	Supervised	Modernized CNN architecture
ViT-Base/16 (IN21K)	768	86.4M	Transformer	Supervised	Global self-attention
DeiT-Base (Distilled)	768	86M	Transformer	Supervised	Data-efficient training
Swin-Base	768	87.8M	Transformer	Supervised	Hierarchical windowed attention
MoCo v3 ViT-B	768	86.6M	Transformer	Self-supervised	Momentum contrast learning
BYOL ResNet-50	2048	25.6M	CNN	Self-supervised	Bootstrap latent learning
MAE ViT-Base	768	112M	Transformer	Self-supervised	High-capacity SSL model
DINOv2-Base	768	86.6M	Transformer	Self-supervised	Strong semantic representations
SimCLR ResNet-50	2048	24M	CNN	Self-supervised	Instance discrimination
CLIP ViT-B/32	512	88M	Transformer	Vision-language contrastive	Multimodal foundation model
EVA02-Base	768	85.8M	Transformer	Self-supervised	Large-scale masked pretraining

Table 2. Summary of the datasets used in our experiments

Dataset	Domain	#Classes	Resolution	Source	Class Balance	Notes
FGVC Aircraft	Natural images	100	Variable	FGVC	Balanced	Fine-grained aircraft recognition
Oxford Flowers-102	Natural images	102	Variable	Oxford	Imbalanced	Fine-grained flower species
CUB-200-2011	Natural images	200	Variable	Caltech-UCSD	Balanced	Fine-grained bird dataset
Food-101	Natural images	101	Variable	ETH Zurich	Balanced	Food category recognition
Oxford-IIIT Pets	Natural images	37	Variable	Oxford-IIIT	Balanced	Pet breed classification
Stanford Dogs	Natural images	120	Variable	Stanford	Balanced	Fine-grained dog breed recognition
Stanford Cars	Natural images	196	Variable	Stanford	Balanced	Fine-grained car model recognition
RetinaMNIST	Medical imaging	5	224 × 224	MedMNIST	Balanced	Retinal OCT images
DTD	Natural images	47	Variable	Oxford	Balanced	Texture recognition

$$\mathrm{GTA} = \int_{k \in \mathcal{K}} A(Y; k)\, \mathrm{d}k, \tag{2}$$

which is approximated in practice via numerical integration over the discrete set $\mathcal{K}$. Normalizing the clustering scales by the sample size ensures that GTA $<$ 1 and that the metric reflects alignment across relative, rather than absolute, resolutions.

This formulation naturally yields a family of metrics, denoted as GTA-N, where N specifies the number of samples used to define the clustering scale (e.g., GTA-500, GTA-1000, GTA-2000). As the number of clusters increases, the normalized conditional entropy $H(Y \mid C_k)$ is non-decreasing, and higher values of GTA indicate stronger and more consistent alignment between the embedding geometry and the ground-truth semantics.

4 Parameter Sensitivity Analysis

Although GTA involves a small number of design parameters, they do not require performance-driven tuning. Instead, each admits a principled selection strategy grounded in statistical or geometric considerations. The primary factors are (1) the use of stratified sampling, (2) the evaluation sample size, and (3) the choice of clustering scales. We evaluate the robustness of GTA to these factors across three datasets and three pretrained models. The details of the evaluated models and datasets are provided in Tables 1 and 2.

4.1 Effect of Stratified Sampling

To assess the effect of stratified sampling, we conduct experiments using three pretrained models (ResNet-50, ViT, MoCo-v3) across three datasets (FGVC-Aircraft, RetinaMNIST, DTD), with sample size fixed at $N = 1{,}000$. We evaluate alignment values over clustering scales $k \in \{0.001, 0.01, 0.05, 0.1, \ldots, 0.9, 1\}$.

Each configuration is repeated ten times, with five runs employing stratified sampling and five runs using non-stratified sampling. Figure 1 summarizes the results. Overall, we observe no statistically significant difference in alignment values between stratified and non-stratified sampling across models, datasets, and clustering scales. This might be because the sample size is large enough for a random sample to be representative.

Nevertheless, for certain clustering scales, non-stratified runs exhibit a tendency toward higher alignment values. This behavior can be attributed to uneven class representation in the sampled subset, where under-represented or missing classes lead to artificially inflated alignment estimates. We note that the three datasets considered differ substantially in their number of classes; detailed statistics are provided in Table 2.

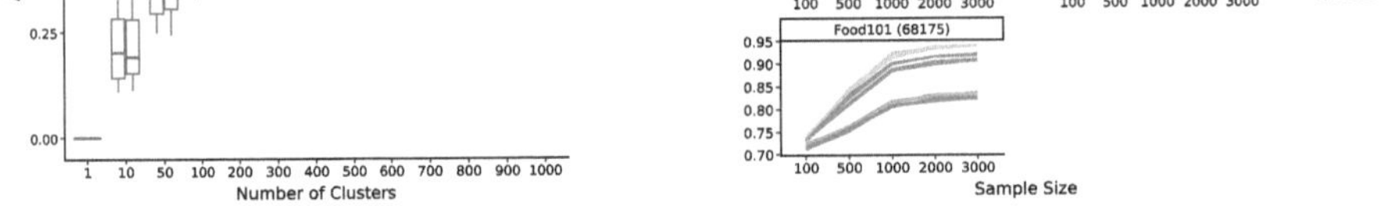

Fig. 1. Left: Alignment values across three datasets and three models while varying sample stratification. Right: GTA values against sample size across models and datasets. Dataset size is mentioned in titles.

4.2 Effect of Sample Size

To assess the effect of sample size, we use five pretrained models (ResNet-50, ViT, MoCo-v3, DINOv2, CLIP) across five datasets (FGVC-Aircraft, DTD, Dogs,

CUB-200-2011, Food101) with stratified sampling. We vary N (100, 500, 1000, 2000, 3000), evaluating over clustering scales $k \in \{0.001, 0.01, 0.05, 0.1, \ldots, 1\}$ for $N \geq 1{,}000$ and $k \in \{0.1, 0.2, \ldots, 1\}$ for $N < 1{,}000$. Each model–dataset pair is repeated five times. Figure 1 shows the resulting GTA values. As noted, GTA is a family of metrics parameterized by N, yielding variants such as GTA-100, GTA-500, and so forth.

The experiment reveals three key observations. First, smaller sample sizes yield lower GTA values, as sparse data produces higher clustering entropy and thus weaker alignment. As N grows, denser data leads to stronger and more stable alignment estimates. Second, larger N produces greater spread across models, enabling finer-grained discrimination between representations. Third, GTA values saturate beyond a certain threshold: even for large datasets, a relatively small subset suffices for accurate estimation. Based on our experiments, we recommend N to be large enough proportional to the number of classes in a given dataset. To have a representative sample, datasets with 1000 classes require a sample much larger than one with 30 classes.

5 Empirical and Theoretical Validation

5.1 Empirical Validation

To evaluate whether GTA truly captures representation alignment with ground truth labels, we conducted a controlled label-noise experiment. Specifically, we progressively injected random noise into the ground-truth labels and monitored how GTA responds. Since the original label distribution reflects meaningful semantic structure, adding random noise effectively degrades the inherent structure that the representation is expected to capture.

For this experiment, we computed GTA on three datasets: CUB-2011, Stanford Dogs, and Food-101, using three backbone models: ViT, ResNet-50, and CLIP. We fixed the sample size to 1,000 instances, computing GTA-1000 in each case. The label noise rate was gradually increased from 0.2 to 0.8.

Figure 2 presents the results. Across all datasets and models, GTA consistently decreases as the noise level increases. This behavior confirms that GTA is sensitive to the semantic alignment between representations and ground-truth

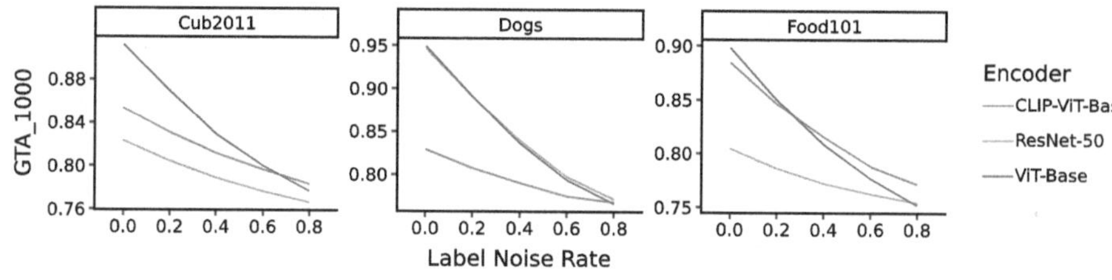

Fig. 2. GTA value degrades as label noise increases, validating that GTA captures representation alignment with ground truth.

labels. As label corruption increases and semantic structure is progressively destroyed, GTA appropriately reflects this degradation by assigning lower scores.

5.2 Theoretical Validation

Theorem 1. *Let Y be a ground-truth label variable over a finite set $\mathcal{Y}$, and let C_k denote the cluster-assignment variable induced by a k-partition of the embedding space. If C_{k_2} is a refinement of C_{k_1} (i.e., $C_{k_1} = \phi(C_{k_2})$ for some deterministic ϕ, as in hierarchical clustering), then:*

$$H(Y \mid C_{k_2}) \leq H(Y \mid C_{k_1}), \tag{3}$$

$$A(Y; k_2) \geq A(Y; k_1). \tag{4}$$

That is, finer clusterings yield lower label entropy and higher alignment.

Proof. Since $C_{k_1} = \phi(C_{k_2})$, the variables form a Markov chain $Y \to C_{k_2} \to C_{k_1}$. By the chain rule:

$$I(Y; C_{k_2}) = I(Y; C_{k_1}) + \underbrace{I(Y; C_{k_2} \mid C_{k_1})}_{\geq 0}, \tag{5}$$

so $I(Y; C_{k_2}) \geq I(Y; C_{k_1})$. Using $I(Y; C) = H(Y) - H(Y \mid C)$ and the fact that $H(Y)$ is constant across clusterings gives $H(Y \mid C_{k_2}) \leq H(Y \mid C_{k_1})$. Finally, substituting $H(Y \mid C_k) = H(Y) - I(Y; C_k)$ into the alignment score definition yields:

$$A(Y; k) = 1 - \frac{H(Y \mid C_k)}{H(Y)} = \frac{I(Y; C_k)}{H(Y)}, \tag{6}$$

and since $H(Y) > 0$ is constant, monotonicity of $I(Y; C_k)$ directly implies $A(Y; k_2) \geq A(Y; k_1)$.

6 Relation to Linear Probing

To investigate the relationship between linear probing and GTA, we train a single-layer classifier on the frozen representations of all models (Table 1) and datasets (Table 2), with layer size matching the representation dimensionality and number of classes. We use a batch size of 64, learning rate of 10^{-3}, and train for 20 epochs, reporting the maximum test accuracy. For GTA, we use GTA-1000 with clustering scales $k \in \{0.001, 0.01, 0.05, 0.1, \ldots, 0.9, 1\}$.

6.1 Global Correlation

The results in Fig. 3 demonstrate a strong positive relationship between GTA and downstream performance, with a Spearman correlation of 0.82 across diverse datasets. This indicates that GTA reliably captures representation quality, as higher GTA values consistently correspond to higher linear probing accuracy. Importantly, GTA not only correlates well with linear probing, but also provides a more faithful and robust assessment in scenarios where linear probing can be misleading.

This behavior is particularly evident for Retinamnist, a medical imaging dataset that lies outside the natural image distribution used to train the encoders. While linear probing yields moderate accuracy on Retinamnist, its GTA score is noticeably lower relative to natural image datasets, correctly reflecting the representation mismatch and reduced semantic alignment. This suggests that GTA is sensitive to distributional shifts and better captures the true transferability of representations, whereas linear probing may overestimate performance due to classifier adaptation.

The converse case also exists: some representations perform poorly under linear probing while still achieving a median GTA value. This can be explained by representation geometry that is not well-suited for linear decision boundaries. In such cases, the intrinsic structure of the embedding space may still preserve meaningful hierarchical organization, even if it is not linearly separable.

Overall, the figure demonstrates that GTA provides a more intrinsic, hyperparameter-free measure of representation quality, while still maintaining strong agreement with established evaluation protocols.

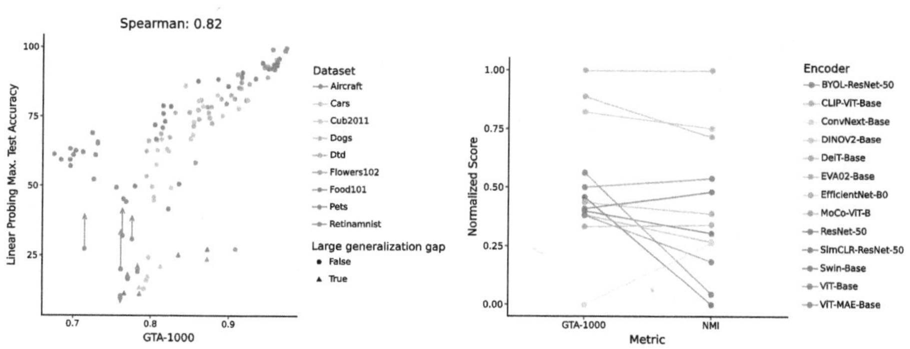

Fig. 3. Left: Correlation between GTA-1000 and linear probing maximum test accuracy across datasets and encoders. The arrows show linear probing performance gain after tuning hyperparameters. Right: Encoder ranks on the Aircraft dataset across metrics.

6.2 Per Dataset Correlation

Linear probing is widely used to assess the task-specific utility of learned representations and is commonly regarded as a proxy for representation quality. In this section, we examine the correlation between GTA scores and linear probing test accuracy, for each dataset. This analysis allows us to evaluate whether higher GTA values are systematically associated with improved downstream performance. If such a relationship is observed, it would indicate that GTA may serve as a training-free and hyperparameter-free alternative for representation evaluation.

Table 3 presents the correlation between linear probing accuracy and the GTA-1000 metric. The results indicate that when the downstream task shares a similar image distribution with the data used during pretraining, GTA correlates strongly with linear probing performance. In contrast, when the distributions differ substantially, this correlation weakens. As shown in Table III, strong correlations are observed across most datasets, with the notable exceptions of DTD and RetinaMNIST, both of which exhibit distribution shifts relative to the pretraining data. These findings suggest that GTA can reliably predict the best-performing encoder when distributional alignment exists between pretraining and downstream data, but its predictive power diminishes under pronounced distribution shifts.

This raises the question of whether this is a limitation of GTA or a caveat of linear probing. We argue that GTA measures the intrinsic alignment between embeddings and task semantics, whereas linear probing extends beyond this by leveraging the adaptability of the classifier.

6.3 Linear Probing Sensitivity

Although it is well established in the literature that linear probing is sensitive to hyperparameter tuning [7], we conducted a small experiment to further examine this effect. Specifically, we considered the Aircraft dataset, where several models had performed worse compared to others.

To investigate whether this behavior could be attributed to suboptimal hyperparameters, we retrained the linear probing classifiers on top of these pretrained models while modifying only the learning rate, increasing it from $1e^{-3}$ to $1e^{-2}$. The evaluated models include ResNet-50, ConvNeXt, ViT-Base, ViT-MAE, SimCLR, BYOL, and MoCo.

The results show that the accuracy of four out of the six evaluated models improved substantially, as indicated by the arrows in Fig. 3.

Furthermore, with the exception of the RetinaMNIST experiments, most cases that deviate from the overall trend, i.e., those exhibiting weak alignment between GTA and linear probing also display a large cumulative generalization gap between the training and testing curves during linear probing.

These observations further highlight the difference between the two evaluation methods: GTA exhibits consistent behavior across settings, whereas linear probing appears considerably more sensitive to hyperparameter choices.

Table 3. Spearman correlation of GTA and Linear Probing per dataset

Dataset	Aircraft	Cars	Cub2011	Dogs	Dtd	Flowers102	Food101	Pets	Retinamnist
Spearman	0.34	**0.95**	**0.96**	**0.93**	0.72	**0.90**	**0.92**	**0.85**	0.39

7 Relation to Retrieval

To investigate GTA suitability more we compared it to retrieval performance on the GPR1200 dataset, we computed GTA-4000 to match the scale of the dataset (1200 class) and used the same clustering scale as in the previous experiment. Retrieval performance was evaluated using mAP@9. The spearman correlation between both metrics was 0.81.

8 Discussion

GTA integrates the alignment score across multiple clustering scales. A natural question arises as to why the alignment is not evaluated at a single clustering scale, for instance by fixing k to the number of ground-truth classes in the dataset.

To investigate this, we conduct an experiment in which we compute the normalized mutual information (NMI) between k-means clustering assignments and ground-truth labels, with k set equal to the number of classes for each dataset. We then compare this metric to GTA. Our results indicate that, on a per-dataset basis, NMI at $k = |\mathcal{Y}|$ exhibits a stronger correlation with linear probing performance than GTA (see Table 4). However, when aggregated across datasets, this correlation decreases (spearman 0.65).

Table 4. Spearman correlation of NMI at k equals number of classes and Linear Probing per dataset

Dataset	Aircraft	Cars	Cub2011	Dogs	Dtd	Flowers102	Food101	Pets	Retinamnist
Spearman	0.85	**0.86**	**0.97**	**0.91**	0.60	**0.93**	**0.95**	**0.75**	0.52

This observation implies that NMI at k equals to number of classes can serve as a reasonable proxy for linear probing within individual datasets. Nevertheless, GTA and NMI produce notably different rankings of encoders, particularly on datasets that deviate from the pretraining distribution. For example, on the Aircraft dataset, GTA tends to favor self-supervised encoders such as SimCLR, MAE, and BYOL over supervised models like ConvNeXt. Figure 3 presents the ranks of the encoders on the Aircraft dataset across both GTA and NMI metrics.

We hypothesize that this behavior stems from structural differences in the learned representations. Self-supervised methods often yield more fragmented or locally separable representations due to their training objectives, whereas supervised methods typically produce more globally integrated feature spaces. As the number of clusters increases well beyond the number of semantic classes, the alignment between clustering assignments and labels for self-supervised representation tends to improve..

In this context, GTA's ability to aggregate alignment across both coarse and fine clustering scales provides a more comprehensive characterization of representation quality, capturing structure that may be overlooked when evaluating at a single scale.

9 Conclusion

This paper introduced Ground-Truth Alignment (GTA), a training-free and hyperparameter-free metric for evaluating learned representations. GTA measures representation quality by integrating normalized conditional label entropy across clustering scales, thereby capturing the extent to which the geometric structure of an embedding space aligns with ground-truth labels. Unlike linear probing, GTA does not require supervised optimization or hyperparameter selection, and instead directly quantifies intrinsic representation–label alignment.

Across 13 pretrained encoders and 9 datasets spanning medical imaging, fine-grained recognition, and general vision, we demonstrated that GTA exhibits strong and consistent correlation with downstream linear probe accuracy and retrieval performance. Sensitivity analysis further confirmed that GTA remains stable with respect to sampling strategy and evaluation sample size, indicating that multiscale entropy integration provides a reliable and principled signal of representation quality.

GTA is designed for classification-oriented evaluation where ground-truth labels are categorical, making it directly applicable to the predominant paradigm of representation learning benchmarking. While regression, segmentation, and non-visual modalities involve fundamentally different label structures, the breadth of evaluated domains already demonstrates meaningful cross-domain generality. Extending GTA to such settings remains a natural direction for future work, alongside exploring its correlation with fine-tuned model performance.

By formalizing representation–label alignment as a scale-integrated entropy measure, this work contributes toward more principled, reproducible, tuning-free, and optimization-free evaluation protocols for learned representations.

Acknowledgments. Special thanks to the Maroun Semaan Faculty of Engineering and Architecture (MSFEA) at the American University of Beirut (AUB) for providing the essential infrastructure needed to conduct this study. A.M owned the research, developed the code base, and performed the experiments. A.C proposed experiments, provided funding and supervision, reviewed the manuscript and gave feedback. GenAI was used to proofread the manuscript.

References

1. Alain, G., Bengio, Y.: Understanding intermediate layers using linear classifier probes (2018). https://arxiv.org/abs/1610.01644
2. Caron, M., Misra, I., Mairal, J., Goyal, P., Bojanowski, P., Joulin, A.: Unsupervised learning of visual features by contrasting cluster assignments (2021). https://arxiv.org/abs/2006.09882

3. Chen, T., Kornblith, S., Norouzi, M., Hinton, G.: A simple framework for contrastive learning of visual representations. In: Proceedings of the 37th International Conference on Machine Learning. ICML'20, JMLR.org (2020)
4. Grill, J.B., et al.: Bootstrap your own latent a new approach to self-supervised learning. In: Proceedings of the 34th International Conference on Neural Information Processing Systems. NIPS '20, Curran Associates Inc, Red Hook, NY, USA (2020)
5. He, K., Fan, H., Wu, Y., Xie, S., Girshick, R.: Momentum contrast for unsupervised visual representation learning. In: 2020 IEEE/CVF Conference on Computer Vision and Pattern Recognition (CVPR), pp. 9726–9735 (2020). https://doi.org/10.1109/CVPR42600.2020.00975
6. Kalibhat, N., Narang, K., Firooz, H., Sanjabi, M., Feizi, S.: Measuring self-supervised representation quality for downstream classification using discriminative features (2023). https://arxiv.org/abs/2203.01881
7. Lee, J.H., Yoon, D., Ji, B., Kim, K., Hwang, S.: Rethinking evaluation protocols of visual representations learned via self-supervised learning (2023). https://arxiv.org/abs/2304.03456
8. Marks, M., et al.: A closer look at benchmarking self-supervised pre-training with image classification. Int. J. Comput. Vision **133**(8), 5013–5025 (2025). https://doi.org/10.1007/s11263-025-02402-w
9. Noroozi, M., Favaro, P.: Unsupervised Learning of Visual Representations by Solving Jigsaw Puzzles. In: Leibe, B., Matas, J., Sebe, N., Welling, M. (eds.) ECCV 2016. LNCS, vol. 9910, pp. 69–84. Springer, Cham (2016). https://doi.org/10.1007/978-3-319-46466-4_5
10. Tian, Y., Sun, C., Poole, B., Krishnan, D., Schmid, C., Isola, P.: What makes for good views for contrastive learning? In: Proceedings of the 34th International Conference on Neural Information Processing Systems. NIPS '20, Curran Associates Inc, Red Hook, NY, USA (2020)
11. Wang, Q., Bruns, B., Scharr, H., Krajsek, K.: Self-supervised learning based on transformed image reconstruction for equivariance-coherent feature representation (2025). https://arxiv.org/abs/2503.18753
12. Wen, L., Wang, X., Liu, J., Xu, Z.: Mveb: self-supervised learning with multi-view entropy bottleneck. IEEE Trans. Pattern Anal. Mach. Intell. **46**(9), 6097–6108 (2024). https://doi.org/10.1109/TPAMI.2024.3380065
13. Zbontar, J., Jing, L., Misra, I., LeCun, Y., Deny, S.: Barlow twins: Self-supervised learning via redundancy reduction (2021). https://arxiv.org/abs/2103.03230

Instance Reweighting for Adversarial Training Helps Only Against the Training Attack

Giovanni Scodeller[1], Dario Lazzato[1,2], Emanuele Ledda[3], Antonio Emanuele Cinà[1], Battista Biggio[3], Giorgio Fumera[3], Davide Anguita[1], Fabio Roli[1,3], and Luca Oneto[1](✉)

[1] University of Genoa, Genoa, Italy
luca.oneto@unige.it
[2] Sapienza University of Rome, Rome, Italy
[3] University of Cagliari, Cagliari, Italy

Abstract. Adversarial training is one of the most effective defenses against adversarial examples, but it is computationally expensive due to the repeated generation of adversarial samples. In practice, efficient yet suboptimal attacks (i.e., PGD) are used during training, motivating approaches that aim to improve AT without increasing its cost. Among these, instance reweighting has gained significant attention. Through a series of experiments across multiple architectures, datasets, and reweighting strategies, we show that: (i) existing instance reweighting methods are often compared under unfair or statistically unsound evaluation protocols, and (ii) their apparent robustness gains typically hold only against the specific attack used during training, failing to generalize to stronger evaluations (i.e., APGD and AutoAttack). These findings indicate that instance reweighting may provide a misleading sense of robustness.

Keywords: Adversarial Training · Instance Reweighting · False Sense of Security

1 Introduction

Adversarial Examples (AEs) are inputs that have been subtly and deliberately perturbed with the goal of inducing Machine Learning (ML) models to produce incorrect predictions [2,29]. To counter this threat, numerous defense strategies have been proposed, including input transformation and purification [27], model obfuscation or masking [21], model regularization [13], model verification [23], and adversarial example detection [30].

Among these approaches, Adversarial Training (AT) has emerged as one of the most effective and extensively studied methods for mitigating vulnerability to AE, consistently demonstrating strong empirical performance across diverse threat models and architectures [22]. This procedure is commonly formulated as

L. Pasa et al. (Eds.): ICANN 2026, LNCS 17089, pp. 506–518, 2027.
https://doi.org/10.1007/978-3-032-38398-3_42

a min-max optimization problem, in which an inner maximization step generates AE, while an outer minimization step updates the model parameters to classify these challenging inputs correctly.

However, the min-max nature of AT makes it computationally demanding, as it involves both model training and the generation of AEs [3,9,18,39]. Regarding the training process, given a fixed model architecture, state-of-the-art theoretical, software, and hardware solutions are already employed, since these strategies for training DNNs are already highly optimized for computational efficiency [4]. In contrast, the generation of AE spans a broad spectrum of methods that vary significantly in both computational cost and attack effectiveness [8]. These methods range from simple, fast, but comparatively less effective methods (e.g., `FGSM` [15]) to more sophisticated, computationally intensive, and highly effective techniques (e.g., `AutoAttack` [11] which is an ensemble of known attacks) [7,8]. For this reason, practical implementations of AT typically rely on an attack procedure that balances effectiveness and computational tractability, with Projected Gradient Descent (`PGD`) [22] often serving as a widely adopted compromise.

The effectiveness of AT can be further enhanced through several complementary strategies. One approach involves designing or employing network architectures that are inherently more robust to adversarial perturbations [25]. Another line of work focuses on enriching the training distribution by generating or leveraging synthetic data, enabling models to better capture adversarially relevant features [32]. Although these techniques can yield substantial improvements, they often introduce considerable additional computational overhead [12], arising from increased model complexity, enlarged training datasets, or more intricate training pipelines. Conversely, several strategies aim to reduce the computational cost of AT, such as selecting only a subset of AE during training [19], decreasing the number of adversarial example generations [37], or, as mentioned earlier, employing lighter attacks [33]. However, these approaches generally result in a marked reduction in AT effectiveness.

For this reason, researchers have proposed several methods to improve the effectiveness of AT without altering its computational requirements, e.g., by changing the AT loss [35] or via Instance Reweighting (IR) [20,28,36,38]. In particular, IR has recently attracted considerable interest as it simply assigns different importance weights to AEs during training, typically giving higher weight during training to more effective (e.g., those with higher misprediction confidence) AEs [28]. A key advantage of IR is that it does not modify the computational cost of AT, as it reweights the same AEs without introducing additional optimization steps or requiring the generation of extra AEs. However, recent work suggests that the gains provided by reweighting may not consistently generalize to stronger evaluations involving attacks not used during training [14].

In this work, we provide a comprehensive and systematic assessment of IR strategies within the context of AT. Our analysis highlights two central findings:

- We show that widely adopted evaluation practices often overlook essential statistical considerations - most notably the use of an appropriate validation

set for model selection - thereby leading to unreliable or misleading comparisons among reweighting methods;
- We demonstrate that many IR strategies fail to generalize beyond the specific attacks used to generate AE during AT: while they may improve robustness against the particular attack employed during training (e.g., `PGD`), they often fail under stronger, more comprehensive adversarial evaluations (e.g., `AutoAttack`).

Taken together, these observations indicate that IR tends to overfit to the training-time attack used to generate AE, providing only narrow and fragile robustness, and thus a false sense of robustness that does not extend to more challenging adversarial scenarios.

2 Preliminaries

Given an input space $\mathcal{X}$ and a finite output space $\mathcal{Y} = \{1, \dots, k\}$, we consider a parametric model family $\mathcal{F}_\Theta$ induced by a parameter space Θ. A common way to represent a multi-class predictor is through a vector-valued scoring function, formally defined as $\boldsymbol{f}_\theta : \mathcal{X} \to \hat{\mathcal{Y}} = \mathbb{R}^k$ where $\boldsymbol{f}_\theta(x)_y$ denotes the score assigned to class $y \in \mathcal{Y}$ (e.g., logits or unnormalized likelihoods). A classifier [4,26] is then obtained by coupling $\boldsymbol{f}_\theta$ with a decision rule, typically the argmax over class scores $\hat{y} = f_\theta(x) = \arg\max_{y \in \mathcal{Y}} \boldsymbol{f}_\theta(x)_y \in \hat{\mathcal{Y}} = \mathcal{Y}$. Moreover, when probability estimates are required, the scores can be converted into a probability distribution over classes via the softmax function [4] $\mathbb{f}_\theta(x) = \text{softmax}(\boldsymbol{f}_\theta(x)) \in \hat{\mathcal{Y}} = [0,1]^k$.

The quality of a classifier can be assessed via a loss function $\ell : \hat{\mathcal{Y}} \times \mathcal{Y} \to \mathbb{R}$, which quantifies the discrepancy between a predicted label $\hat{y}$ and the true label y. In the context of classification, a commonly considered loss is the misclassification error, also referred to as the hard loss function, denoted by $\ell_H(\hat{y}, y)$.

A classifier is typically obtained by a learning algorithm $\mathscr{A}_\mathcal{H} : (\mathcal{X} \times \mathcal{Y})^n \to \mathcal{F}$, characterized by a set of hyperparameters $\mathcal{H}$. Given a dataset of labeled examples $\mathcal{D} = \{(x_i, y_i) \mid i \in \{1, \dots, n\}\}$, the algorithm searches for optimal parameters $\theta \in \Theta$ according to a prescribed learning criterion. The learning algorithm $\mathscr{A}_\mathcal{H}$ implicitly and/or explicitly induces both the functional form of the classifier f_θ and its parameterization θ. Most learning algorithms can be formulated as the solution of the following optimization problem:

$$\min_\theta \sum_{i=1}^n \ell_D(\boldsymbol{f}_\theta(x_i), y_i), \tag{1}$$

that is, the goal is to find the parameters θ that minimize the empirical risk of f_θ over the dataset, using a differentiable surrogate loss ℓ_D that approximates the hard loss ℓ_H.

3 Adversarial Attacks

In this work, we consider the classical indiscriminate white-box test-time bounded-perturbation evasion attack [3,8], hereafter referred to as the simple adversarial attack. In this setting, the attacker has full knowledge of the classifier f_θ and is allowed to perturb the input x within a prescribed attack budget $\delta \in \Delta$, resulting in the generation of an AE $\tilde{x} = x+\delta$ such that $y = f_\theta(x) \neq f_\theta(\tilde{x})$. That is, the attacker aims to induce a misclassification by perturbing the sample x so that the model's prediction differs from the label originally assigned to x, regardless of which incorrect class is predicted.

In order to achieve this goal, the attacker has to solve the following optimization problem [6]:

$$\max_{\delta\in\Delta} \ \ell_H \left(f_\theta(x+\delta), y\right). \tag{2}$$

However, solving this problem poses several challenges. First, defining the admissible perturbations set Δ for the attacker is non-trivial [5] and Δ can be, in general, non-convex. Second, even if Δ is a convex set, Problem (2) is non-convex and non-differentiable due to the ℓ_H and the functional form of DNN-based classifiers.

To address these challenges, several relaxations are commonly adopted in practice. Regarding the first issue, the perturbation set Δ is usually approximated by a norm-bounded set of the form $\Delta = \left\{\delta \in \mathbb{R}^d : \|\delta\|_p \leq \varepsilon\right\}$, that is, the perturbation is constrained within an ℓ_p norm ball of ε-radius. Under this formulation, p is typically selected from $\{1, 2, \infty\}$ [15,20], leading to a convex set, although more sophisticated perturbation sets can also be employed [5]. For the second issue, the hard loss ℓ_H is commonly replaced by some differentiable surrogate loss ℓ_D. Based on these relaxations, Problem (2) can be reformulated as the following tractable optimization problem:

$$\delta^* : \ \arg\max_{\|\delta\|_p\leq\varepsilon} \ \ell_D \left(\boldsymbol{f}_\theta(x+\delta), y\right), \tag{3}$$

where $\tilde{x}^* = x + \delta^*$ is the corresponding AE. In order to solve Problem (3), one of the most widely adopted white-box approaches is the Projected Gradient Descent (`PGD`) attack [11,22], which iteratively updates the perturbation by performing a gradient ascent step on the surrogate loss followed by projection onto the feasible perturbation set. Several variants of `PGD` have been proposed to mitigate its sensitivity to hyperparameter choices, such as the step size and the initialization of the perturbation; a notable example is Auto-PGD (`APGD`) [11], a parameter-free extension that automatically adapts the step size during optimization. Beyond PGD-based approaches, alternative attacks have been introduced, including the Fast Adaptive Boundary (`FAB`) attack [10], which explicitly approximates the classifier's decision boundary to compute minimal adversarial perturbations, and the `Square Attack` [1], a score-based black-box method relying on randomized search rather than gradient information. Finally, methods that combine multiple complementary attacks have also been proposed, most notably `AutoAttack` [11], an ensemble-based evaluation protocol that aggregates the strongest results from `APGD`, `FAB`, and `Square Attack` to provide a reliable and parameter-free benchmark for adversarial robustness evaluation.

4 Adversarial Training

AT extends the standard training by explicitly accounting for AEs. Specifically, AT aims to learn model parameters that minimize the classification error over samples that can be adversarially manipulated within a prescribed perturbation set. Formally, AT is defined as:

$$\min_\theta \sum_{i=1}^n \max_{\delta\in\Delta} \ell_H \left(f_\theta(x_i+\delta), y_i\right). \tag{4}$$

Problem (4) is typically solved via alternating minimization. The outer minimization with respect to θ is handled using the methods developed for Problem (1) using a particular ℓ_D^θ, while the inner maximization over the perturbations δ is addressed using the approaches introduced for Problem (3) with another ℓ_D^δ. In formal terms, Problem (4) takes the following form:

$$\min_\theta \sum_{i=1}^n \ell_D^\theta \left(\boldsymbol{f}_\theta(x_i+\delta_i^*), y_i\right), \quad \delta_i^* \in \arg\max_{\delta\in\Delta} \ell_D^\delta \left(\boldsymbol{f}_\theta(x_i+\delta), y_i\right). \tag{5}$$

In our setting, Vanilla AT (`VAT`) corresponds to Eq. (5), where both ℓ_D^θ and ℓ_D^δ are defined as the standard cross-entropy (CE) loss.

`TRADES` [35], for example, employs a composite loss for the optimization with respect to θ, defined as:

$$\ell_D^\theta \left(\boldsymbol{f}_\theta(x+\delta), y\right) = \mathrm{CE}\left(\boldsymbol{f}_\theta(x+\delta), y\right) + \lambda \mathrm{KL}(\mathbb{f}_\theta(x) | \mathbb{f}_\theta(x+\delta)),$$

which combines the CE loss with a regularization term given by the Kullback-Leibler divergence between the model predictions on clean inputs and their adversarial counterparts balanced by λ. For the inner optimization with respect to the δ, `TRADES` uses $\ell_D^\delta \left(\boldsymbol{f}_\theta(x+\delta), y\right) = \mathrm{KL}(\mathbb{f}_\theta(x) | \mathbb{f}_\theta(x+\delta))$, and typically solves the resulting maximization problem using `PGD`.

`MART` [31] can be viewed as an extension of `TRADES`, where:

$$\ell_D^\theta \left(\boldsymbol{f}_\theta(x+\delta), y\right) = \mathrm{BCE}\left(\boldsymbol{f}_\theta(x+\delta), y\right) + \lambda(1-\mathbb{f}_\theta(x)_y)\mathrm{KL}(\mathbb{f}_\theta(x) | \mathbb{f}_\theta(x+\delta)).$$

The CE is replaced by a boosted cross-entropy (BCE) and the Kullback-Leibler divergence term is reweighted. Instead, the classical CE is used in $\ell_D^\delta \left(\boldsymbol{f}_\theta(x+\delta), y\right) = \mathrm{CE}\left(\boldsymbol{f}_\theta(x+\delta), y\right)$.

5 Instance Reweighting

IR slightly modifies Problem (5) by introducing a weight in front of each training sample [38]. Accordingly, it replaces the uniform average with a weighted objective:

$$\min_\theta \sum_{i=1}^n w_i \ell_D^\theta \left(\boldsymbol{f}_\theta(x_i+\delta_i^*), y_i\right), \quad \text{s.t. } \sum_{i=1}^n w_i = 1, \tag{6}$$

where different strategies have been proposed to set w_i $i \in \{1, \cdots, n\}$.

`GAIRAT` [38] assigns weights with higher importance to AEs that are closer to the model's decision boundary. To quantify this distance, it introduces the least

PGD steps (LPS), defined as the minimum number of PGD iterations required to generate a successful AE.

`MAIL` [20] replaces the discrete and path-dependent LPS with a continuous, path-independent measure of attack strength, named probabilistic margin. The latter is defined as the difference between the predicted probability of the true class and the highest probability assigned to any incorrect class after adversarial perturbation.

`VIR` [14] presents a new instance-wise reweighting approach that explicitly considers both the intrinsic vulnerability of individual training samples under AT and the loss of informative content caused by adversarial perturbations of clean inputs.

`GAIRAT` [38], `MAIL` [20], and `VIR` [14] are state-of-the-art approaches that define w_i in closed form. Other approaches, such as DONE [28] and MEOW [36], instead formulate a meta-optimization problem to learn the weights, which introduces additional optimization overhead and training complexity; therefore, they are not considered in this paper.

6 Objectives and Experimental Setup

In this section, we first restate the objectives of this work and then describe the experimental setup designed to achieve them. In particular, the goals of this study are twofold:

- We aim to compare the considered methods under a unified and controlled evaluation framework, including standardized training procedures, AE generation protocols, hyperparameter selection strategies, and a strict separation between training, validation, and test sets. To ensure full reproducibility, we provide a publicly available implementation[1];
- We investigate whether IR strategies generalize beyond the specific attack used during AT. In particular, AEs are generated using `PGD` during training, while robustness is evaluated using stronger attacks, including `AutoAttack`.

To achieve the above objectives, we adopt the following setup.

Architectures and Datasets. We consider two architectures: `ResNet-18` [16] and `WideResNet-34-10` [34]. All methods are evaluated on `CIFAR-10` [17] and `SVHN` [24]. We use the predefined splits, reserve 10% of the training set for validation, and restrict the test set to 3,000 random samples for computational efficiency.

Training Framework. We evaluate three AT baselines (`VAT`, `TRADES` and `MART`) and three IR methods (`GAIRAT`, `MAIL`, and `VIR`). All experiments are conducted

[1] https://anonymous.4open.science/r/InstanceReweighting-6CE2.

within the original GAIRAT codebase[2], which we adopt as a unified framework for data loading, preprocessing, and AE generation. Each IR method is implemented according to its original formulation and hyperparameters setting if not specified otherwise [14,20,38], ensuring a consistent training pipeline. All models are trained for 100 epochs using mini-batch SGD with momentum 0.9. During training, AEs are generated with `PGD` ($K = 10$ steps, step size $\alpha = 2/255$) under perturbation bound $\epsilon = 8/255$. Moreover, whenever `TRADES` is involved in the training procedure, we follow its reference implementation released by Madry et al.[3], where AEs are generated during training using `PGD` with Kullback-Leibler divergence.

Hyperparameter Configurations. We evaluate all methods under two hyperparameter settings, denoted as `G-HYP` (reported in `GAIRAT` [38]) and `M-HYP` (reported in `MAIL` [20]). All experiments use a batch size of 128. The IR methods are built upon the AT baselines: training initially follows the baseline objective, and the IR objective is introduced after the first learning rate decay.

For `CIFAR-10` under `G-HYP`, the weight decay depends only on the architecture: $5{\cdot}10^{-4}$ for `ResNet-18` and $2{\cdot}10^{-4}$ for `WideResNet-34-10`. The initial learning rate is 0.1, decayed by a factor of 10 at epochs 30 and 60. Instead, for `WideResNet-34-10` with `TRADES`, the decay occurs at epochs 75 and 90.

For `MART`, following the configuration adopted in `GAIRAT`, we adopt the original configuration reported in [31]. Specifically, for `ResNet-18` we use a learning rate of 0.01 with weight decay $3.5 \cdot 10^{-3}$, while for `WideResNet-34-10` we use a learning rate of 0.1 with weight decay $7 \cdot 10^{-3}$. In both cases, the learning rate is decayed by a factor of 10 at epochs 75 and 90.

For `CIFAR-10` under `M-HYP`, `ResNet-18` uses weight decay $3.5{\cdot}10^{-3}$ and initial learning rate 0.01, while `WideResNet-34-10` uses weight decay $7{\cdot}10^{-4}$ and initial learning rate 0.1. In both cases, the learning rate is reduced by a factor of 10 at epochs 75 and 90.

For `SVHN` with `ResNet-18`, both `G-HYP` and `M-HYP` follow the `CIFAR-10` configuration, except that `G-HYP` uses an initial learning rate of 0.01.

As no official `WideResNet-34-10` configuration is provided for `SVHN`, both `G-HYP` and `M-HYP` follow the `CIFAR-10` configuration, with the initial learning rate set to 0.01.

For the `MART-VIR` combination, since the original VIR paper [14] does not provide a specific hyperparameter configuration for `MART`, we adopt the `VIR` hyperparameter configuration used for `TRADES`, given the closer relationship between `MART` and `TRADES`.

Model Selection. We select the checkpoint achieving the highest validation robust accuracy under `PGD-20` ($K = 20$ steps) with $\epsilon = 8/255$ and $\alpha = 2/255$.

[2] https://github.com/zjfheart/Geometry-aware-Instance-reweighted-Adversarial-Training.

[3] https://github.com/yaodongyu/TRADES.

Test-Time Evaluation. Final results are reported in terms of accuracy in percentage on the original samples (NAT) and AEs generated using PGD-100 ($K = 100$ steps), APGD, and AutoAttack with the same $\epsilon = 8/255$ (and $\alpha = 2/255$ for PGD-100) of the in training and validation phases. This allows to compute $\texttt{Acc}_{\texttt{NAT}}$, $\texttt{Acc}_{\texttt{PGD-100}}$, $\texttt{Acc}_{\texttt{APGD}}$ and $\texttt{Acc}_{\texttt{AutoAttack}}$.

7 Experimental Results

In this section, we report the experimental results obtained under the evaluation protocol described in Sect. 6.

Figures 1 and 2 summarize the results for different architectures (ResNet-18 and WideResNet-34-10) and datasets (CIFAR-10 and SVHN, respectively). Each plot reports, on the x-axis, the combinations of AT strategies (VAT, TRADES, and MART), either without instance reweighting (NO_IR) or combined with IR methods (GAIRAT, MAIL, and VIR), together with the considered hyperparameter configurations (G-HYP and M-HYP). The y-axis shows the classification accuracy,in percentage, evaluated on the test set, both on original samples (NAT) and on AEs generated using PGD-100, APGD, and AutoAttack.

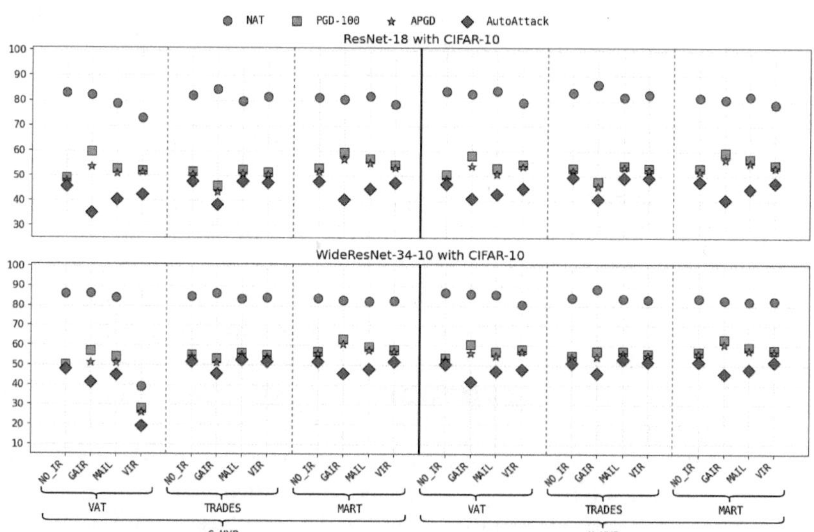

Fig. 1. CIFAR-10: Test accuracy for different architectures (ResNet-18, WideResNet-34-10). The x-axis shows the combinations of AT strategies (VAT, TRADES, MART), either without IR (NO_IR) or with IR methods (GAIRAT, MAIL, VIR), under different hyperparameter configurations (G-HYP, M-HYP). The y-axis reports accuracy on natural samples (NAT) and AEs generated with PGD-100, APGD, and AutoAttack.

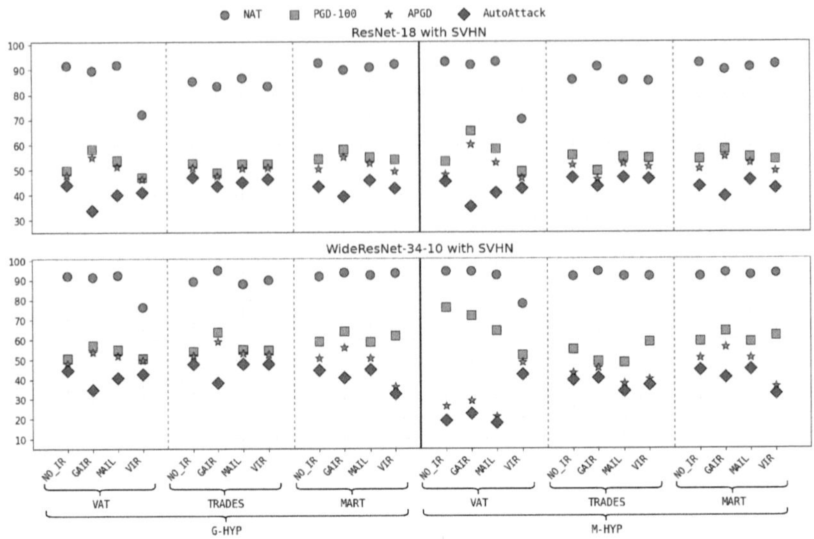

Fig. 2. SVHN: Test accuracy for different architectures (ResNet-18, WideResNet-34-10). The x-axis shows the combinations of AT strategies (VAT, TRADES, MART), either without IR (NO_IR) or with IR methods (GAIRAT, MAIL, VIR), under different hyperparameter configurations (G-HYP, M-HYP). The y-axis reports accuracy on natural samples (NAT) and AEs generated with PGD-100, APGD, and AutoAttack.

From Fig. 1, which reports the results on CIFAR-10, we can draw several observations.

Accuracy on original samples (NAT) appears consistent with the values reported in the original papers across both ResNet-18 and WideResNet-34-10. Robust accuracy under PGD-100 is generally aligned with APGD, with APGD typically yielding slightly lower values, as expected from a stronger first-order attack. Methods achieving the largest improvements under PGD-100 often exhibit a significant drop in performance under stronger attacks (i.e., AutoAttack). This behavior is particularly evident for GAIRAT, especially when combined with VAT and MART, across both architectures. This trend is less pronounced for TRADES, where the gap between PGD-100 and AutoAttack is generally smaller. However, when combined with GAIRAT, TRADES tends to produce lower robustness under stronger attacks, while often achieving the highest clean accuracy. For MART, IR methods consistently improve robustness under PGD-100 compared to the NO_IR baseline, but this improvement is typically accompanied by a degradation under AutoAttack. Results for MART are identical across G-HYP and M-HYP, as the same hyperparameters are used. When combining VIR with MART using hyperparameters originally designed for TRADES, we observe more stable behavior compared to other IR methods, although improvements over the baseline remain limited, and performance often stays close to the corresponding NO_IR model. The combination VAT-VIR under G-HYP shows significantly degraded performance, which is consistent with the fact that this configuration is not defined in the original works, highlighting the sensitivity of IR methods to the chosen hyperparameter setting. Overall, IR methods show a strong dependency on the hyperparameter configuration, which can significantly affect both robustness and generalization, including the epoch at which reweighting is introduced (epoch 60 in `G-HYP` vs epoch 75 in `M-HYP`). Across both architectures, `VIR` generally provides smaller gains compared to `GAIRAT` and `MAIL`, often yielding performance close to the corresponding `NO_IR` baseline, especially under stronger attacks.

From Fig. 2, which reports the results on `SVHN`, we observe a similar behavior, with a few additional considerations.

Comparison with the original works is possible only for the case `ResNet-18`, as `WideResNet-34-10` is not evaluated on this dataset in prior work. In the case of `ResNet-18`, clean accuracy is generally high (above 80%) across most methods, with `M-HYP` often yielding slightly better results. `TRADES`-based models tend to achieve lower clean accuracy compared to `VAT` and `MART`. The combination `VAT-VIR` consistently results in lower clean accuracy, indicating higher sensitivity to the training configuration. Under stronger attacks (i.e., `AutoAttack`), `VIR` tends to degrade less than other IR methods, often achieving better performance than `GAIRAT` and `MAIL`, despite weaker results under `PGD-100`. Differences between `G-HYP` and `M-HYP` are generally limited for `ResNet-18`. However, for `VAT`, `M-HYP` consistently improves robustness under `PGD-100`, particularly when combined with `GAIRAT` and `MAIL`, with limited but consistent transfer to stronger attacks. One possible explanation is that introducing IR at a later stage of training (epoch 75 in `M-HYP`) may reduce over-specialization to the training-time attack. Models trained with `TRADES` generally exhibit lower performance under `PGD-100`, but achieve more stable results under stronger attacks, suggesting a lower degree of over-specialization to the training-time attack. For `WideResNet-34-10`, the most significant differences between `G-HYP` and `M-HYP` are observed for `VAT`-based methods. In particular, `M-HYP` tends to achieve higher robustness under `PGD-100`, but at the cost of reduced generalization to stronger attacks. This behavior suggests a stronger over-specialization to the training-time attack. A possible explanation is that `G-HYP`, which reduces the learning rate at an earlier stage of training, may help mitigate overfitting, whereas the later introduction of IR in `M-HYP` (epoch 75 vs 60) can lead to a more pronounced over-specialization effect. Interestingly, this behavior appears to be partially mitigated when using `VIR` with `VAT` under `M-HYP`. Additional differences between `G-HYP` and `M-HYP` are observed for `TRADES`-based methods, with `G-HYP` generally yielding better performance across both weak and strong attacks.

Overall, the results across both `CIFAR-10` and `SVHN` reveal consistent trends. IR methods consistently improve robustness against attacks similar to those used during training (i.e., `PGD`), often by a significant margin. However, these improvements rarely transfer to stronger evaluations (i.e., `AutoAttack`), where IR methods typically match or underperform the corresponding `NO_IR` baseline. In several cases, the methods achieving the largest gains under `PGD-100` also exhibit the largest degradation under `AutoAttack`, suggesting a form of over-specialization to the training-time attack rather than a genuine improvement in adversarial robustness. Conversely, TRADES-based models tend to provide more stable performance across different attacks, trading off some robustness under PGD-100 for improved generalization to stronger evaluations. Finally, in some cases, the baselines obtained in our experiments achieve higher robustness than those reported in the original works (e.g., for MAIL on ResNet-18). As a result, the relative gains of IR methods may appear smaller in our setting. This discrepancy is likely due to differences in the experimental protocol, including hyperparameter configurations and the use of validation/test splits, which were not handled in the same way in the original works.

8 Conclusions

In this work, we provided a controlled and unified evaluation of IR strategies within AT. By standardizing architectures, hyperparameters, model selection, and evaluation protocols, we enabled a fair comparison between IR methods and strong AT baselines.

Our results highlight two main findings. First, commonly adopted experimental practices, particularly the lack of proper validation-based model selection, can lead to unreliable or inflated robustness claims. When consistent and statistically sound procedures are enforced, the advantages of several IR methods substantially diminish.

Second, IR often improves robustness only against the specific attack used during training. While gains are observed against PGD-based evaluations, they frequently disappear under stronger and more comprehensive attacks such as AutoAttack. This indicates that IR tends to overfit to the training-time attack, providing narrow and fragile robustness rather than genuine generalization within the threat model.

Overall, although computationally appealing, IR does not consistently deliver robust improvements and may create a false sense of security. Our findings stress the importance of rigorous evaluation protocols and attack-agnostic robustness assessment when claiming progress in adversarial defense.

Acknowledgments. This work is partially supported by (i) project ELSA (European Lighthouse on Secure and Safe AI) funded by the European Union's Horizon Europe under the grant agreement No. 101070617, (ii) project SERICS (PE00000014) and FAIR (PE00000013) under the NRRP MUR program funded by the EU - NGEU, and (iii) project FISA-2023-00128 funded by the MUR program "Fondo italiano per le scienze applicate". Lastly, this work was carried out while Dario Lazzaro was enrolled in the Italian National Doctorate on Artificial Intelligence run by the Sapienza University of Rome in collaboration with the University of Genoa, funded by "Unione europea-Next Generation EU, Missione 4 Componente 1 CUP B53C23003580004".

References

1. Andriushchenko, M., Croce, F., Flammarion, N., Hein, M.: Square attack: a query-efficient black-box adversarial attack via random search. In: European Conference on Computer Vision (2020)
2. Biggio, B., et al.: Evasion attacks against machine learning at test time. In: Joint European Conference on Machine Learning and Knowledge Discovery in Databases (2013)
3. Biggio, B., Roli, F.: Wild patterns: ten years after the rise of adversarial machine learning. Pattern Recogn. **84**, 317–331 (2018)
4. Bishop, C.M., Bishop, H.: Deep Learning: Foundations and Concepts. Springer, Cham (2023)
5. Carlini, N., Wagner, D.: Towards evaluating the robustness of neural networks. In: IEEE Symposium on Security and Privacy (2017)
6. Chakraborty, A., Alam, M., Dey, V., Chattopadhyay, A., Mukhopadhyay, D.: A survey on adversarial attacks and defences. CAAI Trans. Intell. Technol. **6**(1), 25–45 (2021)
7. Cinà, A.E., Pintor, M., Demetrio, L., Demontis, A., Biggio, B., Roli, F.: Evaluating the evaluators. In: Trust in Adversarial Robustness Tests. arXiv preprint arXiv:2507.03450 (2025)
8. Cinà, A.E., et al.: Attackbench: evaluating gradient-based attacks for adversarial examples. In: AAAI Conference on Artificial Intelligence (2025)
9. Cinà, A.E., Villani, F., Pintor, M., Schönherr, L., Biggio, B., Pelillo, M.: σ-zero: gradient-based optimization of ℓ_0-norm adversarial examples. In: International Conference on Learning Representations (2025)
10. Croce, F., Hein, M.: Minimally distorted adversarial examples with a fast adaptive boundary attack. In: International Conference on Machine Learning (2020)
11. Croce, F., Hein, M.: Reliable evaluation of adversarial robustness with an ensemble of diverse parameter-free attacks. In: International Conference on Machine Learning (2020)
12. Debenedetti, E., Wan, Z., Andriushchenko, M., Sehwag, V., Bhardwaj, K., Kailkhura, B.: Scaling compute is not all you need for adversarial robustness. arXiv preprint arXiv:2312.13131 (2023)
13. Dingeto, H., Kim, J.: Comparative study of adversarial defenses: adversarial training and regularization in vision transformers and cnns. Electronics **13**(13), 2534 (2024)
14. Fakorede, O., Nirala, A.K., Atsague, M., Tian, J.: Vulnerability-aware instance reweighting for adversarial training. Trans. Mach. Learn. Res. (2023)
15. Goodfellow, I.J., Shlens, J., Szegedy, C.: Explaining and harnessing adversarial examples. arXiv preprint arXiv:1412.6572 (2014)
16. He, K., Zhang, X., Ren, S., Sun, J.: Deep residual learning for image recognition. In: IEEE Conference on Computer Vision and Pattern Recognition (2016)
17. Krizhevsky, A.: Learning multiple layers of features from tiny images, Canada, Toronto (2009)
18. Li, B., Wang, S., Jana, S., Carin, L.: Towards understanding fast adversarial training. arXiv preprint arXiv:2006.03089 (2020)
19. Li, Y., Zhao, P., Lin, X., Kailkhura, B., Goldhahn, R.: Less is more: data pruning for faster adversarial training. arXiv preprint arXiv:2302.12366 (2023)
20. Liu, F., et al.: Probabilistic margins for instance reweighting in adversarial training. In: Neural Information Processing Systems (2021)

21. Liu, X., Hu, T., Yi, P., Li, B., Xie, J., Ma, H.: Gradient semi-masking for improving adversarial robustness. Pattern Recogn. 112433 (2025)
22. Madry, A., Makelov, A., Schmidt, L., Tsipras, D., Vladu, A.: Towards deep learning models resistant to adversarial attacks. arXiv preprint arXiv:1706.06083 (2017)
23. Meng, M.H., et al.: Adversarial robustness of deep neural networks: a survey from a formal verification perspective. IEEE Trans. Dependable Secure Comput. (2022)
24. Netzer, Y., Wang, T., Coates, A., Bissacco, A., Wu, B., Ng, A.Y.: Reading digits in natural images with unsupervised feature learning. In: Neural Information Processing Systems - Workshop on Deep Learning and Unsupervised Feature Learning (2011)
25. Peng, S.Y., et al.: Robust principles: architectural design principles for adversarially robust cnns. arXiv preprint arXiv:2308.16258 (2023)
26. Shalev-Shwartz, S., Ben-David, S.: Understanding Machine Learning: From Theory to Algorithms. Cambridge University Press, Cambridge (2014)
27. Song, Y., Kim, T., Nowozin, S., Ermon, S., Kushman, N.: Pixeldefend: leveraging generative models to understand and defend against adversarial examples. arXiv preprint arXiv:1710.10766 (2017)
28. Sow, D., Lin, S., Wang, Z., Liang, Y.: Doubly robust instance-reweighted adversarial training. arXiv preprint arXiv:2308.00311 (2023)
29. Szegedy, C., et al.: Intriguing properties of neural networks. arXiv preprint arXiv:1312.6199 (2013)
30. Villani, F., Maljkovic, I., Lazzaro, D., Sotgiu, a., Cinà, A.E., Roli, F.: Robust image classification with multi-modal large language models Pattern Recogn. Lett. **194**, 1–7 (2025)
31. Wang, Y., Zou, D., Yi, J., Bailey, J., Ma, X., Gu, Q.: Improving adversarial robustness requires revisiting misclassified examples. In: International Conference on Learning Representations (2019)
32. Wang, Z., Pang, T., Du, C., Lin, M., Liu, W., Yan, S.: Better diffusion models further improve adversarial training. In: International Conference on Machine Learning (2023)
33. Wong, E., Rice, L., Kolter, J.Z.: Fast is better than free: revisiting adversarial training. In: International Conference on Learning Representations (2020)
34. Zagoruyko, S., Komodakis, N.: Wide residual networks. In: British Machine Vision Conference (2016)
35. Zhang, H., et al.: Theoretically principled trade-off between robustness and accuracy. In: International Conference on Machine Learning (2019)
36. Zhang, J., Hong, Y., Zhao, Q.: Memorization weights for instance reweighting in adversarial training. In: AAAI Conference on Artificial Intelligence (2023)
37. Zhang, J., et al.: Attacks which do not kill training make adversarial learning stronger. In: International Conference on Machine Learning (2020)
38. Zhang, J., Zhu, J., Niu, G., Han, B., Sugiyama, M., Kankanhalli, M.: Geometry-aware instance-reweighted adversarial training. arXiv preprint arXiv:2010.01736 (2020)
39. Zhao, W., Alwidian, S., Mahmoud, Q.H.: Adversarial training methods for deep learning: a systematic review. Algorithms **15**(8), 283 (2022)

Distributional Loss for Robust Classification

Kathleen Anderson(✉) and Thomas Martinetz

Institute for Neuro- and Bioinformatics, University of Lübeck, Lübeck, Germany
{k.anderson,martinetz}@uni-luebeck.de

Abstract. This paper proposes a novel loss concept for supervised classification tasks. Rather than enforcing a direct mapping from each input sample to a single assigned label, we define an optimization objective over all classifier outputs as a bimodal Gaussian distribution. This softer target formulation implicitly captures class ambiguity, mitigates overfitting, and encourages the learning of more robust decision boundaries, all without requiring additional label information. Experimental results demonstrate consistent improvements in robustness, with particularly pronounced gains in low-data regimes, while requiring only minimal modifications to standard training pipelines.

Keywords: Supervised Learning · Density Estimation · Small Datasets

1 Introduction

The domain of robust classification strategies is vast and spans a wide range of different approaches, including dataset regularization or augmentation, network architectures, and pretraining. An aspect whose relevance has diminished in recent years is the choice of classification training loss, with cross entropy often serving as the gold standard for any classification task.

In this work, we consider a supervised classification setting in which inputs $x \sim P(x)$ are associated with class labels $c_k \sim P(c_k \mid x)$, where $k = 1, \ldots, K$ denotes the K possible classes to which an input can be assigned. In an idealized case, there are only two values which $P(c_k \mid x)$ may take: if $P(c_k \mid x) = 1$, the input x always belongs to class c_k with certainty. Conversely, if $P(c_k \mid x) = 0$, the input x will never belong to class c_k. These probabilities of 0 and 1 represent extreme cases. In practice, due to noise, class overlap, or shared features among different classes, posterior probabilities are rarely exactly 0 or 1.

Many loss functions, such as the cross-entropy loss, are nevertheless defined with respect to these extremes [33]. In a training dataset $\{(x^{(n)}, c_k^{(n)}), n = 1, ..., N\}$, to each input $x^{(n)}$ is assigned a hard class label $c_k^{(n)} \in \{c_1, \ldots, c_K\}$. This corresponds to a target distribution for which $P(c_k^{(n)} \mid x^{(n)}) = 1$ for the correct class and $P(c_l^{(n)} \mid x^{(n)}) = 0$ for all $l \neq k$. As a result, the classifier is effectively trained to reproduce these degenerate (one-hot) target distributions.

L. Pasa et al. (Eds.): ICANN 2026, LNCS 17089, pp. 519–530, 2027.
https://doi.org/10.1007/978-3-032-38398-3_43

Assigning each sample a single definitive label may fail to capture the intrinsic complexity of the underlying task: in many practical scenarios, samples are not perfectly representative of a single class, but instead exhibit characteristics shared across multiple classes. Using softer, more informative labels, ideally reflecting the true class probabilities, can improve classification robustness [14,24], but such high-quality probabilistic annotations are rarely available in practice. Enforcing a rigid, hard label assignment instead may encourage overfitting by compelling the model to learn overly confident and potentially misleading decision boundaries [12].

This work utilizes the hypothesis that for most natural datasets the distribution of the posterior probability values $p_k(x) = P(c_k \mid x)$ for inputs $x \sim P(x)$, is more likely to follow a bimodal distribution as illustrated in Fig. 1, with modes concentrated near 0 and 1. We propose a *distributional loss* that explicitly leverages this bimodal structure. The training labels now indicate only to which mode of the distribution each sample should be assigned. The proposed training objective encourages the classifier to distribute predictions in accordance with this bimodal behavior.

To achieve this, we employ two complementary differentiable estimators of the Kullback–Leibler divergence. The first is based on a *probabilistic encoder*, as commonly used in variational autoencoders, while the second relies on a non-parametric density estimator derived from pairwise distances between samples. These two loss components complement each other, with one creating a robust base and the other fitting more closely to the true distribution.

Experimental results demonstrate improved robustness, particularly in settings with limited training data. The proposed objective is especially advantageous in regimes where the risk of overfitting is high. The required modifications to the training procedure are minimal and can be readily integrated into virtually any classifier architecture or training pipeline.

2 Related Work

Cross entropy in a combination with a softmax continues to serve as the standard loss for neural network classification in contemporary research [33]. Contradicting its omniscience, previous research has identified various shortcomings of this loss function, and the idea of augmenting or replacing cross entropy has been approached from many angles.

One of the discovered disadvantages is a lacking resilience to noise or outliers. It has been shown that contrary to cross entropy, the absolute distance between network output logits and target label is robust to label noise [8], which motivated the proposal of new loss functions that enable smooth transitions between cross entropy and absolute distance [40], based on a scalar hyperparameter.

Another caveat of cross entropy training is that it has been found to motivate overconfidence - the output of a trained classifier almost always claims a very high probability (i.e. a high confidence) for one class and a very low probability for all others, even when the selected target class is incorrect. An added confidence penalty can improve the classification accuracy [9,25].

The application of softmax implicitly enforces margins between class logits, but depending on the dataset, the resulting margin was shown to be inconsistent or insufficient. Explicitly increasing the margin between class logits can yield more robust classifiers [4,5,21,22,30,32,34].

On a more theoretical basis, another line of research reported findings about a reweighted Taylor-decomposition of standard cross entropy loss [10,20].

Previous research has also, like this paper, pondered on the idea of replacing cross entropy entirely. Common alternatives include variations of cosine similarity [3,17,31] and correntropy [27,29,39]. Correntropy in its basic form denotes a simple kernelized distance between classifier output and target vector: $L_{corr}(x, y) = \kappa(x - y)$, with κ usually defined to be a Gaussian. Empirical exploration yielded other entirely new (and in some cases even counter-intuitive) loss functions [2,7], which also outperform their cross entropy baseline.

Naturally, specialized loss functions have also been employed for a multitude of other purposes, such as the search for robust representations [6,15,35,36] or the regularization of intermediate feature layers [23].

3 Training for a Distribution

Following the general considerations outlined in the introduction, we now turn to a concrete setting. We consider a classifier implemented as a neural network with K output units, one for each class.

When using a softmax layer, the outputs of the classifier are interpreted as posterior probabilities $p_k(x) = P(c_k \mid x)$ for a given input x. Let $y_k(x)$ denote the corresponding pre-softmax activations (logits). In the typical classification scenario, all target probabilities are given as either $p_k(x) = 1$ or $p_k(x) = 0$ (marked red in Fig. 1). While the probabilities $p_k(x)$ are constrained to the interval $(0, 1)$, the logits $y_k(x)$ take values in $(-\infty, \infty)$, with $y_k(x) \to -\infty$ for $p_k(x) \to 0$ and $y_k(x) \to \infty$ for $p_k(x) \to 1$. The scalar values $y_k(x)$ will henceforth be abbreviated as y when the indices are clear.

Our new distributional target is based on the assumption that for most datasets, a softer curve is closer to the ground truth (drawn in green in Fig. 1). Instead of aiming for ∞ and $-\infty$, the logit values y are aimed at a bimodal Gaussian distribution with means $-m$ and m.

Contrary to cross entropy, our new loss does not require its inputs to be probabilities and therefore does not include a softmax layer, but is computed directly on the logits y. The fixed target distribution nonetheless implicitly constrains the range of output values. A notable effect of omitting softmax is that the $y_k(x)$ for one x do not directly influence each other: a high $y_k(x)$ for one class c_k does not necessarily lower the value of the other $y_l(x), l \neq k$.

To minimize the divergence between the newly defined target distribution and the output of the classifier, we use the Kullback-Leibler (KL) divergence and two strategies of computing it: a parametric solution that is robust, but very strict, and a non-parametric estimator which is more sensitive to noise,

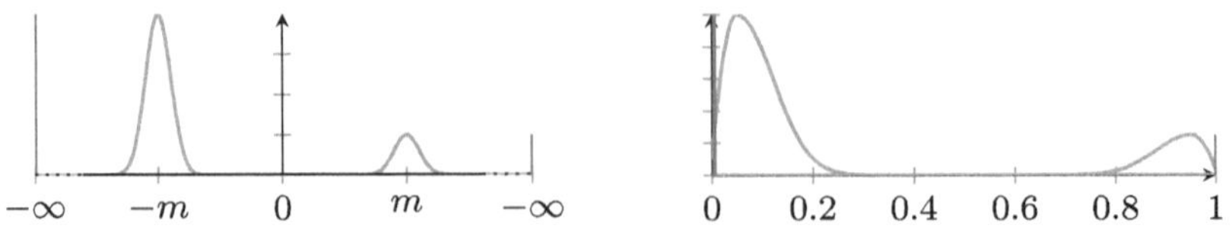

(a) Logits $y_k(x)$ before softmax

(b) Posterior probabilities $p_k(x)$ after softmax

Fig. 1. Target for of a single output unit (i.e. corresponding to a single class), for a standard classification task, defined by softmax cross entropy (red) and our new formula (green).

but adapts more flexibly to a distribution. The two losses are combined with a balancing factor λ as

$$L(y) = \lambda L_{\text{param}}(y) + L_{\text{pointwise}}(y, Y) \tag{1}$$

where the pointwise loss $L_{\text{pointwise}}$ is computed for the logit value y in relation to the set Y of all y.

3.1 Parametric Probabilistic Encoders

A probabilistic encoder is usually used as a component of a variational autoencoder (VAE) [16]. Instead of directly computing the logit $y(x)$ for an input x, the output of a probabilistic encoder f consists of the parameters of a Normal distribution, μ_x and σ_x. Analogously to a VAE, we define the target to be another Normal distribution $\mathcal{N}(\mu_{\text{target}}, \sigma_{\text{target}})$, and compute the loss via the closed form equation for the KL divergence

$$L_{\text{param}}(\mu_x, \sigma_x) = KL(\mathcal{N}(\mu_x, \sigma_x) \mid \mathcal{N}(\mu_{\text{target}}, \sigma_{\text{target}})). \tag{2}$$

The standard deviation σ_{target} is shared between all inputs, while μ_{target} is defined as $+m$ or $-m$ depending on the class assigned to input x, resulting in the bimodal Gaussian distribution depicted in Fig. 1.

3.2 Generating Samples from a Probabilistic Encoder

A probabilistic encoder does not generate logits, but distributions over logits, each defined by their parameters μ_x and σ_x. The pointwise divergence estimation introduced in the next section, however, is based on a fixed set of points. To utilize the logits y itself, one can implement the same reparameterization trick that is also employed by a standard VAE: draw a random $n \in \mathcal{N}(0, 1)$ and compute $y(x) = \sigma_x n + \mu_x$. Gradient descent can be done as if y was directly generated by the encoder f.

The reparameterization trick comes with another advantage: the logits can be augmented by drawing several $y(x)$ per input x. Since the pointwise divergence estimation in the next section requires many data points to work well, this natural dataset expansion can be especially beneficial.

3.3 Non-parametric Pointwise Divergence

The outcome of the probabilistic encoder, after reparameterization, is a set of logit values y drawn from many Normal distributions, with one individual distribution $\mathcal{N}(\mu_x, \sigma_x)$ corresponding to every datapoint x. The KL divergence loss described in the previous section aims to make all of those $\mathcal{N}(\mu_x, \sigma_x)$ match one of two targets, $\mathcal{N}(+m, \sigma_{\text{target}})$ or $\mathcal{N}(-m, \sigma_{\text{target}})$. However, the combined set of all y can reflect a target distribution even if individual points originate from distinct $\mathcal{N}(\mu_x, \sigma_x)$. In fact, it can be beneficial to allow deviations for some x, e.g. to make the classifier respond with a low confidence to a noisy sample.

The second divergence estimator balances the strict goal of the probabilistic encoder with a non-parametric density estimation. Instead of a purely sample-wise evaluation, the estimator approximates the distribution over an entire batch of logits. As before, the scalar logit values y can be split into two categories:

positive when the class c assigned to x corresponds to the output unit, aiming for $\mathcal{N}(+m, \sigma)$. The standard softmax cross entropy formulation would have trained for $y \to \infty$ for all y in this group (see Fig. 1).

negative y are those where the logit dimension does not match the target class, which aim for an overall distribution of $\mathcal{N}(-m, \sigma)$.

This assignment yields a total of $2K$ groups for K classes: one positive and one negative group for each of the K classifier output units. Let $Y_l \subset Y$ be a set of y, all belonging to the same group and aimed at the same target distribution. For each Y_l, we sample a matching set T_l of points from the assigned target distribution[1], $|Y_l| = |T_l|$.

As a minimization objective proportional to the KL–divergence between the distributions $P(Y)$ and $P(T)$, we insert one of the subsequently described divergence measures d into

$$L_{\text{pointwise}}(y, Y) = d(y, Y_l), \tag{3}$$

with $Y_l \subset Y$ referring to the set of all scalar logit values that correspond to the same group as y.

Nearest-Neighbor Estimation. The nearest neighbor density estimation method is based on a simple intuition: for each $y \in Y_l$, estimate the distance between y and its neighbors. A close proximity to many neighbors indicates a high density area. If the distributions of two sets Y_l and T_l align, the average neighbor distances for a point at the location of y will be the same when estimated in either Y_l or T_l. The strategy was first used in the context of entropy estimation [18] and has only recently been applied to gradient based learning [1].

For a value y, let $\{y^{(1)}, ..., y^{(m)}\}$ and $\{t^{(1)}, ..., t^{(m)}\}$ be the m closest neighbors to y in Y_l and T_l respectively, sorted by their distance $\Delta(y, y^{(i)})$. The KL

[1] While the explicit estimation of density values is also possible, we found this more indirect estimation of comparing two sets of points to yield better results.

divergence between $P(Y_l)$ and $P(T_l)$ can be decreased by minimizing

$$d_{\text{nn}}(y, Y_l) = \frac{1}{m}\sum_{i=1}^{m}(\Delta(y, y^{(i)}) - \Delta(y, t^{(i)}))^2. \tag{4}$$

Interestingly, d_{nn} is not only proportional to the true KL divergence, but can be reshaped into an accurate approximation of it, with the approximation error decreasing with the smoothness of the true distribution. Refer to [1] for a more in-depth derivation and analysis.

The hyperparameter m heavily influences the accuracy of the d_{nn} estimator: considering only very close neighbors can make the estimate sensitive to noise and too uneven, choosing a very large m may oversmooth the estimate and obscure divergence. To ease the influence of m, we chose $\Delta(y, y^{(i)}) = e^{-(y-y^{(i)})^2}$. The Gaussian kernel weights close neighbors higher, but still factors distant neighbors into the difference.

Kernel Density Estimation. When computing the distributional divergence on one-dimensional data, the popular kernel density estimation (KDE) [26] is exceedingly similar to nearest-neighbor estimation, the main difference being that the density is estimated as the kernelized sum over all points first, and compared second:

$$d_{\text{kde}}(y) = \left(\sum_{i=1}^{|Y_l|}\Delta_h(y, y^{(i)}) - \sum_{i=1}^{|Y_l|}\Delta_h(y, t^{(i)})\right)^2. \tag{5}$$

This rearrangement makes the ordering of $\{y^{(1)}, ..., y^{(m)}\}$ and $\{t^{(1)}, ..., t^{(m)}\}$ irrelevant. It is also less sensitive: even when point distances diverge significantly for singular neighboring points, the sum over all neighbors may still yield two similar density estimates.

The traditional implementation considers all points as "neighbors" ($m = |Y_l|$), eliminating the hyperparameter m in favor of another hyperparameter: h, the bandwidth of the kernel $\Delta_h(a) = e^{-\frac{a}{h}^2}$. As the name suggests, h determines the width of the kernel, and effectively decides how distant neighbors factor into the result.

The choice of h is critical for the accuracy of KDE [28]. Unlike standard applications, manually tuning h is ineffective for our purpose, as the dataset is subject to constant change over the course of the training. Our simple solution is to combine KDE and nearest neighbor strategies and scale h by the average distance between a point and its closest neighbors.

4 Distributional Loss Compared to Cross Entropy

As demonstrated in numerous prior studies (see Sect. 2), the standard cross-entropy loss exhibits several limitations, particularly in settings with small or low-quality datasets. The proposed distribution-based loss addresses many of these shortcomings, thereby increasing the likelihood that the learned model better approximates the underlying true data-generating distribution.

One beneficial property it shares with many related papers is the ability to explicitly define and enforce margins between classes. Since the target mean for the negative and the positive group can be chosen freely, the margin between the two is an integral part of the loss formulation.

Countering the presumed tendency towards overconfidence [9,25], our target distributions do not only allow, but even enforce logit vectors with a lower confidence. A small percentage of classification results is expected to be uncertain, reflecting the true distribution of most datasets. Noisy samples are not forcibly treated as perfect representatives of their class.

A less obvious advantage is the independence between the dimensions of a logit vector. The classification problem assigns exactly one class to each sample, but a single input might resemble multiple classes (e.g. an image of a distant four legged animal within an animal classification dataset) or none at all (e.g. an unclear image lacking any identifiable animal). The independent logit evaluation allows such cases: a sample can be given a high score for more than one class. Nonetheless, the classifier is trained for a globally normalized solution: the output distribution over the full dataset should remain constant.

Concluding, it can be said that training with the distributional loss introduces a more intricate bias for the labeling of the dataset. Instead of mapping each sample to one class index, the loss defines properties over the entire set of outputs. Our experiments have shown that this target is generally more difficult to train for, demanding more training time and more careful tuning of hyperparameters to locate a satisfactory loss minimum. However, the assumption of a smoother label distribution is usually much closer to the true classification problem. Therefore, once a good training accuracy is achieved, the result is more likely to generalize well.

5 Experiments

We compare our new method against several other loss functions. Softmax cross entropy (CE) serves as a baseline, and results are also presented for *PolyLoss* [20], a popular variation of cross entropy that reweights the first term of its Taylor Expansion, and *T-vMF* Loss [17], which modifies the final network layer to regularize penultimate features using a generalized cosine similarity[2].

Unless stated otherwise, all experiments have been performed using the same standard ResNet18 [13] architecture with initial weights obtained by pretraining on ImageNet. For a clean comparison, we otherwise abstain from using advanced pretraining, augmentation or normalization techniques.

[2] Note that our comparison is limited to other specialized loss functions. Assessing the effect of additional strategies (such as label smoothing) on all compared loss functions remains to be evaluated in future work.

To offer a perspective on different types of image data, experiments are performed for four datasets: Cifar10 [19] is one of the most commonly used general image classification datasets, featuring 60000 32×32 images split into 10 balanced classes. Caltech256 [11] is a more challenging dataset comprising 30607 images from 256 diverse categories. The dataset is not balanced, each class is represented by a minimum of 80 and a maximum of 827 images. Images have been rescaled to 256×256. We also evaluate BloodMNIST: a set of 17091 microscopic images of individual blood cells divided into 8 categories, and PathMNIST: 107180 stained histological images from 9 different types of tissues. BloodMNIST images originate from the MedMNIST dataset [37] and have been scaled to 28×28. PathMNIST was published in an expanded version with larger images [38] and our experiments are performed on image size 128×128. These two medical datasets deviate more pronounced from the pretraining distribution.

For the non-parametric component of the distributional loss (Sect. 3.3), results are shown for the divergence estimate that proved most effective for the respective dataset: KDE for BloodMNIST and Caltech256, and nearest-neighbor densities for Cifar10 and PathMNIST.

The full implementation including all training details and hyperparameters are available in the public github repository[3].

5.1 Classification Accuracy

Our results confirm that changing the training objective function can notably affect the test accuracies of a model. Visible in Table 1, the distributional loss generally increases the test accuracy, achieving the most notable improvements on small datasets. The strongest increase can be observed on BloodMNIST, where the distributional loss outperforms cross entropy by up to 17.2% points. The only dataset not improved by our method is Caltech256, where cross entropy tends to outperform both our newly introduced loss and related work. One can hypothesize that the large 256 dimensional logit vectors fall into a different task category than those of a 10- or 8- class problem. Adapting the distributional loss to this slightly different setting remains future work.

5.2 Overfitting to Random Noise

Theoretically, a neural network with a fixed architecture will always have the same capacity: it is able to represent a specific collection of functions, independently of the training method. In practice, however, the loss function defines the topology of the optimization landscape. A subset of solutions might be theoretically realizable by a network, but not discovered by the training procedure.

As a simple evaluation of the effect of a loss function on the practical complexity of a model, we train a classifier on an artificially generated dataset. Input samples have been generated as Gaussian noise, categorical target labels are drawn from a uniform distribution. There is no statistical relationship between

[3] https://github.com/ka-anderson/distributional-classification-loss.

Table 1. Average classification test accuracies in % for different dataset sizes.

		Samples per Class			
		2	20	200	full dataset
Cifar10	CE	23.50	47.35	67.13	83.28
	Distributional	**27.94**	**55.23**	**71.20**	**84.42**
	PolyLoss	25.12	48.17	67.43	83.49
	T-vMF	**27.90**	52.73	68.92	83.40
Caltech256	CE	37.98	**75.01**		**83.38**
	Distributional	39.11	73.27		80.91
	PolyLoss	38.32	74.08		83.34
	T-vMF	**40.89**	69.37		75.42
BloodMNIST	CE	44.63	73.35	90.06	95.40
	Distributional	**61.82**	**83.14**	**92.32**	**96.71**
	PolyLoss	51.63	78.65	**92.14**	95.29
	T-vMF	55.45	79.90	**92.01**	96.05
PathMNIST	CE	56.29	83.50	94.72	**99.63**
	Distributional	59.76	**90.04**	**96.06**	**99.61**
	PolyLoss	**61.19**	75.35	95.04	99.53
	T-vMF	56.94	88.28	95.44	99.43

inputs and labels, hence the test accuracy is always equivalent to random guessing. Any training accuracy exceeding random chance indicates that the model has memorized a direct sample–label mapping.

The plots in Fig. 2 show the **training** accuracy of a linear classifier, a six layer MLP, and a small Resnet. MLP and linear classifier have been trained to learn binary labels on 20 dimensional random vectors, the Resnets are given samples with shape 30×30, attempting to arrange them into 10 classes.

The distributional loss clearly limits the memorization ability, whereas cross entropy can memorize an almost perfect training score for all dataset sizes. While it does not guarantee an increased test accuracy, a lower likelihood of memorizing random labels makes it more likely that a classifier is forced to learn meaningful structures.

5.3 Ablation Study

To evaluate the relevance of all loss components, we repeat our Cifar10 experiments using (i) only the parametric closed-form loss term, (ii) only the non-parametric pointwise loss term and (iii) the pointwise loss term, with a target standard deviation $\sigma = 0$. This final experiment is limited to the pointwise term, since the parametric KL divergence is not defined for $\sigma = 0$ and highly unstable for very small σ.

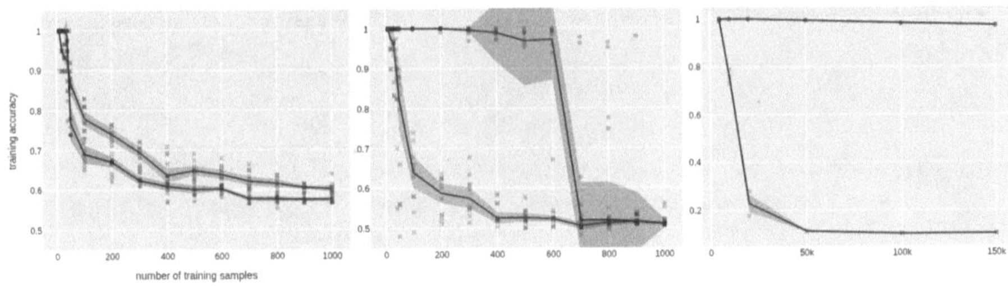

Fig. 2. Training accuracy of a linear classifier, and MLP and a resnet on random noise. Blue circles show models trained with cross entropy, red crosses models trained with the distributional loss. (Color figure online)

The resulting accuracies in Table 2 demonstrate a clear benefit of using the parametric and the non-parametric loss term concurrently. Interestingly, enforcing a non-zero variance is especially important for smaller datasets: when trained on the full Cifar10 training set, the pointwise loss achieves the same accuracies, independent of target variance. However, when using 2 samples per class, the model trained for zero variance barely outperforms the cross entropy baseline. One can assume that large datasets implicitly enforce a smoother target distribution - if a certain subject A can be classified as either class i or class j, a very large dataset will contain multiple views of A, being labeled as either i or j, so that minimal training error is achieved by predicting both i and j with a lower confidence.

As the full distributional loss still outperforms classical cross entropy by 1.1 %pt. on the full dataset, its other properties (such as the explicit margin and the independent logit evaluation) are likely to be beneficial for any dataset size.

Table 2. Average test accuracies on the Cifar10 dataset when using different variations of the distributional loss.

	Samples per Class			
	2	20	200	full dataset
full distributional	27.94	55.23	71.20	84.41
non-parametric	25.93	52.97	68.67	84.13
parametric	25.91	55.10	68.65	84.24
zero variance	23.65	51.35	68.29	84.15
CE baseline	23.50	47.35	67.13	83.28

6 Conclusion

This paper explores the idea of a distributional loss function: a classification loss that makes the distribution over all classifier outputs adhere to a specified structure rather than evaluating each sample and its corresponding label separately. The robustness of this distributional loss is most likely linked to a different bias imposed on the logit output of the network: it is assumed that the distribution of classification labels follows a Gaussian distribution, including samples with varying degrees of representativeness for their class, ranging from typical to atypical examples. This assumption is more reflective of natural classification tasks, and can be shown to reduce overfitting more effectively than cross entropy.

While the proposed loss function comes at a cost of a moderate increase in training time, it constitutes an effective alternative for its intended purpose. In general, it improves classification accuracy, demonstrating superior performance over other recent loss formulations in the majority of evaluated scenarios. In settings with a high risk of overfitting, the test accuracy improved by up to 17 percentage points.

Disclosure of Interests. The authors have no competing interests to declare that are relevant to the content of this article.

References

1. Anderson, K., Martinetz, T.: Nearest-neighbor density estimation for dependency suppression. In: ICPR (2026)
2. Barron, J.: A general and adaptive robust loss function. In: CVPR (2019)
3. Barz, B., Denzler, J.: Deep learning on small datasets without pre-training using cosine loss. In: WACV (2020)
4. Cao, K., Wei, C., Gaidon, A., Arechiga, N., Ma, T.: Learning imbalanced datasets with label-distribution-aware margin loss. In: NeurIPS (2019)
5. Elsayed, G.F., Krishnan, D., Mobahi, H., Regan, K., Bengio, S.: Large margin deep networks for classification. In: NeurIPS (2018)
6. Frosst, N., Papernot, N., Hinton, G.E.: Analyzing and improving representations with the soft nearest neighbor loss. In: ICML (2019)
7. Gajowniczek, K., Chmielewski, L.J., Orłowski, A., Ząbkowski, T.: Generalized entropy cost function in neural networks. In: ICANN (2017)
8. Ghosh: Robust loss functions under label noise for deep neural networks. In: AAAI (2017)
9. Gonzalez, S., Miikkulainen, R.: Improved training speed, accuracy, and data utilization through loss function optimization. In: CEC (2020)
10. Gonzalez, S., Miikkulainen, R.: Optimizing loss functions through multi-variate taylor polynomial parameterization. In: Genetic and Evolutionary Computation Conference (2021)
11. Griffin, G., Holub, A., Perona, P.: Caltech 256 (2022)
12. Guo, C., Pleiss, G., Sun, Y., Weinberger, K.Q.: On calibration of modern neural networks. In: ICML (2017)
13. He, K., Zhang, X., Ren, S., Sun, J.: Deep residual learning for image recognition. In: CVPR (2016)
14. Hinton, G.E., Vinyals, O., Dean, J.: Distilling the knowledge in a neural network. preprint (2015)
15. Khosla, P., et al.: Supervised contrastive learning. In: NeurIPS (2020)
16. Kingma, D.P., Welling, M.: Auto-encoding variational bayes. In: ICLR (2014)

17. Kobayashi, T.: T-vmf similarity for regularizing intra-class feature distribution. In: CVPR (2021)
18. Kozachenko, L.F., Leonenko, N.N.: Sample estimate of the entropy of a random vector. Problemy Peredachi Informatsii (1987)
19. Krizhevsky, A.: Learning multiple layers of features from tiny images. University of Toronto, Technical report (2009)
20. Leng, Z., et al.: Polyloss a polynomial expansion perspective of classification loss functions. In: ICLR (2021)
21. Lezama, J., Qiu, Q., Sapiro, P.M., G.: Ole orthogonal low-rank embedding, a plug and play geometric loss for deep learning. In: CVPR (2018)
22. Liu, W., Wen, Y., Yu, Z., Yang, M.: Large-margin softmax loss for convolutional neural networks. In: ICML (2016)
23. Ma, Y., Jiao, L., Liu, F., Wen, M., Li, L., Ma, W.: Predicting and enhancing the fairness of dnns with the curvature of perceptual manifolds. TPAMI (2024)
24. Nguyen, Q., Valizadegan, H., Hauskrecht, M.: Learning classification models with soft-label information. J. Am. Med. Inf. Assoc. (2014)
25. Pereyra, G., Tucker, G., Chorowski, J., Kaiser, L., Hinton, G.: Regularizing neural networks by penalizing confident output distributions. In: ICLR (2017)
26. Rosenblatt, M.: Remarks on some nonparametric estimates of a density function. Ann. Math. Stat. (1956)
27. Shi, W., Gong, Y., Tao, X., Zheng, N.: Training dcnn by combining maxmargin, max-correlation objectives, and correntropy loss for multilabel image classification. IEEE Trans. Neural Netw. Learn. Syst. (2017)
28. Silverman, B.W.: Density Estimation for Statistics and Data Analysis. Routledge, Abingdon (2018)
29. Singh, A., Pokharel, R., Principe, J.: The c-loss function for pattern classification. Pattern Recogn. (2014)
30. Sokolić, J., Giryes, R., Sapiro, G., Rodrigues, M.R.: Robust large margin deep neural networks. CoRR (2016)
31. Sun, P., Jin, X., Su, W., He, Y., Xue, H., Lu, Q.: A visual inductive priors framework for data-efficient image classification. In: ECCV Workshop (2020)
32. Sun, S., Chen, W., Wang, L., Liu, X., Liu, T.Y.: On the depth of deep neural networks a theoretical view. In: AAAI (2016)
33. Terven, J., Cordova-Esparza, D.M., Romero-González, J.A., Ramírez-Pedraza, A., Chávez-Urbiola, E.A.: A comprehensive survey of loss functions and metrics in deep learning. Artif. Intell. Rev. (2025)
34. Wan, W., Zhong, Y., Li, T., Chen, J.: Rethinking feature distribution for loss functions in image classification. In: CVPR (2018)
35. Wu, M., Mosse, M., Zhuang, C., Yamins, D., Goodman, N.: Conditional negative sampling for contrastive learning of visual representations. In: ICLR (2021)
36. Wu, Z., Efros, A.A., Yu, S.: Improving generalization via scalable neighborhood component analysis. In: ECCV (2018)
37. Yang, J., Shi, R., Ni, B.: Medmnist classification decathlon: a lightweight automl benchmark for medical image analysis. In: IEEE 18th International Symposium on Biomedical Imaging (2020)
38. Yang, J., et al.: Medmnist v2-a large-scale lightweight benchmark for 2d and 3d biomedical image classification. Sci. Data (2023)
39. Yuan, C., Zhou, C., Peng, J., Li, H.: Mixture correntropy-based robust distance metric learning for classification. Knowl.-Based Syst. (2024)
40. Zhang, Z., Sabuncu, M.: Generalized cross entropy loss for training deep neural networks with noisy labels. In: NeurIPS (2018)

Neural-CLS: Mitigating Similarity-Induced Interference via Neural-Inspired Complementary Learning Scheme for Class-Incremental Learning

Pengyue Zhang, Liu Yang(✉), and Qi Shen

Tianjin University, Tianjin, China
{pennyzhang,yangliuyl,shenqi_405}@tju.edu.cn

Abstract. Class-Incremental learning (CIL) aims to learn new classes sequentially while preserving previously learned knowledge. Although pre-trained vision-language models such as CLIP provide transferable representations, they remain vulnerable to similarity-induced interference: semantically similar classes often yield entangled features and conflicting updates, making new classes harder to distinguish while accelerating forgetting of old ones. Motivated by the complementary learning systems (CLS) view from neuroscience, which emphasizes both neocortical representation integration and hippocampal pattern separation, we propose **Neural-CLS** (**Neural**-Inspired **C**omplementary **L**earning **S**cheme), a CLIP-based framework for exemplar-free CIL with (i) a Knowledge Integration module that enriches class prototypes with fine-grained textual descriptions, providing more informative semantic supervision than class names alone, and (ii) a Graph-Driven Dynamic Grouping module that separates highly similar classes via graph coloring, thereby reducing structural interference during incremental learning. These two complementary components are unified by a group-aware knowledge routing mechanism, which performs similarity-conditioned, group-specific feature transformation for both visual and textual embeddings. In this way, Neural-CLS effectively mitigates similarity-induced interference by improving both new-class discrimination and old-knowledge retention. Extensive experiments on standard benchmarks demonstrate the superior performance of Neural-CLS in exemplar-free CIL.

Keywords: Class-Incremental Learning · Continual Learning · CLIP

1 Introduction

Conventional machine learning assumes all data are available simultaneously, yet real-world applications—autonomous driving, robotic perception, medical diagnosis—often encounter sequential data from evolving tasks. This necessitates Class-Incremental Learning (CIL) [14], which aims to continuously learn

L. Pasa et al. (Eds.): ICANN 2026, LNCS 17089, pp. 531–543, 2027.
https://doi.org/10.1007/978-3-032-38398-3_44

new classes while retaining prior knowledge. However, CIL faces a fundamental stability-plasticity dilemma, often leading to catastrophic forgetting (CF), where neural networks overwrite previously acquired knowledge when adapting to new data distributions.

Traditional CIL methods mainly mitigate catastrophic forgetting through rehearsal [2,14], regularization [1,15], or parameter-isolation strategies [4]. More recently, pre-trained models like Transformers [3] and CLIP [13] have emerged as a promising foundation for CIL, with prompt-based methods such as L2P [19], DualPrompt [20], CODA-Prompt [16] and adapter-based methods such as SimpleCIL [23] and RAPF [6] enabling parameter-efficient adaptation.

Meanwhile, recent studies [10,21] highlight that similarity-induced interference, both inter-task and intra-task, plays a critical role in forgetting. As shown in Fig. 1, similar classes often yield entangled representations and conflicting updates during sequential adaptation. Methods such as APD [22] and HALRP [9] attempt to mitigate this issue through parameter decomposition or low-rank adaptation, while theoretical results [10] suggest that reducing similarity among tasks improves generalization and robustness.

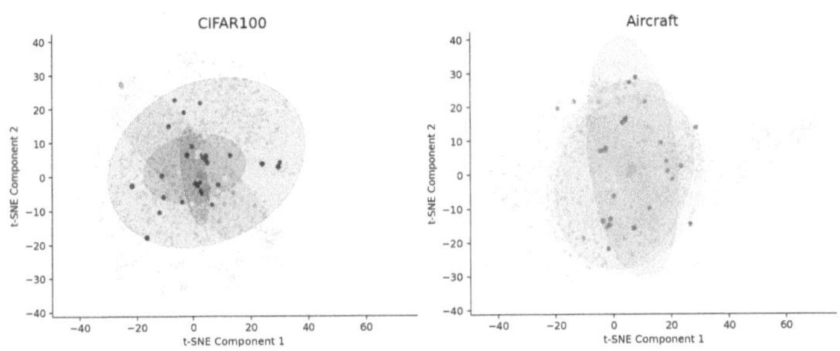

Fig. 1. Each point represents an image sample. Gray points show the overall embedding distribution, while colored points and translucent ellipses highlight representative classes. The overlap among highlighted classes indicates feature entanglement and similarity-induced confusion in the representation space. (Color figure online)

CLIP [13], as a representative pre-trained vision-language model, achieves strong zero-shot performance by aligning images and text in a shared embedding space. However, its supervision in class-incremental learning is often limited to coarse class names. For example, the prompt "a photo of a bird" conveys only a broad class identity, while omitting finer details that are essential for distinguishing between semantically similar subcategories such as different bird varieties. As new classes are introduced, this limitation can yield insufficiently discriminative representations, increasing confusion among semantically similar classes. Under sequential adaptation, such confusion induces conflicting updates that distort the shared representation space, weaken cross-modal alignment, and ultimately aggravate forgetting of previously learned classes. Together, these observations indicate that CIL requires not only preserving prior knowledge and enriching class representations, but also explicitly organizing overlapping classes to reduce interference throughout the incremental process.

In contrast to machine learning systems that remain vulnerable to representational entanglement, humans exhibit remarkable lifelong learning abilities. As illustrated in Fig. 2, the **Complementary Learning Systems**(CLS) framework, as proposed by McClelland et al. [12], provides a neurobiological model for understanding how the brain manages knowledge acquisition and integration over time. In this framework, the neocortex and hippocampus work together in a complementary manner to support lifelong learning. The neocortex integrates and refines representations, enabling generalization and the formation of distributed representations across related concepts while the hippocampus supports pattern separation, encoding new experiences into distinct representations, thereby reducing interference from similar experiences.

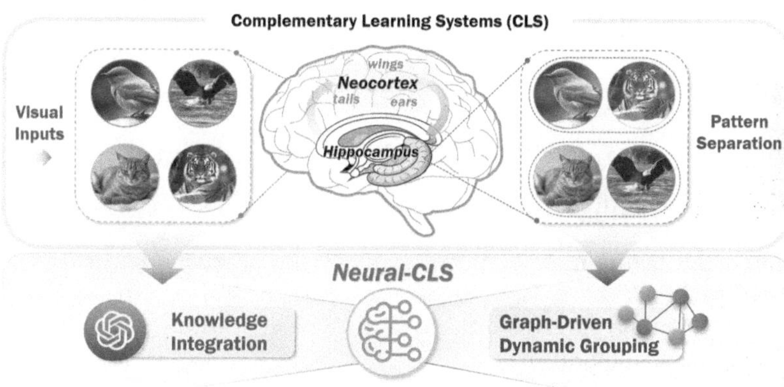

Fig. 2. Conceptual illustration of the CLS perspective underlying Neural-CLS. Neocortex-inspired knowledge integration improves fine-grained representation learning, while hippocampus-inspired graph-driven grouping performs pattern separation to disentangle overlapping classes and reduce incremental interference.

Motivated by this principle, we propose **Neural**-Inspired **C**omplementary **L**earning **S**cheme, a CLIP-based CIL framework inspired by the complementary roles of representation integration and pattern separation. **Neural-CLS** consists of two components: Knowledge Integration, which enhances new-concept representations through richer semantic supervision, and Graph-Driven Dynamic Grouping, which structures inter-class relationships to mitigate similarity-induced interference. Specifically, Neural-CLS leverages GPT-5 to generate fine-grained class descriptions that complement CLIP's coarse class-name supervision, and builds a class similarity graph to partition mutually confusable classes into distinct groups via graph coloring. Instead of applying a shared transformation to all classes, we introduce a Group-aware Knowledge Routing (GKR) module for each group, enabling group-specific feature transformation on both visual and textual branches through residual adapter pairs. As new classes arrive, the model updates existing groups or creates new ones, while optimizing only the corresponding GKR module under a shared feature space. During inference, outputs from all GKRs are aggregated for final prediction. Hence, our contributions are summarized as follows (Fig. 3):

- We present **Neural-CLS**, a novel CLIP-based framework for exemplar-free CIL inspired by the complementary roles of representation integration and pattern separation.
- We propose a **Knowledge Integration** approach that enriches CLIP with fine-grained semantic descriptions beyond coarse class names, improving discriminability for semantically similar classes.
- We develop a **Graph-Driven Dynamic Grouping** module with Group-aware Knowledge Routing mechanism, which applies group-specific feature transformations to mitigate similarity-induced interference during sequential learning, achieving state-of-the-art performance in exemplar-free CIL settings.

2 Method: Neural-CLS for Class-Incremental Learning

2.1 Problem Setting

In exemplar-free class-incremental learning (CIL), a model learns from a sequence of tasks $\{\mathcal{T}_1, \dots, \mathcal{T}_N\}$, where each task introduces disjoint label sets $\mathcal{Y}_t$. At step t, only the current dataset $\mathcal{D}_t = \{(x_i, y_i)\}$ with $y_i \in \mathcal{Y}_t$ is accessible, and the model is evaluated over

$$\mathcal{Y}^{1:t} = \bigcup_{\tau=1}^{t} \mathcal{Y}_\tau. \tag{1}$$

2.2 Knowledge Integration

We build upon CLIP with an image encoder f_v and a text encoder f_t. Given an image x and a class prompt p_c, we obtain

$$z_x = f_v(x), \quad z_c = f_t(p_c), \tag{2}$$

In class-incremental learning, a key challenge arises from similarity-induced interference, where classes with highly similar representations are prone to confusion under sequential updates. In standard CLIP, class representations are derived from template prompts such as

$$p_c^{\text{tmp}} = \text{``}\texttt{a photo of a } c\text{''}, \quad z_c^{\text{tmp}} = f_t(p_c^{\text{tmp}}). \tag{3}$$

These prompts provide only coarse semantic information and may lead to insufficient separation between similar classes. For example, the prompt "a photo

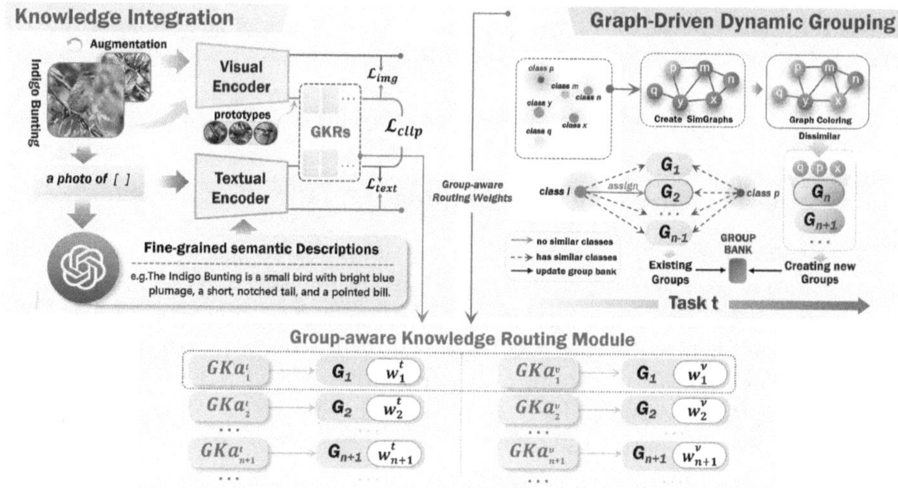

Fig. 3. Illustration of Neural-CLS. At each incremental step, the Knowledge Integration branch enriches CLIP text supervision with fine-grained descriptions, while the Graph-Driven Dynamic Grouping branch separates highly similar classes into different groups. Based on the resulting structure, the Group-aware Knowledge Routing mechanism activates group-aware knowledge adapters for both visual and textual features, reducing similarity-induced interference during class-incremental learning.

of an Indigo Bunting" only encodes the class identity, while a description such as "a small bird with bright blue plumage, a short, notched tail, and a pointed bill" captures more distinctive visual characteristics.

Such additional information helps distinguish classes that would otherwise be closely aligned in the embedding space. To address this issue, we enhance class representations by incorporating descriptive information while preserving the original template-based structure. For each class c, we construct a descriptor pool $\mathcal{D}_c^{\text{text}} = \{d_c^{(1)}, \ldots, d_c^{(K)}\}$, where each description captures a distinct visual characteristic. We then define the description-based feature as

$$z_c^{\text{desc}} = \frac{1}{K} \sum_{k=1}^{K} f_t(c \texttt{ with } d_c^{(k)}). \tag{4}$$

Instead of directly replacing the template representation, we combine the template feature and the description-based feature:

$$z_c^{\text{grp}} = \beta z_c^{\text{tmp}} + (1 - \beta) z_c^{\text{desc}}. \tag{5}$$

To further align the two representations, we enforce consistency between template-based and description-based features:

$$\mathcal{L}_{\text{text}} = -\cos(z_c^{\text{tmp}}, z_c^{\text{desc}}). \tag{6}$$

On the visual side, we adopt a similar strategy using data augmentation. Given an image x, we generate an augmented view $\mathcal{A}(x)$ and enforce feature consistency:

$$\mathcal{L}_{\text{img}} = -\cos(f_v(x), f_v(\mathcal{A}(x))). \tag{7}$$

2.3 Graph-Driven Dynamic Grouping

While the enriched representations improve class discrimination, highly similar classes may still remain close in the embedding space. When such classes are learned within the same update stage, their representations tend to overlap, leading to unstable decision boundaries. To further alleviate this issue, we introduce a Graph-Driven Dynamic Grouping Module that organizes classes in a conflict-aware manner. Instead of grouping similar classes together, we explicitly separate classes with high similarity that are likely to interfere, so that they are processed under different adaptation paths.

Class Similarity. We first compute a prototype for each class. The visual prototype is defined as

$$\mu_c^v = \frac{1}{|\mathcal{D}_c|} \sum_{(x_i, y_i = c)} f_v(x_i), \tag{8}$$

and the textual prototype is given by

$$\mu_c^t = z_c^{\text{grp}}. \tag{9}$$

We then define a fused similarity between two classes:

$$S(c_i, c_j) = (1 - \alpha)\cos(\mu_{c_i}^v, \mu_{c_j}^v) + \alpha \cos(\mu_{c_i}^t, \mu_{c_j}^t). \tag{10}$$

Conflict-Aware Assignment. Given a new class c, we evaluate its compatibility with each existing group g. A group is considered valid only if no member in the group is overly similar to c:

$$\max_{c' \in \mathcal{C}_g} S(c, c') < \tau. \tag{11}$$

Among all valid groups, we assign c to the one with the lowest average similarity:

$$g^*(c) = \arg\min_g \frac{1}{|\mathcal{C}_g|} \sum_{c' \in \mathcal{C}_g} S(c, c'). \tag{12}$$

This strategy ensures that classes within each group remain as dissimilar as possible, thereby reducing potential interference.

Graph-Based Partitioning. If a class cannot be assigned to any existing group, we construct a similarity graph over unresolved classes. Each class is treated as a node, and an edge is formed when

$$S(c_i, c_j) \geq \tau. \tag{13}$$

We then apply Welsh–Powell graph coloring to partition the graph into disjoint subsets, where no connected nodes share the same group. Each subset is instantiated as a new group, ensuring separation among highly similar classes.

Dynamic Update. As new classes arrive, the grouping structure is updated incrementally through the above assignment and partitioning procedures, resulting in a dynamic grouping mechanism that adapts to evolving class distributions while maintaining low intra-group similarity.

Group Prototypes. Finally, we compute group-level prototypes as:

$$\mu_g^v = \frac{1}{|\mathcal{C}_g|} \sum_{c \in \mathcal{C}_g} \mu_c^v, \quad \mu_g^t = \frac{1}{|\mathcal{C}_g|} \sum_{c \in \mathcal{C}_g} \mu_c^t. \tag{14}$$

Group-Aware Knowledge Routing Mechanism. The graph-driven dynamic grouping module provides a structured partition over classes along with group-level prototypes. We leverage this structure to guide feature transformation via a Group-aware Knowledge Routing (GKR) mechanism, containing two group-aware knowledge adapters, corresponding to the visual and textual branches. Each adapter is associated with a group and defines a group-specific transformation. For a feature $z \in \mathbb{R}^d$, the adapter of group g is defined as

$$A_g(z) = z + \lambda W_g^{(2)} \sigma\left(W_g^{(1)} z\right), \tag{15}$$

where $W_g^{(1)} \in \mathbb{R}^{m \times d}$, $W_g^{(2)} \in \mathbb{R}^{d \times m}$, and $\sigma(\cdot)$ is a non-linear activation.

To determine which groups should contribute to the transformation, we compute routing weights based on the similarity between input features and group prototypes. For the visual branch, we define

$$r_g^v = \cos(z_x, \mu_g^v), \quad w_g^v = \frac{\exp(r_g^v / T)}{\sum_{g'} \exp(r_{g'}^v / T)}, \tag{16}$$

where T is a temperature parameter. Similarly, for the textual branch, we define

$$r_g^t = \cos(z_c, \mu_g^t), \quad w_g^t = \frac{\exp(r_g^t / T)}{\sum_{g'} \exp(r_{g'}^t / T)}. \tag{17}$$

The routed representations are obtained by aggregating the outputs of group-specific adapters:

$$z'_x = \sum_g w_g^v A_g^v(z_x), \qquad z'_c = \sum_g w_g^t A_g^t(z_c). \tag{18}$$

Table 1. The table presents the average accuracy ($\bar{A}$) and final-task accuracy (A_l) for Neural-CLS and baseline methods.

Method	CIFAR100				ImageNet-R				UCF101			
	B0 Inc10		B50 Inc10		B0 Inc20		B100 Inc20		B0 Inc10		B50 Inc10	
	$\bar{A}$	A_l	$\bar{A}$	A_l	$\bar{A}$	A_l	$\bar{A}$	A_l	$\bar{A}$	A_l	$\bar{A}$	A_l
L2P	82.68	73.11	80.87	72.98	76.05	66.41	72.74	66.88	86.47	76.28	84.07	76.54
DualPrompt	81.62	72.39	80.13	72.48	76.09	66.78	73.11	67.71	85.33	75.91	84.19	76.44
CODA-Prompt	82.42	73.41	78.7	71.58	77.83	68.81	73.62	68.18	87.65	80.26	82.93	75.12
SimpleCIL	84.27	76.59	80.15	76.53	81.12	74.33	76.91	74.62	90.31	85.79	88.03	85.74
RAPF	85.13	77.62	82.21	77.84	81.18	70.61	76.22	70.11	92.15	80.41	90.42	81.43
Neural-CLS	**86.55**	**79.63**	**83.23**	**79.7**	**85.74**	**81.21**	**83.16**	**81.34**	**93.85**	**89.54**	**92.80**	**89.81**
Method	CUB200				Stanford Cars				Aircraft			
	B0 Inc20		B100 Inc20		B0 Inc10		B50 Inc10		B0 Inc10		B50 Inc10	
	$\bar{A}$	A_l	$\bar{A}$	A_l	$\bar{A}$	A_l	$\bar{A}$	A_l	$\bar{A}$	A_l	$\bar{A}$	A_l
L2P	71.24	58.06	75.62	66.23	76.51	61.95	76.48	65.55	46.87	29.04	44.29	32.17
DualPrompt	69.93	57.49	74.34	64.85	76.38	63.02	76.77	67.43	44.28	25.79	46.08	33.57
CODA-Prompt	73.09	62.88	74.02	62.25	80.09	66.59	75.18	64.07	46.06	27.72	44.97	32.17
SimpleCIL	83.36	77.42	79.67	76.44	92.18	86.74	88.71	86.97	60.24	48.54	54.03	47.82
RAPF	79.15	62.72	72.81	62.91	82.77	62.97	75.95	63.08	50.37	23.49	40.43	25.42
Neural-CLS	**86.68**	**80.15**	**82.98**	**79.86**	**94.29**	**90.45**	**92.09**	**90.69**	**70.07**	**59.17**	**64.82**	**59.86**

2.4 Training Objective

Classification Loss. We compute the CLIP-style contrastive loss using routed features:

$$\mathcal{L}_{\text{clip}} = \text{InfoNCE}(z'_x, z'_y), \tag{19}$$

where z'_y denotes the routed textual embedding corresponding to the ground-truth class y.

Group Regularization. We regularize the group structure by encouraging intra-group compactness and inter-group separation:

$$\mathcal{L}_{\text{group}} = \lambda_{\text{intra}} \sum_{g,c} (1 - \cos(\mu_c^v, \mu_g^v)) + \lambda_{\text{inter}} \sum_{g \neq g'} \max(0, \cos(\mu_g^v, \mu_{g'}^v) - \delta). \tag{20}$$

Thus, we optimize the model with a multi-term objective that aligns routed representations and regularizes group structure:

$$\mathcal{L} = \mathcal{L}_{\text{clip}} + \lambda_t \mathcal{L}_{\text{text}} + \lambda_i \mathcal{L}_{\text{img}} + \lambda_g \mathcal{L}_{\text{group}}. \tag{21}$$

2.5 Incremental Training and Inference

At incremental step t, the model is first trained on the current-task data using Eq. (21). During training, we optionally replay perturbed prototype features to stabilize previously learned knowledge. After training, image prototypes of newly introduced classes are recomputed, and the grouping structure is updated following the procedure in Sect. 2.3.

During inference, we first construct routed textual prototypes for all seen classes. Given a query image, we compute its routed visual feature by first estimating routing weights via Eq. (16), and then aggregating group-specific transformations as in Eq. (18). We then match it against the routed textual features. Optionally, a group-aware logit bias is applied to slightly increase the scores of classes belonging to the most relevant groups.

3 Experiments

3.1 Experimental Setup

Datasets. We evaluated the proposed method on six benchmark datasets that cover both generic and fine-grained class-incremental recognition: CIFAR100 [8], ImageNet-R [5], UCF101 [17], CUB200 [18], Stanford Cars [7], and Aircraft [11]. Following prior CIL works, we adopt "B-m Inc-n" to split the classes, where m and n denote the number of initial and incremental classes.

Baselines. We compare Neural-CLS with representative prompt and adapter-based CIL baselines, including L2P [19], DualPrompt [20], CODA-Prompt [16], SimpleCIL [23] and RAPF [6]. All methods are deployed with the same CLIP as initialization. After each incremental step, the model is evaluated on all classes observed so far.

Implementation Details. We build our framework on top of CLIP with a frozen vision–language backbone. Fine-grained semantic descriptions are generated offline and used for knowledge integration. The model is optimized with SGD for 10 epochs, with an initial learning rate of 0.05, batch size 64 and weight decay 0.05. All experiments were conducted on a single Nvidia RTX 3090 GPU with random seed 1993, 42, 2024, and the mean results are reported.

3.2 Main Results

Table 1 summarizes the overall comparison across five benchmarks. Neural-CLS consistently achieves the best performance by under all evaluated settings, demonstrating the effectiveness of jointly modeling knowledge integration and group-driven dynamic grouping. The gains are particularly notable on fine-grained datasets such as CUB200, Stanford Cars and Aircraft, where classes exhibit strong semantic and visual overlap, demonstrating the effectiveness of our framework in mitigating similarity-induced interference and improving robustness in CIL. Neural-CLS also performs strongly on more diverse domains such as ImageNet-R and UCF101, suggesting that the proposed framework is not limited to a specific visual regime.

On CUB200, ImageNet-R and Stanford Cars, Neural-CLS surpasses all compared methods in both $\bar{A}$ and A_l, indicating that the proposed framework improves not only final-task retention but also the overall incremental learning

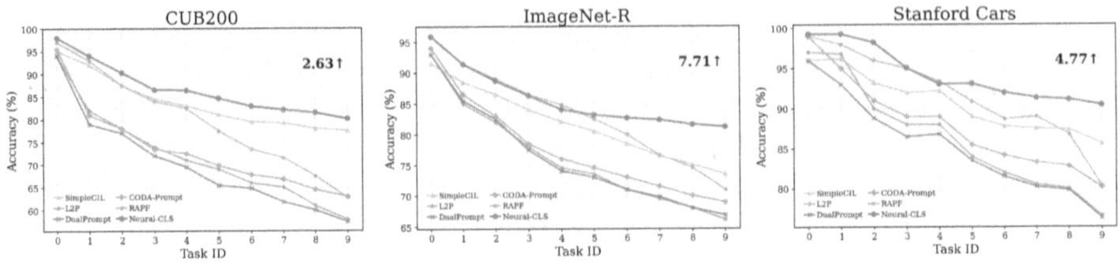

Fig. 4. The task-wise accuracy curves shown in the plot demonstrate the performance over 10 incremental tasks.

trajectory shown in Fig. 4. Overall, these results verify that combining Knowledge Integration with Graph-Driven Dynamic Grouping provides a more robust solution than using prompt tuning or shared adaptation alone.

3.3 Ablation Study

We conduct ablation studies to quantify the contribution of each component in Neural-CLS. The results in Table 2 show that both knowledge integration and group-driven dynamic grouping provide clear gains over the baseline.

Most importantly, combining both components leads to the strongest performance confirming that semantic refinement and structural disentanglement are complementary: the former improves *what* is represented, while the latter regulates *how* similar classes are adapted under sequential interference.

Table 2. Ablation results for CIFAR100 B0Inc10.

Method	$\bar{A}$	A_l
W/o Knowledge Integration	84.98	76.54
W/o Graph-Driven Dynamic Grouping	85.52	77.81
Neural-CLS	**86.55**	**79.63**

3.4 Analysis

We further visualize the class-level similarity structure on CUB200 before and after incremental training under our proposed method. As shown in Fig. 5, when classes are reordered according to their learned distribution, the similarity matrix exhibits a clearer block-wise structure. Highly similar classes are less likely to co-occur within the same group, while more distinct regions emerge across groups, suggesting that Neural-CLS improves not only classification performance but also the underlying relational organization of the representation space. This property is particularly desirable when overlapping semantics constitute a major source of catastrophic forgetting.

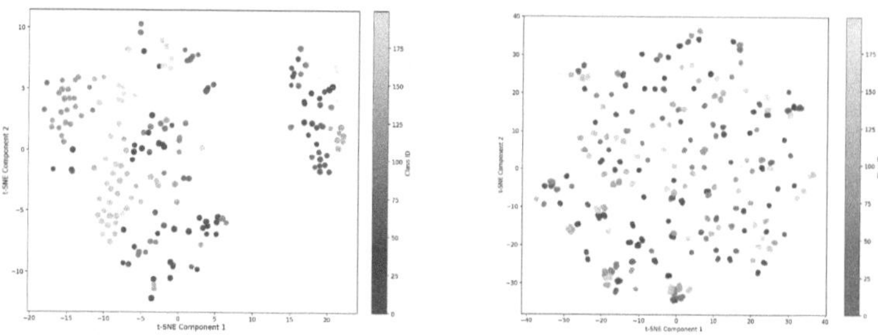

Fig. 5. t-SNE visualization of CUB200 class centers before (left) and after (right) incremental training with our proposed method.

4 Conclusion

We proposed a novel framework for class-incremental learning (CIL) that effectively addresses both catastrophic forgetting by reducing similarity-induced interference. Through incorporating a Knowledge Integration module, which enriches class prototypes with fine-grained textual descriptions, and a Graph-Driven Dynamic Grouping module, which organizes classes based on their similarity structure, Neural-CLS improves both final-task accuracy and the overall incremental learning trajectory. Our extensive experiments on various datasets show that Neural-CLS consistently outperforms existing CIL methods, demonstrating significant improvements in robustness and generalization across tasks. The proposed framework not only enhances the model's ability to retain knowledge over time but also allows it to adapt effectively to new classes, achieving state-of-the-art performance in exemplar-free settings. Future work will explore reducing inference overhead and extend Neural-CLS to more complex incremental scenarios.

Acknowledgements. This work was supported in part by the National Natural Science Foundation of China under Grant 62476194; and in part by the Independent Research Project of State Key Laboratory of Synthetic Biology under Project HCZC-202606A.

References

1. Aljundi, R., Kelchtermans, K., Tuytelaars, T.: Task-free continual learning. In: Proceedings of the IEEE/CVF Conference on Computer Vision and Pattern Recognition (CVPR) (2019)
2. Chaudhry, A., Ranzato, M., Rohrbach, M., Elhoseiny, M.: Efficient lifelong learning with A-GEM. In: Proceedings of the 35th International Conference on Machine Learning (ICML) (2018)
3. Dosovitskiy, A., et al.: An image is worth 16x16 words: transformers for image recognition at scale. In: Proceedings of the 9th International Conference on Learning Representations (ICLR) (2021)

4. He, J., Duan, Z., Zhu, F.: CL-LoRA: continual low-rank adaptation for rehearsal-free class-incremental learning. In: Proceedings of the IEEE/CVF Conference on Computer Vision and Pattern Recognition (CVPR) (2025)
5. Hendrycks, D., et al.: The many faces of robustness: a critical analysis of out-of-distribution generalization. In: Proceedings of the IEEE/CVF International Conference on Computer Vision (ICCV) (2021)
6. Huang, L., Cao, X., Lu, H., Liu, X.: Class-incremental learning with CLIP: adaptive representation adjustment and parameter fusion. In: Proceedings of the European Conference on Computer Vision (ECCV) (2024)
7. Krause, J., Stark, M., Deng, J., Fei-Fei, L.: 3D object representations for fine-grained categorization. In: Proceedings of the IEEE/CVF International Conference on Computer Vision (ICCV) (2013)
8. Krizhevsky, A., Hinton, G.: Learning multiple layers of features from tiny images. Technical report, University of Toronto (2009)
9. Li, J., et al.: Hessian aware low-rank perturbation for order-robust continual learning. IEEE Trans. Knowl. Data Eng. **36**(11), 6385–6396 (2024)
10. Lin, S., Ju, P., Liang, Y., Shroff, N.: Theory on forgetting and generalization of continual learning. In: Scarlett, J., Krause, A., Brunskill, E., Cho, K., Engelhardt, B., Sabato, S., Scarlett, J. (eds.) Proceedings of the 40th International Conference on Machine Learning (ICML). Proceedings of Machine Learning Research, vol. 202, pp. 21078–21100. PMLR (2023)
11. Maji, S., Rahtu, E., Kannala, J., Blaschko, M., Vedaldi, A.: Fine-grained visual classification of aircraft. In: Proceedings of the IEEE/CVF Conference on Computer Vision and Pattern Recognition (CVPR) (2013)
12. McClelland, J.L., McNaughton, B.L., O'Reilly, R.C.: Why there are complementary learning systems in the hippocampus and neocortex: insights from the successes and failures of connectionist models of learning and memory. Psychol. Rev. **102**(3), 419–457 (1995)
13. Radford, A., et al.: Learning transferable visual models from natural language supervision. In: Proceedings of the 38th International Conference on Machine Learning (ICML). Proceedings of Machine Learning Research, vol. 139, pp. 8748–8763. PMLR (2021)
14. Rebuffi, S.A., Kolesnikov, A., Sperl, G., Lampert, C.H.: iCaRL: incremental classifier and representation learning. In: Proceedings of the IEEE Conference on Computer Vision and Pattern Recognition (CVPR) (2017)
15. Shi, Y., et al.: Mimicking the oracle: an initial phase decorrelation approach for class incremental learning. In: Proceedings of the IEEE/CVF Conference on Computer Vision and Pattern Recognition (CVPR), pp. 16722–16731 (2022)
16. Smith, J.S., et al.: CoDA-Prompt: continual decomposed attention-based prompting for rehearsal-free continual learning. In: Proceedings of the IEEE/CVF Conference on Computer Vision and Pattern Recognition (CVPR), pp. 11909–11919 (2022)
17. Soomro, K., Zamir, A.R., Shah, M.: UCF101: a dataset of 101 human action classes from videos in the wild. In: Proceedings of the IEEE Conference on Computer Vision and Pattern Recognition Workshops, CVPRW (2012)
18. Wah, C., Branson, S., Welinder, P., Perona, P., Belongie, S.: The Caltech-UCSD birds-200-2011 dataset. Technical report CNS-TR-2011-001, California Institute of Technology (2011)
19. Wang, Z., et al.: Learning to prompt for continual learning. In: Proceedings of the IEEE/CVF Conference on Computer Vision and Pattern Recognition (CVPR) (2022)

20. Wang, Z., et al.: Dualprompt: complementary prompting for rehearsal-free continual learning. In: Proceedings of the European Conference on Computer Vision (ECCV), pp. 631–648 (2022)
21. Wu, T., et al.: Curriculum-meta learning for order-robust continual relation extraction. In: Proceedings of the AAAI Conference on Artificial Intelligence (AAAI) (2021)
22. Yoon, J., Kim, S., Yang, E., Hwang, S.J.: Scalable and order-robust continual learning with additive parameter decomposition. In: Proceedings of the AAAI Conference on Artificial Intelligence (AAAI) (2019)
23. Zhou, D.W., Cai, Z.W., Ye, H.J., Zhan, D.C., Liu, Z.: Revisiting class-incremental learning with pre-trained models: generalizability and adaptivity are all you need. Int. J. Comput. Vision (2024)

Addressing Class Imbalance in Kernel Thinning

Blanca Cano-Camarero(✉), Ángela Fernández-Pascual, and José R. Dorronsoro

Dpto. Ingeniería Informática, Universidad Autónoma de Madrid, Madrid, Spain
{blanca.cano,a.fernandez,jose.dorronsoro}@uam.es

Abstract. Kernel Thinning (KT) is an efficient sample reduction technique for kernel-based methods, but its application to classification tasks, particularly under class imbalance, may result in the loss of minority class representatives and biased subsets. To address this limitation, we propose several supervised extensions, namely Stratified Kernel Thinning (SKT), Conditional Kernel Thinning (CKT), and Joint Kernel Thinning (JKT), together with variants that incorporate SMOTE-based rebalancing. An extensive experimental study on a diverse collection of imbalanced datasets shows that SKT is the only method that guarantees class preservation by construction and exhibits the most robust and stable behavior across all reduction levels. However, SKT is not always the best-performing approach in terms of balanced accuracy. In particular, SMOTE-based variants such as SSKT and SJKT often achieve superior predictive performance for small to moderate reductions of the datasets. Moreover, our results indicate that on strongly imbalanced datasets, KT and its supervised variants can act as an implicit regularizer for KSVM, in some cases matching or even outperforming the full model. Overall, these findings reveal a trade-off between robustness and accuracy, guiding the choice of thinning strategies in highly skewed classification scenarios.

Keywords: Kernel Thinning · Imbalanced Classification · Kernel Methods · Sample Reduction · Coresets

1 Introduction

Kernel methods remain a cornerstone of modern machine learning due to their solid theoretical grounding and their ability to capture complex, non-linear relationships. Nevertheless, their practical applicability is often limited by their high computational cost, which grows rapidly with the number of training samples [6].

Recently, a novel approach known as Kernel Thinning (KT) [4] has been proposed as a data compression strategy aimed at reducing the effective size of the dataset while preserving its essential properties for kernel-based learning. This line of work opens a promising avenue for mitigating the computational burden of kernel methods. However, most of the subsequent research [5] has

L. Pasa et al. (Eds.): ICANN 2026, LNCS 17089, pp. 544–557, 2027.
https://doi.org/10.1007/978-3-032-38398-3_45

focused on regression tasks, where the integration of such techniques is more direct.

In the context of classification, the situation is more challenging. The thinning process may lead to the complete omission of specific classes in the reduced dataset, which compromises the validity of the learned model [1]. To the best of our knowledge, this issue has not been addressed in the literature, and no general methodology has been established to make these approaches consistently applicable to classification settings. To overcome this limitation, we introduce in this work label-aware variants of KT that incorporate supervision directly through the kernel definition.

We propose a class-stratified variant of KT, termed SKT (Sect. 2.2). We further introduce Conditional KT (CKT) by replacing MMD with its conditional form, CMMD [10] (Sect. 2.3), and Joint KT (JKT) using joint feature–label kernels (Sect. 2.4). These methods, together with standard KT (Sect. 2.1) and SMOTE variants (Sect. 2.5), are evaluated in Sect. 3, demonstrating effectiveness under class imbalance. Conclusions and further work are presented in Sect. 4.

2 Kernel Thinning for Classification Problems

2.1 Kernel Thinning

KT is a recently developed sampling technique that efficiently compresses a probability distribution P while preserving its statistical properties more effectively than standard i.i.d. sampling or classical thinning methods [4]. Given a positive-definite kernel function k, KT reduces an input dataset of n samples drawn from P into a smaller subset of size $\lfloor n/2^m \rfloor$, where m is a user-defined integer parameter controlling the compression level. The goal is to retain a comparable MMD in the associated Reproducing Kernel Hilbert Space (RKHS) [9].

Beyond the base kernel k, KT employs an auxiliary function known as the *square-root kernel*, which enables efficient recursive data splitting. More precisely, let $k : \mathbb{R}^d \times \mathbb{R}^d \to \mathbb{R}$ be a symmetric, positive-definite kernel. A function $k_{\text{rt}} : \mathbb{R}^d \times \mathbb{R}^d \to \mathbb{R}$ is called a *square-root kernel* for k if $k_{\text{rt}}(x, \cdot)$ is square-integrable for all $x \in \mathbb{R}^d$, and

$$k(x, y) = \int_{\mathbb{R}^d} k_{\text{rt}}(x, z)\, k_{\text{rt}}(y, z)\, dz, \quad \forall x, y \in \mathbb{R}^d. \tag{1}$$

For instance, if k is the Gaussian kernel $k(x, y) = \exp(-\gamma\|x - y\|^2)$, a valid square-root kernel is given by $k_{\text{rt}}(x, y) = \left(\frac{4\gamma}{\pi}\right)^{d/4} \exp\left(-2\gamma\|x - y\|^2\right)$.

The KT algorithm consists of two main steps [4]:

- **KT-Split:** Recursively partitions the dataset into 2^m candidate subsets of size approximately $n/2^m$ using randomized halving guided by k_{rt} and MMD. This step ensures that each split remains both balanced and diverse.
- **KT-Swap:** Selects the candidate subset that minimizes MMD w.r.t. the full dataset under the target kernel k, and refines it through a greedy swapping procedure.

Together, these operations produce a reduced coreset of $\lfloor n/2^m \rfloor$ elements with strong theoretical guarantees in terms of MMD preservation. A priori, KT does not incorporate label information, and to the best of our knowledge, only [5] has tackled the problem for regression and kernel ridge regression. In the context of classification, we propose the extensions explained in the following subsections.

2.2 Stratified Kernel Thinning

Stratified Kernel Thinning (SKT) extends KT to supervised settings. The underlying idea can be formalized from a probabilistic perspective. Let X denote the feature space and c the class label. The marginal distribution of the data can be decomposed as $P(X) = \sum_{c=1}^{C} P(X \mid Y = c)P(Y = c)$. KT aims to approximate the overall distribution $P(X)$ using a representative subset of samples. However, when class labels are available, another approximation can be obtained by applying KT to each conditional component $P(X \mid Y = c)$ individually. Formally, SKT approximates each class-conditional distribution and reconstructs the global approximation as $\widehat{P}(X) = \sum_{c=1}^{C} \widehat{P}(X \mid Y = c)\widehat{P}(Y = c)$, where $\widehat{P}(Y = c)$ denotes the empirical proportion of class c. This formulation provides a probabilistic justification for stratification: by preserving or adjusting $\widehat{P}(Y = c)$, the method allows one to maintain the prior class proportions or intentionally rebalance the dataset by changing the prior of each class to another weight, which may improve the performance of subsequent supervised models. Therefore, instead of processing the full dataset jointly, SKT partitions the data according to class labels and applies the KT procedure independently within each class.

Let $\mathcal{D} = (x_i, y_i)_{i=1}^{n}$ denote the dataset, where $y_i \in \{1, \ldots, C\}$ represents the class label. For each class c, a subset $\mathcal{S}_c \subseteq \{x_i : y_i = c\}$ of size $r_c = \left\lfloor \frac{|\{y_i = c\}|}{2^{m}_{c}} \right\rfloor$ is selected by minimizing the MMD within that class:

$$\mathcal{S}_c = \arg \min_{\mathcal{S} \subseteq \{x_i : y_i = c\} \wedge |\mathcal{S}| = r_c} \text{MMD}^2(\widehat{P}_c, \widehat{P}_{\mathcal{S}}), \tag{2}$$

where $\widehat{P}_c$ is the empirical distribution of all samples in class c, and $\widehat{P}_{\mathcal{S}_c}$ is that of the selected subset. The final coreset is obtained by merging all class-specific subsets, $\mathcal{S}\text{KT} = \bigcup_{c=1}^{C} \mathcal{S}_c$.

This class-wise stratification ensures balanced representation and prevents majority classes from dominating the selected subset, while maintaining the computational efficiency and theoretical guarantees of standard KT. Additionally, as ensembles do, it reduces the variance of the method. Moreover, by adjusting different m_c per class, it offers a natural mechanism to re-balance classes and potentially enhance the posterior performance of supervised models.

2.3 Conditional Kernel Thinning

Conditional Kernel Thinning (CKT) extends the Stratified-KT approach by explicitly aligning the conditional feature distributions across classes. Instead of treating each class independently, CKT seeks to minimize the discrepancy between the conditional distributions $P(X|Y = c)$ and $P_S(X|Y = c)$ for all classes $c \in \{1, \dots, C\}$ by the discrepancy metric during KT. Formally, instead of minimizing MMD, CKT minimizes the *Conditional Maximum Mean Discrepancy* (CMMD) [10] between the empirical distributions of the original dataset and the thinned coreset:

$$\mathrm{CMMD}^2 = \sum_{c=1}^{C} w_c \, \mathrm{MMD}^2\big(\widehat{P}(X|Y{=}c),\, \widehat{P}_{\mathcal{S}}(X|Y{=}c)\big), \tag{3}$$

where w_c denotes a class weight (typically $w_c = {\pi_c}^{-1}$, where π_c is the empirical class prior), $\widehat{P}(X|Y{=}c)$ is the conditional distribution of class c in the full dataset, and $\widehat{P}_{\mathcal{S}}(X|Y{=}c)$ is its corresponding distribution in the selected subset.

2.4 Joint Kernel Thinning

Joint Kernel Thinning (JKT) is an alternative that incorporates label information directly into the kernel function, yielding a label-aware version of KT that operates jointly over the feature–label space. Let $(x_i, y_i) \in X \times Y$ denote the samples, where x_i represents the input features and y_i its class label. We define a joint kernel over $X \times Y$ as

$$k\big((x_i, y_i), (x_j, y_j)\big) = k_X(x_i, x_j) k_Y(y_i, y_j), \tag{4}$$

where k_X is a positive-definite kernel over the input space and k_Y is a kernel over the label space. For discrete labels, k_Y can be defined as a Kronecker delta, i.e.

$$k_Y(y_i, y_j) = \mathbb{I}(y_i = y_j), \tag{5}$$

where $\mathbb{I}$ is one if $y_i = y_j$ and zero otherwise. Alternatively, k_Y can be defined as a similarity kernel that captures relations among classes.

Since the product of two positive-definite kernels is itself positive definite, the combined function in (4) defines a valid kernel on the joint space (X, Y). Moreover, its associated square-root kernel can also be defined as the product of the corresponding square-root kernels. Consequently, the corresponding RKHS norm and its associated MMD preserve the same mathematical properties as in standard KT.

2.5 Kernel Thinning in Combination with SMOTE

SMOTE [2] is a synthetic oversampling technique designed to address class imbalance by generating artificial samples of minority classes through interpolation between existing instances. Its computational cost is $\mathcal{O}(n_{\min}^2 d + s \cdot d)$, where $n_{\min}$ denotes the number of samples in the minority class, d the data dimensionality, and s the number of synthetic samples to be generated.

In our setting, SMOTE is applied prior to sample selection in order to balance class sizes and thus prevent underrepresentation or even complete omission of minority classes after coreset selection. This preprocessing step is particularly meaningful when combined with KT and JKT, giving rise to the variants denoted as SSKT and SJKT, respectively. In contrast, applying SMOTE in conjunction with CKT is not meaningful because, when classes are balanced, the CMMD term in CKT reduces to the standard MMD, rendering the class-aware formulation redundant.

3 Experiments: KSVM on Thinned Subsets

The experimental objective is to evaluate how label-aware thinning affects downstream classification. To this end, we trained a KSVM [6] on subsets obtained from *KT*, *SKT*, *CKT*, *JKT*, *SJKT* and *SSKT* and measured the resulting balanced accuracy (BA) on a test set. Additionally, we studied the final coreset class distribution ratio, ρ. Note that BA is defined as $\mathrm{BA} = \frac{1}{C}\sum_{c=1}^{C} \frac{\mathrm{TP}_c}{\mathrm{TP}_c+\mathrm{FN}_c}$, where TP_c and FN_c denote the number of true positives and false negatives of class c.

Given a labeled dataset $\mathcal{D} = \{(x_i, y_i)\}_{i=1}^{n}$, we split the data into train, validation, and test sets using stratified sampling (30/20/50). All methods operate on the same partitions, and features are standardized using min-max scaling. After selecting the best hyperparameters, the training and validation sets are merged.

We repeated each experiment ten times for each model and for $m \in \{1, \ldots, 5\}$, corresponding to dataset reductions to $\frac{100}{2^m}$ (i.e., 50%, 25%, 12.5%, 6.25%, and 3.12%). The common experimentation schema is described in Algorithm 1.

Algorithm 1. Experimental pipeline

Require: Dataset split $\mathcal{D}_{\text{train}}, \mathcal{D}_{\text{val}}, \mathcal{D}_{\text{test}}$
Require: KT method $\mathcal{V}_{KT}$ and m.
Require: $C_{\text{KSVM}} \in \{10^{-2}, \ldots, 10^{3}\}$, $\gamma \in \{10^{-3}, \ldots, 10^{1}\}/d$, where d is the dataset dimension.

$\mathcal{S}_m \leftarrow \mathcal{V}_{KT}(\mathcal{D}_{\text{train}}, m)$
$(C^\star, \gamma^\star) \leftarrow \arg\max_{C,\gamma} \mathrm{BA}\left(f_{C,\gamma}, \mathcal{D}_{\text{val}}\right)$
▷ Where $f_{C,\gamma} \leftarrow \textsc{Train}\big(\mathrm{KSVM}(C,\gamma), \mathcal{S}_m\big)$
$\mathcal{D}_{\text{tv}} \leftarrow \mathcal{D}_{\text{train}} \cup \mathcal{D}_{\text{val}}$
$\tilde{\mathcal{S}}_m \leftarrow \mathcal{V}_{KT}\left(\mathcal{D}_{\text{tv}}, m\right)$
$f \leftarrow \textsc{Train}\big(\mathrm{KSVM}(C^\star, \gamma^\star), \tilde{\mathcal{S}}_m\big)$
$\mathrm{BA}(\mathcal{V}_{KT}, m) \leftarrow BA(f, \mathcal{D}_{\text{test}})$
$\rho(\mathcal{V}_{KT}, m) \leftarrow \frac{\text{number of element in majority class}}{\text{number of elements minority class}}$.

3.1 Implementation Details

All implementations and experiments are available at https://doi.org/10.5281/zenodo.18413030 and build upon the official `goodpoints` library [4], which provides reference implementations of KT-SPLIT and KT-SWAP, as well as efficient MMD computation. This framework underlies the variants described in Sects. 2.2–2.4.

For the stratified variant (*SKT*), the dataset is partitioned by class, KT is applied independently to each subset, and the results are merged into the final coreset, ensuring class balance without modifying the original code.

For the conditional variant (*CKT*), the empirical discrepancy in the selection step is redefined by overriding `squared_emp_rel_mmd` to compute the conditional form (3). Class weights are set inversely proportional to class priors ($w_c = n/n_c$), enabling minimization of class-conditional discrepancies instead of a single marginal objective. For the joint variant, *JKT*, a composite kernel couples features and labels (4) where k_x is a Gaussian kernel on features and k_Y is a label kernel (Kronecker delta defined in (5)). The corresponding square-root kernel used during KT-SPLIT is $k_{\text{rt}}\big((x, y), (x', y')\big) = k_{X_{\text{rt}}}(x, x')\, k_Y(y, y')$, with $k_{X_{\text{rt}}}$ being the square-root kernels of k_X, and k_Y its own square-root kernel, i.e., $k_{Y\,rt} = k_Y$. All other components of the algorithm, including recursive splitting, candidate subset generation, and greedy swapping, remain unchanged to maintain compatibility with the theoretical properties of the original implementation.

Finally, SMOTE was implemented using the `imbalanced-learn` library [8]. Models involving SMOTE first balance the class distribution and then apply the desired kernel thinning strategy, or directly train the KSVM in the case of SMOTE-KSVM. Since SMOTE increases dataset size, we increase the number of splits, m_{SMOTE}, in order to preserve approximately the coreset size $\lfloor \frac{n}{2^m} \rfloor$.

Table 1. Characteristics of the selected datasets.

Name	Samples	Dimension	Ratio	Classes
abalone	4,177	10	9.7:1	7
abalone_19	4,177	10	130:1	19
car_eval_34	1,728	21	11.9:1	3
car_eval_4	1,728	21	25.6:1	4
coil_2000	9,822	85	16:1	2
ecoli	336	7	8.6:1	8
isolet	7,797	617	12:1	2
mammography	11,183	6	42.0:1	2
oil	937	49	21.9:1	3
optical_digits	5,620	64	9.1:1	10
ozone_level	2,536	72	34:1	2
pen_digits	10,992	16	9.4:1	10
protein_homo	145,751	74	111.5:1	2
satimage	6,435	36	9.3:1	6
scene	2,407	294	12.6:1	6
sick_euthyroid	3,163	42	9.8:1	2
solar_flare_m0	1,389	32	19:1	2
spectrometer	531	93	10.8:1	3
thyroid_sick	3,772	52	15.3:1	2
us_crime	1,994	100	12.3:1	2
webpage	34,780	300	33:1	2
wine_quality	4,898	11	26:1	6
yeast_me2	1,484	8	28.1:1	2
yeast_m18	2,417	103	13:1	8

3.2 Datasets

The datasets employed (see Table 1) in this study were obtained from the `imbalanced-learn` library [8], a specialized repository designed for research on imbalanced data problems. This collection provides standardized and widely used benchmark datasets that cover a broad range of domains and imbalance ratios, facilitating a consistent evaluation of algorithms under different imbalance conditions. In Table 1, the ratio represents how many elements of the majority class correspond to one element of the minority class.

3.3 Mean Balanced Accuracy and Ratio

After ten repetitions of the experiment, the mean BA obtained are reported in Table 2. If none of the repetitions produced a valid subset, this is with representations of all classes, a "–" symbol is shown. To determine whether performance differences between models are statistically significant, we follow the proposed methodology for multiple classifier comparisons by [3]. For each pair of models $(\mathcal{V}_i, \mathcal{V}_j)$, we perform a paired Wilcoxon signed-rank test [11] on the run-wise performance differences. To account for multiple comparisons, p-values are adjusted using the Holm–Bonferroni [7] procedure at significance level α. Model $\mathcal{V}_i$ is considered significantly better than model $\mathcal{V}_j$ if it achieves a higher mean performance and the corresponding Holm-adjusted FWER is smaller than $\alpha = 0.05$. The final ranking is obtained using a competition ranking scheme, where the rank of a model is 1 plus the number of models that are significantly better than it. Together with the mean BA, the rank is reported in brackets, along with a superscript indicating the number of coresets that fail to achieve representation of all classes. Note that the baseline models (KSVM and SMOTE-KSVM), which are trained on the full dataset, do not participate in the ranking calculation. A summary of the ranking is shown in Table 3. Overall, SMOTE-based methods (SJKT and SSKT) consistently achieve the best rankings across all compression levels, with statistically indistinguishable performance between them. This confirms the strong impact of rebalancing the class prior. Among the non-oversampling approaches, no single method uniformly dominates.

We first analyze how well the proposed methods approximate KSVM performance on large-scale problems. For datasets with more than 10,000 instances or features—namely `protein_homo`, `webpage`, and `mammography`—most approaches remain competitive with KSVM at low and moderate reduction levels. SMOTE-based variants consistently achieve the best performance, particularly for small and medium values of m, while classical KT-based methods degrade more rapidly as m increases.

In `protein_homo`, SMOTE-based variants dominate across almost all values of m, frequently ranking first and exhibiting stable BA even under aggressive reductions. In `webpage`, all methods provide strong approximations for $m = 1$ and $m = 2$; however, feasibility issues arise for some SMOTE-based variants as m increases, whereas SKT remains among the most robust alternatives. For `mammography`, although some KT-based methods perform competitively at low

Table 2. Mean balanced accuracy, ranking and number of models that omitted classes

Dataset	m	KSVM	SMOTE-KSVM	KT	SKT	CKT	JKT	SJKT	SSKT
protein_homo	1	95.1	95.5 ± 0.5	94.6 ± 0.4(3)	94.7 ± 0.3(3)	94.6 ± 0.4(3)	94.8 ± 0.4(2)	**95.1 ± 0.3(1)**	**94.9 ± 0.5(1)**
protein_homo	2			93.7 ± 0.7(6)	94.4 ± 0.5(3)	94.3 ± 0.6(3)	94.2 ± 0.5(4)	94.9 ± 0.5(2)^{1}	**95.2 ± 0.4(1)**
protein_homo	3			93.8 ± 0.5(3)	93.6 ± 0.8(5)	93.9 ± 0.5(3)	93.6 ± 0.4(6)	**94.4 ± 1.1(1)**	**94.2 ± 0.8(1)**
protein_homo	4			92.7 ± 0.7(2)	92.8 ± 0.2(2)	92.6 ± 0.8(2)	92.3 ± 0.3(6)	92.4 ± 7.0(4)	**94.6 ± 0.4(1)**
protein_homo	5			91.1 ± 0.8(3)	90.0 ± 0.6(5)	91.2 ± 0.1(3)	88.8 ± 6.0(6)	**94.4 ± 0.4(1)**	**94.2 ± 0.6(1)**
webpage	1	92.6	91.4 ± 0.6	90.2 ± 0.5(4)	**91.5 ± 1.1(1)**	90.3 ± 0.6(4)	90.3 ± 0.6(4)	90.7 ± 0.4(2)	90.5 ± 0.4(2)
webpage	2			88.7 ± 2.1(3)	87.8 ± 2.4(4)	88.1 ± 1.3(4)	88.4 ± 1.6(3)	**90.3 ± 0.5(1)**	**90.0 ± 0.8(1)**
webpage	3			**86.4 ± 0.0(1)**	84.0 ± 0.2(4)	84.5 ± 0.7(2)	84.6 ± 0.8(2)	–	49.9 ± 0.3(5)
webpage	4			81.5 ± 6.3(2)	**82.0 ± 3.8(1)**	75.2 ± 11.0(4)	**82.6 ± 3.4(1)**	–	50.0 ± 0.8(5)
webpage	5			**76.5 ± 9.4(1)**	69.5 ± 5.1(3)	**76.0 ± 8.5(1)**	64.5 ± 10.0(4)	–	50.0 ± 0.8(5)
mammography	1	90.1	90.7 ± 0.1	89.2 ± 0.8(5)	**90.4 ± 0.8(1)**	90.0 ± 0.4(3)	89.3 ± 1.7(5)	89.9 ± 1.0(3)	**90.3 ± 0.5(1)**
mammography	2			86.0 ± 5.1(5)	89.5 ± 0.4(3)	81.8 ± 5.6(6)	89.2 ± 1.4(3)	**90.1 ± 0.8(1)**	**89.8 ± 1.2(1)**
mammography	3			81.1 ± 5.4(6)	**87.2 ± 3.9(1)**	86.2 ± 1.8(4)	83.6 ± 8.5(4)	**87.3 ± 4.5(1)**	**88.3 ± 3.2(1)**
mammography	4			81.5 ± 1.5(4)^{1}	84.8 ± 4.4(3)	75.0 ± 16.8(5)	77.3 ± 9.7(5)	**89.9 ± 0.8(1)**	**88.6 ± 4.1(1)**
mammography	5			**68.0 ± 11.8(1)^{1}**	82.1 ± 1.7(3)	**75.5 ± 4.6(1)^{1}**	**63.3 ± 14.0(1)^{1}**	**89.9 ± 0.6(1)**	89.2 ± 0.8(2)
coil_2000	1	67.0	66.3 ± 0.6	65.4 ± 1.1(3)	65.6 ± 1.1(2)	65.3 ± 0.8(4)	65.1 ± 1.1(4)	**66.0 ± 0.9(1)**	**66.1 ± 1.0(1)**
coil_2000	2			62.5 ± 4.5(6)	64.5 ± 1.8(3)	64.7 ± 0.4(3)	64.9 ± 1.0(3)	**66.5 ± 0.8(1)**	66.1 ± 0.8(2)
coil_2000	3			62.8 ± 2.4(5)	64.9 ± 1.2(3)	62.3 ± 4.4(5)	63.9 ± 2.1(4)	**66.2 ± 1.1(1)**	**66.3 ± 1.7(1)**
coil_2000	4			59.6 ± 5.2(5)	62.9 ± 0.5(3)	61.0 ± 4.2(4)	60.9 ± 4.2(4)	64.8 ± 2.3(2)	**65.7 ± 1.4(1)**
coil_2000	5			56.6 ± 2.7(4)	60.1 ± 1.0(3)	–	–	**62.4 ± 1.4(1)**	60.9 ± 3.3(2)
isolet	1	94.9	95.6 ± 0.3	70.1 ± 1.1(5)	67.7 ± 5.8(5)	70.8 ± 1.0(3)	70.5 ± 0.7(4)	**81.8 ± 0.4(1)**	**81.7 ± 0.4(1)**
isolet	2			53.3 ± 2.0(6)	57.1 ± 0.3(4)	58.5 ± 8.4(3)	57.7 ± 2.8(3)	**90.7 ± 0.6(1)**	**90.7 ± 0.6(1)**
isolet	3			53.3 ± 9.6(2)	**57.1 ± 0.1(1)**	50.0 ± 0.0(4)	50.2 ± 0.4(3)	–	–
isolet	4			50.0 ± 0.0(2)	**55.7 ± 3.0(1)**	50.0 ± 0.0(2)	50.0 ± 0.0(2)	–	–
isolet	5			50.0 ± 0.0(2)	**56.4 ± 2.2(1)**	50.0 ± 0.0(2)	50.0 ± 0.0(2)	–	–
satimage	1	90.1	90.3 ± 0.3	87.6 ± 2.4(5)	86.4 ± 0.0(6)	88.6 ± 1.3(3)	88.2 ± 2.0(3)	**90.4 ± 1.5(1)**	**90.3 ± 1.4(1)**
satimage	2			84.7 ± 0.2(4)	85.6 ± 1.2(3)	84.1 ± 1.0(6)	84.6 ± 0.1(5)	87.6 ± 1.1(2)	**89.0 ± 0.9(1)**
satimage	3			83.4 ± 0.3(3)	82.9 ± 0.4(5)	83.4 ± 0.4(3)	82.5 ± 0.1(6)	–	–
satimage	4			77.6 ± 9.9(4)	81.0 ± 0.4(3)	81.1 ± 0.8(3)	77.3 ± 9.8(5)	**84.1 ± 1.1(1)^{7}**	–
satimage	5			**77.2 ± 5.3(1)**	76.9 ± 2.3(3)	71.2 ± 11.6(6)	**78.0 ± 5.4(1)**	79.0 ± 2.2(2)^{3}	**80.0 ± 1.2(1)^{2}**
optical_digits	1	98.4	98.1 ± 0.2	93.4 ± 6.4(6)	95.0 ± 0.6(4)	95.7 ± 0.9(3)	94.5 ± 0.3(5)	97.8 ± 0.3(2)	**97.9 ± 0.3(1)**
optical_digits	2			94.6 ± 0.2(3)	92.7 ± 0.5(6)	94.2 ± 0.9(4)	93.0 ± 0.5(5)	96.4 ± 0.9(2)	**96.8 ± 0.7(1)**
optical_digits	3			90.6 ± 0.8(4)	90.9 ± 0.2(3)	90.7 ± 0.8(4)	90.1 ± 1.2(6)	**94.4 ± 0.1(1)^{6}**	94.0 ± 0.2(2)^{6}
optical_digits	4			92.3 ± 0.9(2)	87.0 ± 0.6(6)	92.2 ± 0.9(2)	87.6 ± 2.4(5)	92.3 ± 1.5(2)	**93.2 ± 0.5(1)**
optical_digits	5			86.8 ± 3.1(3)	84.1 ± 12.2(4)	75.5 ± 17.8(5)	78.5 ± 15.6(4)	91.1 ± 1.4(2)	**92.2 ± 0.7(1)**
wine_quality	1	74.8	74.5 ± 0.5	69.6 ± 3.7(2)	65.8 ± 3.0(6)	68.3 ± 3.1(4)	68.0 ± 4.4(4)	**71.7 ± 2.6(1)**	70.2 ± 2.9(2)
wine_quality	2			62.3 ± 2.9(5)	66.2 ± 3.4(3)	63.4 ± 4.3(5)	65.2 ± 4.3(3)	70.1 ± 3.9(2)	**72.6 ± 1.5(1)**
wine_quality	3			61.3 ± 6.0(4)	63.3 ± 3.7(3)	61.5 ± 5.5(4)	63.0 ± 4.4(3)	70.4 ± 2.0(2)	**71.8 ± 1.7(1)**
wine_quality	4			**51.2 ± 1.7(1)^{8}**	59.9 ± 5.8(3)	**52.5 ± 0.0(1)^{8}**	**49.0 ± 3.9(1)^{7}**	**70.1 ± 2.9(1)**	67.8 ± 4.0(2)
wine_quality	5			–	62.9 ± 4.8(3)	–	–	**70.0 ± 1.6(1)**	**66.7 ± 8.0(1)**
abalone_19	1	71.4	74.5 ± 1.0	64.4 ± 8.3(4)	65.9 ± 10.1(2)	**68.0 ± 7.2(1)**	66.7 ± 6.9(2)	**69.0 ± 5.7(1)**	**70.1 ± 5.5(1)**
abalone_19	2			64.4 ± 7.1(3)^{2}	58.0 ± 5.5(6)	61.3 ± 7.0(4)	60.3 ± 7.0(4)	**72.7 ± 4.8(1)**	69.2 ± 3.9(2)
abalone_19	3			**56.4 ± 7.0(1)^{3}**	60.8 ± 6.7(3)	–	–	**73.7 ± 7.5(1)**	**70.6 ± 9.4(1)**
abalone_19	4			–	64.6 ± 5.6(3)	–	–	**76.4 ± 5.5(1)^{2}**	69.5 ± 1.4(2)^{1}
abalone_19	5			–	62.5 ± 7.7(3)	–	–	**73.9 ± 6.1(1)**	**73.6 ± 5.5(1)^{1}**
abalone	1	80.6	79.9 ± 0.4	78.8 ± 1.0(5)	**80.0 ± 0.9(1)**	79.1 ± 1.2(5)	**80.1 ± 0.8(1)**	**80.1 ± 0.8(1)**	**79.9 ± 0.6(1)**
abalone	2			78.6 ± 1.5(5)	**79.6 ± 0.5(1)**	79.3 ± 1.0(2)	**79.4 ± 0.8(1)**	**79.5 ± 0.5(1)**	78.7 ± 1.4(5)
abalone	3			75.3 ± 3.9(6)	79.1 ± 1.4(3)	78.8 ± 1.2(3)	**79.9 ± 0.6(1)**	**79.6 ± 0.8(1)**	78.7 ± 1.2(3)
abalone	4			76.7 ± 2.0(6)	77.5 ± 1.4(4)	78.6 ± 1.5(2)	**78.7 ± 2.7(1)**	**79.6 ± 0.6(1)**	77.7 ± 1.5(4)
abalone	5			70.1 ± 8.6(5)	72.8 ± 8.5(3)	72.5 ± 8.4(3)	59.8 ± 12.7(6)	**79.0 ± 1.1(1)**	77.7 ± 1.8(2)
thyroid_sick	1	86.3	86.3 ± 0.8	84.7 ± 0.0(2)	**83.7 ± 6.3(1)**	84.7 ± 0.0(3)	84.6 ± 0.0(4)	**86.1 ± 1.4(1)**	84.1 ± 3.2(2)
thyroid_sick	2			**86.0 ± 0.0(1)**	83.7 ± 0.0(5)	83.9 ± 2.7(4)	85.6 ± 2.2(2)	**85.9 ± 1.8(1)**	84.8 ± 2.2(4)
thyroid_sick	3			82.8 ± 0.0(2)	77.1 ± 1.0(5)	79.1 ± 1.6(3)	78.2 ± 2.2(4)	**84.7 ± 2.1(1)**	78.6 ± 13.6(2)
thyroid_sick	4			72.3 ± 7.2(3)	79.0 ± 1.0(2)	67.6 ± 12.2(4)	70.4 ± 7.4(5)	**79.2 ± 3.8(1)**	74.7 ± 4.5(3)
thyroid_sick	5			**64.3 ± 14.0(1)^{1}**	**77.4 ± 1.5(1)**	**63.0 ± 18.3(1)^{8}**	**65.5 ± 14.9(1)^{1}**	–	–
sick_euthyroid	1	90.2	90.3 ± 0.3	89.1 ± 0.4(2)	86.6 ± 0.7(6)	87.7 ± 1.2(4)	87.6 ± 0.5(4)	89.5 ± 2.1(2)	**90.8 ± 0.3(1)**
sick_euthyroid	2			86.1 ± 1.7(6)	87.7 ± 0.4(3)	87.5 ± 1.7(3)	87.3 ± 1.5(3)	**89.6 ± 1.6(1)**	**89.3 ± 2.1(1)**
sick_euthyroid	3			78.5 ± 0.4(5)	83.5 ± 1.8(3)	78.3 ± 0.9(5)	79.2 ± 8.5(4)	**89.2 ± 0.5(1)**	89.0 ± 2.2(2)
sick_euthyroid	4			79.1 ± 9.0(3)	74.5 ± 1.0(6)	78.3 ± 1.3(4)	**87.1 ± 0.0(1)**	84.9 ± 3.1(2)	81.8 ± 5.1(3)
sick_euthyroid	5			67.3 ± 7.4(3)	69.3 ± 1.5(3)	65.3 ± 6.4(4)	62.0 ± 8.9(5)	**73.1 ± 2.1(1)**	**72.8 ± 2.2(1)**

(continued)

Table 2. (*continued*)

ozone_level	1	76.2	76.6 ± 0.6	75.6 ± 2.0(3)	75.4 ± 3.2(2)	75.6 ± 2.0(3)	76.6 ± 0.1(2)	76.1 ± 2.9(2)	**78.9 ± 2.8(1)**
ozone_level	2			73.0 ± 5.8(4)	75.8 ± 2.1(3)	71.4 ± 6.7(4)	72.6 ± 9.5(3)	**77.2 ± 2.2(1)**	**77.7 ± 2.4(1)**
ozone_level	3			**69.8 ± 7.6(1)**^{1}	**77.5 ± 3.1(1)**	**65.7 ± 11.4(1)**^{1}	72.9 ± 5.8(4)	76.8 ± 1.7(2)	76.4 ± 1.4(2)
ozone_level	4			**72.3 ± 0.3(1)**^{6}	62.2 ± 9.1(3)	**71.9 ± 4.5(1)**^{6}	**74.5 ± 1.5(1)**^{4}	**76.9 ± 1.7(1)**	**75.9 ± 2.9(1)**
ozone_level	5			**69.1 ± 11.5(1)**^{8}	63.0 ± 7.8(3)	–	**66.2 ± 9.0(1)**^{7}	**74.2 ± 2.8(1)**	**74.4 ± 1.5(1)**
yeast_ml8	1	61.2	60.7 ± 4.4	54.5 ± 2.8(2)	**55.8 ± 2.8(1)**	54.4 ± 4.0(2)	55.3 ± 2.3(2)	**55.7 ± 4.8(1)**	**57.4 ± 5.5(1)**
yeast_ml8	2			57.6 ± 3.0(2)	52.7 ± 2.2(5)	**59.0 ± 3.7(1)**	52.9 ± 2.2(5)	54.7 ± 3.7(3)	55.2 ± 1.8(3)
yeast_ml8	3			54.7 ± 1.8(2)	51.4 ± 1.1(4)	50.9 ± 1.7(5)	52.2 ± 1.9(3)	–	–
yeast_ml8	4			50.1 ± 0.2(5)	51.3 ± 1.7(3)	51.6 ± 3.1(2)	50.1 ± 0.2(5)	**55.4 ± 3.3(1)**^{5}	52.4 ± 1.4(2)^{5}
yeast_ml8	5			50.0 ± 0.0(3)	49.9 ± 0.2(3)	**50.0 ± 0.1(1)**^{8}	**50.0 ± 0.0(1)**^{3}	**51.9 ± 3.4(1)**^{5}	**50.1 ± 0.2(1)**^{5}
scene	1	72.1	69.1 ± 0.6	70.3 ± 0.5(3)	**70.6 ± 1.1(1)**	69.8 ± 1.1(4)	**71.0 ± 2.1(1)**	68.2 ± 0.7(6)	68.6 ± 1.7(5)
scene	2			**68.9 ± 2.9(1)**	67.7 ± 6.2(3)	**68.8 ± 2.5(1)**	68.3 ± 6.4(3)	66.6 ± 3.9(5)	67.1 ± 4.2(4)
scene	3			64.6 ± 6.3(4)	**65.9 ± 5.4(1)**	**63.9 ± 7.7(1)**	64.4 ± 6.7(4)	65.0 ± 7.9(2)	**65.1 ± 8.0(1)**
scene	4			**55.6 ± 3.8(1)**^{1}	**56.7 ± 3.5(1)**	**58.7 ± 9.6(1)**^{2}	**56.1 ± 3.0(1)**^{1}	51.3 ± 3.9(2)	53.2 ± 6.8(2)
scene	5			–	**53.0 ± 2.6(1)**	–	–	49.7 ± 0.6(3)	51.4 ± 1.2(2)
us_crime	1	83.3	83.6 ± 0.5	**84.8 ± 0.3(1)**	84.2 ± 0.2(3)	83.5 ± 0.8(5)	82.7 ± 1.2(6)	84.4 ± 0.7(2)	84.1 ± 0.9(3)
us_crime	2			82.8 ± 2.3(3)	82.2 ± 0.9(5)	83.4 ± 1.0(3)	81.5 ± 5.7(6)	83.5 ± 2.5(2)	**84.5 ± 0.7(1)**
us_crime	3			80.5 ± 3.5(3)	81.4 ± 1.7(2)	62.1 ± 15.7(6)	74.2 ± 13.2(4)	**82.7 ± 0.8(1)**^{7}	**75.3 ± 16.9(1)**^{6}
us_crime	4			**70.8 ± 7.9(1)**^{1}	52.8 ± 8.8(4)	**66.2 ± 15.7(1)**^{1}	71.5 ± 4.8(3)	**82.7 ± 0.7(1)**^{6}	**82.4 ± 2.1(1)**^{6}
us_crime	5			**74.4 ± 0.6(1)**^{5}	69.8 ± 5.3(3)	**50.0 ± 0.0(1)**^{8}	**69.0 ± 8.4(1)**^{3}	74.5 ± 10.1(2)^{2}	**80.0 ± 1.0(1)**^{6}
car_eval_4	1	99.4	99.4 ± 0.0	98.5 ± 1.2(3)	98.8 ± 0.9(3)	97.8 ± 2.1(5)	98.1 ± 1.2(5)	99.3 ± 0.2(2)	**99.4 ± 0.1(1)**
car_eval_4	2			93.6 ± 3.1(4)	89.9 ± 14.1(4)	**96.7 ± 2.0(1)**	94.5 ± 3.0(4)	**97.3 ± 1.3(1)**	96.8 ± 0.9(2)
car_eval_4	3			74.1 ± 20.8(5)	88.5 ± 4.4(3)	83.9 ± 18.1(2)	67.7 ± 10.2(5)	**93.4 ± 0.7(1)**^{6}	92.2 ± 0.0(2)^{8}
car_eval_4	4			**59.2 ± 12.6(1)**^{5}	75.3 ± 10.2(2)	**74.4 ± 22.3(1)**^{5}	**70.4 ± 18.7(1)**^{7}	**78.2 ± 19.3(1)**	**84.7 ± 13.1(1)**^{1}
car_eval_4	5			**50.0 ± 0.0(1)**^{7}	**68.5 ± 15.9(1)**	**60.5 ± 2.8(1)**^{8}	–	**58.9 ± 17.8(1)**^{6}	50.0 ± 0.0(2)^{7}
car_eval_34	1	99.2	98.9 ± 0.9	93.6 ± 1.7(5)	93.7 ± 1.4(5)	94.4 ± 1.1(4)	96.9 ± 1.5(2)	**97.6 ± 1.0(1)**	**97.3 ± 1.9(1)**
car_eval_34	2			92.2 ± 2.3(3)	91.7 ± 0.9(5)	92.7 ± 3.0(3)	92.6 ± 2.7(3)	**94.6 ± 1.0(1)**	**94.9 ± 1.0(1)**
car_eval_34	3			**88.1 ± 4.0(1)**	85.7 ± 2.9(2)	84.8 ± 12.8(2)	82.3 ± 11.6(3)	–	–
car_eval_34	4			65.4 ± 19.9(4)	80.7 ± 3.4(3)	**76.0 ± 16.4(1)**^{1}	**76.4 ± 15.1(1)**^{1}	**87.5 ± 2.0(1)**^{8}	**87.9 ± 1.0(1)**^{4}
car_eval_34	5			**54.2 ± 10.3(1)**^{4}	**67.7 ± 15.3(1)**	–	**63.4 ± 18.9(1)**^{8}	50.0 ± 0.0(3)^{3}	56.1 ± 10.5(2)^{1}
yeast_me2	1	82.8	82.5 ± 2.5	**82.0 ± 3.3(1)**	**81.3 ± 2.0(1)**	**82.1 ± 3.4(1)**	79.9 ± 4.1(5)	**81.6 ± 2.8(1)**	79.2 ± 4.2(5)
yeast_me2	2			72.3 ± 9.0(5)	78.3 ± 4.5(3)	–	79.3 ± 3.0(2)	79.7 ± 3.0(2)	**81.0 ± 5.3(1)**
yeast_me2	3			72.6 ± 11.6(3)	64.3 ± 10.1(5)	68.3 ± 12.9(4)	74.0 ± 11.6(3)	**80.3 ± 4.5(1)**	**81.6 ± 2.5(1)**
yeast_me2	4			**54.4 ± 4.2(1)**^{8}	67.6 ± 10.3(3)	–	**50.0 ± 0.0(1)**^{7}	**79.1 ± 5.6(1)**	**80.8 ± 2.6(1)**
yeast_me2	5			–	69.8 ± 3.0(2)	–	–	67.5 ± 13.7(2)^{1}	**76.3 ± 10.4(1)**
solar_flare_m0	1	70.9	71.9 ± 1.7	**72.0 ± 1.1(1)**	**70.5 ± 7.3(1)**	**72.2 ± 2.1(1)**	66.6 ± 9.0(4)	68.0 ± 5.4(4)	66.4 ± 8.8(4)
solar_flare_m0	2			67.1 ± 6.0(4)	**68.7 ± 6.9(1)**	56.3 ± 2.6(5)	54.6 ± 4.9(6)	**69.2 ± 4.7(1)**	**69.5 ± 3.7(1)**
solar_flare_m0	3			**59.5 ± 9.9(1)**^{2}	63.2 ± 1.6(2)	–	–	**66.6 ± 8.9(1)**	**69.1 ± 7.0(1)**
solar_flare_m0	4			**50.0 ± 0.1(1)**^{8}	61.7 ± 2.9(3)	–	–	**68.9 ± 4.8(1)**	**70.4 ± 2.5(1)**
solar_flare_m0	5			–	**66.2 ± 2.5(1)**	–	–	61.2 ± 6.6(2)	61.0 ± 5.5(2)^{3}
oil	1	73.8	74.6 ± 1.6	**74.7 ± 4.1(1)**	68.8 ± 1.7(4)	69.7 ± 8.0(4)	67.8 ± 10.1(4)	72.8 ± 4.8(2)	**75.0 ± 2.4(1)**
oil	2			60.4 ± 6.1(4)	64.0 ± 2.5(3)	54.6 ± 7.8(6)	58.8 ± 5.3(4)	**69.1 ± 3.9(1)**	**69.2 ± 4.6(1)**
oil	3			**51.6 ± 3.3(1)**^{4}	56.1 ± 5.1(3)	**58.3 ± 6.4(1)**^{4}	**57.4 ± 7.6(1)**^{4}	61.1 ± 4.8(2)	**64.2 ± 5.2(1)**
oil	4			**53.4 ± 5.8(1)**^{7}	55.6 ± 2.9(2)	–	**49.7 ± 0.5(1)**^{7}	**59.2 ± 8.5(1)**^{2}	**59.9 ± 7.6(1)**
oil	5			–	**58.8 ± 5.1(1)**	–	–	**49.7 ± 0.5(1)**^{8}	**60.4 ± 6.3(1)**
spectrometer	1	94.9	94.6 ± 1.1	92.2 ± 0.3(3)	90.0 ± 6.6(4)	88.3 ± 13.6(3)	92.6 ± 0.4(2)	**95.2 ± 1.2(1)**	91.8 ± 6.9(4)
spectrometer	2			81.6 ± 18.1(3)	86.0 ± 14.4(3)	77.1 ± 16.6(5)	83.6 ± 14.0(4)	**92.8 ± 8.3(1)**	**90.0 ± 10.3(1)**
spectrometer	3			66.2 ± 18.6(3)	61.9 ± 9.7(3)	**81.3 ± 17.5(1)**^{5}	**67.3 ± 8.5(1)**^{4}	**95.5 ± 2.0(1)**	**95.0 ± 1.7(1)**
spectrometer	4			**88.1 ± 0.0(1)**^{8}	62.5 ± 11.3(2)	**70.4 ± 23.6(1)**^{6}	**58.8 ± 11.2(1)**^{6}	**83.1 ± 8.5(1)**^{2}	**93.0 ± 8.0(1)**
spectrometer	5			**71.2 ± 3.4(1)**^{6}	67.2 ± 6.1(3)	**67.7 ± 2.0(1)**^{6}	**68.4 ± 2.3(1)**^{7}	**86.7 ± 5.5(1)**	**86.0 ± 7.5(1)**
ecoli	1	83.6	84.0 ± 1.8	79.1 ± 10.6(6)	83.6 ± 2.2(2)	**85.0 ± 2.4(1)**	82.5 ± 11.8(3)	84.2 ± 2.0(2)	83.4 ± 2.3(3)
ecoli	2			80.7 ± 11.5(3)	**85.9 ± 2.6(1)**	**85.4 ± 2.7(1)**	84.4 ± 2.7(2)	83.1 ± 3.1(4)	83.9 ± 3.9(3)
ecoli	3			**85.2 ± 5.4(1)**^{2}	**84.7 ± 4.3(1)**	**55.7 ± 14.0(1)**^{4}	**85.8 ± 2.1(1)**^{6}	**83.7 ± 3.5(1)**	**84.5 ± 3.9(1)**
ecoli	4			**69.2 ± 22.3(1)**^{6}	76.8 ± 14.5(3)	–	**70.2 ± 28.6(1)**^{8}	**85.9 ± 4.0(1)**	83.8 ± 4.1(2)
ecoli	5			–	68.1 ± 14.1(2)	–	**79.1 ± 19.5(1)**^{6}	**73.3 ± 13.7(1)**^{1}	**78.7 ± 6.9(1)**^{2}

reduction levels, both BA and stability deteriorate markedly under more aggressive reductions.

Table 3. Mean ranking (± standard deviation) of the thinning methods + KSVM classifier across all datasets. Methods in **bold** are not significantly different from the best according to the Nemenyi test.

m	KT	SKT	CKT	JKT	SJKT	SSKT
1	4.2 ± 1.7	3.8 ± 1.7	3.8 ± 1.5	4.3 ± 1.5	$\mathbf{2.2 \pm 1.2}$	$\mathbf{2.7 \pm 1.7}$
2	4.4 ± 1.6	4.0 ± 1.6	4.2 ± 1.5	4.3 ± 0.9	$\mathbf{2.1 \pm 1.3}$	$\mathbf{2.0 \pm 1.3}$
3	4.4 ± 1.3	3.7 ± 1.6	4.5 ± 1.3	4.3 ± 1.5	$\mathbf{2.0 \pm 1.4}$	$\mathbf{2.2 \pm 1.3}$
4	4.5 ± 1.3	4.0 ± 1.4	4.1 ± 1.3	4.4 ± 1.5	$\mathbf{2.0 \pm 1.4}$	$\mathbf{1.9 \pm 1.2}$
5	3.8 ± 0.8	4.0 ± 1.6	5.0 ± 1.0	4.7 ± 1.4	$\mathbf{1.8 \pm 1.4}$	$\mathbf{1.8 \pm 0.6}$

Overall, SMOTE-based variants provide the closest approximation to KSVM in terms of BA on large datasets, but they may still omit minority classes. Moreover, SMOTE introduces non-negligible computational overhead, especially as class imbalance increases, since KT and its variants operate on an expanded dataset prior to reduction. Consequently, although SMOTE-based methods achieve strong rankings, SKT often offers a more favorable trade-off in practice, attaining similar mean ranks with substantially lower computational cost.

For medium and small datasets (below 10,000 features), most methods perform closely to KSVM well and frequently match or outperform it in terms of average ranking. A notable exception is `isolet`, where SMOTE-based approaches perform well for small values of m but degrade sharply for $m > 3$. In these regimes, several methods collapse to a balanced accuracy of 50% in binary classification problems (`isolet` and `us_crime`) and exhibit poor rankings, indicating a pathological behavior in which the classifier predicts a single class. This highlights the intrinsic difficulty of this dataset under aggressive reduction.

For the remaining datasets, the proposed methods often achieve rankings comparable to or better than KSVM, and in several cases one of them consistently rank first. This behavior can be attributed to an implicit regularization effect induced by data reduction or resampling. Such improvements are particularly evident in datasets such as `satimage`, `abalone_19`, `sick_euthyroid`, `ozone_level`, `us_crime`, `car_eval_4`, `solar_flare_m0`, `oil`, `spectrometer`, and `ecoli`, where performance gains become more pronounced as dataset size decreases.

Let's now analyze the missing classes in the experiments performed, the ranking dynamics that appear with the reduction of level m and, finally, let's focus on the class ratios.

Analysis of Missing Classes. Missing results, indicated by superscripts in Table 2, occur when at least one class is omitted after reduction. As expected, no omission occurs at low compression ($m \leq 2$), where the coreset size remains sufficiently large.

As compression increases, omission becomes more frequent for all non-stratified methods, confirming the theoretical prediction that aggressive thinning exacerbates class loss. In particular, KT, CKT, and JKT exhibit rapidly increasing omission rates, with high variability across datasets.

In contrast, SKT completely eliminates omission across all settings, highlighting the effectiveness of enforcing class-wise selection. This confirms that stratification directly addresses the combinatorial source of omission.

SMOTE-based methods (SSKT and SJKT) consistently reduce omission compared to their non-oversampled counterparts, although they do not fully prevent it. This aligns with the theoretical analysis, where increasing π_c mitigates but does not eliminate omission.

Overall, the results illustrate that omission is primarily driven by class imbalance, and that methods explicitly controlling class proportions (SKT, SMOTE) are the most effective in practice. As datasets are ordered from largest to smallest, a clear global trend emerges: missing-class events increase as dataset size decreases (m grows). For $m = 1$, no method omits classes. At $m = 2$, omissions are rare (mean number of omissions: KT = 0.09, CKT = 0.45; others and SKT = 0). At $m = 3$, the effect intensifies: KT = 0.55, JKT = 1.55, SJKT = 3.14, SSKT = 2.7; SKT remains at 0. For larger reductions, gaps widen. At $m = 4$: KT = 3.14, CKT = 3.59, JKT = 3.09; stratified SMOTE variants reduce this to 2.36 (SJKT) and 2.14 (SSKT). At $m = 5$, omissions are frequent for non-stratified methods (KT = 4.64, CKT = 6.32, JKT = 5.27), while SJKT and SSKT limit them to 2.68 and 2.59; SKT never omits a class.

These results confirm that method choice plays a crucial role as dataset size decreases and reduction becomes more aggressive. While SMOTE-based variants often achieve the best average rankings, they may still suffer from missing classes for large m. For intermediate values of m, class-aware methods (SKT, CKT, SJKT, SSKT, SJKT) do not improve upon simple KT; this can be explained by the fact that the target information is implicitly encoded in the features. However, in the extreme case $m = 5$, KT performs considerably worse in rankings than its variants. By construction, SKT provides the only methods with guarantees for class omission prevention, whereas for SMOTE-methods aggressive oversampling followed by reduction can, in certain regimes, exacerbate rather than mitigate class disappearance.

Analysis of Ranking Dynamics across Reduction Scales. Table 2 in parenthesis reports the mean ranking of each method as a function of the reduction level m, allowing us to analyze how relative performance evolves as the reduction becomes more aggressive. If for a m and a dataset has no score, then maximum rank is assumed.

In the majority of tested datasets, the SMOTE-based methods SJKT and SSKT consistently achieve the highest rankings (typically rank 1 or 2). This suggests that data generation improves over the KT, SKT, or CKT methods when a moderate amount of data is available.

As the reduction factor, m increases to 4 or 5, the rankings exhibit two divergent behaviors:

- Robustness: In datasets such as `optical_digits`, `coil_2000`, and `wine_quality`, SJKT and SSKT maintain their rank 1 status despite the severe reduction in samples.
- Convergence Failure: In datasets like `webpage`, `isolet`, and `scene`, SMOTE methods show a performance collapse or missing values at $m = 5$. In these instances, the rank 1 position reverts to the simpler methods like KT or SKT. This indicates that while SJKT and SSKT have a higher performance ceiling, they may require a minimum sample threshold to remain stable.

Finally, the intermediate methods CKT and JKT generally occupy mid-tier rankings (ranks 3 to 5). Their performance is relatively stable but they rarely surpass the SMOTE variants. However, in specific cases like `abalone_19` at $m = 1$, CKT manages to secure a rank 1.

Class Ratios. Additionally, after subsampling, we computed the ratios between the majority and minority class sizes, defined as the number of majority-class samples divided by the number of minority-class samples. These results are omitted from the paper due to space constraints, but can be computed using the provided software. The most stable behavior corresponds to SKT, which by construction preserves a class-wise partition and therefore maintains a nearly constant ratio. Additionally, the SMOTE-based variants exhibit smaller discrepancies between classes (i.e., ratios closer to one); however, the ratio is not exactly one and tends to increase as m grows. For the remaining methods, no clear pattern is observed, as the class ratio may either increase or decrease depending on the dataset, model and the value of m.

4 Conclusions and Further Work

In conclusion, this work examined the limitations of Kernel Thinning when applied to classification tasks under class imbalance and introduced several label aware extensions—SKT, CKT, and JKT—as well as variants incorporating SMOTE (SSKT from SMOTE + KT and SJKT from SMOTE + JKT). Among these, SKT is the only method that strictly guarantees class preservation by construction and exhibits the most robust and stable behavior across all reduction levels. However, it does not always achieve the highest balanced accuracy. In fact, SMOTE-based variants such as SJKT and SSKT often achieve better balanced accuracy rankings, especially for small to moderate size reductions, highlighting the benefit of explicit data rebalancing prior to thinning. Moreover, our results show that on strongly imbalanced datasets, KT and its label aware variants can act as implicit regularizers for KSVM, in some cases matching or even surpassing the performance of the full model. Overall, these findings suggest that while SKT is the safest choice when class coverage must be guaranteed, alternative label aware and SMOTE-based strategies can yield superior predictive performance in highly skewed classification scenarios and in small to medium-sized datasets. Conversely, in large-scale settings it is preferable to use SKT (or JKT or CKT), since performance differences are marginal while the computational cost is lower.

Even though this work focuses on classification problems, JKT remains the standard—and, to our knowledge, the only—method commonly used for regression. However, in classification we observe that other variants can outperform JKT. It would therefore be of interest to investigate whether this advantage extends to regression by adapting these variants accordingly. One possible direction is to extend SKT to regression via stratification based on binning of the target variable, enabling preservation of the response distribution within the reduced subset. This could help determine whether the benefits observed in classification, such as improved stability and reduced performance degradation, also transfer to continuous prediction settings.

More generally, the JKT, CKT, and SKT variants do not exhibit differences as pronounced as might be expected, except on certain datasets. This suggests that their relative performance is driven by underlying dataset characteristics, with advantages emerging only under specific conditions. Understanding these factors and identifying the properties that favor each variant constitutes an important direction for future research.

Additionally, it would be interesting to study the effect of alternative kernel combinations in JKT, as well as different class-weighting strategies in CKT.

Acknowledgments. The authors acknowledge financial support from project PID2022-139856NB-I00 funded by MCIN/AEI/10.13039/ 501100011033/ FEDER, UE and project IDEA-CM (TEC-2024/COM-89) from the CAM, the Cátedra UAM-IIC de Ciencia de Datos y Aprendizaje Automático, and FPI-UAM, as well as computational support from the Centro de Computación Científica Universidad Autónoma de Madrid (CCC-UAM).

References

1. Cano, B., Fernández, Á., Dorronsoro, J.R.: Analysis of kernel thinning for scalable support vector machines. In: International Conference on Hybrid Artificial Intelligence Systems, pp. 309–321. Springer (2025)
2. Chawla, N.V., Bowyer, K.W., Hall, L.O., Kegelmeyer, W.P.: Smote: synthetic minority over-sampling technique. J. Artif. Intell. Res. **16**, 321–357 (2002)
3. Demšar, J.: Statistical comparisons of classifiers over multiple data sets. J. Mach. Learn. Res. **7**, 1–30 (2006)
4. Dwivedi, R., Mackey, L.: Kernel thinning. JMLR **25**(152), 1–77 (2024)
5. Gong, A., Choi, K., Dwivedi, R.: Supervised kernel thinning. arXiv preprint arXiv:2410.13749 (2024)
6. Hastie, T., Tibshirani, R., Friedman, J.: The Elements of Statistical Learning: Data Mining, Inference, and Prediction, 2nd edn. Springer (2009)
7. Holm, S.: A simple sequentially rejective multiple test procedure. Scand. J. Stat. **6**(2), 65–70 (1979)
8. Lemaître, G., Nogueira, F., Aridas, C.K.: Imbalanced-learn: a python toolbox to tackle the curse of imbalanced datasets in machine learning. J. Mach. Learn. Res. **18**(17), 1–5 (2017). http://jmlr.org/papers/v18/16-365.html

9. Sriperumbudur, B.K., Gretton, A., Fukumizu, K., Schölkopf, B., Lanckriet, G.R.: Hilbert space embeddings and metrics on probability measures. J. Mach. Learn. Res. **11**, 1517–1561 (2010)
10. Wang, W., et al.: A unified joint maximum mean discrepancy for domain adaptation. arXiv preprint arXiv:2101.09979 (2021)
11. Wilcoxon, F.: Individual comparisons by ranking methods. Biometrics Bulletin **1**(6), 80–83 (1945)

Sailfish: A Data Efficiency Optimization Approach for Branch Learning via Full Strong Branching

Feifei Xu, Jinze Zhao, and Haizhou Du(✉)

Shanghai University of Electric Power, Shanghai 201306, China
duhaizhou@shiep.edu.cn

Abstract. Imitating strong branching is a mainstream approach for enhancing variable selection in Mixed-Integer Linear Programming (MILP). However, existing methods typically adopt a hybrid strategy combining strong branching and pseudo-cost branching, which yields only a limited number of strong branching state samples. This leads to high dependence on large datasets and sparse coverage of the state space. To address this issue, we propose Sailfish, a data efficiency optimization approach based on full strong branching. By incorporating weighted sampling and pseudo-label loss, it reduces the number of required training instances while directing the model's focus to critical state samples. Furthermore, we define the Normalized Score Gap (NSG) as evaluation metric based on the degree of search space reduction to identify node importance. Extensive experiments demonstrate Sailfish achieves significant improvements on diverse benchmarks. Sailfish can reduce the number of required MILP instances by up to 85% and the total data collection wall-clock time by up to 69%. Moreover, models trained on such efficiently collected data maintain competitive end-to-end solving performance comparable to baselines.

Keywords: Mixed-Integer Linear Programming · Neural Branching · Data Efficiency · Strong Branch · Normalized Score Gap

1 Introduction

Mixed-Integer Linear Programming (MILP) problems, prevalent in various domains [4,9,12,27], are often solved using the Branch-and-Bound (B&B) algorithm [15]. In B&B, variable selection is a critical step [20]. Although expert heuristics like Strong Branching (SB) are effective in generating smaller search trees [1,22], their high computational cost motivate machine learning researchers to train models that emulate SB to reduce this overhead [13].

Recent advances in machine learning have significantly improved MILP solving efficiency. A prominent approach uses Imitation Learning (IL) with Graph Convolutional Neural Networks (GCNNs) [14] to capture MILP problem structure [8]. Further innovations include hybrid graph neural network/multi-layer

L. Pasa et al. (Eds.): ICANN 2026, LNCS 17089, pp. 558–569, 2027.
https://doi.org/10.1007/978-3-032-38398-3_46

perceptron architectures [10], tree-aware transformers [18], and regularization techniques [11]. To mitigate data scarcity, methods such as contrastive learning [17] and block-based instance generation [19] emerge. A comprehensive survey benchmarks machine learning applications across all B&B components [24].

Conventional machine learning studies employ hybrid strategies [1] like Reliability Pseudo-cost Branching (RPB) [2], where SB is used only intermittently to manage computational costs with cheaper heuristics [8,10,17,19,26,28]. This partial sampling approach has two critical flaws. First, it creates discontinuous state trajectories, complicating the learning process [25]. Second, it is extremely data-inefficient, capturing only a few samples per instance [30]. This requires an excessive number of MILP instances for training, which is very time-consuming in practice. A plausibly reasonable solution is to generate training data using Full Strong Branching (FSB) exclusively. However, this method presents its own severe pitfall: it generates a massive number of states, many of which (*e.g.*, deep in the tree) are far less critical to the final solution time than the states near the root node [5,20]. Model training would cause it to over-learn low-impact decisions on this highly imbalanced data, while failing to master the critical, high-impact decisions at the top of the tree.

To address this issue, our core ideas are threefold. First, we increase the sample density per instance by generating the B&B search tree solely with FSB and retaining all state samples, thereby reducing the total number of required MILP instances while ensuring complete state trajectories. Second, we prioritize critical states via weighted sampling and loss computation, achieved by recording the maximum SB score [6] at each step and normalizing these scores within each instance. Third, to handle ties without penalizing any correct selection, we introduce pseudo-labels that combine model predictions with all valid high-score candidates. Building on these ideas, we propose Sailfish, a data-efficient paradigm designed to achieve high model quality with substantially fewer training instances.

Our **main contributions** are summarized as follows:

- **Sailfish Framework**: A data-efficient approach that exclusively utilizes FSB to preserve all decision states. Sailfish addresses the high instance demand and discontinuous state trajectories prevalent in conventional hybrid sampling methods.
- **Importance-Aware Data Strategy**: A dual-component strategy comprising cross-instance score normalization and adaptive weighted sampling. This mechanism resolves objective scaling discrepancies and dynamically prioritizes critical states near the root node, ensuring high-impact samples dominate the learning process.
- **Robust Training and Evaluation Protocol**: A confidence-driven pseudo-label method combined with a new metric, Normalized Score Gap. This protocol resolves action ambiguity during training by considering all optimal candidates and provides a more robust assessment of search-space reduction.
- **Extensive Experimental Validation**: Empirical results demonstrate that Sailfish drastically reduces MILP instance requirements by up to 85% and

data collection time by up to 69%, while producing models with end-to-end solving performance comparable to heavy-data baselines.

2 Related Work

The integration of machine learning into MILP primarily centers on optimizing the B&B process through IL and data-centric augmentation. This research landscape is defined by a dual focus: IL for Branching in MILP and Generating Additional Instances for Training.

2.1 IL for Branching in MILP

IL serves as a primary vehicle for automating branching decisions by emulating expert heuristics like SB. This paradigm typically represents MILP states as bipartite graphs, using GCNNs to extract features and predict branching scores [8]. Current research trends move beyond this basic architecture to address two specific demands: inference efficiency and structural awareness. To facilitate deployment on CPUs, hybrid models combine GCNNs with Multi-Layer Perceptrons or Gradient Boosting Decision Trees to bypass expensive graph convolutions during inference [10,28]. In parallel, to capture the dynamic evolution of search trees, more complex architectures like Temporo-Attentional GNNs [26], gating mechanisms for tree reduction [29], and tree-aware transformers [18] integrate branching history and temporal dependencies into the decision process. Similar temporal dynamics are addressed through target smoothing to handle the lookback phenomenon [11]. Despite these structural refinements, these methods generally rely on partial sampling from hybrid solvers. Such a strategy inherently discards intermediate states, which leads to discontinuous state trajectories and prevents the model from learning the full solving process [25].

2.2 Generating Additional Instances for Training

The heavy reliance of IL on high-quality expert labels makes data scarcity a critical bottleneck. To expand the training pool, existing literature focuses on synthetic instance generation through various augmentation techniques. For instance, contrastive learning frameworks utilize variable shifting to create augmented MILPs, while decomposition-based methods modify block structures to generate new hard instances while preserving feasibility [17,19]. Other approaches integrate reinforcement learning to enhance sample efficiency by augmenting existing trajectories with diverse solver states [30]. While these strategies effectively increase the volume of training data, they fundamentally focus on the diversity of MILP instances rather than the sampling density within each instance. Consequently, these methods require an immense number of instances to cover the state distribution, neglecting the rich information available within a single, fully-explored search tree.

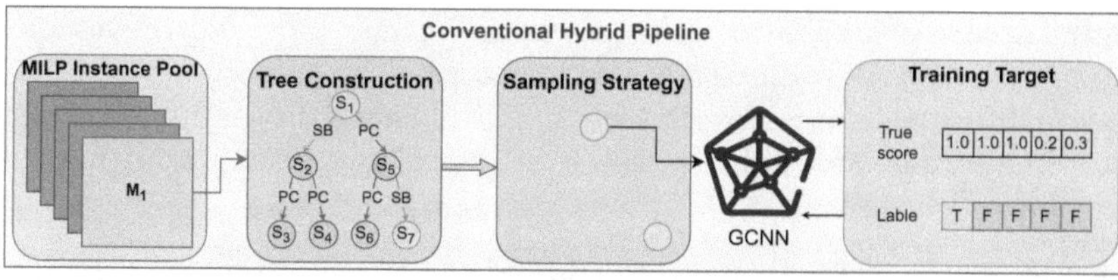

(a) The standard data collection pipeline.

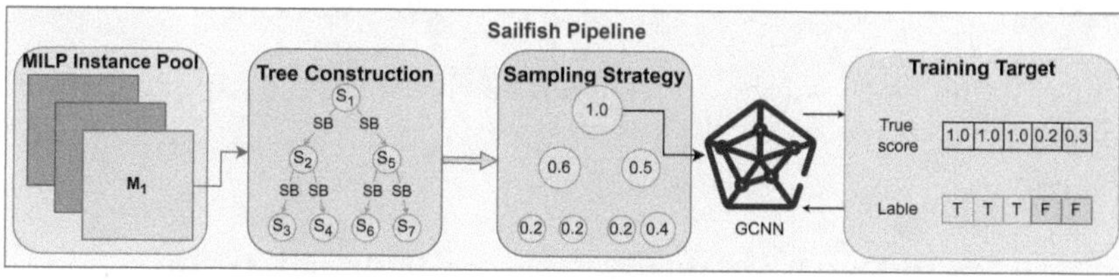

(b) The dense sampling pipeline based on state importance.

Fig. 1. Overview of the data efficiency approach in Sailfish.

3 Methodology

A standard MILP problem can be formally defined as the following optimization model:

$$\min_{\mathbf{x}}\{\mathbf{c}^T\mathbf{x} \mid \mathbf{A}\mathbf{x} \leq \mathbf{b},\, \mathbf{l} \leq \mathbf{x} \leq \mathbf{u},\, \mathbf{x} \in \mathbb{Z}^p \times \mathbb{R}^{n-p}\}, \tag{1}$$

where $\mathbf{c} \in \mathbb{R}^n$, $\mathbf{A} \in \mathbb{R}^{m\times n}$, $\mathbf{b} \in \mathbb{R}^m$, and p denotes the number of integer variables. The B&B algorithm solves this by recursively partitioning the continuous linear programming (LP) relaxation space into a search tree, utilizing upper and lower bounds to prune subtrees and minimize the dual gap. Variable selection determines which fractional variable x_i to split at each node.

Sailfish is designed to maximize training utility while minimizing the required MILP instances. As illustrated in Fig. 1, Sailfish transforms the standard data collection pipeline into a dense sampling process guided by state importance and pseudo-labels. This framework prioritizes high-quality search-space information to ensure robust policy learning with limited data. The following sections detail the core components of Sailfish.

3.1 FSB-Based Data Generation

Traditional hybrid branching methods [1] in operations research interleave SB with Pseudo-Cost (PC) branching to balance optimality and computational efficiency. As implemented in [8], this adopts a fixed 1:19 SB-to-PC ratio, where only SB-generated states (Fig. 1b) are retained as training samples. Although this naive imitation increases sample diversity across MILP instances, it introduces two critical limitations:

- **High Instance Demand**: It requires 1.5–5 times more MILP instances to collect equivalent SB samples compared to Sailfish.
- **Discontinuous State Trajectories**: The sparse SB sampling (1/20 steps) discards intermediate states between SB steps, potentially omitting pivotal branching decisions.

Sailfish employs the FSB strategy to address these limitations by:

- Exclusively using SB throughout the B&B tree (Fig. 1b), preserving all decision states $\{\mathbf{s}_t\}_{t=1}^{T}$.
- Counteracting leaf-node bias through adaptive weighted sampling (Sect. 3.3) and loss reweighting (Sect. 3.4).

3.2 State Importance Quantification and Normalization

To quantify the importance of a state $\mathbf{s}_t$, we first define the SB score $\sigma(\mathbf{s}_t, a)$ for a candidate variable a as the improvement in the dual bound (DB) resulting from branching on a:

$$\sigma(\mathbf{s}_t, a) = |\text{DB}_{\text{after}} - \text{DB}_{\text{before}}|. \tag{2}$$

The importance of state $\mathbf{s}_t$, denoted as $\phi(\mathbf{s}_t)$, is then defined as the maximum score among its candidates: $\phi(\mathbf{s}_t) = \max_a \sigma(\mathbf{s}_t, a)$. However, direct comparison across MILP instances is invalid due to objective scaling discrepancies. Consider two instances:

- **Instance A**: The current state has $\phi = 10$, which happens to be the maximum score in this instance (Relative importance: 100%).
- **Instance B**: The current state has $\phi = 1,000$, but the maximum score in this instance is 11,000 (Relative importance: ~9%).

Although ϕ is 100× larger in Instance B, the relative improvement within its own context suggests that the state in Instance A is actually more critical. We therefore normalize $\phi(\mathbf{s}_t)$ within each instance:

$$\tilde{\phi}(\mathbf{s}_t) = \frac{\phi(\mathbf{s}_t)}{\max_{\mathbf{s}_k \in \mathcal{D}_{\text{instance}}} \phi(\mathbf{s}_k)} \in [0, 1]. \tag{3}$$

This preserves intra-instance importance rankings while enabling cross-instance comparability. In the example above, Instance A's state becomes $\tilde{\phi} = 1.0$, correctly prioritizing it over Instance B's state ($\tilde{\phi} \approx 0.09$).

3.3 Adaptive Weighted Sampling

To prioritize high-importance states while maintaining computational tractability, we design the sampling probability as:

$$p(\mathbf{s}_t) \propto \tilde{\phi}(\mathbf{s}_t)^{\gamma}, \quad \gamma > 0. \tag{4}$$

To ensure that states with normalized scores $\tilde{\phi}(\mathbf{s}_t) \geq 0.9$ constitute a significant portion (*e.g.*, over 20%) of the sampled batch, the exponent γ is typically set between 2 and 3, determined via pilot experiments on each dataset. This contrasts sharply with the baseline uniform sampling strategy [8] where all SB states receive equal weights regardless of their importance scores.

3.4 Pseudo-Label GCNN Training

For states with multiple optimal actions $\mathcal{A}^*(\mathbf{s}_t) = \{a \mid \sigma(\mathbf{s}_t, a) = \phi(\mathbf{s}_t)\}$, we resolve label ambiguity through model-predicted confidence, as shown in the pseudo-label block of Fig. 1a:

$$a^* = \arg \max_{a \in \mathcal{A}^*(\mathbf{s}_t)} \pi_\theta(a|\mathbf{s}_t). \tag{5}$$

This differs from the baseline's rigid rule [8] that selects the first occurrence of the maximum score as the label. Sailfish prevents excessive penalization when the model correctly identifies an alternative optimal action. The weighted cross-entropy loss becomes:

$$\mathcal{L}(\theta) = -\frac{1}{N} \sum_{(\mathbf{s},a^*) \in \mathcal{D}} \tilde{\phi}(\mathbf{s}) \cdot \log \pi_\theta(a^*|\mathbf{s}). \tag{6}$$

3.5 Enhanced Evaluation Protocol

To fairly evaluate the intermediate state, we propose two metrics as follows:

- **Normalized Score Gap (NSG)**: To accurately evaluate performance across instances with varying scales, we calculate the gap using normalized ratios:

$$\text{NSG} = \sum_{\mathbf{s} \in \mathcal{D}_{\text{test}}} \left(1 - \frac{\sigma(\mathbf{s}, \hat{a})}{\phi(\mathbf{s})}\right), \quad \hat{a} = \arg\max \pi_\theta(a|\mathbf{s}). \tag{7}$$

  Here, the term $\frac{\sigma(\mathbf{s},\hat{a})}{\phi(\mathbf{s})}$ represents the ratio of the chosen variable's score to the optimal score (optimal ratio is 1.0).
- **Low-Score Filtering**: To focus on critical decisions, we exclude states with $\tilde{\phi}(\mathbf{s}) < 10^{-4}$ during testing. This removes the vast majority of trivial samples while retaining 99% of the cumulative SB scores.

This protocol aligns with our insight that states exhibit varying solution-critical importance, as quantified in Sect. 3.2.

4 Experiments

We design experimental evaluation to comprehensively assess Sailfish. We aim to answer four key questions:

- **(Q1: Efficiency)** How does our data collection strategy compare to the baseline in terms of sample generation efficiency (*i.e.*, MILP instances consumed and wall-clock time)?
- **(Q2: Policy Quality)** Does the higher-fidelity data generated by Sailfish translate into a superior branching policy, as measured by the NSG?
- **(Q3: Solving Performance)** Crucially, does the policy trained on our efficient data maintain competitive end-to-end MILP solving performance (time, nodes, and wins) compared to the baseline trained on massive datasets?
- **(Q4: Component Contribution)** How do the individual components of Sailfish contribute to the overall performance?

4.1 Experimental Setup

Benchmarks. Following the setup in [8] and the Ecole library [23], Sailfish is thoroughly evaluated on four NP-hard combinatorial optimization problems: Set Covering (SC) [3], Combinatorial Auction (CAU) [16], Maximum Independent Set (MIS) [21], and Capacitated Facility Location (CFL) [7].

Instance Sets. We use two distinct sets of instances for our experiments: **Data Collection Instances:** To generate the training/validation/test samples for the policy network, we generate 10,000 MILP instances for training, 2,000 for validation, and 2,000 for testing for each question. Following [8], these instances are all generated at the *easy* difficulty level. **Solving Performance Instances:** To evaluate end-to-end practical solving performance, we generate a separate, comprehensive testbed comprising 20 *easy*, 20 *medium*, and 20 *hard* instances for each question. Models are trained and tested separately for each benchmark.

Baselines. We compare three distinct branching strategies:

- **RPB (Solver)**: The traditional, non-learning solver heuristic, Reliability Pseudo-cost Branching, which serves as the default in SCIP.
- **Hybrid (Baseline)**: The GCNN model from [8], trained on data collected using their original hybrid branching (partial sampling) method.
- **Sailfish (Ours)**: The exact same GCNN architecture, but trained using our Sailfish data collection and weighting methodology.

This comparison setup effectively isolates the impact of our data generation strategy, as the model architecture remains constant.

Data Collection and Policy Training. To train the learning-based models, we generate two datasets, each containing 100,000 state samples: one using the baseline's Hybrid Branching and one using the proposed Sailfish. For validating and testing the policy network (used for NSG evaluation), we use 20,000 samples for validation and 20,000 for testing. Test samples with normalized scores below 1×10^{-4} are excluded.

Metrics. To evaluate policy quality, we employ the **NSG**, where smaller values indicate closer alignment with the optimal strong branching rules. For practical solving performance, we measure **solving time** (s), **node count** in the branch-and-bound tree, and the **win count** (number of instances where a method achieves the fastest solving time). For time and nodes, smaller values are better; for wins, higher values are better.

4.2 Experimental Results

MILP Instance Consumption Analysis. As summarized in Table 1, Sailfish demonstrates significantly superior efficiency in *both* metrics. It drastically reduces the number of MILP instances required (*e.g.*, from 4,938 to 745 for CFL) and simultaneously achieves a faster total wall-clock time for collecting 100,000 samples. This is because the baseline's partial sampling, while cheap per instance, must process a prohibitively large number of instances, incurring significant overhead.

Table 1. Comparison of sample generation efficiency (to collect 100,000 samples).

Problem	MILP Instances		Time (s)	
	Baseline (Hybrid)	Ours (Sailfish)	Baseline (Hybrid)	Ours (Sailfish)
SC	5,148	**938**	1744.87	**1,241.16**
CAU	2,923	**661**	600.27	**406.25**
MIS	4,076	**2,668**	5,086.24	**5,083.23**
CFL	4,938	**745**	8,033.50	**2,478.98**

Note: Baseline refers to the hybrid branching strategy. Time measured on a 32-core Xeon(R) Platinum 8352V CPU using 30 threads to generate 100k samples.

NSG Comparison. We evaluate the model's policy quality using the NSG, as defined in Sect. 3.5. The NSG metric can handle states with varying importance and multiple optimal actions, providing a more faithful measure of branching decision quality. A lower NSG is better, and the improvement in this metric (*i.e.*, NSG reduction) demonstrates enhanced policy alignment.

Figure 2 illustrates the percentage improvement (reduction) in NSG achieved by Sailfish compared to the baseline across the four benchmarks. The results show a strong, albeit complex, correlation with the data efficiency gains reported in Table 1. For instance, MIS achieves the largest NSG improvement (**17.63%**) alongside significant instance savings (4,076 vs. 2,668). This suggests that the higher-quality training data generated by our Sailfish method translates directly to substantial performance improvements in policy alignment.

Conversely, CAU exhibits minimal NSG reduction (0.96%) despite having the most dramatic instance savings (2,923 vs. 661). This implies that for certain problem-specific structures, the gains from sample efficiency may not proportionally translate to better policy alignment, or that the baseline policy for CAU was already near-optimal under this metric.

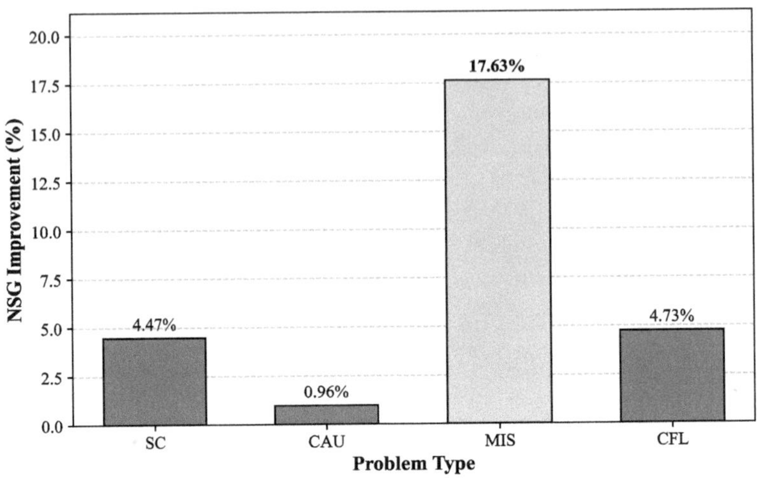

Fig. 2. NSG improvement of Sailfish across four tasks. The most significant gain is observed in the MIS task (17.63%).

Evaluation of Practical Solving Performance. Having established the superior data efficiency (Sect. 4.2) and policy alignment (Sect. 4.2), the compound impact on end-to-end MILP solving performance was subsequently evaluated. We compare the three branching strategies defined in our experimental setup: the traditional **RPB (Solver)**, the baseline **Hybrid** strategy [8], and our **Sailfish** strategy. This comparison effectively isolates the contribution of our data generation strategy, as the underlying model architecture for Hybrid and Sailfish remains identical.

Table 2 presents the detailed solving performance. Both learning-based methods (Hybrid and Sailfish) significantly outperform the traditional RPB baseline. The end-to-end performance also shows a **positive correlation in magnitude** with the NSG results from Sect. 4.2. This is most evident on **MIS**, where a significant NSG improvement (17.63%) translates to a **dramatic reduction** in solving time on hard instances (from 812.77 s to 291.45 s). Conversely, on **SC**, **CFL**, and **CAU**, where the NSG improvements are all in the single-digits, the final solving performance is **highly comparable** to the baseline (*e.g.*, +2.5% time on SC, −1.3% time on CAU).

Crucially, Sailfish achieves this **comparable** solving performance while being trained on a dataset that was collected *far* more efficiently (both in terms of MILP instances and wall-clock time). This validates our core hypothesis: Sailfish represents a superior **data-efficiency paradigm**, successfully producing models with competitive solving capabilities without the prohibitive data collection costs of the conventional hybrid approach.

Table 2. Solving performance comparison (Time, Wins) on test instances for the four problem benchmarks.

Problem	Strategy	Easy		Medium		Hard	
		Time(s)	Wins	Time(s)	Wins	Time(s)	Wins
SC	RPB	4.61	0	44.52	0	459.02	1
	Hybrid	2.67	46	**30.30**	**56**	**397.56**	**66**
	Sailfish	**2.66**	**54**	30.80	44	407.40	33
CAU	RPB	3.02	0	49.23	0	462.54	8
	Hybrid	1.68	39	**32.90**	**60**	433.37	39
	Sailfish	**1.67**	**61**	33.37	40	**427.85**	**53**
MIS	RPB	4.92	8	70.50	0	297.83	26
	Hybrid	**2.94**	**50**	46.85	6	812.77	0
	Sailfish	2.97	42	**25.45**	**94**	**291.45**	**74**
CFL	RPB	43.08	5	142.32	4	379.93	26
	Hybrid	33.33	**48**	**113.50**	42	**370.63**	**50**
	Sailfish	**32.85**	47	123.19	**54**	377.68	24

Note: "Hybrid" denotes the GCNN trained with the baseline data strategy. "Sailfish" denotes the same GCNN trained with our strategy.

4.3 Ablation Study

To understand the individual contribution of Sailfish's key components, we conduct an ablation study (Table 3). We measure the percentage increase in NSG (indicating performance degradation) after removing each of the three components introduced in Sect. 3.2 and 3.4: Pseudo-Labeling, Adaptive Weighted Sampling, and Weighted Loss.

The results confirm that all components are valuable, though their impact is highly problem-dependent. We observe that **Adaptive Weighted Sampling (WS)** is the most critical component, especially for **CFL** (42.86% degradation) and **SC** (21.83%), which suffer from high state importance imbalance. In contrast, **Pseudo-Labeling (PL)** is most crucial for **MIS** (7.24%) and **CAU** (3.57%), problems characterized by high label ambiguity. **Weighted Loss (WL)** acts as a complementary mechanism, most prominent on **SC** (12.56%).

Table 3. Ablation Study: NSG Increase (%) After Component Removal

Component	SC	CAU	MIS	CFL
w/o PL	2.11	3.57	7.24	0.80
w/o WS	21.83	3.01	6.77	42.86
w/o WL	12.56	2.76	2.41	2.55

5 Conclusion

We propose Sailfish, a data efficiency approach that leverages Full Strong Branching to reduce training instance requirements by up to 85%. By maximizing sample density and prioritizing critical states, Sailfish maintains competitive solving performance while significantly lowering data collection cost. This establishes a practical, low-overhead paradigm for training high-quality neural MILP solvers. In future work, reducing the communication latency between GPU and CPU is crucial, as our method runs on GPU while the solver remains on CPU.

References

1. Achterberg, T., Berthold, T.: Hybrid branching. In: van Hoeve, W.-J., Hooker, J.N. (eds.) CPAIOR 2009. LNCS, vol. 5547, pp. 309–311. Springer, Heidelberg (2009). https://doi.org/10.1007/978-3-642-01929-6_23
2. Achterberg, T., Koch, T., Martin, A.: Branching rules revisited. Oper. Res. Lett. (2005)
3. Balas, E., Ho, A.: Set Covering Algorithms Using Cutting Planes, Heuristics, and Subgradient Optimization: A Computational Study. Springer, Heidelberg (1980)
4. Bao, T., Wang, H.: Cyclic scheduling of multi-cluster tools based on mixed integer programming. IEEE Trans. Semicond. Manuf. (2017)
5. Bengio, Y., Lodi, A., Prouvost, A.: Machine learning for combinatorial optimization: a methodological tour d'horizon. Eur. J. Oper. Res. (2021)
6. Chen, D., Batson, R., Dang, Y.: Branch-and-Cut Approach (2011)
7. Cornuéjols, G., Sridharan, R., Thizy, J.: A comparison of heuristics and relaxations for the capacitated plant location problem. Eur. J. Oper. Res. (1991)
8. Gasse, M., Chételat, D., Ferroni, N., Charlin, L., Lodi, A.: Exact combinatorial optimization with graph convolutional neural networks. In: Advances in Neural Information Processing Systems, vol. 32 (2019)
9. Godart, A., Manier, H., Bloch, C., Manier, M.A.: MILP for a variant of pickup & delivery problem for both passengers and goods transportation. In: 2018 IEEE International Conference on Systems, Man, and Cybernetics (SMC) (2018)
10. Gupta, P., Gasse, M., Khalil, E.B., Kumar, M.P., Lodi, A., Bengio, Y.: Hybrid models for learning to branch. branch. In: Advances in Neural Information Processing Systems, vol. 33 (2020)
11. Gupta, P., et al.: Lookback for learning to branch. Trans. Mach. Learn. Res. (2022)
12. Hammad, A., Akbarnezhad, A., Rey, D.: A multi-objective mixed integer nonlinear programming model for construction site layout planning to minimise noise pollution and transport costs. Autom. Constr. (2016)

13. Hoffman, K.L., Padberg, M., Rinaldi, G.: Traveling Salesman Problem. Springer, New York (2013)
14. Kipf, T.N., Welling, M.: Semi-supervised classification with graph convolutional networks. In: 5th International Conference on Learning Representations, ICLR 2017, Toulon, France, 24–26 April 2017, Conference Track Proceedings. OpenReview.net (2017)
15. Land, A.H., Doig, A.G.: An Automatic Method for Solving Discrete Programming Problems. Springer, Heidelberg (2010)
16. Leyton-Brown, K., Pearson, M., Shoham, Y.: Towards a universal test suite for combinatorial auction algorithms. In: Proceedings of the 2nd ACM Conference on Electronic Commerce (2000)
17. Lin, J., Xu, M., Xiong, Z., Wang, H.: CAMBranch: contrastive learning with augmented MILPs for branching. In: The Twelfth International Conference on Learning Representations (2024)
18. Lin, J., Zhu, J., Wang, H., Zhang, T.: Learning to branch with tree-aware branching transformers. Knowl. Based Syst. (2022)
19. Liu, H., et al.: MILP-studio: MILP instance generation via block structure decomposition. In: The Thirty-eighth Annual Conference on Neural Information Processing Systems (2024)
20. Lodi, A., Zarpellon, G.: On learning and branching: a survey. TOP (2017)
21. Lozano, L., Bergman, D., Cire, A.A.: Decision Diagrams for Optimization. Springer, Cham (2020)
22. Patel, J., Chinneck, J.: Active-constraint variable ordering for faster feasibility of mixed integer linear programs. Math. Program. (2007)
23. Prouvost, A., Dumouchelle, J., Scavuzzo, L., Gasse, M., Chételat, D., Lodi, A.: Ecole: a Gym-like library for machine learning in combinatorial optimization solvers, CoRR (2020)
24. Scavuzzo, L., Aardal, K.I., Lodi, A., Yorke-Smith, N.: Machine learning augmented branch and bound for mixed integer linear programming. CoRR (2024)
25. Scavuzzo, L., et al.: Learning to branch with tree MDPs. In: Advances in Neural Information Processing Systems (2022)
26. Seyfi, M., Banitalebi-Dehkordi, A., Zhou, Z., Zhang, Y.: Exact combinatorial optimization with temporo-attentional graph neural networks. In: Machine Learning and Knowledge Discovery in Databases: Research Track (2023)
27. Soylu, A., Oruç, C., Turkay, M., Fujita, K., Asakura, T.: Synergy analysis of collaborative supply chain management in energy systems using multi-period MILP. Eur. J. Oper. Res. (2006)
28. Ye, H., Xu, H., Wang, H., Wang, C., Jiang, Y.: GNN&GBDT-guided fast optimizing framework for large-scale integer programming. In: ICML (2023)
29. Zarpellon, G., Jo, J., Lodi, A., Bengio, Y.: Parameterizing branch-and-bound search trees to learn branching policies. In: AAAI (2021)
30. Zhang, C., et al.: Towards imitation learning to branch for MIP: a hybrid reinforcement learning based sample augmentation approach. In: The Twelfth International Conference on Learning Representations (2024)

Batch Normalization Shapes Representations in Barlow Twins Pretraining

Hans-Oliver Hansen(✉) and Thomas Martinetz

Institute for Neuro- and Bioinformatics, University of Lübeck, Lübeck, Germany
h.hansen@uni-luebeck.de

Abstract. Self-supervised learning (SSL) methods such as Barlow Twins produce feature representations with substantially different internal statistics compared to supervised models. In this work, we analyze the weight distributions of ResNet-50 models pretrained with SSL and supervised learning and observe significantly larger weight magnitudes and BatchNorm statistics in SSL models, suggesting a potential impact on transfer learning. To investigate this, we introduce a simple reparameterization that incorporates BatchNorm running variances into convolutional layers, effectively encoding normalization statistics into the network parameters. While this transformation preserves the functional behavior under fixed statistics, it alters the optimization dynamics during finetuning. Across multiple downstream datasets, we find that this reparameterization consistently degrades full finetuning performance for Barlow Twins-pretrained models, while substantially improving full finetuning performance for supervised ImageNet-pretrained models. In contrast, linear probing with BatchNorm adaptation remains robust across all settings. These results indicate that Barlow Twins-pretrained models are more sensitive to changes in BatchNorm-induced scaling than supervised ImageNet-pretrained models, suggesting that BatchNorm plays a more prominent role in shaping the learned feature distribution under Barlow Twins pretraining.

Keywords: Self-Supervised Learning · Barlow Twins · Batch Normalization · Transfer Learning · Feature Scaling · Representation Learning

1 Introduction

Self-supervised learning (SSL) has become a powerful tool for learning visual representations from unlabeled data. Methods such as SimCLR [2], MoCo [6], and BYOL [5] have demonstrated that representations learned without manual annotations prove to be competitive compared to supervised pretraining on a variety of downstream tasks. Among these approaches, Barlow Twins [16] introduces a correlation-based objective that directly decorrelates feature representations.

L. Pasa et al. (Eds.): ICANN 2026, LNCS 17089, pp. 570–580, 2027.
https://doi.org/10.1007/978-3-032-38398-3_47

A key characteristic of Barlow Twins is the use of a normalized cross-correlation matrix in its loss function, which enforces feature decorrelation for redundancy reduction while maintaining invariance to augmentations to avoid feature collapse. This design has been shown to be essential for successful pretraining. However, the normalization inherent to the objective also alters the internal scaling of feature activations, which may in turn affect the distribution of learned weights.

In this work, we observe that convolutional weights of models pretrained with Barlow Twins exhibit substantially larger variances compared to standard supervised ImageNet [13] pretraining. These differences are consistent across layers and are particularly pronounced in early convolutional stages. This effect is shown in Fig. 1: The left panel shows layer-wise statistics (mean vs. standard deviation), revealing a clear difference between Barlow Twins and ImageNet pretrained models across all layers. The right panel further highlights these differences by plotting the sorted absolute weights on a logarithmic scale, where Barlow Twins exhibits significantly heavier tails, indicating the presence of substantially larger weight magnitudes. Such discrepancies suggest that the internal feature scaling induced during pretraining differs substantially between self-supervised and supervised models, potentially affecting their behavior during downstream adaptation.

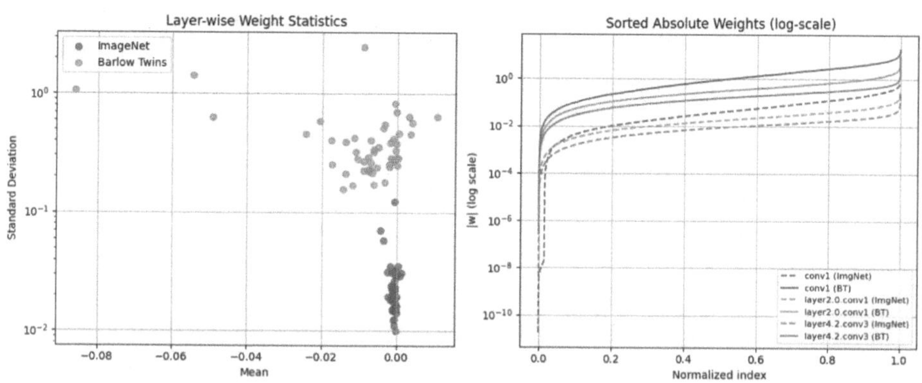

Fig. 1. Weight statistics of convolutional layers for ResNet-50 models pretrained on ImageNet and with Barlow Twins. **Left:** Layer-wise mean and standard deviation of convolutional weights. Barlow Twins exhibits consistently larger variances across all layers compared to supervised pretraining. **Right:** Sorted absolute weights on a logarithmic scale for representative early, intermediate, and late layers. Barlow Twins shows substantially heavier tails, indicating the presence of significantly larger weight magnitudes. These results highlight a pronounced difference in internal feature scaling induced by Barlow Twins and supervised pretraining.

This raises an important question: How do these differences in weight distributions and internal scaling affect downstream adaptation? In particular, Batch

Normalization (BN) plays a central role in controlling activation statistics and stabilizing training. While originally motivated as a mechanism to reduce internal covariate shift [8], it is now widely understood that BN smooths the optimization landscape and enables efficient training [14]. In this work, we focus on its role in controlling feature scaling, which may interact differently with self-supervised and supervised pretraining.

To investigate this, we propose a simple reparameterization that incorporates BatchNorm running variances into the preceding convolutional layers. This transformation preserves the functional behavior of the network at initialization during evaluation, while explicitly modifying the internal scaling of features. By doing so, we encode the activation statistics learned during pretraining into the convolutional weights, allowing us to study how pretraining-derived scaling interacts with downstream optimization.

We evaluate this approach across multiple datasets and transfer learning settings, including full finetuning, linear probing, and linear probing with BatchNorm adaptation. Our results show a consistent pattern: while supervised ImageNet-pretrained models benefit from this reparameterization under full finetuning, Barlow Twins models exhibit a significant degradation in performance, indicating a substantially stronger sensitivity to the original BatchNorm-induced scaling.

Our contributions are as follows:

- We provide an empirical analysis of weight distributions in Barlow Twins and supervised ImageNet pretrained ResNet-50 [7] models, revealing substantial differences in variance and scaling.
- We propose a simple reparameterization that integrates BatchNorm running variances into convolutional layers to study the effect of BatchNorm-induced feature scaling.
- We systematically evaluate the impact of this transformation across multiple datasets and training regimes, including finetuning and linear probing with and without BatchNorm adaptation.
- We show that Barlow Twins representations are significantly more sensitive to internal scaling changes than supervised counterparts, highlighting a pronounced difference in how BatchNorm-induced scaling interacts with Barlow Twins and supervised pretraining.

2 Related Work

Self-supervised learning (SSL) has become a prominent approach for learning rich visual representations from unlabeled data. Early SSL methods relied on *sample-contrastive* objectives, such as SimCLR [2] and MoCo [6], which maximize similarity between augmented views of the same sample while pushing apart features from different samples.

Barlow Twins [16] follows a distinct self-supervised learning approach that is often described as *non-contrastive* or *feature-contrastive*. Rather than explicitly contrasting different samples, the method minimizes the cross-correlation between feature representations of two augmented views, thereby encouraging decorrelation across feature dimensions. Crucially, the loss function operates on a *normalized cross-correlation matrix* rather than a standard covariance matrix. As shown in the ablation study of the original work, this normalization is essential to prevent feature collapse, leading to a significant drop in linear probing performance from 71.4% to 53.4% on ImageNet when removed.

Other non-contrastive approaches, such as VICReg [1] and W-MSE [4], also aim to decorrelate or regularize features, but they rely on additional variance and invariance regularization terms, making the methods conceptually distinct from Barlow Twins' direct correlation minimization. This highlights that the large weight magnitudes observed in Barlow Twins may be an intrinsic consequence of the correlation-based loss rather than a generic property of SSL.

Several studies have examined the role of Batch Normalization (BN) in downstream adaptation. For instance, [15] showed that fine-tuning BN parameters during linear probing can substantially improve performance across multiple SSL frameworks, especially in limited-data regimes. However, their analysis did not include Barlow Twins, leaving open questions about how its unique feature-scaling properties influence transfer learning.

Benchmarking studies [10] systematically compared SSL models across a range of downstream datasets, evaluating both linear probing and full fine-tuning, including the role of BN. These analyses demonstrated that performance differences are influenced not only by the pretraining objective but also by the interaction between convolutional weights and BN statistics. Nevertheless, they did not explicitly investigate how the *intrinsic weight distributions* induced by feature-contrastive pretraining affect downstream adaptation.

Our work builds on these insights by explicitly reparameterizing convolutional and BN layers to study the impact of pretraining-derived running variances on downstream performance. This allows us to directly assess how the internal scaling of features, as established by Barlow Twins, interacts with transfer learning strategies such as full fine-tuning and linear probing with BN adaptation.

3 Method

3.1 Reparameterization of Convolution and BatchNorm Layers

We consider convolutional neural networks in which Batch Normalization (BN) layers are applied after convolutional layers. While the proposed approach is generally applicable to such architectures, we focus on ResNet-50 [7] in our experiments.

Let x denote the input to a convolutional layer with weights W. In modern architectures, convolutional layers typically do not include a bias term when followed by BatchNorm, as the BN offset (β) effectively replaces it. The output of a convolution followed by BatchNorm can then be written as

$$y = \gamma \cdot \frac{Wx - \mu}{\sqrt{\sigma^2 + \epsilon}} + \beta, \tag{1}$$

where μ and σ^2 denote the running mean and variance, that are estimated during training, and γ and β are the learnable affine parameters of the BN layer. Intuitively, the running variance and γ scale the output in a channel-wise fashion.

To analyze the role of BN-induced feature scaling, we reparameterize the convolutional weights by incorporating the running variance into the convolution:

$$W' = \frac{W}{\sqrt{\sigma^2 + \epsilon}}. \tag{2}$$

This yields the transformed output

$$y = \gamma \cdot (W'x - \frac{\mu}{\sqrt{\sigma^2 + \epsilon}}) + \beta. \tag{3}$$

Under fixed BatchNorm statistics and evaluation mode, this reparameterization preserves the functional behavior of the network at initialization. During downstream training, however, BatchNorm uses batch statistics and updates its running estimates depending on the chosen training protocol. Therefore, the transformation should be understood as an intervention on the parameterization and optimization dynamics, rather than as a function preserving transformation throughout training.

3.2 Design Choices

In our reparameterization, we incorporate only the running variance into the convolutional weights, effectively encoding the feature scaling learned from the unlabeled pretraining dataset into the convolutional weights. The running mean and affine parameters (γ, β) remain part of the BatchNorm layer. This design allows us to isolate the effect of feature scaling without altering the learned affine transformation. Intuitively, the running variance σ^2 encodes how the network distributed activations during pretraining, and integrating it into the convolutional weights preserves this internal scaling for downstream adaptation.

BatchNorm layers remain fully trainable during finetuning and linear probing with BN adaptation (LP+BN), enabling them to adjust to the new data distribution. Extending the reparameterization to include affine BN parameters is possible but left for future work.

The reparameterization is applied once after loading pretrained weights and before downstream training. For linear probing without BN adaptation, this transformation does not affect the extracted features.

4 Experiments

4.1 Datasets

We evaluate our approach on four widely used image classification benchmarks: Oxford Flowers 102 [11], Stanford Cars [9], Oxford-IIIT Pets [12], and Describable Textures Dataset (DTD) [3]. These datasets cover a diverse range of visual recognition tasks, including fine-grained classification (Flowers102, Cars, Pets) and texture recognition (DTD). Our selection follows prior work on BatchNorm adaptation in self-supervised learning, enabling a direct comparison to established evaluation protocols [15].

4.2 Pretrained Models

We consider ResNet-50 [7] backbones pretrained using two different paradigms: (i) supervised pretraining on ImageNet [13] and (ii) self-supervised pretraining using Barlow Twins [16]. For both settings, we use publicly available pretrained weights and do not perform additional pretraining.

4.3 Reparameterization

For experiments involving reparameterization, we incorporate the running variance of each BatchNorm layer into the preceding convolutional weights as described in Sect. 3.1. This transformation is applied once after loading the pretrained weights and before downstream training. Models without this modification serve as baselines.

4.4 Training Protocols

We evaluate three common transfer learning strategies:

- **Finetuning:** All network parameters, including the backbone and classification head, are optimized.
- **Linear Probing (LP):** The backbone is frozen and only a linear classification head is trained.
- **Linear Probing with BatchNorm adaptation (LP+BN):** The backbone is frozen except for BatchNorm parameters, which remain trainable alongside the linear head.

For full finetuning, the complete backbone and classification head are optimized in training mode, such that BatchNorm affine parameters and running statistics are updated. For LP, we first extract features with the frozen backbone in evaluation mode using the stored BatchNorm statistics and subsequently train only the linear classifier. For LP+BN, the model is kept in training mode and the optimizer updates the linear classifier together with the BatchNorm affine parameters, while the remaining backbone parameters stay frozen; BatchNorm running statistics are updated during training.

4.5 Implementation Details

All models are trained for 50 epochs using the Adam optimizer with a learning rate of 10^{-3} and no weight decay. We do not use a learning-rate scheduler. The objective is the standard cross-entropy loss. We use a batch size of 32 for all experiments. Each experiment is repeated for five random seeds (0, 42, 1234, 2024, and 9999), and we report the mean and standard deviation of the best validation accuracy achieved during training.

For data preprocessing, all images are resized to 224×224 and normalized using standard ImageNet statistics. During training, we apply random horizontal flipping as the only data augmentation. For evaluation, only deterministic preprocessing (resize, center crop, and normalization) is used.

The code for reproducing our experiments is publicly available.[1]

4.6 Evaluation Protocol

Performance is measured in terms of top-1 validation accuracy. For each run, we record the highest validation accuracy over the course of training, following common practice in transfer learning evaluation.

5 Results

For Barlow Twins pretraining, a consistent trend emerges under full finetuning: the reparameterized variant (BT_norm) performs substantially worse than the original BT model across all datasets. The degradation ranges from 7.2% points on Flowers102 to 31.5 percentage points on Stanford Cars. This indicates that full finetuning of Barlow Twins-pretrained models is highly sensitive to the internal scaling induced by the original BatchNorm statistics.

In contrast, supervised ImageNet-pretrained models exhibit the opposite behavior under full finetuning. The reparameterized variant (ImgNet_norm) consistently outperforms the original ImageNet-pretrained model, with particularly large improvements on Flowers102 and Stanford Cars. Moreover, ImgNet_norm shows substantially lower variance across runs, suggesting that incorporating BatchNorm running variances into the convolutional weights stabilizes downstream optimization for supervised pretraining. Table 1 summarizes the downstream performance across multiple datasets for full finetuning, linear probing (LP), and linear probing with BatchNorm adaptation (LP+BN).

One possible explanation is that the reparameterization changes the effective parameterization of the supervised ImageNet-pretrained model in a way that is favorable for the fixed finetuning protocol used in our experiments. By incorporating BatchNorm running variances into the convolutional weights, the transformation changes the channel-wise scaling of the optimized parameters. This may affect effective learning rates and weight decay interactions during full finetuning, leading to more stable optimization. The fact that this improvement is primarily observed under full finetuning, but not under LP or LP+BN, supports the interpretation that the effect is related to downstream optimization dynamics rather than to a general improvement of the fixed feature representation.

[1] https://github.com/hohansen90/bt-batchnorm-icann2026.

Table 1. Downstream evaluation on multiple datasets. Values indicate mean accuracy ± standard deviation over 5 runs.

Dataset	Pretraining	Finetuning	LP	LP+BN
Flowers102	BT	0.928 ± 0.006	0.920 ± 0.001	0.911 ± 0.002
	BT_norm	0.856 ± 0.010	0.920 ± 0.001	0.911 ± 0.002
	ImgNet	0.664 ± 0.045	0.863 ± 0.002	0.930 ± 0.004
	ImgNet_norm	0.936 ± 0.009	0.863 ± 0.002	0.931 ± 0.005
DTD	BT	0.669 ± 0.005	0.712 ± < 0.001	0.691 ± 0.003
	BT_norm	0.516 ± 0.010	0.712 ± < 0.001	0.691 ± 0.002
	ImgNet	0.497 ± 0.010	0.671 ± 0.003	0.659 ± 0.005
	ImgNet_norm	0.599 ± 0.003	0.671 ± 0.003	0.663 ± 0.002
Oxford-IIIT Pets	BT	0.914 ± 0.002	0.927 ± 0.003	0.927 ± 0.002
	BT_norm	0.794 ± 0.007	0.927 ± 0.003	0.926 ± 0.002
	ImgNet	0.757 ± 0.007	0.936 ± 0.002	0.935 ± 0.004
	ImgNet_norm	0.891 ± 0.010	0.936 ± 0.002	0.935 ± 0.002
Stanford Cars	BT	0.838 ± 0.006	0.515 ± 0.001	0.757 ± 0.003
	BT_norm	0.523 ± 0.086	0.515 ± 0.001	0.757 ± 0.005
	ImgNet	0.496 ± 0.032	0.479 ± 0.003	0.745 ± 0.005
	ImgNet_norm	0.783 ± 0.006	0.479 ± 0.003	0.744 ± 0.009

Linear probing remains largely unaffected by the reparameterization. This is expected, since the transformation preserves the functional behavior of the feature extractor under fixed BatchNorm statistics. Accordingly, LP results are identical for the original and reparameterized variants within each pretraining setting.

LP+BN also remains stable across the reparameterization. For Barlow Twins, adapting BatchNorm layers together with the linear classifier yields nearly identical performance for BT and BT_norm across all datasets. This suggests that BatchNorm adaptation can compensate for the altered internal scaling when the convolutional backbone remains frozen. A similar pattern is observed for supervised ImageNet pretraining, where ImgNet and ImgNet_norm achieve comparable LP+BN performance.

Stanford Cars shows a particularly pronounced difference between LP and LP+BN. Across both pretraining paradigms, LP+BN substantially improves over LP, indicating that BatchNorm adaptation is especially important for this fine-grained classification task. This is consistent with the intuition that fine-grained recognition requires more precise feature scaling and benefits from adapting normalization statistics to the downstream data distribution.

Overall, our findings show that the same reparameterization has opposite effects depending on the pretraining objective. For Barlow Twins, incorporating BatchNorm running variances into the convolutional weights strongly degrades full finetuning performance. For supervised ImageNet pretraining, the same transformation substantially improves and stabilizes full finetuning. These results suggest that Barlow Twins-pretrained models are more sensitive to perturbations of BatchNorm-induced scaling, while supervised models may benefit from a rescaling of the pretrained convolutional weights before downstream adaptation.

At the same time, we do not claim that the observed degradation is solely caused by representational changes. The reparameterization also changes the optimization dynamics during finetuning, for example through altered effective learning rates, weight decay interactions, and the adaptation of BatchNorm statistics. Our results should therefore be interpreted as evidence that Barlow Twins is particularly sensitive to BatchNorm-induced scaling, rather than as a complete causal explanation of the underlying mechanism.

6 Conclusion

In this work, we investigated the impact of internal feature scaling on the transferability of pretrained convolutional neural networks. We found that models pretrained with Barlow Twins exhibit substantially larger weight magnitudes and heavier-tailed distributions compared to supervised ImageNet pretraining.

To investigate these differences, we introduced a simple reparameterization that incorporates BatchNorm running variances into convolutional layers, effectively encoding pretraining statistics into the network weights. While this transformation preserves the functional behavior of the network at initialization under fixed BatchNorm statistics, it changes the internal scaling and optimization dynamics during downstream training.

Our experiments reveal a clear distinction between pretraining approaches: supervised ImageNet-pretrained models benefit from this reparameterization under full finetuning, whereas Barlow Twins-pretrained models consistently lose performance. In contrast, linear probing with BatchNorm adaptation remains largely unaffected, highlighting the importance of adaptive normalization in transfer learning. These findings suggest that Barlow Twins-pretrained models rely more strongly on the BatchNorm-induced scaling established during pretraining. This is consistent with BatchNorm contributing to the shape of the learned feature distribution, but further analyses are required to disentangle representational effects from changes in optimization dynamics, effective learning rates, and BatchNorm statistic adaptation during finetuning.

A limitation of our study is that we focus on a single architecture and one self-supervised learning framework. As future work, it will be important to investigate whether similar scaling effects also appear in other architectures and self-supervised frameworks such as SimCLR, BYOL, MoCo, or VICReg. Another key question is whether the unusually large weight magnitudes observed in Barlow Twins are essential for successful pretraining or rather a byproduct of the learning objective. Further analyses, such as full BatchNorm folding, resetting or freezing BatchNorm statistics, and tracking activation statistics throughout finetuning, may provide deeper insights into how normalization, weight distributions, and learned representations interact in modern self-supervised learning.

References

1. Bardes, A., Ponce, J., LeCun, Y.: VICReg: variance-invariance-covariance regularization for self-supervised learning. In: ICLR 2022 - International Conference on Learning Representations. Online, United States (2022)
2. Chen, T., Kornblith, S., Norouzi, M., Hinton, G.: A simple framework for contrastive learning of visual representations. In: Proceedings of the 37th International Conference on Machine Learning. Proceedings of Machine Learning Research, vol. 119, pp. 1597–1607. PMLR (2020)
3. Cimpoi, M., Maji, S., Kokkinos, I., Mohamed, S., Vedaldi, A.: Describing textures in the wild. In: Proceedings of the IEEE Conf. on Computer Vision and Pattern Recognition (CVPR) (2014)
4. Ermolov, A., Siarohin, A., Sangineto, E., Sebe, N.: Whitening for self-supervised representation learning. In: Proceedings of the 38th International Conference on Machine Learning. Proceedings of Machine Learning Research, vol. 139, pp. 3015–3024. PMLR (2021)
5. Grill, J.B., et al.: Bootstrap your own latent - a new approach to self-supervised learning. In: Larochelle, H., Ranzato, M., Hadsell, R., Balcan, M., Lin, H. (eds.) Advances in Neural Information Processing Systems, vol. 33, pp. 21271–21284. Curran Associates, Inc. (2020)
6. He, K., Fan, H., Wu, Y., Xie, S., Girshick, R.: Momentum contrast for unsupervised visual representation learning. In: 2020 IEEE/CVF Conference on Computer Vision and Pattern Recognition (CVPR), pp. 9726–9735 (2020). https://doi.org/10.1109/CVPR42600.2020.00975
7. He, K., Zhang, X., Ren, S., Sun, J.: Deep residual learning for image recognition. In: 2016 IEEE Conference on Computer Vision and Pattern Recognition (CVPR), pp. 770–778 (2016). https://doi.org/10.1109/CVPR.2016.90
8. Ioffe, S., Szegedy, C.: Batch normalization: accelerating deep network training by reducing internal covariate shift. In: Bach, F., Blei, D. (eds.) Proceedings of the 32nd International Conference on Machine Learning. Proceedings of Machine Learning Research, Lille, France, 07–09 July 2015, vol. 37, pp. 448–456. PMLR (2015). https://proceedings.mlr.press/v37/ioffe15.html
9. Krause, J., Stark, M., Deng, J., Fei-Fei, L.: 3d object representations for fine-grained categorization. In: 2013 IEEE International Conference on Computer Vision Workshops, pp. 554–561 (2013). https://doi.org/10.1109/ICCVW.2013.77
10. Marks, M., et al.: A closer look at benchmarking self-supervised pre-training with image classification. Int. J. Comput. Vision **133**(8), 5013–5025 (2025). https://doi.org/10.1007/s11263-025-02402-w
11. Nilsback, M.E., Zisserman, A.: Automated flower classification over a large number of classes. In: Indian Conference on Computer Vision, Graphics and Image Processing (2008)
12. Parkhi, O.M., Vedaldi, A., Zisserman, A., Jawahar, C.V.: Cats and dogs. In: IEEE Conference on Computer Vision and Pattern Recognition (2012)
13. Russakovsky, O., et al.: ImageNet Large Scale Visual Recognition Challenge (2014)
14. Santurkar, S., Tsipras, D., Ilyas, A., Mądry, A.: How does batch normalization help optimization? In: Proceedings of the 32nd International Conference on Neural Information Processing Systems, NIPS 2018, pp. 2488–2498. Curran Associates Inc., Red Hook (2018)
15. Sirotkin, K., Escudero-Viñolo, M., Carballeira, P., Garcia-Martin, A.: Improved transferability of self-supervised learning models through batch normalization fine-

tuning. Appl. Intell. **54**, 11281–11294 (2024). https://doi.org/10.1007/s10489-024-05758-7

16. Zbontar, J., Jing, L., Misra, I., LeCun, Y., Deny, S.: Barlow twins: self-supervised learning via redundancy reduction. In: Proceedings of the 38th International Conference on Machine Learning. Proceedings of Machine Learning Research, vol. 139, pp. 12310–12320. PMLR (2021)

FAIRVAR: Fair Federated Learning via Variance Regularization

Zahra Kharaghani[1(✉)], Ali Dadras[2], and Tommy Löfstedt[1]

[1] Department of Computing Science, Umeå University, Umeå, Sweden
zahra.kharaghani@slu.se, tommy.lofstedt@umu.se
[2] Department of Mathematics and Mathematical Statistics, Umeå University, Umeå, Sweden
ali.dadras@umu.se

Abstract. Federated learning (FL) allows collaborative training of machine learning models across multiple parties without sharing raw data. However, heterogeneous data can cause some clients to have disproportionate influence on the global model, leading to disparities in their performance. Fairness, understood as reducing these disparities, is therefore a crucial concern in FL and has been addressed in various ways. We studied performance equitable fairness in FL, where the goal is to minimize performance disparities across clients. We evaluated several existing fairness-aware methods and introduce a new gradient-variance-regularized method, implemented in two variants: FairGrad (approximate) and FairGrad* (exact). We theoretically characterize the connections between these methods and, empirically, on heterogeneous benchmarks, show that FairGrad and FairGrad* consistently improve fairness by reducing the variance of client accuracies, while maintaining competitive or improved mean performance compared to existing fairness-aware baselines.

Keywords: Federated Learning · Fairness · Distributed Optimization · Neural Networks

1 Introduction

In federated learning (FL), it is often assumed that individual clients have insufficient data to effectively train machine learning models on their local datasets. By collaborating within the FL framework, clients can jointly train a global model, and by that enhance overall performance without sharing raw data. However, due to data heterogeneity, the global model may perform inconsistently across clients. This disparity poses a significant challenge, particularly for clients with substantially different data distributions or smaller sample sizes. Such clients may experience poor performance when using the global model. To address this issue, it is crucial to ensure that the training process in FL is fair, *i.e.*, for clients to have an incentive to participate, they must benefit from the global model without being disadvantaged by disparities in their local data distributions.

L. Pasa et al. (Eds.): ICANN 2026, LNCS 17089, pp. 581–593, 2027.
https://doi.org/10.1007/978-3-032-38398-3_48

Fairness in FL is generally categorized into three types: collaborative fairness, equitable fairness, and algorithmic fairness. In this work, we focused on equitable fairness, which seeks to minimize disparities in client performances. Within this category, fairness has been further conceptualized in three distinct forms: good-intent equitable fairness [16], Pareto-optimal equitable fairness [2], and performance equitable fairness [13]. Among these research lines, performance equitable fairness has attracted much attention. This approach aims to ensure equal performance across all clients without sacrificing the overall model performance. The local loss functions are usually modified in an attempt to achieve performance equitable fairness. It is, however, still unclear how different proposed loss functions support the goal of performance equitable fairness, making it challenging to determine which approaches are most effective in practical settings. Moreover, most existing methods lack a solid theoretical foundation, making it difficult to justify their use or apply them across diverse FL scenarios.

In this work, we analyzed and evaluated several methods for performance equitable fairness in FL. We distinguished between existing fairness-aware methods, baselines, and our proposed contribution. The existing fairness-aware methods include loss-variance regularization [14] and q-Fair Federated Learning (q-FFL; [13]), which, as we show later, is related to loss-variance regularization. We proposed a new gradient-variance regularization approach for performance equitable fairness, implemented in two variants: FairGrad, an approximate and communication-efficient version, and FairGrad*, an exact version. As baselines, we also included federated averaging (FedAvg; [15]), Agnostic Federated Learning (AFL; [16]), and the more recent AAggFF [5]. Formal definitions of the variance-regularized loss functions are provided in Sect. 3.

Our analysis revealed how these objectives are related under homogeneous and heterogeneous settings. Section 4 provides theoretical insights into the relationships between FedAvg, q-FFL, and two variance-regularized methods, and shows how variance regularization can change the optimization behavior under heterogeneity. Empirical results further demonstrated that the variance-regularized methods generally outperform the baselines in both fairness and overall performance, with the proposed variants, FairGrad and FairGrad*, showing particularly strong results.[1]

2 Related Work

Fairness in FL can be defined in different ways: collaborative fairness [18,19,22], which allocates rewards proportionally to clients' data contributions, algorithmic fairness [1–3,7,20], which ensures that the trained model is not biased against any client, and equitable fairness, which minimizes performance disparities across clients. Equitable fairness includes good-intent fairness [16], performance equitable fairness [13], and Pareto-optimal fairness [2]. Equitable fairness was introduced by Mohri *et al.* [16], where AFL was proposed. They showed that AFL naturally leads to a notion of fairness they refer to as good-intent fairness. This

[1] The code is available at: https://github.com/zahrakhara/FAIRVAR.

method aims to find the best global model by focusing on the worst-case client performance. However, significant performance disparities can still occur under this approach [17], and AFL has been shown to work best with smaller numbers of clients [13]. In contrast to AFL, q-FFL [13] offers greater flexibility and can be fine-tuned to achieve the desired level of fairness. q-FFL utilizes a method called q-FedAvg, which focuses on fairness by dynamically re-weighting clients based on their performance. Clients with lower performance are given higher weights to ensure a more uniform performance distribution across clients. The fairness level is controlled by the parameter q, with larger values placing more emphasis on disadvantaged clients. AFL and FedAvg are special cases of q-FFL, obtained by setting q to a sufficiently large value and $q = 0$, respectively.

AAggFF [5] builds on this line of work by unifying existing fairness-aware aggregation strategies within an online convex optimization framework and enhancing them through sequential decision-making at the server, with variants for both cross-device and cross-silo FL. Another direction studies fairness in FL via bilevel optimization. FedFB [23] selects fair batch representations, while entropy-based aggregation [21] aligns models and gradients through adaptive updates. Some works highlight that data heterogeneity can lead to different levels of irreducible error across clients. Pareto-optimal fairness [2] addresses this by minimizing worst-case loss while preserving overall performance. Building on this idea, methods such as FedMGDA+ [8] and AdaFed [6] extend multi-objective optimization for fair FL.

Variance-regularized methods have been explored in FL for different purposes. For instance, Guo *et al.* [4] proposed FEDIIR, which uses gradient variance regularization to enhance out-of-distribution generalization for non-participating clients. In contrast, we propose an approach that applies gradient variance regularization specifically to improve performance equitable fairness among participating clients in heterogeneous settings, supported by both theoretical analysis and empirical evaluation.

While various directions have been explored, the present work focused on performance equitable fairness, where important gaps remain: The theoretical connections among existing approaches are not well understood, and their performance under heterogeneous data has not been systematically evaluated. In this work, we addressed these gaps and clarified some of the theoretical connections, conceived new variance-regularized methods, and evaluated their effectiveness under heterogeneous data (Sects. 4–6).

3 Statement of the Problem

The standard FL procedure aims to minimize a global average expected risk,

$$\min_{\mathbf{x}\in\mathbb{R}^d} \left\{ \widetilde{F}(\mathbf{x}) := \frac{1}{n}\sum_{i=1}^{n} \tilde{f}_i(\mathbf{x}) = \frac{1}{n}\sum_{i=1}^{n} \mathbb{E}_{\xi_i\sim\mathcal{D}_i}\left[\ell(\mathbf{x},\xi_i)\right] \right\}, \tag{1}$$

where n denotes the number of participants (clients) involved in the federation, $\mathbf{x} \in \mathbb{R}^d$ is the parameters of the machine learning model, $\xi_i \sim \mathcal{D}_i$ is a sample data point from the i-th client's data distribution, $\mathcal{D}_i$, and $\ell(\mathbf{x}, \xi_i)$ is the loss function, indicating the loss of the model with parameters $\mathbf{x}$ on data sample ξ_i. As the true data distribution, $\mathcal{D}_i$, is unknown, clients rely on an empirical set of N_i samples, $\mathcal{S}_i = \{\xi_i^1, \ldots, \xi_i^{N_i}\} \sim \mathcal{D}_i^{N_i}$, and corresponding global empirical risk,

$$\min_{\mathbf{x} \in \mathbb{R}^d} \left\{ F(\mathbf{x}) := \frac{1}{n} \sum_{i=1}^{n} f_i(\mathbf{x}) = \frac{1}{n} \sum_{i=1}^{n} \frac{1}{N_i} \sum_{j=1}^{N_i} \ell(\mathbf{x}, \xi_i^j) \right\}. \tag{2}$$

In the rest of the paper, $F(\mathbf{x})$ denotes the global empirical risk, also referred to as the average loss across clients. Solutions to Eq. (2) may result in significant performance discrepancies among participating clients, particularly when the number of samples, N_i, is small or the client data distributions are highly heterogeneous [24]. To quantify these discrepancies, we outline a fairness criterion for FL, building on but slightly modifying the definitions introduced by Li *et al.* [13] and Nguyen *et al.* [17].

Definition 1 (Performance equitable fairness in FL). *For a set of models, with weights* $\mathcal{X} = \{\mathbf{x} : |F(\mathbf{x}) - \min_{\mathbf{x}'} F(\mathbf{x}')| \leq \varepsilon\}$, *where* $\varepsilon \geq 0$ *is a tolerance for performance degradation, a model* $\mathbf{x}_1 \in \mathcal{X}$ *is said to achieve better* ε*-performance equitable fairness than model* $\mathbf{x}_2 \in \mathcal{X}$ *if the variance of its performance across clients is lower. That is, if* $var((p_i(\mathbf{x}_1))_{i=1}^n) < var((p_i(\mathbf{x}_2))_{i=1}^n)$, *where* $p_i(\mathbf{x})$ *denotes a performance measure (e.g., accuracy) for model* $\mathbf{x}$ *on the i-th client's local data and* $var((p_i(\mathbf{x}))_{i=1}^n)$ *denotes the sample variance across all clients.*

When a model is trained on data from multiple clients, fairness per Definition 1 means that each client has an equal opportunity to benefit from the model. The goal is to achieve equitable performance across all clients, regardless of differences in the data they contribute with. The tolerance parameter, ε, ensures that only models with average loss, $F(\mathbf{x})$, close to the best model (within ε) are considered. This avoids favoring models that achieve low variance simply because they perform poorly on all clients. Instead, the comparison is restricted to high-performing models, among which the fairest model is preferred.

3.1 Evaluated Methods

In this section, we present the objective functions of the methods evaluated in this work for performance equitable fairness in FL. These methods aim to reduce disparities in client performance by addressing differences in client losses or gradients. Each method introduces a tunable parameter that controls the trade-off between average performance and fairness. When set to zero, the objective reduces to the standard finite-sum formulation in Eq. (2); otherwise, it promotes a more uniform distribution of performance across clients, potentially improving outcomes for under-performing clients at the expense of others.

Ling *et al.* [14] proposed to penalize models with high variance in the client performances, which can be implemented using a variance-regularized loss, as

$$\min_{\mathbf{x} \in \mathbb{R}^d} \left\{ L_\lambda(\mathbf{x}) := F(\mathbf{x}) + \frac{\lambda}{2n} \sum_{i=1}^{n} \left(f_i(\mathbf{x}) - F(\mathbf{x})\right)^2 \right\}, \tag{3}$$

where λ controls the degree of fairness and $F(\mathbf{x})$ denotes the average loss, as defined in Equation (2). We propose two new algorithms to solve this objective, detailed in Algorithm 1 and Algorithm 3[2].

We further propose a novel regularization approach aiming to promote performance equity across participating clients in FL by minimizing gradient disparities. This proposed method is formulated as

$$\min_{\mathbf{x}\in\mathbb{R}^d}\left\{J_\gamma(\mathbf{x}) := F(\mathbf{x}) + \frac{\gamma}{2n}\sum_{i=1}^{n}\left\|\nabla f_i(\mathbf{x}) - \nabla F(\mathbf{x})\right\|^2\right\}, \tag{4}$$

where $\nabla F(\mathbf{x}) := \frac{1}{n}\sum_{i=1}^{n}\nabla f_i(\mathbf{x})$ denotes the average gradient, γ controls the degree of fairness, $\|\cdot\|$ denotes the Euclidean norm, and $F(\mathbf{x})$ is defined in Eq. (2), as before. We also propose two new algorithms to solve Eq. (4), that are detailed in Algorithm 2 and Algorithm 4[2].

Li *et al.* [13] proposed q-FFL to improve fairness in FL by prioritizing clients with higher losses through a reweighting mechanism that gives higher weight to under-performing clients. It is formulated as

$$\min_{\mathbf{x}\in\mathbb{R}^d}\left\{H_q(\mathbf{x}) := \sum_{i=1}^{n}\frac{r_i}{q+1}\left(f_i(\mathbf{x})\right)^{q+1}\right\}, \tag{5}$$

where r_i is the proportion of total data samples held by client i, and q is a parameter that controls the degree of fairness.

The FedAvg algorithm [15] aims to minimize the average of local loss functions across clients. The objective, given in Eq. (2), is considered here with uniform weights, $1/n$, for all clients. Eqs. (3), (4), and (5) define the objective functions of the fairness methods considered in this work, however, they are not directly solvable in a federated context because they are not separable with respect to each client's objective. To address this, we propose approximate algorithms for Eqs. (3) and (4) in Sect. 5, with exact variants provided in Kharaghani *et al.* [9]. For q-FFL, we adopt the original optimization algorithm of Li *et al.* [13]. Full proofs of the theoretical results in Sects. 4 and 5 are also available in Kharaghani *et al.* [9].

4 Analysis of Stationary Points

Homogeneous Data: In a homogeneous data environment, the datasets of all n clients are drawn from the same underlying probability distribution, *i.e.*, $\mathcal{D}_i = \mathcal{D}_j$ for all $i, j \in \{1, \ldots, n\}$. As a result, each client's local loss, $f_i(\mathbf{x})$, is an unbiased estimator of the expected average loss, $\mathbb{E}[F(\mathbf{x})]$, and each client's gradient, $\nabla f_i(\mathbf{x})$, is an unbiased estimator of the expected average gradient, $\mathbb{E}[\nabla F(\mathbf{x})]$. This implies that, in homogeneous settings, introducing the regularization terms in Eqs. (3) and (4) does not alter the location of the stationary points on average,

[2] Algorithms 3 and 4 are available in Kharaghani *et al.* [9].

thereby preserving the optimal solutions of the original objective, $F(\mathbf{x})$. Consequently, the regularization terms are unnecessary in such settings, as they do not change the outcome. This conclusion also holds for q-FFL, as the following theorem implies. We show that the expectation of the function $H_q(\mathbf{x})$, when $q = 1$, can be upper bounded by the expectation of the function $L_\lambda(\mathbf{x})$, for a particular value of λ.

Theorem 1. *For $q = 1$ and $r_i = \frac{1}{n}$, we have*

$$\mathbb{E}\big[H_1(\mathbf{x})\big] = \frac{1}{2n}\sum_{i=1}^{n}\mathbb{E}\big[f_i(\mathbf{x})\big]^2 + \frac{n}{n-1}\mathbb{E}\left[\frac{1}{2n}\sum_{i=1}^{n}\left(f_i(\mathbf{x}) - F(\mathbf{x})\right)^2\right].$$

If $\mathbb{E}[f_i(\mathbf{x})] \leq C$ for all i and some constant C, then $\mathbb{E}[H_1(\mathbf{x})] \leq \mathbb{E}[L_\lambda(\mathbf{x})]$, with $\lambda = \frac{n}{C(n-1)}$.

Theorem 1 demonstrates that $H_1(\mathbf{x})$, *i.e.*, q-FFL with $q = 1$, can be expressed in a similar form as $L_\lambda(\mathbf{x})$ for a particular λ. Further, at the stationary points of $F(\mathbf{x})$, the two functions $H_q(\mathbf{x})$ and $L_\lambda(\mathbf{x})$ (for $q = 1$ and $\lambda = \frac{n}{C(n-1)}$) are close, as $F(\mathbf{x})$ attains its minimum value at these points, leading to a small difference between them.

Heterogeneous Data: In the case of heterogeneous data, the datasets of different clients are drawn from different distributions, *i.e.*, $\mathcal{D}_i \neq \mathcal{D}_j$ for some $i, j \in \{1, \ldots, n\}$, resulting in variability in the local objectives and gradients. In this case, the variance terms in the regularized loss functions, $L_\lambda(\mathbf{x})$, $J_\gamma(\mathbf{x})$, and $H_q(\mathbf{x})$ with $q = 1$, are important because they penalize this variability. The regularization encourages the optimization process to find solutions that balance performance across diverse clients, promoting models that generalize better. While the expectation-based arguments in the homogeneous case rely on the unbiasedness of local losses, and local gradients with respect to the global objective, such assumptions do not hold universally in the heterogeneous setting, where client data distributions differ. However, under certain conditions, specifically when client losses or gradients are aligned at the optimum point, similar results can still be obtained.

Theorem 2. *Assume that the f_i's are differentiable. Let $\mathbf{x}^\star \in \arg\min_{\mathbf{x}\in\mathbb{R}^d} F(\mathbf{x})$, $\mathbf{x}^\star_\lambda \in \arg\min_{\mathbf{x}\in\mathbb{R}^d} L_\lambda(\mathbf{x})$, $\mathbf{x}^\star_\gamma \in \arg\min_{\mathbf{x}\in\mathbb{R}^d} J_\gamma(\mathbf{x})$, and $\mathbf{x}^\star_q \in \arg\min_{\mathbf{x}\in\mathbb{R}^d} H_q(\mathbf{x})$, such that*

$$\nabla F(\mathbf{x}^\star) = \nabla L_\lambda(\mathbf{x}^\star_\lambda) = \nabla J_\gamma(\mathbf{x}^\star_\gamma) = \nabla H_q(\mathbf{x}^\star_q) = 0.$$

If either $f_i(\mathbf{x}^\star) = f_j(\mathbf{x}^\star)$ or $\nabla f_i(\mathbf{x}^\star) = \nabla f_j(\mathbf{x}^\star)$ holds for all i, j, then we have $\nabla L_\lambda(\mathbf{x}^\star) = \nabla J_\gamma(\mathbf{x}^\star) = \nabla H_q(\mathbf{x}^\star) = 0$. Similarly, if either $f_i(\mathbf{x}^\star_\lambda) = f_j(\mathbf{x}^\star_\lambda)$ or $\nabla f_i(\mathbf{x}^\star_\lambda) = \nabla f_j(\mathbf{x}^\star_\lambda)$ for all i, j, then $\nabla F(\mathbf{x}^\star_\lambda) = 0$. If the analogous condition holds at $\mathbf{x}^\star_\gamma$ and $\mathbf{x}^\star_q$, then $\nabla F(\mathbf{x}^\star_\gamma) = 0$, and $\nabla F(\mathbf{x}^\star_q) = 0$.

The following theorem investigates when stationary points differ between $F(\mathbf{x})$ and $L_\lambda(\mathbf{x})$. This displacement is useful when stationary points of $F(\mathbf{x})$

lead to unequal client performance, as can occur under heterogeneous data. In such cases, variance regularization can move the solution away from an average-loss optimum toward a point with more balanced performance across clients.

Theorem 3. *Assume that the f_i's are differentiable. Let $\mathbf{x}^\star \in \arg\min_{\mathbf{x}\in\mathbb{R}^d} F(\mathbf{x})$ with $\nabla F(\mathbf{x}^\star) = 0$, and $\mathbf{x}^\star_\lambda \in \arg\min_{\mathbf{x}\in\mathbb{R}^d} L_\lambda(\mathbf{x})$ with $\nabla L_\lambda(\mathbf{x}^\star_\lambda) = 0$. Furthermore, suppose there exist at least two indices $1 \le i, j \le n$ such that $f_i(\mathbf{x}) \neq f_j(\mathbf{x})$ and $\nabla f_i(\mathbf{x}) \neq \nabla f_j(\mathbf{x})$ for $\mathbf{x} \in \{\mathbf{x}^\star, \mathbf{x}^\star_\lambda\}$. Let $\mathcal{G}(\mathbf{x}) = \{\nabla f_i(\mathbf{x})\}_{i=1}^n$ denote the set of gradient vectors at $\mathbf{x}$. Then $\nabla L_\lambda(\mathbf{x}^\star) \neq 0$ and $\nabla F(\mathbf{x}^\star_\lambda) \neq 0$ if at least one of the following conditions holds:*

1. *Subset of clients with distinct gradient directions, others with identical loss values. That is, there exist m linearly independent gradients in $\mathcal{G}(\mathbf{x})$, where $1 < m \le n-1$, and the remaining $n-m$ gradients correspond to loss functions that are equal at $\mathbf{x} \in \{\mathbf{x}^\star, \mathbf{x}^\star_\lambda\}$. Specifically, if $I_{ind} \subseteq \{1, \ldots, n\}$ indexes the m linearly independent gradients, i.e. $\mathcal{G}_{ind}(\mathbf{x}) = \{\nabla f_i(\mathbf{x}) \mid i \in I_{ind}\} \subseteq \mathcal{G}(\mathbf{x})$, and the remaining indices are $I_{rem} = \{1, \ldots, n\} \setminus I_{ind}$, then the corresponding loss values satisfy $f_j(\mathbf{x}) = f_k(\mathbf{x})$ for all $j, k \in I_{rem}$.*
2. *Subset of clients with non-zero gradients, all but one sharing the same loss value. That is, there exist m non-zero gradient vectors with $1 \le m \le n$, and the loss functions corresponding to these gradients are all equal except for one.*

The conditions in Theorem 3 describe when fairness regularization alters the optimization landscape, such that a stationary point of one objective ($\mathbf{x}^\star$ for $F(\mathbf{x})$ or $\mathbf{x}^\star_\lambda$ for $L_\lambda(\mathbf{x})$) is not preserved for the other. Such situations arise naturally in heterogeneous FL. The first condition occurs when clients have substantially different data distributions, leading to distinct gradient directions even when some clients exhibit similar loss values. The second arises when most clients have similar performance, but one client's data distribution results in a noticeably different loss. In both scenarios, the disagreement in either gradient directions or loss values introduces an imbalance that fairness regularization alters, thereby shifting the stationary point between $F(\mathbf{x})$ and $L_\lambda(\mathbf{x})$.

The following theorem shows that even when all clients have identical losses at a stationary point of $F(\mathbf{x})$, small zero-mean perturbations to the client losses can cause the stationary point of $L_\lambda(\mathbf{x})$ to differ from that of $F(\mathbf{x})$.

Theorem 4. *Assume that the f_i's are differentiable. Let $\mathbf{x}^\star \in \arg\min_{\mathbf{x}\in\mathbb{R}^d} F(\mathbf{x})$ such that $f_i(\mathbf{x}^\star) = f_j(\mathbf{x}^\star)$ for all indices i, j. Assume further that, $\nabla f_i(\mathbf{x}^\star) \neq 0$ for at least one index i. Then, there exist non-zero values o_i such that $\bar{o} := \frac{1}{n}\sum_{i=1}^n o_i = 0$ and the function $\hat{L}_\lambda(\mathbf{x})$ defined as $\hat{L}_\lambda(\mathbf{x}) := \hat{F}(\mathbf{x}) + \frac{\lambda}{2n}\sum_{i=1}^n \left(\hat{f}_i(\mathbf{x}) - \hat{F}(\mathbf{x})\right)^2$, where $\hat{f}_i(\mathbf{x}) = f_i(\mathbf{x}) + o_i$, and $\hat{F}(\mathbf{x}) = F(\mathbf{x}) = \frac{1}{n}\sum_{i=1}^n \hat{f}_i(\mathbf{x})$, has a different stationary point than $F(\mathbf{x})$.*

We show that, under certain conditions, the regularized objectives may have stationary points that differ from those of $F(\mathbf{x})$. Stationary points of $L_\lambda(\mathbf{x})$ may reduce local-loss variance, leading to more balanced client performance [9].

5 Algorithms

To minimize the objective in Eq. (3), which aims to minimize both the average and the variance of client loss functions, we propose and evaluate two algorithms: an approximate version (FairLoss; see Algorithm 1) and an exact version (FairLoss*; see Algorithm 3[2]). Similarly, for the objective in Eq. (4), which minimizes the average loss and reduces the variance of the gradients of client loss functions, we introduce two algorithms: an approximate version (FairGrad; see Algorithm 2) and an exact version (FairGrad*; see Algorithm 4[2]). In the approximate variants, clients perform local updates using information from the previous iteration, thereby reducing communication with the server. In the exact variants, clients use information from the current iteration, which provides more accurate updates but at the cost of increased communication. Detailed descriptions of all algorithms are provided in Kharaghani *et al.* [9].

In the following theorem, we show that the surrogate objective and gradient used by each approximate algorithm are consistent with those of its exact counterpart.

Theorem 5. *Let $\bar{a}_\lambda(\mathbf{x}^t)$ and $\bar{a}_\gamma(\mathbf{x}^t)$ denote approximations of the functions $L_\lambda(\mathbf{x}^t)$ and $J_\gamma(\mathbf{x}^t)$, respectively, and let $\bar{\mathbf{g}}_\lambda(\mathbf{x}^t)$ and $\bar{\mathbf{g}}_\gamma(\mathbf{x}^t)$ denote the corresponding gradient approximations at iteration t, as used in Algorithms 1–2. We assume that SGD is used for the updates. Then we have*

$$L_\lambda(\mathbf{x}^t) - \bar{a}_\lambda(\mathbf{x}^t) = -\tfrac{\lambda}{2}\left(F(\mathbf{x}^t) - F(\mathbf{x}^{t-1})\right)^2 \leq 0, \text{ and}$$

$$J_\gamma(\mathbf{x}^t) - \bar{a}_\gamma(\mathbf{x}^t) \leq -\tfrac{\gamma}{2}\left\|\nabla F(\mathbf{x}^t) - \nabla F(\mathbf{x}^{t-1})\right\|^2 \leq 0.$$

Moreover, if each local loss function, $f_i(\mathbf{x})$, is L-smooth and η denotes the learning rate, then $\|\nabla L_\lambda(\mathbf{x}^t)\| \leq \|\bar{\mathbf{g}}_\lambda(\mathbf{x}^t)\| + \lambda L\eta\|\bar{\mathbf{g}}_\lambda(\mathbf{x}^{t-1})\|\,|F(\mathbf{x}^t) - F(\mathbf{x}^{t-1})|$ and

$$\|\nabla J_\gamma(\mathbf{x}^t)\| \leq \|\bar{\mathbf{g}}_\gamma(\mathbf{x}^t)\| + \gamma L^2\eta\|\bar{\mathbf{g}}_\gamma(\mathbf{x}^{t-1})\|.$$

Algorithms 1 and 2 use approximations of the objective functions $L_\lambda(\mathbf{x}^t)$ and $J_\gamma(\mathbf{x}^t)$, together with their gradients $\nabla L_\lambda(\mathbf{x}^t)$ and $\nabla J_\gamma(\mathbf{x}^t)$, denoted by $\bar{a}_\lambda(\mathbf{x}^t)$, $\bar{\mathbf{g}}_\lambda(\mathbf{x}^t)$, and $\bar{a}_\gamma(\mathbf{x}^t), \bar{\mathbf{g}}_\gamma(\mathbf{x}^t)$, respectively. Theorem 5 establishes connections between the approximate and true functions, ensuring that the surrogate functions provide reliable estimates of the original objectives. Specifically, if the sequence $\mathbf{x}^t$ generated by Algorithm 1 satisfies $\lim_{t\to\infty}\mathbf{x}^t = \mathbf{x}^\dagger$ and $\lim_{t\to\infty}\|\bar{\mathbf{g}}_\lambda(\mathbf{x}^t)\| = 0$, then $\mathbf{x}^\dagger$ is a stationary point of L_λ. Analogously, if the sequence $\mathbf{x}^t$ generated by Algorithm 2 satisfies $\lim_{t\to\infty}\mathbf{x}^t = \hat{\mathbf{x}}$ and $\lim_{t\to\infty}\|\bar{\mathbf{g}}_\gamma(\mathbf{x}^t)\| = 0$, then $\hat{\mathbf{x}}$ is a stationary point of J_γ. We present the approximate variants, FairLoss and FairGrad, here. The exact variants, FairLoss* and FairGrad*, along with q-FFL, are provided in Kharaghani *et al.* [9].

6 Experiments

We evaluated the investigated methods on four widely used image datasets: MNIST [12], CIFAR-10, CIFAR-100 [10], and Tiny ImageNet [11], with 50, 50,

Algorithm 1: FairLoss

Input: $\mathbf{x}^0, \lambda, \eta, T$
Initialize: $a^0 = 0$, $\mathbf{g}^0 = 0$
for round $t = 0, 1, \ldots, T-1$ **do**
  — **Client**-level local training ———
  for client $i = 1, \ldots, n$ **do**
    Receives $\mathbf{x}^t, a^t$, and $\mathbf{g}^t$ from the server
    $a_i^t = f_i(\mathbf{x}^t) + \frac{\lambda}{2}\big(f_i(\mathbf{x}^t) - a^t\big)^2$
    $\mathbf{g}_i^t = \nabla f_i(\mathbf{x}^t) + \lambda\big(f_i(\mathbf{x}^t) - a^t\big)\big(\nabla f_i(\mathbf{x}^t) - \mathbf{g}^t\big)$
    $\mathbf{x}_i^{t+1} = \mathbf{x}^t - \eta \cdot \mathbf{g}_i^t$
    Sends $\mathbf{x}_i^{t+1}$, $f_i(\mathbf{x}^t)$, and $\nabla f_i(\mathbf{x}^t)$ to the server
  end for
  — **Server**-level aggregation ———
  Receives all $\mathbf{x}_i^{t+1}$, $f_i(\mathbf{x}^t)$, and $\nabla f_i(\mathbf{x}^t)$
  $\mathbf{x}^{t+1} = \frac{1}{n}\sum_{i=1}^n \mathbf{x}_i^{t+1}$
  $a^{t+1} = \frac{1}{n}\sum_{i=1}^n f_i(\mathbf{x}^t)$
  $\mathbf{g}^{t+1} = \frac{1}{n}\sum_{i=1}^n \nabla f_i(\mathbf{x}^t)$
  Sends $\mathbf{x}^{t+1}$, a^{t+1}, and $\mathbf{g}^{t+1}$ to the clients
end for

Algorithm 2: FairGrad

Input: $\mathbf{x}^0, \gamma, \eta, T$
Initialize: $\mathbf{g}^0 = 0$
for round $t = 0, 1, \ldots, T-1$ **do**
  — **Client**-level local training ———
  for client $i = 1, \ldots, n$ **do**
    Receives $\mathbf{x}^t$ and $\mathbf{g}^t$ from the server
    $\mathbf{g}_i^t = \nabla\big(f_i(\mathbf{x}^t) + \frac{\gamma}{2}\|\nabla f_i(\mathbf{x}^t) - \mathbf{g}^t\|^2\big)$
    $\mathbf{x}_i^{t+1} = \mathbf{x}^t - \eta \cdot \mathbf{g}_i^t,$
    Sends $\mathbf{x}_i^{t+1}$ and $\nabla f_i(\mathbf{x}^t)$ to the server
  end for
  — **Server**-level aggregation ———
  $\mathbf{x}^{t+1} = \frac{1}{n}\sum_{i=1}^n \mathbf{x}_i^{t+1}$
  $\mathbf{g}^{t+1} = \frac{1}{n}\sum_{i=1}^n \nabla f_i(\mathbf{x}^t)$
  Sends $\mathbf{x}^{t+1}$ and $\mathbf{g}^{t+1}$ to the clients
end for

100, and 100 clients, respectively. To simulate varying levels of data heterogeneity among clients, we used the data generation procedure from the FL-bench framework [?] where a Dirichlet distribution is applied with concentration parameters $\alpha \in \{0.05, 0.1, 0.5\}$[3]. Smaller α values result in greater variability in client data distributions and more unequal dataset sizes, thereby increasing the degree of heterogeneity. We compared the performance of the investigated methods FairLoss, FairGrad, FairLoss*, FairGrad*, and q-FFL against the baseline methods FedAvg, AFL, and AAggFF. For each hyperparameter configuration, we trained all methods and selected the epoch that maximized the following validation-based criterion: $\overline{\text{acc}} - t \cdot \sqrt{\frac{\overline{\text{var}}}{m}}$, where the criterion is computed using validation accuracies across clients with $t = 1.96$. For clarity, for each random seed we first computed the average validation accuracy across clients and the variance of validation accuracies across clients. Then, $\overline{\text{acc}}$ denotes the average of these per-seed average accuracies, $\overline{\text{var}}$ denotes the average of these per-seed variances, and m is the number of independent runs (*i.e.*, seeds). Only after selecting this best epoch and its corresponding parameters, the resulting best model was evaluated on the test data, and those test scores are the results that are reported in Tables 1 and 2. Additional results (Tables 3 and 4), together with detailed descriptions of datasets, partitioning strategies, model architectures, training protocols, and hyperparameter selection are provided in Kharaghani *et al.* [9].

[3] https://github.com/KarhouTam/FL-bench.

Table 1. Performance on CIFAR-100 across Dirichlet values (100 clients)

Method	**Dir**(0.05)		**Dir**(0.1)		**Dir**(0.5)	
	Test Accuracy (%)	Test Variance	Test Accuracy (%)	Test Variance	Test Accuracy (%)	Test Variance
FedAvg	66.98 (±0.73)	0.88 (±0.07)	66.69 (±0.21)	0.96 (±0.04)	66.35 (±0.71)	0.92 (±0.07)
AAggFF	66.83 (±0.50)	0.93 (±0.09)	66.68 (±0.46)	0.85 (±0.08)	67.22 (±0.48)	0.89 (±0.06)
AFL	57.50 (±0.36)	4.94 (±0.14)	61.44 (±0.81)	4.01 (±0.11)	67.46 (±0.46)	2.49 (±0.12)
q-FFL	66.36 (±0.32)	0.90 (±0.07)	67.70 (±0.72)	0.86 (±0.02)	67.20 (±0.39)	0.92 (±0.04)
FairLoss	66.46 (±0.78)	0.81 (±0.06)	67.55 (±0.45)	0.88 (±0.06)	66.94 (±0.29)	0.81 (±0.06)
FairLoss*	67.16 (±0.50)	0.88 (±0.05)	67.41 (±0.43)	0.90 (±0.06)	67.26 (±0.33)	0.82 (±0.06)
FairGrad	68.90 (±0.21)	0.82 (±0.03)	69.38 (±0.46)	0.91 (±0.02)	**70.52** (±0.78)	0.86 (±0.04)
FairGrad*	**69.23** (±0.29)	**0.79** (±0.04)	**70.78** (±0.40)	**0.77** (±0.02)	69.67 (±0.16)	**0.78** (±0.03)

Table 2. Performance on Tiny ImageNet across Dirichlet values (100 clients)

Method	**Dir**(0.05)		**Dir**(0.1)		**Dir**(0.5)	
	Test Accuracy (%)	Test Variance	Test Accuracy (%)	Test Variance	Test Accuracy (%)	Test Variance
FedAvg	59.03 (±0.10)	**0.47** (±0.01)	59.39 (±0.27)	**0.53** (±0.01)	59.42 (±0.15)	0.44 (±0.02)
AAggFF	59.15 (±0.23)	0.51 (±0.02)	59.30 (±0.16)	0.57 (±0.02)	59.70 (±0.25)	0.49 (±0.02)
AFL	59.27 (±1.30)	3.67 (±0.17)	58.88 (±1.55)	2.51 (±0.19)	55.16 (±0.52)	1.69 (±0.07)
q-FFL	58.77 (±0.12)	0.56 (±0.02)	59.25 (±0.12)	0.56 (±0.02)	59.71 (±0.01)	0.47 (±0.01)
FairLoss	59.10 (±0.15)	0.49 (±0.03)	59.60 (±0.39)	**0.53** (±0.01)	59.50 (±0.03)	0.46 (±0.03)
FairLoss*	59.15 (±0.31)	0.49 (±0.03)	59.55 (±0.17)	0.54 (±0.04)	59.62 (±0.15)	0.46 (±0.01)
FairGrad	**62.06** (±0.11)	0.52 (±0.01)	62.15 (±0.85)	0.56 (±0.02)	60.01 (±0.02)	0.44 (±0.02)
FairGrad*	59.92 (±0.40)	**0.47** (±0.02)	**63.46** (±0.12)	0.56 (±0.01)	**62.84** (±0.11)	**0.36** (±0.01)

Results and Discussion. For each run, the best epoch and hyperparameters were selected based on the validation set. All reported results are evaluated on the test set and include test accuracy and the variance of test accuracies across clients (the fairness metric). Values in parentheses denote the standard error. Detailed results are presented in Tables 1 and 2; additional results for CIFAR-10 and MNIST (Tables 3 and 4) are provided in Kharaghani *et al.* [9]. In these tables, the reported results correspond to different values of the Dirichlet concentration parameter α. For CIFAR-100 (Table 1), FairGrad* obtained both the best accuracy and lowest variances at $\alpha \in \{0.05, 0.1\}$, and the lowest variance at $\alpha = 0.5$, where FairGrad reached the highest accuracy. The regularized methods generally improved over the baselines in either accuracy, variance, or both. For Tiny ImageNet (Table 2), FairGrad* achieved the lowest variance at $\alpha \in \{0.05, 0.5\}$ and the best accuracy at $\alpha = 0.5$, while FairGrad and FairGrad* attained the highest accuracy at $\alpha = 0.05$ and $\alpha = 0.1$, respectively. Similar to the results on CIFAR-100, the regularized methods improved fairness or performance compared to the baselines. For CIFAR-10, FairGrad consistently outperformed the other methods, achieving the lowest variances (indicating higher fairness) across all α, and the highest accuracies at $\alpha \in \{0.1, 0.5\}$. For MNIST, all methods performed similarly, with regularized variants offering slightly better fairness; in particular, FairGrad* achieved the lowest variance across all α values.

7 Conclusion

In this work, we focused on fairness-aware FL algorithms that explicitly reduce client disparities through variance-based regularization of client losses or gradients. We evaluated several methods in this category, including a gradient variance regularization approach, proposed here for the first time in the context of performance equitable fairness, and implemented in two variants: An approximate version that uses the average gradient computed at the previous iteration, and an exact version that uses the average gradient from the current iteration across clients. The exact variant can be useful when the task is more challenging or client gradients vary substantially, whereas the approximate version is preferable when communication efficiency is the main concern.

The theoretical analysis indicates that, under the sufficient conditions identified in our analysis, the regularized objectives have different stationary points compared to the unregularized ones. Empirical results confirm that, in heterogeneous settings with more clients and complex datasets, the regularized methods improve both fairness and overall model performance, with the gradient alignment variants consistently delivering strong results. In contrast, in homogeneous settings, the theoretical results suggest minimal to no differences between these methods, which aligns with the empirical observations. The experiments focused on image classification benchmarks with full client participation; to evaluate other tasks and partial client participation is an important direction for future work. Overall, this work improves our understanding of variance-based fairness regularization in general and in particular highlights the proposed gradient alignment regularizer as a principled and effective approach to improve fairness in FL.

Acknowledgments. This work was supported by the Swedish Cancer Society (grant number 22 2428 Pj) and Lion's Cancer Research Foundation in Northern Sweden (grant numbers LP 22-2319, LP 24-2367, and AMP 26-1265). The computations were enabled by resources provided by the National Academic Infrastructure for Supercomputing in Sweden (NAISS) at C3SE and HPC2N, partially funded by the Swedish Research Council through grant agreement no. 2022-06725. Ali Dadras was supported by the Wallenberg AI, Autonomous Systems and Software Program (WASP), funded by the Knut and Alice Wallenberg Foundation, and by the Swedish Research Council under grant number 2023-05476. We are also grateful for valuable discussions with Alp Yurtsever.

Disclosure of Interests. The authors have no competing interests to declare that are relevant to the content of this article.

References

1. Chen, H., Zhu, T., Zhou, W., Zhao, W.: AFed: algorithmic fair federated learning. IEEE Trans. Neural Netw. Learn. Syst. **36**(8) (2025)
2. Cui, S., Pan, W., Liang, J., Zhang, C., Wang, F.: Addressing algorithmic disparity and performance inconsistency in federated learning. Adv. Neural. Inf. Process. Syst. **34** (2021)
3. Ezzeldin, Y.H., Yan, S., He, C., Ferrara, E., Avestimehr, A.S.: FairFed: enabling group fairness in federated learning. In: AAAI Conference on Artificial Intelligence, vol. 37 (2023)
4. Guo, Y., Guo, K., Cao, X., Wu, T., Chang, Y.: Out-of-distribution generalization of federated learning via implicit invariant relationships. In: Proceedings of the International Conference on Machine Learning, pp. 11905–11933 (2023)
5. Hahn, S.J., Kim, G.S., Lee, J.: Pursuing overall welfare in federated learning through sequential decision making. In: Proceedings of the 41st International Conference on Machine Learning, PMLR, vol. 235, pp. 17246–17278 (2024)
6. Hamidi, S.M., Yang, E.H.: AdaFed: fair federated learning via adaptive common descent direction. Trans. Mach. Learn. Res. (2024)
7. He, Z., Wang, Z., Dong, X., Sun, P., Ren, J., Ren, K.: Towards fair federated learning via unbiased feature aggregation. IEEE Trans. Depend. Secure Comput. (2025)
8. Hu, Z., Shaloudegi, K., Zhang, G., Yu, Y.: Federated learning meets multi-objective optimization. IEEE Trans. Netw. Sci. Eng. **9**(4) (2022)
9. Kharaghani, Z., Dadras, A., Löfstedt, T.: FAIRVAR: Fair federated learning via variance regularization (2026). arXiv:2508.12042
10. Krizhevsky, A.: Learning multiple layers of features from tiny images. Tech. rep. Department of Computer Science, University of Toronto (2009)
11. Le, Y., Yang, X.: Tiny ImageNet visual recognition challenge. Technical report, Stanford University (2015)
12. LeCun, Y., Bottou, L., Bengio, Y., Haffner, P.: Gradient-based learning applied to document recognition. Proc. IEEE **86**(11) (1998)
13. Li, T., Sanjabi, M., Beirami, A., Smith, V.: Fair resource allocation in federated learning. In: International Conference on Learning Representations (ICLR) (2020)
14. Ling, X., Fu, J., Wang, K., Li, H., Cheng, T., Chen, Z.: FedFDP: fairness-aware federated learning with differential privacy. In: International Conference on Applied Cryptography and Network Security (ACNS), pp. 129–159. Springer (2026)
15. McMahan, B., Moore, E., Ramage, D., Hampson, S., Aguera y Arcas, B.: Communication-efficient learning of deep networks from decentralized data. In: International Conference on Artificial Intelligence and Statistics (2017)
16. Mohri, M., Sivek, G., Suresh, A.T.: Agnostic federated learning. In: Proceedings of the International Conference on Machine Learning (2019)
17. Nguyen, L.M., Hoang, T.N., Chen, P.Y.: Federated Learning: Theory and Practice. Academic Press (2024)
18. Rashid, M.M., Xiang, Y., Uddin, M.P., Tang, J., Sood, K., Gao, L.: Trustworthy and fair federated learning via reputation-based consensus and adaptive incentives. IEEE Trans. Inf. Forensics Secur. **20** (2025)
19. Sim, R.H.L., Zhang, Y., Chan, M.C., Low, B.K.H.: Collaborative machine learning with incentive-aware model rewards. In: Proceedings of the International Conference on Machine Learning (2020)

20. Su, C., Yu, G., Wang, J., Li, H., Li, Q., Yu, H.: Multi-dimensional fair federated learning. In: AAAI Conference on Artificial Intelligence, vol. 38, no. 13 (2024)
21. Tan, J., Wang, X.: FL-bench: a federated learning benchmark for solving image classification tasks (2023). https://github.com/KarhouTam/FL-bench
22. Wang, L., Wang, Z., Shi, Y., Karimireddy, S.P., Tang, X.: Entropy-driven fair and effective federated learning. (2023). arXiv:2011.10464
23. Zeng, Y., Chen, H., Lee, K.: Improving fairness via federated learning (2021). arXiv:2110.15545
24. Zhao, Y., Li, M., Lai, L., Suda, N., Civin, D., Chandra, V.: Federated learning with non-IID data (2018). arXiv:1806.00582

Bi-DKD: Bidirectional Decoupled Knowledge Distillation

Yu Feng[1], Bo Li[1], Fen Wang[1(✉)], Yuan Liu[1], Baoyi Xu[1,2], and Long Yang[1,2]

[1] Hangzhou Institute of Technology, Xidian University, Hangzhou, China
wangfen@xidian.edu.cn

[2] State Key Laboratory of Integrated Services Networks, Xidian University, Xi'an, China

Abstract. Knowledge distillation (KD) transfers knowledge from a large teacher model to a compact student to improve performance. A recent benchmark study, Decoupled Knowledge Distillation (DKD), reveals that the core value of logit distillation lies in non-target-class distillation (NCKD). Meanwhile, related theoretical research indicates that the success of distillation stems from the transfer of "multi-view" features that the student model fails to capture. Inspired by these perspectives, we propose that NCKD serves as the primary carrier for these crucial feature views. To fully learn these feature views, in this paper, we propose Bidirectional Decoupled Knowledge Distillation (Bi-DKD), a cyclical teacher-student learning framework built upon DKD to facilitate the exchange of their complementary feature views. Specifically, Bi-DKD feeds the knowledge learned by the student from the first-round training back to the teacher. This process constructs an enhanced teacher model with a broader ensemble of views. Subsequently, this enhanced teacher guides the training of the final student model. Additionally, to avoid learning stagnation, we introduce an Adaptive Dynamic Weight (ADW) strategy. By dynamically adjusting the NCKD coefficient, this strategy prioritizes the transfer of missing feature views. Extensive experiments on various datasets demonstrate that Bi-DKD significantly outperforms standard DKD and surpasses most existing methods, proving its effectiveness in learning a broader ensemble of feature views.

Keywords: Knowledge Distillation · Decoupled Knowledge Distillation · Bidirectional Decoupled Knowledge Distillation

1 Introduction

Knowledge distillation (KD) [6] has established itself as a pivotal technique for model compression, which transfers the generalization capability of a powerful teacher model to a lightweight student model. By aligning their output representations or intermediate features, KD enables the deployment of computationally efficient models without significant performance degradation, which has become

L. Pasa et al. (Eds.): ICANN 2026, LNCS 17089, pp. 594–606, 2027.
https://doi.org/10.1007/978-3-032-38398-3_49

Fig. 1. Grad-CAM visualizations of the original teacher (ResNet-110), the distilled student (ResNet-32), and the reverse-distilled teacher on CIFAR-100.

increasingly important for on-device application scenarios such as mobile computing [22] and real-time perception [25].

Existing KD approaches primarily fall into two categories: feature-based and logit-based methods. Feature-based methods [4,14] align intermediate features between teacher and student models. In contrast, logit-based methods [7,23] operate directly on the output layer, offering notable advantages in terms of architectural flexibility and lower computational overhead. This simplicity has cemented their popularity in both academia and industry.

A recent benchmark study, Decoupled Knowledge Distillation (DKD) [30], decomposes the standard KD loss into target class knowledge distillation (TCKD) and non-target class knowledge distillation (NCKD), revealing that the core efficacy of logit distillation lies in NCKD. Furthermore, recent theoretical research [1] on the mechanics of distillation indicates that the effectiveness of distillation fundamentally stems from the transfer of "multi-view" features: a single neural network often learns only a limited subset of valid features (views), and the distillation process enables the student model to acquire complementary features that it initially lacked.

Synthesizing these perspectives, we argue that NCKD serves as the primary carrier for these crucial feature views. Based on this insight, explicitly promoting the bidirectional exchange of complementary feature views between the teacher and student models can help both acquire their missing feature views. As illustrated in Fig. 1, the Grad-CAM visualizations on the CIFAR-100 dataset (using ResNet-110 as the teacher and ResNet-32 as the student) corroborate our core hypothesis. The original teacher (denoted by T_0) focuses on specific parts of the object. After the initial distillation, the student (denoted by S_1) not only inherits the teacher's knowledge but also possesses novel, complementary feature views. Through reverse distillation, the enhanced teacher (denoted by T_1) successfully assimilates these new views.

To fully capture these feature views, in this paper, we propose Bidirectional Decoupled Knowledge Distillation (Bi-DKD), a cyclical teacher-student learning framework built upon DKD. Unlike prior unidirectional self-distillation methods [3,26], Bi-DKD explicitly facilitates the mutual exchange of complementary feature views. Specifically, Bi-DKD feeds the multi-view knowledge of a student from the first round of training back to the teacher. This constructs an enhanced teacher model with a broader ensemble of views, which subsequently guides the final student model.

The main contributions of this paper are summarized as follows:

- We propose Bi-DKD, a novel bidirectional cyclical training framework based on the multi-view theory, demonstrating that knowledge transfer can be optimized in a cyclical process to benefit both teacher and student models.
- We introduce an Adaptive Dynamic Weight (ADW) strategy. By dynamically adjusting the NCKD coefficient based on teacher confidence, it prioritizes the transfer of missing feature views and maintains the stability of the bidirectional cycle.
- The entire bidirectional process converges very fast within reasonable computational overhead, building upon pre-trained student/teacher models rather than from scratch.

2 Related Work

Knowledge Distillation (KD) [6] is an effective technique for model compression technique that transfers "dark knowledge" from a large teacher to a lightweight student by mimicking the teacher's soft logits, yielding better performance than hard labels alone. Existing KD methods fall into two main categories: logit-based [10,23,24,30,31] and feature-based distillation [2,4,11,12,16,17].

In recent years, researchers have extensively explored and refined logit-based distillation methods to leverage the teacher's knowledge more efficiently. Existing logit-based distillation methods have been improved mainly in two aspects: knowledge decoupling and knowledge representation optimization. For knowledge decoupling, DKD [30] decomposes the Kullback-Leibler (KL) divergence into distillation of TCKD and NCKD, additionally SDD [23] decouples logit outputs into consistent and complementary terms for supervision. These methods isolate different knowledge types but introduce extra hyperparameters and overhead. For knowledge representation optimization, NKD [24] normalizes non-target logits to emphasize inter-class relations. CTKD [10] adopts an adaptive temperature to modulate the learning difficulty, and WTTM [31] uses an asymmetric temperature scaling strategy to refine the teacher's knowledge. These approaches improve the distillation quality, but generally require task-specific tuning and have limited generalization.

Recently, several works have explored diverse knowledge transfer mechanisms. For instance, Born-Again Neural Networks (BAN) [3] demonstrates that a student parameterized identically to its teacher can significantly outperform the teacher through sequential self-distillation. Similarly, Tf-KD [26] reveals that even a poorly trained teacher can improve a student. It interprets KD as a learned label smoothing regularization, pioneering the concepts of teacher-free and reverse distillation. Building upon the teacher-free concept, Tf-FD [9] proposes a self-regulated feature learning framework that reuses intra-layer and inter-layer features within the student itself to provide teacher-like guidance, completely eliminating the need for an external model. While these methods effectively explore self-distillation and diverse knowledge transfer, they remain confined to unidirectional guidance or knowledge transfer within a single network. This limitation causes them to miss the opportunity for bidirectional enhancement between different models. Departing from these conventional approaches, our proposed Bi-DKD leverages NCKD as an explicit carrier to facilitate a bidirectional, cyclical transfer of complementary feature views between the teacher and student networks, explicitly enabling both the teacher and student models to acquire their respective missing feature views.

3 Method

In this section, we first review DKD as the foundation of our work. Then, we introduce the overall pipeline of the proposed Bi-DKD framework. Recognizing that the teacher and student models inherently capture distinct feature views, Bi-DKD transforms the classical unidirectional distillation into a bidirectional paradigm between two different networks, explicitly facilitating the cyclical exchange of their complementary feature views. Subsequently, we present the ADW strategy, a key component that enables effective bidirectional knowledge transfer in Bi-DKD.

3.1 Review Decoupled Knowledge Distillation

Traditional logit-based KD transfers knowledge by minimizing the KL divergence $\mathcal{L}_{\text{KD}}$ between teacher and student output distributions. Zhao et al. [30] decompose this KL into TCKD and NCKD, which are coupled in classical KD. They argue that standard KD suppresses valuable inter-class similarity information, particularly when the teacher is highly confident.

Specifically, let the logit vectors be defined as $\mathbf{z} = [z_1, z_2, \ldots, z_C] \in \mathbb{R}^{1\times C}$, where C represents the number of classes. The output probability distribution is denoted as $\mathbf{p} = [p_1, p_2, \ldots, p_C] \in \mathbb{R}^{1\times C}$. Here, p_i represents the predicted probability for the i-th class, obtained by applying the softmax function to z_i. To separate the probability of the target class from those of other targets, $\mathbf{b} = [p_t, p_{\setminus t}] \in \mathbb{R}^{1\times 2}$ is defined to represent the binary probabilities for the target class (p_t) and other classes ($p_{\setminus t} = 1 - p_t$). Similarly, $\hat{\mathbf{p}} = [\hat{p}_1, \ldots, \hat{p}_{t-1}, \hat{p}_{t+1}, \ldots, \hat{p}_C] \in \mathbb{R}^{1\times (C-1)}$ is defined to independently model the probabilities among non-target classes, where $\hat{p}_i = \frac{p_i}{\sum_{i\neq t} p_i}$. Let T and S denote the teacher and the student, respectively. The classical KD loss function can then be reformulated as follows:

$$\mathcal{L}_{\mathrm{KD}} = \mathrm{KL}(\mathbf{p}^T \| \mathbf{p}^S) = \mathrm{KL}(\mathbf{b}^T \| \mathbf{b}^S) + (1 - p_t^T)\mathrm{KL}(\hat{\mathbf{p}}^T \| \hat{\mathbf{p}}^S) \tag{1}$$

As shown in Eq. (1), the KD loss is reformulated as a weighted sum of two terms. $\mathrm{KL}(\mathbf{b}^T \| \mathbf{b}^S)$ represents the similarity between the binary probabilities of the teacher and student for the target class. Therefore, it is named Target Class Knowledge Distillation (TCKD). Meanwhile, $\mathrm{KL}(\hat{\mathbf{p}}^T \| \hat{\mathbf{p}}^S)$ represents the similarity of probabilities among non-target classes between the teacher and student, termed Non-Target Class Knowledge Distillation (NCKD). Consequently, Eq. (1) can be rewritten as:

$$\mathcal{L}_{\mathrm{KD}} = \mathrm{TCKD} + (1 - p_t^T)\mathrm{NCKD} \tag{2}$$

This reformulation reveals a shortcoming of classical KD: as the teacher's confidence p_t^T increases, the multiplicative weight $(1 - p_t^T)$ on NCKD decreases, thereby suppressing prominent "dark knowledge" transfer from well-predicted samples [30]. To alleviate this issue, DKD decouples the components into:

$$\mathcal{L}_{\mathrm{DKD}} = \alpha\mathrm{TCKD} + \beta\mathrm{NCKD} \tag{3}$$

where hyperparameters α and β flexibly balance the two components for effective transfer.

3.2 Bidirectional Decoupled Knowledge Distillation(Bi-DKD) Framework

Built upon DKD and the multi-view feature theory, we propose the Bi-DKD framework. As depicted in Fig. 2, the framework consists of three main stages designed to cyclically exchange and expand complementary feature views.

Stage 1: Initial Forward Distillation (Fig. 2(a)). We first train the student S_0 using the standard DKD loss to imitate the softened probability distribution of a large pre-trained teacher T_0:

$$\mathcal{L}_{S_1} = \mathcal{L}_{\mathrm{CE}}(\mathbf{p}^{S_0}, Y) + \mathcal{L}_{\mathrm{DKD}}(\mathbf{p}^{T_0}, \mathbf{p}^{S_0}) \tag{4}$$

where $\mathcal{L}_{\mathrm{CE}}$ denotes the cross-entropy loss with the ground-truth label Y. During this stage, the student model effectively captures the teacher's specific feature views. Meanwhile, due to differences in network architectures and random initialization, the student model S_1 often possesses novel views that the original teacher lacks.

Stage 2: Reverse Distillation with ADW (Fig. 2(b)). To allow the teacher to learn these missing views, we now reverse the teacher-student roles: the strong student S_1 (obtained from Stage 1) serves as the teacher, while the original large teacher T_0 is treated as the student. Importantly, T_0 is not retrained from scratch but is further fine-tuned on top of its original pretrained weights. Reverse

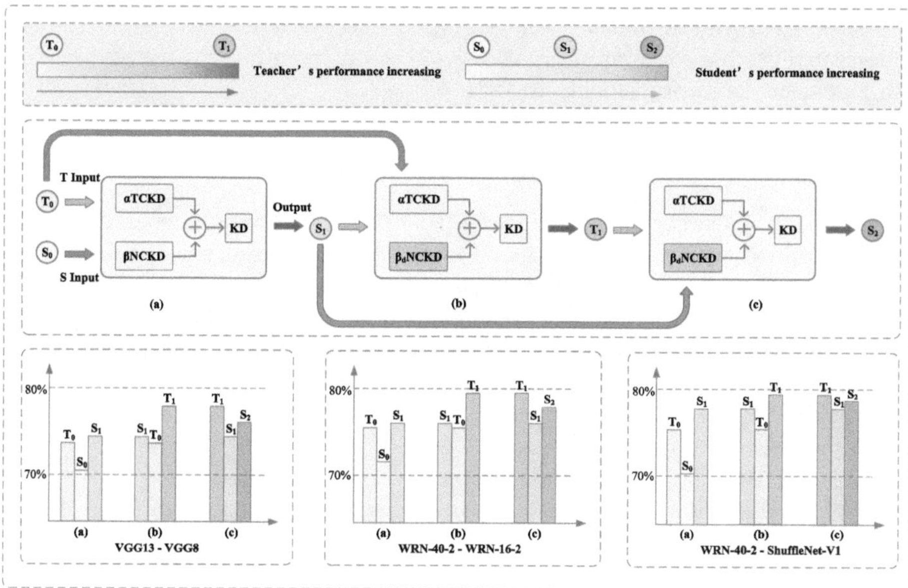

Fig. 2. Bidirectional Decoupled Knowledge Distillation (Bi-DKD) framework.

distillation is performed using a modified DKD loss equipped with ADW that dynamically adjusts the contribution of NCKD:

$$\mathcal{L}_{T_1} = \mathcal{L}_{\text{CE}}(\mathbf{p}^{T_0}, Y) + \alpha\mathcal{L}_{\text{TCKD}}(\mathbf{p}^{S_1}, \mathbf{p}^{T_0}) + \beta_d\mathcal{L}_{\text{NCKD}}(\mathbf{p}^{S_1}, \mathbf{p}^{T_0}) \tag{5}$$

where β_d is determined by the ADW strategy described in Sect. 3.3. Since NCKD is the primary carrier for multi-view features, this reverse process explicitly transfers the student's unique feature views back to the teacher. This yields an enhanced teacher T_1 with a significantly broader ensemble of feature views.

Stage 3: Second Forward Distillation with ADW (Fig. 2(c)). We perform a final round of forward distillation, where the enhanced teacher T_1 further refines the student S_1. Similarly, S_1 is not retrained from scratch but is further optimized on top of its already well-trained weights from Stage 1. This second forward Training follows the same DKD method but again applies ADW on the NCKD term:

$$\mathcal{L}_{S_2} = \mathcal{L}_{\text{CE}}(\mathbf{p}^{S_1}, Y) + \alpha\mathcal{L}_{\text{TCKD}}(\mathbf{p}^{T_1}, \mathbf{p}^{S_1}) + \beta_d\mathcal{L}_{\text{NCKD}}(\mathbf{p}^{T_1}, \mathbf{p}^{S_1}) \tag{6}$$

This phase fully exploits the mutually enhanced knowledge. The student is guided by a teacher that now possesses a much richer union of multi-view features, exposing the student to more comprehensive learning signals.

As shown in Fig. 2, Bi-DKD is applied to three representative teacher-student pairs: VGG13 → VGG8, WRN-40-2 → WRN-16-2, and WRN-40-2 → ShuffleNetV1. The accuracy curves further reveal consistent performance gains across all stages for both the teacher ($T_0 \rightarrow T_1$) and the student ($S_0 \rightarrow S_1 \rightarrow S_2$), demonstrating that bidirectional knowledge transfer and iterative refinement bring measurable improvements to both models.

3.3 Adaptive Dynamic Weight (ADW) Strategy for NCKD

In the Bi-DKD framework, we observe that if a fixed weight coefficient β is directly applied to NCKD in Stages 2 and 3, even if the enhanced teacher T_1 outperforms T_0, the final student S_2 might still underperform S_1. This phenomenon occurs because a static β fails to accommodate the dynamically changing learning requirements across different stages. For instance, when the teacher is highly confident, the student should focus more on the fine-grained knowledge within NCKD, whereas an uncertain teacher requires the student to place greater emphasis on TCKD.

To tackle the above-mentioned problem, we propose the ADW strategy to adjust the weight during bidirectional training. First, for each training sample, we compute the teacher network's confidence margin $z_t - z_{\max}$, where z_t is the logit of the ground-truth target class and $z_{\max}$ is the maximum logit among all non-target classes. Second, we discretize the confidence margin into several intervals. A large number of exploratory experiments consistently reveal that, among the tested β candidates, the best-performing β exhibits a clear monotonically increasing trend with the confidence margin. Guided by this empirical observation, we evaluate multiple candidate β values for each interval and select the one achieving the highest performance as the fixed β for that interval. Finally, during training, each sample is assigned the β of its corresponding interval according to $z_t - z_{\max}$, and the batch-level weight β_d is obtained by averaging these per-sample β values:

$$\beta_d = \frac{1}{N} \sum_{i=1}^{N} \beta_i \tag{7}$$

where N is the batch size and β_i is the adaptive weight for the i-th sample. The ADW strategy ensures that the transfer of missing feature views is adaptively prioritized. It leverages the NCKD carrier efficiently, preventing knowledge degradation and maintaining stable, continuous view expansion throughout the bidirectional cycle.

4 Experiments

4.1 Experimental Settings

Datasets. We conduct experiments on two widely-used benchmark datasets: CIFAR-100 [8], which comprises 50k training images and 10k test images across 100 classes, and ImageNet [18], a large-scale dataset containing over 1.2 million training images and 50k validation images from 1k classes.

Network Architectures. To evaluate the generalization capability of Bi-DKD, we employ diverse network architectures, including ResNet [5], WideResNet (WRN) [27], VGGNet [20], ShuffleNet [13,28], and MobileNetV2 [19]. We construct comprehensive teacher-student pairs with similar model architectures and different model architectures.

Baselines. We compare our method with the state-of-the-art (SOTA) KD approaches, including feature-based methods (e.g., CRD [21], ReviewKD [2]) and logit-based methods (e.g., DKD [30], NKD [24], WTTM [31]).

Implementation Details. All experiments are implemented using PyTorch on NVIDIA RTX 5090 GPUs. For CIFAR-100, models are trained for 240 epochs using an SGD optimizer with a momentum of 0.9, a weight decay of 5×10^{-4}, and a batch size of 128. The distillation temperature $\mathcal{T}$ is set to 4 for all methods. For ImageNet, we follow the standard training protocol as in DKD [30].

4.2 Main Results on CIFAR-100

To comprehensively evaluate the effectiveness of our proposed Bi-DKD method, we conduct extensive experiments on the CIFAR-100 dataset. We consolidate the evaluation across diverse teacher-student pairs, encompassing both similar and different network architectures, to demonstrate the universal applicability of our bidirectional framework. We present the progressive performance change across three distillation stages and the performance comparison between Bi-DKD method and other popular KD approaches, in Table 1 and Table 2, respectively.

Progressive Performance Improvement. As shown in Table 1, our bidirectional framework yields consistent performance gains across different stages for all architectural combinations. After the initial DKD stage, the student S_1 obtains performance improvements across all pairs. Then, the reverse distillation yields the enhanced teacher T_1 with substantial improvements (e.g., from 74.64% to 76.93% for the VGG8 → VGG13 pair, and from 74.31% to 76.34% for the ResNet32 → ResNet110 pair). Subsequently, T_1 is leveraged as the teacher model for S_1 to produce the final student S_2 with further gains. For instance, S_2 brings a +0.46% improvement over S_1 for the WRN-40-2 → ShuffleNet-V1 pair, and a +0.44% gain for the ResNet32×4 → ResNet8×4 pair.

Comparison with Popular Methods. We further evaluate Bi-DKD against state-of-the-art distillation methods. As shown in Table 2, compared with the baseline and vanilla KD, Bi-DKD achieves consistent improvements across all teacher-student pairs. Our method yields a 1∼3% performance boost for both similar architecture and different architecture pairs, attaining the best results on multiple pairs. Specifically, Bi-DKD improves upon vanilla KD by nearly 3% on the ResNet32x4 → ResNet8x4, VGG13 → MobileNetV2, and ResNet32x4 → ShuffleNetV2 pairs. These results demonstrate the effectiveness of the proposed Bi-DKD method in handling teacher and student models with diverse network architectures and layers.

Table 1. Performance comparison of Bi-DKD's progressive stages across various teacher-student pairs on CIFAR-100.

	To Similar Architecture					To Different Architecture			
Teacher (T_0)	WRN-40-2	VGG 13	ResNet 110	ResNet 32×4	WRN-40-2	WRN-40-2	ResNet 32×4	VGG 13	ResNet 32×4
Accuracy	75.61	74.64	74.31	79.42	75.61	75.61	79.42	74.64	79.42
Student (S_0)	WRN-16-2	VGG 8	ResNet 32	ResNet 8×4	WRN-40-1	ShuffleNet-V1	ShuffleNet-V1	MobileNet-V2	ShuffleNet-V2
Accuracy	73.26	70.36	71.14	72.50	71.98	70.50	70.50	64.60	71.82
DKD (S_1)	75.90	74.79	74.01	75.93	74.87	76.83	76.39	69.94	77.00
E-Teacher (T_1)	78.14	76.93	76.34	79.73	78.05	78.16	80.94	75.79	81.02
Bi-DKD (S_2)	76.30	75.02	74.17	76.37	75.43	77.29	76.56	70.11	77.15
S_2-S_1	+0.40	+0.23	+0.16	+0.44	+0.56	+0.46	+0.17	+0.17	+0.15

Table 2. Evaluation of the top-1 accuracy (%) compared with state-of-the-art distillation methods on the CIFAR-100 validation set. △ denotes the improvement of our Bi-DKD over the vanilla KD baseline.

		To Similar Architecture					To Different Architecture			
Teacher	Publication	WRN-40-2	VGG 13	ResNet 110	ResNet 32×4	WRN-40-2	WRN-40-2	ResNet 32×4	VGG 13	ResNet 32×4
Accuracy		75.61	74.64	74.31	79.42	75.61	75.61	79.42	74.64	79.42
Student	Year	WRN-16-2	VGG 8	ResNet 32	ResNet 8×4	WRN-40-1	ShuffleNet-V1	ShuffleNet-V1	MobileNet-V2	ShuffleNet-V2
Accuracy		73.26	70.36	71.14	72.50	71.98	70.50	70.50	64.60	71.82
FitNet [17]	ICLR'15	73.58	71.02	71.06	73.50	72.24	73.73	73.54	63.16	73.54
RKD [16]	CVPR'19	73.35	71.48	71.82	71.90	72.22	72.21	72.28	64.43	73.21
CRD [21]	ICLR'20	75.48	73.94	73.48	75.51	74.14	76.05	75.11	69.11	75.65
ReviewKD [2]	CVPR'21	76.12	74.84	**73.89**	75.63	75.09	77.14	77.45	70.37	77.78
NORM [12]	ICLR'23	75.65	73.95	73.67	76.49	74.82	77.03	77.42	68.94	78.07
FCFD [11]	ICLR'23	**76.34**	**74.86**	-	76.80	**75.43**	**77.81**	78.12	**70.67**	78.20
CAT-KD [4]	CVPR'23	75.60	74.65	73.62	**76.91**	74.82	77.35	**78.26**	69.13	**78.41**
KD [6]	NeurIPS'14	74.92	72.98	73.08	73.33	73.54	74.83	74.07	67.37	74.45
CTKD [10]	AAAI'23	75.45	73.52	73.25	73.70	73.93	75.78	74.48	68.46	75.31
NKD [24]	ICCV'23	75.43	74.86	73.50	76.35	74.25	76.46	76.31	**70.67**	76.92
SDD [23]	CVPR'24	75.86	73.91	73.97	75.09	74.53	76.65	76.30	68.79	76.67
SDD [23]	CVPR'24	75.86	73.91	73.97	75.09	74.53	76.65	76.30	68.79	76.67
WTTM [31]	ICLR'24	**76.37**	74.44	74.13	76.06	74.58	75.42	74.37	69.16	76.55
TeKAP [7]	ICLR'25	75.21	74.00	73.42	74.79	73.80	76.75	74.92	67.39	75.43
Bi-DKD	Ours	76.30	**75.02**	**74.17**	**76.37**	**75.43**	**77.29**	**76.56**	70.11	**77.15**
△	Ours	+1.38	+2.04	+1.09	+3.04	+1.89	+2.46	+2.49	+2.74	+2.70

These results demonstrate that our Bi-DKD framework effectively enables bidirectional knowledge transfer. By cyclically optimizing the knowledge transfer process, the progressive improvements presented across the three stages validate the effectiveness of each component. Specifically, the reverse distillation successfully optimizes the teacher model, while the second forward distillation further boosts the final student model.

Table 3. Evaluate the top-1 and top-5 accuracy (%) of student on the ImageNet-1K validation set.

ResNet34 (teacher): 73.31% Top-1, 91.42% Top-5. ResNet18 (student): 69.75% Top-1, 89.07% Top-5.											
	Features				Logits						
	CRD [21]	ReviewKD [2]	FCFD [11]	CAT-KD [4]	KD [6]	DKD [30]	SDD [23]	WTTM [31]	TeKAP [7]	Bi-DKD	△
Top-1	71.17	71.61	**72.24**	71.26	70.66	71.70	71.44	**72.19**	71.35	72.06	+1.40
Top-5	90.13	90.51	**90.74**	90.45	89.88	90.31	90.05	-	90.54	**90.64**	+0.76

4.3 Validation on ImageNet

We evaluate Bi-DKD on the ResNet34→ResNet18 pair using ImageNet. The experimental results are presented in Table 3. Bi-DKD achieves 72.06% Top-1 and 90.64% Top-5 accuracy. In terms of Top-1 accuracy, it outperforms the standard KD by 1.40% and the baseline DKD by 0.36%. For Top-5 accuracy, it surpasses standard KD by 0.76% and DKD by 0.33%. Additionally, compared to the vanilla student (69.75% Top-1 and 89.07% Top-5), our method yields substantial gains of +2.31% and +1.57%, respectively. These results further confirm the scalability and effectiveness of our proposed Bi-DKD framework on large-scale datasets.

4.4 Computational Cost Analysis

We evaluated computational overhead using WRN-40-2 → WRN-40-1 on CIFAR-100 (RTX 5090). While standard DKD (240 epochs) took 92 min, Bi-DKD added only 29 min (∼31% increase), as Stages 2 and 3 both leverage warm-start initialization from pre-trained weights and require just 35 epochs each to converge. Extending training duration for better performance is common in knowledge distillation [3,7,15,29]. For example, TeKAP [7] incurs a 12.5% overhead via noise injection. Although Bi-DKD required slightly more time than TeKAP, it consistently outperformed TeKAP across all benchmarks. Given that the primary goal for KD is obtain a high-performing student model, this modest additional cost is fully acceptable.

5 Conclusion

In this paper, we propose Bidirectional Decoupled Knowledge Distillation (Bi-DKD), a cyclical training framework grounded in the multi-view feature theory.

Recognizing NCKD as the primary carrier for crucial feature views, Bi-DKD facilitates bidirectional knowledge exchange between teacher and student models. By feeding the unique multi-view features captured by an initially distilled student back to the teacher, we construct an enhanced teacher with a broader ensemble of views to guide the final student. Furthermore, to adaptively prioritize the transfer of reliable missing feature views, we introduce the Adaptive Dynamic Weight (ADW) strategy based on teacher confidence. Extensive experiments on various datasets demonstrate that Bi-DKD continuously expands the decision boundaries of both models, surpassing standard DKD and most existing methods. We hope our view-oriented bidirectional framework provides new insights into the mechanism and potential of knowledge distillation.

Acknowledgement. This work was supported in part by "The Fundamental Research Funds for the Central Universities" under Grant XJSJ25031, the Fundamental Research Funds for the Central Universities and the Innovation Fund of Xidian University under Grant No.YJSJ26004, the National Natural Science Foundation of China (NSFC) under Grant No.62401513, and the Shaanxi High-Level Talents Special Support Program under Grant H025200001.

References

1. Allen-Zhu, Z., Li, Y.: Towards understanding ensemble, knowledge distillation and self-distillation in deep learning. In: The Eleventh International Conference on Learning Representations (2023)
2. Chen, P., Liu, S., Zhao, H., Jia, J.: Distilling knowledge via knowledge review. In: Proceedings of the IEEE/CVF Conference on Computer Vision and Pattern Recognition, pp. 5008–5017 (2021)
3. Furlanello, T., Lipton, Z., Tschannen, M., Itti, L., Anandkumar, A.: Born again neural networks. In: International Conference on Machine Learning, pp. 1607–1616. PMLR (2018)
4. Guo, Z., Yan, H., Li, H., Lin, X.: Class attention transfer based knowledge distillation. In: Proceedings of the IEEE/CVF Conference on Computer Vision and Pattern Recognition, pp. 11868–11877 (2023)
5. He, K., Zhang, X., Ren, S., Sun, J.: Deep residual learning for image recognition. In: Proceedings of the IEEE Conference on Computer Vision and Pattern Recognition, pp. 770–778 (2016)
6. Hinton, G., Vinyals, O., Dean, J.: Distilling the knowledge in a neural network. arXiv preprint arXiv:1503.02531 (2015)
7. Hossain, M.I., Akhter, S., Hong, C.S., Huh, E.N.: Single teacher, multiple perspectives: teacher knowledge augmentation for enhanced knowledge distillation. In: The Thirteenth International Conference on Learning Representations (2025)
8. Krizhevsky, A., Hinton, G.: Learning multiple layers of features from tiny images (2009)
9. Li, L.: Self-regulated feature learning via teacher-free feature distillation. In: Avidan, S., Brostow, G., Cissé, M., Farinella, G.M., Hassner, T. (eds.) ECCV 2022. LNCS, vol. 13686, pp. 347–363. Springer, Cham (2022). https://doi.org/10.1007/978-3-031-19809-0_20

10. Li, Z., et al.: Curriculum temperature for knowledge distillation. In: Proceedings of the AAAI Conference on Artificial Intelligence, vol. 37, pp. 1504–1512 (2023)
11. Liu, D., Kan, M., Shan, S., Chen, X.: Function-consistent feature distillation. In: The Eleventh International Conference on Learning Representations (2023)
12. Liu, X., Li, L., Li, C., Yao, A.: NORM: knowledge distillation via n-to-one representation matching. In: The Eleventh International Conference on Learning Representations (2023)
13. Ma, N., Zhang, X., Zheng, H.-T., Sun, J.: ShuffleNet V2: practical guidelines for efficient CNN architecture design. In: Ferrari, V., Hebert, M., Sminchisescu, C., Weiss, Y. (eds.) Computer Vision – ECCV 2018. LNCS, vol. 11218, pp. 122–138. Springer, Cham (2018). https://doi.org/10.1007/978-3-030-01264-9_8
14. Miles, R., Mikolajczyk, K.: Understanding the role of the projector in knowledge distillation. In: Proceedings of the AAAI Conference on Artificial Intelligence, vol. 38, pp. 4233–4241 (2024)
15. Mirzadeh, S.I., Farajtabar, M., Li, A., Levine, N., Matsukawa, A., Ghasemzadeh, H.: Improved knowledge distillation via teacher assistant. In: Proceedings of the AAAI conference On Artificial Intelligence, vol. 34, pp. 5191–5198 (2020)
16. Park, W., Kim, D., Lu, Y., Cho, M.: Relational knowledge distillation. In: Proceedings of the IEEE/CVF Conference on Computer Vision and Pattern Recognition, pp. 3967–3976 (2019)
17. Romero, A., Ballas, N., Kahou, S.E., Chassang, A., Gatta, C., Bengio, Y.: Fitnets: hints for thin deep nets, CoRR (2014)
18. Russakovsky, O., et al.: Imagenet large scale visual recognition challenge. Int. J. Comput. Vision **115**(3), 211–252 (2015)
19. Sandler, M., Howard, A., Zhu, M., Zhmoginov, A., Chen, L.C.: Mobilenetv 2: inverted residuals and linear bottlenecks. In: Proceedings of the IEEE Conference on Computer Vision and Pattern Recognition, pp. 4510–4520 (2018)
20. Simonyan, K., Zisserman, A.: Very deep convolutional networks for large-scale image recognition. arXiv preprint (2014). arXiv:1409.1556
21. Tian, Y., Krishnan, D., Isola, P.: Contrastive representation distillation. In: International Conference on Learning Representations (2020)
22. Vasu, P.K.A., Gabriel, J., Zhu, J., Tuzel, O., Ranjan, A.: Mobileone: an improved one millisecond mobile backbone. In: Proceedings of the IEEE/CVF Conference on Computer Vision and Pattern Recognition, pp. 7907–7917 (2023)
23. Wei, S., Luo, C., Luo, Y.: Scale decoupled distillation. In: 2024 IEEE/CVF Conference on Computer Vision and Pattern Recognition, pp. 15975–15983 (2024)
24. Yang, Z., Zeng, A., Li, Z., Zhang, T., Yuan, C., Li, Y.: From knowledge distillation to self-knowledge distillation: a unified approach with normalized loss and customized soft labels. In: Proceedings of the IEEE/CVF International Conference on Computer Vision, pp. 17185–17194 (2023)
25. Yeo, S., Jang, Y., Sohn, J.Y., Han, D., Yoo, J.: Can we find strong lottery tickets in generative models? In: Proceedings of the AAAI Conference on Artificial Intelligence, vol. 37, pp. 3267–3275 (2023)
26. Yuan, L., Tay, F.E., Li, G., Wang, T., Feng, J.: Revisiting knowledge distillation via label smoothing regularization. In: Proceedings of the IEEE/CVF Conference on Computer Vision and Pattern Recognition, pp. 3903–3911 (2020)
27. Zagoruyko, S., Komodakis, N.: Wide residual networks, CoRR (2016)
28. Zhang, X., Zhou, X., Lin, M., Sun, J.: Shufflenet: an extremely efficient convolutional neural network for mobile devices. In: Proceedings of the IEEE Conference on Computer Vision and Pattern Recognition, pp. 6848–6856 (2018)

29. Zhang, Y., Xiang, T., Hospedales, T.M., Lu, H.: Deep mutual learning. In: Proceedings of the IEEE Conference on Computer Vision and Pattern Recognition, pp. 4320–4328 (2018)
30. Zhao, B., Cui, Q., Song, R., Qiu, Y., Liang, J.: Decoupled knowledge distillation. In: Proceedings of the IEEE/CVF Conference on Computer Vision and Pattern Recognition, pp. 11953–11962 (2022)
31. Zheng, K., Yang, E.H.: Knowledge distillation based on transformed teacher matching. In: The Twelfth International Conference on Learning Representations (2024)

Data Augmentation as an Engine: A Multidimensional Few-Shot Analysis

Maria Eduarda Mota(✉) and Cleber Zanchettin

Informatics Center, Federal University of Pernambuco (UFPE), Recife, Brazil
{mebm,cz}@cin.ufpe.br

Abstract. This study presents a multidimensional analysis of LLM-based text data augmentation (DA) in few-shot learning (FSL), investigating the dynamics among augmentation strategies, model architectures, and label granularity. We evaluate four FSL paradigms (Linear Probe, Prototypical Networks, SetFit, and LoRA) across five benchmarks and demonstrate that DA can yield over 70% performance gains in noisy scenarios, though it is limited by a trade-off between lexical innovation and semantic fidelity. Our findings reveal a qualitative shift in DA utility, suggesting its role as a boundary stabilizer—anchoring decision frontiers in sparse vector spaces— at $k = 1$ and trending toward a precision tool as k increases within our evaluation. To mitigate semantic drift in higher-k regimes, we introduce a Boundary-Aware GPT3Mix configuration to demonstrate that anchoring synthetic samples within decision frontiers significantly improves model calibration (ECE). Finally, we identify an architectural trade-off where adaptive models offer superior performance potential while static models provide scalability and resilience to synthetic noise, concluding that robust few-shot systems must dynamically adapt augmentation constraints to specific tasks and architectural requirements.

Keywords: Few-Shot Learning · Text Classification · Data Augmentation

1 Introduction

In an era where data drives strategic decision-making, automatic text classification is a fundamental tool for transforming unstructured data into organized taxonomies, such as mapping customer queries to high-cardinality intent spaces or categorizing product descriptions [4]. Despite their success across sectors such as healthcare and finance, the efficacy of state-of-the-art models remains heavily dependent on large labeled datasets [1,4]. However, in specialized domains, expert annotation is often unfeasible due to the prohibitive costs and time constraints [1,4,25], creating a significant bottleneck for high-capacity architectures, such as Deep Learning models, that require substantial training volumes to achieve competitive performance [4,28].

L. Pasa et al. (Eds.): ICANN 2026, LNCS 17089, pp. 607–619, 2027.
https://doi.org/10.1007/978-3-032-38398-3_50

To address these limitations, Few-Shot Learning (FSL) has emerged as a means of adapting text classification models using only a handful of examples per class [1,4]. While FSL offers a scalable alternative, its success depends on the model's ability to extract robust features—domain-invariant predictors that resist overfitting— in low-data regimes [4,13]. In this scenario, Data Augmentation (DA) has gained traction as a strategy to expand the information density of these sets [8]. By generating synthetic variations of the initial real-world samples, DA enables a more comprehensive mapping of the underlying distribution for underrepresented classes [28].

Despite advancements in Large Language Models (LLMs), critical challenges persist, particularly regarding the risk of semantic drift in fine-grained contexts [16,19]. Balancing semantic fidelity with lexical diversity is thus essential for model stability and generalization [7,17]. Driven by these limitations, this work presents a systematic analysis investigating how various DA approaches behave across different label granularities and domains, exploring the trade-off between semantic fidelity and structural variance. We evaluate how this balance influences few-shot performance and calibration. Through a multidimensional analysis of semantic density—defined as the geometric compactness of class embeddings—and dataset complexity, this study provides a holistic evaluation of DA, extending beyond simple performance gains to investigate its impact on model reliability.

2 Related Works

Although Deep Learning has achieved remarkable success in classification tasks, its reliance on large volumes of labeled data remains a significant bottleneck, especially in niche domains [1]. Few-Shot Learning (FSL) and Data Augmentation (DA) address this scarcity by enhancing generalization in low-data regimes [3,21,28]. This section reviews FSL's evolution, the transition of DA toward LLMs, and current metrics for assessing the quality and diversity of synthetic data, highlighting the research gaps our framework aims to address.

2.1 Few-Shot Learning

Methodologies have evolved from static linear probing on frozen embeddings to distance-based reasoning, exemplified by Prototypical Networks [18], which define class boundaries using centroids in a latent space. Subsequently, research transitioned toward more sophisticated adaptation strategies, such as meta-learning and prompt-based tuning [7,11].

With the rise of LLMs, the focus shifted toward efficient adaptation. Parameter-Efficient Fine-Tuning (PEFT) techniques, such as LoRA [10], enable specialization with minimal trainable parameters. More recently, frameworks like SetFit [20] have gained traction by combining contrastive learning with sentence transformers, offering a robust alternative for classification in low-data regimes, validated across benchmarks ranging from coarse-grained topic labeling (AG News) [26] to high-cardinality intents (Banking77) [5].

2.2 Data Augmentation

Data Augmentation (DA) involves synthesizing novel variations while preserving the semantic integrity of the original referenced data [8]. In text classification, this remains a non-trivial challenge, as it requires generating sufficient linguistic variability without compromising the underlying language structure [28]. Traditional strategies, such as Easy Data Augmentation (EDA) [22], utilize rule-based lexical modifications (e.g., synonym replacement). However, these surface-level transformations often fail to capture complex contextual nuances [28].

Recent literature leverages LLMs for generating semantically rich samples [7, 28], introducing a fundamental balance between linguistic diversity and semantic fidelity [7, 17]. In high-granularity datasets such as Banking77, even minor shifts can trigger semantic drift, in which generated samples inadvertently cross narrow boundaries [16]. This risk underscores the need for robust evaluation frameworks to ensure DA effectively contributes to dataset reliability.

2.3 Evaluation Metrics for DA

Assessing synthetic data requires a shift from lexical to semantic metrics. Traditional indicators like BLEU and Self-BLEU are limited by their reliance on exact lexical overlap, often penalizing beneficial linguistic variations while failing to guarantee semantic integrity [2, 17].

Recent studies propose representation-based metrics to address these gaps. Fréchet BERT Distance (FBD) [9] evaluates distributional alignment between real and synthetic samples in latent space [2], while Embedding Similarity captures semantic fidelity [7]. Additionally, MS Jaccard serves as a noise-resistant alternative for assessing the balance between diversity and quality. [2]. Despite these metrics, the correlation between synthetic quality and probabilistic reliability remains under-explored [8, 27]. This work addresses this by integrating these indicators into a multidimensional diagnostic framework to investigate how semantic density dictates DA effectiveness.

3 Proposed Approach

To separate baseline text augmentation paradigms from our specific architectural design adaptation, we structure these strategies into three semantic levels. This approach addresses distinct learning vulnerabilities in few-shot scenarios:

Lexical Level - Synonym Substitution (WordNet) [Baseline Paradigm]: This rule-based technique replaces specific words with WordNet synonyms [22] to increase lexical variety. Decoupling class intent from specific keywords prevents overfitting and promotes generalization to broader semantic concepts [19].

Contextual Level - Context Swapping [Baseline Paradigm]: An LLM baseline designed to increase intra-class variance by replacing primary entities or shifting scenarios within the same domain through generative prompting. This prevents subject-specific biases, forcing the model to capture generalized class intent regardless of the actors or events involved [6].

Semantic Level: We focus on capturing the core meaning of each class to improve model generalization through two distinct approaches. **(i) Semantic Deletion [Baseline Paradigm]** generates "Hard Positives" by instructing an LLM to omit explicit discriminative keywords. By using implicit context and parallel descriptions, the classifier is forced to look beyond semantic shortcuts and capture the global intent [7,27]. **(ii) Generative In-Context Learning (ICL)** leverages LLMs for semantic interpolation. To manage high label cardinality and mitigate prompt confusion, we propose a Dynamic Context Retrieval strategy via SBERT that narrows the LLM's focus to the target class's semantic neighborhood through two specialized variants: **Dynamic GPT3Mix** refines the framework by [24] to handle high-cardinality datasets by dynamically filtering the Prompt Header to include only the top-N most similar labels. When combined with Reference Anchors (ground-truth examples), this mechanism ensures generative relevance and significantly reduces hallucinations and prompt noise. Complementarily, we further extend the approach into **Boundary-Aware GPT3Mix (BA-G3M)**. While Dynamic GPT3Mix only exposes adjacent class names, BA-G3M injects boundary text examples from the nearest neighboring classes, guiding the LLM to synthesize discriminative samples that respect fine-grained boundaries and prevent label overlap.

This integrated approach facilitates the synthesis of high-quality, context-aware samples that leverage LLM internal knowledge while maintaining high fidelity to the original data distribution.

3.1 Experimental Strategy and Benchmarking

To validate the robustness of our augmentation paradigms, we evaluate them across diverse label granularities, domains, and structural complexities. This ensures that our findings generalize across varying semantic densities rather than a single distribution.

Evaluation Datasets: We select AG News [26] and Banking77 [5] to address the contrast between coarse-grained topics and high-cardinality intent detection, respectively. In addition, to explicitly mitigate data contamination risks in LLMs, we integrate contemporary benchmarks, AlleNoise [14], and FinBen-FPB [23], ensuring that our evaluation reflects genuine generative reasoning rather than the retrieval of memorized patterns. Beyond their temporal recency, AlleNoise introduces hierarchical structural complexity through real-world e-commerce product titles. At the same time, FinBen-FPB assesses domain expertise, requiring the model to capture sentiment tied to market expectations and fiscal implications rather than explicit emotional cues.

Downstream Models: These classifiers represent four major FSL paradigms [12,28]: (i) shallow feature extraction (Linear Probe), (ii) metric learning (Prototypical Networks) [18], (iii) contrastive fine-tuning (SetFit) [20], and (iv) parameter-efficient adaptation (LoRA) [10]. This strategic selection allows us to observe DA impacts across models with distinct inductive biases, from static feature extraction to adaptive fine-tuning, ensuring our findings are architecture-agnostic and robust.

3.2 Multidimensional Diagnostic Framework

To evaluate the impact of our augmentation strategies, we propose a diagnostic framework that transcends simple predictive performance. Our premise is that a robust augmentation strategy must satisfy three fundamental pillars:

Generative Quality (Fidelity vs. Diversity): We investigate the trade-off between intent preservation (fidelity) and linguistic variation (diversity). By employing SBERT Cosine Similarity to validate semantic fidelity and the Multi-Scale Jaccard (MS-Jaccard) index to quantify lexical innovation, we ensure that augmentation expands decision boundaries without introducing noise [2,7].

Statistical Naturalness: We evaluate whether the generative process respects the original domain's distribution. To this end, we use the Fréchet BERT Distance (FBD) to detect potential distributional shifts or out-of-distribution (OOD) issues that could compromise model stability [2].

Predictive Reliability: We analyze how synthetic data affects probabilistic calibration. Using Expected Calibration Error (ECE) and the Brier Score, we verify that performance gains do not lead to overconfident models that fail on ambiguous instances.

Additionally, we examine the Contextual Sensitivity of these pillars, investigating how varying shot sizes ($k \in \{1, 5, 16\}$) and dataset granularities affect the behavior of each augmentation paradigm.

4 Experiments

4.1 Datasets

Following the experimental design in Subsect. 3.1, we evaluate our approach across five benchmarks: (i) **AG News** [26], a topic classification task with 4 categories (World, Sports, Business, and Sci/Tech) where we concatenate 'Title' and 'Description' fields for richer context; (ii) **Banking77** [5], a fine-grained intent detection dataset with 77 banking-specific classes characterized by high inter-class similarity; (iii) **AlleNoise** [14], a hierarchical e-commerce benchmark of product titles evaluated at **Level 2** (top 12 classes) and **Level 3** (top 50), using expert-corrected labels (clean_category_id) and filtering underrepresented classes to ensure class balance and stability in our few-shot setup; and (iv) **FinBen-FPB** [23], a 3-class sentiment task (Positive, Negative, Neutral) focused on specialized financial terminology.

4.2 Baseline Models

We evaluate our proposed augmentation strategies against four baseline paradigms detailed in Subsect. 3.1, using the *all-mpnet-base-v2* encoder [15]. All models follow the optimization protocol described in Subsect. 4.3, with hyperparameters selected based on validation Macro-F_1 performance.

Linear Probe: A Logistic Regression classifier where the regularization parameter $C \in \{0.01, 0.1, 1, 10, 100\}$ was optimized via grid search.

Prototypical Network: Class prototypes are computed as mean embedding vectors, with classification determined by squared Euclidean distance [18].

SetFit: Following the framework by [20], this baseline involves contrastive fine-tuning of the encoder followed by classification head optimization.

PEFT (LoRA): Low-Rank Adaptation [10] applied to backbone attention/dense layers. We explored rank values $r \in \{8, 16, 32, 64\}$ with a scaling factor $\alpha = 2r$.

4.3 Implementation Details

Data Augmentation Procedure: For LLM-based approaches, synthetic samples were generated via GPT-4o-mini with prompt structures tailored to each method's intent. We generated $n = 3$ candidates per original seed example at a temperature of 0.7 to balance lexical creativity and semantic integrity. To establish expansion factors of **100%** and **200%**, we performed a semantic filtering step: for each seed, we calculated the SBERT-based Cosine Similarity between the original example and its $n = 3$ synthetic variants, selecting only the top-1 and top-2 most similar candidates from that specific pool, respectively. This process prioritizes synthetic data that strictly preserves the original class intent.

Experimental Protocol: Under a k-shot regime ($k \in \{1, 5, 16\}$), we randomly sampled k training and validation instances per class, utilizing the latter for hyperparameter optimization. To balance statistical variance and computational cost, we ran 20 independent episodes for low-cardinality datasets (< 13 classes) and 10 for high-cardinality benchmarks (≥ 13 classes). Results report the mean Macro-F_1 and standard deviations. Experiments were conducted on an HPC cluster with a single NVIDIA A100 GPU, 4 CPU cores, and 16GB of RAM.

4.4 Evaluation Metrics

Following the diagnostic framework in Subsect. 3.2, we evaluate models across four technical dimensions:

Predictive Performance: We report Macro-F_1 as our primary metric due to the fine-grained nature of some datasets. Additionally, we compute per-class F_1 and mean confusion matrices (averaged over n episodes) to identify misclassification clusters and assess the impact of each strategy on overlapping intents.

Probabilistic Reliability: Model calibration is evaluated via the Brier Score and Expected Calibration Error (ECE). We implement a per-class ECE ($M = 10$ bins) to monitor the alignment between confidence and accuracy and detect potential overconfidence shifts introduced by the augmentation process.

Intrinsic Generative Quality: We evaluate samples across three axes: **(i) Lexical Diversity (MS-Jaccard)** measures structural innovation across n-gram scales ($n \in \{1, 2\}$) via local pairwise calculations between seeds and variants [2]; **(ii) Semantic Fidelity (SBERT Similarity)** validates intent preservation via Cosine Similarity between seed and synthetic embeddings using the *all-mpnet-base-v2* model [7], and **(iii) Naturalness (Fréchet BERT Distance (FBD))** quantifies distributional alignment via the Fréchet distance between Gaussian approximations of latent representations [2].

Computational Efficiency: To examine the practical viability of the methods, we record the total Training Time across all experimental trials.

5 Results and Discussion

Preliminary analysis across all benchmarks indicated that doubling the augmentation scale from 100% to 200% yielded marginal performance gains for most architectures, with LoRA being the notable exception. Consequently, to prioritize a focused evaluation of augmentation strategies, the following discussions center on the 100% scale, which represents the optimal trade-off between computational overhead and classification accuracy.

Table 1. Intrinsic Quality: **MS-J** (diversity) ↓/**SBERT** (fidelity) ↑/**FBD** (distributional health) ↓. Datasets: AG News (AG), Banking77 (B77), AlleNoise L2 (A2), AlleNoise L3 (A3), and FinBen (FB). **Bold** indicates the best result per metric column.

Method	AG	B77	A2	A3	FB
WordNet	0.49/**0.90**/**0.20**	0.41/0.67/0.30	0.45/**0.84**/**0.23**	0.44/**0.84**/**0.14**	0.33/**0.80**/**0.40**
Context Sw.	0.10/0.39/1.09	0.17/0.55/0.35	0.04/0.42/0.76	0.04/0.45/0.40	0.13/0.41/1.08
Sem. Del.	0.06/0.61/0.76	**0.06**/0.61/0.29	0.05/0.58/0.63	0.05/0.57/0.41	0.09/0.69/0.60
Dynam. G3M	**0.01**/0.16/1.38	0.07/0.66/0.18	**0.00**/0.22/0.93	**0.01**/0.30/0.48	**0.01**/0.15/1.34
BA-G3M	0.03/0.32/1.07	0.12/**0.74**/**0.14**	0.14/0.55/0.43	0.16/0.59/0.22	0.07/0.39/0.88

5.1 Intrinsic Quality Assessment: Diversity Vs. Fidelity

The intrinsic metrics in Table 1 highlight a clear trade-off between lexical innovation (MS-Jaccard), semantic fidelity (SBERT), and distributional alignment (FBD). Our analysis identifies distinct behavioral archetypes:

Conservative Stability: Rule-based WordNet exhibits the highest semantic stability (SBERT). Still, it exhibits limited lexical innovation, as evidenced by high MS-Jaccard scores, which may limit effectiveness in complex architectures.

Targeted Structural Expansion: Intermediate transformative approaches, such as Semantic Deletion and Context Swapping, provide a more sophisticated expansion of the feature space. Notably, Semantic Deletion often outperforms Context Swapping in diversity (e.g., $MS\text{-}J = 0.06$ vs. 0.17 in Banking77) by forcing structural rephrasing without key discriminative tokens. Yet, it still struggles with fine-grained semantic drift compared to contrastive methods.

Unconstrained Innovation: Dynamic GPT3Mix demonstrates high creative potential (near-zero MS-Jaccard scores), but at the cost of severe semantic drift. This collapse is evident in AG News and FinBen-FPB, where high FBD and low fidelity (SBERT) indicate that synthetic samples lose the nuanced sentiment required. Interestingly, this effect is less pronounced in high-granularity Banking77, where constrained intent vocabulary likely limits the LLM's hallucination range (lower FBD). Nevertheless, the sharp drop in fidelity in these cases still introduces out-of-intent noise, disrupting the few-shot training process.

The Contrastive Stabilizer: Our proposed Boundary-Aware GPT3Mix acts as a structural anchor. By leveraging boundary examples, it recovers semantic fidelity (e.g., $SBERT = 0.74$ on Banking77, surpassing even WordNet) while preserving high levels of innovation. It significantly reduces FBD, achieving a robust equilibrium between conservative stability and generative innovation.

5.2 Overall Performance Analysis

Results in Table 2 confirm that DA is pivotal in extreme low-data settings ($k \in \{1, 5\}$) by mitigating coarse misclassifications between semantically distant categories. While WordNet ensures semantic fidelity, its lexical stagnation yields only marginal gains relative to generative methods. Notably, simpler architectures (Linear Probe/ProtoNet) appear as the most stable choices at $k = 1$, whereas adaptive paradigms (SetFit/LoRA) exhibit lower baseline performance or high sensitivity to the choice of augmentation.

Crucially, Dynamic GPT3Mix often outperforms Boundary-Aware GPT3Mix at $k = 1$, suggesting that unconstrained latent exploration provides a richer feature set for initial mapping. In contrast, contrastive constraints may be overly restrictive when support seeds are extremely scarce. However, as the baseline anchors primary boundaries (e.g., at $k = 16$), unconstrained DA risks introducing obstructive noise or semantic drift, underscoring the need for boundary-aware precision once a minimal real-world signal is established. Furthermore, while

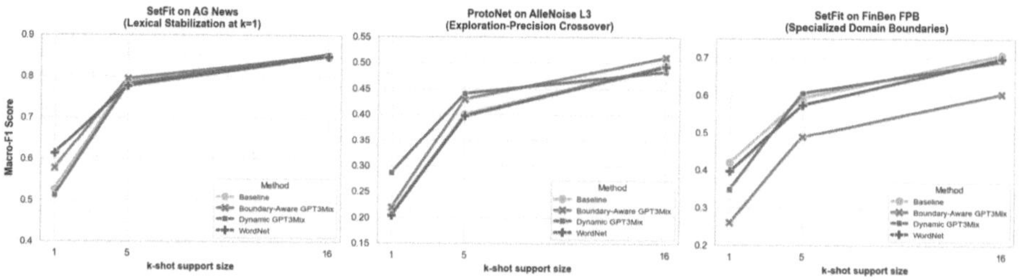

Fig. 1. Downstream Macro-F1 across support sizes (k). Left: WordNet is observed to stabilize contrastive regimes in general domains at $k = 1$. Center: Boundary precision (BA-GPT3Mix) triggers a crossover effect as k increases. Right: Specialized domains appear more sensitive to generative shifts, rendering conservative methods more robust.

DA efficacy is strictly bound by dataset characteristics, extremely specialized domains such as FinBen-FPB expose the limits of synthetic expansion (Fig. 1, Right). While LLM-based 'hallucinations' can be destructive in technical niches where vocabulary precision is non-negotiable, WordNet provides a safer alternative for expert jargon by anchoring the exact lexical markers that generative models often compromise through over-creative paraphrasing. This reinforces that the semantic drift and class-misalignment discussed in Subsect. 5.1 are the primary drivers of performance degradation in specialized domains.

5.3 Model Architecture vs Data Augmentation

Model architecture reveals distinct sensitivities to DA (Table 2). LoRA yields the highest relative F1 gains from synthetic volume, thereby stabilizing its gradient updates at $k = 1$, while its reference baseline remains fragile. Alternatively, Linear Probe and Prototypical Networks (Protonet) act as stable but rigid alternatives. Their restricted learning spaces (frozen embedding representations or centroid-based aggregation) offer resilience against semantic drift but prevent the models from fully adapting to the specific nuances of the augmented data.

In contrast, SetFit exhibits a more volatile and non-linear response to DA. Its reliance on contrastive triplet mining makes it uniquely sensitive to sample fidelity. At $k = 1$, even a minor semantic drift in a synthetic sample creates a "false positive" that distorts the model's entire latent space. This misalignment of embedding, illustrated by SetFit's volatility under generative noise on AG News (Fig. 1, Left), highlights an inherent sensitivity in contrastive regimes with minimal support. This reveals an architectural trade-off: while LoRA requires sufficient data to stabilize, ProtoNet/Linear Probe prioritizes robustness, and SetFit demands high semantic precision to achieve its peak performance.

5.4 Impact of Label Granularity on Robustness

The AlleNoise hierarchy (Subsect. 4.1) provides a controlled environment to isolate label resolution from textual complexity. By using samples from the same

Table 2. Macro F1 performance across all datasets and models for $k \in \{1, 5, 16\}$. Values: Mean F1 (±% Relative Gain) compared to baseline (**Bsl**). WN: WordNet; CS: Context Swapping; D-G3M: Dynamic GPT3Mix; BA-G3M: Boundary-Aware GPT3Mix; SD: Semantic Del. Best results in **bold**. * indicates significant improvement over Bsl ($p < 0.05$, Wilcoxon test). ↑/↓ indicate best/worst model peak F1 for each k.

Dataset	k	Model	Bsl	WN	CS	D-G3M	BA-G3M	SD
AG News	1	Lin. Probe	0.521	0.522 (+0.3%)	0.533 (+2.3%)	**0.548 (+5.2%)**	0.539 (+3.4%)	0.529 (+1.5%)
		LoRA ↓	**0.394**	0.346 (-12.0%)	0.334 (-15.1%)	0.346 (-12.1%)	0.354 (-10.0%)	0.343 (-13.0%)
		ProtoNet	0.521	0.521 (+0.0%)	0.528 (+1.3%)	**0.567 (+8.8%)**	0.524 (+0.5%)	0.520 (-0.3%)
		SetFit ↑	0.526	**0.613 (+16.5%)***	0.574 (+9.1%)	0.510 (-3.0%)	0.577 (+9.7%)*	0.518 (-1.4%)
	5	Lin. Probe	0.720	0.720 (-0.1%)	0.729 (+1.2%)	**0.732 (+1.7%)**	0.732 (+1.6%)	0.725 (+0.7%)
		LoRA ↓	0.608	0.630 (+3.6%)	0.637 (+4.9%)	**0.653 (+7.6%)**	0.645 (+6.2%)	0.640 (+5.4%)
		ProtoNet	0.711	0.712 (+0.2%)	0.738 (+3.8%)*	**0.752 (+5.8%)***	0.738 (+3.8%)*	0.726 (+2.2%)
		SetFit ↑	0.786	0.776 (-1.3%)	0.785 (-0.2%)	0.780 (-0.8%)	0.793 (+0.9%)	**0.801 (+1.8%)**
	16	Lin. Probe ↓	0.814	0.815 (+0.1%)	0.816 (+0.3%)	0.814 (-0.0%)	**0.817 (+0.4%)**	0.817 (+0.3%)
		LoRA	0.827	**0.847 (+2.4%)**	0.845 (+2.1%)	0.845 (+2.2%)	0.842 (+1.8%)	0.844 (+2.1%)
		ProtoNet	0.815	0.817 (+0.2%)	**0.824 (+1.1%)***	0.824 (+1.0%)	0.823 (+0.9%)	0.821 (+0.8%)
		SetFit ↑	0.851	0.848 (-0.5%)	0.840 (-1.3%)	0.852 (+0.1%)	0.849 (-0.2%)	**0.859 (+0.9%)**
AllenoiseL2	1	Lin. Probe ↑	0.258	0.256 (-1.0%)	0.305 (+18.3%)*	**0.327 (+26.7%)***	0.276 (+7.0%)*	0.281 (+8.7%)*
		LoRA ↓	0.123	0.159 (+29.1%)*	0.208 (+69.1%)	**0.211 (+71.8%)***	0.193 (+56.9%)*	0.190 (+55.0%)
		ProtoNet	0.251	0.246 (-2.0%)	0.296 (+17.9%)*	**0.319 (+27.2%)***	0.258 (+2.6%)	0.263 (+4.7%)*
		SetFit	0.258	**0.263 (+1.9%)**	0.060 (-76.5%)	0.185 (-28.1%)	0.234 (-9.3%)	0.153 (-40.6%)
	5	Lin. Probe	0.415	0.410 (-1.2%)	0.457 (+10.2%)*	**0.469 (+13.1%)***	0.450 (+8.5%)*	0.437 (+5.4%)*
		LoRA	0.374	0.401 (+7.2%)	0.445 (+18.8%)	**0.455 (+21.5%)**	0.441 (+17.8%)	0.430 (+14.8%)
		ProtoNet ↑	0.425	0.422 (-0.6%)	0.472 (+11.0%)*	**0.475 (+11.8%)***	0.456 (+7.3%)*	0.448 (+5.3%)*
		SetFit ↓	**0.443**	0.442 (-0.1%)	0.429 (-3.1%)	0.437 (-1.3%)	0.394 (-11.1%)	0.438 (-1.0%)
	16	Lin. Probe	0.507	0.505 (-0.5%)	**0.533 (+5.1%)***	0.531 (+4.7%)*	0.533 (+5.0%)*	0.520 (+2.4%)*
		LoRA ↑	0.500	0.527 (+5.5%)*	0.557 (+11.5%)	0.558 (+11.7%)*	**0.558 (+11.7%)***	0.555 (+11.2%)
		ProtoNet ↓	0.512	0.509 (-0.5%)	**0.529 (+3.4%)***	0.514 (+0.4%)*	0.528 (+3.1%)*	0.522 (+2.0%)*
		SetFit	0.538	0.503 (-6.4%)	0.537 (-0.2%)	0.503 (-6.4%)	**0.539 (+0.3%)**	0.518 (-3.7%)
AllenoiseL3	1	Lin. Probe ↑	0.211	0.210 (-0.7%)	0.269 (+27.3%)*	**0.290 (+37.2%)***	0.235 (+11.4%)*	0.230 (+9.0%)*
		LoRA	0.132	0.172 (+30.5%)	0.234 (+77.8%)	**0.245 (+85.8%)**	0.199 (+50.8%)	0.214 (+62.6%)
		ProtoNet	0.206	0.203 (-1.4%)	0.263 (+27.6%)*	**0.286 (+38.9%)***	0.218 (+6.1%)*	0.221 (+7.2%)*
		SetFit ↓	0.211	0.211 (-0.1%)	0.039 (-81.6%)	0.013 (-93.6%)	**0.214 (+1.5%)**	0.115 (-45.5%)
	5	Lin. Probe ↑	0.393	0.390 (-0.9%)	0.443 (+12.7%)*	**0.450 (+14.3%)***	0.428 (+8.7%)*	0.411 (+4.5%)*
		LoRA	0.306	0.362 (+18.3%)	**0.449 (+46.8%)***	0.430 (+40.5%)*	0.422 (+38.0%)*	0.416 (+35.8%)
		ProtoNet	0.400	0.396 (-1.0%)	**0.441 (+10.2%)***	0.440 (+9.9%)*	0.429 (+7.2%)*	0.418 (+4.4%)*
		SetFit ↓	0.386	0.373 (-3.5%)	0.364 (-5.8%)	0.364 (-5.7%)	**0.390 (+0.9%)**	0.363 (-5.9%)
	16	Lin. Probe	0.511	0.509 (-0.4%)	0.533 (+4.2%)	0.531 (+3.8%)*	**0.536 (+4.8%)***	0.519 (+1.4%)
		LoRA ↑	0.507	0.529 (+4.4%)	0.546 (+7.7%)*	0.544 (+7.4%)*	**0.548 (+8.1%)***	0.536 (+5.8%)
		ProtoNet	0.495	0.493 (-0.5%)	0.502 (+1.4%)	0.483 (-2.5%)	**0.510 (+2.9%)***	0.495 (-0.0%)
		SetFit ↓	0.480	0.208 (-56.7%)	0.485 (+1.1%)*	0.473 (-1.3%)	**0.506 (+5.5%)***	0.485 (+1.0%)
Banking77	1	Lin. Probe	0.545	0.539 (-1.2%)	0.537 (-1.6%)	**0.575 (+5.5%)**	0.561 (+2.9%)	0.540 (-1.0%)
		LoRA ↓	0.443	0.430 (-2.9%)	0.394 (-11.0%)	**0.481 (+8.6%)**	0.467 (+5.5%)	0.414 (-6.4%)
		ProtoNet ↑	0.535	0.532 (-0.5%)	0.547 (+2.3%)	**0.599 (+11.8%)***	0.573 (+7.1%)*	0.542 (+1.3%)*
		SetFit	0.431	**0.483 (+12.2%)***	0.307 (-28.7%)	0.449 (+4.3%)	0.418 (-3.0%)	0.360 (-16.4%)
	5	Lin. Probe ↑	0.810	0.808 (-0.2%)	0.802 (-0.9%)	**0.816 (+0.8%)**	0.814 (+0.6%)	0.805 (-0.6%)
		LoRA ↓	0.710	0.743 (+4.6%)	0.706 (-0.6%)	0.747 (+5.2%)	**0.757 (+6.7%)**	0.741 (+4.3%)
		ProtoNet	0.790	0.788 (-0.2%)	0.789 (-0.2%)	**0.805 (+2.0%)***	0.800 (+1.3%)*	0.789 (-0.1%)
		SetFit	0.781	0.774 (-0.8%)	0.703 (-10.0%)	0.784 (+0.4%)	**0.787 (+0.8%)**	0.786 (+0.7%)
	16	Lin. Probe ↑	0.892	0.892 (-0.1%)	0.885 (-0.8%)	0.892 (+0.0%)	**0.893 (+0.1%)**	0.890 (-0.3%)
		LoRA	0.887	0.887 (-0.0%)	0.864 (-2.7%)	**0.888 (+0.1%)**	**0.888 (+0.1%)**	0.884 (-0.4%)
		ProtoNet ↓	0.848	0.846 (-0.2%)	0.845 (-0.3%)	0.848 (+0.0%)*	**0.849 (+0.2%)**	0.846 (-0.1%)
		SetFit	0.850	0.860 (+1.2%)*	0.829 (-2.5%)	**0.868 (+2.2%)***	0.860 (+1.2%)*	0.858 (+1.0%)
FinbenFPB	1	Lin. Probe	**0.417**	0.411 (-1.5%)	0.336 (-19.5%)	0.320 (-23.2%)	0.336 (-19.4%)	0.405 (-2.9%)
		LoRA ↓	0.335	0.332 (-1.1%)	0.300 (-10.5%)	0.306 (-8.9%)	0.319 (-4.9%)	**0.337 (+0.6%)**
		ProtoNet	**0.417**	0.398 (-4.5%)	0.310 (-25.6%)	0.309 (-25.7%)	0.286 (-31.4%)	0.387 (-7.2%)
		SetFit ↑	**0.420**	0.395 (-5.9%)	0.316 (-24.8%)	0.348 (-17.1%)	0.259 (-38.3%)	0.337 (-19.8%)
	5	Lin. Probe	**0.555**	0.530 (-4.5%)	0.415 (-25.2%)	0.384 (-30.9%)	0.397 (-28.5%)	0.529 (-4.7%)
		LoRA ↓	0.425	**0.454 (+6.9%)***	0.343 (-19.2%)	0.309 (-27.3%)	0.341 (-19.8%)	0.450 (+5.8%)
		ProtoNet	**0.575**	0.558 (-2.9%)	0.402 (-30.1%)	0.355 (-38.2%)	0.390 (-32.1%)	0.544 (-5.4%)
		SetFit ↑	0.593	0.573 (-3.3%)	0.565 (-4.6%)	**0.605 (+2.1%)**	0.488 (-17.6%)	0.574 (-3.1%)
	16	Lin. Probe ↓	**0.627**	0.604 (-3.7%)	0.449 (-28.4%)	0.432 (-31.1%)	0.445 (-29.1%)	0.599 (-4.5%)
		LoRA	0.638	**0.696 (+9.0%)***	0.470 (-26.4%)	0.437 (-31.5%)	0.431 (-32.5%)	0.673 (+5.5%)
		ProtoNet	**0.647**	0.631 (-2.5%)	0.424 (-34.5%)	0.364 (-43.7%)	0.396 (-38.7%)	0.604 (-6.7%)
		SetFit ↑	**0.708**	0.697 (-1.5%)	0.666 (-5.9%)	0.690 (-2.5%)	0.603 (-14.8%)	0.684 (-3.4%)

underlying distribution at Level 2 (12 classes) and Level 3 (50 classes), we directly assess how granularity affects DA efficacy.

AlleNoise gains observed in Table 2 confirm that higher granularity amplifies DA utility at $k = 1$, where synthetic samples provide critical coverage of the high-resolution label space. Conversely, as k increases, relative gains in L2 often surpass those in L3. In denser environments (L3), class proximity heightens sensitivity to neighbor-class noise, causing generative benefits to plateau earlier than in sparser L2 space (Fig. 1, Center). Granularity also exposes distinct architectural vulnerabilities. While Linear Probe and Prototypical Networks eventually saturate, SetFit faces a systemic representational breakdown in dense spaces. Our Boundary-Aware GPT3Mix method mitigates this via boundary-aware prompting: by prioritizing semantic precision over raw diversity, it prevents SetFit's collapse, maintaining baseline stability (+1.5%) at $k = 1$ (L3) where Dynamic GPT3Mix suffers severe degradation (−93.6%).

5.5 Reliability and Computational Efficiency

Scaling the support size (k) consistently reduces Brier Scores, enhancing probabilistic certainty across all architectures. While DA-driven calibration fluctuates at low k, it enhances reliability in high-granularity contexts. Notably, Boundary-Aware GPT3Mix significantly recovers SetFit's calibration in Banking77 (Table 3), reducing its ECE from 0.411 to 0.163 (ΔECE = −0.248), a critical gain for deploying adaptive models in dense label spaces.

Efficiency analysis reveals a pronounced divide between paradigms: static models (e.g., ProtoNet) maintain sub-second training times, whereas adaptive fine-tuning (LoRA, SetFit) incurs significant overhead, with latencies orders of magnitude higher. This gap underscores a critical trade-off between the performance potential of adaptive models and the scalability of feature-based methods.

Table 3. Efficiency vs. Reliability trade-off (k = 16, Banking77). Bold indicates the best result per metric.

Model Architecture	Time (s) ↓	Brier Score ↓	ECE (Bsl) ↓	ECE (BA-G3M) ↓	Δ ECE ↓
Linear Probe (Static)	25.65 ± 0.50	0.201 ± 0.05	**0.199 ± 0.03**	0.195 ± 0.03	−0.003
Prototypical Net (Static)	**0.44 ± 0.00**	0.280 ± 0.00	0.209 ± 0.00	0.206 ± 0.00	−0.002
LoRA (Adaptive)	185.68 ± 19.96	**0.182 ± 0.02**	0.204 ± 0.01	0.213 ± 0.01	+0.009
SetFit (Adaptive)	500.02 ± 6.53	0.422 ± 0.01	0.411 ± 0.00	**0.163 ± 0.01**	**−0.248**

6 Conclusion

This study provided a multidimensional analysis of data augmentation (DA) in few-shot learning, revealing that a critical trade-off between lexical innovation and semantic fidelity governs DA efficacy across architectures and granularities. Our findings indicate a qualitative shift in DA utility across the evaluated scenarios: while augmentation serves as a vital boundary stabilizer at $k = 1$, its role tends toward precision as k increases. In higher-k regimes, where unconstrained synthesis risks inducing semantic drift, our proposed Boundary-Aware GPT3Mix framework maintains reliability by anchoring synthetic samples in boundary-aware contexts, effectively improving model calibration (ECE).

Architecturally, contrastive models like SetFit demand extreme semantic precision in low-k regimes, whereas static alternatives offer greater resilience against noise in dense spaces. As the support size increases to $k = 16$, however, investing in synthetic data quality is more cost-effective than increasing model complexity. With high-quality synthetic data generation, simpler architectures achieve peak performance while avoiding the prohibitive computational overhead of adaptive fine-tuning. This underscores that robust few-shot systems must dynamically adapt augmentation constraints to the specific needs of the task, ensuring that generative expansion enhances, rather than compromises, empirical robustness.

References

1. Ahmadnia, S.: Active Few-Shot Learning for Text Classification. In: NAACL, pp. 6677–6694 (2025)
2. Alihosseini, D., Montahaei, E., Baghshah, M.S.: Jointly measuring diversity and quality in text generation models. In: NAACL, pp. 90–98 (2019)
3. Aljehani, A.: A BERT-based prototypical networks for few-shot Arabic short-text topic detection. J. Electrical Systems **20**(10s), 549–563 (2024)
4. Aljehani, A., Hasan, S.H., Khan, U.A.: Advancing text classification: a systematic review of few-shot learning approaches. Comput. Digit. Syst. **17**(1), 1–21 (2025)
5. Casanueva, I.: Efficient intent detection with dual sentence encoders. In: Proceedings of the NLP for Conversational AI, pp. 38–45 (2020)
6. Chai, Y., Xie, H., Qin, J.S.: Text data augmentation for large language models: a comprehensive survey. Artif. Intell. Rev. 35 (2026)
7. Dai, H., et al.: AugGPT: leveraging ChatGPT for text data augmentation. IEEE Trans. Big Data **11**(3), 907–918 (2025)
8. Feng, S.Y.: A survey of data augmentation approaches for NLP. In: Findings of the Association for Computational Linguistics: ACL-IJCNLP, pp. 968–988 (2021)
9. Heusel, M.: Gans trained by a two time-scale update rule converge to a local Nash equilibrium. Adv. Neural Inf. Process. Syst. **30** (2017)
10. Hu, E.J., et al.: Low-rank adaptation of large language models. ICLR **1**(2), 3 (2022)
11. Lei, S.: Tart: improved few-shot text classification using task-adaptive reference transformation. In: Proceedings of Association for Computational Linguistics, vol. 1, pp. 11014–11026 (2023)
12. Liu, H.: Label-enhanced prototypical network with contrastive learning for multi-label few-shot aspect category detection. In: Proceedings of SIGKDD, pp. 1079–1087 (2022)
13. Liu, X., Gao, Y., Zong, L., Xu, B.: Improve meta-learning for few-shot text classification with all you can acquire from the tasks. In: EMNLP, pp, 223–235 (2024)
14. Rączkowska, A.: Allenoise: large-scale text classification benchmark dataset with real-world label noise. arXiv preprint arXiv:2407.10992 (2024)

15. Reimers, N., Gurevych, I.: Sentence-BERT: sentence embeddings using Siamese BERT-networks. In: EMNLP-IJCNLP, pp. 3982–3992 (2019)
16. Sahu, G.: Data augmentation for intent classification with off-the-shelf large language models. In: Proceedings of NLP for Conversational AI, pp. 47–57 (2022)
17. Shypula, A.: Evaluating the diversity and quality of LLM generated content. arXiv preprint arXiv:2504.12522 (2025)
18. Snell, J., Swersky, K., Zemel, R.: Prototypical networks for few-shot learning. Adv. Neural Inf. Process. Syst. **30** (2017)
19. Suresh, V., Ong, D.: Not all negatives are equal: label-aware contrastive loss for fine-grained text classification. In: EMNLP, pp, 4381–4394 (2021)
20. Tunstall, L.: Efficient few-shot learning without prompts. arXiv preprint arXiv:2209.11055 (2022)
21. Wang, Y.: Generalizing from a few examples: a survey on few-shot learning. ACM Comput. Surv. 1–34 (2020)
22. Wei, J., Zou, K.: Eda: easy data augmentation techniques for boosting performance on text classification tasks. In: EMNLP-IJCNLP, pp. 6382–6388 (2019)
23. Xie, Q., et al.: Finben: a holistic financial benchmark for large language models. Adv. Neural. Inf. Process. Syst. **37**, 95716–95743 (2024)
24. Yoo, K.M.: GPT3Mix leveraging large-scale language models for text augmentation. In: EMNLP, pp, 2225–2239 (2021)
25. Yuan, Y., Liu, Y., Cheng, L.: A multi-faceted evaluation framework for assessing synthetic data generated by large language models (2024). arXiv:2404.14445
26. Zhang, X., Zhao, J., LeCun, Y.: Character-level convolutional networks for text classification. Adv. Neural Inf. Process. Syst. **28** (2015)
27. Zhou, J., et al.: Flipda: effective and robust data augmentation for few-shot learning. In: Proceedings of Association for Computational Linguistics, vol. 1, pp. 8646–8665 (2022)
28. Zhou, Y.: A survey on data augmentation in large model era (2024). arXiv:2401.15422

The Gentle Collapse: Distributional Metrics for Continual Learning

Ahmed Anwar[1,2](✉), Andreas Wagner[3], Federico Raue[1], Tobias Nauen[1,2], and Andreas Dengel[1,2]

[1] German Research Center for Artificial Intelligence (DFKI), Kaiserslautern, Germany
{ahmed.anwar,federico.raue,tobias.nauen,andreas.dengel}@dfki.de
[2] RPTU University Kaiserslautern-Landau, Kaiserslautern, Germany
[3] Hochschule Karlsruhe University of Applied Sciences, Karlsruhe, Germany
andreas.wagner@h-ka.de

Abstract. Accuracy degradation is the standard metric for Catastrophic Forgetting (CF), however, it records only whether forgetting occurred or not. It saturates at the extremes and collapses discretely at task boundaries, hiding the internal structure of what is being forgotten. We introduce six softmax-derived metrics spanning true-label rank (TLR), predictive confidence, and distributional divergence that characterize forgetting continuously, each normalized to $[0, 1]$ with no modification to training. On CIFAR-100, these metrics carry information where accuracy does not: at 0% accuracy, the Confusion Margin spans an IQR of $[0.32, 0.50]$ across classes that accuracy treats identically. We demonstrate that this richer signal is actionable in mitigating catastrophic forgetting. Per-sample metric scores used as loss weights reduce forgetting by 1.3% points over uniform experience replay (ER) on CIFAR-100. Furthermore, the slope of a metric over a small window provides a stable sampling criterion: at a small-window size (e.g. 3 epochs), accuracy-trend degrades to $34.79 \pm 2.32\%$ while log-TLR achieves $41.07 \pm 0.57\%$. This gap is structural since reliable small-window trend estimation requires a continuous signal. On TinyImageNet, log-TLR trend sampling reduces forgetting by 7.7% points over the ER baseline.

Keywords: Continual Learning · Catastrophic Forgetting · Evaluation

1 Introduction

Continual learning addresses a fundamental constraint in modern machine learning: neural networks suffer from Catastrophic Forgetting (CF) of previously learned tasks when trained on new ones [13]. As models grow larger and deployment settings more dynamic, this constraint becomes increasingly expensive, since full retraining is often computationally infeasible. The field has responded with a substantial body of work on replay [16], regularization [8,21], and architectural adaptation [18]. Progress across all of it is measured almost universally by a single quantity: classification accuracy and its degradation across tasks.

L. Pasa et al. (Eds.): ICANN 2026, LNCS 17089, pp. 620–632, 2027.
https://doi.org/10.1007/978-3-032-38398-3_51

Accuracy is a binary outcome defined at the decision boundary, however, forgetting does not happen as a binary sudden shift at the decision boundary. It happens throughout the output distribution: confidence in the correct class erodes, the true label falls in the predicted ranking, and probability mass shifts toward newly learned categories. When accuracy does register a drop in performance, it encodes no information about whether a class is marginally forgotten or entirely displaced. Our experiments validate this: in our CIFAR-100 experiment, in 31 observations where the class accuracy is exactly 0% shown in Fig. 3, the Confusion Margin spans an IQR of $[0.32, 0.50]$ and True-Label Rank a median of 0.65, both metrics continuing to distinguish degrees of forgetting that accuracy collapses to a single value. Two models that accuracy declares indistinguishable may differ substantially in how much residual knowledge they retain and how amenable that knowledge is to recovery.

We introduce six metrics derived from the softmax output distribution that characterize forgetting at a finer resolution than accuracy allows. They fall into three families: true-label rank metrics (TLR, Log-TLR, Reciprocal-TLR), which track the position of the correct class in the predicted ranking as forgetting progresses; confidence and confusion metrics (True-Label Confidence, Confusion Margin), which quantify predictive certainty and proximity to misclassification; and a distribution-matching metric (Normalized KL), which measures global output-distribution shift relative to the ideal one-hot encoded target. All six are normalized to $[0, 1]$, interpretable alongside accuracy, and require no additional forward passes or architectural changes.

The argument of this paper is three-part. First, the metrics expose distributional structure that accuracy cannot see: they carry information at the extremes where accuracy saturates and change gradually where accuracy steps. Second, this structure is actionable at the sample level, since metric scores assign a meaningful gradient of forgetting severity to individual replay samples, and using them to re-weight loss contributions improves performance over uniform replay. Third, this structure is actionable at the temporal level: the slope of a metric over a short trailing window is a stable and informative signal for replay prioritization. At a window of three epochs on CIFAR-100, accuracy-trend sampling collapses to $34.79 \pm 2.32\%$ while log-TLR trend sampling achieves $41.07 \pm 0.57\%$, a gap that reflects the structural impossibility of reliable short-window trend estimation on a binary signal. Continuous distributional metrics do not share this limitation.

Our contributions are as follows. We introduce six normalized, softmax-derived metrics for continual learning evaluation and show empirically that they reveal forgetting dynamics invisible to accuracy-based analysis. We show that per-sample metric scores are effective loss weights for replay, with consistent results on CIFAR-100 and TinyImageNet. We show that metric trend over a short window is an effective and stable sampling criterion, that accuracy-trend sampling exhibits catastrophic instability at short windows that continuous metrics do not, and that this asymmetry is structural not quantitative. Taken together, these results argue that the field's reliance on accuracy has not only limited evaluation; it has limited what interventions are possible for mitigating CF.

2 Related Work

Continual Learning Methods. Catastrophic forgetting [13] is addressed by three families of methods [5,20]. Regularization-based methods protect past knowledge by penalising changes to important parameters (EWC [8], SI [21]) or by self-distillation on new-task data (LwF [9]). Replay-based methods store a buffer of past examples and interleave them with current training. Standard ER [16] uses uniform sampling; GEM [11] and A-GEM [3] add gradient constraints; iCaRL [15] pairs replay with a herding-based exemplar selector; DER/DER++ [2] replays stored logits alongside inputs, combining cross-entropy with a logit-matching distillation term to form a strong baseline; and X-DER [1] further revises stored logits across tasks. Architecture-based methods grow separate capacity per task [18] or isolate sub-networks [19], avoiding interference at the cost of scalability.

Evaluation of Continual Learning. CL evaluation has long centred on the accuracy matrix $A = \{a_{k,j}\}$ [11], from which forward/backward transfer and forgetting measures are derived. Díaz-Rodríguez et al. [6] were among the first to challenge this, arguing that memory overhead, compute cost, and positive transfer deserve equal standing, and proposing a composite CLScore. The CLEVA-Compass [14] formalised multi-dimensional evaluation as a visual framework promoting transparency across diverse scenarios. Despite these efforts, all introduced frameworks still tie catastrophic forgetting to the binary classification accuracy signal. Most relevant to our work, De Lange et al. [4] showed through per-iteration evaluation that all major CL methods suffer from a stability gap, a sharp transient performance drop immediately after a new task begins, which end-of-task evaluation misses entirely. Their metrics quantify worst-case accuracy; ours quantify the degree of forgetting continuously through the full output distribution, and further show that this richer signal enables principled intervention at both the sample and temporal level.

Output-Based Metrics in Classification. Our metrics are all derived from the softmax distribution rather than the top-1 decision. True Label Confidence and Normalized KL are grounded in calibration research [7], where the probability assigned to the true class is recognised as a proper scoring signal independent of the hard prediction. The TLR family extends rank-based evaluation from information retrieval (MRR, nDCG [12]) and top-k accuracy [17] into a continuous scalar in $[0, 1]$, tracking how far the true label is pushed down the ranked list as forgetting progresses. The Confusion Margin is related to inference-time decision margins [10], but used here purely as an evaluation signal rather than a training objective.

3 Methodology

Standard continual learning evaluation centres on the accuracy matrix $A = \{a_{k,j}\}$ [14,20], where $a_{k,j} \in [0,1]$ is the accuracy of task j after learning task k. Aggregate statistics such as average accuracy, forgetting measure, and backward/forward transfer are all derived from this matrix [11]. Because every entry $a_{k,j}$ is a hard decision-boundary outcome, all of these measures inherit the same coarseness: two models that produce identical accuracy matrices can differ substantially in how much residual knowledge they retain.

To see why this matters, consider a classifier trained on four classes (0–3) in two tasks. After task 2, models A and B both misclassify a test sample with true label $y = 0$, giving $a_{2,1} = 0$ in both cases. Yet their softmax outputs tell a very different story:

$$f_2^A(x) = [\mathbf{0.48},\ 0.49,\ 0.02,\ 0.01], \qquad f_2^B(x) = [\mathbf{0.05},\ 0.10,\ 0.70,\ 0.15].$$

Model A has nearly recovered the true class, while model B has almost entirely forgotten it, a distinction that no accuracy-based measure can express.

3.1 Proposed Metrics

We focus on a classification problem with C classes learned over T tasks. Let $f_k(x_i^j)[c]$ denote the c-th softmax output of model f_k, after training on task k, on sample x_i^j from task j, and let y_i^j be its true label. All six metrics below are defined on a test set D_j of size n_j, take values in $[0,1]$, and are oriented so that higher is better, matching the convention of accuracy.

Confusion Margin (CM). The margin between the top predicted probability and the probability assigned to the true label directly measures proximity to misclassification:

$$m_{k,j} = \frac{1}{n_j} \sum_{i=1}^{n_j} \Big(1 + f_k(x_i^j)[y_i^j] - \max_c f_k(x_i^j)[c]\Big).$$

A value of 1 corresponds to a perfectly confident correct prediction; lower values indicate increasing confusion.

True-Label Rank Family (TLR, LTLR, RTLR). These metrics track the position of the true class in the ranked softmax output, where rank 1 denotes a correct classification. Formally, $\mathrm{rank}(y_i^j, f_k(x_i)) = 1 + \sum_{c \neq y_i^j} \mathbb{1}(f_k(x_i)[c] > f_k(x_i)[y_i^j])$. Rank-based evaluation extends established information-retrieval practice (MRR, nDCG [12]) and top-k accuracy [17] into a continuous scalar.

The three variants trade off sensitivity at different parts of the rank distribution:

$$r_{k,j} = \frac{1}{n_j}\sum_{i=1}^{n_j}\left(1 - \frac{\text{rank}(y_i^j, f_k(x_i^j)) - 1}{C-1}\right),$$

$$l_{k,j} = \frac{1}{n_j}\sum_{i=1}^{n_j}\left(1 - \frac{\ln\left(\text{rank}(y_i^j, f_k(x_i^j))\right)}{\ln C}\right),$$

$$s_{k,j} = \frac{1}{n_j}\sum_{i=1}^{n_j}\frac{C - \text{rank}(y_i^j, f_k(x_i^j))}{(C-1)\cdot\text{rank}(y_i^j, f_k(x_i^j))}.$$

TLR (r) penalises all rank displacements equally (linearly); Log-TLR (l) applies a logarithmic compression that softens penalties for distant errors; Reciprocal-TLR (s) concentrates sensitivity at the top ranks, making it the strictest measure for near-correct predictions.

True Label Confidence (CTL). Mean predicted probability of the true class is the simplest distributional signal, directly grounded in calibration research [7]:

$$q_{k,j} = \frac{1}{n_j}\sum_{i=1}^{n_j} f_k(x_i^j)[y_i^j].$$

Normalized KL (NKL). The KL divergence from the model's output to the Dirac distribution on the true class equals $-\ln f_k(x_i)[y_i]$, confirming that CTL corresponds to $1-\exp(\text{KL})$. For a complementary normalization that caps divergence at the uniform-distribution baseline $\ln C$, we define:

$$k_{k,j} = \frac{1}{n_j}\sum_{i=1}^{n_j}\left(1 - \frac{\min(-\ln f_k(x_i^j)[y_i^j],\ \ln C)}{\ln C}\right).$$

NKL captures global output-distribution shift and is the most sensitive metric to severe forgetting where the model assigns near-zero probability to the true class.

4 Experiments

4.1 Experimental Setup

Diagnostic experiments (Sect. 4.2) use Sequential CIFAR-100 with a 5-task class-incremental split (20 classes per task). We train a ResNet-18 backbone with a replay buffer of size 2000, batch size 256, learning rate 0.03, and 50 epochs per task, using Experience Replay (ER) with uniform sampling as the base method. Metrics are computed per class at every epoch, enabling fine-grained analysis across the full training trajectory. Intervention experiments (Sects. 4.3–4.4) additionally include TinyImageNet (200 classes, 10 tasks) to assess generalisation to a harder benchmark. All intervention results are reported as mean ± std across three independent runs. The window size $w = 10$ was selected based on the sensitivity analysis in Sect. 4.4.

4.2 Metrics Reveal Forgetting Dynamics

A Gradual Signal Where Accuracy is Binary. Figure 1 shows mean ± std across Task-1 classes for accuracy, CM, and TLR centred on the Task 1→2 boundary. When the new task begins, accuracy collapses sharply, dropping to near zero within a few epochs with no intermediate signal. The continuous metrics respond to the same event differently: they decay gradually over many epochs, exposing a continuum of forgetting severity that the binary accuracy signal compresses into a single step. At any point after the task boundary, the continuous metrics still carry information about how severely a class has been forgotten, while accuracy has already saturated for most classes.

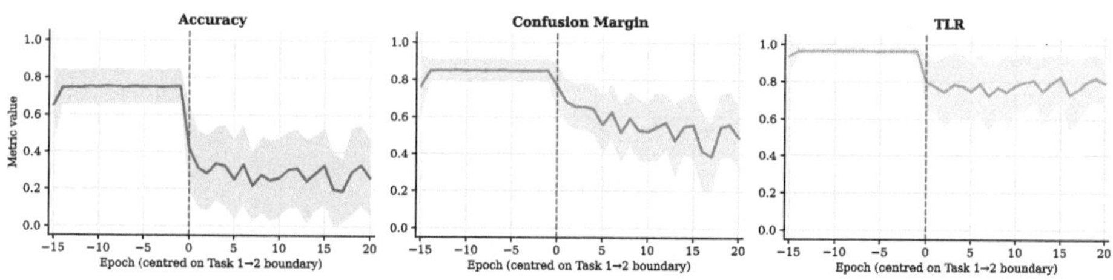

Fig. 1. Mean ± std across Task-1 classes for accuracy, confusion margin, and true-label rank centred on the Task 1→2 boundary (red dashed line). Continuous metrics decay gradually while accuracy collapses sharply, exposing a graded forgetting signal that accuracy cannot provide. (Color figure online)

Figure 2 quantifies this by separating epochs into transition epochs (within ±5 steps of a task boundary) and stable epochs (mid-task). Accuracy produces extreme outlier jumps at boundaries, with the upper whisker exceeding 0.35, and these spikes are absent for all continuous metrics. The continuous metrics show a moderate and consistent elevation at transitions. When a task boundary occurs, accuracy undergoes a structural discontinuity; the continuous metrics register a moderate, trackable increase in change rate. This smoothness is not merely diagnostic: it is precisely what makes short-window trend estimation tractable, as we show in Sect. 4.4.

Information Where Accuracy is Saturated. When accuracy reaches zero for a class, it provides no further signal. Figure 3 shows the distribution of each metric restricted to the n epochs where class-accuracy is in the corresponding accuracy bin ±0.5%. At bin $acc = 0\%$ (when accuracy reaches zero) all samples are indistinguishable accuracy-wise, however CM spans an IQR of $[0.32, 0.50]$

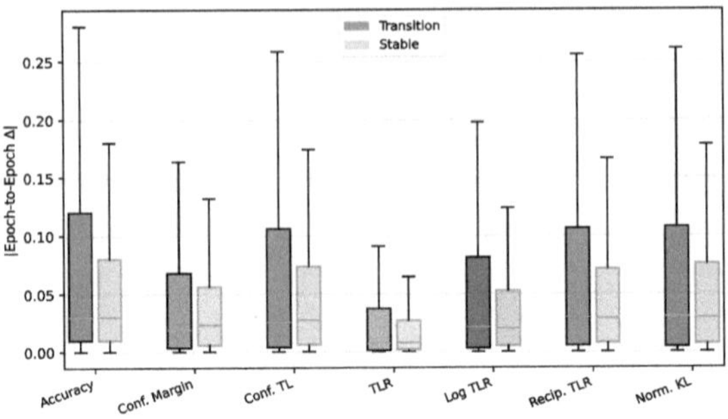

Fig. 2. Epoch-to-epoch absolute change split by phase: transition epochs (±5 steps around each task boundary) versus stable mid-task epochs. Accuracy produces extreme jumps at transitions (upper whisker > 0.35). CTL, RTLR, and NKL are more volatile compared to CM, TLR, and Log-TLR.

with values reaching 0.85, and TLR has a median of 0.65 with IQR $[0.60, 0.76]$, which gives these metrics a discriminative power that can be exploited. A class with CM = 0.85 has its true label nearly at the top of the predicted ranking despite being classified incorrectly; a class with CM = 0.18 has been pushed far from the decision boundary. Accuracy reports both identically nevertheless, which further highlights the drawbacks of relying on accuracy alone in understanding CF.

CTL and NKL show very little spread at 0% accuracy; IQR $[0.00, 0.01]$ and $[0.02, 0.04]$ respectively: at zero accuracy, the model systematically assigns low probability to the true class, so confidence-based measures saturate near zero alongside accuracy. This pattern holds across fixed accuracy levels from 0% to 75%: rank-based and margin-based metrics remain informative at the extremes where confidence-based measures do not as shown in Fig. 3. The complementarity of rank-based and confidence-based metrics guides metric selection for the intervention experiments that follow: NKL and CTL, which carry strong signal in mid-range forgetting, are suited to sample-level weighting; TLR and Log-TLR, which remain discriminative at severe forgetting, are the more robust choice for trend estimation on harder tasks.

4.3 Metric-Weighted Loss

The distributional metrics assign a continuous score to each sample reflecting the current severity of forgetting for its class. We investigate whether concentrating gradient contributions on more-forgotten samples, by using per-sample metric scores as loss weights, produces measurable gains over uniform ER.

At each replay step, each sample is assigned a weight inversely proportional to its current metric score, so that more-forgotten samples exert a larger influence on the parameter update. Table 1 reports final accuracy and forgetting on Sequential CIFAR-100 and TinyImageNet for each metric-weighted variant. On CIFAR-100, Log-TLR produces the highest accuracy ($36.98 \pm 0.55\%$) and

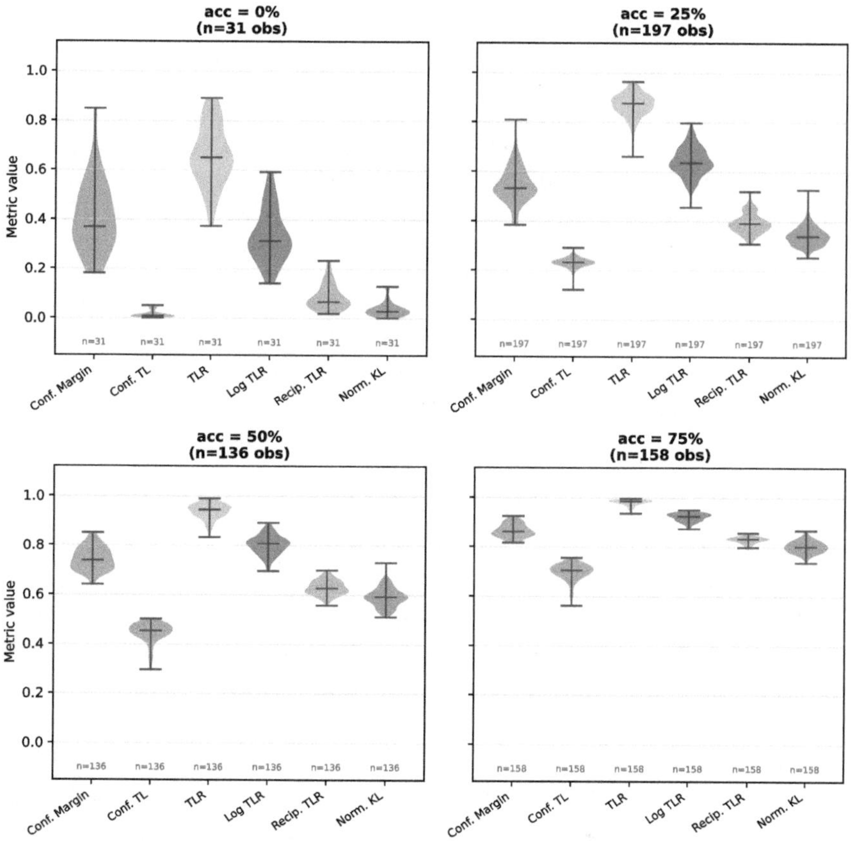

Fig. 3. Distribution of each metric at fixed accuracy levels (±0.5% windows). At 0% accuracy, CM and TLR retain wide distributions while CTL and NKL are more concentrated. Across all intermediate levels, the metrics carry substantial variation beyond what accuracy encodes.

Table 1. Metric-weighted loss on Sequential CIFAR-100 and TinyImageNet ($n = 3$ seeds). Higher accuracy and lower forgetting are better.

Weight Metric	CIFAR-100		TinyImageNet	
	Acc. (%)↑	Forg. (%) ↓	Acc. (%)↑	Forg. (%)↓
ER uniform (baseline)	36.33 ± 0.55	53.20 ± 0.59	23.67 ± 0.81	54.98 ± 0.28
TLR	36.40 ± 0.68	52.30 ± 0.31	23.51 ± 0.38	55.00 ± 1.06
Log-TLR	$\mathbf{36.98 \pm 0.55}$	51.92 ± 0.35	23.72 ± 0.19	$\mathbf{53.38 \pm 1.58}$
Reciprocal-TLR	35.42 ± 0.49	52.55 ± 0.49	23.04 ± 0.30	54.67 ± 0.12
Confusion Margin	36.43 ± 0.40	52.54 ± 0.16	$\mathbf{23.76 \pm 0.60}$	54.05 ± 1.08
Normalized KL	35.56 ± 0.55	$\mathbf{51.49 \pm 0.22}$	23.30 ± 0.46	53.85 ± 0.12
Confidence TL	35.42 ± 0.54	52.10 ± 0.51	22.78 ± 0.90	53.89 ± 0.72
Accuracy	35.21 ± 0.58	52.21 ± 0.84	22.19 ± 0.87	54.54 ± 0.60

a forgetting reduction of 1.3% points over the ER baseline ($51.92 \pm 0.35\%$ vs $53.20 \pm 0.59\%$). Confusion Margin and TLR perform similarly ($36.43 \pm 0.40\%$ and $36.40 \pm 0.68\%$ respectively), while confidence-based metrics trail by roughly 1% point. On TinyImageNet, accuracy improvements are within noise across all variants, but the forgetting reduction is consistent: Log-TLR reduces forgetting from $54.98 \pm 0.28\%$ to $53.38 \pm 1.58\%$, and NKL achieves $53.85 \pm 0.12\%$ with lower variance. Accuracy-weighted loss ($35.21 \pm 0.58\%$ on CIFAR-100, $22.19 \pm 0.87\%$ on TinyImageNet) serves as a confirmation that sample reweighting helps in principle, but being a binary signal it fails to outperform the baseline, while all continuous distributional variants do.

Table 2. Metric trend sampling on Sequential CIFAR-100 and TinyImageNet ($w = 10$, $\gamma = 2.0$, $n = 3$ seeds). Bold marks the best result per column.

Trend metric	CIFAR-100		TinyImageNet	
	Acc. (%)↑	Forg. (%)↓	Acc. (%)↑	Forg. (%)↓
ER uniform (baseline)	36.33 ± 0.55	53.20 ± 0.59	23.67 ± 0.81	54.98 ± 0.28
TLR	39.93 ± 0.94	46.10 ± 1.99	23.14 ± 0.72	46.79 ± 1.21
Log-TLR	41.71 ± 0.73	44.52 ± 0.57	$\mathbf{24.74 \pm 0.81}$	47.26 ± 1.14
Reciprocal-TLR	$\mathbf{41.88 \pm 0.60}$	44.73 ± 1.07	24.16 ± 1.34	49.11 ± 1.37
Confusion Margin	40.67 ± 1.14	46.30 ± 1.31	23.27 ± 0.13	49.39 ± 3.34
Normalized KL	41.13 ± 0.70	45.95 ± 0.77	21.92 ± 4.23	46.54 ± 3.95
Confidence TL	37.30 ± 1.34	51.21 ± 2.02	23.43 ± 1.15	48.32 ± 0.91
Loss	41.44 ± 0.52	$\mathbf{43.73 \pm 0.52}$	24.00 ± 1.37	$\mathbf{46.23 \pm 3.53}$
Accuracy	37.75 ± 0.94	50.65 ± 1.35	24.30 ± 0.27	49.11 ± 1.74

4.4 Metric Trend Sampling

Sample-level metric scores capture the current degree of forgetting for each class. A complementary question is whether the rate of change of a metric over a recent window provides additional information for replay prioritization. The sampling weight is computed as $\text{softmax}(\gamma \cdot slope)$ over classes, where $\gamma = 2.0$ controls the sharpness of prioritization and slope is estimated by linear regression over the trailing window of w epochs. A class whose metric score is declining rapidly is actively being forgotten; a class whose score is stable may not benefit as urgently from additional replay even if its current score is low.

Window Sensitivity on CIFAR-100. Figure 4 shows final accuracy on Sequential CIFAR-100 as a function of window size for five trend signals: Log-TLR, Loss, Normalized KL, Confusion Margin, and Accuracy.

At $w = 3$, accuracy-trend sampling collapses to $34.79 \pm 2.32\%$, with error bands reaching below 33%. Every continuous signal, Log-TLR, Loss, Normalized

KL, and Confusion Margin, is stable at $w = 3$, clustering between 39.8% and 41.1%. As the window grows, accuracy partially recovers, reaching 40.33% at $w = 15$ before falling to 39.14% at $w = 20$, and its variance remains elevated throughout.

Accuracy at the class level is a step function; it moves in discrete jumps as samples cross the decision boundary, and the slope of a step function over a short window is dominated by the noise of those transitions. A longer history is required to average that noise out. This is exactly what Fig. 4 shows: accuracy needs $w \geq 15$ before approaching the performance that continuous metrics achieve at $w = 3$.

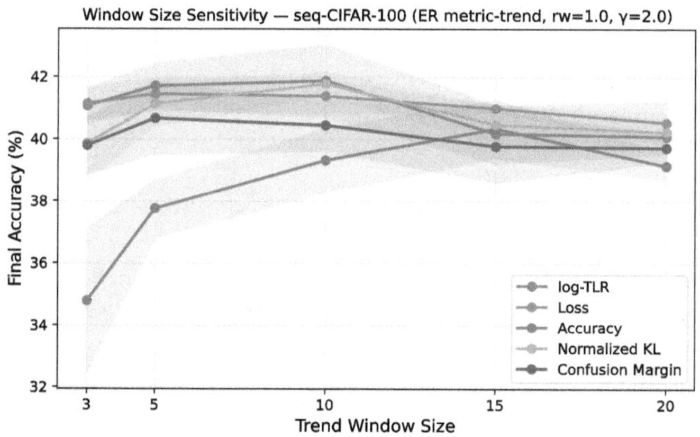

Fig. 4. Final accuracy on Sequential CIFAR-100 as a function of trend window size ($\gamma = 2.0$, $n = 3$ seeds, shaded regions show ± 1 std). At $w = 3$, accuracy-trend sampling collapses to $34.79 \pm 2.32\%$, more than six percentage points below every continuous metric at the same window. All continuous signals are stable across all window sizes. The instability of accuracy-trend is structural: slope estimation on a binary signal is dominated by the noise of discrete threshold crossings.

The key distinction is between continuous and binary signals, not between our proposed metrics and all alternatives. Loss is also a continuous signal and is correspondingly stable across windows ($41.14 \pm 0.53\%$ at $w = 3$ through $40.53 \pm 0.68\%$ at $w = 20$). Table 2 shows that at $w = 10$, Loss achieves the best forgetting on CIFAR-100 ($43.73 \pm 0.52\%$), while Reciprocal-TLR leads on accuracy ($41.88 \pm 0.60\%$); the continuous metrics are broadly competitive, and all substantially outperform accuracy-trend. In settings where the trend window cannot be tuned, rank-based metrics are the safer default, given their additional stability on harder tasks (see below).

Furthermore, Table 2 shows results on TinyImageNet at $w = 10$. Log-TLR achieves a 7.7 percentage point reduction in forgetting relative to the ER baseline ($47.26 \pm 1.14\%$ vs $54.98 \pm 0.28\%$), with a corresponding improvement in final accuracy and low variance across seeds. Loss-trend matches log-TLR on forgetting mean (46.23%) but with considerably higher variance ($\pm 3.53\%$ vs $\pm 1.14\%$), suggesting that rank-based metrics provide a more stable trend signal on harder tasks. Accuracy-trend is competitive on final accuracy ($24.30 \pm 0.27\%$) but substantially weaker on forgetting ($49.11 \pm 1.74\%$), reflecting the partial stabilisation of the binary signal at $w = 10$ noted in the window analysis.

Normalized KL and CM show elevated variance ($\pm 4.23\%$ and $\pm 3.34\%$). While CM performed well in metric-weighted loss, its spread at 0% accuracy visible in Fig. 3 results in sensitivity in a metric trend scenario. Normalized KL, like loss and CTL, is less robust under trend estimation than rank-based metrics, likely because its sensitivity to low-probability tails amplifies noise on the harder task.

5 Conclusion

Accuracy reduces model behaviour to a binary signal at the decision boundary, it is therefore blind to the gradual distributional shifts that precede and accompany forgetting. The lack of this nuance hinders its application to fine-grained interventions that could mitigate CF.

We introduced six softmax-derived metrics spanning true-label rank, confidence and confusion, and distribution matching. On Sequential CIFAR-100, these metrics expose structure that accuracy cannot: they carry substantial information where accuracy is saturated, distinguishing degrees of forgetting that accuracy collapses to a single value, and they evolve continuously at task boundaries where accuracy undergoes discontinuities. We showed that this richer characterisation is actionable in two ways. Per-sample metric scores used as loss weights reduce forgetting by 1.3 percentage points over uniform ER on CIFAR-100, with consistent forgetting improvements on TinyImageNet despite modest accuracy gains. Metric slope over a short trailing window reduces forgetting on TinyImageNet by 7.7 percentage points and improves accuracy under log-TLR. The temporal actionability depends on the signals being continuous: at a window of three epochs, accuracy-trend sampling collapses to $34.79 \pm 2.32\%$ while log-TLR maintains $41.07 \pm 0.57\%$, a gap that reflects the structural impossibility of reliable short-window trend estimation on a binary signal.

The metrics impose no additional training cost and are complementary to existing evaluation frameworks. More broadly, the granularity of an evaluation signal determines the granularity of the interventions it can support. Extending metric-guided interventions to stronger replay methods such as DER++ [2] and iCaRL [15], where the interaction between logit-matching distillation and distributional metric-weighting is non-trivial, is a natural direction for future work.

Acknowledgments. This work was supported by the BMFTR project Albatross (funding code 16IW24002). All compute was done thanks to the Pegasus cluster at DFKI Kaiserslautern.

Disclosure of Interests. The authors have no competing interests to declare that are relevant to the content of this article.

References

1. Boschini, M., Bonicelli, L., Buzzega, P., Porrello, A., Calderara, S.: Class-incremental continual learning into the eXtended DER-verse. IEEE Trans. Pattern Anal. Mach. Intell. **45**(5), 5497–5512 (2023)
2. Buzzega, P., Boschini, M., Porrello, A., Abati, D., Calderara, S.: Dark experience for general continual learning: a strong, simple baseline. In: Advances in Neural Information Processing Systems, vol. 33, pp. 15920–15930 (2020)
3. Chaudhry, A., Ranzato, M., Rohrbach, M., Elhoseiny, M.: Efficient lifelong learning with A-GEM. In: International Conference on Learning Representations (ICLR) (2019)
4. De Lange, M., van de Ven, G.M., Tuytelaars, T.: Continual evaluation for lifelong learning: identifying the stability gap. In: International Conference on Learning Representations (ICLR) (2023)
5. De Lange, M., et al.: A continual learning survey: defying forgetting in classification tasks. IEEE Trans. Pattern Anal. Mach. Intell. **44**(7), 3366–3385 (2022)
6. Díaz-Rodríguez, N., Lomonaco, V., Filliat, D., Maltoni, D.: Don't forget, there is more than forgetting: new metrics for continual learning. In: NeurIPS Continual Learning Workshop (2018)
7. Guo, C., Pleiss, G., Sun, Y., Weinberger, K.Q.: On calibration of modern neural networks. In: International Conference on Machine Learning (ICML). Proceedings of Machine Learning Research, vol. 70, pp. 1321–1330 (2017)
8. Kirkpatrick, J., et al.: Overcoming catastrophic forgetting in neural networks. Proc. Natl. Acad. Sci. **114**(13), 3521–3526 (2017)
9. Li, Z., Hoiem, D.: Learning without forgetting. IEEE Trans. Pattern Anal. Mach. Intell. **40**(12), 2935–2947 (2018)
10. Liu, W., Wen, Y., Yu, Z., Yang, M.: Large-margin softmax loss for convolutional neural networks. In: International Conference on Machine Learning (ICML), vol. 2, p. 7 (2016)
11. Lopez-Paz, D., Ranzato, M.: Gradient episodic memory for continual learning. In: Advances in Neural Information Processing Systems, vol. 30 (2017)
12. Manning, C.D., Raghavan, P., Schütze, H.: Introduction to Information Retrieval. Cambridge University Press (2008)
13. McCloskey, M., Cohen, N.J.: Catastrophic interference in connectionist networks: the sequential learning problem. Psychol. Learn. Motiv. **24**, 109–165 (1989)
14. Mundt, M., Lang, S., Delfosse, Q., Kersting, K.: CLEVA-compass: a continual learning EValuation assessment compass to promote research transparency and comparability. In: International Conference on Learning Representations (ICLR) (2022)
15. Rebuffi, S.A., Kolesnikov, A., Sperl, G., Lampert, C.H.: iCaRL: Incremental classifier and representation learning. In: IEEE Conference on Computer Vision and Pattern Recognition (CVPR), pp. 2001–2010 (2017)
16. Rolnick, D., Ahuja, A., Schwarz, J., Lillicrap, T., Wayne, G.: Experience replay for continual learning. In: Advances in Neural Information Processing Systems, vol. 32 (2019)
17. Russakovsky, O., et al.: ImageNet large scale visual recognition challenge. Int. J. Comput. Vision **115**(3), 211–252 (2015)
18. Rusu, A.A.: Progressive neural networks. arXiv preprint arXiv:1606.04671 (2016)
19. Serrà, J., Surís, D., Miron, M., Karatzoglou, A.: Overcoming catastrophic forgetting with hard attention to the task. In: International Conference on Machine

Learning (ICML). Proceedings of Machine Learning Research, vol. 80, pp. 4548–4557 (2018)
20. Wang, L., et al.: A comprehensive survey of continual learning: theory, method and application. IEEE Trans. Pattern Anal. Mach. Intell. **46**(8), 5362–5383 (2024)
21. Zenke, F., Poole, B., Ganguli, S.: Continual learning through synaptic intelligence. In: International Conference on Machine Learning (ICML). Proceedings of Machine Learning Research, vol. 70, pp. 3987–3995 (2017)

Multiclass Learning on the Simplex

Edoardo Legnaro[1(✉)], Sabrina Guastavino[1], and Francesco Marchetti[2]

[1] Department of Mathematics, University of Genova, Genova, Italy
edoardo.legnaro@edu.unige.it

[2] Department of Mathematics "Tullio Levi-Civita", University of Padova, Padova, Italy

Abstract. Binary classification naturally separates scoring from decision making: a model outputs a score, and a threshold turns it into a label. By contrast, in multiclass classification, softmax outputs are usually followed by argmax, which fixes one decision rule in advance. In this paper, we present a simplex-based view in which multiclass prediction is interpreted as choosing a family of class regions on the probability simplex S_m over m classes. A threshold vector $\tau \in S_m$ induces such regions on the simplex, with argmax recovered at the barycenter, and threshold tuning becomes post-hoc selection of better decision regions for a target score. The same geometric framework also supports threshold-aware training. We formalize this through a common score functional defined on probability measures over threshold vectors: Dirac measures recover fixed decision rules such as argmax and post-hoc threshold tuning, while general measures recover threshold-aware training. In this view, multiclass classification is not only prediction followed by argmax, but prediction on the simplex together with a choice of decision regions.

Keywords: multiclass classification · simplex geometry · threshold tuning · score-oriented learning

1 Introduction

In binary classification, scoring and decision making are naturally separated: a trained model produces a scalar score, and a threshold converts that score into a label. Performance can then be optimized post hoc by selecting the threshold that best matches the target metric [4]. In multiclass classification, this separation is not as straightforward, as softmax outputs are interpreted probabilistically and prediction is reduced to the argmax rule [13]. The resulting pipeline fixes the decision rule in advance and offers no native analogue of threshold tuning.

On the other hand, softmax outputs live in the simplex, so multiclass prediction has a geometric domain. Yet, the standard argmax decision rule uses only one specific family of decision regions on that domain. Existing work approaches this scoring-decision gap from several directions, including reductions to binary classification, calibration, cost-sensitive decision making, metric-oriented learning, and alternatives to argmax. Section 2 discusses these connections and clarifies the position of the present framework.

L. Pasa et al. (Eds.): ICANN 2026, LNCS 17089, pp. 633–644, 2027.
https://doi.org/10.1007/978-3-032-38398-3_52

In this work, we show that threshold tuning and MultiSOL [12] can be expressed within a single optimization framework over threshold-induced decision-region collections. MultiSOL is the multiclass score-oriented training construction in which the loss is evaluated under a probability law on simplex thresholds, so training accounts for a family of threshold-induced decision rules. We identify the common mathematical object behind fixed argmax inference, post-hoc threshold optimization, and threshold-aware training. We formalize this by introducing a score functional on probability measures over threshold vectors: Dirac measures correspond to deterministic threshold-induced decision-region collections, while general measures correspond to distributions over such collections. Within this formulation, argmax is the barycentric Dirac case, threshold tuning is post-hoc optimization over Dirac measures, and MultiSOL is training under a non-degenerate threshold law.

This unified formulation provides three concrete benefits. First, it makes explicit that multiclass decision making can be optimized over simplex decision regions rather than being tied to argmax. Second, it places post-hoc decision-rule selection and score-oriented training in the same functional language, clarifying how their optimization variables differ: the threshold law is optimized after training in threshold tuning, whereas the network parameters are optimized under a prescribed threshold law in MultiSOL. Third, it separates the geometric decision-rule component from the chosen performance score, so that accuracy, macro F1, and other confusion-matrix-based criteria can be discussed within the same decision-region framework.

This paper is structured as follows. Section 2 reviews related approaches. Section 3 introduces multiclass prediction on the simplex and revisits argmax through its simplex decision regions. Section 4 presents threshold-induced regions and their decision-rule interpretation. Section 5 formulates post-hoc threshold tuning as score-oriented optimization on the simplex. Section 6 then places deterministic and distributional threshold selection within a single functional over threshold-induced decision-region collections. Section 7 specializes the distributional case to MultiSOL. Section 8 concludes and points to the experimental validation in [8,12].

2 Related Work

Classical multiclass learning often reduces a multiclass problem to several binary tasks. One-vs-rest, one-vs-one, and coding-based reductions are unified in the margin-classifier framework of [1]. These approaches modify the training or aggregation pipeline, whereas the present paper studies the decision regions induced directly on the probability simplex by a multiclass score vector. Post-hoc calibration methods provide another way to act after training by improving the probabilistic interpretation of classifier outputs [5,14,21]. Calibration is complementary to the present framework: calibrated scores may be used as simplex points, but the threshold-induced decision rule still determines how those points are converted into labels.

Decision rules that account for asymmetric costs or target metrics are also closely related. Cost-sensitive learning formalizes classification when different errors have different penalties [2], and metric-oriented methods optimize objectives such as F1 or other confusion-matrix-based criteria directly [7,15]. In neural-network training, SOL functions define losses from probabilistic confusion matrices for binary classification [11], with subsequent theoretical development for weighted metrics [10]. The AnyLoss construction similarly derives differentiable losses from matrix-based binary metrics by smoothing the thresholding operation [6]. These works share the goal of aligning training or decision making with the evaluation score; the simplex framework considered here makes the multiclass decision regions themselves the common object, so that post-hoc threshold tuning and threshold-distribution training can be expressed through the same functional.

Several recent works also question the default role of argmax in multiclass prediction. Geometry-aware adaptation replaces argmax with a label-geometry-based rule when additional metric information on labels is available [16], while the inflated argmax modifies the argmax decision step to improve stability [17]. In contrast, threshold-induced simplex regions do not require an external label metric and retain a single-label decision rule; the classical argmax is recovered as the barycentric threshold. The threshold-based formulation introduced in [8] uses this geometry for post-hoc multiclass threshold tuning. The MultiSOL construction of [12] then uses distributions over simplex thresholds to train score-oriented multiclass models. The contribution of the present work is to place these deterministic and distributional uses of threshold-induced simplex regions within one optimization functional over probability measures.

3 Multiclass Prediction on the Simplex

Let $\Omega \subset \mathbb{R}^d$ be the input space and let $\{C_1, \dots, C_m\}$ be $m > 2$ classes, represented by one-hot vectors $\{\mathbf{e}_1, \dots, \mathbf{e}_m\}$. A classifier (e.g. a neural network) with parameters θ, evaluated at an input point $x \in \Omega$, outputs

$$\hat{\mathbf{y}}_\theta(x) = (\hat{y}_\theta^1(x), \dots, \hat{y}_\theta^m(x)),$$

typically after a softmax layer. Hence,

$$\sum_{j=1}^{m} \hat{y}_\theta^j(x) = 1, \qquad \hat{y}_\theta^j(x) \geq 0,$$

so each prediction lies in the simplex

$$S_m = \left\{ \mathbf{z} \in \mathbb{R}^m \;\middle|\; \sum_{j=1}^{m} z_j = 1,\; z_j \geq 0 \right\}. \tag{1}$$

The standard multiclass decision rule is

$$j^\star(x) = \underset{j=1,\dots,m}{\arg\max}\, \hat{y}_\theta^j(x).$$

Equivalently, argmax defines the decision regions

$$A_j = \{\mathbf{z} \in S_m \mid z_j \geq z_k,\ \forall k \neq j\}, \qquad j = 1, \ldots, m,$$

which form a partition of S_m up to tie sets of measure zero. Thus, in standard multiclass inference, the classifier maps each input to a point in the simplex, and argmax turns that point into a class label via a single fixed family of decision regions. The rest of the paper examines what changes when that family of decision regions is allowed to vary.

4 Threshold Geometry and Decision Rules

A binary threshold separates the interval $(0, 1)$ into decision regions. The simplex analogue is a threshold vector $\boldsymbol{\tau} = (\tau_1, \ldots, \tau_m) \in S_m$, which induces class regions

$$R_j(\boldsymbol{\tau}) = \{\mathbf{z} \in S_m \mid z_j - z_k > \tau_j - \tau_k,\ \forall k \neq j\}, \qquad j = 1, \ldots, m. \tag{2}$$

or equivalently

$$R_j(\boldsymbol{\tau}) = \{\mathbf{z} \in S_m \mid z_j - \tau_j > z_k - \tau_k,\ \forall k \neq j\}. \tag{3}$$

The sets $R_j(\boldsymbol{\tau})$ leave only tie hyperplanes undecided. On finite data, we complete the rule by a fixed deterministic tie-breaking convention, equivalently by selecting one element of

$$\underset{j=1,\ldots,m}{\arg\max} \left(\hat{y}^j_\theta(x) - \tau_j \right). \tag{4}$$

We write $h_{\boldsymbol{\tau},\theta}(x)$ for this completed decision rule. This convention affects only exactly tied shifted scores and does not change the geometric statements below. Each threshold therefore defines class regions, and hence a multiclass decision rule up to tie sets of measure zero: given $\hat{\mathbf{y}}_\theta(x)$, predict class j when $\hat{\mathbf{y}}_\theta(x) \in R_j(\boldsymbol{\tau})$. The inequalities in (2) are coupled across classes: changing one component of $\boldsymbol{\tau}$ shifts every boundary between that class and the others. As a result, decisions are adjusted through simplex-wide trade-offs rather than through independent one-vs-rest cutoffs. The regions in (2) are pairwise disjoint, and the barycentric threshold

$$\boldsymbol{\tau}_{\text{arg}} = \left(\frac{1}{m}, \ldots, \frac{1}{m} \right) \tag{5}$$

recovers the argmax rule. In fact, if $\mathbf{z} \in R_j(\boldsymbol{\tau}) \cap R_k(\boldsymbol{\tau})$ with $j \neq k$, then

$$z_j - z_k > \tau_j - \tau_k \quad \text{and} \quad z_k - z_j > \tau_k - \tau_j,$$

which is impossible. If $\boldsymbol{\tau} = \boldsymbol{\tau}_{\text{arg}}$, it follows $\tau_j - \tau_k = 0$ for all j, k, so (2) becomes $z_j > z_k$ for all $k \neq j$, which is precisely the argmax decision rule.

Figure 1 illustrates this geometry for $m = 3$.

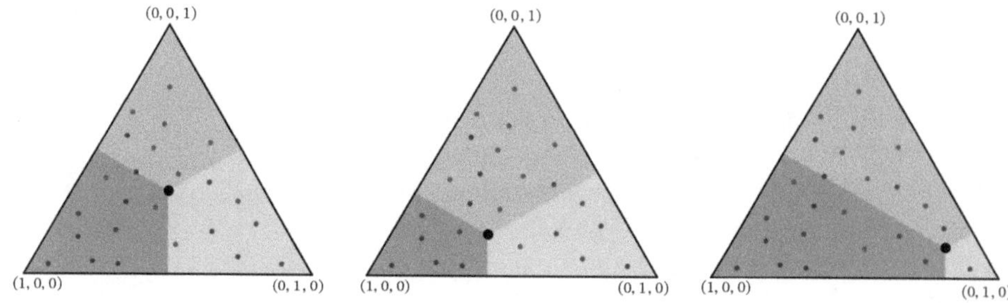

Fig. 1. Threshold vectors induce different decision regions on the 2-simplex and, consequently, different decision rules. The barycenter corresponds to argmax. Figure reproduced from [8].

5 Threshold-Based Optimization

Let $\{(x_i, y_i)\}_{i=1}^n$ be the set of n samples, where $x_i \in \Omega$ denotes the input sample and $y_i \in \{\mathbf{e}_1, \dots, \mathbf{e}_m\}$ its one-hot encoded class label. Once the decision rule depends on $\boldsymbol{\tau}$, performance becomes a function of $\boldsymbol{\tau}$ as well. For each class j, define the one-vs-rest confusion matrix

$$\mathrm{CM}_j(\boldsymbol{\tau}, \theta) = \begin{pmatrix} \mathrm{TN}_j(\boldsymbol{\tau}, \theta) & \mathrm{FP}_j(\boldsymbol{\tau}, \theta) \\ \mathrm{FN}_j(\boldsymbol{\tau}, \theta) & \mathrm{TP}_j(\boldsymbol{\tau}, \theta) \end{pmatrix},$$

with

$$\mathrm{TP}_j(\boldsymbol{\tau}, \theta) = \sum_{i=1}^n \mathbf{1}_{\{y_i = \mathbf{e}_j\}} \mathbf{1}_{\{\hat{\mathbf{y}}_\theta(x_i) \in R_j(\boldsymbol{\tau})\}},$$

$$\mathrm{FP}_j(\boldsymbol{\tau}, \theta) = \sum_{i=1}^n \mathbf{1}_{\{y_i \neq \mathbf{e}_j\}} \mathbf{1}_{\{\hat{\mathbf{y}}_\theta(x_i) \in R_j(\boldsymbol{\tau})\}},$$

$$\mathrm{FN}_j(\boldsymbol{\tau}, \theta) = \sum_{i=1}^n \mathbf{1}_{\{y_i = \mathbf{e}_j\}} \mathbf{1}_{\{\hat{\mathbf{y}}_\theta(x_i) \notin R_j(\boldsymbol{\tau})\}},$$

$$\mathrm{TN}_j(\boldsymbol{\tau}, \theta) = \sum_{i=1}^n \mathbf{1}_{\{y_i \neq \mathbf{e}_j\}} \mathbf{1}_{\{\hat{\mathbf{y}}_\theta(x_i) \notin R_j(\boldsymbol{\tau})\}}.$$

Let s be a binary one-vs-rest score such as precision, recall, F1, or binary accuracy. The corresponding macro one-vs-rest multiclass score is obtained by averaging over classes:

$$s(\boldsymbol{\tau}, \theta) = \frac{1}{m} \sum_{j=1}^m s(\mathrm{CM}_j(\boldsymbol{\tau}, \theta)). \tag{6}$$

This classwise construction should be distinguished from standard multiclass accuracy. When accuracy is used in its usual multiclass sense, as in the left panel of Fig. 2, we compute

$$\mathrm{Acc}(\boldsymbol{\tau}, \theta) = \frac{1}{n} \sum_{i=1}^{n} \mathbf{1}_{\{h_{\boldsymbol{\tau},\theta}(x_i) = \arg\max_{j=1,\dots,m} y_i^j\}}, \tag{7}$$

with the same fixed tie-breaking convention. The same grid-search threshold-tuning procedure can be applied to either type of validation score. In the formal development below, however, $s(\boldsymbol{\tau}, \theta)$ denotes a macro one-vs-rest score of the form (6); standard accuracy is used in Fig. 2 as an illustrative validation score. For a trained model with parameters $\theta^\star$, threshold tuning is then the post-hoc optimization problem

$$\boldsymbol{\tau}^\star \in \arg\max_{\boldsymbol{\tau} \in S_m} s(\boldsymbol{\tau}, \theta^\star). \tag{8}$$

Equation (8) is the multiclass analogue of binary threshold selection. In practice, (8) is solved by evaluating a grid or a random sample of thresholds on the simplex and selecting the best validation score [8]. Since the barycentric threshold belongs to the search space, the tuned rule cannot underperform argmax on the validation set for the chosen score, and whether that gain transfers beyond the validation split depends on the usual requirement that the validation split be informative for the test distribution.

The computational cost of threshold tuning is governed by the number of threshold vectors evaluated. For a validation set of size n, m classes, and G candidate thresholds, a direct evaluation of (8) requires applying the threshold-induced decision rule to all validation predictions for each candidate, with cost on the order of $O(Gnm)$ after the network outputs have been computed. A uniform grid on the simplex becomes rapidly expensive as m grows, because the number of grid points at a fixed resolution increases combinatorially with the simplex dimension $m-1$. For this reason, exhaustive grids are mainly practical in low-dimensional visualizations such as the three-class examples below, whereas random sampling, adaptive search, or structured families of thresholds are more suitable for larger numbers of classes. Once a threshold has been selected, inference remains cheap: predicting with a fixed $\boldsymbol{\tau}$ amounts to applying argmax to the shifted scores $\hat{y}_\theta^j(x) - \tau_j$.

Equation (8) already optimizes over threshold-induced decision-region collections, because each threshold vector determines one such collection. Before passing to probability measures on threshold-induced decision-region collections, Fig. 2 makes the role of $\boldsymbol{\tau}$ explicit for a fixed trained predictor.

The example uses the SOLAR-STORM1 validation set, a three-class dataset of solar active-region cutouts [3,18]. The predictor is a trained Data Efficient Image Transformer classifier [19] with about 86.6M parameters, namely the `deit_base_patch16_224` model considered in [9]. For each threshold vector $\boldsymbol{\tau} \in S_3$, the associated simplex decision rule is applied post hoc to the same network outputs, and the resulting standard multiclass validation accuracy and macro F1 are displayed in the heatmaps. Both scores vary substantially across the simplex, and their maximizers lie relatively far from the barycenter. This highlights how in multiclass classification performance depends not only on the predictor but also on the threshold-induced decision regions used to turn its outputs into labels.

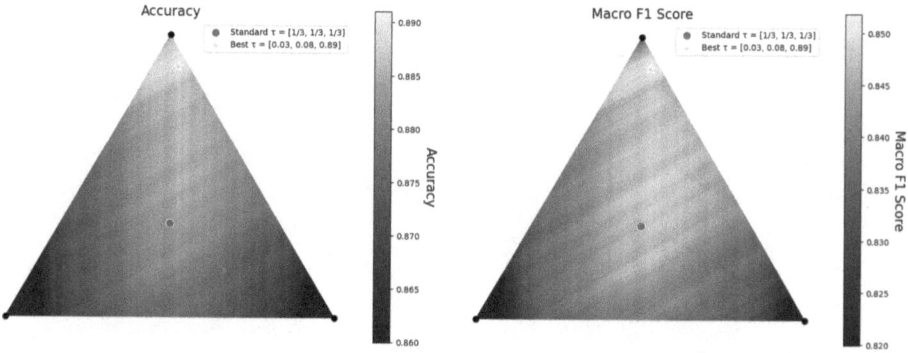

Fig. 2. Accuracy (left) and macro F1 (right) on the SOLAR-STORM1 validation set as functions of the simplex threshold for the trained `deit_base_patch16_224` model. The red circle marks the barycentric argmax threshold, and the green star marks the best validation threshold for the displayed score. In this example, the optima are $(0.00, 0.06, 0.94)$ for accuracy and $(0.01, 0.08, 0.91)$ for macro F1, computed over a grid of 20,301 thresholds. Figure reproduced from [8].

6 Optimization Over Threshold-Induced Decision Regions

The threshold-tuning problem (8) treats a single threshold vector as the optimization variable. The same geometry admits a broader formulation in which the optimization variable is a probability measure over threshold vectors, equivalently over threshold-induced decision-region collections. Let

$$\mathcal{R}(\boldsymbol{\tau}) = \{R_1(\boldsymbol{\tau}), \dots, R_m(\boldsymbol{\tau})\}$$

denote the collection of decision regions associated with $\boldsymbol{\tau}$, and let $\mathcal{P}(S_m)$ be the set of Borel probability measures on S_m. For $\mu \in \mathcal{P}(S_m)$ and each class j, define the aggregated confusion matrix

$$\overline{\mathrm{CM}}_j(\mu, \theta) = \int_{S_m} \mathrm{CM}_j(\boldsymbol{\tau}, \theta)\, d\mu(\boldsymbol{\tau}) = \mathbb{E}_{\boldsymbol{\tau}\sim\mu}[\mathrm{CM}_j(\boldsymbol{\tau}, \theta)]\,, \tag{9}$$

where the integral is taken componentwise. This yields the score functional

$$\Phi_s(\mu, \theta) = \frac{1}{m}\sum_{j=1}^{m} s\big(\overline{\mathrm{CM}}_j(\mu, \theta)\big)\,. \tag{10}$$

Here scores such as precision, recall, and F1 are evaluated on nonnegative real-valued confusion-matrix entries through their usual algebraic formulas. Thus, instead of choosing a single threshold, one may optimize over deterministic or distributional selections of threshold-induced decision-region collections.

Proposition 1. *Let $\delta_{\boldsymbol{\tau}}$ denote the Dirac measure at $\boldsymbol{\tau} \in S_m$. Then, for every score s and parameter vector θ:*

1. $\Phi_s(\delta_{\boldsymbol{\tau}}, \theta) = s(\boldsymbol{\tau}, \theta)$;
2. $\Phi_s(\delta_{\boldsymbol{\tau}_{\text{arg}}}, \theta) = s(\boldsymbol{\tau}_{\text{arg}}, \theta)$, *i.e., the score obtained under argmax;*
3. *for a trained model $\theta^\star$, the threshold-tuning problem (8) is equivalent to*

$$\mu^\star \in \operatorname*{arg\,max}_{\mu \in \{\delta_{\boldsymbol{\tau}} : \boldsymbol{\tau} \in S_m\}} \Phi_s(\mu, \theta^\star);$$

4. *for any threshold law $P \in \mathcal{P}(S_m)$, the distributional objective underlying MultiSOL is $-\Phi_s(P, \theta)$.*

Proof. By (9), one has

$$\overline{\mathrm{CM}}_j(\delta_{\boldsymbol{\tau}}, \theta) = \mathrm{CM}_j(\boldsymbol{\tau}, \theta),$$

so item 1 follows immediately from (6). Item 2 is then obtained by evaluating item 1 at the barycentric threshold $\boldsymbol{\tau}_{\text{arg}}$ and using the argument in Sect. 4. Item 3 is a direct restatement of (8) through item 1. Finally, item 4 follows from (10), since for a general law P the functional is computed from the confusion matrices averaged over thresholds, which is precisely the quantity on which MultiSOL is based.

This proposition isolates the common mathematical object used by the three decision schemes discussed in this paper. Argmax corresponds to a fixed Dirac measure at the barycenter, threshold tuning optimizes over Dirac measures after training, and MultiSOL keeps a general threshold law and optimizes the model parameters instead. When the score s is affine in the entries of the confusion matrix, (10) coincides with the expected macro score under μ. In general, for nonlinear scores such as F1, $\Phi_s(\mu, \theta)$ does not coincide with the expected macro score under μ; it remains a score-oriented functional on aggregated confusion matrices.

7 Learning with Threshold Distributions

Section 6 shows that threshold tuning is the Dirac-measure case of the functional Φ_s. MultiSOL uses the same functional in the distributional setting: a probability law $P \in \mathcal{P}(S_m)$ is fixed on the simplex, and the model parameters are trained through the loss

$$\ell_s(\theta) = -\Phi_s(P, \theta) = -\frac{1}{m} \sum_{j=1}^{m} s(\mathbb{E}_{\boldsymbol{\tau} \sim P}[\mathrm{CM}_j(\boldsymbol{\tau}, \theta)])\,. \tag{11}$$

In other words, while threshold tuning changes only the decision rule applied to already computed network outputs, MultiSOL changes the learned representation itself by training under a distribution of decision rules, extending the binary SOL construction presented in [11] to the multiclass simplex. For scores that are affine in the entries of the confusion matrix, minimizing (11) coincides exactly with maximizing the expected macro score under P; for nonlinear scores, it yields a score-oriented surrogate with the same decision-geometric interpretation [10].

In practice, (11) is approximated by Monte Carlo samples $\boldsymbol{\tau}_1, \ldots, \boldsymbol{\tau}_N \sim P$. Since the condition $\hat{\mathbf{y}}_\theta(x) \in R_j(\boldsymbol{\tau}_r)$ is non-differentiable, this is replaced by the smooth quantity

$$\pi_j(\hat{\mathbf{y}}_\theta(x), \boldsymbol{\tau}_r) = \prod_{k \neq j} \sigma\Big(\lambda(\hat{y}^j_\theta(x) - \hat{y}^k_\theta(x) - \tau_{r,j} + \tau_{r,k})\Big), \qquad (12)$$

where σ is the sigmoid and $\lambda > 0$ controls the steepness of the relaxation. Rather than considering a single threshold-induced decision-region collection, the resulting loss averages soft membership over thresholds. Dirichlet priors are especially natural here because they are supported on the simplex and allow one to control how strongly training emphasizes thresholds near or far from the barycenter.

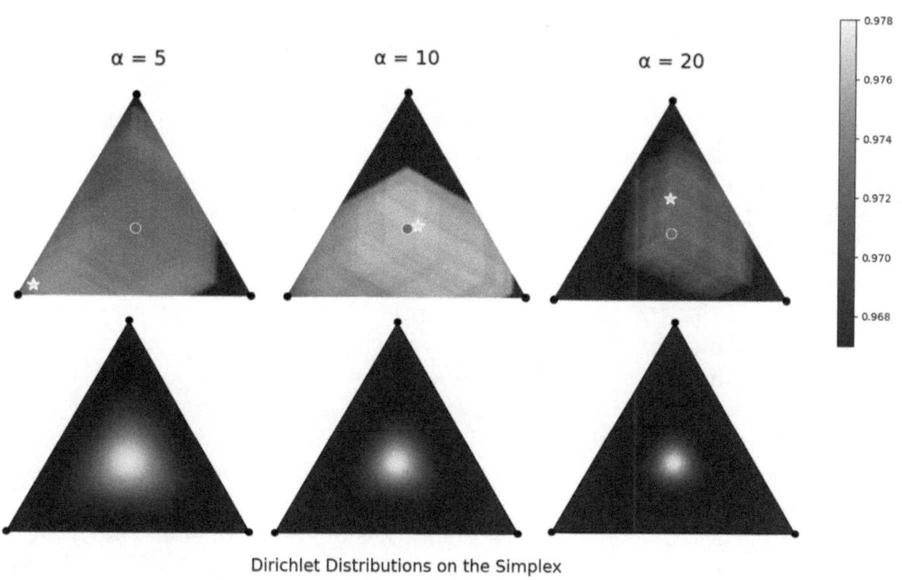

Fig. 3. Changing the Dirichlet prior over thresholds changes the post-hoc accuracy landscape on the simplex for a ResNet10t trained on the three-class FashionMNIST subset T-shirt/top, Trouser, and Pullover. Figure reproduced from [12].

Figure 3 illustrates this dependence on the threshold prior for a three-class FashionMNIST problem restricted to T-shirt/top, Trouser, and Pullover [20].

The trained predictor is a ResNet10t with approximately 4.9M parameters, as in [12]. The bottom row shows Dirichlet priors on the simplex for increasing concentration parameters, while the top row shows the corresponding post-hoc accuracy landscapes after training. The location and extent of the high-accuracy regions change with the prior. In other words, training with a distribution of thresholds shapes which threshold-induced decision regions become effective at inference time.

This dependence also highlights that MultiSOL defines a family of threshold-aware training objectives indexed by the threshold law P and the smoothing parameter λ. These choices determine which threshold-induced decision-region collections are emphasized during optimization and how sharply region membership is approximated.

8 Conclusion

The simplex viewpoint provides a geometric interpretation of multiclass classification: a network predicts a point in the simplex, and the decision rule chooses class regions on that simplex. Within this interpretation, argmax is recovered from the barycentric threshold, threshold tuning is the post-hoc optimization of a single threshold-induced decision-region collection for a fixed predictor, and MultiSOL is optimization during training under a distribution of such collections. Relative to [8], which introduced the threshold-based multiclass decision rule and its post-hoc tuning, and to [12], which introduced MultiSOL as a score-oriented training construction on the simplex, the contribution of the present paper is to place these two developments within a single optimization framework over threshold-induced decision-region collections. More precisely, the paper identifies a common functional on probability measures over such collections: argmax is recovered by evaluating that functional at the barycentric Dirac measure, threshold tuning corresponds to optimizing it over Dirac measures for a fixed predictor, and MultiSOL corresponds to optimizing model parameters under a prescribed threshold law. Thus both constructions follow from the same functional by fixing different arguments: post-hoc threshold tuning fixes the trained parameters $\theta^\star$ and optimizes the measure variable within the class of Dirac measures, whereas MultiSOL fixes a threshold law P and optimizes the model parameters θ during training. This perspective makes threshold geometry a common language for multiclass decision making, post-hoc score optimization, and score-oriented learning. The present framework is intended for single-label classifiers whose outputs lie in S_m and whose decisions can be written as shifted-score comparisons. It directly covers macro one-vs-rest scores through classwise confusion matrices; micro-averaged scores can be handled by replacing the outer class average with a global confusion-matrix functional, whereas hierarchical or structured losses require a task-specific score functional. In problems with many classes, exhaustive threshold grids are limited by the combinatorial growth of the simplex search space, so practical use should rely on random, adaptive, or structured threshold families and on held-out validation to control overfitting, especially under class imbalance. Empirical validation of the threshold-tuning and MultiSOL constructions can be found in [8,12].

References

1. Allwein, E.L., Schapire, R.E., Singer, Y.: Reducing multiclass to binary: a unifying approach for margin classifiers. J. Mach. Learn. Res. **1**, 113–141 (2000)
2. Elkan, C.: The foundations of cost-sensitive learning. In: Proceedings of the Seventeenth International Joint Conference on Artificial Intelligence, pp. 973–978 (2001)
3. Fang, Y., Cui, Y., Ao, X.: Deep learning for automatic recognition of magnetic type in sunspot groups. Adv. Astron. **2019**(1), 9196234 (2019)
4. Fawcett, T.: An introduction to ROC analysis. Pattern Recogn. Lett. **27**(8), 861–874 (2006). https://doi.org/10.1016/j.patrec.2005.10.010
5. Guo, C., Pleiss, G., Sun, Y., Weinberger, K.Q.: On calibration of modern neural networks. In: Proceedings of the 34th International Conference on Machine Learning. Proceedings of Machine Learning Research, vol. 70, pp. 1321–1330 (2017)
6. Han, D., Moniz, N., Chawla, N.V.: Generation of loss functions from matrix-based binary classification metrics. ACM Trans. Knowl. Discov. Data **20**(5), 1–40 (2026). https://doi.org/10.1145/3805705
7. Joachims, T.: A support vector method for multivariate performance measures. In: Proceedings of the 22nd International Conference on Machine Learning, pp. 377–384 (2005). https://doi.org/10.1145/1102351.1102399
8. Legnaro, E., Guastavino, S., Marchetti, F.: Multiclass threshold-based classification and model evaluation. CoRR abs/2511.21794 (2025). https://doi.org/10.48550/arXiv.2511.21794
9. Legnaro, E., Guastavino, S., Piana, M., Massone, A.M.: Deep learning for active region classification: a systematic study from convolutional neural networks to vision transformers. Astrophys. J. **981**(2), 157 (2025)
10. Marchetti, F., Guastavino, S., Campi, C., Benvenuto, F., Piana, M.: A comprehensive theoretical framework for the optimization of neural networks classification performance with respect to weighted metrics. Optim. Lett. **19**(1), 169–192 (2025). https://doi.org/10.1007/s11590-024-02112-1
11. Marchetti, F., Guastavino, S., Piana, M., Campi, C.: Score-oriented loss (SOL) functions. Pattern Recogn. **132**, 108913 (2022). https://doi.org/10.1016/j.patcog.2022.108913
12. Marchetti, F., Legnaro, E., Guastavino, S.: The multiclass score-oriented loss (MultiSOL) on the simplex. CoRR abs/2511.22587 (2025). https://doi.org/10.48550/arXiv.2511.22587
13. Murphy, K.P.: Machine Learning: A Probabilistic Perspective. MIT Press, Cambridge (2012)
14. Niculescu-Mizil, A., Caruana, R.: Predicting good probabilities with supervised learning. In: Proceedings of the 22nd International Conference on Machine Learning, pp. 625–632 (2005). https://doi.org/10.1145/1102351.1102430
15. Puthiya Parambath, S.A., Usunier, N., Grandvalet, Y.: Optimizing F-measures by cost-sensitive classification. In: Advances in Neural Information Processing Systems, vol. 27, pp. 2123–2131 (2014)
16. Roberts, N., et al.: Geometry-aware adaptation for pretrained models. Adv. Neural. Inf. Process. Syst. **36**, 49628–49658 (2023)

17. Soloff, J., Barber, R., Willett, R.: Building a stable classifier with the inflated argmax. Adv. Neural. Inf. Process. Syst. **37**, 70349–70380 (2024)
18. Tianchi Datasets: SOLAR-Automatic Recognition of Magnetic Type in Sunspot Groups Dataset (2020). https://tianchi.aliyun.com/dataset/74779. Accessed 30 Mar 2026
19. Touvron, H., Cord, M., Douze, M., Massa, F., Sablayrolles, A., Jégou, H.: Training data-efficient image transformers & distillation through attention. In: International Conference on Machine Learning, pp. 10347–10357. PMLR (2021)
20. Xiao, H., Rasul, K., Vollgraf, R.: Fashion-MNIST: a novel image dataset for benchmarking machine learning algorithms. arXiv preprint arXiv:1708.07747 (2017)
21. Zadrozny, B., Elkan, C.: Transforming classifier scores into accurate multiclass probability estimates. In: Proceedings of the Eighth ACM SIGKDD International Conference on Knowledge Discovery and Data Mining, pp. 694–699 (2002). https://doi.org/10.1145/775047.775151

Understanding Fisher SAM: Scale-Invariant Sharpness, Generalization, and Convergence

Yingcong Zhou[1], Li Wang[2], Pingfan Wu[1], and Zhiguo Fu[1](✉)

[1] School of Information Science and Technology, AI4S Center, Northeast Normal University, Changchun, Jilin 130117, China
{zhouyc821,wupingfan994,fuzg432}@nenu.edu.cn

[2] Key Laboratory of Education Blockchain and Intelligent Technology, Ministry of Education, Guangxi Normal University, Guilin 541004, Guangxi, China
wangl028@gxnu.edu.cn

Abstract. Sharpness-aware minimization (SAM) improves generalization by seeking flat minima via local perturbations, but its definition of sharpness is limited to the Euclidean space, which may fail to capture the intrinsic geometry of overparameterized models. To address this issue, recent work introduces Fisher information to construct geometry-aware perturbations and propose Fisher SAM. However, the theoretical properties of such Fisher-based sharpness remain insufficiently understood, limiting our understanding of its role in generalization. In this paper, we provide a theoretical analysis of Fisher-based sharpness from an statistical manifolds perspective. We show that this sharpness is inherently scale-invariant with respect to parameter rescaling, without requiring any explicit normalization. We further derive a generalization bound under weaker assumptions, which reveals an additional structural factor involving the product of the parameter norm and the trace of the Fisher information matrix, highlighting their joint influence on generalization beyond sharpness alone. Building on this insight, we develop a sharpness-aware optimization method with a theory-driven regularization term that is absent in Fisher SAM, and establish convergence guarantees for the proposed algorithm. Experiments on benchmark datasets demonstrate improved generalization performance and consistently flatter minima.

Keywords: Generalization Theory · Fisher Sharpness-Aware Minimization · Scale Invariance · Optimization

1 Introduction

Modern deep neural networks are typically highly overparameterized and achieve remarkable generalization performance across a wide range of domains [12,19].

Our code is public: https://github.com/pingfanw/Sharpness-Aware-Minimization-on-Statistical-Manifold.

Supplementary Information The online version contains supplementary material available at https://doi.org/10.1007/978-3-032-38398-3_53.

L. Pasa et al. (Eds.): ICANN 2026, LNCS 17089, pp. 645–657, 2027.
https://doi.org/10.1007/978-3-032-38398-3_53

Despite their ability to fit training data with near-zero error, different solutions with similar training loss can exhibit significantly different generalization behaviors. A growing body of empirical and theoretical studies suggests that this phenomenon is closely related to the geometry of the loss landscape, where flatter minima tend to generalize better than sharper ones [1,9,16]. Motivated by this connection, Sharpness-aware minimization (SAM) [3] have been proposed and achieved state-of-the-art results on various tasks. However, a fundamental issue arises from the definition of sharpness in SAM it is sensitive to parameter rescaling, raising concerns about its reliability [2].

To address this scale dependence, several recent studies have been carried out [7,10,20]. For example, the invariant sharpness measure proposed by [17] is obtained by compensating for the changes in the loss Hessian through multiplication with the scalar products of weight matrices. Specifically, [13] introduced the concept of Adaptive Sharpness-Aware Minimization(ASAM), which adds the normalization operator in the parameter space that cancels out the effect of any scaling operator. However, this normalization technique forcibly changes the shape of the parameter space to adapt to the scale changes, thus destroys the true structure of the parameter space.

Existing sharpness measures are typically defined in the Euclidean space. However, many deep learning tasks such as image classification use the cross-entropy loss or negative log-likelihood as the loss function, which require us to measure the difference of the distribution of the parameters on the statistical manifolds rather than the Euclidean norm of the parameters themselves. In this setting, Euclidean perturbations may fail to capture the intrinsic geometry of the model. To address this mismatch, [11] defines sharpness in a geometry-aware manner using Fisher information matrix (FIM), and develops Fisher SAM, which has been shown to improve generalization.

While this algorithm better aligns with the intrinsic structure of the model, the theoretical properties remain insufficiently understood. In particular, it is unclear whether Fisher-based sharpness resolves scale dependence, how it relates to generalization performance under weaker conditions, and whether the Fisher SAM admits rigorous convergence guarantees. In this paper, we provide the theoretical analysis of Fisher SAM to address these questions. The main contributions are summarized as follows,

- We derive an explicit formulation of Fisher-based sharpness, providing a concrete characterization that facilitates theoretical analysis.
- We show that Fisher-based sharpness is inherently invariant to parameter rescaling, without requiring explicit normalization or reparameterization, providing a theoretical justification that is absent in prior work.
- We derive a generalization error bound under weak conditions, and reveal an additional structural factor involving the product of the parameter ℓ_2 norm and the trace of the Fisher matrix. This result shows that generalization is governed not only by sharpness, but also by their joint interaction.
- Motivated by the above analysis, we develop a sharpness-aware optimization method with a novel regularization term induced by the generalization bound, and establish its theoretical convergence guarantees.

- We validate the proposed method on standard benchmark datasets, demonstrating improved generalization and consistently flatter minima.

2 Preliminaries

Consider a model $f : \mathcal{X} \to \mathcal{Y}$ parametrized by a weight vector $\theta \in \Theta \subseteq \mathbb{R}^k$ and a loss function $l : \Theta \times \mathcal{X} \times \mathcal{Y} \to \mathbb{R}_+$. Given a training data set $S = \{(x_1, y_1), (x_2, y_2), \cdots, (x_n, y_n)\}$ drawn i.i.d from the distribution $\mathcal{D}$, the training loss can be defined as $L_\mathcal{S}(\theta) = \sum_{i=1}^n l(y_i, f(x_i, \theta))/n$. The expected loss of a model is $L_\mathcal{D}(\theta) = \mathbb{E}_{(x,y)\sim\mathcal{D}}[l(y_i, f(x_i, \theta))]$, and the generalization performance of the model can be measured by the generalization gap $L_\mathcal{D}(\theta) - L_\mathcal{S}(\theta)$.
Sharpness-Aware Minimization (SAM). SAM measures sharpness of a solution θ via the worst-case loss increase within a Euclidean neighborhood [3]:

$$\max_{\|\epsilon\|_p \leq \rho} L_\mathcal{S}(\theta + \epsilon) - L_\mathcal{S}(\theta), \tag{1}$$

where $p \geq 1, \rho > 0$ and ϵ is the perturbation of the parameters. However, this definition relies on Euclidean perturbations and is sensitive to parameter rescaling, lacking scale invariance, i.e.

$$\max_{\|\epsilon\|_p \leq \rho} L_\mathcal{S}(\theta + \epsilon) \neq \max_{\|\epsilon\|_p \leq \rho} L_\mathcal{S}(A\theta + \epsilon),$$

where A is a scaling operator on the weight space that does not change the loss function, i.e. $L_\mathcal{S}(A\theta) = L_\mathcal{S}(\theta)$.
Adaptive Sharpness-Aware Minimization (ASAM). In order to overcome the scale dependency of (1), [13] introduced the adaptive sharpness which adds the normalization operator in the parameter space that cancels out the effect of A. In details, for any invertible scaling operator A, $T_{A\theta}^{-1} A = T_\theta^{-1}$ holds, where T_θ^{-1} is a normalization operator of θ, $\{T_\theta, \theta \in \mathbb{R}^k\}$ is a family of invertible linear operators on $\mathbb{R}^k$. And the adaptive sharpness is defined by

$$\max_{\|T_\theta^{-1}\epsilon\|_p \leq \rho} L_\mathcal{S}(\theta + \epsilon) - L_\mathcal{S}(\theta),$$

where $1 \leq p \leq \infty$. Although it achieves scale invariance, [11] points out that ASAM's parameter magnitude-scaled neighborhood choice is rather ad hoc, not fully reflecting the underlying geometry of the parameter manifold.
Fisher Sharpness-Aware Minimization (Fisher SAM). To better capture the underlying structure of deep models, [11] consider the distribution of the parameters, perform a perturbation on statistical manifolds of this distribution, and then use the KL divergence to quantify the difference between the original and perturbed model. The result can be stated by the following lemma.

Lemma 1. *[11] Given the distribution of the parameters $q(y|x, \theta)$, the KL divergence between $q(y|x, \theta)$ and its perturbation model $q(y|x, \theta + \epsilon)$ is*

$$\mathrm{KL}(q(y|x, \theta) \parallel q(y|x, \theta + \epsilon)) \approx \frac{1}{2}\epsilon^\top F(\theta)\epsilon, \tag{2}$$

where $F(\theta) = \mathbb{E}_{q(y|x,\theta)}\left[\nabla_\theta \log q(y|x,\theta) \nabla_\theta \log q(y|x,\theta)^\top\right]$ *is the Fisher information matrix (FIM).*

The sharpness in terms of the FIM is defined as follows.

Definition 1. *[11] The sharpness metric of the parameter* θ *is defined by*

$$\max_{\epsilon^T F(\theta)\epsilon \leq \rho^2} L_{\mathcal{S}}(\theta + \epsilon) - L_{\mathcal{S}}(\theta), \tag{3}$$

where the optimal perturbation is given by

$$\epsilon^* = \rho \frac{F(\theta)^{-1} \nabla_\theta L_{\mathcal{S}}(\theta)}{\sqrt{\nabla_\theta L_{\mathcal{S}}(\theta)^T F(\theta)^{-1} \nabla_\theta L_{\mathcal{S}}(\theta)}}. \tag{4}$$

Accordingly, Fisher SAM solves the following minimax optimization problem

$$\min_\theta \max_{\epsilon^T F(\theta)\epsilon \leq \rho^2} L_{\mathcal{S}}(\theta + \epsilon), \tag{5}$$

While Fisher SAM provides a geometry-aware formulation, a systematic theoretical understanding of its properties is still lacking. This motivates the theoretical analysis in this paper.

3 Fisher-Based Sharpness: Formulation and Scale Invariance

Geometric Motivation. To better understand the necessity of geometry-aware perturbations, we provide an intuitive example illustrating the discrepancy between Euclidean distance and statistical manifold distance.

Example 1. In Fig. 1, we compare distances measured by the Euclidean metric and the statistical manifold metric. Under the statistical manifold distance, data points can appear substantially farther apart, as it captures the underlying data geometry and distributional structure. In contrast, the Euclidean distance only accounts for coordinate-wise differences and ignores correlations across dimensions, which may lead to misleading proximity relationships. This example highlights that, in deep learning, measuring perturbations in terms of distributional divergence is more appropriate than relying solely on Euclidean distance. More examples can be found in Appendix A.1.

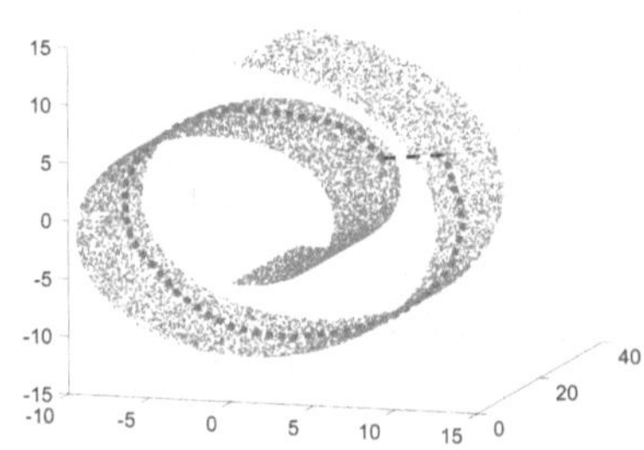

Fig. 1. Comparison of the Euclidean distance (blue line) and the statistical manifold distance (orange curve). The data are generated by mapping uniformly sampled points in a rectangular region onto a nonlinear "Swiss Roll" surface embedded in 3D space. (Color figure online)

Give the weighted Euclidean norm (also known as the Mahalanobis distance) $\|x\|_A = \sqrt{x^T A x}$, which measures the length or the magnitude of x in the "norm" induced by the matrix A. Recall that Definition 1 essentially considers the robustness of the distribution of the parameters. We give the explicit formulation of Fisher-based sharpness.

Lemma 2. *The sharpness can be computed by*

$$\max_{\epsilon^T F(\theta)\epsilon \leq \rho^2} L_{\mathcal{S}}(\theta+\epsilon) - L_{\mathcal{S}}(\theta) = \rho \, \|\nabla_\theta L_{\mathcal{S}}(\theta)\|_{F(\theta)^{-1}}. \tag{6}$$

The Lemma 2 (proof is shown in Appendix A.2) shows that this metric is governed by the weighted Euclidean norm of $\nabla_\theta L_{\mathcal{S}}(\theta)$, which induced by the FIM.

Comparison with SAM. Compared with the perturbation introduced by [3], which lies in a rigid region with a fixed radius $\epsilon^* = \rho \frac{\nabla_\theta L_{\mathcal{S}}(\theta)}{\|\nabla_\theta L_{\mathcal{S}}(\theta)\|_2}$. We show that, in particular, if $F(\theta)^{-1}$ is the identity matrix, then

$$\|\nabla_\theta L_{\mathcal{S}}(\theta)\|_{F(\theta)^{-1}} = \nabla_\theta L_{\mathcal{S}}(\theta)^T \nabla_\theta L_{\mathcal{S}}(\theta) = \|\nabla_\theta L_{\mathcal{S}}(\theta)\|_2,$$

which is the standard Euclidean norm. The Fisher-based sharpness recovers the sharpness of SAM as a special case. By Cramér-Rao low bound, we know that in the one-dimensional case, a larger value of $F(\theta)$ corresponds to a smaller variance. Thus, [3] actually shows that the sharpness metric is based on an isotropic covariance matrix, that is, it has the same perturbation in any dimension. However, the sharpness in [11] is based on an anisotropic covariance matrix, which means that dimensions with high variance have a higher weight, while those dimensions with low variance have a lower weight. Thus Fisher-based metric can suit the parameter rescaling. More importantly, in the context of deep learning, the importance of each dimension of the parameter is not consistent. The Fisher-based sharpness takes into account the correlations between different dimensions of the parameter space by the Fisher matrix and can therefore capture the underlying structure of the parameter space more accurately.

Scale Invariance of Fisher-based Sharpness. We now show that the sharpness described in Definition 1 is inherently scale-invariant.

Theorem 1. *For any invertible scaling operator A, which does not change the loss function, i.e. $L_{\mathcal{S}}(A\theta) = L_{\mathcal{S}}(\theta)$, the values of the sharpness at θ and $A\theta$ are the same as*

$$\max_{\epsilon^T F(\theta)\epsilon \leq \rho^2} L_{\mathcal{S}}(\theta+\epsilon) - L_{\mathcal{S}}(\theta) = \max_{\epsilon^T F(A\theta)\epsilon \leq \rho^2} L_{\mathcal{S}}(A\theta+\epsilon) - L_{\mathcal{S}}(A\theta).$$

The proof can be found Appendix A.3.

Compared with ASAM. Compared with the perturbation in [13], which lies in an adaptive ellipsoidal region $\epsilon^* = \rho \frac{T_\theta^2 \nabla_\theta L_{\mathcal{S}}(\theta)}{\|T_\theta \nabla_\theta L_{\mathcal{S}}(\theta)\|_p}$. We show that, in particular, if $p = 2$ and T_θ is set to $F(\theta)^{-\frac{1}{2}}$, the Fisher-based sharpness aligns with [13]. Essentially, the FIM acts as a normalization operator, adaptively adjusting the perturbation region. However, imposing the normalization T_θ, intuitively, will forcibly alter the shape of the parameter space to accommodate scale changes, and ultimately undermine the true structure of the parameter space and loss of the original correlation between different dimensions of the parameters. The Fisher-based sharpness does not require any transformation of the parameter space and its scale invariance is inherent. We give a simple but persuasive example to verify that the Fisher-based sharpness is scale-invariant (Appendix A.4).

4 Fisher Geometry, Sharpness, and Generalization

In Sect. 3, we prove the Fisher-based sharpness is inherently scale-invariant. In this section, we establish its connection to generalization by deriving a novel generalization bound under weaker assumptions.

Generalization Bound. Existing analyses in [11] establish generalization guarantees under relatively restrictive assumptions, and primarily attribute generalization performance to sharpness. The following result extends this connection to a more general setting.

Theorem 2. *For the training set $S \sim \mathcal{D}$, the number of samples $n \in \mathbb{N}$, $\sigma > 0$ and $\delta > 0$, if $L_\mathcal{D}(\theta) \leq \mathbb{E}_\epsilon[L_\mathcal{D}(\theta+\epsilon)]$, where $\epsilon \sim \mathcal{N}(0, \sigma^2 F(\theta)^{-1})$, then with probability $1-\delta$, we have the following generalization bound*

$$L_\mathcal{D}(\theta) \leq \max_{\epsilon^T F(\theta)\epsilon \leq \rho^2} L_\mathcal{S}(\theta+\epsilon) + \frac{1}{\sqrt{n}} + \sqrt{\frac{\|\theta\|^2 \cdot \operatorname{tr}(F(\theta)) \left(1+\sqrt{\log n}\right)^2 / \rho^2 + kc + 2\log\frac{n}{\delta}}{4(n-1)}}, \tag{7}$$

where k is the number of parameters, c is a positive constant, $\rho = \sigma\left(1+\sqrt{\ln(n)}\right)$.

The proof is an extension of [3,13]. The condition $L_\mathcal{D}(\theta) \leq \mathbb{E}_\epsilon[L_D(\theta+\epsilon)]$ is a general assumption which means that adding Gaussian perturbation $\epsilon \sim \mathcal{N}(0, \sigma^2 F(\theta)^{-1})$ should not decrease the test error. More details can be found in Appendix A.5. We now provide the following simplified form of Theorem 2.

Corollary 1. *Under the same conditions as Theorem 2, the generalization bound satisfies*

$$L_\mathcal{D}(\theta) \leq \max_{\epsilon^T F(\theta)\epsilon \leq \rho^2} L_\mathcal{S}(\theta+\epsilon) + h\left[\|\theta\|^2 \cdot \operatorname{tr}(F(\theta))\right], \tag{8}$$

where $h(\cdot)$ is a strictly increasing function with respect to $\|\theta\|^2 \cdot \operatorname{tr}(F(\theta))$.

Remark on the Generalization Error Bound. To better understand the role of sharpness in Theorem 2, we rewrite the bound in (8) as:

$$L_\mathcal{D}(\theta) \leq \underbrace{\left[\max_{\epsilon^T F(\theta)\epsilon \leq \rho^2} L_\mathcal{S}(\theta+\epsilon) - L_\mathcal{S}(\theta)\right]}_{\text{sharpness}} + L_\mathcal{S}(\theta) + h\left[\|\theta\|^2 \cdot \operatorname{tr}(F(\theta))\right].$$

This decomposition highlights that generalization is controlled by three factors: the empirical loss, the sharpness, and a complexity term. Importantly, compared with the generalization bounds in [11], Theorem 2 reveals an additional factor $\|\theta\|^2 \cdot \mathrm{tr}(F(\theta))$ that is absent in prior analyses. This result indicates that sharpness alone is insufficient to fully characterize generalization. From a geometric perspective, while the sharpness term captures local sensitivity of the loss landscape, this additional term depends jointly on the parameter scale and the global curvature encoded by the FIM.

Empirical Validation. To support the theoretical findings, we analyze the correlation between sharpness and generalization in Fig. 2. The results show that the Fisher-based sharpness metric exhibits a stronger and more consistent correlation with generalization compared to SAM and ASAM. In particular, lower sharpness values correspond more reliably to better generalization performance, supporting the theoretical advantage of the Fisher-based sharpness in capturing the relationship between flatness and generalization.

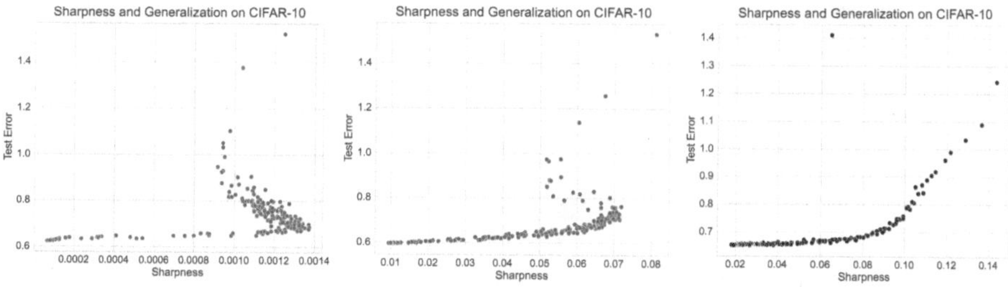

Fig. 2. The correlations between sharpness and generalization. The sharpness values are generated by SAM (left), ASAM (center), and SMSAM (right), respectively. These experiments are conducted on CIFAR-10 for WideResNet-28-6, and the experimental settings align with those described in Sect. 6.1

5 SMSAM: Statistical Manifolds-Based Sharpness-Aware Minimization

Motivated by Corollary 1, we propose a new algorithm, termed Statistical Manifold Sharpness-Aware Minimization (SMSAM), which integrates both Fisher-based sharpness and the complexity term $\|\theta\|^2 \cdot \mathrm{tr}(F(\theta))$. To explicitly control this factor, we incorporate it as a regularizer ($\lambda > 0$ is a regularization parameter) and formulate the following minimax problem:

$$\min_{\theta} \max_{\epsilon^T F(\theta)\epsilon \leq \rho^2} L_{\mathcal{S}}(\theta + \epsilon) + \frac{\lambda}{2}\|\theta\|^2 \cdot \mathrm{tr}\left(F(\theta)\right). \tag{9}$$

Compared with Fisher SAM, which only focuses on geometry-aware perturbations, SMSAM additionally incorporates an explicit complexity regularization

Algorithm 1. SMSAM algorithm

Input: Loss function $l : \mathcal{W} \times \mathcal{X} \times \mathcal{Y} \rightarrow \mathbb{R}_+$, training dataset $\mathcal{S} := \cup_{i=1}^{n}\{(x_i, y_i)\}$, Mini-batch size b, Perturbation radius ρ, Regularization parameter λ, Learning rate η.
Output: Trained weight θ.
Initialize weight θ_0, $t = 0$.

while not converged **do**
  Compute gradient $\nabla_\theta L_\mathcal{B}(\theta_t)$ of the batch's training loss;
  Compute ϵ_t per equation (10);
  Compute mini-batch gradient approximation for the SAM objective: $g(\theta_t) = \nabla_\theta \left[L_\mathcal{B}(\theta_t + \epsilon_t) + \frac{\lambda}{2}\|\theta_t\|^2 \cdot \|\nabla_\theta L_\mathcal{B}(\theta)\|^2\right]$;
  Update weights: $\theta_{t+1} = \theta_t - \eta g(\theta_t)$;
  $t = t + 1$;
end while
return θ_t.

term derived from the generalization bound, leading to a fundamentally different optimization objective. Then, the two-step procedure of SMSAM is as follows,

$$\epsilon_t = \rho \frac{F(\theta_t)^{-1}\nabla_\theta L_\mathcal{S}(\theta_t)}{\sqrt{\nabla_\theta L_\mathcal{S}(\theta_t)^T F(\theta_t)^{-1}\nabla_\theta L_\mathcal{S}(\theta_t)}}, \tag{10}$$

$$\theta_{t+1} = \theta_t - \eta\nabla_\theta \left[L_\mathcal{S}(\theta_t + \epsilon_t) + \frac{\lambda}{2}\|\theta_t\|^2 \cdot \mathrm{tr}\left(F(\theta_t)\right)\right]. \tag{11}$$

The main computational challenges of SMSAM are computing $F(\theta)^{-1}$ in (10) and $\mathrm{tr}(F(\theta))$ in (11). Regarding the computation of tr(F(θ)), it can be expressed as $\mathrm{tr}(F(\theta)) = \mathbb{E}_{x\sim\mathcal{X},\hat{y}\sim p(y|x;\theta)}\left[\|\frac{\partial}{\partial\theta}L(x,\hat{y})\|_2^2\right]$. In the preliminary experiments, [8] found that we can use the norm of the expected gradient rather than the expected norm of the gradient to give a more direct approximations of $\mathrm{tr}(F(\theta))$, i.e., $\mathrm{tr}(F_\mathcal{B}(\theta)) \approx \left\|\frac{1}{\mathcal{B}}\sum_{i=1}^{\mathcal{B}} \nabla l_i(\theta)\right\|^2 = \|\nabla_\theta L_\mathcal{B}(\theta)\|^2$. To address the computation of $F(\theta)^{-1}$, we utilize the well-known Kronecker-Factored Approximate Curvature method (K-FAC) proposed by [15]. In details, it is derived by approximating various large blocks of the FIM (corresponding to entire layers) as being the Kronecker product of two much smaller matrices. Then, we design an algorithm to compute ϵ_t (see details and computational overhead in Appendix A.6).

Now we can gives the pseudo-code for the full SMSAM algorithm as shown in Algorithm 1. The base optimizer used in this algorithm is SGD. We note that each SAM weight update requires two back propagation operations, i.e., one to compute ϵ_t and another to compute the final gradient. To provide an intuitive understanding of the advantages and disadvantages of SMSAM, we compare it with SAM and ASAM on a toy model and visualize the optimization trajectories of all three methods. Detailed descriptions are provided in Appendix A.8.

Convergence Analysis. Next, we establish a theoretical convergence guarantee for SMSAM under standard assumptions (see Appendix A.7), which are common for nonconvex stochastic optimization.

Theorem 3. *Let $\nabla L_{\mathcal{S}}(\theta_t)$ and $\nabla L_{\mathcal{B}}(\theta_t)$ satisfy the assumptions in Appendix A.7. Choosing $\eta_t = \frac{\eta_0}{\sqrt{T}} \leq \frac{2}{3\beta}$ and $\rho_t = \frac{\rho_0}{\sqrt{T}}$, then Algorithm 1 ensures that*

$$\frac{1}{T}\sum_{t=0}^{T-1} \mathbb{E}[\|\nabla L_{\mathcal{S}}(\theta_t)\|^2] \leq O\left(\frac{\tilde{\sigma}^2}{\sqrt{T}}\right), \quad \frac{1}{T}\sum_{t=0}^{T-1} \mathbb{E}[\|\nabla L_{\mathcal{S}}(\theta_t + \epsilon_t)\|^2] \leq O\left(\frac{\tilde{\sigma}^2}{\sqrt{T}}\right).$$

We provide the proof in Appendix A.7. Theorem 3 shows that Algorithm 1 achieves the convergence rate $\mathcal{O}(1/\sqrt{T})$. To clarify the relationship between SMSAM and Fisher SAM, we summarize their key differences in Table 1.

Table 1. Comparison between Fisher SAM and the proposed SMSAM.

Aspect	Fisher SAM	SMSAM (Ours)
Sharpness formulation	Fisher-based	Same (revisited and analyzed)
Scale invariance	Not analyzed	Explicitly established
Generalization bound	sharpness	sharpness and $\|\theta\|^2 \cdot \mathrm{tr}(F(\theta))$
Regularization term	Not included	$\|\theta\|^2 \cdot \mathrm{tr}\,(F(\theta))$
Fisher approximation	Empirical diagonal minibatch	K-FAC approximation
Convergence guarantee	Not provided	$\mathcal{O}(1/\sqrt{T})$ established

6 Experimental Results

In this section, we validate the effectiveness of SMSAM through numerical experiments on various rectifier neural networks and datasets. Firstly, we evaluate the impact on the model generalization performance and demonstrate that SMSAM effectively enhances the generalization performance. Secondly, we employ visualization techniques to analyze the loss landscapes of models trained using SMSAM. The results illustrate that SMSAM can lead to flatter minima.

6.1 Standard Training

To evaluate generalization performance, we compare the best test accuracies of SAM, ASAM, FisherSAM, and SMSAM on CIFAR-10/100 across various rectifier networks, including ResNet [5], WideResNet [21], DenseNet [6], PyramidNet [4], and VGG [18]. The results are presented in Table 2. Experimental details and additional experimental results are see Appendix A.9.1 and A.10, respectively.

6.2 Transfer Learning

Transfer learning shows the generalization of models when trained on sufficient labeled data and finetuned on a novel dataset. We use SGD, ASAM, FisherSAM and SMSAM to train ViTs (pre-trained on ImageNet 1K). Results are shown in Table 3, with details in Appendix A.9.2.

Table 2. Comparison of test accuracy (%) on CIFAR-10/100 datasets.

CIFAR-10	SAM	ASAM	FisherSAM	SMSAM
ResNet-20	89.41$\pm$0.11	90.07$\pm$0.07	89.18$\pm$0.03	**90.32**$\pm$0.03
ResNet-56	90.54$\pm$0.09	91.18$\pm$0.11	91.08$\pm$0.03	**91.28**$\pm$0.06
WideResNet-28-6	94.25$\pm$0.08	93.94$\pm$0.04	93.81$\pm$0.02	**94.44**$\pm$0.11
WideResNet-58-2	92.67$\pm$0.05	93.04$\pm$0.08	93.09$\pm$0.03	**93.41**$\pm$0.14
DenseNet-121	76.19$\pm$0.08	72.14$\pm$0.26	80.88$\pm$0.1	**84.25**$\pm$0.15
VGG19-BN	87.91$\pm$0.14	90.07$\pm$0.11	**91.44**$\pm$0.09	91.32$\pm$0.04
Pyramidnet-164	91.71$\pm$0.19	92.46$\pm$0.22	92.44$\pm$0.13	**93.29**$\pm$0.17
CIFAR-100	SAM	ASAM	FisherSAM	SMSAM
ResNet-20	68.66$\pm$0.16	69.25$\pm$0.09	69.31$\pm$0.04	**69.89**$\pm$0.22
ResNet-56	74.13$\pm$0.06	74.12$\pm$0.26	74.32$\pm$0.03	**75.29**$\pm$0.14
WRN-28-6	80.61$\pm$0.12	81.07$\pm$0.08	81.10$\pm$0.05	**81.97**$\pm$0.12
WRN-58-2	77.47$\pm$0.16	77.04$\pm$0.18	77.60$\pm$0.08	**78.01**$\pm$0.10
DenseNet-121*	**72.15**$\pm$0.06	69.91$\pm$0.14	70.19$\pm$0.17	70.26$\pm$0.14
VGG19-BN*	72.18$\pm$0.15	71.87$\pm$0.19	73.33$\pm$0.11	**74.49**$\pm$0.26
Pyramidnet-164	69.17$\pm$0.12	68.99$\pm$0.17	71.83$\pm$0.23	**74.35**$\pm$0.06

Table 3. Transfer learning results on Tiny-ImageNet and Food-101.

Tiny-ImageNet	SGD	SAM	ASAM	FisherSAM	SMSAM
Vit-tiny	78.84$\pm$0.07	79.45$\pm$0.03	79.17$\pm$0.02	79.59$\pm$0.02	**79.77**$\pm$0.04
Vit-small	85.85$\pm$0.03	86.48$\pm$0.12	86.76$\pm$0.04	86.57$\pm$0.1	**87.15**$\pm$0.07
Deit-tiny	76.07$\pm$0.08	77.42$\pm$0.03	77.02$\pm$0.08	77.43$\pm$0.05	**77.62**$\pm$0.07
Deit-small	84.69$\pm$0.04	85.60$\pm$0.09	86.45$\pm$0.07	86.53$\pm$0.1	**86.86**$\pm$0.06
Food-101	SGD	SAM	ASAM	FisherSAM	SMSAM
Vit-tiny	83.45$\pm$0.2	83.87$\pm$0.12	84.31$\pm$0.15	84.77$\pm$0.05	**85.04**$\pm$0.16
Deit-tiny	82.07$\pm$0.2	82.24$\pm$0.1	82.25$\pm$0.1	82.58$\pm$0.06	**83.05**$\pm$0.07

6.3 Visualization of Landscapes

We visualize the loss landscapes of the models trained with SAM, ASAM and SMSAM through the visualization technique suggested by [14]. The experimental settings are consistent with Sect. 6.1. Figure 3 shows that comparing SAM and ASAM, SMSAM is able to find the flatter minima.

6.4 Robustness to the Hyperparameter ρ

We evaluate SMSAM, SAM, and ASAM under varying radii on MNIST, FashionMNIST, and CIFAR-10 (see Appendix A.9.3). SMSAM maintains higher test accuracy over a wider range of ρ, indicating improved robustness (see Fig. A.6

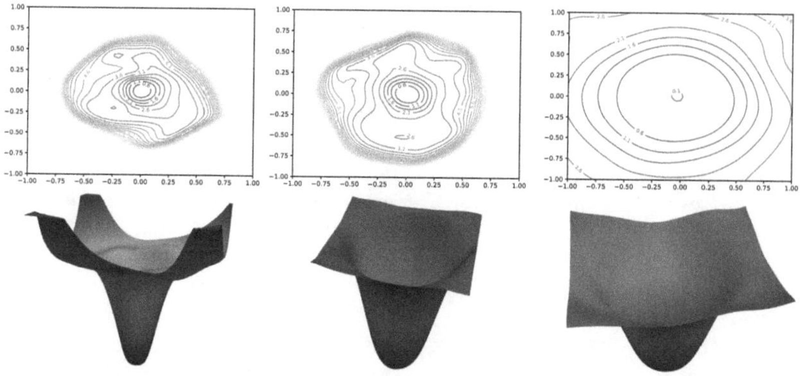

Fig. 3. The 2D (row 1) and 3D (row 2) visualization of loss landscape for SAM (left), ASAM (center) and SMSAM (right) on CIFAR-10 for WideResNet-28-6.

of Appendix A.9.3). This can be attributed to the well-designed perturbation neighborhood in SMSAM, which respects the intrinsic geometry of the parameter space and effectively controls the parameter norm and the trace of the FIM.

7 Conclusion

In this paper, we revisit sharpness-aware minimization from the perspective of statistical manifolds. We show that the Fisher-based sharpness is inherently scale-invariant, providing a principled explanation for geometry-aware perturbations. We establish a connection between Fisher sharpness and generalization by deriving the novel generalization bound, which reveals that generalization depends not only on sharpness but also on a complexity term involving the parameter norm and the FIM trace. Motivated by this insight, we propose SMSAM, a bound-driven optimization method that integrates both effects and admits a convergence guarantee. Experiments show that SMSAM improves generalization and promotes convergence to flatter solutions. Future work includes developing more efficient approximations of the FIM, exploring alternative geometry-aware distances beyond KL divergence, and deepening the theoretical understanding of the optimization dynamics and implicit bias of SAM-type methods.

Acknowledgments. This study was funded by the National Natural Science Foundation of China (Grant No. 62472082) and Guangxi Normal University 2025 Joint Cultivation Project of the National Natural Science Foundation of China (Grant No. RZ2500003784).

References

1. Chaudhari, P., et al.: Entropy-SGD: biasing gradient descent into wide valleys. In: International Conference on Learning Representations (2017)

2. Dinh, L., Pascanu, R., Bengio, S., Bengio, Y.: Sharp minima can generalize for deep nets. In: International Conference on Machine Learning, pp. 1019–1028 (2017)
3. Foret, P., Kleiner, A., Mobahi, H., Neyshabur, B.: Sharpness-aware minimization for efficiently improving generalization. In: International Conference on Learning Representations (2021)
4. Han, D., Kim, J., Kim, J.: Deep pyramidal residual networks. In: Proceedings of the IEEE Conference on Computer Vision and Pattern Recognition, pp. 5927–5935 (2017)
5. He, K., Zhang, X., Ren, S., Sun, J.: Deep residual learning for image recognition. In: Proceedings of the IEEE Conference on Computer Vision and Pattern Recognition, pp. 770–778 (2016)
6. Huang, G., Liu, Z., Van Der Maaten, L., Weinberger, K.Q.: Densely connected convolutional networks. In: Proceedings of the IEEE Conference on Computer Vision and Pattern Recognition, pp. 4700–4708 (2017)
7. Ibayashi, H., Hamaguchi, T., Imaizumi, M.: Minimum sharpness: scale-invariant parameter-robustness of neural networks. In: International Conference on Machine Learning (2021)
8. Jastrzebski, S., et al.: Catastrophic fisher explosion: early phase fisher matrix impacts generalization. In: International Conference on Machine Learning (2020)
9. Jiang, L., Huang, D., Liu, M., Yang, W.: Beyond synthetic noise: deep learning on controlled noisy labels. In: International Conference on Machine Learning, vol. 119, pp. 4804–4815 (2020)
10. Karakida, R., Akaho, S., Amari, S.: The normalization method for alleviating pathological sharpness in wide neural networks. In: Advances in Neural Information Processing Systems, pp. 6403–6413 (2019)
11. Kim, M., Li, D., Hu, S.X., Hospedales, T.M.: Fisher SAM: information geometry and sharpness aware minimisation. In: International Conference on Machine Learning, pp. 11148–11161 (2022)
12. Kolesnikov, A., et al.: Big transfer (bit): general visual representation learning. In: Computer Vision-ECCV 2020: 16th European Conference, pp. 491–507 (2020)
13. Kwon, J., Kim, J., Park, H., Choi, I.K.: ASAM: adaptive sharpness-aware minimization for scale-invariant learning of deep neural networks. In: International Conference on Machine Learning, vol. 139, pp. 5905–5914 (2021)
14. Li, H., Xu, Z., Taylor, G., Studer, C., Goldstein, T.: Visualizing the loss landscape of neural nets. In: Advances in Neural Information Processing Systems, vol. 31 (2018)
15. Martens, J., Grosse, R.B.: Optimizing neural networks with kronecker-factored approximate curvature. In: International Conference on Machine Learning, pp. 2408–2417 (2015)
16. Neyshabur, B., Bhojanapalli, S., McAllester, D., Srebro, N.: Exploring generalization in deep learning. In: Advances in Neural Information Processing Systems, pp. 5947–5956 (2017)
17. Petzka, H., Kamp, M., Adilova, L., Sminchisescu, C., Boley, M.: Relative flatness and generalization. In: Advances in Neural Information Processing Systems, vol. 34, pp. 18420–18432 (2021)
18. Simonyan, K., Zisserman, A.: Very deep convolutional networks for large-scale image recognition. In: International Conference on Learning Representations (2015)
19. Tan, M., Le, Q.: EfficientNet: rethinking model scaling for convolutional neural networks. In: International Conference on Machine Learning, pp. 6105–6114 (2019)

20. Tsuzuku, Y., Sato, I., Sugiyama, M.: Normalized flat minima: exploring scale invariant definition of flat minima for neural networks using PAC-Bayesian analysis. In: International Conference on Machine Learning, vol. 119, pp. 9636–9647 (2020)
21. Zagoruyko, S., Komodakis, N.: Wide residual networks. In: Proceedings of the British Machine Vision Conference (2016)

Special Session: Complexity, Robustness, and Accuracy of Neural Networks

Top-Theta Attention: Sparsifying Transformers by Compensated Thresholding

Konstantin Berestizshevsky(✉), Renzo Andri, and Lukas Cavigelli

Computing Systems Laboratory, Huawei Research, Zurich, Switzerland
{konstantin.berestizshevsky,renzo.andri,lukas.cavigelli}@huawei.com

Abstract. We present Top-Theta (Top-) Attention, a training-free method for sparsifying transformer attention during inference. Our key insight is that static, per-head thresholds can be calibrated to retain the desired constant number of significant elements per attention row. This approach enables content-based sparsity without retraining, and it remains robust across data domains. We further introduce compensation techniques to preserve accuracy under aggressive sparsification, establishing attention thresholding as a practical and principled alternative to top-k attention. We provide extensive evaluation on natural language processing tasks, showing that Top- achieves 3–10× reduction in $\boldsymbol{V}$-cache usage and up to 10× fewer attention elements during inference while degrading no more than 1% in accuracy.

Keywords: Large Language Models · Self-attention · Sparsity

1 Introduction

The self-attention mechanism is central to the success of modern large language models [25], yet it presents a fundamental scalability challenge: its quadratic complexity and the linear growth of the KV cache with sequence length [15] make memory bandwidth the dominant bottleneck during inference [9]. Among promising, scalable solutions is the attention sparsification [7,27], restricting each token to attend only to the most relevant subset of the context.

Existing sparsification approaches each carry significant limitations. Fixed-pattern methods (e.g., StreamingLLM [29], DuoAttention [28]) are computationally cheap but cannot capture dynamic, long-range dependencies. Content-based eviction methods (H_2O [31], SnapKV [18], PyramidKV [3], FastGen [10]) select tokens dynamically but permanently discard KV entries, incurring irreversible information loss. The most principled content-based approach, Top-k attention [12], retains all KV data and selects the k most relevant tokens per row, but requires a full-row sort, which is hardware-unfriendly, has super-linear complexity, and is incompatible with the tiled, pipelined execution of modern GPU kernels. SparQ [21] mitigates accuracy loss via value-mean compensation

L. Pasa et al. (Eds.): ICANN 2026, LNCS 17089, pp. 661–671, 2027.
https://doi.org/10.1007/978-3-032-38398-3_54

but inherits the same sorting bottleneck. LTP [16] avoids sorting, but requires retraining.

We propose Top- , which retains the content-based selectivity of Top-k while replacing the expensive sort with a simple elementwise comparison against a calibrated scalar threshold θ. Elements below θ are excluded from the softmax and $\boldsymbol{V}$-matrix multiplication, reducing memory reads without evicting any KV entries. Since each attention head exhibits a characteristic and stable score distribution, a static threshold suffices to retain $\approx k$ elements per row in $O(n)$ time with no row dependency, making Top- fully compatible with tiled and distributed inference kernels. To further preserve accuracy under aggressive sparsification, we introduce softmax denominator compensation (SDC) and mean $\boldsymbol{V}$-row compensation (VMC). Our central finding is:

Central Finding: Static per-head thresholds, calibrated **once** for a given model on a few hundred samples, reliably sparsify attention to $\approx k$ elements per attention row and generalize across data domains.

Based on our central finding, our Top- method delivers the following advantages:

1. *Efficiency* - 3–10× fewer $\boldsymbol{V}$-rows and up to 10× fewer attention elements on LLaMA2/3 with ≤1% accuracy drop.
2. *Tiling compatible* - our elementwise thresholding carries no row dependency, enabling pipelined and distributed kernels.
3. *No retraining* - one-time calibration on a few hundred samples suffices.
4. *Domain robust* - thresholds encode model structure rather than data statistics, transferring across domains without recalibration.

2 Background

2.1 Self-attention and Sparsity

Multi-head self-attention (MHA) projects an input sequence $\boldsymbol{X} \in \mathbb{R}^{n\times D}$ into queries, keys, and values $\boldsymbol{Q}, \boldsymbol{K}, \boldsymbol{V} \in \mathbb{R}^{n\times d}$, computes pre-softmax scores $\boldsymbol{A} = \boldsymbol{Q}\boldsymbol{K}^T$, and produces output $\boldsymbol{P}$:

$$\boldsymbol{S} = \text{row\_softmax}(\boldsymbol{A}) \triangleq \left[\frac{e^{A_{ij}}}{\sum_{k=1}^{n} e^{A_{ik}}}\right]_{\substack{1\le i\le n\\ 1\le j\le n}}, \qquad \boldsymbol{P} = \boldsymbol{S}\boldsymbol{V} \tag{1}$$

where n is the sequence length, D is the hidden dimension, and d is the head dimension. The post-softmax matrix $\boldsymbol{S} \in [0,1]^{n\times n}$ values resemble a log-normal distribution [20], with values concentrated near 0 and entropy decreasing in deeper layers [26]. Since $\boldsymbol{S}$ is an order-preserving transformation of $\boldsymbol{A}$, small values in column i diminish the contribution of $\boldsymbol{V}$-row i to $\boldsymbol{P}$, *enabling ranking and pruning of attention elements (either* $\boldsymbol{A}$ *or* $\boldsymbol{S}$*) by significance.*

Generative Decoding and KV-Cache. During decoding [22], only the current token's $\boldsymbol{q}, \boldsymbol{k}, \boldsymbol{v} \in \mathbb{R}^d$ are computed while the full $\boldsymbol{K}, \boldsymbol{V}$ are loaded from cache, reducing attention to vectors $\boldsymbol{a}, \boldsymbol{s} \in \mathbb{R}^n$ and output $\boldsymbol{p} = \boldsymbol{s}\boldsymbol{V}$. This is heavily memory-bandwidth-bound: all n rows of $\boldsymbol{K}$ and $\boldsymbol{V}$ are loaded for a single multiply-add each, so *reducing memory reads yields a direct speedup.* Grouped query attention (GQA) [1] shares one $\boldsymbol{K}, \boldsymbol{V}$ pair across g attention heads, reducing bandwidth by $g\times$ with negligible accuracy cost, yet the memory bandwidth bottleneck remains.

2.2 Top-k Attention and Its Limitations

The related work of Top-k attention [12] retains the k largest elements per attention row, discarding the rest. However, exact top-k selection requires full-row aggregation with at best $O(\log n)$ complexity [30], imposing a global row dependency that is incompatible with tiled kernels. In pipelined kernels such as paged-attention [17], each KV block is processed sequentially through $\boldsymbol{Q}\boldsymbol{K}^T$, softmax, and $\boldsymbol{S}\boldsymbol{V}$; a global top-$k$ would require materializing all partial scores before pruning, defeating pipelining. Elementwise thresholding, by contrast, makes pruning decisions locally per block with no cross-block synchronization. *We conjecture that top-k search can be replaced by comparison to a carefully calibrated threshold.*

3 Top-θ Method

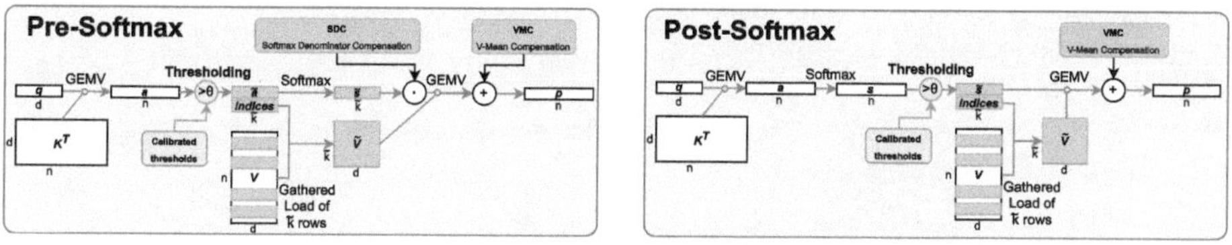

Fig. 1. Two variants of Top- attention for inference at generative decoding.

We propose Top- attention, which compares each attention vector against a calibrated threshold, pruning elements below it from softmax and $\boldsymbol{V}$-multiplication. Since each attention head exhibits a characteristic score distribution, a static threshold suffices to retain k elements per row on average, replacing the $O(n \log n)$ top-k sort with an $O(n)$ elementwise operation without row dependency. Section 3.1 explains our offline threshold calibration, and Sect. 3.2 presents the efficient inference that employs the calibrated thresholds.

3.1 Threshold Calibration

Top- requires a short calibration procedure. To this end, Algorithm 1 calibrates a threshold $\theta_r(k)$ per attention row r, of a given transformer layer and head by collecting the $\frac{n-k}{n}$-quantile of each attention row over a calibration set, then averaging across samples (l. 19). A "top-k at calibration" step (ll. 8, 15) ensures downstream activations reflect the sparsification pattern, improving threshold stability. We found that a uniform parameter k per layer suffices, and we provide the optional offset α for conservative user cases, when thresholding (that selects $\tilde{k} \approx k$) should not violate the expected k at inference. Thresholds for a"Post-softmax" variant of Top- can be obtained by applying the quantile to $\boldsymbol{S}$ instead of $\boldsymbol{A}$. Rows $r < k$ are skipped due to causal masking.

As shown on Fig. 2a, few hundred calibration samples suffice: while larger sets improve fidelity of keeping the desired k elements per row, the downstream accuracy is weakly affected. Stored thresholds are negligible in size: for LLaMA-3-70B (80 layers, 64 heads, 1200 rows, float16) the total is only 11.8 MB (0.0008% of model size).

Figure 2b illustrates that threshold values vary per head and row (sequence length), justifying per-(head, row) individual thresholds. Pre-softmax thresholds converge to a constant, whereas post-softmax thresholds decrease with length due to softmax normalization.

Algorithm 1 Calibrate($\mathcal{C}, k, \alpha$) - threshold calibration of a single attention head

Require: $\mathcal{C}$ (Calibration set of inputs)
Require: $k \in \mathbb{N}$ (elements to keep per atten. row)
Require: $\alpha \in \mathbb{R}$ (calibration offset in std_devs)
1: $\Theta_r = \emptyset, \forall r$ {empty sets of observed thresholds}
2: **for** $\boldsymbol{X} \in \mathcal{C}$ **do**
3: **if** is_prefill($\boldsymbol{X}$) **then**
4: $\boldsymbol{A} = \boldsymbol{Q}(\boldsymbol{X})\boldsymbol{K}^T(\boldsymbol{X})$
5: $n = \text{num\_rows}(\boldsymbol{A})$
6: **for** $r = k$ to $n-1$ **do**
7: $\Theta_r = \Theta_r \cup \{\text{quantile}_{\frac{n-k}{n}}(\boldsymbol{A}_r)\}$
8: $\boldsymbol{A}_r = \text{top}_k(\boldsymbol{A}_r)$
9: **end for**
10: $\boldsymbol{S} = \text{row\_softmax}(\boldsymbol{A})$
11: **else** {generative decoding, $X \in \mathbb{R}^d$}
12: $\boldsymbol{a} = \boldsymbol{Q}(\boldsymbol{X})\boldsymbol{K}^T(\boldsymbol{X})$
13: $n = \text{length}(\boldsymbol{a})$
14: $\Theta_n = \Theta_n \cup \{\text{quantile}_{\frac{n-k}{n}}(\boldsymbol{a})\}$
15: $\boldsymbol{a} = \text{top}_k(\boldsymbol{a})$
16: $\boldsymbol{s} = \text{softmax}(\boldsymbol{a})$
17: **end if**
18: **end for**
19: **return** $\theta_r = \text{mean}(\Theta_r) + \alpha \cdot \text{std\_dev}(\Theta_r), \forall r$

3.2 Top-θ Inference

At inference (Fig. 1), each attention row n is compared against θ_n; sub-threshold elements are discarded and only the $\tilde{k}$ selected $\boldsymbol{V}$-rows are loaded into a compact $\tilde{\boldsymbol{V}}$ for the final product $\boldsymbol{p}$. Two practical details: (i) if a sequence length was not seen during calibration, the nearest calibrated threshold is used; (ii) $\tilde{k}$ varies around k due to input-dependent distributions, capping at k degraded generative performance, so we instead control variability via calibration set size.

Softmax Denominator Compensation (SDC). Post-softmax sparsification outperformed pre-softmax in our experiments, which motivated us to develop SDC, that emulates post-softmax accuracy during pre-softmax sparsification by rescaling. Let $\mathcal{I} \subseteq \{0, \ldots, n-1\}$ denote a set of indices that we intend to keep during the sparsification, and let us establish the relation between post-softmax sparsified vector $\tilde{\boldsymbol{s}}$ and the pre-softmax sparsified vector that underwent softmax

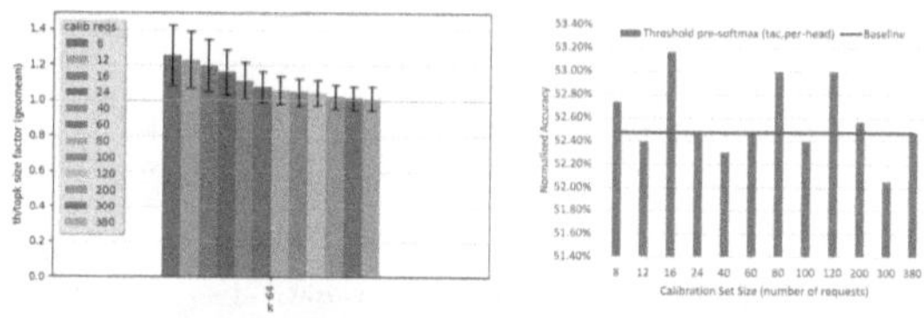

(a) **Calibration set size impact** – LLaMA2-7B on ARC-C. **Left**: average $\tilde{k}/k$ ratio (closer to 1 indicates better Top-k fidelity) for calibration set sizes ranging from 8 to 380. **Right**: test accuracy as a function of calibration set size.

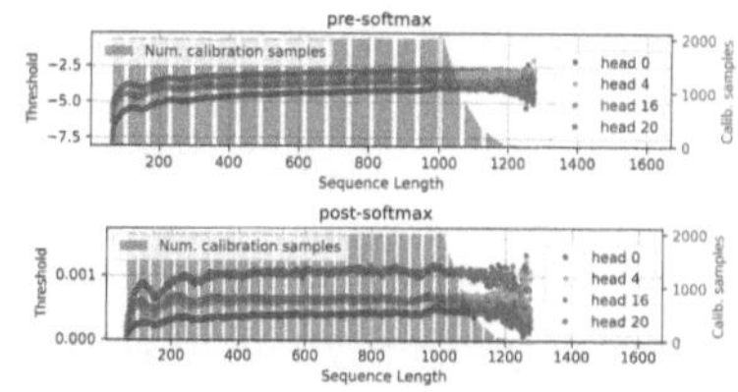

(b) **Threshold values vs. sequence length** – LLaMA2-7b, 11th layer, k=64. Scatter at long sequence lengths reflects calibration data scarcity (low purple bars).

Fig. 2. Calibration properties. Left: effect of calibration set size on Top-k-approximation fidelity and downstream accuracy. Right: pre- and post-softmax threshold values as a function of sequence length. (Color figure online)

(softmax($\tilde{\boldsymbol{a}}$)). Let R and E denote the sums of exponents of the selected and discarded elements, respectively.

$$\forall i \in \mathcal{I}, \tilde{\boldsymbol{s}}_i = \text{softmax}(\boldsymbol{a})_i = \frac{e^{a_i - \max_l a_l}}{\sum_{j=0}^{n-1} e^{a_j - \max_l a_l}} = \frac{e^{a_i - \max_l a_l}}{\sum_{j\in\mathcal{I}} e^{a_j - \max_l a_l} + \sum_{j\notin\mathcal{I}} e^{a_j - \max_l a_l}}$$
$$= \frac{e^{a_i - \max_l a_l}}{R+E} = \frac{e^{a_i - \max_l a_l}}{R} \cdot \frac{R}{R+E} = \text{softmax}(\tilde{\boldsymbol{a}})_i \cdot \frac{R}{R+E} \qquad (2)$$

hence, by performing pre-softmax sparsification and multiplying by $R/(R+\tilde{E})$, we approximate the desired $\tilde{\boldsymbol{s}}$, where $\tilde{E} \approx E$ is estimated via one of three strategies:

1. `offline-calibrated`: calibrate a static value $\tilde{E}$ per (layer, head, row); requires extra calibration and memory.
2. `exp-threshold`: $\tilde{E} = \gamma(n-\tilde{k})e^{\theta}$, γ=0.05; 3 scalar ops, no extra memory or calibration. This approximation is motivated by the fact that threshold is the upper-bound of all discarded elements.
3. `exact`: $\tilde{E} = E$ computed directly; costs $2(n-\tilde{k})$ scalar ops, no extra memory or calibration.

The correction factor $R/(R+\tilde{E})$ can be applied after softmax or after the $\boldsymbol{SV}$ product, analogously to FlashAttention [6], to not impede tiling.

V-Mean Compensation (VMC). When attention rows do not sum to 1 (post-softmax or pre-softmax+SDC), the contribution of discarded tokens is missing from $\boldsymbol{p}$. VMC approximates this missing contribution using the mean $\boldsymbol{V}$-row $\boldsymbol{\mu}$:

$$\boldsymbol{\mu} = \frac{1}{n}\sum_{i=0}^{n-1} \boldsymbol{V}_i, \qquad \beta = 1 - \sum_{i=0}^{\tilde{k}-1} \tilde{\boldsymbol{s}}_i, \qquad \hat{\boldsymbol{p}} = \tilde{\boldsymbol{s}}\tilde{\boldsymbol{V}} + \beta\boldsymbol{\mu} \qquad (3)$$

The approximation improves with longer sequences as $\boldsymbol{\mu}$ averages over more tokens. During generative decoding, $\boldsymbol{\mu}$ is maintained as a running mean.

4 Evaluations

Experiments use the LM Evaluation Harness [8] for normalized accuracy on multiple-choice Q&A benchmarks (standard few-shot settings), the official HumanEval harness [5] for pass@1 on 164 code generation tasks, and Long-Bench [2] (qmsum, gov_report) for long-sequence summarization. Experiments on 70B models require 192 GB GPU memory; 7B/8B results on non-LongBench tasks reproduce on a single 24 GB GPU. Full source code and reproducibility scripts are provided[1].

We compare three attention variants: (i) **Baseline** – full attention; (ii) **Top-k** – exact top-k elements per row; (iii) **Top-θ** – thresholding to $\tilde{k} \approx k$ elements. Compensations (SDC, VMC) are applied to Top-k and Top- for a fair comparison.

4.1 Top-θ Overall Performance

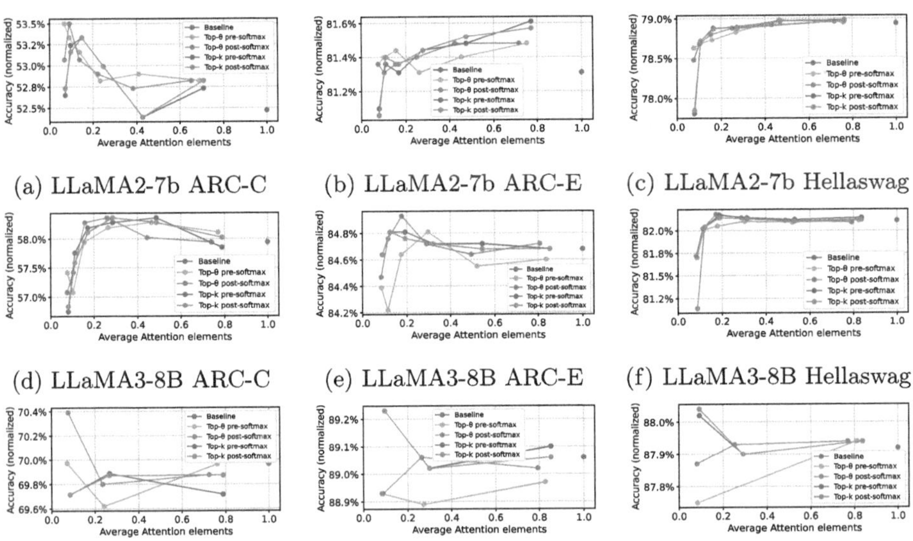

(a) LLaMA2-7b ARC-C (b) LLaMA2-7b ARC-E (c) LLaMA2-7b Hellaswag

(d) LLaMA3-8B ARC-C (e) LLaMA3-8B ARC-E (f) LLaMA3-8B Hellaswag

(g) LLaMA3-70B ARC-C (h) LLaMA3-70B ARC-E (i) LLaMA3-70B Hellaswag

Fig. 3. Prefill tasks – accuracy vs. fraction of active attention elements. Post-softmax Top-k/Top- use VMC; pre-softmax variants use VMC+SDC. Both methods match or exceed baseline accuracy at 2–10× sparsity.

Figure 3 evaluates LLaMA2 [24] and LLaMA3 [11] on **prefill-heavy** Q&A tasks with $k \in [32, 512]$ (k=512 for the first two layers; Top- calibrated on 10% of the training/validation set). Across all models and datasets, both Top-k and Top- *increase* accuracy by 0.2–1% over the baseline while reducing active attention elements by 2–5×. Post-softmax consistently outperforms pre-softmax.

[1] https://github.com/huawei-csl/top-theta-attention.

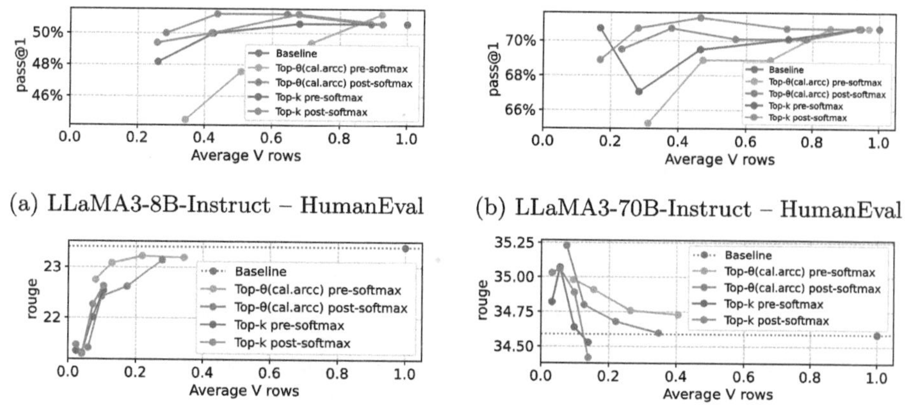

(a) LLaMA3-8B-Instruct – HumanEval

(b) LLaMA3-70B-Instruct – HumanEval

(c) LLaMA3.1-8B-Instruct – LongBench-qmsum

(d) LLaMA3.1-8B-Instruct – LongBench-gov_report

Fig. 4. Generative tasks – accuracy vs. fraction of required $\boldsymbol{V}$-rows. Top- thresholds calibrated on ARC-C and transferred to Human-eval and LongBench. Post-softmax Top-k/Top- use VMC; pre-softmax variants use VMC+SDC.

Figure 4 evaluates **generative** tasks, with Top- thresholds cross-domain transferred from ARC-C. On Human-eval ($k \in [32, 512]$, greedy decoding), post-softmax Top- preserves pass@1 within 1% of baseline at 3× $\boldsymbol{V}$-row reduction for 8B, and 5× for 70B at 1% cost. On LongBench (∼15k token prompts, $k \in [128, 768]$), reductions reach 10× despite GQA, with Rouge-L within 1% of baseline and often +0.5%. Overall, Top- and Top-k perform comparably.

4.2 Ablation Studies

Pre vs Post-softmax Thresholding. We evaluate how different the impact of pre-softmax attention matrix sparsification (applied on matrix $\boldsymbol{A}$) is compared to post-softmax attention matrix sparsification ($\boldsymbol{S}$). Figure 5a depicts how the different thresholdings impact the accuracy of LLaMA2-7b on the Hellaswag dataset. For comparison, the Top-k approach is also evaluated alongside Top- . Overall, *post-softmax preserves more accuracy compared to pre-softmax thresholding.*

Thresholding Different Layers. We explored the impact of thresholding different layers with a different target k. For the LLaMA models, we have observed that targeting a slightly higher k in the first layers is crucial. As a best practice, we found keeping 2 initial layers at $k = 512$, whereas the rest of the layers could be sparsified more aggressively. Figure 5b shows the LLaMA2-7b model accuracy on Hellaswag, comparing Top-k and Top- using higher k in first 2 layers against using equal k in all layers. All variants do not perform any compensations. *The*

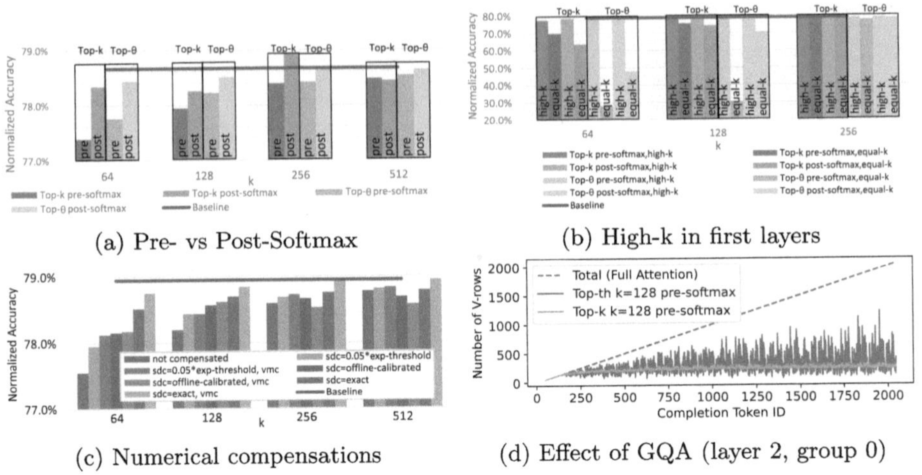

(a) Pre- vs Post-Softmax

(b) High-k in first layers

(c) Numerical compensations

(d) Effect of GQA (layer 2, group 0)

Fig. 5. Ablation studies on Hellaswag (LLaMA2-7b) and Human-eval (LLaMA3-8B-Instruct). **(a)** Post-softmax thresholding preserves higher accuracy than pre-softmax. **(b)** Higher k in the first two layers substantially recovers accuracy. **(c)** SDC + VMC compensation closes the accuracy gap to baseline. **(d)** Despite GQA group size $g = 4$, only $\approx$250–300 $\boldsymbol{V}$-rows are needed at k=128, indicating strong inter-head agreement.

conclusion is that preserving denser attention matrices in the early layers is beneficial for the accuracy of a downstream task, which is aligned with other works on sparsity [23] and on quantization [13].

Numerical Compensations. We evaluate the proposed numerical compensation methods SDC and VMC, finding that more explicit SDC variants (exp-threshold, exact) substantially recover degraded accuracy on the challenging Hellaswag task with LLaMA2-7b especially with Top- attention, as shown in Fig. 5c; combining SDC with VMC further improves results, effectively closing the accuracy gap between the baseline and the non-compensated Top- attention. However, on the ARC-C and ARC-E datasets, Top- attention alone already outperforms the baseline by reducing attention noise, so applying compensations in these cases offers little benefit and may even reduce accuracy back to baseline levels by reintroducing noise. *Overall, SDC and VMC compensations almost entirely recover the accuracy.*

Impact of Grouped Query Attention (GQA). In GQA, multiple attention heads share the same $\boldsymbol{V}$ matrix; for LLaMA-3-8B models with a group size of $g = 4$, this means that when applying Top-k, up to $4 \cdot k$ rows of $\boldsymbol{V}$ matrix will be needed if heads select completely different tokens. However, as shown in Fig. 5d, both Top-k and Top- with $k = 128$ typically select only 250–300 $\boldsymbol{V}$ rows per group, indicating substantial agreement among heads - a pattern observed across many layers.

While our method uses independent per-head selections, GQA sparsification could be improved by considering two alternatives: (1) discarding $\boldsymbol{V}$-rows requested by few attention heads, or (2) augmenting selections for all heads with the group's union. Discarding (1) maintains a stricter budget but adds complexity (e.g., vote thresholds) and risks dropping crucial tokens. Conversely, augmenting (2) is simpler to implement via a mask union. Our limited evaluation showed only a minor accuracy uplift using augmenting (2), which we attribute to the added tokens being less important to the individual heads.

Distribution Shifts. To examine how domain-sensitive the calibrated threshold is, we evaluate on Human-eval the two following Top- variants: 1) calibrated on Q&A dataset of ARC-C, and 2) calibrated on the first 10% of Human-eval tasks. As seen in Fig. 6a, the Top- post-softmax is even more accurate when using thresholds calibrated on a different dataset. For pre-softmax, there is a benefit of calibrating on the same dataset. We also evaluate using MedMCQA, where we compare the Top- calibrated on ARC-C with Top-k on Fig. 6b - showing that both sparsification methods perform equally well. *Overall, the thresholds show resilience towards distribution shift, suggesting that they are strongly associated with the model rather than with the data, allowing a single calibration per model.*

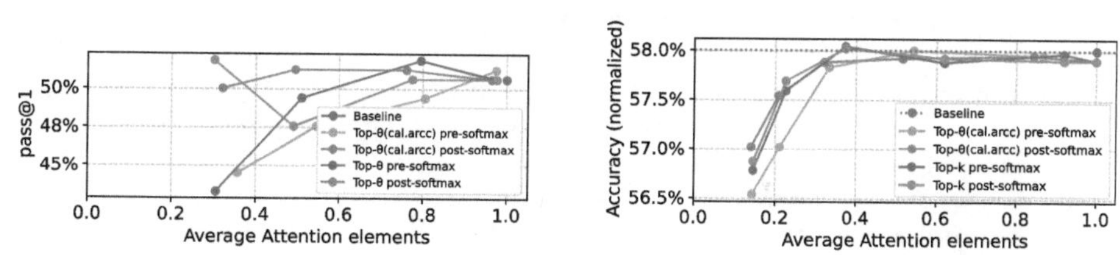

(a) LLaMA3-8B-Instruct - Human-eval (b) LLaMA3-8B MedMCQA

Fig. 6. Distribution shift - Top- calibrated on different task (labeled with *cal.arcc*) shows comparable accuracy and attention reduction compared to Top- calibrated on the same task (Fig. 6a), and compared to Top-k (6b).

4.3 Top-Theta Kernel

We implemented a prototype Top- MHSA kernel on an Ascend NPU (24 Davinci cores, 376 TOPS/sec FP16, 1.8 TB/sec memory bandwidth), achieving a 1.17× speedup over the optimized `IncrementalFlashAttentionV4` decoding kernel [14] at 48 heads, D=128, k=128, and n=2048. Sparse kernel efficiency is inherently hardware-dependent, relying on low-level DMA, bandwidth, and parallel execution capabilities. To fully unlock the potential of sparsity-based attention, future memory subsystems with native gather support and compact KV-cache layouts would translate the theoretical bandwidth reductions of Top- more directly into wall-clock speedups, a trend already anticipated in the memory systems literature [4, 19].

5 Conclusion

We have presented Top- , a new sparse attention algorithm based on fixed, offline-calibrated thresholds. At inference time, it boils down to a simple elementwise operation, which overcomes the limitation of full-row dependent algorithms, such as Top-k, and thereby unconstrains tiling and distributed inference implementations. Moreover, only a small subset of calibration samples is required to calibrate the thresholds, which we showed to be resilient to input distribution shifts. Therefore, a short calibration is needed once per model. Furthermore, we show minor to no performance degradation using 10× less attention elements at the computationally-bound prefill phase and using 3× (for shorter sequence tasks) to 10× less $\boldsymbol{V}$ matrix rows (for longer sequence, decoding-heavy tasks) at the memory bandwidth-bound generative decoding phase. These reductions unlock promising speedups for inference.

References

1. Ainslie, J., Lee-Thorp, J., de Jong, M., Zemlyanskiy, Y., Lebrón, F., Sanghai, S.: GQA: Training generalized multi-query transformer models from multi-head checkpoints. arXiv:2305.13245 (2023)
2. Bai, Y., et al.: LongBench: a bilingual, multitask benchmark for long context understanding. In: Proceedings of the 62nd Annual Meeting of the Association for Computational Linguistics (Volume 1: Long Papers), Bangkok, Thailand, pp. 3119–3137. Association for Computational Linguistics (Aug 2024)
3. Cai, Z., et al.: PyramidKV: dynamic KV cache compression based on pyramidal information funneling (2025)
4. Chatterjee, N., O'Connor, M., Loh, G.H., Jayasena, N., Balasubramonia, R.: Managing dram latency divergence in irregular GPGPU applications. In: SC'14: Proceedings of the International Conference for High Performance Computing, Networking, Storage and Analysis, pp. 128–139. IEEE (2014)
5. Chen, M., et al.: Evaluating large language models trained on code (2021)
6. Dao, T., et al.: FlashAttention: fast and memory-efficient exact attention with io-awareness. In: Oh, A. (ed.) Advances in Neural Information Processing Systems, vol. 35, pp. 16344–16359. Curran Associates, Inc (2022)
7. Fuad, K.A.A., Chen, L.: A survey on sparsity exploration in transformer-based accelerators. Electronics **12**(10), 2299 (2023)
8. Gao, L., et al.: A framework for few-shot language model evaluation (2024)
9. Ge, S., Zhang, Y., Liu, L., Zhang, M., Han, J., Gao, J.: Model tells you what to discard: adaptive KV cache compression for LLMs (2024)
10. Ge, S., Zhang, Y., Liu, L., Zhang, M., Han, J., Gao, J.: Model tells you what to discard: adaptive KV cache compression for LLMs (2024). https://doi.org/10.48550/arXiv.2310.01801, arXiv:2310.01801 [cs]
11. Grattafiori, A., et al.: The llama 3 herd of models (2024). arXiv:2407.21783
12. Gupta, A., Dar, G., Goodman, S., Ciprut, D., Berant, J.: Memory-efficient transformers via Top-k attention. CoRR abs/2106.06899 (2021)
13. Huang, X., Shen, Z., Dong, P., Cheng, K.T.: Quantization variation: a new perspective on training transformers with low-bit precision. Trans. Mach. Learn. Res. (2024)

14. Huawei: CANNCommunityEdition. Computer Software (2024). https://www.hiascend.com/en/software/cann/community
15. Keles, F.D., Wijewardena, P.M., Hegde, C.: On the computational complexity of self-attention. In: International Conference on Algorithmic Learning Theory, pp. 597–619. PMLR (2023)
16. Kim, S., et al.: Learned token pruning for transformers. In: Proceedings of the 28th ACM SIGKDD Conference on Knowledge Discovery and Data Mining, pp. 784–794 (2022)
17. Kwon, W., et al.: Efficient memory management for large language model serving with paged attention. In: Proceedings of the 29th Symposium on Operating Systems Principles, pp. 611–626 (2023)
18. Li, Y., et al.: SnapKV: LLM knows what you are looking for before generation. Adv. Neural. Inf. Process. Syst. **37**, 22947–22970 (2024)
19. Miao, H., Jeon, M., Pekhimenko, G., McKinley, K.S., Lin, F.X.: Streambox-HBM: stream analytics on high bandwidth hybrid memory. In: Proceedings of the Twenty-Fourth International Conference on Architectural Support for Programming Languages and Operating Systems, pp. 167–181 (2019)
20. Nahshan, Y., Kampeas, J., Haleva, E.: Linear log-normal attention with unbiased concentration (2024)
21. Ribar, L., Chelombiev, I., Hudlass-Galley, L., Blake, C., Luschi, C., Orr, D.: SparQ attention: bandwidth-efficient LLM inference. arXiv:2312.04985 (2023)
22. Shi, L., Zhang, H., Yao, Y., Li, Z., Zhao, H.: Keep the cost down: a review on methods to optimize LLM's KV-cache consumption (2024)
23. Tang, J., Zhao, Y., Zhu, K., Xiao, G., Kasikci, B., Han, S.: Quest: query-aware sparsity for efficient long-context LLM inference. arXiv:2406.10774 (2024)
24. Touvron, H., et al.: Llama 2: open foundation and fine-tuned chat models. arXiv:2307.09288 (2023)
25. Vaswani, A., et al.: Attention is all you need. In: Advances in Neural Information Processing Systems (2017)
26. Vig, J., Belinkov, Y.: Analyzing the structure of attention in a transformer language model (2019)
27. Wang, X., Salmani, M., Omidi, P., Ren, X., Rezagholizadeh, M., Eshaghi, A.: Beyond the limits: a survey of techniques to extend the context length in large language models (2024)
28. Xiao, G., et al.: DuoAttention: efficient long-context LLM inference with retrieval and streaming heads. arXiv (2024)
29. Xiao, G., Tian, Y., Chen, B., Han, S., Lewis, M.: Efficient streaming language models with attention sinks (2024)
30. Zhang, J., Naruse, A., Li, X., Wang, Y.: Parallel Top-K algorithms on GPU: a comprehensive study and new methods. In: Proceedings of the International Conference for High Performance Computing, Networking, Storage and Analysis. SC '23, New York, NY, USA. Association for Computing Machinery (2023)
31. Zhang, Z., et al.: H2O: heavy-hitter oracle for efficient generative inference of large language models.https://doi.org/10.48550/arXiv.2306.14048, arXiv:2306.14048 [cs] (2023)

Adaptive Gradient Reversal for Efficient Backdoor Unlearning with Minimal Poisoned Samples

Yunfeng Zhang(✉), Yuyuan Sun, Zhankai Li, Tao Liu, Kailong Zhu, and Xuehu Yan

National University of Defense Technology, Hefei, China
{zhangyunfeng24,zhukailong,yanxh17}@nudt.edu.cn

Abstract. Backdoor attacks, as critical security threats in deep learning, have garnered increasing research attention. Existing mitigation approaches often fail to achieve true unlearning due to the absence of backdoor knowledge, merely marginalizing backdoor patterns. We propose the Backdoor Gradient Reversal and Adaptive Scaling (GRAS) method, which leverages a small set of poisoned samples to enhance backdoor unlearning. Extensive experiments demonstrate that GRAS maintains model's benign accuracy while significantly reducing backdoor attack success rate and improving the classification of poisoned samples. Notably, our experiments reveal that even a minimal number of poisoned samples contain most discriminative features of backdoor patterns, with a single poisoned sample enabling effective mitigation.

Keywords: Deep Neural Networks · Backdoor Defense · Backdoor Unlearning · Adaptive Gradient Reversal

1 Introduction

The widespread adoption of deep learning across critical domains such as autonomous driving [21] and security systems [17] highlights its transformative potential. Resource-constrained edge environments (e.g., IoT devices) necessitate smaller Deep Neural Networks (DNNs) for practical deployment, maintaining significant security concerns. Backdoor attacks–via data poisoning or model modification–remain a critical threat, enabling attackers to implant stealthy triggers that activate under specific conditions while preserving normal functionality on clean inputs [1]. Backdoor attacks via data poisoning exhibit high adaptability and concealment, making it particularly perilous in adversarial settings where attackers may poison exposed datasets to manipulate inference outcomes.

As research on backdoor attacks advances, trigger design has evolved from human-perceptible, fixed-pattern approaches to sample-specific, imperceptible manipulations. Contemporary poisoned datasets often contain samples with

---

K. Zhu and X. Yan are co-corresponding authors.

L. Pasa et al. (Eds.): ICANN 2026, LNCS 17089, pp. 672–682, 2027.
https://doi.org/10.1007/978-3-032-38398-3_55

increasingly stealthy backdoor triggers [4,6,19,20]. However, not all poisoned samples can maintain effective concealment. With the advancement of backdoor detection methods [7,8,11,25,29], it becomes feasible to capture a limited number of poisoned samples through approaches such as anomaly analysis, manual-assisted detection, and threat intelligence. Especially in the field of backdoor sample detection, some existing state-of-the-art methods [7,8] require first training a backdoor model on a poisoned dataset, and then performing re-training or fine-tuning after the detection to obtain a clean model. Therefore, exploring efficient and stable backdoor unlearning methods can effectively complement the existing backdoor defense system, reduce the cost of re-training or fine-tuning, and lower the risk of the backdoor being reactivated.

Existing works have explored different methods to mitigate backdoor attacks based on data poisoning in various scenarios. The latest defense methods have achieved outstanding defense effects under specific conditions and scenarios [3,13,22,27]. However, due to the lack of knowledge about backdoors, many mitigation methods do not truly make the model forget the backdoor patterns. Some studies have shown that residual backdoor in purified models can still be reactivated using specific methods [15,26].

In this paper, we propose a novel fine-tuning-based backdoor unlearning method named Gradient Reversal and Adaptive Scaling (GRAS), which enables backdoor unlearning using limited benign samples and a minimal number of poisoned samples, without requiring explicit trigger extraction. We conducted comprehensive evaluations by comparing GRAS against six representative backdoor mitigation methods across three benchmark datasets under eight canonical backdoor attacks. Extensive experimental results demonstrate that GRAS effectively eliminates backdoor patterns while preserving model performance on benign data. The enhanced poisoned sample restoration accuracy and t-SNE visualization results further confirm that GRAS achieves trigger features unlearning rather than mere patterns isolation.

Our key finding demonstrates that even a single poisoned sample contains sufficient discriminative features for effective backdoor unlearning. Statistical evidence shows that GRAS can achieve acceptable defense performance when only one poisoned sample is available, highlighting GRAS's remarkable efficiency in eliminating backdoor patterns from minimal poisoned data.

2 Related Work

2.1 Backdoor Attacks

Backdoor attacks establish trigger-target label mappings via data poisoning, by injecting malicious samples to associate triggers with attacker-defined labels. BadNet [6] employs a colored patch as a trigger. Blended [4] integrates a special pattern (e.g., Hello Kitty) into training images. SIG [2] and LC [18] advance clean-label attacks that avoid explicit label modification, thereby enhancing concealment. Recent methods achieve imperceptible, sample-dependent triggers: Refool [12] utilizes reflection phenomena, LF [24] and FTrojan [19] operate in frequency domain, BPP [20] employs image quantization and dithering to embed hidden triggers.

2.2 Backdoor Defenses

Backdoor mitigation techniques aim to reduce the impact of backdoor attacks through model refinement. Basic fine-tuning leverages DNNs' natural forgetting mechanism. FT-SAM [27] enhances fine-tuning by incorporating Sharpness-Aware Minimization (SAM). RNP [10] unlearns benign features at the neuron level and reconstructs them at the filter level. FST [14] reinitializes the fully connected layer weights and restores the model's benign performance using an additional penalty term based on the dot product of the original poisoned and recovered weights, forcing features to shift. ULRL [13] refines FST by leveraging statistical features during the forgetting of clean samples to locate backdoor-related neurons, replacing the penalty term with cosine similarity.

While effective at reducing attack success rates, these methods lack backdoor knowledge, limiting accurate poisoned sample restoration. Dunnett et al. [5] utilize poisoned samples for mitigation, improving poisoned sample recognition but relying on the often impractical assumption that triggers can be fully reversed through Backdoor Trigger Inversion (BTI). A more realistic scenario is that a small number of poisoned samples can be captured via backdoor detection techniques [7,8,11,25,29], manual screening, or threat intelligence.

3 Proposed Method

3.1 Threat Model

Adversary Capabilities. We assume attackers have successfully implanted a backdoor into a DNN model $N(\theta_p)$, where θ_p denotes the backdoor-infected parameters. For clean sample x_c with correct label y_c, the model maintains normal performance: $\mathbb{E}_{x_c \sim P}[\mathcal{L}(N(x_c;\theta_p), y_c)] < \varepsilon_1$. When embedded with a predefined trigger through transformation $p(x)$, the model outputs the attacker-specified backdoor label y_p: $\mathbb{E}_{x_c \sim P}[\mathcal{L}(N(p(x_c);\theta_p), y_p)] < \varepsilon_2$. Here P represents the benign data distribution, $\mathcal{L}$ denotes the loss function (e.g., cross-entropy), and ε_1, ε_2 are small thresholds.

Defender Capabilities. The defender possesses: (1) the backdoor-infected model architecture and parameters $N(\theta_p)$, (2) limited clean dataset $\mathcal{D}_c$ with $|\mathcal{D}_c| \leq 5\% \times |\mathcal{D}_{train}|$, and (3) a small poisoned dataset $\mathcal{D}_p$ with $|\mathcal{D}_p| \leq 100$.

Defender Objective. Find parameters θ_c such that $N(\theta_c)$ minimizes backdoor attack success rate while maintaining accuracy on benign tasks.

3.2 Vanilla Backdoor Unlearning

Backdoor learning can be formulated as a dual-task optimization process that simultaneously operates on both the clean training subset $\mathcal{D}_{t_c}$ and the poisoned training subset $\mathcal{D}_{t_p}$:

$$\arg\min_{\theta}[\underbrace{\mathbb{E}_{(x_c,y_c)\in D_{t_c}}\mathcal{L}(N(x_c;\theta),y_c)}_{Benign\ Task} + \underbrace{\mathbb{E}_{(x_p,y_p)\in D_{t_p}}\mathcal{L}(N(x_p;\theta),y_p)}_{Backdoor\ Learning}]. \tag{1}$$

As illustrated in Fig. 1a, the red arrow indicates parameter updates toward backdoor learning. Backdoor unlearning reverses this process to maximize backdoor attack risk.

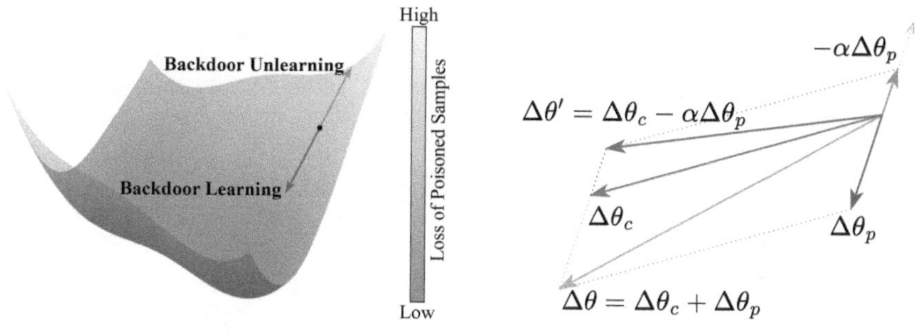

(a) Backdoor Learning and Unlearning (b) Vanilla Backdoor Gradient Reversal

Fig. 1. The Principle of Vanilla Backdoor Unlearning.

We define $\Delta\theta$ as the parameter update direction, decomposed into $\Delta\theta_c$ (benign task) and $\Delta\theta_p$ (backdoor learning) (Fig. 1b). $\Delta\theta_c$ and $\Delta\theta_p$ can be derived from reverse gradients on clean and poisoned samples. The unlearning update direction becomes:

$$\Delta\theta' = \Delta\theta_c - \alpha\Delta\theta_p, \tag{2}$$

where α balances unlearning strength and model stability.

Our experiments demonstrate that fixed α fails to balance benign performance and backdoor unlearning. As shown in Fig. 2, larger α (red-circle) rapidly reduces attack success rate (ASR) but causes gradient explosion in later fine-tuning, leading to catastrophic degradation. Conversely, smaller α (blue-square) preserves benign performance but yields insufficient backdoor unlearning. This indicates that it is challenging to find a suitable value of α that simultaneously ensures stable performance and effective backdoor unlearning. Moreover, different attacks exhibit varying degrees of sensitivity to α.

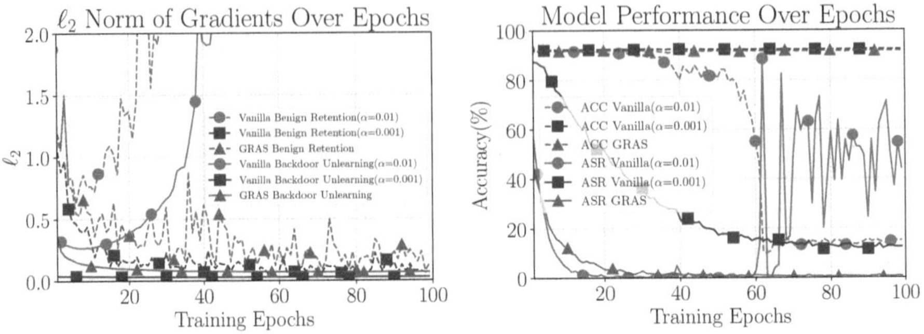

(a) ℓ_2 Norm Dynamics (b) ACC & ASR Dynamics

Fig. 2. Scaling Effects on Unlearning Blended Attack. (Color figure online)

3.3 Gradient Reversal and Adaptive Scaling (GRAS)

We propose an adaptive scaling mechanism for backdoor unlearning based on the current attack risk level, defined as:

$$\Delta\theta = \begin{cases} \Delta\theta_c - \dfrac{\alpha}{\mathcal{L}_p}\Delta\theta_p, & \mathcal{L}_p \geq 1 \\ \Delta\theta_c - \alpha\Delta\theta_p, & \mathcal{L}_p < 1 \end{cases}, \tag{3}$$

where $\mathcal{L}_p = \frac{1}{|D_p|}\sum_{(x_p,y_p)\in D_p}\mathcal{L}(N(x_p;\theta),y_p)$ represents the average loss on poisoned samples.

This formulation dynamically adjusts unlearning intensity by inversely scaling with $\mathcal{L}_p$, achieving dual objectives: 1) The piecewise function prevents numerical instability when $\mathcal{L}_p \approx 0$ by eliminating division by small values; 2) Maintains unlearning efficacy while avoiding catastrophic forgetting through risk-aware gradient scaling.

As illustrated in Fig. 2, the green-triangle curve demonstrates our GRAS, where the adaptive scaling factor modulates update magnitude based on current attack risk. This contrasts with the fixed α approach shown in red-circle/blue-square curves, highlighting the method's ability to balance unlearning speed and model stability.

4 Experiments

4.1 Experimental Setup

We evaluated our approach on three benchmark datasets: 1) CIFAR-10, 2) GTSRB, and 3) Tiny-ImageNet. Baseline backdoored models were trained on the PreActResNet-18 architecture using the stochastic gradient descent (SGD) optimizer with cosine annealing (multi-stage learning rate reduction for Tiny-ImageNet), with an initial learning rate of 0.01 and momentum of 0.9. The training epochs were set as: 1) 100 epochs for CIFAR-10, 2) 50 epochs for GTSRB, 3) 200 epochs for Tiny-ImageNet.

Attack Setup. We validated our method against eight backdoor attacks: 1) BadNet [6], 2) Blended [4], 3) BPP [20], 4) FTrojan [19], 5) LF [24], 6) Refool [12], 7) SIG [2], and 8) WaNet [16]. All attacks used 10% poisoning ratio targeting class 0, following default parameters from *BackdoorBench* [23]. SIG was excluded for GTSRB/Tiny-ImageNet due to insufficient dataset size for poisoning requirements.

Defense Setup. We compared with six state-of-the-art backdoor mitigation methods: 1) NAD [9], 2) FT-SAM [27], 3) NPD [28], 4) RNP [10], 5) FST [14], and 6) ULRL [13]. Each defense was provided with 5% clean samples from training sets, with parameter settings following original papers and *Backdoor-Bench*. RNP used 1% clean samples due to sensitivity to excessive data [10].

Our GRAS method maintained consistent purification settings across all experiments. We set hyperparameter $\alpha = 0.1$, randomly selected 50 poisoned samples for unlearning, and performed 20-epoch fine-tuning using SGD optimizer with cosine annealing (initial learning rate 0.01). To mitigate random fluctuations, each defense method was executed five times with distinct random seeds for every backdoored model, and results reported as the average of five runs.

Evaluation Metrics. We measured defense effectiveness through three metrics: 1) Benign Accuracy (ACC): model performance on clean test samples, 2) Attack Success Rate (ASR): proportion of poisoned samples misclassified to the target label, and 3) Restoration Accuracy (RA): accuracy of poisoned samples classified correctly to their original labels.

4.2 Experimental Results

The experimental results in Table 1 show that:

GRAS's Performance in Unlearning Backdoor Patterns. While maintaining benign model performance, GRAS significantly reduces the average attack success rate on the CIFAR-10 dataset from 97.8% to 3.24%, while increasing the poisoned sample restoration accuracy from 2.12% to 81.31%. On the GTSRB dataset, GRAS drops the average attack success rate from 96.45% to 8.73%, with poisoned sample restoration accuracy rising from 3.46% to 77.14%. For the Tiny-ImageNet dataset, GRAS declines the average attack success rate from 99.44% to 0.57%, while increasing the poisoned sample restoration accuracy from 0.43% to 44.91%.

Limitations of Existing Approaches in Restoring Poisoned Samples. We perform t-SNE dimensionality reduction on the output features of purified models. As shown in Fig. 3, after Blended backdoor purification, some methods still exhibit concentrated cluster distributions for poisoned samples after purification, while GRAS achieves sparse distribution patterns for poisoned samples.

Table 1. Comparative Evaluation of Defense Performance Across Three Datasets. **Note:** Defense failure (ASR⩾20% or ACC drop⩾20%) is *gray-shaded*, best results are **bolded**, with benign accuracy (ACC) in **blue***, attack success rate (ASR) in **red**†, and restoration accuracy (RA) in **green**‡.

Dataset	Attack	No Defense			NAD			FT-SAM			NPD		
		ACC	ASR	RA	ACC	ASR	RA	ACC	ASR	RA	ACC	ASR	RA
CIFAR-10	BadNet	91.82	93.79	5.90	89.60	1.46	89.10	**91.79***	1.67	**90.81**‡	*64.79*	*0.00*	*52.93*
	Blended	93.69	99.76	0.24	*92.25*	*65.08*	*27.49*	**92.55***	16.34	59.27	*57.29*	*0.00*	*34.74*
	BPP	94.13	99.96	0.03	92.45	1.79	**87.61**‡	92.91	3.72	86.40	*61.79*	*0.00*	*58.85*
	FTrojan	93.66	99.99	0.01	92.14	1.81	69.19	**92.61***	1.57	80.71	80.77	**0.00**†	66.44
	LF	93.11	98.97	0.97	*91.71*	*32.54*	*59.87*	**92.29***	4.63	**83.83**‡	*64.39*	*1.97*	*46.18*
	Refool	92.37	93.23	6.54	*91.41*	*22.10*	*65.00*	**92.13***	4.82	80.49	*60.91*	*0.08*	*49.00*
	SIG	84.49	97.87	2.12	89.82	4.26	56.45	**91.19***	1.69	61.36	*75.01*	*33.06*	*39.80*
	WaNet	92.92	98.81	1.17	92.68	0.71	**91.20**‡	**92.83***	0.92	91.08	*66.91*	*0.00*	*56.67*
	AVG	92.02	97.80	2.12	91.51	16.22	68.24	**92.29***	4.42	79.24	*66.48*	*4.39*	*50.58*
GTSRB	BadNet	97.60	94.36	5.47	**98.94***	0.31	**98.51**‡	*98.07*	*40.77*	*58.75*	80.92	13.60	67.26
	Blended	98.72	99.99	0.01	**98.26***	17.51	**42.73**‡	*98.22*	*97.74*	*2.12*	*98.47*	*96.11*	*2.58*
	BPP	97.66	99.82	0.16	**99.07***	0.00	92.83	98.28	0.00	**92.97**‡	*73.76*	*0.00*	*38.57*
	FTrojan	98.48	100.00	0.00	98.03	0.00	**97.53**‡	*97.75*	*25.26*	*65.00*	*98.15*	*80.00*	*10.48*
	LF	98.04	99.55	0.44	**98.27***	2.59	75.99	*97.88*	*93.61*	*5.61*	*97.90*	*87.35*	*3.35*
	Refool	96.41	83.39	16.21	*98.43*	*34.27*	*56.76*	*97.14*	*76.32*	*22.39*	*89.49*	*21.44*	*44.03*
	WaNet	97.55	98.01	1.95	**99.11***	0.00	**98.83**‡	98.44	0.05	97.68	78.06	17.16	47.06
	AVG	97.78	96.45	3.46	**98.59***	7.81	**80.45**‡	*97.97*	*47.68*	*49.22*	*88.10*	*45.09*	*30.47*
Tiny-ImageNet	BadNet	56.12	99.90	0.10	50.98	0.16	50.13	**55.44***	0.15	**54.90**‡	*10.66*	*0.00*	*10.40*
	Blended	56.49	99.67	0.22	*51.15*	*97.17*	*1.92*	*55.32*	*99.31*	*0.49*	*21.65*	*0.00*	*7.63*
	BPP	57.92	99.91	0.08	52.33	0.34	47.55	**57.95***	0.14	49.48	*20.22*	*0.00*	*13.85*
	FTrojan	56.19	100.00	0.00	*51.02*	*49.31*	*22.98*	*55.02*	*95.07*	*3.23*	*25.73*	*0.19*	*14.94*
	LF	55.98	98.59	1.05	*51.02*	*95.04*	*3.32*	*54.95*	*95.67*	*2.90*	*23.99*	*0.00*	*16.03*
	Refool	55.79	98.51	1.19	*50.92*	*85.89*	*9.21*	*55.09*	*94.61*	*3.88*	*14.96*	*14.87*	*6.58*
	WaNet	57.26	99.49	0.38	52.67	**0.02**†	40.80	**57.52***	0.04	42.68	*26.89*	*0.00*	*17.75*
	AVG	56.54	99.44	0.43	*51.44*	*46.85*	*25.13*	*55.90*	*55.00*	*22.51*	*20.59*	*2.15*	*12.46*

Dataset	Attack	RNP			FST			ULRL			**GRAS (Ours)**		
		ACC	ASR	RA	ACC	ASR	RA	ACC	ASR	RA	ACC	ASR	RA
CIFAR-10	BadNet	92.54	**0.00**†	35.62	88.74	0.60	87.84	90.78	2.04	89.50	90.52	1.04	90.44
	Blended	91.12	1.31	49.00	91.29	**0.31**†	41.13	92.41	3.38	51.53	91.62	4.83	**69.09**‡
	BPP	*91.37*	*39.06*	*42.95*	91.36	8.40	65.86	**93.13***	7.04	71.02	92.20	**0.75**†	87.56
	FTrojan	92.54	**0.00**†	35.62	91.02	**0.00**†	29.41	92.42	0.02	36.53	92.24	1.18	**86.90**‡
	LF	91.27	4.43	54.69	90.96	**1.03**†	65.76	92.27	1.66	69.77	91.54	13.63	78.17
	Refool	*89.92*	*28.42*	*55.57*	90.20	**3.12**†	69.84	91.68	4.20	75.93	91.41	4.08	**82.96**‡
	SIG	*88.80*	*52.66*	*32.58*	89.66	**0.00**†	44.42	90.30	0.24	53.98	85.77	0.16	**66.94**‡
	WaNet	91.91	1.82	80.81	90.42	0.36	83.08	92.72	0.51	89.94	90.77	**0.22**†	88.41
	AVG	90.64	16.01	54.54	90.46	**1.73**†	60.92	91.96	2.39	67.28	90.76	3.24	**81.31**‡
GTSRB	BadNet	97.68	6.14	86.87	97.81	**0.00**†	96.32	98.02	**0.00**†	96.58	97.79	**0.00**†	97.82
	Blended	*98.06*	*86.73*	*4.79*	98.04	3.25	30.37	98.10	**0.01**†	20.82	*96.27*	*34.21*	*39.55*
	BPP	*98.02*	*90.38*	*6.41*	98.36	**0.00**†	45.27	98.15	**0.00**†	23.39	97.90	**0.00**†	75.69
	FTrojan	**98.14***	3.22	1.35	97.54	1.14	19.04	97.83	**0.00**†	9.85	97.90	4.64	92.46
	LF	97.75	15.67	3.36	97.92	0.05	23.34	97.81	**0.00**†	12.76	97.15	2.56	**84.39**‡
	Refool	96.82	7.09	42.75	**97.45***	0.02	**58.75**‡	97.12	**0.01**†	50.16	96.44	19.68	56.09
	WaNet	*97.96*	*69.78*	*13.74*	98.27	**0.00**†	90.06	98.23	**0.00**†	57.10	97.53	**0.00**†	93.98
	AVG	*97.78*	*39.86*	*22.75*	97.91	0.64	51.88	97.89	**0.00**†	38.67	97.28	8.73	77.14
Tiny-ImageNet	BadNet	47.87	0.68	45.24	*48.98*	*29.93*	*29.64*	51.48	**0.00**†	38.33	53.30	0.01	53.05
	Blended	47.31	0.38	14.55	*48.62*	*20.16*	*15.66*	**50.76***	**0.00**†	16.37	48.81	0.10	**30.72**‡
	BPP	49.17	0.71	7.01	50.33	11.44	22.64	52.73	0.01	22.34	54.95	**0.00**†	**49.55**‡
	FTrojan	47.25	0.43	2.51	*49.23*	*81.97*	*4.89*	50.83	**0.00**†	5.76	**52.34***	0.45	**48.40**‡
	LF	*46.69*	*25.19*	*17.01*	*48.44*	*27.53*	*21.97*	50.88	**0.00**†	26.15	**51.63***	0.01	**45.72**‡
	Refool	47.05	1.68	18.01	*48.79*	*28.52*	*23.42*	51.11	**0.00**†	26.93	**51.82***	2.17	**39.24**‡
	WaNet	48.57	0.11	28.40	50.48	0.16	39.58	*53.51*	*55.79*	*24.49*	54.00	1.27	**47.70**‡
	AVG	47.70	4.17	18.96	*49.27*	*28.53*	*22.54*	51.61	7.97	22.91	**52.41***	**0.57**†	**44.91**‡

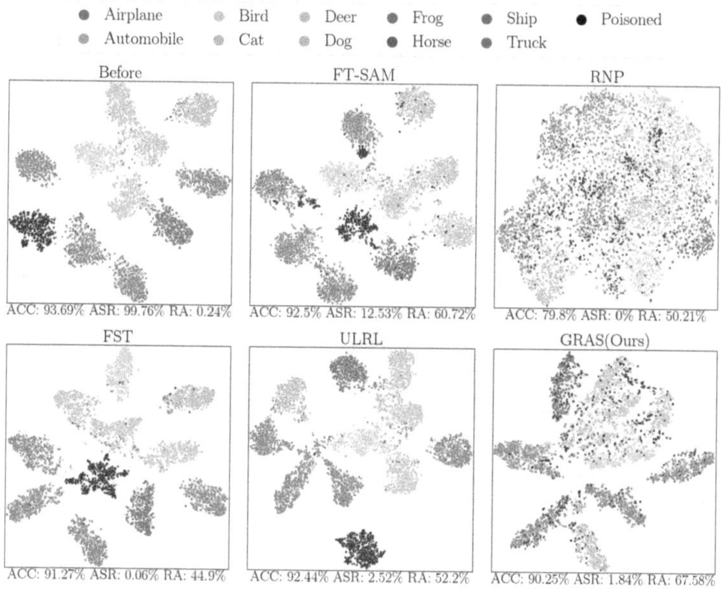

Fig. 3. t-SNE Visualization of Model Output Features after Different Defenses.

4.3 Robustness and Generalization Analysis of GRAS

Effect of Varying the Number of Poisoned Samples We examined the impact of varying the number of backdoor samples utilized during unlearning on defense performance, maintaining consistency with the main experimental setup Sect. 4.1 for all other parameters. As illustrated in Fig. 4a, GRAS demonstrates effective defense across three datasets. Each data point denotes the mean performance calculated from 24 independent runs for CIFAR-10 (and 21 for GTSRB/Tiny-ImageNet), encompassing all attack configurations over 3 distinct random seeds. Notably, GRAS maintains acceptable defense efficacy even when exposed to a single poisoned sample.

Impact of Fine-Tuning Epochs. As shown in Fig. 4b, across three datasets, GRAS achieves satisfactory average defense performance after the first 20 fine-tuning epochs. As fine-tuning progresses, GRAS maintains stable defense performance without significant degradation or fluctuations.

Generalization Across Architectures. We validated GRAS's generalization to different model architectures by testing on VGG-19 and ResNet-18. Table 2 summarizes the results (other settings follow Sect. 4.1). This demonstrates GRAS's architecture-agnostic effectiveness.

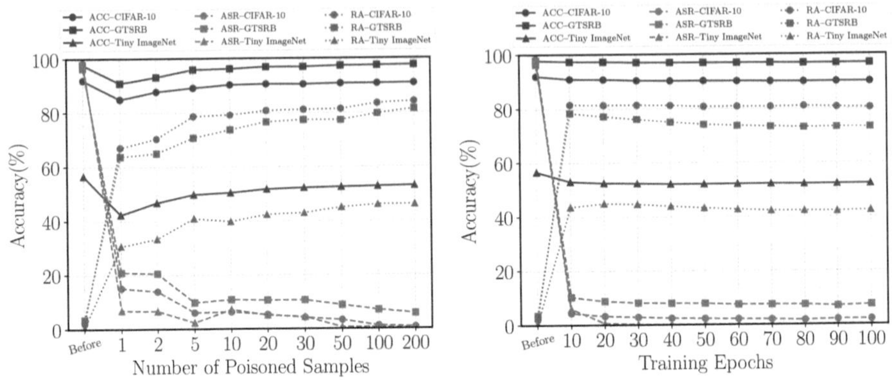

(a) Different Poisoned Samples

(b) Different Fine-tuning Epochs

Fig. 4. Impact of Poisoned Sample Quantity and Epochs.

Table 2. Defense Performance Across Different Architectures on CIFAR-10.

Model	Attack	No Defense			GRAS(Ours)		
		ACC	ASR	RA	ACC	ASR	RA
VGG-19	BadNet	88.37	96.18	3.57	86.00	0.68	76.99
	Blended	90.29	99.28	0.69	79.95	0.58	36.53
	BPP	91.86	99.69	0.31	86.17	2.23	79.64
	FTrojan	82.81	14.12	81.10	85.36	1.26	85.58
	Refool	88.04	89.74	8.80	84.52	1.43	66.91
	WaNet	89.85	95.22	4.52	83.09	2.36	80.18
	AVG	88.54	82.37	16.50	84.18	1.42	70.97
ResNet-18	BadNet	83.63	89.50	9.29	82.98	2.54	81.35
	Blended	84.71	98.82	0.97	81.82	12.90	63.94
	BPP	85.30	80.51	17.84	81.93	2.24	80.08
	FTrojan	79.95	8.31	78.26	81.66	1.73	81.38
	Refool	83.18	86.84	10.29	81.86	2.60	72.04
	WaNet	82.41	91.73	7.20	81.78	1.22	79.01
	AVG	83.20	75.95	20.64	82.01	3.87	76.30

5 Conclusion

We propose GRAS, an adaptive gradient reversal method for backdoor unlearning that dynamically adjusts unlearning intensity to maintain benign performance, effectively reducing attack success rates, and improving restoration accuracy of poisoned samples. Although our current work demonstrates effective backdoor mitigation using minimal poisoned samples, several critical challenges remain. Future work will explore adaptive sampling strategies for acquiring clean and poisoned samples tailored to real-world deployment constraints, as well as theoretical guarantees for unlearning efficacy.

References

1. Bai, Y., et al.: Backdoor attack and defense on deep learning: a survey. IEEE Trans. Comput. Soc. Syst. (2024)
2. Barni, M., Kallas, K., Tondi, B.: A new backdoor attack in CNNs by training set corruption without label poisoning. In: 2019 IEEE International Conference on Image Processing (ICIP), pp. 101–105. IEEE (2019)
3. Chen, C., Hong, H., Xiang, T., Xie, M.: Anti-backdoor model: a novel algorithm to remove backdoors in a non-invasive way. IEEE Trans. Inf. Forensics Secur. (2024)
4. Chen, X., Liu, C., Li, B., Lu, K., Song, D.: Targeted backdoor attacks on deep learning systems using data poisoning. arXiv preprint arXiv:1712.05526 (2017)
5. Dunnett, K., Arablouei, R., Miller, D., Dedeoglu, V., Jurdak, R.: Unlearning backdoor attacks through gradient-based model pruning. In: 2024 54th Annual IEEE/IFIP International Conference on Dependable Systems and Networks Workshops (DSN-W), pp. 46–54. IEEE (2024)
6. Gu, T., Liu, K., Dolan-Gavitt, B., Garg, S.: BadNets: evaluating backdooring attacks on deep neural networks. IEEE Access **7**, 47230–47244 (2019)
7. Hou, L., Feng, R., Hua, Z., Luo, W., Zhang, L.Y., Li, Y.: IBD-PSC: input-level backdoor detection via parameter-oriented scaling consistency. In: Proceedings of the 41st International Conference on Machine Learning, pp. 18992–19022 (2024)
8. Hou, L., Luo, W., Hua, Z., Chen, S., Zhang, L.Y., Li, Y.: Flare: towards universal dataset purification against backdoor attacks. IEEE Trans. Inf. Forensics Secur. (2025)
9. Li, Y., Lyu, X., Koren, N., Lyu, L., Li, B., Ma, X.: Neural attention distillation: erasing backdoor triggers from deep neural networks. In: 9th International Conference on Learning Representations, ICLR 2021 (2021)
10. Li, Y., et al.: Reconstructive neuron pruning for backdoor defense. In: International Conference on Machine Learning, pp. 19837–19854. PMLR (2023)
11. Liu, X., et al.: Detecting backdoors during the inference stage based on corruption robustness consistency. In: Proceedings of the IEEE/CVF Conference on Computer Vision and Pattern Recognition, pp. 16363–16372 (2023)
12. Liu, Y., Ma, X., Bailey, J., Lu, F.: Reflection backdoor: a natural backdoor attack on deep neural networks. In: Vedaldi, A., Bischof, H., Brox, T., Frahm, J.-M. (eds.) ECCV 2020. LNCS, vol. 12355, pp. 182–199. Springer, Cham (2020). https://doi.org/10.1007/978-3-030-58607-2_11
13. Min, N.M., Pham, L.H., Sun, J.: Unified neural backdoor removal with only few clean samples through unlearning and relearning. IEEE Trans. Inf. Forensics Secur. (2025)
14. Min, R., Qin, Z., Shen, L., Cheng, M.: Towards stable backdoor purification through feature shift tuning. Adv. Neural. Inf. Process. Syst. **36**, 75286–75306 (2023)
15. Min, R., Qin, Z., Zhang, N.L., Shen, L., Cheng, M.: Uncovering, explaining, and mitigating the superficial safety of backdoor defense. Adv. Neural. Inf. Process. Syst. **37**, 77934–77961 (2024)
16. Nguyen, T.A., Tran, A.T.: WANet - imperceptible warping-based backdoor attack. In: International Conference on Learning Representations (2021). https://openreview.net/forum?id=eEn8KTtJOx

17. Samed, A., Sagiroglu, S.: Explainable artificial intelligence models in intrusion detection systems. Eng. Appl. Artif. Intell. **144**, 110145 (2025)
18. Turner, A., Tsipras, D., Madry, A.: Label-consistent backdoor attacks. arXiv preprint arXiv:1912.02771 (2019)
19. Wang, T., Yao, Y., Xu, F., An, S., Tong, H., Wang, T.: An invisible black-box backdoor attack through frequency domain. In: Avidan, S., Brostow, G., Cissé, M., Farinella, G.M., Hassner, T. (eds.) ECCV 2022. LNCS, vol. 13673, pp. 396–413. Springer, Cham (2022). https://doi.org/10.1007/978-3-031-19778-9_23
20. Wang, Z., Zhai, J., Ma, S.: BPPAttack: Stealthy and efficient trojan attacks against deep neural networks via image quantization and contrastive adversarial learning. In: Proceedings of the IEEE/CVF Conference on Computer Vision and Pattern Recognition (CVPR), pp. 15074–15084 (2022)
21. Wei, L., Yang, J., Jin, H., Cui, J., Li, J., He, D.: Sustainable learning-based intrusion detection system for VANets. IEEE Trans. Intell. Transp. Syst. (2025)
22. Wei, S., Zhang, M., Zha, H., Wu, B.: Shared adversarial unlearning: Backdoor mitigation by unlearning shared adversarial examples. Adv. Neural. Inf. Process. Syst. **36**, 25876–25909 (2023)
23. Wu, B., et al.: Backdoorbench: a comprehensive benchmark and analysis of backdoor learning. Int. J. Comput. Vis. 1–88 (2025)
24. Zeng, Y., Park, W., Mao, Z.M., Jia, R.: Rethinking the backdoor attacks' triggers: a frequency perspective. In: Proceedings of the IEEE/CVF International Conference on Computer Vision (2021)
25. Zhang, H., et al.: Test-time backdoor detection for object detection models. In: Proceedings of the Computer Vision and Pattern Recognition Conference, pp. 24377–24386 (2025)
26. Zhu, M., Liang, S., Wu, B.: Breaking the false sense of security in backdoor defense through re-activation attack. Adv. Neural. Inf. Process. Syst. **37**, 114928–114964 (2024)
27. Zhu, M., Wei, S., Shen, L., Fan, Y., Wu, B.: Enhancing fine-tuning based backdoor defense with sharpness-aware minimization. In: Proceedings of the IEEE/CVF International Conference on Computer Vision, pp. 4466–4477 (2023)
28. Zhu, M., Wei, S., Zha, H., Wu, B.: Neural polarizer: a lightweight and effective backdoor defense via purifying poisoned features. Adv. Neural. Inf. Process. Syst. **36**, 1132–1153 (2023)
29. Zhuang, Z., Ding, Y., Zheng, S., Shao, J., Zhu, D., Wang, H.: Cracking the code of backdoor attacks with confidence consistency. IEEE Trans. Inf. Forensics Secur. (2026)

Robust Loss Functions for RBF Networks and Their Derivatives

Jan Kalina(✉) and Kristýna Dvořáková

The Czech Academy of Sciences, Institute of Computer Science, Pod Vodárenskou věží 2, 182 00 Prague 8, Czech Republic
{kalina,dvorakova}@cs.cas.cz
http://www.cs.cas.cz/staff/kalina/en

Abstract. This paper explores highly robust methods for training Radial Basis Function (RBF) networks, focusing on derivatives of their loss functions. RBF networks are well-known for their ability to model complex nonlinear relationships, but their sensitivity to outliers and anomalies poses a challenge in real-world applications. To address this, we recall and build upon several recently proposed highly robust versions of RBF networks, which are inspired by robust parameter estimation techniques used in linear regression. We provide a detailed derivation of derivatives of the fitted values and loss functions, which are critical for optimizing the network and enhancing its robustness to noisy data. Furthermore, the paper discusses the significance of these derivatives in the context of training and model performance, emphasizing their role in improving the network's stability and robustness. The results presented here contribute to the development of more reliable and efficient RBF networks, especially for tasks involving complex and noisy datasets.

Keywords: Nonlinear regression · RBF networks · outliers · robust estimation · loss function · derivatives

1 Introduction

Radial Basis Function (RBF) networks are a class of artificial neural networks widely used for nonlinear regression tasks. They model complex relationships between inputs and outputs by transforming the input data into a higher-dimensional space using radial basis functions, which allows them to capture intricate patterns and nonlinearities in the data [1]. The aim of this paper is to derive the derivatives of the loss function for several highly robust versions of RBF networks, inspired by robust regression estimators commonly used in robust statistics [5].

The robustness properties of RBF networks have been explored in studies such as [2]. Addressing outliers, anomalies, and adversarial examples is crucial for various neural network versions. Such outliers may arise from measurement errors, transmission noise, faulty sensors, or incorrectly recorded observations,

L. Pasa et al. (Eds.): ICANN 2026, LNCS 17089, pp. 683–694, 2027.
https://doi.org/10.1007/978-3-032-38398-3_56

while adversarial examples correspond to intentionally manipulated inputs that induce erroneous network predictions. In practical applications, outlier removal in RBF networks is commonly performed, though this is typically done without relying on robust parameter estimation. A robust approach to handling outliers in RBF networks was introduced in [10] through simple yet effective modifications to the traditional RBF structure. The proposed robust RBF network incorporates an outlier detection mechanism, enabling adaptation to noisy or erroneous data points. This is achieved by modifying the distance metric used to compute the RBF centers and by adjusting the weights to reduce the influence of outliers during training.

Further advancing the concept, a multi-layer RBF network architecture was proposed in [8], which simplifies out-of-distribution (OOD) detection by effectively combining classification and OOD detection into a single step. Adversarial noise propagation (ANP) for robust neural networks was also studied [13]. Additionally, the automatic design of robust RBF regression networks, integrating conventional pattern recognition techniques, evolutionary approaches, and recent advancements in computational and statistical learning theory, was explored in [12]. Further references on recent robustness issues in RBF networks can be found in Sect. 2 or in the references cited therein.

The main contribution of this paper is the analytical derivation of gradients of robust regression losses for RBF networks, while the loss functions themselves are taken from existing robust statistics literature. Despite recent advancements in robust RBF networks, existing methods often do not fully address the challenges of robust parameter estimation, especially in the context of multi-layer networks and their application to noisy or outlier-prone data, which motivates the need for a more comprehensive approach. In particular, explicit gradient formulas are derived for LTA- and LWS-based robust estimators (least trimmed absolute deviations and least weighted squares) within the RBF framework.

After we recall the nonlinear regression model in Sect. 1.1, Sect. 2 overviews the RBF network model and robust methods for training. Derivatives of their loss functions are formulated in Sect. 3. Section 4 brings conclusions.

1.1 Model

Model. *We consider the nonlinear regression model*

$$Y_i = \varphi(X_i) + e_i, \quad i = 1, \dots, n, \tag{1}$$

with an unknown function φ. *The continuous random variable* $Y_i \in \mathbb{R}$ *represents the response variable and the vectors* $X_i = (X_{i1}, \dots, X_{ip})^T \in \mathbb{R}^p$, *which are* p*-dimensional random vectors available for the total number of* n *measurements, play the role of predictors (regressors). The random errors* $e_1, \dots, e_n$ *are independent (not necessarily identically distributed) and independent on the predictors. While* φ *is unknown, we consider an approximation of* (1) *by means of the model*

$$Y_i = f(X_i, \theta) + e_i, \quad i = 1, \dots, n, \tag{2}$$

with a known nonlinear function f, which is determined up to the parameter $\theta \in \mathbb{R}^d$.

2 Radial Basis Function Networks

Let us start by recalling the RBF network model for the nonlinear regression Model 1.1. The robust estimators of its parameters are then recalled in Sect. 2.1.

Radial basis function (RBF) networks contain an input layer with p inputs, a single hidden layer with the total number N of RBF units (neurons), and a linear output layer. Let us use the standard notation exploited e.g. in [1]. The user chooses the value of N together with a radially symmetric differentiable function (kernel)

$$\rho : \mathbb{R}_0^+ \to \mathbb{R}_0^+, \quad \text{where } \mathbb{R}_0^+ := \{x \in \mathbb{R};\ x \geq 0\}. \tag{3}$$

The model for the RBF network has the form

$$f(x) = \sum_{j=1}^{N} a_j \rho\left((x - c_j)^\top (x - c_j)\right), \quad x \in \mathbb{R}^p, \tag{4}$$

with parameters $c_1, \ldots, c_N \in \mathbb{R}^p$ and $a_1, \ldots, a_N \in \mathbb{R}$, and possibly with other parameters corresponding to the (typically Gaussian) kernel ρ. The analysis in this paper is developed for the Gaussian kernel, although several steps may extend to other radial kernels under additional regularity conditions. We do not pursue this generalization here.

We will assume that ρ is the Gaussian kernel in the form

$$\rho(x) = \exp\left\{-\frac{x^2}{2\sigma^2}\right\}, \quad x \in \mathbb{R}, \quad \sigma > 0, \tag{5}$$

and we will need its derivatives

$$\frac{\delta \rho(x)}{\delta x} = \exp\left\{-\frac{x^2}{2\sigma^2}\right\}\left(-\frac{x}{2\sigma^2}\right) = -\frac{x}{\sigma^2}\rho(x), \quad x \in \mathbb{R}. \tag{6}$$

The task of training an RBF network is the task to find the estimates $\hat{a}_1, \ldots, \hat{a}_N$ and $\hat{c}_1, \ldots, \hat{c}_N$ of the parameters $a_1, \ldots, a_N$ and $c_1, \ldots, c_N$, respectively; σ^2 plays the role of a nuisance parameter. We will use the notation $\theta \in \mathbb{R}^d$ for all the parameters of an RBF network; if the Gaussian kernel is used, then we have

$$\theta = (a_1, \ldots, a_N, c_1, \ldots, c_N, \sigma)^T \in \mathbb{R}^d \tag{7}$$

with $d = 2N + 1$. Throughout the paper, estimates of the parameter θ will be denoted as $\hat{\theta}$.

The standard (plain) approach to training RBF networks is based on

$$\underset{\tilde{\theta} \in \mathbb{R}^d}{\operatorname{argmin}} \sum_{i=1}^{n} u_i^2(\tilde{\theta}), \tag{8}$$

where $u_1(\hat{\theta}), \ldots, u_n(\hat{\theta})$ are residuals for a given estimate $\hat{\theta} \in \mathbb{R}^d$. Alternatively, we could refer to the basic RBF networks as LS-RBF to emphasize that the loss function resembles the least squares (LS) of residuals, although this notation is not commonly used in the literature. Further, we denote the fitted values by

$$\hat{Y}_i = \hat{Y}_i(\hat{\theta}) = \sum_{j=1}^{N} \hat{a}_j \rho \left(||X_i - \hat{c}_j|| \right), \quad i = 1, \ldots, n, \tag{9}$$

where $\hat{a}_j \in \mathbb{R}$ and $\hat{c}_j \in \mathbb{R}^p$ for $j = 1, \ldots, N$.

2.1 Robust Training of RBF Networks

The standard training of RBF networks suffers from the presence of outliers in the data. Thus, several robust alternatives, which are resistant to severe outliers, were proposed. The definition of some of the robust versions exploits the concept of weight functions

$$\psi : (0,1) \rightarrow [0, +\infty), \tag{10}$$

which have to be non-increasing and continuous. Let us use the notation

$$u^2_{(1)}(\hat{\theta}) \leq \cdots \leq u^2_{(n)}(\hat{\theta}) \tag{11}$$

for squared residuals arranged in ascending order.

The robust RBF networks based on the least weighted squares (LWS) estimator are obtained by

$$\operatorname*{argmin}_{\tilde{\theta} \in \mathbb{R}^d} \sum_{i=1}^{n} \psi \left(\frac{i - 1/2}{n} \right) u^2_{(i)}(\tilde{\theta}). \tag{12}$$

We refer to the loss function in (12) as the nonlinear least weighted squares (LWS). If we take

$$w_i = \psi \left(\frac{i - 1/2}{n} \right), \quad i = 1, \ldots, n, \tag{13}$$

the definition (12) can be equivalently expressed as

$$\operatorname*{argmin}_{\tilde{\theta} \in \mathbb{R}^d} \sum_{i=1}^{n} w_i u^2_i(\tilde{\theta}). \tag{14}$$

LTS-RBF network, which is based on the least trimmed squares estimator (LTS) (cf. [11]), is obtained by

$$\operatorname*{argmin}_{\tilde{\theta} \in \mathbb{R}^d} \sum_{i=1}^{h} u^2_{(i)}(\tilde{\theta}) \tag{15}$$

for a given $h \in [n/2, n)$.

LTA-RBF networks, inspired by the least trimmed absolute value estimator (LTA) studied in [18], are obtained by

$$\operatorname*{argmin}_{\tilde{\theta}\in\mathbb{R}^d} \sum_{i=1}^{h} |u(\tilde{\theta})|_{(i)}, \quad h \in [n/2, n). \tag{16}$$

The LTA estimator can be generalized to the least weighted absolute value estimator (LWA); we are not aware of any applications of this estimator to real data. In an analogous spirit, LWA-RBF networks inspired by the least weighted absolute value estimator (LWA) may be defined by

$$\operatorname*{argmin}_{\tilde{\theta}\in\mathbb{R}^d} \sum_{i=1}^{n} \psi\left(\frac{i-1/2}{n}\right) |u(\tilde{\theta})|_{(i)} \tag{17}$$

with a given weight function ψ, where we use the notation

$$|u(\hat{\theta})|_{(1)} \leq \cdots |u(\hat{\theta})|_{(n)} \tag{18}$$

for absolute values of residuals arranged in ascending order.

A tool for a suitable choice of the trimming constant for robust NNs were investigated in [6]; this tool (sequential outlier detection) may be exploited for estimating the contamination level in given data.

3 Derivatives for Robust RBF Networks

This section is devoted to derivatives of highly robust RBF networks for Model 1.1. Although the robust objective functions are directly adopted from robust regression, the derivation of their gradients in the nonlinear RBF network setting requires separate analytical treatment and constitutes the main contribution of this paper. Before we formulate the derivatives of their loss functions, we present the following preparatory results for partial derivatives of the fitted values.

3.1 Derivatives of the Fitted Values

First, let us recall

$$||X_i - \hat{c}_j|| = \sqrt{(X_i - \hat{c}_j)^T (X_i - \hat{c}_j)} = \sqrt{\sum_{k=1}^{p} (X_{ik} - \hat{c}_{jk})^2}, \tag{19}$$

where again the hats denote (any) given estimates of the unknown parameters.

We proceed to formulating partial derivatives of the fitted values with respect to the vectors $\hat{c}_1, \ldots, \hat{c}_N$. Let us prepare

$$\begin{aligned} \frac{\delta}{\delta \hat{c}_{lt}} \sum_{k=1}^{p} (X_{ik} - \hat{c}_{jk})^2 &= \frac{\delta}{\delta c_{lt}} \left((X_{it} - \hat{c}_{lt})^2 \right) \mathbb{1}[j = l] \\ &= -2(X_{it} - \hat{c}_{lt}) \end{aligned} \tag{20}$$

for $i = 1, \ldots, n,\ j = 1, \ldots, N,\ l = 1, \ldots, N,$ and $t = 1, \ldots, p$. We will also need

$$\begin{aligned}
&\frac{\delta}{\delta \hat{c}_{lt}} \sqrt{\sum_{k=1}^{p} (X_{ik} - \hat{c}_{jk})^2} \\
&= -\frac{1}{2||X_i - \hat{c}_j||} \frac{\delta}{\delta \hat{c}_{lt}} \left(\sum_{k=1}^{p} (X_{ik} - \hat{c}_{jk})^2 \right) \\
&= -\frac{1}{2||X_i - \hat{c}_j||} (-2)(X_{it} - \hat{c}_{lt}) \mathbb{1}[j = l] \\
&= \frac{X_{it} - \hat{c}_{lt}}{||X_i - \hat{c}_j||} \mathbb{1}[j = l]. \qquad (21)
\end{aligned}$$

We use these preparatory results to derive the following result, which holds for any choice of the loss function for the RBF network.

Lemma 1. *We consider the model (2), use the notation of Sect. 1.1, and assume the Gaussian kernel (5). Then, it holds that*

1.

$$\frac{\delta \hat{Y}_i}{\delta \hat{a}_j} = \rho\left(||X_i - \hat{c}_j||\right), \quad i = 1, \ldots, n, \quad j = 1, \ldots, N, \tag{22}$$

2.

$$\frac{\delta \hat{Y}_i}{\delta \hat{c}_{lt}} = -\frac{a_l}{\hat{\sigma}^2} (X_{it} - \hat{c}_{lt}) \rho\left(||X_i - \hat{c}_l||\right) \tag{23}$$

for $i = 1, \ldots, n,\ l = 1, \ldots, N,$ *and* $t = 1, \ldots, p$.

3.2 Derivatives of Robust Loss Functions in Linear Regression

The derivatives for the robust RBF networks will be inspired by corresponding derivatives in the linear regression model. We recall the derivatives for the LTS and LWS estimators and derive original formulas for the derivatives also for the LTA and LWA estimators.

Let us recall the standard linear regression model

$$Y_i = \beta_0 + \beta_1 X_{i1} + \cdots + \beta_p X_{ip} + e_i, \quad i = 1, \ldots, n. \tag{24}$$

The robust estimators investigated here represent analogs of the methods of Sect. 2, if considered in the linear model (24). Each of the considered robust estimators is obtained as

$$\hat{\beta} = \operatorname*{argmin}_{\tilde{\beta} \in \mathbb{R}^p} \xi(\tilde{\beta}), \tag{25}$$

where $\xi(\tilde{\beta})$ is the loss function depending on $\tilde{\beta} \in \mathbb{R}^p$ and the residuals will be denoted as $u_1(\hat{\beta}), \ldots, u_n(\hat{\beta})$.

The loss function of the LTS estimator in (24) is defined as

$$\xi_{\text{LTS}}(\hat{\beta}) = \frac{1}{h}\sum_{i=1}^{h} u_{(i)}^2(\hat{\beta}) = \sum_{i=1}^{n} u_i^2(\hat{\beta})\mathbb{1}\left[u_i^2(\hat{\beta}) \leq u_{(h)}^2(\hat{\beta})\right]. \tag{26}$$

To express its derivatives, we may recall the following result given on p. 7 of [15] stating that

$$\frac{\partial \xi_{\text{LTS}}(\hat{\beta})}{\partial \hat{\beta}} = -\frac{2}{n}\sum_{i=1}^{n}\left[u_i(\hat{\beta})X_i\mathbb{1}[u_i^2(\hat{\beta}) \leq u_{(h)}^2(\hat{\beta})]\right] \tag{27}$$

almost everywhere, where $\mathbb{1}$ denotes an indicator function. The expression (27) contains $p+1$ particular derivatives for individual elements of $\hat{\beta}$. In (27), the i-th squared residual is compared with the h-th largest squared residual.

The LWS estimator in the model (24) has the loss function

$$\xi_{\text{LWS}}(\hat{\beta}) = \sum_{i=1}^{n} \psi\left(\frac{i-1/2}{n}\right) u_{(i)}^2(\hat{\beta}) \tag{28}$$

exploits a specified weight function ψ. The set of derivatives of the loss function in (24) has the form

$$\frac{\partial \xi_{\text{LWS}}}{\partial \hat{\beta}} = -2\sum_{i=1}^{n} X_i u_i \psi\left(\hat{F}^{(n)}\left(|u_i(\hat{\beta})|\right)\right), \tag{29}$$

where $\hat{F}^{(n)}$ denotes the empirical distribution function

$$\hat{F}^{(n)}(r) = \frac{1}{n}\sum_{j=1}^{n} \mathbb{1}[|u_j(\hat{\beta})| < r], \quad r \in \mathbb{R}; \tag{30}$$

a detailed proof was given on p. 183 of [16].

The LWA estimator has the loss function

$$\xi_{\text{LWA}}(\hat{\beta}) = \sum_{i=1}^{n} \psi\left(\frac{i-1/2}{n}\right) |u(\hat{\beta}|_{(i)} = \sum_{i=1}^{n} |u_i|\psi\left(\hat{F}_n(|u_i(\hat{\beta})|)\right), \tag{31}$$

where the residuals depend on $\hat{\beta}$. The derivatives of the loss function have the form

$$\begin{aligned}\frac{\delta \xi_{\text{LWA}}(\hat{\beta})}{\delta \hat{\beta}_j} &= \sum_{i=1}^{n} sgn(u_i(\hat{\beta}))\psi\left(\hat{F}_n(|u_i(\hat{\beta})|)\right)\frac{\delta u_i(\hat{\beta})}{\delta \hat{\beta}_j}\\ &= -\sum_{i=1}^{n} sgn(u_i(\hat{\beta}))\psi\left(\hat{F}_n(|u_i(\hat{\beta})|)\right)\frac{\delta}{\delta \hat{\beta}_j}\hat{Y}_i,\\ &= -\sum_{i=1}^{n} sgn(u_i(\hat{\beta}))X_{ij}\psi\left(\hat{F}_n(|u_i(\hat{\beta})|)\right), \quad j = 1,\dots,p, \quad u_i \neq 0.\end{aligned} \tag{32}$$

The LTA estimator has the loss function

$$\xi_{\text{LTA}}(\hat{\beta}) = \sum_{i=1}^{h} |u(\hat{\beta})|_{(i)} = \sum_{i=1}^{n} |u_i(\hat{\beta})| \, \mathbb{1}\left[|u(\hat{\beta})|_{(i)} \leq |u(\hat{\beta})|_{(h)}\right]. \tag{33}$$

Its derivatives have the form

$$\begin{aligned}\frac{\delta \xi_{\text{LTA}}(\hat{\beta})}{\delta \hat{\beta}_j} &= \sum_{i=1}^{n} sgn(u_i(\hat{\beta}))\mathbb{1}\left[|u(\hat{\beta})|_{(i)} \leq |u(\hat{\beta})|_{(h)}\right] \frac{\delta}{\delta \hat{\beta}_j} u_i(\hat{\beta}) \\ &= -\sum_{i=1}^{n} sgn(u_i(\hat{\beta}))\mathbb{1}\left[|u(\hat{\beta})|_{(i)} \leq |u(\hat{\beta})|_{(h)}\right] \frac{\delta}{\delta \hat{\beta}_j} \hat{Y}_i \\ &= -\sum_{i=1}^{n} sgn(u_i(\hat{\beta}))X_{ij}\mathbb{1}\left[|u(\hat{\beta})|_{(i)} \leq |u(\hat{\beta})|_{(h)}\right], \end{aligned} \tag{34}$$

where $j = 1, \ldots, p$ and $u_i(\hat{\beta}) \neq 0$.

3.3 Derivatives of Loss Functions of Robust RBF Networks

Back to Model 1.1, we investigate the loss functions denoted as

$$\xi_{\text{RBF}}(\hat{\theta}) = \frac{1}{n}\sum_{i=1}^{n} u_i^2(\hat{\theta}) = \frac{1}{n}\sum_{i=1}^{n}(Y_i - \hat{Y}_i)^2, \tag{35}$$

$$\xi_{\text{LTS-RBF}}(\hat{\theta}) = \frac{1}{h}\sum_{i=1}^{h} u_{(i)}^2(\hat{\theta}), \tag{36}$$

$$\xi_{\text{LWS-RBF}}(\hat{\theta}) = \sum_{i=1}^{n} \psi\left(\frac{i-1/2}{n}\right) u_{(i)}^2(\hat{\theta}), \tag{37}$$

$$\xi_{\text{LTA-RBF}}(\hat{\theta}) = \frac{1}{h}\sum_{i=1}^{h} |u(\hat{\theta})|_{(i)}, \tag{38}$$

and

$$\xi_{\text{LWA-RBF}}(\hat{\theta}) = \sum_{i=1}^{n} \psi\left(\frac{i-1/2}{n}\right) |u(\hat{\theta})|_{(i)} \tag{39}$$

considered for a given estimate $\hat{\theta} \in \mathbb{R}^d$ of $\theta \in \mathbb{R}^d$.

Lemma 2. *We consider the model (2), use the notation of Sect. 1.1, and assume the Gaussian kernel (5). Then, it holds that*

1.

$$\frac{\delta}{\delta \hat{a}_j}\xi_{RBF}(\hat{\theta}) = \frac{2}{n}\sum_{i=1}^{n}(Y_i - \hat{Y}_i)\rho(||X_i - \hat{c}_j||), \quad j = 1, \ldots, n, \tag{40}$$

2.

$$\frac{\delta}{\delta \hat{c}_{lt}} \xi_{RBF}(\hat{\theta}) = -\frac{2\hat{a}_l}{n\hat{\sigma}^2} \sum_{i=1}^{n} (Y_i - \hat{Y}_i)(X_{it} - \hat{c}_{lt}) \rho(||X_i - \hat{c}_j||), \quad (41)$$

where $l = 1, \ldots, N$ *and* $t = 1, \ldots, p$,

3.

$$\frac{\delta}{\delta \hat{a}_j} \xi_{LTS\text{-}RBF}(\hat{\theta}) = -2 \sum_{i=1}^{n} u_i(\hat{\theta}) \mathbb{1} \left[u_i^2(\hat{\theta}) \leq u_{(h)}^2 \right] \rho(||X_i - \hat{c}_j||) \ a.e. \quad (42)$$

where $j = 1, \ldots, N$,

4.

$$\frac{\delta}{\delta \hat{c}_{lt}} \xi_{LTS\text{-}RBF}(\hat{\theta}) = \frac{2\hat{a}_l}{\hat{\sigma}^2} \sum_{i=1}^{n} u_i \left[u_i^2 \leq u_{(h)}^2 \right] (X_{it} - \hat{c}_{lt}) \rho(||X_i - \hat{c}_j||) \ a.e., \quad (43)$$

where $l = 1, \ldots, N$ *and* $t = 1, \ldots, p$,

5.

$$\frac{\delta}{\delta \hat{a}_j} \xi_{LWS\text{-}RBF}(\hat{\theta}) = -2 \sum_{i=1}^{n} \psi \left(\frac{i - 1/2}{n} \right) \rho(||X_i - \hat{c}_j||), \quad j = 1, \ldots, N, \quad (44)$$

6.

$$\frac{\delta}{\delta \hat{c}_{lt}} \xi_{LWS\text{-}RBF}(\hat{\theta}) = \frac{2\hat{a}_l}{\hat{\sigma}^2} \sum_{i=1}^{n} u_i \psi \left(\frac{i - 1/2}{n} \right) (X_{it} - \hat{c}_{lt}) \rho(||X_i - \hat{c}_j||), \quad (45)$$

where $l = 1, \ldots, N$ *and* $t = 1, \ldots, p$,

7.

$$\frac{\delta}{\delta \hat{a}_j} \xi_{LTA\text{-}RBF}(\hat{\theta}) = -\sum_{i=1}^{n} sgn(u_i(\hat{\theta})) \rho(||X_i - \hat{c}_j||) \mathbb{1} \left[u_i^2(\hat{\theta}) \leq u_{(h)}^2(\hat{\theta}) \right], \quad (46)$$

where $j = 1, \ldots, N$ *and* $u_i(\hat{\theta}) \neq 0$,

8.

$$\frac{\delta}{\delta \hat{c}_{lt}} \xi_{LTA\text{-}RBF}(\hat{\theta}) = \frac{\hat{a}_l}{\hat{\sigma}^2} \sum_{i=1}^{n} sgn(u_i(\hat{\theta}))(X_{it} - \hat{c}_{lt}) \rho(||X_i - \hat{c}_j||) \mathbb{1} \left[u_i^2(\hat{\theta}) \leq u_{(h)}^2(\hat{\theta}) \right], \quad (47)$$

where $l = 1, \ldots, N$ $t = 1, \ldots, p$, *and* $u_i(\hat{\theta}) \neq 0$,

9.

$$\frac{\delta}{\delta \hat{a}_j} \xi_{LWA\text{-}RBF}(\hat{\theta}) = -\sum_{i=1}^{n} sgn(u_i(\hat{\theta})) \rho(||X_i - \hat{c}_j||) \psi \left(\hat{F}_n(|u_i(\hat{\theta})|) \right), \quad (48)$$

where $j = 1, \ldots, N$ *and* $u_i \neq 0$,

10.

$$\frac{\delta}{\delta \hat{c}_{lt}} \xi_{LWA\text{-}RBF}(\hat{\theta}) = \frac{\hat{a}_l}{\hat{\sigma}^2} \sum_{i=1}^{n} sgn(u_i(\hat{\theta}))(X_{it} - \hat{c}_{lt})\rho(||X_i - \hat{c}_j||)\psi\left(\hat{F}_n(|u_i(\hat{\theta})|)\right), \tag{49}$$

where $l = 1, \ldots, N$, $t = 1, \ldots, p$, *and* $u_i(\hat{\theta}) \neq 0$.

The proofs are short using the preparatory results of Sect. 3.1 and the derivatives of the loss functions of Sect. 3.2. Here, the steps used for the linear model in Sect. 3.2 are applied to the nonlinear model appropriately.

For the LTS-based versions, the derivatives exist almost everywhere (a.e.). We can say that the presented derivatives are elegant and straightforward to compute for real data. Deriving these derivatives is straightforward using derivatives of the inner function. We can see that the derivatives exist everywhere (or almost everywhere) for all the considered robust versions of RBF networks.

3.4 Regularized Versions and Relation to Approximate Neurocomputing

Regularization is a commonly employed and powerful technique in neural networks to mitigate overfitting and enhance generalization. For non-robust RBF networks, regularization methods have been suggested to prevent overfitting, especially when the dataset has a smaller number of samples (n), particularly if p is not too small [3]. The most widely used regularization approach applies weight regularization (weight decay) using L_2 or L_1 norms, focusing specifically on the values $a_1, \ldots, a_N$ in (4). However, the integration of robustness with regularization in RBF networks has not been explored in the existing literature. Let $\xi(\hat{\theta})$ denote the chosen loss function of an RBF network, where $\hat{\theta}$ includes, among other estimates, the values $\hat{a}_1, \ldots, \hat{a}_N$ and $\hat{c}_1, \ldots, \hat{c}_N$.

Consider the L_2-regularization given by the following optimization problem:

$$\underset{\tilde{\theta} \in \mathbb{R}^d}{\text{argmin}} \left[\xi(\tilde{\theta}) + \lambda \sum_{j=1}^{N} \tilde{a}_j^2 \right] \tag{50}$$

where ξ represents one of the loss functions mentioned earlier, and the regularization parameter $\lambda > 0$ is chosen through cross-validation. In contrast, L_1-regularization, which encourages sparsity in the parameters, uses the penalization term $\lambda \sum_{j=1}^{N} |\hat{a}_j|$.

Penalization-based regularization enables RBF networks to be applied to high-dimensional data where $n < p$, and it can also be extended to other robust variants of RBF networks. While regularization enhances the local robustness of the standard LS-RBF [17], it does not improve robustness to outliers. This motivates the use of robust, regularized versions of RBF networks, such as the model in (50), for handling contaminated data.

If regularized version of the robust RBF networks are used, then derivatives of the commonly used penalization terms can be obtained simply e.g. in the form

$$\frac{\delta}{\delta \hat{a}_j}\left(\lambda \sum_{k=1}^{N} \hat{a}_k^2\right) = 2\lambda \hat{a}_j, \quad j = 1, \dots, N, \tag{51}$$

$$\frac{\delta}{\delta \hat{c}_{lt}}\left(\lambda \sum_{k=1}^{N} \sum_{s=1}^{N} \hat{c}_{ks}^2\right) = 2\lambda \hat{c}_{lt}, \quad l = 1, \dots, N, \ t = 1, \dots, N, \tag{52}$$

$$\frac{\delta}{\delta \hat{a}_j}\left(\lambda \sum_{k=1}^{N} |\hat{a}_k|\right) = \lambda sgn(\hat{a}_j), \quad j = 1, \dots, N, \ \hat{a}_j \neq 0, \tag{53}$$

$$\frac{\delta}{\delta \hat{c}_{lt}}\left(\lambda \sum_{k=1}^{N} \sum_{s=1}^{N} |\hat{c}_{ks}|\right) = \lambda sgn(\hat{c}_{lt}), \quad l = 1, \dots, N, \ t = 1, \dots, N, \ \hat{c}_{lt} \neq 0. \tag{54}$$

Regularization ensures local robustness to small perturbations e.g. due to rounding or quantization effects. At the same time, regularization may be interpreted as a Bayesian approach, which corresponds to estimating (training) the parameters with a possible contamination by small errors. For this reason, regularized estimation is important for the field of **approximate neurocomputing**, where the effect of rounding the parameters or computing in a low-precision arithmetic are of interest [14].

4 Discussion and Conclusion

This paper introduced robust training methods for RBF networks based on derivatives of robust regression losses. We derived analytical expressions for derivatives of both fitted values and loss functions, enabling robust optimization in the presence of outliers.

The derived formulas can be used in the computation of robust RBF networks, where derivatives play a central role in gradient-based optimization and backpropagation. They may also be useful for hypothesis testing in regression models [9]. For LWS- and LWA-based estimators, derivatives exist everywhere, while for LTS- and LTA-based estimators they exist almost everywhere. Their availability can improve optimization efficiency and convergence behavior in training robust RBF networks.

Experimental evaluations of robust RBF networks can be found in [7]. While derivatives for plain (non-robust) quantiles are straightforward, those for robust quantiles require more complex treatment, relying on the derivatives of robust estimators. These directions present promising opportunities to extend the framework of robust neural networks [4].

Acknowledgements. The work is supported by the grant 25-15490S (LEDNeCo: Low Energy Deep Neurocomputing) of the Czech Science Foundation. The author discussed the methods with Petra Vidnerová, who is supported by the program of the Strategy AV 21 "AI: Artificial Intelligence for Science and Society".

References

1. Blokdyk, G.: Artificial neural network: A complete guide. CreateSpace Independent Publishing Platform, Scotts Valley (2017)
2. Gu, T., Want, J., Tang, D., Wang, J., Guo, T.: Reconstruction of measurement data with multiple outliers using novel domain-based RBF. Mech. Syst. Signal Process. **214**, 111385 (2024)
3. Han, Z., Qian, X., Huang, H., Huang, T.: Efficient design of multicolumn RBF networks. Neurocomputing **450**, 253–263 (2021)
4. Kalina, J.: On robust information extraction from high-dimensional data. Serbian J. Manage. **9**, 131–144 (2014)
5. Kalina, J.: A robust pre-processing of BeadChip microarray images. Biocybern. Biomed. Eng. **38**, 556–563 (2018)
6. Kalina, J.: From robust neural networks towards robust nonlinear quantile estimation. Seq. Anal. **44**, 326–350 (2025)
7. Kalina, J., Tumpach, J., Holeňa, M.: On combining robustness and regularization in training multilayer perceptrons over small data. In: Proceedings IJCNN 2022, pp. 1–8 (2022)
8. Khanna, A., Ling, C., Everett, D., Raff, E., Inkawhich, N.: Multi-layer radial basis function networks for out-of-distribution detection (2025). arXiv:2501.02616v1
9. Mandel, F., Barnett, I.: Permutation-based hypothesis testing for neural networks (2023). ArXiv:2301.11354v1
10. Qian, C., He, N., He, L., Li, R., Cheng, F.: A weighted Gaussian process regression model based on improved local outlier factor and its application in state of health estimation of lithium-ion battery. Engineering Applications of Artificial Intelligence 138, Part A, 109314
11. Rousseeuw, P.J., Van Driessen, K.: Computing LTS regression for large data sets. Data Min. Knowl. Disc. **12**, 29–45 (2006)
12. Sánchez, V.D.: Searching for a solution to the automatic RBF network design problem. Neurocomputing **42**, 147–170 (2002)
13. Seghouane, A.K., Shokouhi, N.: Adaptive learning for robust radial basis function networks. IEEE Trans. Cybern. **51**, 2847–2856 (2021)
14. Sze, V., Chen, Y.B., Yang, T.J., Emer, J.S.: Efficient Processing of Deep Neural Networks. Morgan and Claypool, San Rafael (2020)
15. Víšek, J.Á.: The least trimmed squares. Part I: consistency. Kybernetika **42**, 1–36 (2006)
16. Víšek, J.Á.: Consistency of the least weighted squares under heteroscedasticity. Kybernetika **47**, 179–206 (2011)
17. Werner, T.: Quantitative robustness of instance ranking problems. Ann. Inst. Stat. Math. **75**, 335–368 (2021)
18. Wilcox, R.R.: Introduction to robust estimation and hypothesis testing, 4th edn. Academic Press, London (2017)

Learning to Extrapolate in Industrial Data-Scarce Scenarios via Differential Ensemble Neural Networks

Hao Huang(✉), Jinbo Cao, and John Karigiannis

GE Vernova Advanced Research, Niskayuna, NY, USA
{hao.huang1,caoj,john.karigiannis}@gevernova.com

Abstract. Neural networks typically fail when test data deviate from training data—a fundamental limitation for robust industrial deployment. This challenge is compounded in data-scarce industrial settings where learning from few examples with uncertainty quantification is essential. We introduce DELTANET, a framework that achieves robust Out-of-Distribution (OOD) extrapolation by reformulating regression as differential ensemble prediction using training instances as reference anchors. Rather than learning direct input-output mappings, DELTANET learns differences between samples, yielding N^2 pairwise training combinations from N training samples—a combinatorial expansion that enables effective learning from limited data. We investigate how specific architecture choices—activation functions, Hadamard gating, and Y-aware conditioning—influence extrapolation accuracy and robustness, providing empirical guidance on network design for OOD generalization. DELTANET further integrates learned per-prediction uncertainty with precision-weighted ensemble aggregation, decomposing uncertainty into epistemic and aleatoric components for informative confidence estimates. Evaluated on several industrial data-scarce scenarios, DELTANET consistently outperforms baselines in both prediction accuracy and uncertainty calibration across feature-space and target-space extrapolation scenarios.

Keywords: Data-Scarcity · OOD Prediction · Uncertainty Quantification

1 Introduction

Machine learning excels when 1) the training set is sufficiently large and 2) the test data follow a similar distribution, but many real-world applications require extrapolation—predicting outside the training distribution. This is particularly challenging in industrial domains with **data scarcity** and **distribution shift**: collecting comprehensive data is often expensive or impossible, yet critical predictions are needed precisely where data cannot be collected. We consider two structured extrapolation scenarios, which we term X-OOD and Y-OOD to denote out-of-distribution in the extrapolation sense: (1) X-OOD (Feature-Space Extrapolation), where test features lie beyond the training feature range, and (2) Y-OOD (Target-Space Extrapolation), where test targets lie beyond the

L. Pasa et al. (Eds.): ICANN 2026, LNCS 17089, pp. 695–707, 2027.
https://doi.org/10.1007/978-3-032-38398-3_57

training target range. In this work, we do not aim to address arbitrary distribution shifts, but rather explicit range extrapolation driven by underlying physical mechanisms.

Unlike prior anchor-based or transductive extrapolation methods [16,17,23], our work uniquely combines nonlinear differential encoders, Y-aware conditioning on reference targets, and learned per-prediction uncertainty with weighted ensemble aggregation—enabling both X-OOD and Y-OOD extrapolation with calibrated uncertainty in a single framework (see Sect. 2 for detailed comparison). Such challenges manifest across industrial applications, such as nozzle emission prediction (X-OOD) where emissions above 3200F are vital for compliance, but testing is costly; therefore, models must extrapolate from low temperature data (<3000F); or battery capacity prediction (Y-OOD), in which the training data rarely covers late-stage degradation or long cycles (multi-year lifespan). In such cases, accurate forecasts from limited data are necessary to enable virtual validation without costly testing.

Traditional methods fail in the above scenarios because: (1) neural networks produce unreliable OOD predictions, (2) tree-based methods cannot extrapolate by design, and (3) Gaussian Processes revert to uninformative priors in regions far from training distribution. We introduce DELTANET, a neural framework with three key innovations:

Differential Learning: Instead of learning $f : \mathbf{x} \rightarrow y$, DELTANET predicts differences: given test sample $\mathbf{x}_1$ and reference $(\mathbf{x}_2, y_2)$, predict $y_\Delta = y_1 - y_2$, then get final prediction by $\widehat{y_1} = y_2 + \widehat{y_\Delta}$. It enables extrapolation by anchoring to known references. The pairwise formulation naturally constructs up to N^2 training combinations from N training samples—a combinatorial expansion (not independent augmentation, as many pairs share samples) that exposes the model to a rich space of inter-sample differential patterns, acting as a structural regularizer particularly valuable in data-scarce settings.

Y-Aware Encoding: The model conditions on reference output y_2, enabling context-dependent predictions crucial for extrapolation.

Precision-Weighted Ensemble: At inference, DELTANET generates predictions from all references in the training set, selects top-K with lowest uncertainty, and aggregates via precision-weighted averaging.

2 Related Work

Extrapolation in Machine Learning and Industrial Applications. Standard neural networks fail when test inputs lie outside the training convex hull [26], partly because ReLU networks partition input space into linear regions [3]. [18] and [27] show DNNs exhibit low-frequency bias that hinders extrapolation; [10] shows DNNs revert to constant predictions on OOD inputs. [13] provide causal conditions for extrapolation, while Engression [23] models full conditional distributions but requires expensive generative sampling. These challenges are acute in industrial domains where data collection is expensive or dangerous: battery degradation [2,22] requires forecasting from limited early-life data, emission prediction at extreme temperatures is prohibitive to test physically, and [17] and [21] study prediction under distribution shift in materials science. DELTANET addresses both X-OOD and Y-OOD across these domains using efficient discriminative differential learning with ensemble aggregation.

Uncertainty Quantification. Bayesian neural networks [4,6] provide principled but expensive uncertainty; deep ensembles [14] are practical but require multiple models. Heteroscedastic models [9] may not generalize OOD; conformal prediction [20] assumes exchangeability; [21] identify "known unknowns" for abstention. DELTANET combines learned per-prediction uncertainty with ensemble calibration.

Differential and Relative Learning. Learning from differences appears in Siamese [12] and prototypical networks [24]. Our work also relates to instance-based learning [1] and anchoring [15]. Most related, [16] propose transductive extrapolation via bilinear embeddings with training anchors. DELTANET differs in its use of deep nonlinear encoders, Y-aware conditioning, and learned uncertainty.

DELTANET Positioning. Table 1 contrasts DELTANET with the most related approaches. No prior method combines all three of deep nonlinear differential encoding, Y-aware conditioning, and learned per-prediction uncertainty. DELTANET uniquely integrates these components with precision-weighted ensemble aggregation, enabling both X-OOD and Y-OOD extrapolation with calibrated uncertainty in a single framework.

Table 1. Comparison of DELTANET with related methods along key dimensions. ✓: explicitly designed for; (✓): applicable but not specifically designed for; ✗: not supported.

Method	Diff. Encoding	Y-Aware Cond.	Learned Uncert.	X-OOD	Y-OOD
Transductive [16]	Bilinear	✗	✗	✓	(✓)
ExtrapolationNN [17]	✗	✗	✗	(✓)	✓
Engression [23]	✗	✗	Generative	✓	✓
DELTANET (Ours)	**Deep NN**	✓	✓	✓	✓

3 Problem Setting and High-Level Approach Description

$\mathcal{D}_{\text{train}} = \{(\mathbf{x}_i, y_i)\}_{i=1}^N$ denotes the training dataset with $\mathbf{x}_i \in \mathbb{R}^d$ and $y_i \in \mathbb{R}$. We address two structured extrapolation scenarios. Formally: **Y-OOD:** $y_j \notin \text{range}(\{y_i\}_{i=1}^N)$, **X-OOD:** $x_j^{(k)} \notin \text{range}(\{x_i^{(k)}\}_{i=1}^N)$ for some feature dimension k. DELTANET tackles both via differential learning under the assumption that the underlying differential mechanism remains invariant across the extrapolation regime (see Sect. 4.4).

Feature Scaling Assumption. Because DELTANET learns from feature differences $\mathbf{x}_\Delta = \mathbf{x}_1 - \mathbf{x}_2$, the magnitudes of individual features directly influence the learned differential representation. To ensure that no single feature dominates due to its scale, all input features are standardized to zero mean and unit variance using statistics computed only from the training set. Target values are similarly normalized. This preprocessing is applied consistently across all methods (including baselines) for fair comparison. Below are some key notations used in the paper:

$\mathbf{x}_\Delta = \mathbf{x}_1 - \mathbf{x}_2$: Feature difference; $y_\Delta = y_1 - y_2$: Target difference to predict;
$\mathbf{x}_{\text{ref}} = \mathbf{x}_2$: Reference features; $y_{\text{ref}} = y_2$: Reference target.

At a high level, DELTANET consists of three phases:

Phase 1: Pairwise Reformulation. The core insight of DELTANET is to reframe regression, not as direct mapping $f : \mathbf{x} \rightarrow y$, but as differential prediction:

$$\text{DELTANET: } (\mathbf{x}_1 - \mathbf{x}_2, \mathbf{x}_2, y_2) \xrightarrow{g_\theta} \widehat{y_\Delta}, \quad \widehat{y_1} = y_2 + \widehat{y_\Delta} \tag{1}$$

This formulation yields N^2 training combinations from N training samples. While these combinations are not statistically independent, this expansion exposes the model to a rich space of inter-sample differential patterns and acts as a structural regularizer.

Phase 2: Model Training. Train network to predict $\widehat{y_\Delta}$ and uncertainty σ.

Phase 3: Ensemble Prediction. Generate predictions $\{\widehat{y_\Delta}, \sigma\}$ from all references, select top-K with lowest uncertainty, aggregate via precision-weighted ensembling.

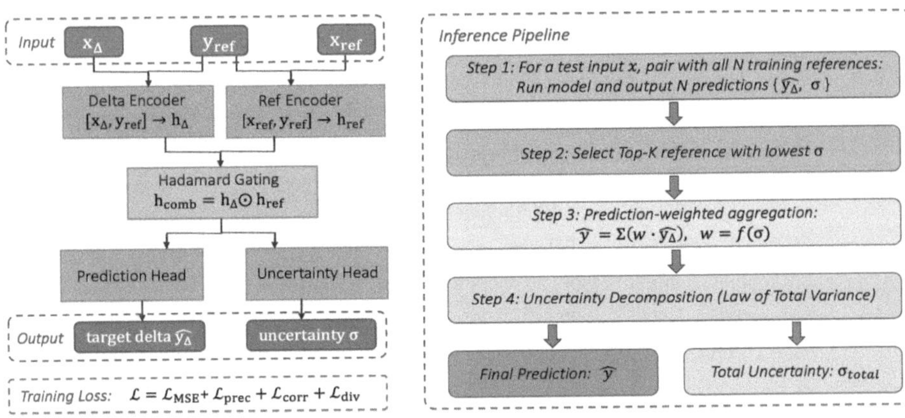

Fig. 1. DELTANET complete pipeline. Left: Training phase – Delta and Reference Encoders combine via Hadamard gating; dual heads predict target difference $\widehat{y_\Delta}$ and uncertainty σ, trained with a four-term loss. Right: Inference – N reference-anchored predictions are filtered by top-K uncertainty selection and aggregated via precision-weighted ensembling with epistemic/aleatoric decomposition.

4 Detailed Description of DELTANET

4.1 Model Architecture

Figure 1 shows the main components and complete pipeline of DELTANET.

Delta Encoder. The delta encoder processes the concatenation of the feature difference $\mathbf{x}_\Delta$ and the reference output y_{ref} through a series of LeakyReLU activations and linear layers, producing embedding $\mathbf{h}_\Delta$. LeakyReLU is beneficial for extrapolation: it preserves gradient information across the entire input range, enabling the network to extend learned trends into unseen regions (ablation in Sect. 5 confirms this benefit). The inclusion of y_{ref} enables **Y-aware encoding**: the model learns that the same feature difference may have different implications depending on the reference (ablation study in Sect. 5 confirms this benefit).

Reference Encoder. The reference encoder processes the concatenation of reference features $\mathbf{x}_{\text{ref}}$ and reference output y_{ref} through an input projection (linear layer followed by ReLU, without skip connection) and subsequent ResNet blocks where ReLU is applied after the residual addition, producing embedding $\mathbf{h}_{\text{ref}}$. The use of ReLU in the reference encoder serves a different purpose than LeakyReLU in the delta encoder: the output $\mathbf{h}_{\text{ref}}$ is guaranteed non-negative, which produces sparse activations that act as a natural gating signal for the downstream Hadamard interaction.

Hadamard-Product Gating. We use element-wise Hadamard-product multiplication to combine the delta and reference embeddings: $\mathbf{h}_{\text{combined}} = \mathbf{h}_{\Delta} \odot \mathbf{h}_{\text{ref}}$. Since $\mathbf{h}_{\text{ref}} \geq 0$ element-wise, this implements a selection/suppression mechanism: dimensions where the reference encoder outputs large values are amplified, while near-zero dimensions are suppressed. Unlike additive combination, this gating design enables the reference to control which delta dimensions are passed through. The combined representation $\mathbf{h}_{\text{combined}}$ is shared by both the prediction head and uncertainty head.

Prediction Head. The prediction head is a deep residual network that takes the multiplicative interaction as input to predict $\widehat{y_{\Delta}}$. Each residual block consists of two linear layers with ReLU activation and dropout, plus a skip connection.

Uncertainty Head. A parallel head predicts uncertainty σ from the concatenation of the Hadamard-product output $\mathbf{h}_{\text{combined}}$ and the reference output y_{ref}, ensuring that uncertainty estimates are derived from input-space representations. The final output goes through a softplus activation with a small positive offset ($\sigma = \text{softplus}(\cdot) + \varepsilon$, $\varepsilon = 0.01$), ensuring $\sigma > 0.01$. The learned σ is trained to correlate with prediction errors for precision-weighted aggregation (refer to Sect. 4.2).

4.2 Training Phase: Loss Function

We train DELTANET with a multi-objective loss that balances prediction accuracy, uncertainty calibration, and uncertainty diversity within each training batch with size B:

MSE Loss. The primary prediction loss $\mathcal{L}_{\text{MSE}} = \frac{1}{B}\sum_{i=1}^{B}(y_{\Delta}^{(i)} - \widehat{y_{\Delta}}^{(i)})^2$.

Precision-Weighted Loss. For uncertainty calibration, we use a precision-weighted squared error loss that encourages the model to assign lower uncertainty to more accurate predictions. We define precision as $p^{(i)} = 1/(\sigma^{(i)})^2$, where $\sigma^{(i)}$ is the predicted uncertainty for the i-th sample. The loss is: $\mathcal{L}_{\text{prec}} = \sum_{i=1}^{B} w^{(i)}(y_{\Delta}^{(i)} - \widehat{y_{\Delta}}^{(i)})^2$ where $w^{(i)} = p^{(i)}/\sum_j p^{(j)}$. Note that $\mathcal{L}_{\text{prec}}$ alone could admit a trivial solution ($\sigma \to \infty$, driving all weights to zero). This degeneracy is explicitly prevented by the auxiliary losses below: $\mathcal{L}_{\text{corr}}$ forces σ to be rank-correlated with actual prediction errors (making uniform or inflated σ suboptimal), and $\mathcal{L}_{\text{diversity}}$ penalizes low variance in σ across the batch (directly opposing constant or collapsed uncertainty). Together, these losses ensure that the uncertainty head produces informative, well-spread estimates.

Correlation Loss. To ensure uncertainty is informative, we maximize the correlation between predicted uncertainty and actual errors: $\mathcal{L}_{\text{corr}} = 1 - \text{Pearson}(\sigma, |y_{\Delta} - \widehat{y_{\Delta}}|)$.

Diversity Loss. To encourage the model to produce diverse uncertainty estimates rather than collapsing to constant values, we include a diversity term: $\mathcal{L}_{\text{diversity}} = -\text{std}(\sigma)$.
Loss Weights. The full loss is: $\mathcal{L} = \lambda_1 \mathcal{L}_{\text{MSE}} + \lambda_2 \mathcal{L}_{\text{prec}} + \lambda_3 \mathcal{L}_{\text{corr}} + \lambda_4 \mathcal{L}_{\text{diversity}}$, with $\lambda_1 = 5, \lambda_2 = 100, \lambda_3 = 5$, and $\lambda_4 = 1$. The large λ_2 compensates for the small gradient magnitude of the precision-weighted loss (which uses normalized weights summing to 1). A **full sensitivity analysis** is provided in supplementary Section G.

4.3 Inference Phase: Precision-Weighted Ensemble Prediction

At inference time, DELTANET generates N predictions for each test sample (one per training reference), then selects the top-K references with the lowest predicted uncertainty (we use $K = 50$; sensitivity analysis in supplementary Section H confirms robustness for $K \in [20, 100]$). The ensemble aggregates these selected predictions using precision-weighted averaging. For each selected reference j, we obtain $\widehat{y_j}$ via Eq. 1, compute precision as $p_j = 1/\sigma_j^2$, normalize to obtain weights $w_j = p_j / \sum_{k=1}^{K} p_k$, and compute the final prediction as $\widehat{y} = \sum_{j=1}^{K} w_j \widehat{y_j}$.

4.4 Theoretical Motivation

The effectiveness of Eq. 1 can be understood through Taylor expansion. For a smooth function f: $f(\mathbf{x}_1) - f(\mathbf{x}_2) \approx \nabla f(\mathbf{x}_2)^\top (\mathbf{x}_1 - \mathbf{x}_2)$. This first-order approximation suggests that differences are locally linear, even when f is non-linear.
For X-OOD: When $\mathbf{x}_1$ lies outside the training feature range but $\mathbf{x}_1 - \mathbf{x}_2$ resembles training differences in direction and magnitude, DELTANET can generalize more effectively than other popular models. The gradient $\nabla f(\mathbf{x}_2)$ is estimated at training points, and if the function is smooth, this gradient extrapolates reasonably to nearby X-OOD regions. **For Y-OOD:** When y_1 lies outside the training target range, the differential approach helps because the model learns relationships between feature differences and target differences. If the relationship $y_\Delta \sim g(\mathbf{x}_\Delta, \mathbf{x}_{\text{ref}}, y_{\text{ref}})$ is consistent across the domain, predictions can extrapolate to unseen target values.
Core Assumption (Invariant Differential Mechanism). DELTANET rests on an assumption inspired by causal extrapolation [13] and compositional generalization [25]: there exists a function g such that for all $\mathbf{x}_1, \mathbf{x}_2$ in the domain: $y_1 - y_2 = g(\mathbf{x}_1 - \mathbf{x}_2, \mathbf{x}_2, y_2) + \epsilon$. where ϵ is small noise. Crucially, this assumes the causal mechanism relating input differences to output differences is invariant across training and OOD regions, while the conditioning on $(\mathbf{x}_2, y_2)$ enables context-aware adaptation. When g models stable physical laws (e.g., thermodynamics of emissions or electrochemistry of degradation), learning it from training data enables valid extrapolation to OOD regimes governed by the same laws. We emphasize that this is an empirical assumption whose validity depends on the smoothness and physical consistency of the specific domain—the Taylor expansion motivation supports this assumption for near-boundary extrapolation of smooth functions, but does not provide formal guarantees for arbitrary extrapolation distances. Consistent outperformance across all four datasets (Sect. 5) provides empirical evidence that this assumption holds in the tested industrial domains.

When DELTANET May Fail. The core assumption reveals when DELTANET may fail to extrapolate reliably: 1) *Phase transitions:* When the underlying mechanism changes outside the training regime (e.g., material phase changes, regime shifts), the learned g no longer applies. 2) *Strong non-additive interactions:* When factors interact non-compositionally such that y_Δ depends on complex higher-order interactions not captured by the differential structure. 3) *Discontinuities:* When the function f has discontinuities or sharp transitions, the smooth differential approximation breaks down.

5 Experiments

We evaluate DELTANET across four industrial domains, each presenting distinct and realistic OOD extrapolation challenges. For comprehensive analysis, we also report in-distribution (ID) performance, denoted as X-ID and Y-ID in the tables.

5.1 Datasets, Problem Settings, and Baselines

BATTERY-1: An internal lithium battery aging dataset with 444 samples from Battery Energy Storage Systems with extended operational lifespans. Total features include the number of cycles, temperature, state of charge, cycling rate, and depth of discharge etc. The goal is to predict long-term normalized capacity $\in [0.75, 1.0]$ from limited early-stage cycling data (Days in Test spans ~1–650). **BATTERY-2:** NASA Li-ion Battery Aging Dataset [7] (504 samples from four 18650 cells cycled to failure) with features including voltage, current, charge/discharge cycles, temperature and load measurements. The target is remaining capacity (range $[1.15, 2.04]$; time span ~2780–3700 s).

Nozzle Test on NO_x Emission (X-OOD). An internal proprietary combustion nozzle dataset (687 samples) for predicting NO_x emissions. Features include design parameters (e.g. tube diameter, injection angle, discharge coefficients) and operating conditions (e.g. flame temperature, hydrogen content, compressor discharge pressure/temperature). Flame temperature spans 1815–3500F; testing above 3000F is prohibitively expensive, so models are trained on temperature $\leq T_{threshold}$ and tested on $> T_{threshold}$, with $T_{threshold} \in \{3000, 3200, 3400\}$F.

Light-Duty Vehicle CO_2 Emission (X-OOD). The Canadian vehicle fuel consumption dataset [5] (7,385 vehicles) with features including engine size (1.0–8.4L), cylinders, and fuel consumption metrics. Models predict emissions for large-engine vehicles (>6L) using only the ~1,017 vehicles with engine ≤6L as training data—a naturally data-scarce X-OOD scenario evaluating whether DELTANET can learn emission scaling despite nonlinear thermodynamic effects.

Baselines. We compare DELTANET against four neural network models (three of which are extrapolation-oriented) and five ML baselines. *NN baselines:* (1) **Transductive** [16]—uses training anchors and bilinear differences to extrapolate beyond input support; (2) **ExtrapolationNN** [17]—matches input in latent space for reliable extrapolation beyond training label ranges; (3) **Engression** [23]—models full conditional distributions via distributional regression for extrapolation; (4) **ResNet-MLP**—residual network with skip connections as a typical deep learning baseline. *Classical ML baselines:* (5) **SVR**—RBF kernel regression; (6) **Random Forest**—decision tree ensemble; (7) **Gradient Boosting**—sequential ensemble fitting residuals; (8) **Ridge Regression**—L2-regularized linear model; (9) **Gaussian Process (GP)** [19]—exact GP regression with RBF kernel and automatic relevance determination, providing principled Bayesian uncertainty. **Hyperparameter Tuning:** All baselines were tuned via grid search with 5-fold cross-validation (full search ranges in supplementary Section E).

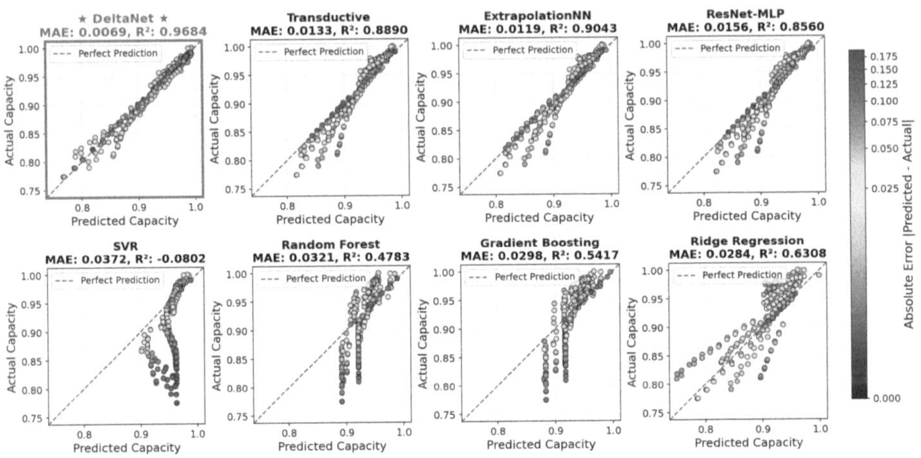

Fig. 2. Predicted vs. actual BATTERY-1 capacity for Days in Test > 180 (X-OOD), shown for a single seed. Points are colored by absolute error $|\widehat{y} - y|$ (blue = low, red = high; power normalization $\gamma = 0.25$). DELTANET's predictions cluster tightly around the diagonal with predominantly blue points, while baselines show greater scatter. (Color figure online)

5.2 Implementation Details

Key settings for DELTANET: Adam optimizer [11] with learning rate 5×10^{-5}, weight decay 10^{-4}, batch size 32, early stopping (patience 20), embedding dimension 120, dropout 0.1. Loss weights: λ_1=5, λ_2=100, λ_3=5, λ_4=1, selected via grid search over validation sets from all four datasets, optimizing average MAE and uncertainty-error correlation (details in supplementary Section D). A full sensitivity analysis (supplementary Section G) confirms robustness within a 2× range. Top-K=50 references used at inference; sensitivity to K is analyzed in supplementary Section H. Models converge within 80–150 epochs. Full architectural details and training procedure are provided in the supplementary Section D. All experiments use PyTorch 2.1 on an NVIDIA A100 GPU and are repeated over 20 random seeds, where each seed controls both the train/validation split (80%/20%) and model initialization; all methods share the same split for each seed to enable fair comparison. A detailed complexity analysis and comparison with baselines (model size, training cost, inference cost including wall-clock times per test sample) is provided in supplementary Section J.

5.3 Results on Battery Capacity Prediction

A predicted-vs-actual battery capacity visualization is shown in Fig. 2. Table 2 presents MAE results on Battery Capacity Prediction for NN baselines (R^2 in supplementary Section A.2; classical ML results in supplementary Section F).

Time-based extrapolation (X-OOD): DELTANET achieves 42% lower MAE (8.7 vs. 15.0, $\times 10^{-3}$) than Transductive for Days in Test > 180. Engression (12.8) outperforms Transductive but remains 47% worse than DELTANET. Notably, even with only 61 training samples (Days in Test > 60), DELTANET still outperforms all baselines.

Capacity-based extrapolation (Y-OOD): For degraded capacity < 0.90 (trained on ≥ 0.90), DELTANET reduces MAE by 20% vs. Engression and 25% vs. ExtrapolationNN. Engression is the second-best method on Y-OOD tasks, consistent with its distributional regression design, but still lacks the anchor-based precision of DELTANET.

Classical ML limitations: SVR, RF, GB, and GP produce MAE 2–4× worse than DELTANET on OOD tasks. Notably, GP (RBF kernel) achieves competitive ID performance (e.g., MAE 3.2 on BATTERY-1 Y-ID) but degrades substantially on OOD (e.g., MAE 48.2 on Y-OOD, all $\times 10^{-3}$), confirming that GP predictions revert toward uninformative priors far from training data (supplementary Section F).

Table 2. Battery Capacity Prediction MAE ($\times 10^{-3}$) Results (mean ± std over 20 rounds). n_{tr}: training set size (for ID rows, reduced because some in-distribution samples are held out as test). **Bold** indicates best performance.

Dataset	Extrap. Condition	DELTANET	Transductive	ExtrapolationNN	Engression	ResNet-MLP
Time-based Extrapolation (X-OOD: Training on early cycles, testing on later cycles)						
BATTERY-1	Days in Test > 180 (n_{tr} = 185)	**8.7 ± 1.4**	15.0 ± 3.0	11.1 ± 1.5	12.8 ± 2.2	14.2 ± 2.0
BATTERY-1	Days in Test > 120 (n_{tr} = 122)	**17.0 ± 1.7**	19.8 ± 1.3	19.1 ± 1.5	19.5 ± 1.9	20.1 ± 1.9
BATTERY-1	Days in Test > 60 (n_{tr} = 61)	**26.4 ± 6.1**	30.0 ± 7.3	31.4 ± 7.1	30.8 ± 7.9	32.9 ± 10.4
BATTERY-2	time > 3200 (n_{tr} = 315)	**32.5 ± 6.5**	39.7 ± 9.1	44.5 ± 10.0	41.8 ± 9.5	50.8 ± 14.6
BATTERY-2	time > 3000 (n_{tr} = 189)	**46.3 ± 2.2**	55.2 ± 2.4	57.5 ± 2.9	53.8 ± 3.1	52.5 ± 9.7
Capacity-based Evaluation (Training on high-capacity ≥0.90, testing on Y-OOD and Y-ID)						
BATTERY-1	cap < 0.90, Y-OOD (n_{tr} = 362)	**16.4 ± 3.4**	22.5 ± 2.0	21.9 ± 1.8	20.5 ± 2.5	22.4 ± 2.3
BATTERY-1	cap ≥ 0.90, Y-ID (n_{tr} = 300)	**2.9 ± 0.1**	2.9 ± 0.2	3.0 ± 0.2	3.1 ± 0.2	3.0 ± 0.2
BATTERY-2	cap < 1.4, Y-OOD (n_{tr} = 401)	**16.1 ± 1.4**	18.9 ± 1.3	21.3 ± 4.2	19.8 ± 2.8	23.5 ± 5.4
BATTERY-2	cap ≥ 1.4, Y-ID (n_{tr} = 321)	**7.5 ± 0.6**	9.9 ± 1.3	8.0 ± 0.6	8.5 ± 0.8	11.6 ± 1.5

Table 3. Emission Prediction MAE (mean±std over 20 rounds). n_{tr}: training set size (for ID rows, reduced because some in-distribution samples are held out as test).

Dataset	Extrap. Condition	DELTANET	Transductive	ExtrapolationNN	Engression	ResNet-MLP
Temperature-based Extrapolation (X-OOD: Training on low temp., testing on high temp.)						
NOZZLE	temp > 3000 (n_{tr} = 521)	**4.928 ± 0.929**	7.724 ± 1.015	7.061 ± 0.923	7.892 ± 1.134	9.154 ± 1.576
NOZZLE	temp > 3200 (n_{tr} = 664)	**11.034 ± 0.943**	16.134 ± 0.799	15.192 ± 1.588	16.547 ± 1.402	18.280 ± 1.176
NOZZLE	temp > 3400 (n_{tr} = 680)	**13.843 ± 1.237**	20.769 ± 1.658	20.239 ± 2.337	19.875 ± 1.893	19.560 ± 1.334
Engine Size-based Evaluation (Training on ≤6L, testing on X-OOD (>6L) and X-ID (≤6L))						
Vehicles	engine > 6, X-OOD (n_{tr} = 1017)	**9.213 ± 1.869**	11.549 ± 2.421	12.719 ± 1.890	12.083 ± 2.156	13.296 ± 2.073
Vehicles	engine ≤ 6, X-ID (n_{tr} = 813)	**4.252 ± 1.116**	6.826 ± 1.402	5.056 ± 1.097	5.438 ± 1.215	9.289 ± 1.310

5.4 Results on Emission Prediction

Table 3 presents MAE results for NN baselines on emission prediction (R^2 results in supplementary Section A.1; classical ML baseline results in supplementary Section F).
High-Temperature Extrapolation: DELTANET consistently outperforms across all temperature thresholds, with MAE increasing monotonically as the threshold rises—despite n_{tr} also increasing ($521 \rightarrow 664 \rightarrow 680$). This contrasts with the battery results (Table 2), where a higher threshold reduces n_{tr} and MAE degrades primarily due to data scarcity. For the nozzle, higher thresholds retain most training data but restrict the test set to increasingly extreme temperatures where the emission response becomes highly nonlinear, so *extrapolation distance* dominates over training set size. At >3200F, DELTANET achieves MAE of 11.03, a 27% reduction vs. ExtrapolationNN and 32% reduction vs. Transductive. At the most extreme threshold (>3400F), DELTANET maintains MAE of 13.84 while baselines degrade more sharply (Transductive: 20.77, ExtrapolationNN: 20.24). Engression (16.55–19.88) performs comparably to Transductive but consistently worse than DELTANET.
Large-Engine Extrapolation: For engines > 6L, DELTANET achieves MAE of 9.21 vs. Transductive (11.55), Engression (12.08), and ExtrapolationNN (12.72)—a 20–28% reduction—while also achieving the best ID performance (MAE 4.25 on $\leq$ 6L).
Classical ML Baselines (supplementary Section F) fail on OOD tasks with MAE 2–3× worse than DELTANET, despite GP achieving competitive ID results.
Statistical Significance. We apply paired t-tests and Wilcoxon signed-rank tests on the raw per-seed MAE scores (with Holm-Bonferroni correction). Against all NN base-

Table 4. Ablation study on key DELTANET components (mean±std over 20 rounds). MAE (↓), R^2 (↑), and uncertainty-error correlation (↑). **Bold** = full model.

Variant	**NOZZLE** (>3200) MAE	R^2	**BAT-1** (cap<.90) MAE($\times 10^{-3}$)	R^2	**Corr.** (BAT-1)
Full DELTANET	**11.03±0.94**	**.836±.021**	**16.4±3.4**	**.684±.041**	**.701±.038**
w/o Differential Encoding	14.52±1.53	.701±.035	24.8±5.1	.387±.067	.528±.054
w/o Y-Aware Encoding	12.18±1.12	.793±.028	21.3±4.2	.502±.053	.634±.045
w/o Precision-Weighted Ens.	12.74±1.21	.776±.031	19.6±3.9	.581±.048	.453±.051
w/o LeakyReLU	11.62±1.03	.812±.024	17.5±3.6	.654±.044	.672±.041
w/o Hadamard Gating	12.09±1.08	.798±.026	18.5±3.8	.618±.046	.649±.043

Table 5. Uncertainty quality on Battery Capacity Prediction: error correlation (Corr, ↑) and Prediction Interval Coverage Probability at 68%/95% nominal levels (PICP, closer to nominal is better). **Bold** = best. Full reliability analysis in supplementary Section K.

Data	DELTANET Corr	P68	P95	Transduct. Corr	P68	P95	ExtrapNN Corr	P68	P95	Engression Corr	P68	P95	ResNet Corr	P68	P95
BAT-1	**.70**	**.63**	**.91**	.48	.42	.74	.57	.51	.82	.54	.49	.80	.52	.45	.77
BAT-2	**.68**	**.61**	**.89**	.52	.44	.76	.55	.48	.79	.51	.46	.78	.49	.43	.75

lines, all conditions are significant at corrected $p < 0.05$ level under both tests—except for the Transductive method on the BATTERY-1 Y-ID (capacity ≥ 0.90, in-distribution) setting, where performance is nearly identical. Crucially, **all OOD conditions are statistically significant** against every baseline. Full per-condition t-statistics, Wilcoxon W-statistics, and corrected p-values are provided in supplementary Section I.

5.5 Ablation Study and Uncertainty Quantification

To quantify each component's contribution, we remove or replace one at a time while keeping all other settings identical (same pairwise training, same N^2 combinations, same network size). Table 4 reports results on NOZZLE (X-OOD, temp > 3200F) and BATTERY-1 (Y-OOD, cap < 0.90). For "w/o Diff. Enc." we replace $(\mathbf{x}_1 - \mathbf{x}_2, \mathbf{x}_2, y_2)$ with $(\mathbf{x}_1, \mathbf{x}_2, y_2)$, keeping the pairwise framework otherwise identical.
Key Findings. Differential Encoding is the most critical component—removing it degrades R^2 from 0.684 to 0.387 on Y-OOD and increases MAE by 51%, despite the model retaining the same pairwise training and size. This confirms that the explicit difference computation provides a structural prior the network cannot easily recover through learned representations alone. Y-Aware Encoding primarily benefits Y-OOD tasks (30% MAE increase when removed). Precision-Weighted Ensemble affects both accuracy (15–20% MAE increase) and calibration (correlation drops from 0.70 to 0.45). LeakyReLU and Hadamard Gating each provide moderate 5–13% MAE gains.

DELTANET decomposes total uncertainty via the **Law of Total Variance** [8, 14] into epistemic (reference disagreement) and aleatoric (per-prediction) components: $\sigma_{\text{total}} = \sqrt{\sigma^2_{\text{epi}} + \sigma^2_{\text{ale}}}$, where $\sigma_{\text{epi}} = \sqrt{\sum_k w_k(\widehat{y_1}^{(k)} - \widehat{y_1})^2}$ and $\sigma_{\text{ale}} = \sqrt{\sum_k w_k \sigma_k^2}$. Baselines use MC Dropout [6] (NN methods) or method-specific estimates (supplementary Section C). Table 5 shows DELTANET achieves the highest uncertainty-error correlation (0.70 on BATTERY-1, 0.68 on BATTERY-2) and the closest-to-nominal PICP at both 68% and 95% levels (e.g., 63%/91% vs. nominal 68%/95% on BATTERY-1), outperforming all baselines. Classical ML uncertainty results and full reliability diagrams are provided in supplementary Sections C and K.

6 Conclusion

We introduced DELTANET, a framework for accurate prediction under structured distribution shift with limited training data. By reformulating regression as differential ensemble prediction, DELTANET contributes to several open questions: **Robustness to structured distribution shift:** DELTANET learns differential relationships that remain stable beyond training support, addressing both X-OOD and Y-OOD shifts. **Learning from few examples in high-dimensional settings:** The pairwise reformulation yields N^2 training combinations from N samples, acting as structural regularization for data-scarce industrial scenarios. **Influence of architecture choices on accuracy and robustness:** Our ablation study reveals how asymmetric activations (LeakyReLU vs. ReLU), Hadamard gating, and Y-aware conditioning each contribute to extrapolation robustness, providing empirical guidance for OOD generalization. **Understanding and modeling uncertainties:** DELTANET integrates learned per-prediction uncertainty with precision-weighted ensemble aggregation, decomposing total uncertainty into epistemic and aleatoric components for well-ordered uncertainty estimates under distribution shift. Future work will explore conformal prediction guarantees for provable coverage under distribution shift, adaptive reference selection, and extensions to multi-output or time series extrapolation.

References

1. Aha, D.W., Kibler, D., Albert, M.K.: Instance-based learning algorithms. Mach. Learn. **6**(1), 37–66 (1991)
2. Attia, P.M., et al.: Closed-loop optimization of fast-charging protocols for batteries with machine learning. Nature **578**(7795), 397–402 (2020)
3. Balestriero, R. and LeCun, Y. (2021). Learning in high dimension always amounts to extrapolation. *arXiv preprint* arXiv:2110.09485
4. Blundell, C., Cornebise, J., Kavukcuoglu, K., and Wierstra, D. (2015). Weight uncertainty in neural networks. In *ICML*, pages 1613–1622
5. Government of Canada (2024). Fuel consumption ratings dataset. https://open.canada.ca/data/en/dataset/98f1a129-f628-4ce4-b24d-6f16bf24dd64
6. Gal, Y., Ghahramani, Z.: Dropout as a Bayesian approximation: Representing model uncertainty in deep learning. In: ICML, (2016)
7. GitHub: alg-x (2024). Battery capacity prediction using regression. https://github.com/alg-x/Battery-Capacity-Prediction-Using-Regression
8. Hoeting, J.A., Madigan, D., Raftery, A.E., Volinsky, C.T.: Bayesian model averaging: A tutorial. Stat. Sci., (1999)
9. Kendall, A., Gal, Y.: What uncertainties do we need in Bayesian deep learning for computer vision? NeurIPS, p. 30. (2017)
10. Kang, K., Gupta, A., Levine, S.: Deep neural networks tend to extrapolate predictably. In: International Conference on Learning Representations (ICLR) (2024)
11. Kingma, D.P., Ba, J.: Adam: A method for stochastic optimization. In: International Conference on Learning Representations (ICLR) (2015)
12. Koch, G., Zemel, R., Salakhutdinov, R.: Siamese neural networks for one-shot image recognition. In: ICML Deep Learning Workshop, (2015)
13. Kong, L., et al.: Towards understanding extrapolation: A causal lens. In: Advances in Neural Information Processing Systems (NeurIPS), (2024)
14. Lakshminarayanan, B., Pritzel, A., Blundell, C.: Simple and scalable predictive uncertainty estimation using deep ensembles. NeurIPS, p. 30. (2017)
15. Narayanaswamy, V., et al.: On the use of anchoring for training vision models. In: Advances in Neural Information Processing Systems, (2024)
16. Netanyahu, A., Gupta, A., Simchowitz, M., Zhang, K., Agrawal., P.: Learning to extrapolate: A transductive approach. In: ICLR, (2023)
17. Pan, M., et al.: Towards extrapolation in deep material property regression. In: International Conference on Learning Representations, (2024)
18. Rahaman, N., et al. (2019). On the spectral bias of neural networks. In *ICML*, pages 5301
19. Rasmussen, C.E., Williams, C.K.I.: Gaussian Processes for Machine Learning. MIT Press (2006)
20. Romano, Y., Patterson, E., Candes, E.: Conformalized quantile regression. NeurIPS, p. 32. (2019)
21. Segal, N., et al.: Known unknowns: Out-of-distribution property prediction in materials and molecules. Nature Communications Materials (8), 6 (2025)

22. Severson, K.A., et al.: Data-driven prediction of battery cycle life before capacity degradation. Nat. Energy **4**(5), 383–391 (2019)
23. Shen, X., Simon, N., Sahoo, D., Peters, J.: Engression: Extrapolation through the lens of distributional regression. Biometrika, (2024)
24. Snell, J., Swersky, K., Zemel, R.: Prototypical networks for few-shot learning. NeurIPS, p. 30. (2017)
25. Wiedemer, T., Kirsch, L., Steenbrugge, X., Schmidhuber, J.: Compositional generalization from first principles. In: Advances in Neural Information Processing Systems, (2023)
26. Xu, K., et al.: Neural networks fail to learn periodic functions. NeurIPS, 33 (2020)
27. Xu, Z.Q.J., Zhang, Y., Luo, T., Xiao, Y., Ma, Z.: Frequency principle: Fourier analysis sheds light on deep neural networks. Communications in Computational Physics **28**(5), 1746–1767 (2021)

Sparse Neural Networks with Architectural Constraints for Feature Selection in Tabular Data

Lars Fluri(✉)

Faculty of Business and Economics, University of Basel, Basel, Switzerland
lars.fluri@unibas.ch

Abstract. Sparse neural networks are often studied in highly overparameterized regimes, yet many tabular regression problems involve only a moderate number of correlated predictors. In this setting, it is unclear whether sparsity-inducing training alone is sufficient or whether it should be coupled with additional architectural heuristics such as pyramid-style hidden-layer resizing. This paper presents a controlled empirical ablation of whether pyramid-style hidden-layer resizing adds predictive value once input sparsity is already imposed. We compare dense baselines, one-shot sparse training, and sparse-to-sparse training under matched optimization settings, sparsity budgets, and data splits. Across synthetic and benchmark regression tasks, sparsity alone accounts for most of the observed generalization gains. Pyramid-style resizing yields smaller and inconsistent changes in the tested tabular MLP regimes, including dataset-specific anomalies that do not translate into robust average improvements. These results motivate treating resizing as a design choice to validate under the evaluated sparse-training settings rather than as a default companion to sparse feature selection.

Keywords: Sparse neural networks · feature selection · pruning · tabular regression · multilayer perceptrons

1 Introduction

Sparse neural networks have become an increasingly popular tool for regression on tabular data, where the number of predictors is moderate, features are often correlated, and sample sizes are limited. In this setup, researchers care not only about predictive accuracy, but also about robustness, stability, and the ability to identify a compact set of relevant input variables. As a result, a wide range of methods has been proposed to introduce sparsity into neural models, including feature-level gating mechanisms, weight pruning, and dynamic sparse training procedures.

Alongside these sparsity mechanisms, it is common practice to impose additional architectural heuristics, most notably pyramid-style hidden-layer resizing. Pyramid-style width rules, including geometric decreases across hidden lay-

L. Pasa et al. (Eds.): ICANN 2026, LNCS 17089, pp. 708–718, 2027.
https://doi.org/10.1007/978-3-032-38398-3_58

ers, have long been used as simple architecture heuristics for feed-forward networks [9]. Such rules are often motivated by the intuition that reduced input dimensionality should be accompanied by reduced hidden capacity, both to control overfitting and to improve interpretability. Consequently, pyramid-style resizing is frequently treated as a natural complement to sparsity, rather than as an independent modeling choice requiring empirical validation.

The main question in this paper is therefore not whether sparse training can work, but whether pyramid-style resizing contributes meaningful marginal value once a sparse feature budget is already enforced. Sparsity and architectural shape represent distinct inductive biases: sparsity directly targets irrelevant or redundant predictors, whereas resizing alters model capacity in a coarse way that may help, harm, or simply duplicate the effect of sparsity. Whether that second ingredient remains useful under matched sparse training conditions is an empirical question.

This paper contributes a controlled empirical ablation of sparse neural architecture design in tabular regression. We compare dense baselines, one-shot sparse training, and dynamic sparse-to-sparse methods, each evaluated with and without pyramid-style resizing, while holding optimization procedures, sparsity budgets, and data splits fixed. The contribution is therefore empirical and methodological rather than algorithmic: the experiments isolate the marginal effect of pyramid-style resizing instead of proposing a new pruning method or a new feature-selection family. Across the reported tasks, most generalization gains are explained by sparsity itself, while resizing produces smaller and less consistent effects. Section 2 positions this question relative to sparse training and neural feature selection, Sect. 3 describes the experimental design, Sect. 4 summarizes the datasets and evaluation protocol, Sect. 5 reports the results, and Sect. 6 discusses anomalies and scope limitations.

2 Related Work

This paper connects two lines of research that are often discussed separately: (i) *neural feature selection* for tabular prediction, which targets an explicit, global subset of inputs; and (ii) *sparse training and pruning*, which targets sparse parameterizations and may or may not yield stable feature-level sparsity. A third axis is *architectural heuristics*, especially pyramid-style hidden-layer resizing, that are frequently bundled with sparsity but rarely isolated as an independent design choice. The present work focuses on that separation and positions its contribution as an empirical test of one commonly bundled heuristic.

2.1 Neural Feature Selection for Tabular Models

A large body of work enforces feature selection by introducing *input gates* whose activations indicate whether a predictor is used. A widely used approach is differentiable ℓ_0 regularization via stochastic gates and the hard-concrete distribution, which optimizes an expected ℓ_0 objective jointly with model weights

[8]. Stochastic Gates (STG) is closely related and provides a probabilistic relaxation for feature indicators with gradient-based learning of selection variables [10]. Another prominent approach, LassoNet, introduces a linear skip connection and hierarchical constraints such that a feature can participate in nonlinear layers only if its skip representative is active, producing a regularization path of sparse solutions and explicit global selection [6]. In contrast to purely gate-based methods, this hierarchy tries to stabilize selection and improve interpretability by tying nonlinear usage to a main-effect proxy. In addition to global gating, some tabular architectures incorporate sequential or instance-wise feature selection. TabNet, for example, uses sequential attention masks to choose features at each decision step, yielding interpretable attributions but not necessarily a single global subset shared across samples [1]. Because the present study concerns global feature selection under a fixed feature budget, such attention-based approaches are complementary rather than directly substitutable.

2.2 Pruning, Sparse Training, and Dynamic Sparse Training

Sparse training and pruning methods traditionally aim to obtain sparse networks with minimal accuracy loss. The ℓ_0-regularized gating framework can also be viewed as inducing sparsity through learned stochastic masks [8]. A separate and influential direction is *dynamic sparse training* (DST), which maintains sparsity throughout training while allowing connectivity to change. RigL is a canonical DST method that periodically prunes connections by magnitude and regrows new connections based on gradient information, enabling training with a fixed parameter budget [2]. A key conceptual development is that DST can provide "over-parameterization over time" even when the instantaneous network is sparse. In-Time Over-Parameterization formalizes this perspective and argues that exploring many sparse connectivities across training can close the expressivity gap between sparse and dense training [7]. Layerwise sparsity allocation has also emerged as an important design variable in pruning performance; LAMP proposes a layer-adaptive magnitude score for global pruning to avoid ad hoc layerwise heuristics [5]. More broadly, surveys and benchmark studies emphasize that simple sparsification baselines can be surprisingly strong and that complex pruning strategies are often inconsistent [3,4].

2.3 Architectural Heuristics Versus Sparsity Mechanisms

Across these threads, architectural heuristics are frequently bundled with sparsity mechanisms and treated as a default regularization choice. However, prior work does not directly answer the question studied here: whether pyramid-style resizing, implemented here through a geometric width schedule, provides additional benefit once input-level sparsity is already enforced under matched optimization and data conditions. The present experiments address that narrower question by separating feature-level sparsity from pyramid-style resizing within otherwise comparable sparse neural training pipelines.

3 Methods

3.1 Model Class and Problem Setup

We study supervised regression using multilayer perceptrons (MLPs) with explicit sparsity at the input layer. Let $\mathbf{x} \in \mathbb{R}^p$ denote the input vector and $y \in \mathbb{R}$ the target variable. All neural models share a common base architecture consisting of fully connected hidden layers with nonlinear activation functions and a linear output layer. Differences between methods arise solely from how sparsity is imposed, maintained, and whether sparsity is coupled to pyramid-style resizing during training.

Table 1. Neural network hyperparameters for the fixed-budget experiments. The symbol K denotes the target number of active input features, while p denotes the preprocessed input dimensionality.

Hyperparameter	**Value**
Training regimes	sparse-to-sparse, one-shot, dense
Target active features (K)	dataset-specific
Hidden layers	3
Initial hidden widths	$\lfloor p/2 \rfloor, \lfloor p/4 \rfloor, \lfloor p/8 \rfloor$
Pyramid scaling factor	2 ($r = 0.5$)
Optimizer	Adam
Epochs	200
Activation function	ReLU
Learning rate	0.01
Batch size	32
Early stopping patience	20
Feature importance criterion	first-layer ℓ_1 norm
Neuron removal criterion	ℓ_1 norm
Re-addition criterion	gradient-based
Sparse rewiring interval	10 epochs
Feature elimination cutoff	0.5

3.2 Study Design

The study is designed as a controlled empirical comparison of sparse architecture choices. The central quantity of interest is the marginal value of pyramid-style resizing once sparsity is already enforced. We therefore compare matched model pairs that differ only in whether hidden-layer widths are resized during training. Optimization settings, data splits, feature budgets, and evaluation protocols are otherwise held fixed, so the comparison isolates the architectural heuristic rather than broader pipeline differences.

3.3 Sparse-to-Sparse Training

We focus on sparse-to-sparse training strategies, in which sparsity is enforced throughout optimization. At any training step, only a subset of $K < p$ input features is active, and feature selection is integrated directly into training via iterative pruning and optional rewiring. The target sparsity level K is fixed *ex ante*, reflecting applied settings where interpretability or robustness motivates a constrained number of predictors. Feature budgets are enforced by retaining the highest-ranked first-layer ℓ_1 feature scores and masking all other inputs; sparse-to-sparse runs revisit the active set every 10 epochs using a 10% feature reshuffle and first-layer rewiring fraction. Features with low importance scores are removed, and reintroduced features are chosen using gradient-based criteria. One-shot variants apply the same feature-ranking rule once at the feature-elimination cutoff, after which the selected support is held fixed. Performance is evaluated across multiple values of K, allowing us to analyze the trade-off between model capacity, feature selection, and generalization.

3.4 Pyramid-Style Resizing Ablation

To isolate the role of architectural constraints, we introduce an explicit ablation. Let K_t denote the number of active input features at training step t, and let $h_\ell(t)$ denote the width of hidden layer ℓ. We consider two architectural regimes:

$$h_\ell(t) = \begin{cases} h_\ell^{(0)}, & \text{(fixed width)}, \\ \max\{1, \lfloor K_t \cdot r^\ell \rfloor\}, & \text{(pyramid-style resizing)}, \end{cases} \tag{1}$$

where $h_\ell^{(0)}$ is the initial hidden-layer width and $r = 0.5$ is the fixed ratio between successive hidden layers. With three hidden layers, the initial fixed-width networks use $\lfloor p/2 \rfloor$, $\lfloor p/4 \rfloor$, and $\lfloor p/8 \rfloor$ units. The reported ablation keeps this ratio fixed and tests the presence or absence of resizing rather than tuning r itself. In the pyramid-style regime, downstream layers are progressively reduced as input sparsity increases, forming a geometric width schedule. In the fixed-width regime, hidden-layer widths remain unchanged throughout training. This formulation enables a direct comparison between sparsity alone and sparsity coupled with architectural resizing under otherwise identical training dynamics.

3.5 Baselines and Experimental Control

In addition to sparse neural models, we consider several baselines to contextualize performance. A dense MLP trained without input sparsity serves as a reference model under the same optimization protocol. We further include Group Lasso regularization applied at the input layer, which induces structured sparsity via convex penalties, and a hard-concrete gating model that learns stochastic binary feature masks through an ℓ_0-inspired relaxation. All neural models are trained with Adam using the same learning rate, batch size, data splits, and early-stopping rule (Table 1), so that performance differences can be attributed to sparsity mechanisms and architectural constraints rather than unrelated training changes.

We evaluate the following models:

- **Dense MLP.** A standard multilayer perceptron trained on all p input features without sparsity.
- **Sparse NN (Fixed).** A sparse-to-sparse MLP constrained to exactly K active input features with fixed hidden-layer widths.
- **Sparse NN (Pyramid).** The same sparse model with hidden-layer widths resized according to the pyramid-style rule.
- **One-Shot Sparse NN (Fixed).** A model in which feature selection is applied once at a predefined training epoch, followed by continued training with fixed sparsity and architecture.
- **One-Shot Sparse NN (Pyramid).** The one-shot variant combined with pyramid-style resizing after feature selection.
- **Group Lasso.** A neural model with group-lasso regularization applied to the input layer.
- **Hard-Concrete Gating.** A neural feature selection model using stochastic binary gates trained via a hard-concrete relaxation.

4 Experiments

We evaluate sparse neural network models on both synthetic and real-world regression tasks. The experimental design is intended to disentangle the effects of input-level sparsity, pyramid-style resizing, and training strategy under a fixed feature budget. Experiments are conducted on one synthetic regression dataset with known ground-truth sparsity and two standard tabular benchmark datasets (`accounting` and `ames`). Synthetic data allow controlled assessment of generalization and overfitting behavior under sparsity constraints, while real datasets assess empirical performance in realistic settings. We compare iterative sparse training (DST-style), one-shot sparse training, and classical regularization baselines alongside a dense MLP reference. For sparse neural models, we explicitly study the effect of enabling or disabling pyramid-style resizing. Each dataset uses a deterministic 70/15/15 train/validation/test split with split seed 42; continuous preprocessing standardizes predictors and targets using training-split statistics only. Reported summaries use 30 model seeds (0–29) unless otherwise stated. Predictive performance is summarized using out-of-sample R^2 and, for the synthetic task, complementary MSE summaries. For real datasets, we also report paired performance differences between pyramid variants at a fixed sparsity level, approximate confidence intervals, and paired win-rates (Table 2).

Table 2. Overview of datasets used in the experiments. Feature counts refer to the final preprocessed input dimensionality.

Dataset	Task	n	p	K
Accounting Data	Reg.	8740	47	15
Ames Housing	Reg.	2930	79	15
Friedman-1	Reg.	5000	100	5

Table 3. Synthetic regression results (mean ± 95% CI over $n = 30$ runs).

Model	Train R^2	Val R^2	Test R^2	Train MSE	Val MSE	Test MSE
Sparse (DST, fixed)	0.999 ± 0.000	0.465 ± 0.019	0.505 ± 0.014	0.016 ± 0.006	8.979 ± 0.324	9.493 ± 0.269
Sparse (DST, pyramid)	0.998 ± 0.000	0.403 ± 0.023	0.455 ± 0.016	0.027 ± 0.005	10.018 ± 0.390	10.451 ± 0.302
One-shot (fixed)	0.523 ± 0.006	0.417 ± 0.007	0.419 ± 0.009	8.409 ± 0.109	9.787 ± 0.122	11.133 ± 0.167
One-shot (pyramid)	0.158 ± 0.081	0.147 ± 0.081	0.136 ± 0.070	14.829 ± 1.434	14.317 ± 1.364	16.553 ± 1.344
Group Lasso	0.409 ± 0.005	0.406 ± 0.005	0.372 ± 0.004	10.410 ± 0.087	9.980 ± 0.087	12.025 ± 0.077
Hard Concrete	0.450 ± 0.008	0.461 ± 0.011	0.414 ± 0.012	9.687 ± 0.144	9.055 ± 0.187	11.237 ± 0.238
Dense MLP	0.990 ± 0.002	0.469 ± 0.009	0.477 ± 0.007	0.171 ± 0.036	8.913 ± 0.145	10.013 ± 0.135

5 Results

5.1 Synthetic Regression Results

Table 3 reports predictive performance on the synthetic regression task for sparse neural models, classical sparsity baselines, and a dense MLP. Sparse neural networks without pyramid-style resizing achieve strong out-of-sample performance under a strict feature budget. Iterative sparse training with fixed hidden widths attains the highest test R^2 among the sparse models and performs comparably to the dense MLP while using far fewer active inputs. This is especially visible in the Friedman-1 design, where sparse DST without resizing outperforms both classical sparsity baselines and its pyramid-style counterpart.

Pyramid-style resizing degrades average synthetic performance for both iterative and one-shot sparse training in this setup. It slightly reduces in-sample error in some cases but yields worse validation and test performance, suggesting that the additional architectural constraint can add bias without improving generalization in this controlled setting. The degradation is strongest for one-shot training, where the pyramid-style variant shows a sharp drop in test R^2 and a large increase in test MSE. The dense MLP attains near-perfect in-sample fit but does not outperform sparse DST on held-out data, reinforcing the interpretation that input-level sparsity provides the dominant regularization effect under the evaluated sparse-training settings.

5.2 Fixed-Budget Performance and Pyramid Ablation

Table 4 summarizes test performance at the fixed feature budgets used in the paired comparisons. Across the `accounting` and `ames` datasets, pyramid-style resizing does not improve average test performance for the sparse neural models. Differences in R^2 are negative for both sparse DST and one-shot training, and

Table 4. Fixed-budget test performance and paired pyramid-style ablation summary. For ablation-capable methods, Off and On denote fixed-width and pyramid-style variants, Δ is the paired difference (On - Off), CI is the approximate 95% half-width, and Win is the fraction of paired seeds with On > Off. Methods without pyramid-style resizing are reported in the final column only.

Dataset	Method	K	Off	On	Δ	CI	Win	No ablation
Accounting	Group Lasso	15	—	—	—	—	—	0.063 ± 0.018
Accounting	Hard Concrete	15	—	—	—	—	—	-0.151 ± 0.059
Accounting	One-shot	15	0.180 ± 0.013	0.136 ± 0.012	-0.0433	± 0.016	0.13	—
Accounting	Sparse (DST)	15	0.190 ± 0.017	0.183 ± 0.013	-0.0068	± 0.020	0.50	—
Ames Housing	Group Lasso	15	—	—	—	—	—	0.582 ± 0.026
Ames Housing	Hard Concrete	15	—	—	—	—	—	0.634 ± 0.026
Ames Housing	One-shot	15	0.865 ± 0.005	0.761 ± 0.093	-0.1040	± 0.094	0.23	—
Ames Housing	Sparse (DST)	15	0.909 ± 0.003	0.900 ± 0.005	-0.0090	± 0.005	0.27	—
Friedman-1	Group Lasso	5	—	—	—	—	—	0.833 ± 0.018
Friedman-1	Hard Concrete	5	—	—	—	—	—	0.700 ± 0.020
Friedman-1	One-shot	5	0.927 ± 0.002	0.843 ± 0.102	-0.0845	± 0.103	0.83	—
Friedman-1	Sparse (DST)	5	0.875 ± 0.003	0.860 ± 0.012	-0.0146	± 0.012	0.33	—

the paired confidence intervals remain close to or overlap zero only for the already small sparse DST effects. The corresponding win-rates are also below 0.5 except for a single tie-like sparse DST case on `accounting`. These results indicate that any marginal effect of resizing is weak relative to the contribution of sparsity itself.

The Friedman-1 comparison is less uniform. Table 4 still shows a negative mean paired difference for the one-shot pyramid-style variant, yet its win-rate is 0.83. Taken together, these summaries indicate high variability rather than a stable advantage: pyramid-style resizing is better in many paired runs, yet the mean effect remains negative because a smaller number of poor runs dominate the average. We return to this anomaly in Sect. 6.

5.3 Feature-Selection Stability

Table 5 reports feature-selection stability through the Kuncheva index. These values measure overlap among selected subsets under a fixed feature budget; they do not certify recovery of a uniquely correct feature set. Moderate or low stability can arise even when predictive performance is similar across runs, especially in the presence of correlated predictors and near-substitutable variables. Under that interpretation, the main pattern is that pyramid-style resizing can change which features are selected without delivering a corresponding predictive gain. For example, sparse DST on `accounting` shows slightly higher Kuncheva stability under resizing, whereas sparse DST on `ames` shows slightly lower stability, yet neither case translates into better average test performance. The stability results therefore support a cautious practical conclusion: pyramid-style resizing affects subset overlap, but it does not provide reliable evidence of improved model quality or better real-data feature recovery.

Table 5. Feature-selection stability measured by the Kuncheva index (mean ± 95% CI across $\binom{n}{2}$ seed pairs) at the dataset-specific feature budgets used in the paired comparisons. For methods without pyramid resizing, only the no-ablation estimate is reported.

Dataset	Method	Off	On	No ablation
Accounting	Group Lasso	—	—	0.398 ± 0.012
Accounting	Hard Concrete	—	—	0.428 ± 0.014
Accounting	One-shot	0.265 ± 0.014	0.265 ± 0.014	—
Accounting	Sparse (DST)	0.228 ± 0.014	0.279 ± 0.014	—
Ames Housing	Group Lasso	—	—	0.632 ± 0.008
Ames Housing	Hard Concrete	—	—	0.683 ± 0.009
Ames Housing	One-shot	0.702 ± 0.007	0.702 ± 0.007	—
Ames Housing	Sparse (DST)	0.603 ± 0.005	0.577 ± 0.005	—
Friedman-1	Group Lasso	—	—	0.331 ± 0.008
Friedman-1	Hard Concrete	—	—	0.328 ± 0.008
Friedman-1	One-shot	0.325 ± 0.008	0.325 ± 0.008	—
Friedman-1	Sparse (DST)	0.098 ± 0.010	0.113 ± 0.009	—
Synthetic	Group Lasso	—	—	0.838 ± 0.005
Synthetic	Hard Concrete	—	—	0.688 ± 0.005
Synthetic	One-shot	0.681 ± 0.005	0.678 ± 0.005	—
Synthetic	Sparse (DST)	0.257 ± 0.007	0.258 ± 0.006	—

6 Discussion

The empirical picture is consistent with a narrow claim. For the tested MLP regimes on tabular regression problems, most of the performance benefit comes from enforcing input-level sparsity, while pyramid-style resizing shows no robust marginal advantage under matched training conditions. This does not imply that architectural resizing is never useful. It means only that the present experiments do not justify treating it as a default companion to sparse feature selection in the setting studied here.

One unresolved exception is the Friedman-1 one-shot comparison. Table 4 reports a negative mean paired difference for the pyramid-style variant even though the same variant wins in most paired runs. That combination suggests a high-variance, dataset-specific exception in which occasional large losses outweigh more frequent small gains. The current evidence is insufficient to identify the mechanism. In particular, the present experiments do not determine whether the pattern arises from the one-shot ranking step, sensitivity to the pyramid-style schedule, or properties of the synthetic design itself.

The stability results reinforce the same point. Kuncheva indices are useful for comparing seed-to-seed overlap, but they do not by themselves establish better recovery of a true feature set, especially on real datasets where the ground truth is unknown and multiple correlated subsets may support similar predictions. In this study, pyramid-style resizing sometimes nudges the stability score upward and sometimes downward, but those shifts are not matched by systematic improvements in test performance. Stability is therefore informative as a descriptive property of the selected subsets, not as a stand-alone validation of feature relevance.

The ablation does not fully isolate architectural shape from total parameter count or effective capacity. Pyramid variants may compress later hidden layers more aggressively, especially at small initial widths, so some observed differences may reflect capacity changes rather than shape alone. Parameter-matched fixed-width comparisons, larger-width ablations, and broader width-ratio sweeps are therefore natural extensions for a longer journal version rather than camera-ready additions.

The scope of the conclusions is limited to MLPs for tabular regression, the sparsity mechanisms reported here, and the feature budgets used in the experiments. The results should not be generalized to classification, broader neural architectures, or theoretical claims about sparse training in general. The paired differences and win-rates provide empirical evidence over repeated seeds, but they should not be read as formal dominance guarantees.

7 Conclusion

Across synthetic and real-world tabular regression tasks, the reported experiments show that input-level sparsity is the primary driver of the observed generalization gains. Once sparsity is imposed, pyramid-style hidden-layer resizing provides smaller and less consistent effects, with no robust average improvement and one unresolved high-variance anomaly on Friedman-1. For the evaluated regimes, fixed-width sparse MLPs are therefore a competitive and often safer default than coupling sparse feature selection with pyramid-style resizing.

A Dataset Summaries

- **Accounting Data**: a collection of 47 financial statement and indicator variables where the target is free cash flow one year ahead in a cross-sectional prediction task.
- **Ames Housing**: the Ames housing price dataset with 79 variables and 2930 observations.
- **Friedman-1**: a synthetic regression dataset with $n = 5000$, $p = 100$, and noise level 1.

References

1. Arik, S.Ö., Pfister, T.: TabNet: attentive interpretable tabular learning. In: Proceedings of the AAAI Conference on Artificial Intelligence. vol. 35, pp. 6679–6687 (2021)
2. Evci, U.: Rigging the lottery: making all tickets winners. In: Proceedings of the 37th International Conference on Machine Learning, pp. 2943–2952 (2020)
3. Gale, T.: The state of sparsity in deep neural networks. arXiv preprint arXiv:1902.09574 (2019)
4. Hoefler, T., Alistarh, D., Ben-Nun, T., Dryden, N., Peste, A.: Sparsity in deep learning: pruning and growth for efficient inference and training in neural networks. J. Mach. Learn. Res. **22**(241), 1–124 (2021)
5. Lee, N., Ajanthan, T., Torr, P.H.S.: Layer-adaptive magnitude-based pruning for deep neural networks. In: International Conference on Learning Representations (2021)
6. Lemhadri, I., Ruan, F., Abraham, L., Tibshirani, R.: LassoNet: a neural network with feature sparsity. J. Mach. Learn. Res. **22**(127), 1–29 (2021)
7. Liu, S., Chen, T., Atashgahi, Z., Li, Z.: In-time over-parameterization: on the global convergence of dynamic sparse training. In: Proceedings of the 38th International Conference on Machine Learning, pp. 7038–7048 (2021)
8. Louizos, C., Welling, M., Kingma, D.P.: Learning sparse neural networks through ℓ_0 regularization. In: International Conference on Learning Representations (2018)
9. Masters, T.: Practical neural network recipes in C, Morgan Kaufmann (1993)
10. Yamada, Y., Kubo, M., Yamane, K.: Feature selection using stochastic gates. In: Proceedings of the 37th International Conference on Machine Learning, pp. 10648–10659 (2020)

Architecture-Aware Structured Pruning for Parameter-Efficient Dense Small-Object Detection

Rafael Subo A. Yap and Arren Matthew C. Antioquia(✉)

Center for Computational Imaging and Visual Innovations, De La Salle University, Manila, Philippines
{rafael_yap,arren.antioquia}@dlsu.edu.ph

Abstract. Dense small-object detection and counting are essential in applications such as mosquito surveillance, agricultural phenotyping, and digital pathology, where large numbers of small instances must be detected accurately in crowded scenes. While modern object detectors achieve strong performance in these tasks, they often rely on parameter-heavy architectures that limit deployment in resource-constrained environments. This study provides a systematic empirical benchmarking of architecture-aware structured pruning for dense small-object detection across three domains using the Aedes aegypti egg dataset, the CoNSeP histopathology dataset, and the Sorghum-Grain-Count dataset. YOLOv4, YOLOv8n, and RT-DETR are analyzed under single-module and multi-module pruning strategies targeting backbone, neck, head, and transformer components. Experimental results show that backbone layers are consistently the most resilient to pruning, enabling significant parameter reduction and inference speed improvements with minimal impact on detection performance. In contrast, feature aggregation modules such as YOLO necks and RT-DETR transformer layers are considerably more sensitive and produce larger degradation in detection and counting accuracy when pruned. These findings establish practical guidelines for designing parameter-efficient detectors capable of maintaining reliable performance in extreme-density small-object detection scenarios.

Keywords: Structured Pruning · Small-Object Detection · Model Compression

1 Introduction

Dense small-object counting and detection are critical tasks across public health, agriculture, and medical diagnostics. Mosquito-borne diseases infect hundreds of millions of people annually, making vector surveillance essential for disease prevention [10]. Monitoring *Aedes* mosquito populations relies on counting eggs

L. Pasa et al. (Eds.): ICANN 2026, LNCS 17089, pp. 719–730, 2027.
https://doi.org/10.1007/978-3-032-38398-3_59

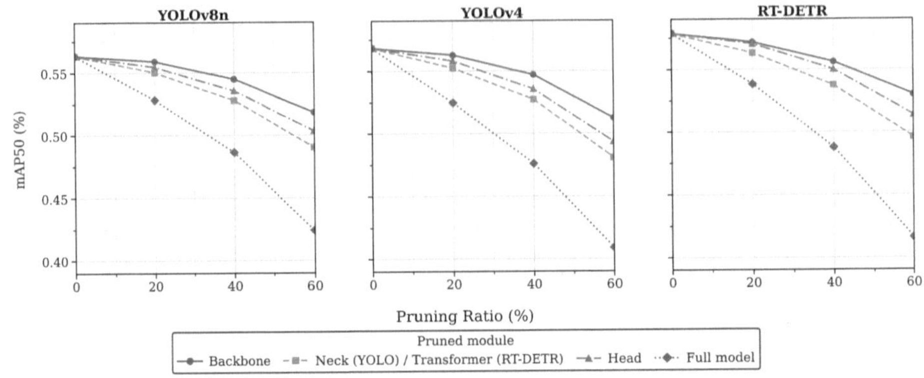

Fig. 1. Module-wise sensitivity to structured pruning for YOLOv4, YOLOv8n, and RT-DETR. Each plot shows mAP50 versus pruning ratio, averaged across the Aedes aegypti Egg, CoNSeP, and Sorghum-Grain-Count datasets. Backbone pruning results in the smallest performance degradation, while neck and transformer layers exhibit higher sensitivity. Pruning multiple modules simultaneously produces the largest performance drops.

collected from ovitraps to estimate breeding site activity and guide control interventions [14]. Similarly, in crop science, accurate counting of sorghum grains and seedlings enables early yield forecasting and accelerates breeding programs, where manual counting remains labor-intensive and time-consuming [5,15]. In digital pathology, automated detection of microscopic features in duodenal biopsies supports celiac disease diagnosis, where pathologists must identify subtle histologic patterns across numerous tissue patches [16,20]. While these applications demand high accuracy, their reliance on parameter-heavy architectures motivates systematic investigation into pruning for parameter reduction.

Recent work demonstrates domain-specific detectors with strong performance, but all employ large-scale models requiring substantial memory and compute. EggCountAI [10] achieves 98.88% egg counting accuracy using Mask R-CNN with millions of parameters. Sorghum grain counting [1,7,15] employs heavyweight YOLO variants trained on GPU workstations. Recent YOLO-based approaches reach expert-level performance for multi-class cell detection and counting in celiac histology patches [17]. Compression techniques like pruning [6,21] have accelerated generic detection networks, yet their applicability to small-object counting, where dense targets stress detection capacity, remains underexplored across domains. No systematic study examines how structured pruning across backbone, neck, and head modules impacts small-object detection and counting performance across diverse domains.

Rather than proposing a novel pruning algorithm, this study provides a systematic empirical benchmarking of standard architecture-aware structured pruning to evaluate module-wise sensitivity in small-object detection. Unlike prior work that is often limited to a single model, dataset, or pruning strategy, this study investigates both single-module and multi-module pruning configurations across multiple detection architectures, including YOLO-based models [11] and RT-DETR [22], as shown in Fig. 1. The approach is evaluated on multiple extreme-density datasets spanning diverse application domains such as entomology, agriculture, and pathology, enabling a comprehensive analysis of pruning behavior under varying conditions. The results identify consistent patterns in module sensitivity, reveal optimal pruning configurations that achieve significant parameter savings with minimal performance degradation, and provide practical guidelines for designing parameter-efficient small-object detectors across different domains.

2 Related Work

Dense small-object detection [19,23] have been extensively studied across entomology, agriculture, and digital pathology, where automated quantification replaces labor-intensive manual inspection. In mosquito surveillance, EggCountAI [10] employs Mask R-CNN to achieve high egg-counting accuracy under controlled imaging conditions, while subsequent studies extend detection to field-collected ovitrap images using CNN-based object detection frameworks [8,14,18]. In agricultural phenotyping, YOLO-based pipelines have been widely adopted for seedling and grain counting from drone [5] and smartphone imagery [15], and large-scale benchmarks such as Sorghum-Grain-Count (SGC) [1] encourage the use of high-capacity detection networks capable of handling thousands of densely packed instances per image. Reconstruction-based approaches [7] further improve structural accuracy in complex panicles. In computational pathology, automated detection of histologic structures supports celiac disease diagnosis [16,20], while detection-based counting frameworks such as DeGPR [17] achieve expert-level performance in crowded microscopic tissue patches.

Despite differences in imaging modality and visual variability, these domains share a common technical challenge: densely packed small objects occupy limited pixel area, suffer from occlusion and overlap, and place significant strain on detector localization capacity.

Modern object detectors address small-object sensitivity through multi-scale feature aggregation and architectural optimization. YOLO-based models, including YOLOv4 [2] and YOLOv8n [11], employ backbone–neck–head decompositions with feature pyramid designs that enhance small-object localization while maintaining real-time performance. Transformer-based detectors such as DETR [4] reformulate detection as a set prediction problem using global attention mechanisms and eliminate hand-crafted post-processing such as non-maximum suppression. Real-time variants such as RT-DETR [22] demonstrate that transformer-based approaches can match or surpass YOLO models in real-time detection scenarios, offering alternative trade-offs between backbone feature extraction, global reasoning through attention, and prediction heads. These architectural differences motivate module-specific analysis when optimizing detection efficiency.

Although architectural innovation has improved detection accuracy and inference speed, most domain-specific counting systems prioritize predictive performance over parameter efficiency. Heavyweight YOLO variants, Mask R-CNN backbones, and transformer-based detectors are typically deployed on GPU workstations, limiting applicability in field-based or edge settings. Model compression strategies [9,21], including pruning, quantization, and lightweight architectural redesign, aim to reduce parameter count and computational cost. Structured filter pruning methods [12,13], which remove entire convolutional channels based on importance criteria, have been shown to produce hardware-friendly reductions while preserving performance after fine-tuning. In detection-specific contexts, pruning [6] has been applied to reduce YOLO model size and computational load, and anchor-level pruning strategies [3] have been proposed to improve efficiency in object detection pipelines. However, these studies primarily evaluate generic detection benchmarks and rarely consider dense small-object counting scenarios characterized by extreme instance density.

Furthermore, prior compression research seldom investigates pruning sensitivity across distinct architectural modules such as backbone feature extractors, multi-scale neck modules, transformer encoders/decoders, or prediction heads. The interaction between pruning location and dense small-object counting fidelity remains underexplored, particularly across heterogeneous domains with significantly different visual characteristics. This gap motivates a unified, architecture-aware structured pruning study across various datasets from different domains, which enables systematic analysis across multiple modules, models, and datasets.

3 Method

3.1 Datasets

Experiments use three small-object detection datasets with splits following their source papers. The Aedes aegypti egg dataset [18] contains 434 field and laboratory images with 12,513 total egg instances using a 70/15/15 train/validation/test split. The CoNSeP dataset [17], processed following DeGPR methodology, comprises H&E stained colorectal adenocarcinoma histology images originally at 1000×1000 ($40\times$ magnification) containing 24,319 multi-class nuclei annotations, which are converted from segmentation masks to bounding boxes. Following DeGPR's preprocessing, each image is split into four 500×500 tiles yielding 108 training and 56 test tiles. The Sorghum-Grain-Count (SGC) dataset [1] provides manually annotated panicle and quadrant images from 316 genotypes of the Sorghum Association Panel, with fixed splits consisting of 74 training panicles from 70 genotypes, 25 development panicles, and 25 test panicles. This results in a total of 124 annotated panicles from 100 genotypes, where each panicle contains 30–3500 grains (average $\sim$1400) with bounding box annotations around individual grains. All images are resized to 640×640 while strictly preserving each paper's original train/validation/test splits.

3.2 Baseline Training

The YOLOv4 [2], YOLOv8n [11], and RT-DETR [22] models are initialized from COCO pretrained weights and fine-tuned on each dataset using standardized hyperparameters selected through systematic tuning: learning rate 0.001, momentum 0.95, weight decay 0.0001, SGD optimizer, and 100–300 training epochs. Tuning explored LR {0.001, 0.005, 0.01}, momentum {0.9, 0.95, 0.97}, weight decay {0.0001, 0.0005, 0.001}, and epochs {100, 150, 200, 250, 300}. The configuration yielding the best validation performance was consistently selected for all models and datasets. All experiments and FPS measurements were conducted on a single NVIDIA RTX 4050 GPU.

3.3 Architecture-Aware Structured Pruning

Structured pruning is applied according to the architectural decomposition of each model family to analyze the sensitivity of different modules within the detection pipeline.

YOLO-Based Models. For YOLOv4 and YOLOv8n, pruning is applied to three primary components: the backbone, neck, and head. The backbone extracts visual representations, the neck aggregates multi-scale information, and the head performs final predictions.

Channel-level structured pruning is used to remove entire convolutional filters based on L_2-norm magnitude ranking, which identifies less important channels within the network. Pruning ratios of 20%, 40%, and 60% are evaluated to study the effects of increasing compression levels. After pruning, each configuration is fine-tuned for 10 epochs using the same hyperparameter settings as the baseline training to recover performance.

RT-DETR. RT-DETR follows a transformer-based detection architecture and does not use the traditional YOLO-style neck-head separation. Instead, pruning regions are defined based on its architectural components: the convolutional backbone, the transformer encoder-decoder layers responsible for global contextual reasoning, and the final prediction head used for classification and bounding box regression.

Structured pruning removes entire channels in convolutional layers and reduces hidden dimensions in linear layers based on magnitude ranking. Similar to the YOLO-based models, pruning ratios of 20%, 40%, and 60% are evaluated. Each pruned model is then fine-tuned for 10 epochs using the same training configuration as the baseline model.

3.4 Multi-module Pruning

In addition to single-module pruning, multi-module pruning is applied to evaluate the combined effects of removing parameters from multiple architectural components simultaneously. For YOLO-based models, pruning combinations include

Table 1. Baseline detection, counting, and efficiency performance of YOLOv8n, YOLOv4, and RT-DETR across the Aedes aegypti Egg, CoNSeP, and Sorghum-Grain-Count (SGC) datasets.

Dataset	Model	Params (M)	mAP50	MAE	Pearson's r	FPS
Aedes aegypti Eggs [18]	YOLOv8n [11]	43.63	0.325	7.10	**0.996**	**21.38**
	YOLOv4 [2]	62.00	0.310	6.80	0.991	14.92
	RT-DETR [22]	81.00	**0.431**	**6.20**	0.987	18.20
CoNSeP [17]	YOLOv8n [11]	3.01	**0.518**	**4.63**	**0.925**	32.00
	YOLOv4 [2]	6.00	0.505	4.85	0.918	28.00
	RT-DETR [22]	31.80	0.495	5.10	0.912	**38.00**
SGC [1]	YOLOv8n [11]	45.00	0.847	**347.2**	0.740	**85.00**
	YOLOv4 [2]	68.00	**0.890**	400.0	0.740	65.00
	RT-DETR [22]	88.00	0.810	430.0	**0.742**	20.00

backbone and neck, backbone and head, neck and head, as well as the full combination of backbone, neck, and head. For RT-DETR, the evaluated combinations include backbone and transformer layers, backbone and prediction head, transformer and head, and the full combination of backbone, transformer, and head.

Each multi-module configuration is pruned using the same ratios of 20%, 40%, and 60% and subsequently fine-tuned for 10 epochs under the same hyperparameter settings used for baseline training. This setup allows consistent comparison of pruning sensitivity across architectural components and model families.

4 Experimental Results

We first establish baseline performance for YOLOv4 [2], YOLOv8n [11], and RT-DETR [22] across the three dense small-object datasets. Each model is trained using the best hyperparameter configuration identified during tuning. Table 1 reports detection performance, counting accuracy, and efficiency metrics, serving as reference for subsequent pruning analysis.

As shown in Table 1, performance varies across datasets. RT-DETR achieves the best results on Aedes aegypti Eggs [18], while YOLOv8n leads on CoNSeP [17] and YOLOv4 attains the highest mAP50 on SGC [1]. Overall, YOLO-based detectors provide greater efficiency, whereas detection and counting performance do not always align in dense scenes.

Table 2 shows average detection performance drop, *i.e.,* ΔmAP50, under structured pruning. The results represent the average performance drop across pruning ratios of 20%, 40%, and 60% for each evaluated model. Backbone pruning consistently results in the smallest degradation, with YOLOv8n decreasing by 4.13%, YOLOv4 by 4.90%, and RT-DETR by 4.79%. In contrast, YOLO necks and RT-DETR transformer layers are more sensitive, with performance reductions of up to 8.50% for neck components and 8.54% for transformer layers.

Table 2. Average ΔmAP50 performance drop under structured pruning across architectural modules. Results are averaged over pruning ratios of 20%, 40%, and 60% for each configuration across the Aedes aegypti Egg, CoNSeP, and Sorghum-Grain-Count (SGC) datasets.

Model	Pruned Module	ΔmAP50 (%)			
		Aedes a. Eggs [18]	**CoNSeP** [17]	**SGC** [1]	**Avg. ΔmAP50 (%)**
YOLOv8n [11]	Backbone	−4.72%	−3.86%	−3.82%	−4.13%
	Neck	−7.90%	−7.01%	−7.01%	−7.31%
	Head	−6.56%	−5.53%	−5.47%	−5.86%
	Backbone+Neck	−10.26%	−10.30%	−9.72%	−10.09%
	Backbone+Head	−8.92%	−8.75%	−8.23%	−8.63%
	Neck+Head	−11.69%	−12.81%	−11.18%	−11.89%
	Backbone+Neck+Head	−14.36%	−16.41%	−14.09%	−14.95%
YOLOv4 [2]	Backbone	−4.95%	−4.95%	−4.79%	−4.90%
	Neck	−8.49%	−8.58%	−8.43%	−8.50%
	Head	−7.10%	−6.93%	−6.85%	−6.96%
	Backbone+Neck	−11.51%	−11.88%	−11.72%	−11.70%
	Backbone+Head	−10.32%	−10.76%	−10.04%	−10.37%
	Neck+Head	−12.90%	−13.66%	−13.78%	−13.45%
	Backbone+Neck+Head	−15.91%	−17.69%	−17.60%	−17.07%
RT-DETR [22]	Backbone	−5.57%	−5.39%	−3.42%	−4.79%
	Transformer (Enc/Dec)	−10.52%	−8.75%	−6.34%	−8.54%
	Head	−6.96%	−6.67%	−4.81%	−6.15%
	Backbone+Transformer	−16.09%	−13.47%	−9.51%	−13.02%
	Backbone+Head	−13.15%	−11.52%	−7.90%	−10.85%
	Transformer+Head	−16.32%	−13.67%	−10.00%	−13.33%
	Backbone+Transformer+Head	−21.81%	−18.86%	−13.17%	−17.94%

Multi-module pruning, especially combinations including necks or transformers, leads to compounded drops, which shows that performance degradation is amplified when sensitive components are pruned together. These patterns suggest that targeted pruning should focus on less sensitive modules to preserve detection performance while achieving compression.

Counting error (ΔMAE) analysis in Table 3 follows similar trends. Results represent the average error increase across pruning ratios of 20%, 40%, and 60% for each evaluated model. Backbone pruning minimally affects counting, with errors of 11.43% for YOLOv8n, 13.12% for YOLOv4, and 11.44% for RT-DETR, whereas pruning necks or transformer layers significantly increases error, reaching 18.12% for the YOLOv8n neck and 19.87% for the RT-DETR transformer. Multi-module pruning further amplifies counting errors, exceeding 40% in some cases. Overall, these results reinforce that necks and transformers are critical for both detection and counting. Thus, pruning strategies must carefully preserve these components to maintain model reliability.

Table 4 summarizes the efficiency–accuracy tradeoff under structured pruning. Values are averaged across all single-module and multi-module pruning configurations and across the three datasets for each model and pruning ratio. As the pruning ratio increases, models achieve substantial parameter reduction, reaching up to 45.68% for YOLOv8n, 44.83% for YOLOv4, and 36.04% for RT-DETR, along with faster inference, including up to a 42.4% increase in frames per second for RT-DETR. Detection performance decreases gradually with higher pruning

Table 3. Average ΔMAE increase indicating counting error sensitivity under structured pruning across architectural modules. Results are averaged over pruning ratios of 20%, 40%, and 60% for each configuration across the Aedes aegypti Egg, CoNSeP, and Sorghum-Grain-Count (SGC) datasets.

Model	Pruned Module	ΔMAE (%)			
		Aedes a. Eggs [18]	**CoNSeP** [17]	**SGC** [1]	**Avg. ΔMAE (%)**
YOLOv8n [11]	Backbone	9.16%	10.87%	14.25%	11.43%
	Neck	14.18%	16.63%	23.56%	18.12%
	Head	12.11%	14.33%	18.95%	15.13%
	Backbone+Neck	19.25%	24.91%	33.64%	25.93%
	Backbone+Head	16.67%	22.53%	28.84%	22.68%
	Neck+Head	22.16%	29.59%	40.65%	30.80%
	Backbone+Neck+Head	28.87%	38.95%	52.65%	40.16%
YOLOv4 [2]	Backbone	9.81%	9.97%	19.58%	13.12%
	Neck	15.20%	15.12%	30.42%	20.24%
	Head	12.74%	13.74%	26.25%	17.58%
	Backbone+Neck	21.47%	23.02%	42.92%	29.14%
	Backbone+Head	18.97%	21.31%	38.08%	26.12%
	Neck+Head	24.51%	25.77%	50.00%	33.43%
	Backbone+Neck+Head	31.86%	33.68%	64.58%	43.37%
RT-DETR [22]	Backbone	12.10%	12.00%	10.23%	11.44%
	Transformer (Enc/Dec)	20.43%	20.33%	18.84%	19.87%
	Head	14.95%	16.20%	15.04%	15.40%
	Backbone+Transformer	31.99%	29.67%	28.91%	30.19%
	Backbone+Head	27.15%	25.33%	24.42%	25.63%
	Transformer+Head	32.96%	30.13%	30.47%	31.19%
	Backbone+Transformer+Head	45.43%	41.67%	41.08%	42.73%

ratios. The tradeoff highlights that moderate pruning provides a favorable balance between efficiency gains and minimal accuracy loss.

To visually illustrate the effects of structured pruning on detection and counting performance, Fig. 2 presents qualitative results for YOLOv4 across the datasets. For the Aedes aegypti Egg dataset, the ground truth count of 128 objects is reasonably approximated by the baseline prediction of 115, but pruning progressively reduces the predicted count to 101, 87, and 65 at 20%, 40%, and 60% pruning, respectively. Similar trends appear in the CoNSeP and SGC datasets. Visually, higher pruning levels introduce a greater number of missed detections, particularly in densely populated regions where small objects overlap. These qualitative examples support the quantitative results by showing that while moderate pruning retains much of the spatial detection structure, aggressive pruning significantly reduces detection completeness and counting accuracy.

Overall, the experimental results reveal consistent patterns across detection, counting, and efficiency metrics. Backbone layers are the most resilient to pruning, exhibiting only minor performance degradation even at higher pruning ratios, likely due to the redundancy of features learned in convolutional backbones. In contrast, feature aggregation modules, such as the YOLO neck and RT-DETR transformer layers, are highly sensitive to pruning because they are responsible for multi-scale feature fusion and contextual information processing, making their performance more susceptible to channel removal. This sensitivity is further amplified when multiple modules are pruned simultaneously. These

Table 4. Average efficiency–accuracy tradeoff under structured pruning. Values are averaged across all single-region and multi-region pruning configurations and across the Aedes aegypti Egg, CoNSeP, and Sorghum-Grain-Count (SGC) datasets for each model and pruning ratio. Parameter reduction and FPS increase are computed relative to the average unpruned baseline of each model.

Model	Ratio	Params (M)	Params ↓ (%)	FPS ↑ (%)	ΔmAP50 (%)
YOLOv8n [11]	20%	25.67	15.96%	8.0%	−2.50%
YOLOv8n [11]	40%	20.91	31.54%	15.0%	−7.69%
YOLOv8n [11]	60%	16.59	45.68%	20.1%	−17.41%
YOLOv4 [2]	20%	37.96	16.26%	10.9%	−3.29%
YOLOv4 [2]	40%	31.45	30.62%	21.6%	−9.47%
YOLOv4 [2]	60%	25.01	44.83%	28.0%	−18.58%
RT-DETR [22]	20%	59.50	11.11%	11.4%	−3.66%
RT-DETR [22]	40%	50.86	24.01%	25.0%	−9.52%
RT-DETR [22]	60%	42.81	36.04%	42.4%	−17.91%

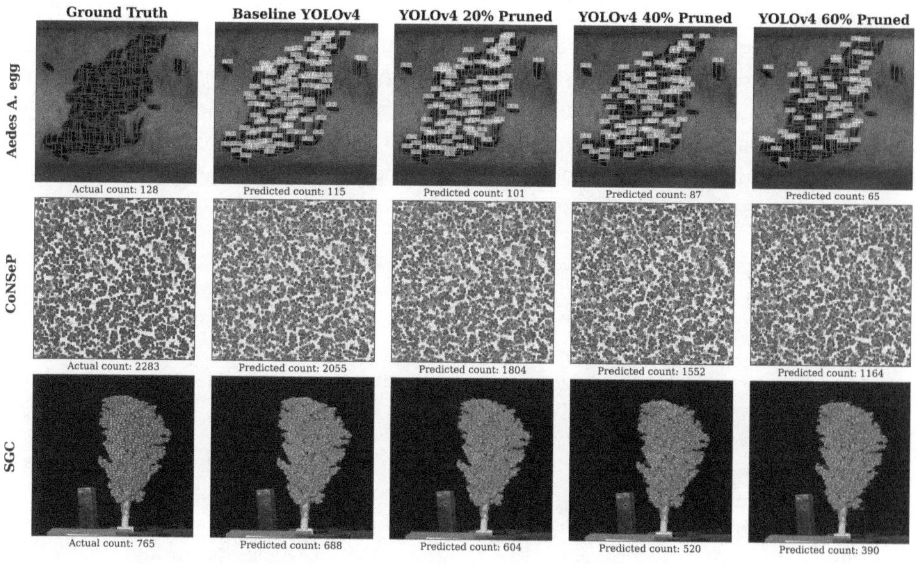

Fig. 2. Qualitative detection and counting results for YOLOv4 under increasing structured pruning levels. Rows correspond to the Aedes aegypti Egg, CoNSeP, and Sorghum-Grain-Count (SGC) datasets, while columns show ground truth annotations, the baseline YOLOv4 model, and YOLOv4 pruned at 20%, 40%, and 60%. Green bounding boxes represent detected objects counted by the model, while red bounding boxes indicate ground-truth instances that were missed by the detector. As the pruning ratio increases, the number of missed detections (red boxes) grows and the overall detection density decreases, illustrating the impact of aggressive pruning on dense small-object detection and counting.

results underscore that targeted pruning strategies focusing on less sensitive regions are essential to achieve a favorable balance between model compression, inference efficiency, and detection reliability.

5 Conclusion

This study presents a systematic evaluation of architecture-aware structured pruning for dense small-object detection and counting across three diverse domains: mosquito egg surveillance, histopathology cell detection, and agricultural grain counting. Results demonstrate that substantial parameter reductions can be achieved while largely preserving detection performance when pruning is applied strategically.

Across all datasets and models, consistent architectural sensitivity patterns emerge. Backbone layers exhibit the greatest resilience to pruning, allowing significant parameter reduction and inference speed improvements. In contrast, feature aggregation components, including YOLO neck modules and RT-DETR transformer layers, are considerably more sensitive. Pruning these regions produces the largest degradation in both detection accuracy and counting reliability, particularly in dense scenes where precise instance localization is critical. Furthermore, simultaneous pruning of multiple regions amplifies these effects. Pruning strategies should prioritize compression of backbone feature extractors while preserving feature fusion and contextual reasoning components. Future work may explore complementary compression approaches, including knowledge distillation, quantization, and neural architecture search, to further improve efficiency while maintaining robustness in extreme-density detection scenarios.

References

1. Bakshi, A., et al.: Sorghum-grain-count: a large image dataset and benchmark for sorghum grain count estimation. Smart Agricul. Technol. **12**, 101218 (2025). https://doi.org/10.1016/j.atech.2025.101218, https://www.sciencedirect.com/science/article/pii/S2772375525004496
2. Bochkovskiy, A., Wang, C.Y., Liao, H.Y.M.: YOLOv4: Optimal speed and accuracy of object detection. arXiv preprint arXiv:2004.10934 (2020). https://arxiv.org/abs/2004.10934
3. Bonnaerens, M., Freiberger, M., Dambre, J.: Anchor pruning for object detection. Comput. Vis. Image Underst. **221**, 103445 (2022). https://doi.org/10.1016/j.cviu.2022.103445, https://www.sciencedirect.com/science/article/pii/S1077314222000601
4. Carion, N., et al.: End-to-end object detection with transformers. In: Vedaldi, A., Bischof, H., Brox, T., Frahm, J.-M. (eds.) ECCV 2020. LNCS, vol. 12346, pp. 213–229. Springer, Cham (2020). https://doi.org/10.1007/978-3-030-58452-8_13
5. Chen, H., et al.: Estimation of sorghum seedling number from drone image based on support vector machine and yolo algorithms. Front. Plant Sci. **15 - 2024** (2024). https://doi.org/10.3389/fpls.2024.1399872, https://www.frontiersin.org/journals/plant-science/articles/10.3389/fpls.2024.1399872

6. Endrawati, D.N., Ibad, S.I., Syafalni, I., Sutisna, N., Mulyawan, R., Adiono, T.: YOLOv3- tiny's weight size reduction using pruning and quantization. In: 2021 15th International Conference on Telecommunication Systems, Services, and Applications (TSSA), pp. 1–5 (2021). https://doi.org/10.1109/TSSA52866.2021.9768258
7. Freeman, H., Schneider, E., Kim, C.H., Lee, M., Kantor, G.: 3D reconstruction-based seed counting of sorghum panicles for agricultural inspection. In: 2023 IEEE International Conference on Robotics and Automation (ICRA), pp. 9594–9600 (2023). https://doi.org/10.1109/ICRA48891.2023.10161400
8. Hamesse, C., et al.: Ovitrap monitor - online application for counting mosquito eggs and visualisation toolbox in support of health services. Eco. Inform. **75**, 102105 (2023). https://doi.org/10.1016/j.ecoinf.2023.102105, https://www.sciencedirect.com/science/article/pii/S1574954123001346
9. He, J., Jiang, J., Zhang, C.: A survey of lightweight methods for object detection networks. Array **29**, 100589 (2026). https://doi.org/10.1016/j.array.2025.100589, https://www.sciencedirect.com/science/article/pii/S2590005625002164
10. Javed, N., López-Denman, A.J., Paradkar, P.N., Bhatti, A.: EggCountAI: a convolutional neural network-based software for counting of Aedes aegypti mosquito eggs. Parasites Vectors **16**(1), 341 (2023). https://doi.org/10.1186/s13071-023-05956-1
11. Jocher, G., Qiu, J., Chaurasia, A.: Ultralytics YOLO (2023). https://github.com/ultralytics/ultralytics
12. Li, H., Kadav, A., Durdanovic, I., Samet, H., Graf, H.P.: Pruning filters for efficient convnets. In: International Conference on Learning Representations (2017)
13. Luo, J.H., Wu, J., Lin, W.: ThiNet: a filter level pruning method for deep neural network compression. In: 2017 IEEE International Conference on Computer Vision (ICCV). pp. 5068–5076. IEEE Computer Society, Los Alamitos, CA, USA (2017). https://doi.org/10.1109/ICCV.2017.541, https://doi.ieeecomputersociety.org/10.1109/ICCV.2017.541
14. Naranjo-Alcazar, J., Grau-Haro, J., Zuccarello, P., Almenar, D., Lopez-Ballester, J.: Automatic counting and classification of mosquito eggs in field traps. arXiv (2024). https://arxiv.org/abs/2405.20656
15. Santiago, G.N., Cisdeli Magalhaes, P.H., Carcedo, A.J., Marziotte, L., Mayor, L., Ciampitti, I.A.: Deep learning methods using imagery from a smartphone for recognizing sorghum panicles and counting grains at a plant level. Plant Phenomics **6**, 0234 (2024). https://doi.org/10.34133/plantphenomics.0234, https://www.sciencedirect.com/science/article/pii/S2643651524002310
16. Stoleru, C.A., Dulf, E.H., Ciobanu, L.: Automated detection of celiac disease using Machine Learning Algorithms. Sci. Rep. **12**(1), 4071 (2022). https://doi.org/10.1038/s41598-022-07199-z
17. Tyagi, A.K., Mohapatra, C., Das, P., Makharia, G., Mehra, L., AP, P., Mausam: DeGPR: deep guided posterior regularization for multi-class cell detection and counting. In: 2023 IEEE/CVF Conference on Computer Vision and Pattern Recognition (CVPR), pp. 23913–23923. IEEE Computer Society, Los Alamitos, CA, USA (2023). https://doi.org/10.1109/CVPR52729.2023.02290, https://doi.ieeecomputersociety.org/10.1109/CVPR52729.2023.02290
18. Vicente, M.N.d.O., et al.: Aedes aegypti egg counting with neural networks for object detection (2024). https://arxiv.org/abs/2403.08016
19. Wang, L., Zhou, Z., Shi, G., Guo, J., Liu, Z.: Small object detection based on bidirectional feature fusion and multi-scale distillation. In: Wand, M., Malinovská, K., Schmidhuber, J., Tetko, I.V. (eds.) Artificial Neural Networks and Machine Learning - ICANN 2024, pp. 200–214. Springer Nature Switzerland, Cham (2024)

20. Wei, J.W., Wei, J.W., Jackson, C.R., Ren, B., Suriawinata, A.A., Hassanpour, S.: Automated detection of celiac disease on duodenal biopsy slides: a deep learning approach. J. Pathology Inform. **10**(1), 7 (2019). https://doi.org/10.4103/jpi.jpi_87_18, https://www.sciencedirect.com/science/article/pii/S2153353922003704
21. Wei, Y., Pan, X., Qin, H., Ouyang, W., Yan, J.: Quantization mimic: towards very tiny CNN for object detection. In: Ferrari, V., Hebert, M., Sminchisescu, C., Weiss, Y. (eds.) ECCV 2018. LNCS, vol. 11212, pp. 274–290. Springer, Cham (2018). https://doi.org/10.1007/978-3-030-01237-3_17
22. Zhao, Y., et al.: DETRs beat YOLOs on real-time object detection. In: 2024 IEEE/CVF Conference on Computer Vision and Pattern Recognition (CVPR), pp. 16965–16974. IEEE Computer Society, Los Alamitos, CA, USA (2024). https://doi.org/10.1109/CVPR52733.2024.01605, https://doi.ieeecomputersociety.org/10.1109/CVPR52733.2024.01605
23. Zhou, M., Yi, C., Li, M., Wan, H., Li, G., Han, D.: Transforming limitations into advantages: improving small object detection accuracy with SC-AttentionIoU loss function. In: Iliadis, L., Papaleonidas, A., Angelov, P., Jayne, C. (eds.) Artificial Neural Networks and Machine Learning - ICANN 2023, pp. 221–232. Springer Nature Switzerland, Cham (2023)

Author Index

L. Pasa et al. (Eds.): ICANN 2026, LNCS 17089, pp. 731–733, 2027.
https://doi.org/10.1007/978-3-032-38398-3

FSC
www.fsc.org
MIX
Papier | Fördert
gute Waldnutzung
FSC® C083411

Kolibri 360 GmbH
Ferdinand-Jühlke-Straße 7
99095 Erfurt, Deutschland
produktsicherheit@kolibri360.de